Marian Bubak   Hamideh Afsarmanesh
Roy Williams   Bob Hertzberger (Eds.)

High Performance Computing and Networking

8th International Conference, HPCN Europe 2000
Amsterdam, The Netherlands, May 8-10, 2000
Proceedings

 Springer

Series Editors

Gerhard Goos, Karlsruhe University, Germany
Juris Hartmanis, Cornell University, NY, USA
Jan van Leeuwen, Utrecht University, The Netherlands

Volume Editors

Marian Bubak
University of Mining and Metallurgy (AGH)
Institute of Computer Science and Academic Computer Center
CYFRONET
al. Mickiewicza 30, 30-059 Cracow, Poland
E-mail: bubak@uci.agh.edu.pl

Hamideh Afsarmanesh
Bob Hertzberger
Universiteit van Amsterdam
Faculteit der Natuurwetenschappen, Wiskunde en Informatica
1098 SJ Amsterdam, The Netherlands
E-mail:{bob,hamideh}@science.uva.nl

Roy Williams
California Institute of Technology
Caltech 158-79
Pasadena, CA 91125, USA
E-mail:roy@cacr.caltech.edu

Cataloging-in-Publication Data applied for

Die Deutsche Bibliothek - CIP-Einheitsaufnahme

High performance computing and networking : 8th international
conference ; proceedings / HPCN Europe 2000, Amsterdam, The
Netherlands, May 8 - 10, 2000. Marian Bubak ... (ed.). - Berlin ;
Heidelberg ; New York ; Barcelona ; Hong Kong ; London ; Milan ; Paris
; Singapore ; Tokyo : Springer, 2000
 (Lecture notes in computer science ; Vol. 1823)
 ISBN 3-540-67553-1

CR Subject Classification (1991): C.2.4, D.1-2, E.4, F.2, G.1-2, J.1-2, J.3, J.6

ISSN 0302-9743
ISBN 3-540-67553-1 Springer-Verlag Berlin Heidelberg New York

This work is subject to copyright. All rights are reserved, whether the whole or part of the material is
concerned, specifically the rights of translation, reprinting, re-use of illustrations, recitation, broadcasting,
reproduction on microfilms or in any other way, and storage in data banks. Duplication of this publication
or parts thereof is permitted only under the provisions of the German Copyright Law of September 9, 1965,
in its current version, and permission for use must always be obtained from Springer-Verlag. Violations are
liable for prosecution under the German Copyright Law.

Springer-Verlag is a company in the BertelsmannSpringer publishing group.
© Springer-Verlag Berlin Heidelberg 2000
Printed in Germany

Typesetting: Camera-ready by author, data conversion by Christian Grosche, Hamburg
Printed on acid-free paper   SPIN 10721098   06/3142   5 4 3 2 1 0

Preface

This volume contains the proceedings of the international HPCN Europe 2000 event which was held in the Science and Technology Centre Watergraafsmeer, Amsterdam, the Netherlands, May 8-10, 2000.

HPCN (*High Performance Computing and Networking*) Europe event was organized for the first time in 1993 in Amsterdam as the result of several initiatives in Europe, the United States of America, and Japan. Succeeding HPCN events were held in Munich (1994), Milan (1995), Brussels (1996), and Vienna (1997), returning to Amsterdam in 1998 to stay.

The HPCN event keeps growing and advancing every year, and this year the event consisted of the scientific conference, focused workshops, and several associated events. The plenary lectures were presented by six renowned speakers:

- Henk van der Vorst, University of Utrecht, The Netherlands: *Giant Eigenproblems within Reach*,
- Wolfgang Gentzsch, CTO, Gridware Inc., Germany: *The Information Power Grid is Changing our World*,
- Bernard Lecussan, SupAero and ONERA/CERT/DTIM, France: *Irregular Application Computations on a Cluster of Workstations*,
- Miguel Albrecht, European Southern Observatory, Garching, Germany: *Technologies for Mining Terabytes of Data*,
- Hans Meinhardt, Max-Planck-Institut, Germany: *The Algorithmic Beauty of Sea Shells*, and
- Ingo Augustin, CERN, Geneva, Switzerland: *Towards Multi-petabyte Storage Facilities*.

The conference consisted of parallel tracks presenting 52 selected papers, and one track presenting 25 posters. The areas covered in the conference include: Industrial and General End-User Applications of HPCN, Computational and Computer Sciences, and this year the scope of the conference was further expanded by an additional area to emphasize the information management aspects, and the importance of the web-based cooperative application infrastructures.

In the area of *Web-Based Cooperative Applications* presented papers addressed: virtual enterprises and laboratories, cooperation coordination, as well as advanced web-based tools for tele-working. The area of *Industrial and End-User Applications of HPCN* consisted of papers focused on parallelisation of industrial codes, data-mining, and network applications. The papers presented in the area of *Computational Science* were dedicated to problem solving environments, metacomputing issues, load balancing and partition techniques, and new parallel numerical algorithms. In the area of *Computer Science Research in HPCN* the following subjects were presented: Java in HPC, cluster computing, monitoring and performance, as well as compilation and low-level algorithms.

The newly emerging domains and applications of HPCN were covered within five thematic workshops and three associated events. The *Java in High Performance Computing* workshop (chaired by Vladimir Getov) focused on the use of Java in simulations, distributed resource management, on-line processing, data-intensive applications, and other emerging research topics that combine distributed object technology with networking. The *LAWRA* workshop (chaired by Jerzy Waśniewski) is devoted to the new, recursive formulation of basic algorithms in numerical software packages. Recursion leads automatically to better utilization of memory, offers very concise program structures, and results in significant speedup on modern SMP processors. Several challenging requirements of the virtual laboratory environments such as the problem solving and computing issues, data mining, and the collaborative work in emerging scientific and engineering domains were addressed within the *Virtual Laboratory* workshop (chaired by Bob Hertzberger). The main goal of the *Cluster Computing* workshop (chaired by Mark Baker and Wolfgang Gentzsch) is to find out how clusters, built with commodity-off-the-shelf hardware components and free or commonly used software, may redefine the concept of high performance and availability computing. At the *EuroStore* workshop (chaired by Fabrizio Gagliardi) efficiency, reliability, and manageability of very large storage systems (Multi-PB) were discussed. These problems, being of great importance for industrial applications, have been observed in high energy physics.

The three associated events of the HPCN 2000 conference were: the MPR event – *Massive Parallel Computing* (organized by Job Kleuver), the NCF event – *Dutch Super Computing* (organized by Jaap Hollenberg), and the symposium on *Modeling and Simulation of Morphogenesis and Pattern Formation in Biology* (organized by Jaap Kaandorp). This symposium addresses the investigation of self-organization and emergent behavior in biological systems with particle-based techniques.

The conference proceedings reflect the state of the art in several main areas of research, within the wide spectrum of HPCN. It is worth mentioning that the deadline for contributed papers was January 18, 2000. All the accepted papers and posters, as well as a selection of some papers presented at the workshops, are included in the proceedings. We thank all contributors for their cooperation, and we are pleased to observe the high quality of the submitted contributions. The best conference papers will also be selected later for publication in a special issue of the North-Holland journal *Future Generation Computer Systems*.

The selection of papers for HPCN 2000 would not have been possible without the support and careful evaluation of all the submissions by the members of the HPCN 2000 program committee, and their associated reviewers. The organizing committee is grateful for all the invaluable suggestions and the cooperation that we received from the reviewers. Their help made it possible to get at least three referee reports for each paper.

We would like to express our high gratitude to the members of the local organizing committee and the conference secretariat. Our sincere thanks go to Lodewijk Bos and Rutger Hamelynck. We greatly appreciate all the personal efforts and dedication of Anne Frenkel for both creating the HPCN Europe web pages and helping with the organization of paper distribution and review results, and those of Berry van Halderen for both setting up the on-line paper submission software and preparing papers for the proceedings.

We would like to thank the computer support groups, the FdNWI faculty of the University of Amsterdam, headed by Gert Poletiek, for the electronic communication support, and SARA in Amsterdam, headed by Jaap Hollenberg, for the distribution of the program and participation calls for HPCN 2000.

The organizers acknowledge the support of the DUTCH HPCN foundation, and the help of the University of Amsterdam for making its facilities available for this event.

March 2000

Marian Bubak
Hamideh Afsarmanesh
Bob Hertzberger
Roy Williams

Organization

Event Chairman:

Bob Hertzberger, University of Amsterdam, NL

Scientific Organization:

Marian Bubak, University of Mining and Metallurgy (AGH), PL
    Conference Chair
Hamideh Afsarmanesh, University of Amsterdam, NL
    Conference Co-chair
Roy Williams, California Institute of Technology, USA
    Conference Co-chair

Program Committee

Hamideh Afsarmanesh
Dan Aharoni
Dick van Albada
Vassil Alexandrov
Farhad Arbab
Jan Astalos
Amnon Barak
Ammar Benabdelkader
Siegfried Benkner
Marian Bubak
Luis M. Camarinha-Matos
Paolo Cremonesi
Przemyslaw Czerwinski
Miroslav Dobrucky
Asuman Dogac
Jack Dongarra
Iain Duff
Dick Epema
Murat Ezbiderli
Martin Frey

Wlodzimierz Funika
Cesar Garita
Wolfgang Gentzsch
Alexandros Gerbessiotis
Vladimir Getov
Luc Giraud
Alexander Godlevsky
Forouzan Golshani
Ted Goranson
Andrzej M. Goscinski
Ralf Gruber
Necip Hamali
Alfons Hoekstra
Vasyl Horodisky
Cengiz Icdem
Peter Kacsuk
Ersin C. Kaletas
Nikos Karacapilidis
Erwin Laure
Heather Liddell

Bob Madahar
Tomàs Margalef
Vladimir Marik
Eduard Mehofer
Hans Moritsch
Zsolt Nemeth
Gustaf Neumann
Deniz Oguz
George A. Papadopoulos
Norbert Podhorszki
Kees van Reeuwijk
Alexander Reinefeld
Dirk Roose
Erich Schikuta

Giuseppe Serazzi
Viera Sipkova
Henk J. Sips
Krzysztof Sowa
Yusuf Tambag
Arif Tumer
Henk A. van der Vorst
Roland Wagner
Willy Weisz
Roy Williams
Kam-Fai Wong
Zahari Zlatev

Workshop Chairs:

Mark Baker (Cluster Computing)
Fabrizio Gagliardi (Eurostore)
Vladimir S. Getov (Java in High Performance Computing)
Bob Hertzberger (Virtual Laboratory)
Jerzy Waśniewski (LAWRA - Linear Algebra with Recursive Algorithms)

Associated Event Chairs:

Jaap Kaandorp (Modeling and Simulation of Morphogenesis and Pattern Formation in Biology)
Job Kleuver (MPR - Massive Parallel Computing)
Jaap Hollenberg (NCF - Dutch Super Computing)

Local Organization:

Lodewijk Bos
Rutger Hamelynck, Conference Office, University of Amsterdam
Anne Frenkel, University of Amsterdam
Berry van Halderen, University of Amsterdam
Joost Bijlmer, University of Amsterdam

Table of Contents

I Computational Science Track

Session 1 - Problem Solving Environments

A Problem Solving Environment Based on Commodity Software 3
    D.J. Lancaster, J.S. Reeve

DOVE: A Virtual Programming Environment for High Performance Parallel Computing 12
    H.D. Kim, S.H. Ryu, C.S. Jeong

Session 3 - Metacomputing

The Problems and the Solutions of the Metacomputing Experiment in SC99... 22
    S. Pickles, F. Costen, J. Brooke, E. Gabriel, M. Müller, M. Resch, S. Ord

Grid Computing on the Web Using the Globus Toolkit 32
    G. Aloisio, M. Cafaro, P. Falabella, C. Kesselman, R. Williams

Data Futures in DISCWorld 41
    H.A. James, K.A. Hawick

Session 6 - Partitioners / Load Balancing

Algorithms for Generic Tools in Parallel Numerical Simulation 51
    D. Lecomber, M. Rudgyard

Dynamic Grid Adaption for Computational Magnetohydrodynamics 61
    R. Keppens, M. Nool, P.A. Zegeling, J.P. Goedbloed

Parallelization of Irregular Problems Based on Hierarchical Domain Representation 71
    F. Baiardi, S. Chiti, P. Mori, L. Ricci

Dynamic Iterative Method for Fast Network Partitioning................ 81
    C.S. Jeong, Y.M. Song, S.U. Jo

Session 9 - Numerical Parallel Algorithms

ParIC: A Family of Parallel Incomplete Cholesky Preconditioners 89
    M. Magolu monga Made, H.A. van der Vorst

A Parallel Block Preconditioner Accelerated by Coarse Grid Correction ... 99
   C. Vuik, J. Frank

Towards an Implementation of a Multilevel ILU Preconditioner on
Shared-Memory Computers 109
   A. Meijster, F.W. Wubs

Session 11 - Numerical Parallel Algorithms

Application of the Jacobi–Davidson Method to Spectral Calculations in
Magnetohydrodynamics 119
   *A.J.C. Beliën, B. van der Holst, M. Nool, A. van der Ploeg,
   J.P. Goedbloed*

PLFG: A Highly Scalable Parallel Pseudo-random Number Generator for
Monte Carlo Simulations 127
   C.J.K. Tan, J.A. Rod Blais

parSOM: Using Parallelism to Overcome Memory Latency in Self-Organizing
Neural Networks 136
   Ph. Tomsich, A. Rauber, D. Merkl

II Web-Based Cooperative Applications Track

Session 2 - Virtual Enterprises / Virtual Laboratories

Towards an Execution System for Distributed Business Processes in a
Virtual Enterprise 149
   L.M. Camarinha-Matos, C. Pantoja-Lima

Towards a Multi-layer Architecture for Scientific Virtual Laboratories 163
   *H. Afsarmanesh, A. Benabdelkader, E.C. Kaletas, C. Garita,
   L.O. Hertzberger*

Session 4 - Cooperation Coordination

Modelling Control Systems in an Event-Driven Coordination Language ... 177
   T.A. Limniotes, G.A. Papadopoulos

Ruling Agent Motion in Structured Environments 187
   M. Cremonini, A. Omicini, F. Zambonelli

Dynamic Reconfiguration in Coordination Languages 197
   G.A. Papadopoulos, F. Arbab

Session 7 - Advanced Web-Based Tools for Tele-working

Developing A Distributed Scalable Enterprise JavaBean Server. 207
    Y. Guo, P. Wendel

CFMS - A Collaborative File Management System on WWW 217
    S. Ruey-Kai, C. Ming-Chun, C. Yue-Shan, Y. Shyan-Ming, T. Jensen, H. Yao-Jin, L. Ming-Chih

Adding Flexibility in a Cooperative Workflow Execution Engine 227
    D. Grigori, H. Skaf-Molli, F. Charoy

A Web-Based Distributed Programming Environment 237
    K.F. Aoki, D.T. Lee

III   Computer Science Track

Session 5 - Monitoring and Performance

Performance Analysis of Parallel N-Body Codes....................... 249
    P. Spinnato, G.D. van Albada, P.M.A. Sloot

Interoperability Support in Distributed On-Line Monitoring Systems 261
    J. Trinitis, V. Sunderam, T. Ludwig, R. Wismüller

Using the SMiLE Monitoring Infrastructure to Detect and Lower the
Inefficiency of Parallel Applications.................................. 270
    J. Tao, W. Karl, M. Schulz

Session 8 - Monitoring and Performance

Run-Time Optimization Using Dynamic Performance Prediction 280
    A.M. Alkindi, D.J. Kerbyson, E. Papaefstathiou, G.R. Nudd

Skel-BSP: Performance Portability for Skeletal Programming 290
    A. Zavanella

Self-Tuning Parallelism 300
    O. Werner-Kytölä, W.F. Tichy

A Novel Distributed Algorithm for High-Throughput and Scalable
Gossiping 313
    V. De Florio, G. Deconinck, R. Lauwereins

Session 13 - Low-Level Algorithms

Parallel Access to Persistent Multidimensional Arrays from HPF
Applications Using *Panda*... 323
    P. Brezany, P. Czerwinski, A. Swietanowski, M. Winslett

High Level Software Synthesis of Affine Iterative Algorithms onto Parallel
Architectures 333
    A. Marongiu, P. Palazzari, L. Cinque, F. Mastronardo

Run-Time Support to Register Allocation for Loop Parallelization of Image
Processing Programs 343
    N. Zingirian, M. Maresca

A Hardware Scheme for Data Prefetching 353
    S. Manoharan, K. See-Mu

Session 15 - Java in HPC

A Java-Based Parallel Programming Support Environment 363
    K.A. Hawick, H.A. James

A Versatile Support for Binding Native Code to Java 373
    M. Bubak, D. Kurzyniec, P. Łuszczek

Task Farm Computations in Java 385
    M. Danelutto

Session 16 - Clusters

Simulating Job Scheduling for Clusters of Workstations 395
    J. Santoso, G.D. van Albada, B.A.A. Nazief, P.M.A. Sloot

A Compact, Thread-Safe Communication Library for Efficient Cluster
Computing 407
    M. Danelutto, C. Pucci

EPOS and Myrinet: Effective Communication Support for Parallel
Applications Running on Clusters of Commodity Workstation 417
    A.A. Fröhlich, G.P. Tientcheu, W. Schröder-Preikschat

Distributed Parallel Query Processing on Networks of Workstations 427
    C. Soleimany, S.P. Dandamudi

IV   Industrial and End-User Applications Track

Session 10 - Parallelisation of Industrial Applications

High Scalability of Parallel PAM-CRASH with a New Contact Search
Algorithm 439
    J. Clinckemaillie, H.G. Galbas, O. Kolp, C.A. Thole, S. Vlachoutsis

Large-Scale Parallel Wave Propagation Analysis by GeoFEM 445
    K. Garatani, H. Nakamura, H. Okuda, G. Yagawa

Explicit Schemes Applied to Aeroacoustic Simulations: The RADIOSS-
CFD System 454
    D. Nicolopoulos, A. Dominguez

Session 12 - Data Analysis and Presentation

Creating DEMO Presentations on the Base of Visualization Model 460
    E.V. Zudilova, D.P. Shamonin

Very Large Scale Vehicle Routing with Time Windows and Stochastic
Demand Using Genetic Algorithms with Parallel Fitness Evaluation 467
    M. Protonotarios, G. Mourkousis, I. Vyridis, T. Varvarigou

Extracting Business Benefit from Operational Data 477
    T.M. Sloan, P.J. Graham, K. Smyllie, A.D. Lloyd

Session 14 - Miscellaneous Applications

Considerations for Scalable CAE on the SGI ccNUMA Architecture 487
    S. Posey, C. Liao, M. Kremenetsky

An Automated Benchmarking Toolset 497
    M. Courson, A. Mink, G. Marçais, B. Traverse

Evaluation of an RCube-Based Switch Using a Real World Application ... 507
    E.C. Kaletas, A.W. van Halderen, F. van der Linden, H. Afsarmanesh,
    L.O. Hertzberger

MMSRS - Multimedia Storage and Retrieval System for a Distributed
Medical Information System 517
    R. Słota, H. Kosch, D. Nikolow, M. Pogoda, K. Breidler, S. Podlipnig

V Posters

Web-Based Cooperative Applications

Dynamically Transcoding Data Quality for Faster Web Access 527
    C. Chi-Huing, L. Xiang, A. Lim

Easy Teach & Learn[(R)]: A Web-Based Adaptive Middleware for Creating
Virtual Classrooms 531
    T. Walter, L. Ruf, B. Plattner

Industrial and End-User Applications

A Beowulf Cluster for Computational Chemistry 535
    K.A. Hawick, D.A. Grove, P.D. Coddington, H.A. James,
    M.A. Buntine

The APEmille Project 539
  E. Panizzi, G. Sacco

A Distributed Medical Information System for Multimedia Data -
The First Year's Experience of the PARMED Project 543
  H. Kosch, R. Słota, L. Böszörményi, J. Kitowski, J. Otfinowski,
  P. Wójcik

Airport Management Database in a Simulation Environment 547
  A. Pasquarelli, T. Hruz

Different Strategies to Develop Distributed Object Systems at University
of La Laguna 551
  A. Estévez, F.H. Priano, M. Pérez, J.A. González, D.G. Morales,
  J.L. Roda

DESIREE: DEcision Support System for Inuandation Risk Evaluation and
Emergencies Management 555
  G. Adorni

Database System for Large-Scale Simulations with Particle Methods 558
  D. Kruk, J. Kitowski

Computational Science

Script Wrapper for Software Integration Systems 560
  J. Fischer, A. Schreiber, M. Strietzel

Implementation of Nested Grid Scheme for Global Magnetohydrodynamic
Simulations of Astrophysical Rotating Plasmas 564
  T. Kuwabara, R. Matsumoto, S. Miyaji, K. Nakamura

Parallel Multi-grid Algorithm with Virtual Boundary Forecast Domain
Decomposition Method for Solving Non-linear Heat Transfer Equation 568
  G. Qingping, Y. Paker, Z. Shesheng, D. Parkinson, W. Jialin

High Performance Computing on Boundary Element Simulations 572
  J.M. Cela, A. Julià

Study of Parallelization of the Training for Automatic Speech Recognition 576
  E.M. Daoudi, A. Meziane, Y.O. Mohamed El Hadj

Parallelization of Image Compression on Distributed Memory Architecture 580
  E.M. Daoudi, E.M. Jaâra, N. Cherif

Parallel DSMC on Shared and Hybrid Memory Multiprocessor Computers 584
  G.O. Khanlarov, G.A. Lukianov, D.Yu. Malashonok, V.V Zakharov

Population Growth in the Penna Model for Migrating Population 588
    A.Z. Maksymowicz, P. Gronek, W. Alda, M.S. Magdoń-Maksymowicz,
    M. Kopeć, A. Dydejczyk

Use of the Internet for Distributed Computing of Quantum Evolution 592
    A.V. Bogdanov, A.S. Gevorkyan, A.G. Grigoryan, E.N. Stankova

Computer Science

Debugging MPI Programs with Array Visualization 597
    D. Kranzlmueller, R. Kobler, R. Koppler, J. Volkert

An Analytical Model for a Class of Architectures under Master-Slave
Paradigm 601
    Y. Yalçınkaya, T. Steihaug

Dynamic Resource Discovery through MatchMaking 605
    O.F. Rana

A New Approach to the Design of High Performance Multiple Disk
Subsystems: Dynamic Load Balancing Schemes 610
    A.I. Vakali, G.I. Papadimitriou, A.S. Pomportsis

Embarrassingly Parallel Applications on a Java Cluster 614
    B. Vinter

A Revised Implicit Locking Scheme in Object-Oriented Database Systems . 618
    W. Jun, K. Kim

Active Agents Programming in HARNESS 622
    M. Migliardi, V. Sunderam

VI Workshops

LAWRA Workshop

LAWRA Workshop: Linear Algebra with Recursive Algorithms:
http://lawra.uni-c.dk/lawra/ 629
    F. Gustavson, J. Waśniewski

Java in HPC Workshop

Communicating Mobile Active Objects in Java 633
    F. Baude, D. Caromel, F. Huet, J. Vayssière

A Service-Based Agent Framework for Distributed Symbolic Computation 644
    R.D. Schimkat, W. Blochinger, C. Sinz, M. Friedrich, W. Küchlin

Performance Analysis of Java Using Petri Nets 657
  O.F. Rana, M.S. Shields

Cluster Computing Workshop

A Framework for Exploiting Object Paralellism in Distributed Systems ... 668
  W. Chen, M.T. Yong

Cluster SMP Nodes with the ATOLL Network: A Look into the Future of
System Area Networks.. 678
  *L. Rzymianowicz, M. Waack, U. Brüning, M. Fischer, J. Kluge,
  P. Schulz*

An Architecture for Using Multiple Communication Devices in a
MPI Library 688
  H. Pedroso, J. Gabriel Silva

Results of the One-Year Cluster Pilot Project 698
  *K. Koski, J. Heikonen, J. Miettinen, H. Niemi, J. Ruokolainen,
  P. Tolvanen, J. Mäki, J. Rahola*

Clusters and Grids for Distributed and Parallel Knowledge Discovery 708
  M. Cannataro

Author Index 717

Track I
Computational Science Track

A Problem Solving Environment Based on Commodity Software

David Lancaster and J.S. Reeve

Electronics & Computer Science
University of Southampton, Southampton SO17 1BJ, U.K.
djl@ecs.soton.ac.uk

Abstract. Following the common use of of commodity hardware to build clusters, we argue that commodity software should be harnessed in a similar way to support Scientific and Engineering work. Problem Solving Environments (PSE) provide the arena where commodity software technology can modernise the development and execution environment. We describe a PSE prototype based on the standard software infrastructure of Java Beans and CORBA that illustrates this idea and provides advanced PSE functionality at minimum effort for medium sized heterogeneus platforms.

1 Introduction

In recent years, scientists and engineers in poorly funded University departments have realised that clusters of PC's provide a cost-effective computing platform for their computationally intensive needs. These clusters [1] are constructed from cheap, off-the-shelf, commodity hardware components and usually run free operating systems such as Linux.

The same forces that have led to the flood of cheap hardware components have also led to the availability of cheap commodity software. It is natural to propose that this software should also be harnessed towards the needs of academic scientists. G. Fox [2] has been a proponent of this idea and has identified the software supporting distributed information systems as being most relevant. In particular, the standardised software that supports web applications, such as CORBA[3], Java[4] and XML[5], is widely available, often freely or for nominal sums.

The proposed role for this commodity software is to modernise the software environment within which scientists and engineers develop and execute applications. The typical software environment for scientific work has changed little over many years, and often remains at the level of command line driven compilers and debuggers. To compile and execute code, the detailed idiosyncracies of each platform in a typically heterogeneus system must be recalled. Problem Solving Environments (PSE's) have been proposed [6] as a more modern environment. A PSE is a software environment that provides support for all stages in both the development and execution of problem solving code. Some PSE's provide further functionality but the essential ingredients are:

- **Development**: The PSE should follow modern software trends by being modular and allowing graphical programming. The components or modules perform well-defined computational functions relevant to the target field of the PSE: for example in engineering they may be solvers of various kinds or grid generators. The solution to a problem is achieved by composing diverse components into a full application that addresses the problem. Composition, which consists in wiring up the components may be done using graphical programming which is widely used in commercial fields and is generally held to improve programmer productivity. This approach enables novel ideas to be tested by rapidly prototyping complex combinations of components.
- **Execution**: A distributed hardware system is assumed, and the the PSE must be able to run each component (typically requiring substantial computation) on remote machines in a transparent manner. The hardware environment available to the users we have in mind is heterogeneous and there is considerable scope for improvements in efficiency by scheduling components to appropriately chosen machines. This is a time-consuming management task, and one gain from using a PSE will be that it avoids the need for the user to be concerned with details of the respective platforms. Infrastructure supporting seamless computing of this kind is provided by various projects such as Globus[7] and Unicore[8].

In this paper we show how a basic PSE with these features can rapidly and simply be put together using commodity software technologies: JavaBeans[4] and CORBA[3]. JavaBeans enable the PSE to use sophisticated graphical tools for development and CORBA addresses the requirements of transparently executing code on heterogeneous platforms.

As befits a prototype system we emphasise simplicity of design. The PSE is designed to use the commmodity layers in such a way as to minimize the additional infrastructure code that must be written and should be regarded as an experimental architecture. In this first step, we have tried to include all the basic functionality but we have not been deeply concerned about the performance. Even so we have incorporated some techniques, notably in data transfer and non-blocking CORBA calls, in order to hide latency.

In the remaining sections of this paper we give a more detailed description of the commodity software technologies, introduce the prototype PSE and explain the roles that the software layers play in this PSE. The next section describes the relevant features of JavaBeans and of CORBA before we proceed to a fuller description of the PSE prototype architecture in section 3. After a brief overview of related work, the conclusion summarises our experiences and mentions some of the directions for further research.

2 Commodity Software Technologies

The commodity technologies related to distributed computing that we have in mind include CORBA, Microsoft's DCOM, Java based software and proposals put forward by the World Wide Web Consortium. It is not apparent which, if any, of these will survive in the future. Nonetheless they do have common characteristic features: they are cheap (often with free implementations), widely available and standardized. As a result of strong competition, implementations are robust and of high quality. These are the features needed in designing new software environments for Scientific/Engineering purposes. The resultant systems are intended to combine the necessary perfomance needed for computationally intensive applications with the rich functionality of commodity systems. In this paper we emphasise the functionality, but in the conclusion we will return to discuss the performance.

Our prototype PSE uses JavaBeans and CORBA. These are appropriate for a prototype that is intended to be used with a heterogeneous collection of medium sized machines such as PC clusters, small shared memory machines and individual workstations that run a variety of operating systems. A more highly featured PSE that was intended to be used with more substantial computing resources would also need software layers that connect to queuing systems and databases. To better organize the computational components and allow them to be easily shared, a large PSE might also benefit from software standards such as XML.

2.1 JavaBeans

JavaBeans are reusable software components that can be manipulated visually in a Builder tool such as the Beanbox[9]. The beans are written in standard Java but conform to certain naming patterns that the builder tool uses to discover their properties.

Java beans fulfill the need for a sophisticated user interface that allows graphical programming by connecting computational components. In the prototype PSE, each bean represents a component that performs some computational function. The development stage of the PSE consists of selecting beans from a palette and wiring them together in a builder tool in such a way as to compose the computational components into an application that provides the means of solving the problem at hand. An approach of this sort should be familiar to anyone who has used tools such as AVS [10].

A fully functional PSE would have its own visual tool that provides all the builder functionality to wire up beans besides supporting the organisation of the palette of beans, various pieces of run-time support along with help and guidance tools. However, the aim of this work is to demonstrate how far one can proceed without such proprietary development, and although commercial builder tools may not provide all the functionality desired, they are interchangeable and support the majority of usage scenarios.

2.2 CORBA

CORBA provides a framework for distributed computing. The features that endear it to industry are that it is object oriented, that it is inter-operable across platforms and a variety of languages and that it is already a long-lived and well supported standard [3]. To enable the interoperablity, interfaces are defined in a language known as IDL and communication takes place transparently through an ORB. There are many suppliers of ORB's [11] such as Sun who include an ORB in their JDK1.2.

CORBA provides a platform independent way of distributing the computations performed by the PSE so as to take advantage of high performance machines and software implementations. The components, or computational modules of the PSE are presented to the user in the form of the JavaBeans described above. These beans are only used to allow graphical composition, the actual computational code is not written in Java as part of the bean, but is in some other language (possibly legacy code) that resides on a different machine. The Java beans are hollow and merely make CORBA calls to the computational code.

For a small PSE with few components, it might be possible to decide in advance where each piece of computational code should run. Then, with the help of the CORBA nameservice the appropriate machine could be selected at runtime. In more sophisticated PSE's a scheduler would be used to select a machine to run the code depending on load and other factors.

3 Prototype

The discussion above should already have clarified the overall architecture of our prototype and the roles of the JavaBeans and CORBA. Before proceeding to describe the prototype in more detail, it may be worth being more explicit about one aspect of the overall architecture: the control flow. The whole PSE is based on a data flow model which is intuitive in this context and is made manifest in the way that the beans are wired together in the builder tool. Indeed, the information about the control flow has its origin in the act of wiring, and remains encapsulated in the builder tool. As soon as one component has completed its task and generated its data, control passes to the next bean which dispatches a request to the CORBA object that implements the computational work of the component. We emphasise this only to contrast it with the case where the control flow information in the builder tool is first abstracted and passed to a separate dispatcher (often combined with scheduler). Our choice of architecture is made for simplicity and to minimize the additional software that has to be written. It turns out that this architecture is extremely flexible and well adapted for computational steering.

To illustrate the use of the PSE prototype and to make the following discussion more concrete we have implemented some simple components that perform operations to diagonalize a matrix. The method of diagonalization is standard and described in [12, 13]. Below we list and briefly describe these components. Figure 1 shows how the components would typically be wired together in order to solve a certain problem in solid state physics. However, one of the main purposes of a PSE is that the components can be reused to solve other problems.

- **Hamiltonian Matrix**: Generate a matrix. The matrix corresponds to the Hamiltonian for a simple problem in solid state physics. It is symmetric and has zeros on the diagonal.
- **Random Diagonal**: Add random terms to the diagonal. This corresponds to a disordered medium in the solid state problem. The strength of the random terms is a variable parameter.
- **Householder**: Turn a symmetric matrix into tridiagonal form using a householder transformation.
- **QL**: Diagonalize a tridiagonal matrix using QL decomposition.
- **Visualize**: Display a matrix.

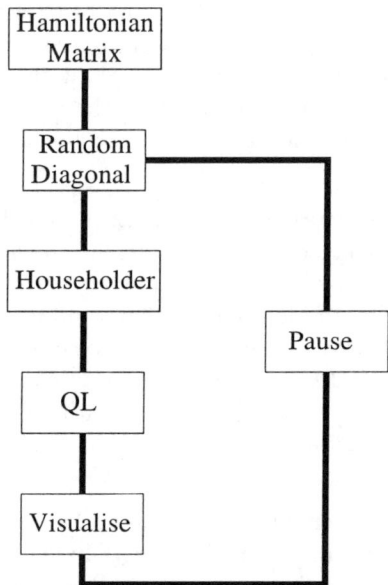

Fig. 1. Components wired in a Builder tool to diagonalize a matrix appearing in a problem of disordered media. The function of each component is described in the text. The direction of control flow is clockwise around the loop in this example.

The beans representing these components have interfaces described in the beaninfo class that prescribe how they can be wired together in the builder tool. Besides listing any parameters, this class lists the type and form of the inputs and outputs that the component expects. The JavaBean code is essentially limited to desribing these interfaces and the CORBA client code needed to call the computational module residing on a server. For this reason we call these beans "hollow". In order to avoid unnecessary data movement and thereby improve performance, the form of the inputs/outputs are not explicit matrices, but are CORBA references to data objects that contain the matrices.

The CORBA part of the prototype development requires an IDL for the system including the data objects for the matrices. This IDL is too long and detailed to include in a paper but is available on http://gather.ecs.soton.ac.uk/PSE/Docs. The CORBA servers have evolved since the beginning of this project and now provide facilities to manage the life-cycle of computational components.

We employ a performance enhancement that makes non-blocking CORBA calls to hide latency following a standard CORBA technique. Consider for example the Householder component which accepts a symmetric matrix and generates a tridiagonal matrix. When this bean starts to run, it first contacts the scheduler which provides an object reference on some remote machine to the computational class that implements the householder transformation. The bean then uses this reference to submit the object reference of the symmetric matrix data class and control then passes back to the bean which then waits. Meanwhile, the Householder object uses the matrix reference to transfer the matrix data to its local address space and then proceeds with the computation, finally creating a new matrix object in which to store the resulting tridiagonal matrix. The object reference to this tridiagonal matrix object is returned to the bean indicating completion of the job.

For the purposes of this work we implemented a scheduler that allows several versions of each component to exist on different machines and selects which one to use on request. This scheduler is dumb in the sense that the algorithm for selection is extremely simple, being either random or based on prior user choices. It does however provide the same kind of interface expected in a more sophisticated version. CORBA initialization employs a nameserver which is used to register servers that are started by hand on whatever remote machines are available. The scheduler contacts the nameserver to obtain all the information about which components are available on which machines.

The flow of control between one component and another takes place using the standard bean event mechanism and PropertyChangeEvents trigger this flow. For example, when the householder bean in the example above, obtains the reference to the tridiagonal matrix object, it fires a PropertyChangeEvent that is picked up by the next bean, in this case the QL bean, that has been wired to listen for such events.

As shown in the example in figure 1, there is no need to avoid loops when wiring the components. In this case there is no feedback of data up the loop, merely control flow. We have shown an explicit pause component that delays firing the PropertyChangeEvent and that allows the program to be controlled directly from the builder tool. In the situation shown in the figure, a Hamiltonian matrix of given size is generated once, and then on subsequent traversals of the loop different random terms are added to the diagonal and the resulting matrix is diagonalized. This corresponds to a typical need in the study of disordered systems where many realizations of the disorder must be averaged. A full implementation would contain components that perform the averaging automatically.

As befits a design based on CORBA, the prototype works on a wide variety of platforms, and we have used both workstations and PC's running various flavours of UNIX and Windows NT. We have used this project to test several ORBS with different functionality, including the JDK1.2 ORB, Orbacus[14] and JacORB[15]. Because we needed support for JDK1.2 at the start of this project, the user interface has been based on the beanbox builder tool. However, we expect that the components we have developed can be used in any commercial builder tool.

4 Related Work

Although work on PSE's has been continuing for some years, before the recent flood of commodity software technology these efforts tended to use proprietary or specially developed infrastructure and have remained isolated.

The work closest to our own both in spirit and in use of commodity software is Webflow. The Webflow PSE [16] has a three tier architecture consisting of a Webflow editor front end, CORBA middle layer and Globus back end. The commodity middle layer has replaced an special purpose layer in a previous version of this tool. The main thrust of this work has been to allow meta-applications to be created simply at a high level. The connection to the Globus back end enables the PSE to target more substantial computational resources than our prototype. This work is now being used in a major PSE project called the Gateway [17].

Other PSE projects that use commodity software technology include JACO3 which couples simulations with CORBA and Java [18] and PSEware which provides a toolkit for building PSE's [19].

5 Conclusion

We have demonstrated the ease with which commodity software technologies can be harnessed to provide advanced problem solving environments targeted for Scientists and Engineers. Our prototype is a proof-of-concept of this approach adapted for medium sized heterogeneous platforms.

As a first step, our demonstration has concentrated on providing the necessary functionality but the main challenge remains to obtain good performance out of systems built from commodity software. This challenge should not be confused with the difficulties in obtaining good performance from any PSE: in general there is a tradeoff between perfomance and flexibility and PSE's are certainly intended to be more flexible than the old-fashioned environment. The difficulty arises in deciding the level of granularity of the components: if the components are large, they will tend to be efficient but inflexible and *vice versa*. In the prototype we have described, CORBA is used to allow each computational component to be implemented in whatever language is appropriate (Fortran can be accomodated even though it has no CORBA binding [20]) and is thus expected to be efficient. The performance problems with commodity based PSE's arise from communication between distributed objects. Work in this area, including high-performance CORBA implementations and comparisons between ORB's may be found in [21]. We have not provided any performance figures for the example included in the prototype because it was intended for illustrative purposes and is not an appropriate benchmark.

With the takeup of commodity software in this area, the range of possible PSE architecures is becoming more restricted so we may expect some convergence leading to the possibility of standardisation. Most useful would be standardisation of the form of the components, and work in this area is already in progress [22].

We also mention the improvements that would make this prototype PSE a more useful and robust tool. To allow the freer import and sharing of new components they should be described in some language such as XML along with tools that use this description to automatically create most of the bean and beaninfo code as well as providing a skeleton for the computational part of the code [23]. The other major area for improvement is in the scheduler, which should monitor load and provide a more sophisticated selection algorithm.

Acknowledgments

David Lancaster would like to acknowledge discussions with Peter Lockey and Matthew Shields. This work was done in the context of a UK EPSRC grant entitled *Problem Solving Environments for Large Scale Simulations*.

References

1. D. Ridge, D. Becker, P. Merkey and T. Sterling. *Beowulf: Harnessing the Power of Parallelism in a Pile-Of-PC's.* Proc. 1997 IEE Aerospace Conference.
   See the Beowulf Project page at CESDIS:
   http://cesdis.gsfc.nasa.gov/linux/beowulf/beowulf.html
2. G. Fox, W. Furmanski, T. Haupt, E. Akarsu and H. Ozdemir. *HPcc as High Performance Commodity Computing on top of Integrated Java, CORBA, COM and Web Standards.* Proc. of Europar 1998, Springer. 55-74, 1998.
3. The CORBA specification is controlled by the Object Management Group. http://www.omg.org/
4. Sun's central site for information about Java is at: http://java.sun.com/
5. World Wide Web Consortium, *Extensible Markup Language (XML)* http://www.w3.org/XML/
6. E. Gallopoulos, E. Houstis and J.R. Rice. *Problem Solving Environments for Computational Science.* IEEE Comput. Sci. Eng., **1**, 11-23, 1994.
   E. Gallopoulos, E. Houstis and J.R. Rice. *Workshop on Problem Solving Environments: Findings and Recommendations.* ACM Comp. Surv., **27**, 277-279, 1995.
7. I. Foster and C. Kesselman. *Globus: A Metacomputing Infrastructure Toolkit.* Int. J. Supercomp. Appl., **11**, 115-128, 1997. http://www.globus.org
8. Uniform Access to Computing Resources. http://www.fz-juelich.de/unicore
9. The Beanbox is a free reference builder tool from Sun. Commercial tools are listed at: http://java.sun.com/beans/tools.html

10. AVS is produced by Advanced Visual Systems Inc. http://www.avs.com/
11. A list of ORB's is given at:
    http://adams.patriot.net/~valesky/freecorba.html/
12. W.H. Press, B.P.Flannery, S.A.Teukolsky and W.T.Vetterling. *Numerical Recipes.* In various languages. CUP.
13. J. Reeve and M. Heath *An Efficient Parallel Version of the Householder-QL Matrix Diagonalization Algorithm.* Par. Comp. **25**, 311-319, 1999.
14. http://www.ooc.com/ob.html
15. http://www.multimania.com/dogweb/
16. T. Haupt, E. Akarsu and G. Fox. *WebFlow: A Framework for Web Based Metacomputing.* Proc. of HPCN 1999, Springer. 291-299, 1999.
17. http://www.osc.edu/~kenf/theGateway
18. http://www.arttic.com/projects/JACO3/default.html
19. http://www.extreme.indiana.edu/pseware/
20. Besides well known techniques of wrapping Fortran, some work on an F90 binding was done in the context of the Esprit PACHA project.
21. http://www.cs.wustl.edu/~schmidt/corba-research-performance.html
22. R. Armstron, D. Gannon, A. Geist, K. Keahey, S. Kohn, L. McInnes, S. Parker and B. Smolinski. *Toward a Common Component Architecture for High-Performance Scientific Computing.* Available from the CCA Forum: http://z.ca.sandia.gov/~cca-forum/
23. O.F. Rana, M. Li, D.W. Walker and M.Shields. *An XML based Component Model for Generating Scientific Applications and Performing Large Scale Simulations in a Meta-Computing Environment.* Available from: http://www.cs.cf.ac.uk/PSEweb/

DOVE: A Virtual Progamming Environment for High Performance Parallel Computing

Hyeong Do Kim, So-Hyun Ryu, and Chang Sung Jeong*

Department of Electronics Engineering, Korea University, Anamdong 5-ka
Sungbuk-ku, Seoul 136-701, Korea
FAX: +82-2-926-7620, Tel: +82-2-3290-3229
kimhd@snoopy.korea.ac.kr
csjeong@charlie.korea.ac.kr
messias@snoopy.korea.ac.kr

Abstract. In this paper, we present a new parallel programming environment, which is called DOVE(Distributed Object-oriented Virtual computing Environment), based on distributed object model. A parallel program is built as a collection of concurrent and autonomous objects interacting with one another via method invocation. It appears to a user logically as a single virtual computer for a set of heterogeneous hosts connected by a network as if objects in remote site reside in one virtual computer. The main goal of DOVE is to provide users with easy-to-use programming environment while supporting efficient parallelisms encapsulated and distributed among objects. Efficient parallelisms are supported by various types of method invocation and multiple method invocation to the object group. For the performance evaluation purpose of DOVE, we have developed two parallel applications both on DOVE and PVM. Our experiment shows that DOVE has better performance than PVM and provides an efficient and easy-to-use parallel programming environment for a set of heterogeneous and clustered computers.

1 Introduction

During the last decade, increases in computing power of individual machines and advances in high-speed computer networks make clustered computer more attractive and considerable trend in high performance computing paradigm[1]. Currently, various programming environments have been developed for clustered computing environments. Most of the parallel programing environments for clustered computing are based on distributed shared memory model and message passing model.

Recently, distributed object models such as OMG CORBA[2], JAVA/RMI[3] and DCOM[4] have been introduced to tackle the problems inherent in distributed computing on a heterogeneous environment. Distributed object model consists of objects which interact via method invocation, while message passing

---

* This work has been supported by KISTEP and KOSEF under contract 99-NF-03-07-A-01 and 981-0926-141-2.

model consists of processes which communicate with each other via message passing. Distributed object model has several benefits over message passing model. It provides an easy programming environment by supporting transparency of distributed objects, plug and play of software as well as the advantages of object oriented programming such as reusability, extensibility, and maintainability through abstraction, encapsulation and inheritance. However, it lacks some functionalities for parallel applications, since they are based on client-server model.

In this paper, we present a new parallel programming environment, which is called DOVE(Distributed Object-oriented Virtual computing Environment), for clustered computing based on a distributed object model. The main goal of DOVE is to provide users with easy-to-use programming environment while supporting efficient parallelisms encapsulated and distributed among objects. An easy-to-use and transparent parallel programming environment is provided by heterogeneity achieved through automatic data marshalling and unmarshalling, stub and skeleton objects generated automatically by DOVE IDL compiler, and object life management and naming service offered by object manager. Thus, in DOVE, the code of a parallel program has little difference with the sequential one. For the performance evaluation purpose of DOVE, we have developed two parallel programs with applications to ray-tracing and genetic simulated annealing respectively both on DOVE and PVM.

The outline of our paper is as follows: In section 2, we describe previous works which are related to our work. In section 3, we present the details of DOVE object model, and in section 4, describe object manager. In section 5 and 6, we present an architecture of DOVE run-time system and its implementation respectively. In section 7, we present a performance comparison with PVM for two applications. Finally, a conclusion will be given in section 8.

2 Related Work

In this section, we describe several parallel programming environments for clustered computing which are related to our work. MPI[5] is a single standard programming interface mainly designed for developing high performance parallel applications with emphasis on a variety of communication patterns and communication topology. However, MPI lacks in functionalities such as process control, resource management and fault-tolerance. PVM[6] is one of the most widely used distributed and parallel computing systems based on the message passing model, and connects together separate physical machines into a virtual computer by providing process control, simple message passing and packing constructs and dynamic process group management. In PVM, a daemon process, which runs on each host of a virtual machine, is used not only as process controller but also message router, which may result in communication bottleneck as all tasks heavily depend on daemon processes.

Legion[7] is an architecture based on a distributed object model, and designed to build system service which provides a virtual machine using shared object, shared name space, fault-tolerance. Legion uses data flow model as par-

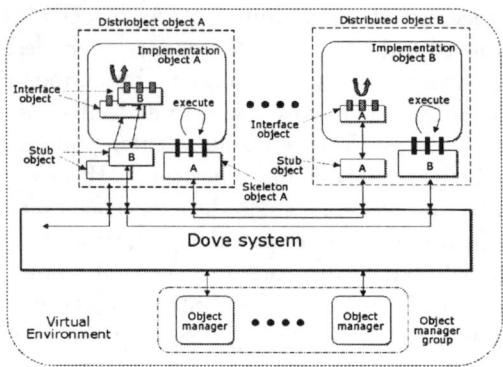

Fig. 1. DOVE object model

allel computation model, and parallelisms are implicitly supported by the underlying runtime system. However, management of data dependency graph for every invocation as well as scheduling nodes of the graph may incur additional computation overhead, and no support for object group may cause communication inefficiency. CORBA[2] is a vendor-independent standard which aims at interoperability and portability of distributed applications. CORBA defines a distributed object model for accessing distributed objects. It includes an Interface Description Language, and a specification for the functionality of run-time systems that enable access to objects. But CORBA is based on client-server model rather than a parallel computing model, and hence it is not adequate to provide a parallel programming environment. It does not support group operations, and has some difficulties in implementing efficient parallelisms by using asynchronous communications.

3 DOVE Object Model

DOVE is based on distributed object model which consists of several distributed objects interacting with each other using method invocation mechanism. Distributed object is divided into *interface object* and *implementation object*. (See figure 1.) The interface object is distributed to applications which are to use the distributed object, and it provides interaction point to its corresponding implementation object. Users can issue a method invocation to the distributed object by invoking the method of its interface object, a local representative of the distributed object. Method invocation to the interface object is converted to the invocation message by *stub object*, and sent to the corresponding implementation object by *DOVE run-time system*. On the opposite side, the receiver's run-time system unmarshals the invocation message, and invokes the appropriate method of the target implementation object through *skeleton object*. A reply message is sent from the implementation object back to the interface object, and returned like a normal function call. This mechanism, which is called *remote method invocation*(RMI), allows transparent access to the object irrespective of whether it resides in local or remote site. In DOVE, distributed object behaves either as client or server to interact with other distributed objects. In other words, it executes the implementation object for incoming RMI requests, while during the execution of the implementation object, it generates a remote invocation to the other distributed object using the interface object as a client. In DOVE, *object manager* exists per host and it provides a set of indispensable services, such as object creation, deletion and naming services, to build a transparent and easy-to-use clustered computing environment. A set of object managers constitutes a single object group which determines the domain of the virtual parallel environment that might encompass a huge number of machines and networks.

3.1 Remote Method Invocation

DOVE provides three kinds of RMI to support various synchronization modes during data exchange between remote objects: *synchronous, deferred synchronous* and *asynchronous* RMIs. In synchronous RMI, sender is blocked until the corresponding reply is arrived. In deferred synchronous call, the sender can do other work immediately without awaiting the reply of the RMI, but later at some point must wait for the reply in order to use the return values. In asynchronous RMI, the sender can proceed without awaiting the reply similarly as in deferred synchronous one, but the corresponding *upcall* method is invoked on the arrival of its reply. These communication types are not the new ones, and often used to acquire high performance in distributed system.

In DOVE, these types of RMI are provided with more ease and intuitive way. Synchronous RMI has no constraint on its return type, and is identical to the general method invocation to local object. With deferred synchronous and asynchronous call, their return type is **void**. A value of type *Waiter* is returned for both of them, and used for synchronization between two threads one of which issues the RMI, and the other returns its reply respectively. In deferred synchronous RMI, sender uses the value of type *Waiter* to wait for the arrival of the corresponding reply. In asynchronous mode, an *upcall* method, specified when a RMI is issued, is invoked using a new thread generated automatically on the arrival of its reply. In this case, the value of type *Waiter* is used for waiting for the end of *upcall* method, not for the arrival of the reply.

3.2 Object Group

In distributed system, group communication pattern is often used, since it provides simple and powerful abstraction. In DOVE, group communication mechanism is supported by introducing a new construct, *object group*, as means of grouping objects and naming them as one unit for RMIs. An interface object can be bound to its corresponding object group, and a RMI issued on the interface object is transparently multicast to each implementation object in the group. Interface object to the object group has the same interfaces as the one to the single object, and provides an interaction point with multiple objects in the object group so that user treats it just like a single object. Therefore, the concept of object group allows users to do more simple programming, and to have chance to get better performance if the underlying communication layer supports multicasting facilities.

An object group can be created at any time by creating its interface object with **GROUP** flag. It is identified by its user-defined name and class name of the interface object. It is implicitly deleted according to their life time policy which is specified when the group is created. Distributed objects may join or leave their groups at any time by issuing *join()* or *leave()* methods on the interface object that is bound to the group. Each object group may consist of objects with different type, but should have one common type from which they are inherited, and only the methods inherited from the common type can be invoked.

3.3 Multiple Method Invocation

A remote method invocation issued to an object group, called *multiple method invocation*(MMI), is transparently multicast to each object in the object group. Since the MMI to the object group may return one reply message per each member object in the group, the user needs to decide what replies to select from them. In DOVE, three types of MMI are supported: *multicast/select, multicast/gather* and *scatter/gather*. *Multicast/select* invocation is returned with only one reply which is obtained by applying operations such as minimum, maximum, and sum to the replies of objects in the group. *Multicast/select* and *multicast/gather* invocations multicast the same request to each member in the object group, while *scatter/gather* invocation multicasts the different requests to each member by storing them in sequences. Each sequence is an array of variable size. Both of *multicast/gather* and *scatter/gather* MMIs are returned with sequences which store each reply from the members in the object group. These MMIs can be also invoked in synchronous, deferred synchronous and asynchronous modes.

4 DOVE Object Manager

In order to provide users with a consistent and uniform view of the distributed clustered system, we need to provide additional services which can supports distribution transparencies. For these purpose, we design a special distributed object which is called *object manager*. Object manager is responsible for object life control such as creation, location and deletion of objects on its local host, naming service for binding object name to its physical address and event propagation to other object managers.

4.1 Object Life Management

Object is created as a process either by a user at console or by an object manager on its local machine. When an interface object is newly created, its local object manager creates its corresponding implementation object on a remote site by cooperating with the other object manager at the remote site. After the implementation object is created, it is connected to the interface object.

An object is created with three different life time: *transient*, *fixed* and *permanent*. A distributed object with transient life time automatically deletes itself after its creator has been disconnected from it. This mechanism efficiently keeps the system resource from being wasted by unused objects without any additional overhead. An independent and long-term object such as an object manager can be created with fixed or permanent life time. The life time of the object group is similar to that of object. A transient group is deleted when it is empty.

4.2 Name Service

An object may have a name given by user, i.e. an alias represented by user-defined string when it is created. It is much easier and more user friendly to use an alias for object instead of its physical address. A name of the implementation object may be specified at the creation of the interface object. Then, the interface object is connected to the implementation object with the name immediately if it already exists; otherwise, after the implementation object with the name is created by object manager. Internally, the object reference is obtained from the name service of object manager, and used to connect to the existing implementation object. Naming service makes remote objects appear to users as the ones in one virtual computer by encapsulating binding operations from users, and hence provides an easy-to-use programming environment. Each object has its local cache to store binding information about its currently accessing objects, and first consult its local cache for binding information. If it fails, it requests the binding information to its local object manager. If the object manager does not contain the binding information, it multicasts the requests for the binding information to the object manager group. This kind of hierarchical naming service where the binding information is distributed in the local cache, local object manager and remote object managers, makes DOVE scalable.

5 DOVE Run-Time System

DOVE system is built as a multilayered architecture to provide the system modularity and extensibility as shown in figure 2. It fully supports various synchronization mechanisms and object group for RMI and MMI as well as heterogeneity through automatic marshalling and unmarshalling. The brief description of each layer is given below.

Application Layer: Application layer contains a user program which interacts with DOVE through stub and skeleton objects. Distributed objects are defined using *Interface Description Language(IDL)*. An IDL compiler is developed to automatically generate codes for stub and skeleton objects, which include codes for marshal and unmarshal methods. User can make use of different interfaces for the same operation to support diverse RMIs and MMIs. The calling semantics of method invocations are determined by their parameters. A method invoked

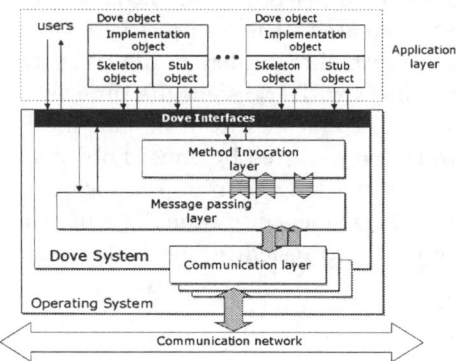

Fig. 2. Multi-layered Architecture of DOVE System

through stub object is marshalled into an invocation structure which is then passed to method invocation layer.

Method Invocation Layer: Method invocation layer has a responsibility to provide the application layer with functionalities of the various types of RMIs in more ease and intuitive way. It marshals each invocation structure into an invocation message, and then passes it to a message passing layer after storing the calling semantic of the RMI into an invocation table to support various synchronization schemes. When the method invocation layer receives a reply message, it unmarshals the message and then performs proper actions such as signaling to the application layer or creating a new thread for execution of the *upcall* method, according to the semantic of the RMI.

Message Passing Layer: Message passing layer carries out object name binding to physical address, and delivers messages to distributed object using one of the underlying communication layers. Each communication layer can be implemented using any distributed protocol which makes use of different interfaces and naming schemes. The main purpose of message passing layer is to encapsulate method invocation layer from communication layer, and to provide the method invocation layer with uniform interfaces for message delivery and object group membership. Also, the message passing layer stores information about group membership, and takes care of MMI to the object group by using multicasting function if the communication layer supports it; otherwise by iterative execution of unicast function in the communication layer.

Communication Layer: In order to achieve the full functionality of DOVE system, the communication layer should be equipped with reliable unicast, reliable multicast and process group management. The minimal requirement for communication layer is reliable unicast likes TCP/IP. Currently, DOVE has installed two communication layers, one for reliable unicast using TCP/IP and another for reliable and ordered group communication using IP multicast.

6 DOVE Implementation

DOVE is implemented in C++ class library with a set of interfaces which are independent on the underlying platform including operating system and communication system. Each layer of DOVE system is implemented as an independent class with its own thread of control, and interacts with other layers through uniform and invariable interfaces in the library. An IDL compiler is also designed and implemented so that by compiling IDL, stub and skeleton objects are automatically generated in C++ codes using the DOVE class library. Currently, DOVE system is installed and tested on Microsoft's Windows NT/98 platform, and various types of Unix machine including Solaris, Irix and Linux.

7 Evaluation

For the performance evaluation purpose of DOVE, we have developed two parallel applications: ray-tracing and genetic simulated annealing. They are implemented both on DOVE and PVM, and the experimental environment consists of 17 heterogeneous machines, two Ultrasparc1, two SGI O2, a SGI Octane, 9 Pentium II PCs running Linux and three Pentium II PCs running Windows NT/98 connected by 100Mbps ethernet. Since each machine has different computing powers, we have measured the relative performance with respect to Linux machine by comparing the execution time of the identical sequential program on each machine. Then, the expected speed up is computed as a sum of each relative performance of the participating machines. The efficiency of the parallel program is defined as the ratio of the actual speed up to the expected speed up.

7.1 Parallel Ray-Tracing

Ray tracing[8] is a simple and powerful technique for rendering realistic images by tracing all the rays from eye through each pixel in the view plane. Since it requires high computation time for the calculation of intersections between rays and objects as well as the tracing of all the reflected and refracted rays, it has been considered as an adequate application for parallel computation. A parallel ray-tracing software is designed in master/slave scheme where one master task manager object interacts with ray tracer objects. Master/slave paradigm suits well in the clustered computing environment, since its dynamic nature of job scheduling can reduce the unbalance of computational power among the heterogeneous machines. Further, the task parallelism in master/slave scheme is very similar to the object parallelism in DOVE. One master task manager object is running on a Ultrasparc1 workstation for job distribution, and each of ray tracer objects on the other machine. Task manager object schedules the assignment of pixels to ray tracer objects which in turn calculate the value of each pixel. The relative performance of the machines obtained by executing the identical sequential ray-tracing program is shown in table 1. Table 2 shows the execution time, speed up and efficiency of ray-tracing according to the number

Table 1. Measurement of relative performance for ray-tracing

machines	M_1	M_2	M_3	M_4	M_5	M_6	M_7
OS (spec.)	Linux (PII-300)	Win98 (PII-300)	Win98 (PII-366)	WinNT (PII-300)	Solaris2.5 (USparc1)	IRIX6.3 (O_2)	IRIX6.5 (Octane)
running time	427.475	312.460	253.670	307.602	858.340	781.636	466.224
relative perf.	1.0	1.368	1.685	1.390	0.498	0.547	0.917

Table 2. Performance results of parallel ray-tracing (Parenthesis in number of ray tracers represents a machine used for running an additional ray tracer object.)

number of ray tracers		$1(M_1)$	$2(M_1)$	$4(M_{3,5})$	$8(M_{1,1,4,6})$	$12(M_{1,2,6,7})$	$16(M_{1,1,1,1})$
expected speedup		1.0	2.0	4.183	8.120	11.952	15.952
DOVE	time(sec)	427.475	217.300	104.298	54.935	37.790	29.630
	speedup	1.0	1.967	4.099	7.781	11.312	14.931
	efficiency(%)	100.0	98.36	97.98	95.83	94.64	93.6
PVM	time(sec)	427.475	227.622	107.803	56.217	38.566	29.336
	speedup	1.0	1.878	3.965	7.604	11.084	14.572
	efficiency(%)	100.0	93.900	94.796	93.646	92.740	91.347

of ray tracers. As the number of ray tracer increases, DOVE shows an almost linear speedup and better performance than PVM. This results from the fact that in DOVE, computation and communication between objects can be efficiently overlapped using asynchronous RMI, and that DOVE executes RMIs directly to the object, while in PVM all messages from different processes are routed through daemon process which runs in each host, producing overhead in daemon process.

7.2 Parallel Genetic Simulated Annealing

GSA(Genetic Simulated Annealing)[9] is a hybrid method which exploits the genetic algorithm and the local selection strategy of simulated annealing. This approach eliminates the processing bottleneck of global selection by making the selection decisions locally.

A parallel implementation of GSA for TSP(Traveling Salesman Problem) is built both on DOVE and PVM. TSP is to find a path touring all of the cities, visiting each exactly once, and returning to the originating city, such that the total distance traveled is minimized. For the experiment of parallel GSA solver, 52 cities with optimal path of 7544.37 distance is used, total 240 solutions are maintained, and 5000 iterations are made. Initially, the solutions are evenly and statically distributed over GSA solver objects. At every iteration step of GSA, each GSA solver object takes a solution from other object, and executes *crossover* and *mutate* operations on the two solutions, one from itself and the other from the different object, to generate new solutions. Among the solutions, one solution is selected using the simulated annealing algorithm at each of GSA solvers. Table 3 shows the experimental performance of parallel GSA solver built both on DOVE and PVM. Since each of parallel GSA solvers exchanges its current solution at every iteration, both of DOVE and PVM show performance degradation as the number of machines increases. However, DOVE shows better speed up than PVM as the number of machines increases.

Table 3. Performance results of parallel GSA.

number of machines		1	2	4	8
expected speedup		1.0	2.0	4.0	8.0
DOVE	time (sec)	1318.050	679.740	364.763	207.497
	speedup	1.0	1.939	3.613	6.352
	efficiency (%)	100.0	96.95	90.34	79.40
PVM	time (sec)	1318.050	687.918	386.072	264.043
	speedup	1.0	1.916	3.414	5.357
	efficiency (%)	100.0	95.80	85.35	66.96

8 Conclusion

In this paper, we have presented a new object-oriented parallel computing environment called DOVE which is based on a distributed object model, and designed to provide a easy-to-use programming environment for clustered or networked parallel computer. Efficient parallelism is supported by diverse RMI, MMI for object group and multi-layered architecture of DOVE system. An easy-to-use transparent programming environment is provided by stub and skeleton objects generated by DOVE IDL compiler, and object management and naming service offered by object manager.

Currently, DOVE is implemented in C++ class library, and user can develop a parallel or distributed applications on a heterogeneous environment using the class libraries as if he resides in one virtual computer. With DOVE, user can easily build a parallel program as a collection of distributed objects over networks. Thus, the code of parallel program has little difference with the sequential one. For the experiment of DOVE, two parallel softwares for ray tracing and genetic simulated annealing have been developed for both of DOVE and PVM, and it is shown that DOVE has better performance than PVM.

References

1. R. Buyya, High Performance Cluster Computing: Systems and Architecture, volumn 1, Prentice Hall PTR, New Jersey, 1999.
2. Object Management Group Inc. , The Common Object Request Broker: Architecture and Specification, OMG Document Revision 2.2, February 1998.
3. T. B. Downing, Java RMI : Remote Method Invocation, IDG Books worldwide, 1998.
4. E. Frank and III. Redmond, DCOM : Microsoft Distributed Component Object Model, IDG Books worldwides, 1997.
5. MPI Forum, MPI: A Message-Passing Interface Standard, International Journal of Supercomputer Application 8, No. 3, 1994.
6. A. Geist, A. Beguelin and et. al., PVM 3 User's guide and Reference manual, ORNL/TM-12187, September 1994.
7. M. Lewis and A. Grimshaw, The Core Legion Object Model, University of Virginia Computer Science Technical Report CS-95-35, August 1995.
8. T. Whitted, An Improved Illumination Model for Shaded Display, Communication of ACM 23, No. 6, June 1980.
9. H. Chen, N. S. Flann, and D. W Watson, Parallel Genetic Simulated Annealing: A Massively Parallel SIMD Algorithm, IEEE Transactions on Parallel and Distributed Systems 9, No. 2, February 1998.

The Problems and the Solutions of the Metacomputing Experiment in SC99

Stephen Pickles[1], Fumie Costen[1], John Brooke[1], Edgar Gabriel[2],
Matthias Müller[2], Michael Resch[2], and Stephen Ord[3]

[1] Manchester Computing, The University of Manchester
Oxford Road, Manchester, M13 9PL, United Kingdom
[2] High Performance Computing Center Stuttgart
Allmandring 30, D-70550 Stuttgart, Germany
[3] N.R.A.L., Jodrell Bank
Macclesfield, Cheshire, SK11 9DL, United Kingdom

Abstract. An intercontinental network of supercomputers spanning more than 10000 miles and running challenging scientific applications was realized at the Supercomputing'99 (SC'99) conference in Portland, Oregon, USA using PACX-MPI and ATM PVCs. In this paper, we describe how we constructed the heterogeneous cluster of supercomputers, the problems we confronted in terms of multi-architecture and the way several applications handled the specific requirements of a metacomputer.

1 Overview of the SC99 Global Network

During Supercomputing'99 a network connection was set up that connected systems in Europe, the US and Japan. For the experiments described in this paper 4 supercomputers were linked together.

	latency	bandwidth
HLRS-CSAR	20ms	0.8 MBit/s
HLRS-PSC	78ms	1.6 MBit/s
CSAR-PSC	58ms	0.5 MBit/s

Table 1. Bandwidth and latency of the network connections between the different sites.

- A Hitachi SR8000 at Tsukuba/Japan with 512 processors (64 nodes)
- A Cray T3E-900/512 at Pittsburgh/USA with 512 processors
- A Cray T3E-1200/576 at Manchester/UK with 576 processors
- A Cray T3E-900/512 at Stuttgart/Germany with 512 processors

Together this virtual supercomputer had a peak performance of roughly 2.1 Teraflops. The network was based on

The Problems and the Solutions of the Metacomputing Experiment in SC99   23

- National high speed networks in each partner country
- The European high speed research network TEN-155
- A transatlantic ATM connection between the German Research Network (DFN) and the US research network Abilene
- A transpacific ATM research network (TransPAC) that was connecting Japanese research networks to STAR-TAP at Chicago

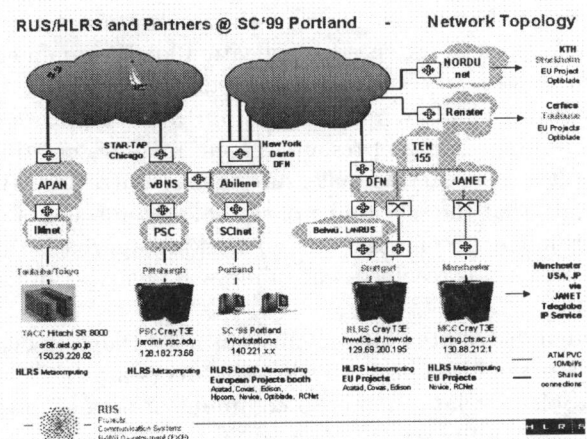

Fig. 1. Network configuration as set up during Superomputing'99.

Details of the network configuration can be seen in figure 1. Using ATM, PVCs were set up with a bandwidth of 10 MBit/s in the European networks. On all other connections bandwidth was shared with other users. The sustained bandwidth as well as the latency is described in table 1.

2   Problems of Multi-architecture

Heterogeneous metacomputing introduces some problems that are similar to those well known from cluster computing and some that are very specific [2]. The most important ones are

- Different data representation on each system
- Different processor speed of each system
- Different communication speed for internal messages of each system
- Different communication speed of messages internal to a system and messages between systems
- Lack of a common file system
- Lack of resource management

The problem of load imbalance due to different processor speed has to be tackled on the application level. So far there is no standard way for an application to determine the speed of a processor. And especially with respect to a given application it is difficult to get information about the sustained performance. An approach that is currently investigated is to start the application with a standard distribution. The application would then collect information about the amount of load imbalance. Based on these data and on the characteristics of the network connection the application would then decide how much effort should be spent to redistribute the load.

Such an approach, however, is based on the assumption that the basic parameters (processor load, network speed, application load) remain constant. Allowing these parameters to vary–such as is necessary if working in a non-dedicated environment or using adaptive methods–will further complicate the problem.

The problem of scheduling of resources is an issue of scientific research. For metacomputing it is necessary to make sure that resources are available concurrently. That means that the scheduling of several systems and of networks would have to be coordinated. So far there is no working approach that would allow this. Until this problem is resolved scheduling will be a task of the user and will not be automatic.

The problem of distributed file systems was not tackled in our approach. We assumed that data would be in place whenever necessary.

All other problems relate mainly to communication and have to be tackled by the message passing library that is used. To specially focus on the problems of metacomputing HLRS has developed an MPI library called PACX-MPI [1, 3] that allows to couple big systems into a single resource from the communication point of view. Based on the experience of several other projects the library relies on three main concepts:

- Single MPI_COMM_WORLD for an application: When starting an application with PACX-MPI a local MPI_COMM_WORLD is created on each system in the metacomputer. These are combined by PACX_MPI to form a single MPI_COMM_WORLD.
- Usage of communication daemons: On each system in the metacomputer two daemons take care of communication between systems. This allows bundling of communication and to avoid having thousands of open connections between processes. In addition it allows handling of security issues centrally.
- Splitting of internal and external communication: For all internal communication PACX-MPI makes use of the native library of the system. It only implements communication for external messages–i.e. messages that have to be sent between systems. This has the advantage that optimized communication inside a system can be combined with portable communication protocols between systems.

In figure 2 global numbering denotes the global MPI_COMM_WORLD while local numbering corresponds to the local communicators. Processes 0 and 1 on each system are acting as daemons for external communication.

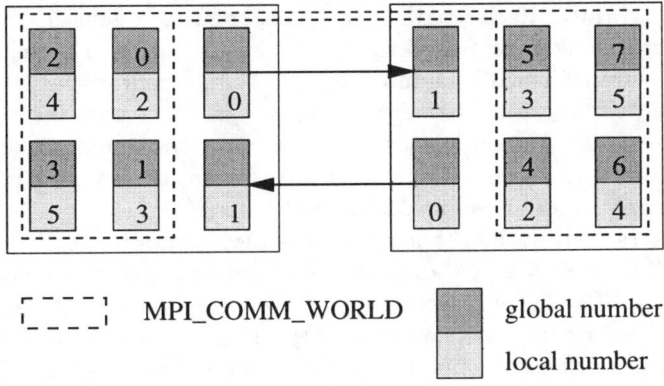

Fig. 2. Basic concept of PACX-MPI. Two systems integrated into one single MPI_COMM_WORLD.

3 Use of Experimental Data

In this section we describe a novel use of metacomputing, namely the processing of data from an experimental facility (in this case the Jodrell Bank radio telescope). This presents particular and severe challenges to our metacomputer since the distribution of the data around the metacomputer makes severe demands on the intercontinental bandwidth. This experiment involved coupling of the three T3E machines only.

3.1 The Jodrell Bank De-dispersion Pulsar Search Code[6]

The application involved the search for pulsars in the radio signal from a globular cluster. The observed radio signal from pulsars manifests itself at the telescope as a periodic increase in broadband radio noise. It order to observe the pulse with a high signal to noise ratio we need to observe across a wide band of radio frequencies. As the space between the earth and the pulsar (the interstellar medium) is slightly charged it is dispersive and therefore different frequencies propagate at different velocities. To reconstruct the pulsar signal it is necessary to first determine the relationship between a given radio frequency and the time lag in the arrival of the pulse (called the dispersion measure DM) and then use this to reconstruct the signal. Since a globular cluster is a compact object, the DM will be approximately constant for all pulsars in the cluster and this is an effective method of searching.

3.2 Implementation on the Metacomputer

Data Handling and Transmission. The total data from a continuous sequence of measurements from the telescope is first broken up into lengths which have to be of length at least twice as long as the dispersive delay time across the observing bandwidth. Each length is then subjected to a Fourier transform and multiplied by the Fourier transform of the inverse filter. The resultant data length is then transformed back into the time domain and processed to produce a de-dispersed time series. The next length is then processed, and the resultant series are then concatenated together. In practical terms the input data-set only has to be forward transformed once, but has to transformed back into the time domain as many times as we have trial DMs.

This process defines the challenge for the metacomputer. If the data is read onto one T3E (in this case at Manchester) then subsets of the data have to be sent to the remote T3Es via a connection that is two orders of magnitude slower than the inter-processor connection within a T3E. This might seem to invalidate the whole metacomputing concept, however the requirement that many trial DMs be applied to each chunk of data means that we can peform sufficient processing on the remote machines to cover the cost of data transmission. Moreover since our ultimate goal is real-time processing of experimental data (i.e. processing directly after the observations via network links to a large computing facility) we would find that the remote host has far more processing power than the machines onto which the data is initially read. In this context it is important that we operate with subsets of the observations, so that the FFT processing can be applied to such chunks independently and the remote machines can begin processing as soon as they receive a chunk of data that is suffiently large to satisfy the constraint imposed by the maximum dispersive delay time.

In the SC99 experiment the data was preloaded onto the Manchester T3E. PACX-MPI does not have automatic startup on Cray-T3E systems since only one user process is permitted per Application PE, thus spawning processes is not possible. The three T3Es were therefore started manually together, however the data was not preloaded on the remote T3Es since the transmission of data is essential to this application.

Work Distribution Model. The original intention was to adopt a master-slave model in which one (or more) master processes read the disk and distribute work to slave processes. This approach is highly attractive because it naturally furnishes an effective method of load-balancing. The main drawback is that, in the absence of complicated and time-consuming preventative steps, no one processor would get two contiguous chunks of the global timeseries, thus the occurence of gaps between work units would be maximized, a fact which would detract from the integrity of the results. Instead we adopted a client-server model. A handful of server nodes poll for requests for data from clients, and the bulk of intelligence resides with the clients who do the actual work. By allocating a single, contiguous portion of the global data set to each worker, gaps occur only a processor boundary, just as in the original code. By letting each processor keep track of where it is up to in the global data set, the impact on the logic and structure of the code is minimized. The disadvantage of the client-server model adopted here is that we are compelled to tackle the load-balancing problem statically, taking into account *estimates* of the MPI bandwidth, processor speed, disk performance and so on. The algorithm used is described below in general terms to permit the analysis to be extended to metacomputing clusters of differing numbers and types of host machines.

Static Load Balancing Algorithm. Let N be the number of hosts, and let n_i be the number of client processors to use on host i, discounting the extra 2 processors required by PACX-MPI and those processors on host 1 which have been assigned to server duties. Denote the bandwidth, in megabytes/sec, from host 1 (where the input data resides) to host i by bw_i. The rate, in megabytes/sec at which data can be read from disk on host 1 is denoted by r; it is assumed that this rate is approximately independent of the number of processors accessing the disk at any one time.

The size of one record is denoted by w. The computational time required to process one record of data on host 1 is determined experimentally and denoted by t_τ. The time to process the same quantity of data on other hosts is estimated by multiplying t_τ by the ratio perf (1) : perf (i) where perf (i) is the peak speed in Mflops of the processors on host i.[1] This approximation is justified in a homogeneous metacomputer, but is likely to overestimate (underestimate) slightly the compute time on processors slower (faster) than those of host 1.

The amount of processing per record can be determined by the parameter n_{slopes} (each dispersion measure gives a slope on a time-lag versus frequency plot). t_τ is now to be re-interpreted as the average compute time per record per unit slope in the regime where n_{slopes} is large enough that the compute time per megabyte can be well approximated by $t_\tau \times n_{\text{slopes}}$.[2]

Any job will process a total of v records. The load balancing problem is to determine the proportion $v_1 : \ldots : v_n$, $\sum_{i=1}^{N} v_i = v$ in which to distribute data to the hosts.

The elapsed wall-clock time t_i to process v_i records on host i is estimated by: $t_i = v_i t_{\text{proc}}(i)/n_i + n_i t_{\text{wait}}(i)$ where $t_{\text{wait}}(i) = (w/\text{bw}_i + w/r)$ is the time that a client processor has to wait for a single work unit, and $t_{\text{proc}}(i) = n_{\text{slopes}} t_\tau \text{perf}(1)/\text{perf}(i)$ is the time that it takes the client to process it. The first term in the expression for t_i is the time required computation. The second term is essentially twice the start up time, i.e. the time elapsed until all processors on the host have received their first work unit.

This term will be dominated by communications bandwidth on remote hosts and by disk access speed on the local host. A similar cost is paid at the end (*rundown time*) as some processors will finish their allocated work before others.

The condition used to balance the work load is that all hosts finish at the same time. $t_1 = t_2 = \ldots = t_N$ Using these equations leads to a linear system with $N+1$ equations and $N+1$ unknowns $(v_1, \ldots, v_n, t)$.

---

[1] This assumes that all the processors on any given host are clocked at the same speed, as was the case in the SC99 demonstrations.

[2] The forward FFT's are computed only once, but the inverse FFT's must be computed for each slope.

$$\begin{pmatrix} a_1 & 0 & \cdots & 0 & -1 \\ 0 & a_2 & \cdots & 0 & -1 \\ \vdots & \vdots & \ddots & \vdots & \vdots \\ 0 & 0 & \cdots & a_N & -1 \\ 1 & 1 & \cdots & 1 & 0 \end{pmatrix} \begin{pmatrix} v_1 \\ v_2 \\ \vdots \\ v_N \\ t \end{pmatrix} = \begin{pmatrix} b_1 \\ b_2 \\ \vdots \\ b_N \\ v \end{pmatrix}$$

Here $a_i = (t_\tau \mathrm{perf}(1))/(n_i \mathrm{perf}(i))$ and $b_i = n_i w (w/\mathrm{bw}_i + w/r)$ The validity of this method depends on the implicit assumption that no client processor experiences dead time waiting for other clients to receive their data. A global condition which expresses this is the inequality $t_i > v_i (w/\mathrm{bw}_i + w/r)$. More carefully, we may define $t_{\mathrm{else}}(i) = (n_i - 1) t_{\mathrm{wait}}(i)$ as the time that it takes all the other processors on host i to receive their work units. Then the dead time $t_{\mathrm{dead}}(i)$ that is lost between work units is expressed like this: $t_{\mathrm{dead}}(i) = t_{\mathrm{else}}(i) - t_{\mathrm{proc}}(i)$ for $t_{\mathrm{else}}(i) > t_{\mathrm{proc}}(i)$ or else $t_{\mathrm{dead}}(i) = 0$. Fig. 3 shows the time relationship of time to request/transfer/process data and data flow. The processors have dead time to wait for the next record in case 1 because n_{slopes} is too small. In case 2, n_{slopes} is sufficiently large so that processors spend sufficiently long on one record of data so that the next record can be supplied with no dead time. A

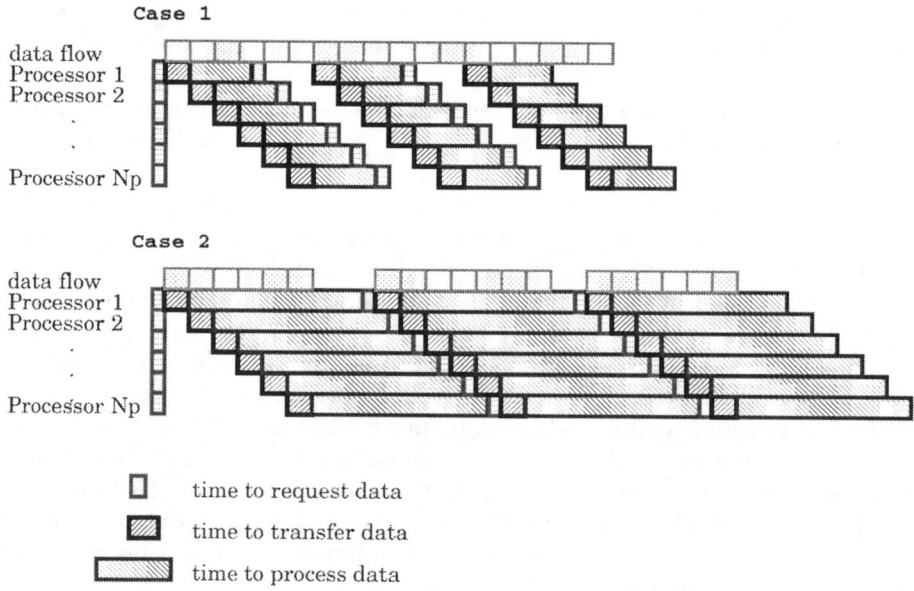

Fig. 3. Data flow. In case 1, processors have dead time to wait for the next processing. In case 2, processors carry out the successive work quickly without dead time.

drawback of this approach, when coupled with the FIFO processing of requests by servers, is that in practice the run down time is usually longer than the start up time. Typically there will be some clients on a given host with one more work unit than other clients on the same host, but there is no guarantee that the more heavily loaded clients will be served before the others. Consequently, some clients may start their last work unit before other clients have receive their second last one. These end effects can be expected to be more pronounced when the number of records per client is small, or equivalently, when the start up time is comparable to the total elapsed time. Indeed, this was confirmed in the metacomputing experiments at SC99; the estimates of elapsed time in longer runs with 10 or so records per client were entirely acceptable.

4 Computational Fluid Dynamics

Another application used during sc99 was a CFD-code called URANUS (Upwind Algorithm for Nonequilibrium Flows of the University of Stuttgart) [4]. This program has been developed to simulate the reentry phase of a space vehicle in a wide altitude velocity range. The reason why URANUS was tested in such a metacomputing environment is, that the nonequilibrium part has been finished and will be parallelized soon. For this version of URANUS, the memory requirements for a large configuration exceeds the total memory which can be provided by a single machine today.

URANUS consists mainly of three parts. The first part (preprocessing) reads the input data and distributes all information to the application nodes. Since the code handles different blocks in the mesh, we could distribute in an ideal case each block on a machine. Thus, the preprocessing step remained local on each host [3]. The second part is the processing part. For the single block code, different strategies were developed to hide latency [4], but these methods were not yet adopted to the multiblock code. Finally, the third part is the postprocessing, including some calculations and the writing of the results to disk. For this part, the same techniques could be used like in the preprocessing part.

During sc99 we were able to simulate the Crew-Rescue Vehicle (X-38) of the new international space station using a four block mesh with a total of 3.6 Million cells. For this, we used 1536 nodes on the Cray T3E's in Manchester, Stuttgart and Pittsburgh.

5 Molecular Dynamics

A further application that was adapted for metacomputing is a molecular dynamic program for short range interaction[5] simulating granular matter. The parallel paradigm applied here is domain decomposition and message passing with MPI (message passing interface). Therefore every CPU is responsible for a part of the domain and has to exchange informations about the border of its domain with its adjacent neighbors. Static load balancing can be implemented by assigning domains of different size to each processor. To reduce the amount

---

[3] This did not work for large simulations, since we had to distribute four blocks on three Cray T3E

of load imbalance between the T3Es and the SR8000, the nodes of the SR8000 were used as eight processors.

Latency Hiding. As described in section 1 the latency of a metacomputer is in the range of milliseconds and thus several orders of magnitudes larger than in todays parallel computers. For a tightly coupled algorithm like a MD simulation latency hiding has to be implemented to get reasonable performance. To achieve this it was decided to group the communication in two sections. First the communication for the first dimension is initiated by calls to MPI_Isend. In a second stage this communication is completed and the calls for all remaining dimensions are performed, allowing a partial overlap of communication with computation. In this application the force calculation for particles that interact only with particles of the core domain is performed between the first and the second stage. If the calculation for one particle takes about $100\mu s$ one needs around 750 particles to hide the latency. As soon as latency hiding is implemented this is no major restriction.

Bandwidth. As a first approach latency hiding also helps to hide the increased communication duration due to a small bandwidth. Increasing the number of particles does however not only increase computing time but also the amount of data to be sent. For a latency of 74ms and a bandwidth of 5MBit/s in both directions even with particles consisting of only 24 bytes data (e.g. three doubles representing a three dimensional position) there is a break even between latency and communication duration above 15000 particles, a value easily reached in real applications. In the problem treated here, a particle consisted of 10 floating point numbers: three vectors position, velocity and the force acting on the particle plus one scalar variable for the particles radius. The force however is local information, because it is calculated on every processor. It is therefore not necessary to transmit this information. In addition, the force is a function of the position and size of the particles. Thus, the velocity only needs to be transmitted whenever a particles crosses the domain between two processors, and not if the information about the particle is only needed to calculate the interaction. With this considerations the amount of required data per particle can be reduced from ten to about four floating point numbers. In addition to the compression techniques offered by PACX-MPI it reduces the bandwidth requirement significantly and is a key point for efficient metacomputing.

6 Conclusion and Future Plan

We have described the realization of a global metacomputer which comprised supercomputers in Japan, the USA, Germany and the UK. The network spanning 10000 miles and several research networks was optimized by using either dedicated PVCs or improved routes. During sc99 the usability of this metacomputer has been demonstrated by several demanding applications. This takes the concept of a global metacomputer forward in two major ways. Firstly we have shown that it is possible to link together different types of machine, in this case Cray-T3E and Hitachi SR8000. Secondly we have shown that even data-intensive applications, such as processing output from experimental facilities, can be successfully undertaken on such a metacomputer.

In the future we hope to extend the cooperation between the participating sites to incorporate the experiences gained into the future development of the PACX-MPI library and the implementation in the application codes. We also intend to test other libraries designed for metacomputing such as Stampi[7], which is an implementation of MPI-2 for heterogeneous clusters. Furthermore, we are planning to extend the Jodrell Bank De-dispersion Pulsar Search Code so that we can process experimental data as the experiment is in progress rather than transfering to storage media for post-processing. This could be vital; if the experiment needs recalibration or intervention, this can be detected immediately rather than wasting precious time on a valuable resource. Also if results can be interpreted as they are being produced the experimental regime can be adapted accordingly, again maximising the usage made of the facility.

References

1. Edgar Gabriel, Michael Resch, Thomas Beisel, Rainer Keller, 'Distributed Computing in a heterogenous computing environment' in Vassil Alexandrov, Jack Dongarra (Eds.) 'Recent Advances in Parallel Virtual Machine and Message Passing Interface', Lecture Notes in Computer Science, Springer, 1998, pp 180-188.
2. Michael Resch, Dirk Rantzau and Robert Stoy 'Metacomputing Experience in a Transatlantic Wide Area Application Testbed.', Future Generation Computer Systems (15)5-6 (1999) pp. 807-816.
3. Matthias A. Brune, Graham E. Fagg, Michael Resch, 'Message-Passing Environments for Metacomputing', Future Generation Computer Systems (15)5-6 (1999) pp. 699-712.
4. Thomas Bönisch, Roland Rühle, 'Adapting a CFD Code for Metacomputing' pp. 119-125 in Parallel Computational Fluid Dynamics, Development and Applications of Parallel Technology, editors C.A. Lin et al., Elsevier 1999
5. M. Müller,' Molecular Dynamics with C++. An object oriented approach', Workshop on Parallel Object-Oriented Scientific Computing , Lissabon/Portugal, 1999
6. S. Ord, I. Stairs, and F. Camilo, 'Coherent De-dispersion Observations at Jodrell Bank' in Proceedings of I.A.U. colloquium 177: 'Pulsar astronomy 2000 and beyond' - editors M. Kramer, N Wex, and R. Wielebinski. - in press
7. H. Koide, et al, 'MPI based communication library for a heterogeneous parallel computer cluster, Stampi', ,Japan Atomic Energy Research Institute, http://ssp.koma.jaeri.go.jp/en/stampi.html

The authors would like to thank Tsukuba Advanced Computing Center, Pittsburgh Supercomputing Center, Manchester Computing Center, High Performance Computing Center Stuttgart, BelWü, DFN, JANET, TEN-155, Abilene, vBNS, STAR-TAP, TransPAC, APAN and IMnet for their support in these experiments.

Grid Computing on the Web Using the Globus Toolkit

Giovanni Aloisio[1], Massimo Cafaro[1], Paolo Falabella[1],
Carl Kesselman[2], and Roy Williams[3]

[1] Dept. of Innovation Engineering
University of Lecce, Italy
{giovanni.aloisio,massimo.cafaro,paolo.falabella}@unile.it
[2] Information Sciences Institute
University of Southern California
carl@isi.edu
[3] Center for Advanced Computing Research
California Institute of Technology
roy@caltech.edu

Abstract. In this paper we present and discuss an architecture that allows transparent access to remote supercomputing facilities from a web gateway. The implementation exploits the Globus toolkit and provides users with fast, secure and reliable access to parallel applications. We show the usefulness of our approach in the context of Digital Puglia, an active digital library of remote sensing digital data.

1 Introduction

This paper describes a design pattern that enables a user to run applications on a Computational Grid [1] using a standard Java-enabled browser. We define a Grid application as consisting of one or more programs, located on geographically distributed machines; these programs access data stored on distributed databases. Among the most important requirements is the capability to find the right programs and to run them with the right data by means of high level requests. The user should be also allowed to check the status of the remote execution, get the output of the application and, in case of an interactive application, steer it from the browser.

In order to integrate Grid applications and the Web, we advise the use of a three tier architecture: (1) the client web browser, (2) middleware running on a secure webserver, (3) the supercomputers of the Computational Grid.

The issues related to Computational Grids and metacomputing have been abundantly described in literature and, in our context, we address essentially the same problems: how do we locate resources, how do we secure these resources and the transactions involving them, how do we start executing an application on a remote machine and so forth. Instruments to help solving these problems have been developed in recent years [2]; among them we chose to use the Globus toolkit which by this time has been released in its version 1.1. The rest of the paper is organized as follows:

section 2 presents the issues related to Grid applications; we discuss them in sections 3 – 7 describing the solutions provided by our framework. Section 8 shows how we exploit the framework in the context of Digital Puglia, an active digital library of remote sensing data and section 9 concludes this work.

2   Web – Grid Integration

In this section we discuss the issues arising from the integration of Computational Grid applications with the Web; in particular, here we address the following:
1. Security: The Globus tools use X.509v3 certificates to mutually authenticate a user and the Grid machines. We propose a two-step mechanism to enable user – Grid authentication from a web browser: a login and password from the browser to the web-server and a certificate from there to the Grid.
2. Resource management: An application running on the Grid could exploit distributed, heterogeneous resources. If we want our application to find the resources dynamically we must store somewhere the information needed to locate the machines, files, applications and so on. This is exactly the aim of the Grid Information Service, an LDAP [3] Directory Service which complements the Globus Toolkit. Applications may need a broker to translate high level requests made by the users in low level system requests.
3. Remote start and status: Commands are provided in the Globus toolkit to run interactive applications or to submit batch jobs on a remote machine taking care of the authentication issues and to check the status of the execution. These commands can be included in CGI scripts or Java servlets to be run on a web server hosted on the second tier of our proposed architecture.
4. Steering: A Java applet can be used to provide a user friendly interface to the Grid application.
5. Result retrieval: The final output produced by the application is held on a remote machine or on a collection of remote machines. The choice of an appropriate strategy is needed to let the user have the results on her web browser in the minimum time.

In order to implement the approach proposed in this paper, the Grid has to be set up in the following way:
- The first tier needs just a Java-enabled browser: this is the basic functionality that all workstations, laptops, and operating systems can provide;
- Globus has to be installed on the machines on the second tier and on the third tier;
- A web server must be installed on the machines on the second tier: we strongly suggest it to be a *secure* web server.

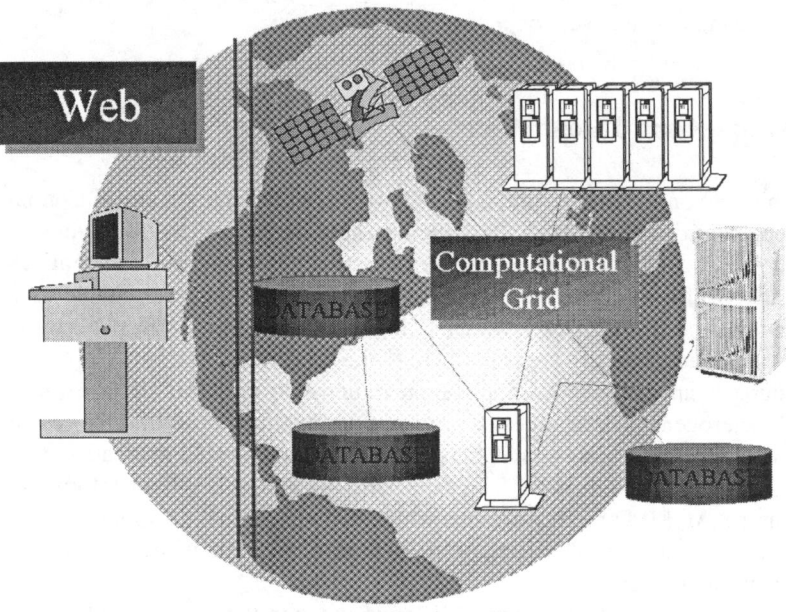

Fig. 1. Accessing the power of a Computational Grid via Web

3 Security

A Computational Grid consists of a set of valuable resources like parallel supercomputers, workstations, databases and smart instruments, so in order to prevent unauthorized access to the Grid we need a strong authentication mechanism.

We propose a two step mechanism: the user must authenticate herself from the client browser to the second tier and also from the second tier to the Grid. A list of trusted users is held on the second tier, and authentication via web is done through a login and password scheme. To guarantee the safety of the password we strongly suggest the use of a secure web server to avoid transmitting the password in clear over an insecure channel.

The second step of the authentication may depend on different security policies. We can decide to map all the users on the same Globus certificate (a "guest" account) or have each trusted user store her individual certificate and private key on the second tier. If we use just one certificate and private key for all the users, we shall not need to create an account for all the users on every machine of the Grid and to store every user's certificate and private key on the second tier. This approach is much easier to manage and it scales better with an increasing number of machines on the third tier.

On the other hand, we might need to use the more complex one to one mapping if the application needs to know which user is making a certain request: for instance in case of personalized options, of different privileges, or to capture accurate accounting information from the supercomputing resources.

Before the machine on the second tier can run any Globus command on the Grid, a Globus proxy must be created with the "grid-proxy-init" command. If the mapping is one to one, the password the user has to enter to access the system must be the PEM pass phrase that protects her private key.

4 Resource Management

Applications running in a Grid environment should be resource aware and adaptive, i.e., capable of handling heterogeneity, and able to adjust their behavior dynamically in response to changes in the Grid. When the user formulates a high-level query the application should be capable of runtime discovery of computing resources (that can be geographically distributed), resource reservation through allocation and/or co-allocation and remote execution on the available pool of machines; one or more executables may be involved in the computation, and a dataset collected from a distributed database. To achieve this goal there should be:

1. A way to store and retrieve information about machines, executables and data; the Grid Information Service, formerly known as Meta Directory Service, is an LDAP based directory service used to provide uniform access to structure and state information about entities composing the Grid.
2. One or more *Resource Brokers* in charge of processing high-level requests formulated by the users and translating them in low level requests understandable to the Grid, if necessary exploiting the information available in the Grid Information Service.
3. Grid software infrastructure tools like the Globus toolkit middleware that provides a bag of core services, including a low level scheduler API, the Grid Information Service, multimethod communication and QoS management, single sign-on and key management, remote file access and Grid status monitoring.

5 Remote Start and Status of the Execution

Once the machine on the second tier has located the executables, the Globus commands or API can be used to run it remotely. A Java servlet or CGI script can be easily written to incorporate the *globusrun* (or *globus-job-submit*) command.

The essential parts of this CGI script are shown below (PERL language is used in this example):

```
# environment variables needed by Globus
$ENV{HOME} = "/users/home/jdoe";

$ENV{GLOBUS_INSTALL_PATH} = "/usr/local/globus";
# creation of a grid proxy: if we are using a one to
# one mapping, certname and keyname will point to the
# certificate and private key we are using for
# everybody. The pwstdin option allows us to read the
# PEM pass phrase from the stdin and thus to use
# redirection
system(grid-proxy-init -pwstdin -cert $certname -key \
$keyname < $PEM_file);
# execution of the remote program
system(globus-job-submit $remote_machine $executable);
```

The status of execution can be checked with the utility *globus-job-status*, using the job-id returned by *globus-job-submit* and which uniquely identifies the submitted job.

The application may require command-line arguments and/or input files. An applet or web form can assist the users during the phase of input generation providing a friendly GUI that can be used to derive visually input parameters; a CGI or a servlet on the second tier will then process the request generating command line arguments, files or both.

The *globus-job-submit* and *globusrun* commands both allow passing command-line arguments to the application, so it is quite straightforward to add them to the sample code snippet presented above.

The *globus-rcp* command mimics on the Grid the functionalities provided by the standard *cp* and *rcp* Unix commands and can thus be used to transfer input files on the machines belonging to the third tier before starting the remote execution.

6 Steering

Not all of the Grid applications will be batch jobs that can be submitted to a pool of Grid machines, so that steering of an interactive application is an important concern here.

Let us assume that we have to develop an interactive application. We advise writing a multithreaded application in which one of the threads will be responsible for handling the flow of control instructions via sockets, and the use of Java applets to provide the users with a friendly GUI that can be used to steer the application at runtime.

The interaction involves bi-directional network communication over sockets, one channel is used to transmit steering instructions, the other one to report diagnostic output about commands executed.

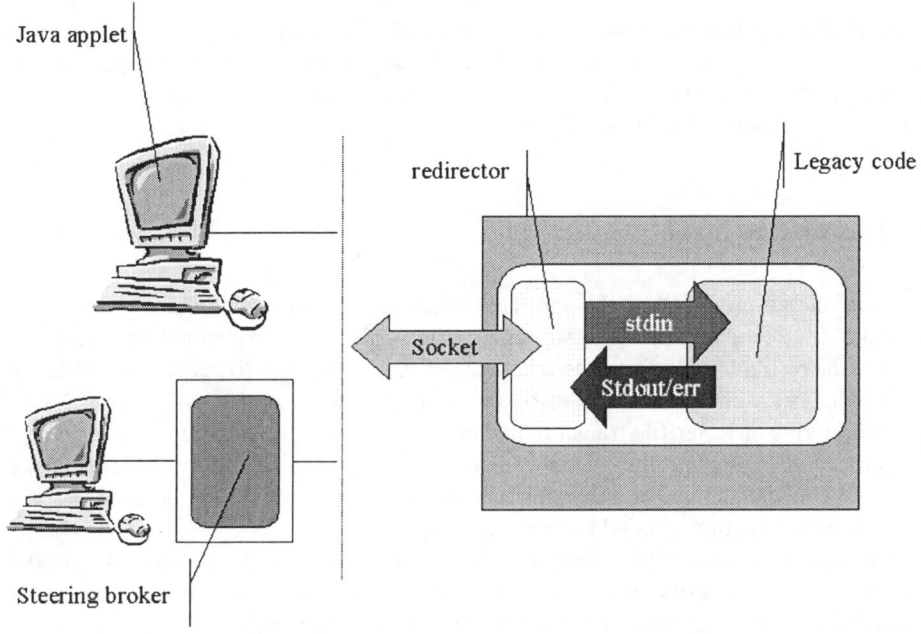

Fig. 2. The use of a *redirector* allows the steering of interactive application without having to change the code. The user interface is in an applet. We might have to use a *steering broker* to connect the applet to the application if there is no possibility to install a web server on the back-end machine.

The next steps in the design depend on the possibility to install a web server on the machines belonging to the third tier. This in turn depends on the local policies in use at the different sites composing the Grid. As an example, it may not be feasible to install a web server on a supercomputer because it can compromise the security of the system. Anyway, if system managers allow it, the easiest thing to do is to have the applet come directly form the third tier to the browser.

Otherwise, the applet must be installed on the middle tier web server. Since unsigned applets can't communicate over a network connection with arbitrary machines, we can address the problem either exploiting signed applets or by means of a *steering broker* whose aim is to dispatch to Grid machines the commands forwarded to it from the applet and to the applet the diagnostic stream output generated by the remote machines. The process of applet signing is not straightforward and differs considerably if the applet is to be signed for use with different client browsers, so we feel confident that the steering broker approach is to be preferred.

Communication between the steering broker and the machines on the Grid may use XTI, TLI or the Nexus communication library instead of sockets while communication with the applet is restricted to sockets.

Now, let us suppose that we have at our disposal an interactive legacy application we want to make available on the Web using the steering mechanism described above.

In addition to the steering broker we need to add a *redirector*, that is, a small piece of code that runs on the same machine as our legacy application with the purpose of redirecting the standard input, output and error of the application to sockets connected to the steering broker (fig 2).

7 Result Retrieval

The final output produced by the application is held on a remote machine or on a collection of remote machines. Two strategies are possible here. We can choose either to retrieve the result directly from the machines of the third tier or to gather the results on the middle tier web server. It is difficult to predict which option will give the best results in terms of faster file transfers lacking dynamic information about the speed of network connections and the available bandwidth that may be reduced due to heavy traffic. We advise the use of a dynamic reconfiguration procedure to choose the route that minimizes the time needed by file transfers.

If accelerating data output movement is not a critical issue, the designers may decide to trade the benefits arising from a dynamic reconfiguration with the ease of a static design, in which one of the available options is simply chosen randomly.

If the output has to be retrieved directly from the third tier, Globus provides a useful tool, i.e., the *globus-gass-server*, that can be exploited to transfer files using the HTTP protocol directly from the client browser. Otherwise, if the output has to be migrated on the second tier, we suggest the use of a manager – worker scheme, designating the machine on the middle tier as the manager and the others belonging to the Grid as the workers.

The manager machine is in charge of gathering the output generated by the application, so that the user will retrieve all the files from the manager machine; the task can be carried out using the globus-rcp command to put the output files on a directory accessible from the web.

8 Web Access to the Grid in Digital Puglia

Digital Puglia [4-7] is an active digital library of remote sensing data which allows interactive browsing and parallel post processing on the Grid using the Web as a gateway. A set of trusted user was defined and we provided them with a couple of parallel applications, a supervised bayesan classifier and a Principal Component Analysis.

Since our users are remote sensing experts, not computer scientists, we decided to map all of them to a single Globus certificate. The users access Grid applications from the University of Lecce Digital Puglia web site, filling in an authentication form. Then, an applet allows them to interactively derive the input parameters, as needed by the applications. A couple of CGI scripts start and monitor execution progress on the remote machine, an HP Exemplar machine at Caltech.

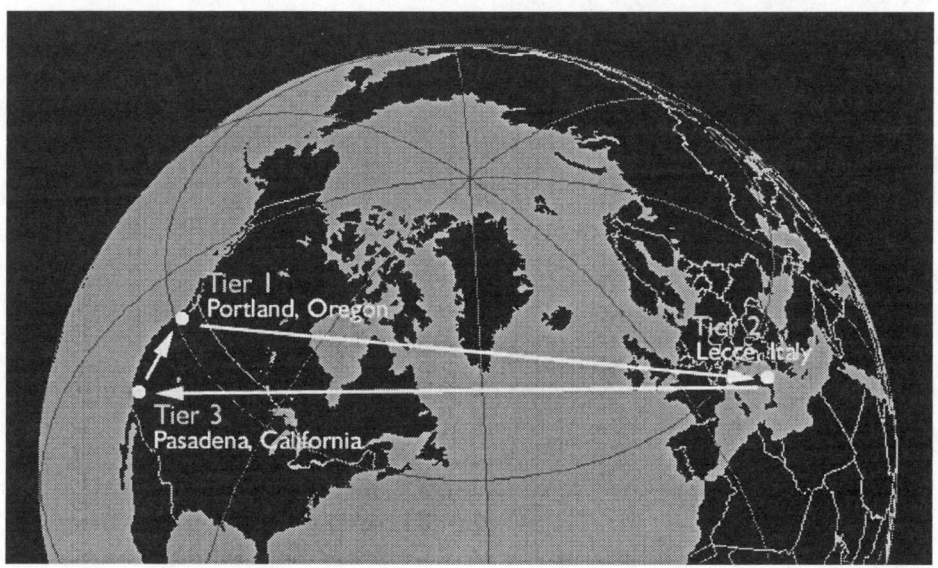

Fig. 3. In the Digital Puglia active library, presented at SuperComputing'99 the possibility to access a worldwide Computational Grid via a web browser was demonstrated

The final output, which is an image, is then retrieved directly from the browser where it can be visualized and saved.

An example session with the supervised classifier was shown in the NPACI exhibit at SuperComputing 1999 to demonstrate the feasibility of interactive web access to Grid applications.

9 Conclusions

We have presented and discussed a framework to build Grid applications that can be transparently started and steered from a Java-enabled web browser. The architecture is based on the Globus toolkit to provide users with fast, secure and reliable access to applications on the Computational Grid. We have showed an example of the use of this framework in the context of Digital Puglia, an active digital library of remote sensing digital data.

References

1. Ian Foster, Carl Kesselman: The Grid. Blueprint for a new computing infrastructure. Morgan Kaufmann, San Francisco (1999)
2. http://www.gridforum.org
3. LDAP v3 Protocol (RFC 2251)
4. R. D. Williams, G. Aloisio, M. Cafaro, P. Messina, J. Patton, "SARA: The Synthetic Aperture Radar Atlas", http://www.cacr.caltech.edu/sara/
5. G. Aloisio, M. Cafaro, P. Messina, R. Williams, "A Distributed Web-Based Metacomputing Environment", Proc. HPCN Europe 1997, Vienna, Austria, Lecture Notes In Computer Science, Springer-Verlag, n.1225 (1997) 480-486.
6. G. Aloisio, G. Milillo, R.D. Williams, "An XML Architecture for High Performance Web-Based Analysis of Remote Sensing Archives", Future Generation Comp. Sys., 16 (1999) 91-100.
7. G. Aloisio, M. Cafaro, R. Williams, " The Digital Puglia Project: an active digital library of remote sensing data", Lect. Notes in Comp.Sci. (Springer) 1593 (1999) 563-572

Data Futures in DISCWorld

H.A. James and K.A. Hawick

Distributed & High Performance Computing Research Group
Department of Computer Science, University of Adelaide
SA 5005, Australia
{heath,khawick}@cs.adelaide.edu.au
Tel +61 8 8303 4519, Fax: +61 8 8303 4366

Abstract. Data futures in a metacomputing system refer to data products that have not yet been created but which can be uniquely named and manipulated. We employ data flow mechanisms expressed as high level task graphs in our DISCWorld metacomputing system. Nodes in these task graphs can themselves be expanded into further task graphs or can be represented as futures in the processing schedule. We find this this a generally useful approach to the dynamic and optimal execution of task graphs. In this paper we describe our DISCWorld Remote Access Mechanism for Futures (DRAMFs) and its implementation using Java technology. DRAMFs embody these ideas and allow data products to be lazily de-referenced and to be manipulated as first class objects in the object space of the metacomputing system. Our system aids in the provision against partial failure or network disruptions in a distributed system such as DISCWorld.
Keywords: data futures; task graphs; Java; metacomputing.

1 Introduction

It is important to provide a flexible yet powerful mechanism to express task graphs in a metacomputing or problem solving environment. We have developed a rich data pointer mechanism in our DISCWorld [6, 7] metacomputing system, and in this paper we present an extension to this mechanism for controlling data **futures**. We show how this can be used to enable lazy and eager scheduling placement of sub components in a task graph in a metacomputing environment.

The remote data access problem is not readily implemented by existing mechanisms in network-oriented programming systems such as Java's networking package, Java RMI or RPC. In such systems, all results are returned to the user upon completion of their request. What is needed, and is not supported by either Java or RPC, is a method whereby the user is returned a pointer to a result that can be accessed directly if they wish to inspect the data that the pointer represents, or can be passed to another server for further processing. DISCWorld Remote Access Mechanisms (DRAMs) provide for this need. The problem of global addresses for pointers is tackled by the construction and use of canonical names for data. This is possible under the general DISCWorld framework. There is no concept of the pointer referring to physical address space on any particular machine in the distributed infrastructure; this enables DRAMs to be scalable, realisable and failure resistant. The differences between DRAMs and other systems' mechanisms for representing high-level data pointers are discussed in [5].

When a DRAM is used to construct a complex processing request, the object to which the DRAM refers is not actually used until it is demanded by the DISCWorld daemon. Most other technologies do not allow for on-demand or "lazy" processing; the value that the object references is evaluated when the reference is returned. To aid construction of processing requests, DRAMs contain *metadata* describing the objects to which they refer. The Java definition of a DISCWorld DRAM is shown in figure 1. The DRAM definition mandates sufficient metadata or "recipe" information that will allow reconstruction of the pointed-to object in case of a restart. The DRAM mechanism allows a remote node the option of transferring the object, to which a DRAM refers, to any other DISCWorld node.

```
public abstract class DRAM implements Serializable
{
    private String publicName;      // descriptive name for GUI use
    private String globalName;      // internal ID
    private Icon icon;              // associated icon eg thumbnail image
    private String description;     // long free textual description
    private String className;       // query-able search-able text
                                    //   representation of class
    private String remoteServer;    // location of the Data to which
                                    //   the DRAM points
    private int objectMobility;     // whether the object being pointed to
                                    // can be transferred across
                                    // the network
    private int objectSize;         // size of the object being pointed
                                    // to, in bytes
}
```

Fig. 1. DRAM Java base class. The DRAM provides a high-level pointer to an object. Metadata describing the object to which this DRAM refers is recorded to enable efficient scheduling and placement decisions in a wide-area, high-latency environment.

Using a global naming mechanism to refer to DRAMs, we allow equivalence between DRAMs to be established. This naming scheme is independent of the DRAM mechanism and we assume its existence as does CORBA [12]. We can use a simple implementation of the unique naming scheme provided by either Java/RMI or by CORBA. These are usually based on "mangled" strings derived from host names or URLs. DRAMs from different nodes can have the *same* name, implying that they are copies of the same object, even if they were created on different DISCWorld servers.

There are two direct subtypes of DRAM: a pointer to data, DRAMD; and a pointer to metacomputing services, DRAMS. Since a DRAM is a Java object, it can have extra functionality programmed in for use at the client end. This functionality could be graphical semantics or data filter/compression services [8].

For example, data to be transferred could be explicitly compressed at the server and uncompressed at the client end by the DRAM framework. This would not be possible without the bytecode portability provided by Java.

2 DRAM Futures

In a high-level distributed system, user requests are represented as process networks of metacomputing services linked together by the sharing of data. Such sharing of data may be due to the services' reliance on the same input data, or the sharing may be more explicit, as in a producer-consumer relationship. Such process networks may be specified and their services statically assigned to hosts in the distributed system. This approach can lead to optimal solutions in the case where the characteristics of the distributed system are known in advance, but if all characteristics are *not* known in advance, the dynamic allocation of services to hosts and the optimisation of execution schedules can produce the best suboptimal solutions [8].

To implement the optimisation of process networks, we have extended the DRAM concept to include the notion of pointers to data that have not yet been created. The extended DRAMs can be passed between clients and servers. We term these pointers to not-yet-created data DRAM Futures, or DRAMFs. DRAMFs are created with the name of the data that they *will* represent, and as such, are equivalent to DRAMs in the operations that can be performed on them [5]. DRAMFs are assigned an approximate size for the data they point to, based on the mean historical size of the previous data products of the creating service on that node. However, unlike DRAMs, DRAMFs are only able to point to data, not metacomputing services. This restriction is common sense, as it is not sensible to have a pointer to a service which does not yet exist. When a DRAM is *inspected*, the data to which the object points is retrieved; in the case of a DRAMF, a request is sent to the server that is to create the data. When the data has been created, a DRAMD of the same name as the DRAMF is returned containing the data. Thus, the DRAMF is replaced with a DRAMD, which contains the exact size of the data item, instead of the DRAMF's original estimate.

The implementation of DRAMFs extends the DRAM Java base class. This implements the Serializable interface and embeds an integer estimate of when data will be available from the time that the DRAMF is dereferenced, without any execution optimisations. A typical example of a DRAMF's contents is shown in figure 2. For simplicity, the `estimatedTimeAvailable` field is used as an offset from the time that the holder of the DRAMF requests the data be created. This value consists of the current waiting time for the host node start the service, as well as the mean run time of the service that will create the data. If the data represented by the DRAMF uses any other DRAMFs as inputs to its service, the `estimatedTimeAvailable` field also includes the estimated time of the DRAMF upon which this service needs to wait, and the estimated time to transfer the resulting data between DISCWorld daemons. We consider the time taken to estimate the data's arrival time to be small when compared with the time taken to transfer necessary data and perform the services that the DRAMF represents.

Instance Variable	Value
publicName	Future for GMS5 1998122500:IR1 (0,0), (2291,2291) 100% zoom
globalName	ERIC:SingleImage:GMSImage:IR1(Integer:00,Integer:25, Integer:12,Integer:1998,Integer:0,Integer:2291, Integer:0,Integer:2291,Integer:100)
icon	*thumbnailed representation of data*
description	Future GMS5 image 00Hrs 25DEC1998 IR1 channel, Original Dimensions, 100% zoom
className	image.satellite.GMS5
remoteServer	cairngorm.cs.adelaide.edu.au:1965
objectMobility	2
objectSize	5248681
estimatedTimeAvailable	102

Fig. 2. Example contents of a DRAMF expressed as name/value pairs. The globalName specifies the "recipe" by which the data can be reconstructed in the event that the remoteServer is temporarily unavailable and then later restored.

In principle, DRAM Futures are similar to programming-language level Futures [14], Wait-by-necessity [1] and Promises [10] mechanisms. Whereas Futures and Promises were designed to hide the latency in RPC-based systems, DRAMFs are intended to be used as a high-level pointer to data, in exactly the same way as DRAMDs and DRAMSs. The main difference between DRAMFs and the other mechanisms is their granularity, and the operations which can be performed on them. The Futures and Promises mechanisms are fine-grained. Because the Futures actually represent a memory location in the requesting program, they are tied to the instance of that program. They are not able to be sent between programs. DRAMFs are higher granularity than the other systems, abstracting away from any instance of a creating program. Their ability to be referenced outside the scope a particular instance of the DISCWorld server allows them to be treated as first-class objects.

DRAMFs can be sent between servers, and may also be used by clients in the composition of new processing requests. What this means is that a DRAMF may be copied and moved to many other servers. Subsequently, it may be used as an input to a service by a scheduler, and only when that schedule is *executed*, will the data to which that DRAMF points be copied to a remote machine. Thus, by using the DRAM Futures mechanism, and more generally, the DRAM mechanism, unnecessary bulk data transfers are prevented. User process networks are represented as networks of DRAMs. This may be done either explicitly, with the user constructing a graph of nodes (as discussed in section 3), or implicitly by the user selecting a pre-designed process network from an available library. When the user has created their process network, it is sent to a DISCWorld daemon for processing.

The normal behaviour of the placement mechanism in DISCWorld is to assume that the server will be willing to execute a service or network of services sent to it. If the server is *not* willing to execute the service, either because it is too heavily loaded or the user who owns the job is not allowed to execute services on this machine, or for some other reason, the onus of selecting a new daemon to host the service(s) is on the local daemon. Although beyond the scope of this paper, relocation of the service requires re-placement of the affected services to other daemons and transmitting the modified process network to those daemons which have been designated to produce data that the services were to consume. Nothing needs to be sent to those daemons that will consume the data that is produced by the services on this node, as they will receive a DRAMF from the daemon that has accepted the execution request. The case where no daemon can be found to execute the affected service(s) has not been considered, nor has the case in which a number of daemons continually attempt to assign the services to each other in an infinite loop.

When a daemon creates DRAMFs to some future data, the service does not automatically begin executing. The agreement by a daemon to execute a service with given parameters is, however, binding. The daemon has acknowledged that if requested, it will perform the computation and return the result to any requesters. If the input data required by a service is to be created by a previous service on perhaps a remote machine, then the DRAMFs for the current service cannot be made until the input DRAMFs are received. They are needed in order to estimate the time at which the data, to which the DRAMFs point, should become available. It is in this way that a complex processing request may be set up. This allows, essentially, resource reservations to be made for daemons that will participate in the processing of a request. Thus, while a processing request may be expressed as a network of services connected by data transfer, culminating in the production of some desired data, its execution is constructed in reverse order. Unlike other models of resource reservation, multiple services can be reserved on a single DISCWorld node simultaneously.

In the DISCWorld model, the partial results of processing requests are deemed to be as significant as the final results. Therefore, in addition to sending the DRAMFs to the nodes that may use them, they are also sent to the daemon or client which submitted the processing request. In the case of the user client, DRAMFs corresponding to all the partial products are returned – if the user wishes to view the contents of the DRAMF, they can request the DRAM's contents. As a DRAMF represents the result of a remote computation that has not begun execution, when the results are requested the service is started, and the result is returned when available. When the DRAMF is inspected by the user client or any other daemon, a request is sent to the producing daemon, which initiates the computation.

Optimisations are allowed on the services used to satisfy a processing request. The server may be aware of one or more servers that already possess the data which is to be created, or have supplied DRAMFs to the data sought. If the data is available, and the time spent transferring it will not exceed the time spent waiting until the DRAMF can be satisfied and the data returned, then the decision may be made to transfer the data from an alternative source. By the same reasoning, if there exists another DRAMF pointing to a different data source,

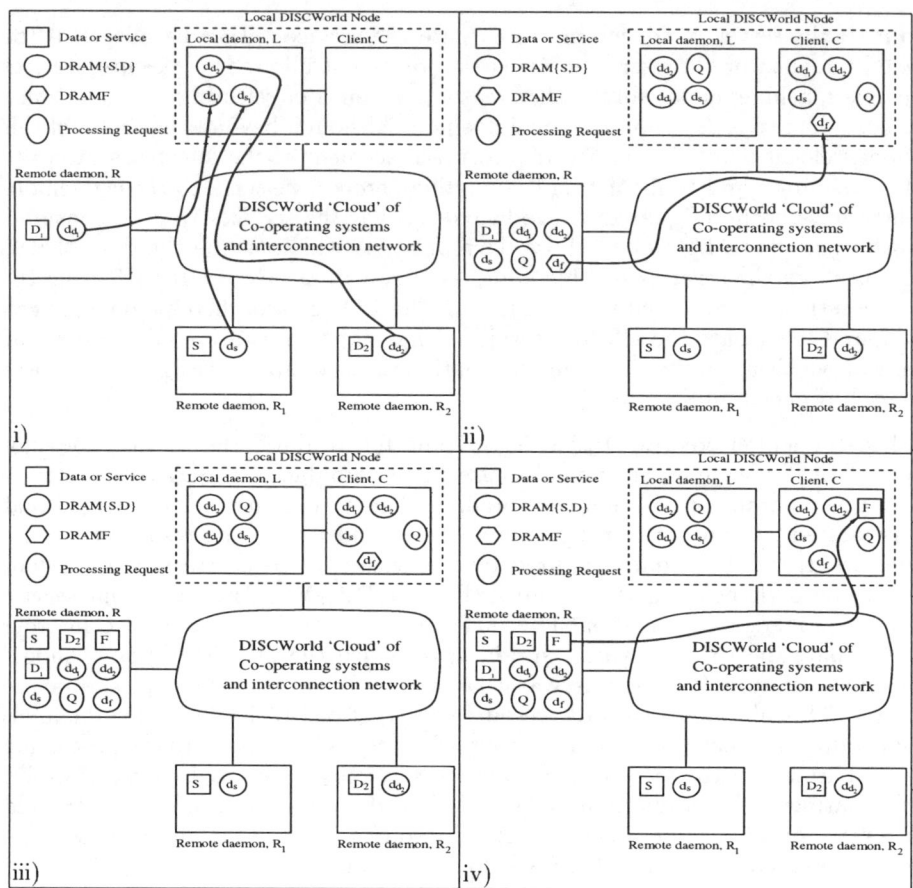

Fig. 3. Execution sequence of process networks within DISCWorld. These show: i) DRAM representations of available data and services are distributed around the system; ii) user logs on to client and submits a processing request, causing a future to be returned which represents the resultant data; iii) the future is inspected, causing execution of the processing request to begin; iv) the client's request for the bulk data item F is satisfied.

and the estimated time is lower than that returned, the alternative DRAMF may be dereferenced in the expectation that either the computation has already been requested by another holder of the DRAMF, or the data should be available sooner due to the lower estimated arrival time. The method by which daemons are made aware of DRAMs on different servers is beyond the scope of this paper [13]. Using partial or final data products of previous processing requests allows the possibility of pruning the execution schedule for the current processing request. Thus, if there is no need to re-compute a data product, then the system will avoid it if possible. Of course, the situation may arise where a user has inspected a DRAMF corresponding to a final data product, causing the computation to be started, and at the same time inspecting a DRAMF to a partial data product. If the DRAMF that is an input to the service which is to produce the final data product has an estimated time greater than a cheaper source of the same output data, then the data may be retrieved from the alternate source, even though by inspecting the partial product DRAMF, the user has dramatically reduced the time it would take to retrieve the data. This case is not addressed in the current implementation – we simply make a best-estimate of the execution times.

In the prototype version of the daemon, there is no maximum time limit between when a DRAMF is returned from a server and the execution must be started. In future implementations, depending on the cost that the user is willing to incur, multiple requests may be sent out to competing data sources, in order to achieve the fastest possible turnaround time for the processing request. Using DRAMFs to represent the data allows the system to judge whether it is more economically feasible to wait until a result that is to be computed at one node is available than retrieve the data from another node, where it already exists but may be expensive to access. The retrieval of data from another node may have one of two effects: either the data the node was to produce now becomes redundant and a waste of processing cycles, or the data is scheduled for use by a subsequent service (after the service has found the data from another source). If the result is no longer needed, depending on the node's management policy, and usually based on the load of the node and its storage capacity, it may choose to proceed with the computation on the chance the data will be needed within the time frame that the node is willing to store results for, or may decide to cancel the production of the result. In turn, in the event of cancelling the production of the result can be propagated up through the list of servers scheduled to create the data that is no longer needed. Of course, the creation of data is only possible if there are no other servers that use the data to be created. Thus, the execution tree is pruned so that only the services that *create data* and those that create the data that is *required*, are actually executed.

The steps by which a user processing request is executed are shown in figure 3. Figure 3i) shows DRAM representations of the data and services being sent to the daemon local to that from which the client submits the request. For example, the DRAM corresponding to data D_1 on remote daemon R is named d_{d_1} and the DRAMS corresponding to service S on remote daemon R_1 is named d_{d_s}.

When a client C connects to the local DISCWorld daemon L a copy of each of the DRAMS is sent to C. Submission of a processing request Q from C to L causes a DRAMF, d_f, representing the output of Q to be returned to C. This is shown in figure 3ii). It can be seen from this figure that the submission of the processing request has also caused a number of DRAMs to be copied around the system. Until the data (or program) represented by each of the DRAMs is needed, it is not transferred between daemons. When the DRAMF is inspected, the DISCWorld framework sends a message from the current holder of the DRAMF (C) to the DRAMF's originating server (R), causing creation of the DRAMF's data to begin. Shown in figure 3 iii), DRAMF d_f represents the result of processing request Q, which is executed to completion. It is at this point in the execution that optimisation can be performed. The daemon attempts to achieve service execution in the quickest amount of time. If the data that is to be used as input is available from another node, or is already present on the local node, or the current node has a DRAMF for the same data, with a sooner estimate of availability, the daemon may decide to use the alternate data source. One of the assumptions of the DISCWorld model is that the size of a data item is the same no matter where it is created. Thus, if a daemon has a DRAMD and a DRAMF to the same data, the estimated size of the DRAMF can be updated to the (exact) size of the DRAMD. When the data to which the DRAMF points is actually created, F, the DISCWorld system converts the local DRAMF d_f to a DRAMD. This is done because the data is actually available now and the estimation time information contained within the DRAMF is no longer relevant. Finally, the outstanding request for the DRAMF's contents is satisfied and the bulk data item F is transferred to the requester C. At C the DRAMF is converted to a DRAMD and the processing request is satisfied. This is shown in figure 3 iv).

3 Discussion

It is also convenient to use DRAMs as a mechanism for deferred delivery of computations. A DRAM may be delivered to a client program with an indication that the remote data to which it refers will only be available within certain times. This allows the token to be redeemed at some time in the future when a long computation has completed. The concept of a deferred delivery mechanism is also useful where a user or client program receives a *token* to some data that may not yet be created or directly available. This allows the user or client to submit a request, and upon a subsequent login to the DISCWorld environment, receive the results. The token is used both as a placeholder for the request result, but also as proof of authorisation to request the status of the pending request. DRAMs are used to implement the deferred delivery mechanism of DISCWorld, and are the equivalent of data tokens. Common mechanisms for notifying users of deferred delivery products in other systems include the creation of output files in pre-defined directories, and the receipt of electronic mail bearing an authentication key to be used to gain access to the results at a server. We believe that while these methods are indeed efficient and perform well for their intended applications, the DISCWorld philosophy is to abstract away from the user's file system and to use a more scalable, portable mechanisms such as embodied by DRAMs.

Although our original concept for a DRAM was that of a remote data pointer in an object-oriented framework. It was convenient to extend this idea to allow DRAMs to have a special graphical or iconic representation in a DISCWorld client environment. Our present implementation allows for a DRAM to be attached to a graphical icon in the client screen area, and to be manipulated using the usual mouse semantics. These include being able to drag and drop iconic DRAMs onto other applications and other sensitive screen areas. We are presently exploring collaborating graphical clients that allow two interactive users to share a graphical workspace and to exchange data using DRAMs. Graphical User Interface builder environments such as AVS and JavaStudio make use of similar graphical semantics to move around iconic representations of code and data items and allow these to be interconnected. Our DISCWorld service infrastructure lends itself particularly well to such semantics and we are also building this capability into the DISCWorld graphical client environment. DRAMs have a direct analogy to the way in which user queries are submitted to the DISCWorld metacomputing environment. Service DRAMs and Data DRAMs are able to be combined together into a process network which depicts the manner, or "recipe" in which a result is constructed. The process network is then either serialised, and sent to the DISCWorld server for processing, or is converted into a textual form called Distributed Job Placement Language (DJPL) [8] before being sent to the DISCWorld server.

4 Summary and Conclusions

We have introduced the DISCWorld Remote Access Mechanism Future (DRAMF), which builds on our base DRAM entity. In order to facilitate the implementation of large-scale distributed service execution, we have extended the DRAM concept to allow data that has not yet been created to be manipulated. Future data is encapsulated in the DRAM Future (DRAMF). DRAMFs are used extensively in DISCWorld to optimise service execution time and cost, especially as there are no guarantees on how accurate a node's global system state information is. Optimisation is achieved through data and service re-use on a per-node basis.

We have shown how DRAMFs' ability to be independently named and manipulated makes them first-class objects in the DISCWorld system. Their status as DRAMs allows their contents to be referenced even in the presence of temporary network and server disruptions. We have demonstrated how data futures can be implemented in a Java environment making use of services provided by our DISCWorld system. We believe a futures mechanism like DRAMFs is a useful feature for any metacomputing environment to enable proper control and optimal placement of directed acyclic task graphs.

References

1. Denis Caromel. Toward a Method of Object-Oriented Concurrent Programming. In *Communications of the ACM*, 36(9):90–102, September 1993.
2. P. D. Coddington, K. A. Hawick and H. A. James. *Web-Based Access to Distributed High-Performance Geographic Information Systems for Decision Support*, Proc. of Hawai'i International Conference on System Sciences (HICSS-32), Maui, January 1999.
3. K. A. Hawick and P. D. Coddington. *Interfacing to Distributed Active Data Archives*, International Journal on Future Generation Computer Systems, Special Issue on Interfacing to Scientific Data Archives, 16(1999) 73–89. Editor, Roy Williams.

4. K. A. Hawick and H. A. James. Distributed High-Performance Computation for Remote Sensing, Proc. of Supercomputing '97, San Jose, November 1997
5. K. A. Hawick, H. A. James and J. A. Mathew. Remote Data Access in Distributed Object-Oriented Middleware, *To appear in Parallel and Distributed Computing Practices*, 1999.
6. K. A. Hawick, H. A. James, A. J. Silis, D. A. Grove, K. E. Kerry, J. A. Mathew, P. D. Coddington, C. J. Patten, J. F. Hercus and F. A. Vaughan. DISCWorld: An Environment for Service-Based Metacomputing. Invited article *Future Generation Computing Systems (FGCS)*, (15)623–635, 1999.
7. K. A. Hawick, H. A. James, C. J. Patten and F. A. Vaughan. *DISCWorld: A Distributed High Performance Computing Environment*, Proc. of HPCN Europe '98, Amsterdam, April 1998.
8. Heath A. James. *Scheduling in Metacomputing Systems*, PhD Thesis, The University of Adelaide, July 1999.
9. H. A. James and K. A. Hawick. *Resource Descriptions for Job Scheduling in DISCWorld*, Proc 5th IDEA Workshop, Fremantle, Feb 1998.
10. B. Liskov and L. Shrira. Promises: Linguistic Support for Efficient Asynchronous Procedure Calls in Distributed Systems. In Proc. SIGPLAN'88 Conf. Programming Language Design and Implementation, pp260–267, June 1988.
11. J. A. Mathew, A. J. Silis and K. A. Hawick. *Inter Server Transport of Java Byte Code in a Metacomputing Environment*, Proc. TOOLS Pacific (Tools 28) - Technology of Object-Oriented Languages and Systems, Melbourne, 1998.
12. Object Management Group. *CORBA/IIOP 2.2 Specification*, July 1998, Available from http://www.omg.org/corba/cichpter.html.
13. Andrew Silis and K. A. Hawick. The DISCWorld Peer-To-Peer Architecture. In Proc. Fifth IDEA Workshop, February 1998.
14. Edward F. Walker, Richard Floyd and Paul Neves. Asynchronous Remote Operation Execution in Distributed Systems. In Proc. Tenth Int. Conf. Distributed Computing Systems, pages 253–259, May 1990.

Algorithms for Generic Tools in Parallel Numerical Simulation*

David Lecomber and Mike Rudgyard**

Oxford University Computing Laboratory
Oxford, UK
{David.Lecomber,Mike.Rudgyard}@comlab.ox.ac.uk

Abstract. COUPL+ is a simple and relatively complete environment for applications that make use of unstructured and hybrid grids for numerical simulations. The package automates parallelization of applications by handling the partitioning of data and dependent data. Primitives are provided to maintain halo interfaces and ensure copy coherency. This paper explores some of the algorithms behind the COUPL+ library, analysing the performance on a cluster of P-II-450 workstations. A multi-level partitioning algorithm for skewed data is presented, involving solving the multi-set median-finding problem. Partitioning elements over a set of pre-partitioned nodes is explored and a novel solution is found reducing communication requirements of the resulting distribution.

1 Introduction

For the majority of data-parallel applications there is a need to partition not only data, but indirection lists (or *connectivities*) that relate the distinct *sets* that the data span. Once the data and connectivities are partitioned, such applications then need to maintain data coherency across the processors using message-passing or (virtual) shared-memory constructs.

This situation is typical of parallel PDE simulations based on unstructured or hybrid grids, such as those occurring in computational fluid dynamics, electromagnetics or structural analysis. Connectivities define relationships between the elements, faces, edges and nodes of the mesh, or the edges or nodes of bounding surfaces. For example, most finite-element techniques primarily make use of element-to-node pointers, while finite-volume techniques require edge-to-node or face-to-element pointers. To solve resulting linear or non-linear equations, sparse matrix representations may also be required, and connectivities may be transformed into edge-based or irregular graph-based structures.

These problems require the management of partitioning of nodes, data and elements. For example, having partitioned the nodes a partitioning of the elements must be induced, or vice-versa. The connectivities, described by lists

---

* This work was partly funded by the European Community as part of the JULIUS project under contract ESPRIT EP25050
** Smith Institute Research Fellow

of nodes, must then have these nodes relabelled to point to the new location of nodes following partitioning. Indeed, by creating a halo containing copies of boundary nodes from adjacent partitions it is possible to relabel connectivities so that all nodes mentioned appear local. The system ensures the halo is consistent.

Such tasks are non-trivial, and arise frequently in many codes, it is therefore essential that tools be made to manage this for the programmer. COUPL+ [RLS98,RSD96][1] is an efficient parallel library developed to meet this need. As well as offering a range of integrated grid-manipulation functions and iterative methods, it provides tools for parallel I/O, including distributed algorithms for partitioning data and connectivities, and a simple parallel loop syntax that may be used to ensure that copies of interface data are maintained. The package is configurable for PVM [GBD+94], MPI [SOHL+95] and BSP [Val90] machines.

The present paper explores some of the inner workings of the COUPL+ library, describing how a new bulk-synchronous approach to these tasks has been implemented with optimal performance. Graph-partitioning tools are included in the COUPL+ but are not detailed in this paper.

2 Performance Prediction

The programming model that is adopted for the COUPL+ library is BSP. In this model the communication mechanism is only guaranteed to have completed pending remote writes and reads at barrier synchronization points where all the processors synchronize. The system is parameterized as follows

1. p – the number of processors.
2. g – where the time required to effect h-relations in continuous message traffic is gh time units. A h-relation is a message pattern in which no processor receives, nor sends, more than h words.
3. l – the minimum time between successive synchronization operations.

BSP predicts the time between successive synchronization points to be directly related to the maximum communication traffic, h, and computation, w, seen at any of the processors by the formula $t = w + gh + l$. Algorithms have been written in this *bulk-synchronous* style for portability and may be analysed effectively.

3 Geometric Partitioning

Optimal bisection of a mesh is at least NP-hard. A heuristic is to split the 3D–space by some plane: this provides the smallest boundary surface area for a bounded region if oriented correctly. The first objective is to determine the orientation of the plane, a number of alternatives have been previously extolled; surveys of the area are found in [Sim91,Far88]. COUPL+ presently supports

---

[1] See http://www.comlab.ox.ac.uk/oucl/work/david.lecomber

1. Inertial bisection
   (a) Compute eigenvector of inertial matrix for data [Far88].
   (b) Bisect partition at mid-points.
   (c) Repeat (a) and (b) until number of partitions is p.
2. Slicing
   (a) Compute eigenvector of entire set as above, once.
   (b) Recursively bisect along this axis.
3. xyz-cutting
   (a) Bisect along x-axis, then y, then z until complete.
4. User provided.

To obtain a perfectly balanced partitioning, in terms of the number of nodes, the plane must evenly bisect the points. Consequently if we have the normal to the cutting plane, we must find the median distance along this normal relative to some origin. Median finding is significantly faster than sorting. Randomized algorithms for parallel median finding are well established [GS96,JáJ92] with running time $O(\frac{n}{p})$ with high probability for all practical parallel machines. In the case of inertial bisection the computation of the inertial matrix and eigenvector is also $O(\frac{n}{p})$ for such machines.

It is not desirable to move data between processors at each level of the recursion: this imposes $O(\frac{gn \log p}{p})$ communication cost in the worst case where each element is moved at each level before finding the destination processor. At each stage of the algorithm elements merely move locally – data is sent to the ultimate destination when this has been finally computed.

Definition 1. *A partitioning $\mathcal{P}$ of a set S is defined to be a sequence $\langle P_0, \ldots P_{k-1} \rangle$ of disjoint sets such that $\bigcup_{0 \leq i < k} P_i = S$.*

Note 1. The subset of elements of a set P that reside on processor j is written P^j. Furthermore for any P, $\bigcup_{0 \leq i < p} P^i = P$.

The optimality of median finding and the first stage of inertial bisection require perfect balance in the initial data. This is true for the initial bisection into two sets: each processor holds $\frac{n}{p}$ elements of the one initial set. Thereafter, if the bisection algorithm is applied to each element of existing partitioning $\mathcal{P}$ to create partitioning $\mathcal{Q}$ then we risk skew in P_i for each i – even though for each processor j we have $\sum_{0 \leq i < |\mathcal{P}|} |P_i^j| = \frac{n}{p}$ because data is not moved until after the final partitioning. It is possible, for example, to have $|P_i^j| = \delta_{i,(j/2)} \frac{n}{p}$ after the penultimate phase of bisecting[2]. This problem arises in the situation where data has already been partitioned using the same method, for example.

Hence obvious algorithms require amendment; this is our first contribution.

---
[2] δ_{ij} is 0 if $i \neq j$ and 1 otherwise

3.1 p-Way Partitioning

An algorithm for computing a partitioning of S into p partitions where $p = 2^k$ is presented, this is easily generalized to any $p \in \mathbb{N}^+$.

1. Initially $\mathcal{P} := \langle P_0 \rangle$.
2. Repeat the following until $|\mathcal{P}| = p$.
   (a) Globally, compute eigenvectors for each $P_i \in \mathcal{P}$.
   (b) On each processor j, for each node $v \in P_i^j$, compute its scalar value using the eigenvector for P_i.
   (c) Use a parallel multi-median finder to determine the median scalar of each P_i concurrently.
   (d) On each processor j re-order the elements of each P_i^j into sets Q_{2i}^j and Q_{2i+1}^j consisting of those elements with scalar value less than and respectively greater than the median of P_i.
   (e) $\mathcal{P} := \langle Q_0, \ldots \rangle$.
3. Communicate data to destination in parallel: for $0 \leq i, j < p$, processor j sends P_i^j to processor i.

Step 2(a) is performed in optimal $O(n/p)$ time for each method provided. It is sufficient that the "eigenvector" for each disjoint subset in $\mathcal{P}$ can be determined from (i) an associative global reduction, (ii) a constant-time operation on the result of this reduction. Thus local contributions to the reduction for every P_i are calculated concurrently without synchronization. As the total number of locally-held elements is n/p for each processor, this is perfectly balanced. A single synchronization is required to gather these partial results by partition to single processors. The processor in charge of the result for partition P_i, say processor i, reduces the p partial results to a single value and computes the appropriate vector in $O(1)$ time. The results are broadcast to all processors using only one synchronization. The total cost of step 2(b) is hence $O(l + gp + n/p)$, or $O(n/p)$ provided $l + gp = O(n/p)$, which is a valid assertion. Step 2(c), parallel multi-median finding, is described later. Step 2(d) is perfectly balanced as the total number of elements moved around locally is n/p for each processor.

Now, provided parallel multi-median finding is $O(n/p)$ then total complexity of the entire method is $O(\frac{n}{p}(g + \log p))$. The g-term is incurred in the final movement of data to the destination processor.

3.2 Parallel Multi-median Finding

Definition 2. *The parallel multi-median finding problem is to find the median of k sets $P_0 \ldots P_{k-1}$ each of total size n/k and distributed such that for each $0 \leq j < p$, $|\bigcup_{0 \leq i < k} P_i^j| = n/p$.*

In [GS96] the median of a single evenly distributed set is computed with computation $2n/p + o(n/p)$ and communication time of $(1 + o(1))n/p$ provided $g = o(n^\varepsilon / \log^{1/3} n)$ for a fixed $0 \leq \varepsilon < 1/3$. Their algorithm attains these bounds with high probability (ie. $1 - O(n^{1-\rho})$ for some $\rho > 1$).

The crux of the algorithm in [GS96], and others in [JáJ92], is the random sampling of a small subset of data. The sample is then sorted. It can be shown that the median of the whole set lies between two predetermined elements, l and u, of the sorted sample with probability $1 - O(n^{1-\rho})$. The number, k, of elements of the original data less than the lower bound, l, is then computed. The original data lying between these l and u are then selected. The size of this subset is shown to be small, and hence this data is now sorted. The median can then immediately be picked out as the element at position $n/2 - k$ of this set.

The novel algorithm used by COUPL+ applies the technique *concurrently* to all partitions in a partitioning to remove imbalance. Let $t = 2(\frac{n}{k})^{2/3}(3\rho \log n)^{1/3}$ for some $\rho > 1$. Let $\mu = (n/k)^{1/3}(3\rho \log n)^{2/3}$ and define $l = t/2 - \mu$ and $u = t/2 + \mu$. These parameter values follow from [GS96] routinely.

1. Each processor broadcasts the sizes of each subset held locally.
2. For each $0 \leq i < k$, a subset Q_i of size t is chosen with each processor choosing a set of size t/p. Processor, j, randomly selects elements from P_i – using the sizes returned in step 1 to determine the processors required element lie on. The elements are exchanged in one aggregated communication.
3. Using any efficient deterministic sorting algorithm, sort the sample sets concurrently – ie. do the appropriate computation and communication for each set before synchronizing in each phase of the algorithm.
4. Broadcast the l and u elements (l_i and u_i) of each sorted sample Q_i, from the processors holding these elements to all processors.
5. For each of the processor j and set i compute l_i^j and u_i^j, the number of elements in P_i^j less than l_i and greater than u_i respectively.
6. Concurrently reduce for each i sum the l_i^j and u_i^j. Determine whether, for all i, the medium of P_i is between l_i and u_i.
7. On the assumption of success in the previous step, let M_i be the values held between l_i and u_i for each i. Locally calculate M_i^j for each i, j.
8. Balance these sets by randomly distributing values across processors. Using any efficient deterministic sorting algorithm, sort these sets concurrently.
9. For each i select the required $\frac{n}{2k} - |l_i|$-th element from the sorted sets M_i.

The cost of these steps is now detailed. Steps 1 and 6 are broadcasts of k and $2k$ values from each processor. BSP reduction of a vector of width c can be done in time $(cg + c)mp^{1/m} + lm$ for any $m > 1$. In practice $m = 1$ is selected and this reduction takes one synchronization.

In step 2 each processor j determines the t/p elements that it needs from P_i by choosing elements uniformly from P_i; to do this requires knowledge of the local sizes P_i^l for each l (hence step 1). This ensures a random distribution to the elements in the sample. Near perfect balance for the sourcing of elements is obtained, but only over the union of $P_0 \ldots P_{k-1}$. Consequently aggregating this step into one synchronization yields complexity of $(1 + o(1))ktg/p$.

In steps 3 and 8 the sorting is performed in $O((kt \log t)/p)$[GS96]. In the latter case the result (only) holds with high probability. The initial balancing phase of step 8 requires *at most* ktg time. Step 4 is certainly performed in time no greater than $2kgp + l$. Steps 5 and 7 may actually be performed together. Furthermore, because each processor holds a total of n/p elements these steps are completed in total time n/p. Step 9 is dominated by preceding steps.

If p and g are independent of n it is now observed that $ktg = o(n/p)$, and also that $O((tk \log t)/p) = o(n/p)$ as $tk = o(n^{1-\varepsilon})$ for any $\varepsilon < 1/3$. Consequently the asymptotically significant stages of this algorithm are step 5 and 7. These two stages require no more than two comparisons per element and thus the algorithm has complexity $2n/p + o(n/p)$,

It is noted that the individual failure probability for each median finding is certainly $O(n^{1-\rho})$, but since there are no more than $p/2$ medians to find then the failure of at least one is $O(pn^{1-\rho})$ but this is $O(n^{1-\varepsilon})$ for any $\varepsilon < \rho$ provided p is $o(\log n)$. For fixed p then, naturally, this is $O(n^{1-\rho})$.

3.3 Data Tracking

It is important to keep track of data as it is moved around the COUPL+ system, for I/O purposes and in order to import values or elements associated with the data. Partitioning information for the nodes is distributed amongst the processors – each processor holds an array A in which $A[i]$ gives the original location of the i node current stored at this processor. It is also possible to store an inverse of this *mapping* which records where a node has been moved to, rather than where it has come from. The inverse may be computed with a communication cost of $O(gn/p)$ if space is limited.

4 Importing a Connectivity

Having determined locations for the nodes of a mesh, the tetrahedra or other such elements must be partitioned consistently with this distribution.

Definition 3. *If E represents a set of elements then define $n(E)$ to be the set of nodes pointed to by E.*

Assuming that the inverse mapping for nodes is available, the algorithm proceeds as follows. Let E be the set of elements.

1. Each processor j computes $n(E^j)$, the set of nodes required by the locally held portion of E.
2. Nodes in sets $n(E^j)$ are sent to the original hosting processors of the nodes.
3. For each node received by processor j from processor k in the previous step, send back the new destination information to k.
4. Locally, for each j and for each element of E^j, relabel the nodes pointed to by using the destination information for $n(E^j)$ received.
5. Locally, for each j and for each element e of E^j determine a destination for e according to the following rule:
   (a) If the majority of nodes are held by processor k, send e to k.
   (b) Otherwise, choose a random processor from the set of processors holding nodes of e.
6. Move the elements to their destinations.

It should be noted that the penultimate step is carried out locally. This method raises a problem: a locally favourable move for an element is made independently of the elements surrounding it. If, for example, elements are simply edges and the task is to minimize the number of remote nodes that are communicated between processors, Fig. 1 shows the possible mix between partitions at a boundary (indicated with lightly dashed line). Most edges require the import of a different boundary node. In contrast, suppose boundary edges were grouped so that in the

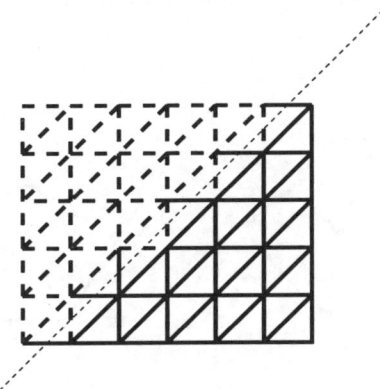

Fig. 1. Example partitioning of simple mesh of edges

top half of the figure the edges were assigned to the (dashed) left-hand partition and the bottom half to the right hand. In this arrangement each boundary node shares its remote node with another boundary edge held on the same processor: The communication requirement is halved without a balance penalty. For triangle-based meshes non-optimal sawtooth-like results are often obtained.

4.1 Smoothing an Imported Graph

Evidently the distribution of elements must be improved. Only boundary elements need be considered – those having at least one node on another processor – the other elements are already perfectly placed. Standard graph-based techniques such as Kernighan-Lin [KL70,KK96,Hen98] are inapplicable directly. These move nodes in order to improve edge-cut. Our innovation is to use these edge-refinement algorithms on the *dual* graph of the boundary elements. The dual graph expresses dependencies between elements in the graph: two elements are linked if they require the same node.

Definition 4. *Let* $G = (V, E)$ *be a graph with vertices* $v \in V$ *and elements* $e \in E$. *The dual graph is defined by* $D(G) = (V', E')$ *where* $V' = E$ *and*

$$E' = \{(e_1, e_2) | \ e_1, e_2 \in E \bullet n(e_1) \cap n(e_2) \neq \emptyset\}.$$

Minimizing edge-cut in the dual corresponds to minimizing communication between elements in the source graph. There is added complication in that the problem is not strictly isomorphic to edge-cut refinement problems: freedom to move elements is less as this task cannot be executed in isolation to the location knowledge of the nodes to which the elements point.

It is possible for a partitioning of the nodes to lead to imbalance when used to import elements. This balance can be important to solvers performing computation on elements and nodes. So-called multi-constraint refinement methods may be used, a generalization of Kernighan-Lin. This is an area of future interest.

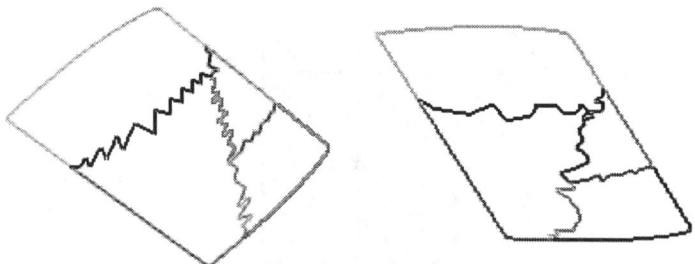

Fig. 2. Imported graph (i) without smoothing and (ii) with smoothing

COUPL+ currently uses a greedy algorithm for refining the dual graph of elements. Elements are moved if doing so improves the edge cut of the dual graph locally – thus an element is moved to a processor that already requires some of the the nodes which the element requires. Table 1 shows that a 10% improvement in edge cut is obtained with one pass of this algorithm. Figure 2(ii) shows the improvement obtained over the initial partitioning of edges in a 4-way partitioning of the M6 wing.

Mesh	Nodes	Elements	Processors	Distinct remote nodes per processor	
				Initial partitioning	Post-smoothing
F15	690,991	4,247,412	8	6881	6235
			4	5905	5357
			2	3942	3559
M6 Coarse	10,812	57,564	8	550	491
			4	600	526
			2	464	399
Falcon	155,932	847,361	8	2188	1926
			4	2294	2005
			2	1817	1592

Table 1. Set communication requirements

5 Experimental Results

The performance shown in the tables is that for an 8-node P-II-450 cluster using a standard PVM library. The nodes of substantial mesh, the F15, can be partitioned into 8 sets in under 4 seconds.

Mesh	Nodes	Processors	Time (s)
F15	690,991	8	3.61
		7	3.44
		4	3.70
		2	3.97
M6 Coarse	10,812	8	0.84
		4	0.26
		2	0.12
Falcon	155,932	8	1.47
		7	1.33
		4	1.10
		2	0.99

Table 2. Node Partitioning times on P-II-450 cluster

Partitioning considered as a function of p is atypical of parallel algorithms – in this case as p increases so does the work as we ask it to partition. The test of using only seven processors is included to show applicability. Performance for the Falcon and F15 aircraft meshes is good. The M6 wing is too small to hide the latency when partitioned and consequently poor performance is obtained.

Mesh	Elements	Processors	Time (s)
F15	4,247,412	8	20.69
		4	24.26
		2	33.68
M6 Coarse	57,564	8	0.32
		4	0.37
		2	0.52
Falcon	847,361	8	3.46
		4	4.56
		2	5.81

Table 3. Edge Importing times on P-II-450 cluster

For comparison loading the elements of the F15 grid onto the system takes 81 seconds. In excess of half of the above running time is associated with the movement of data to the appropriate home, therefore little improvement on the above would seem possible.

Mesh	Processors	Time (s)
F15	8	20.04
	4	13.63
	2	5.88
M6 Coarse	8	1.50
	4	1.37
	2	0.68
Falcon	8	6.05
	4	5.05
	2	2.63

Table 4. Interface smoothing times on P-II-450 cluster

Interface smoothing is a costly operation. Most of the cost is incurred in constructing the dual graph, rather than the refinement, and presently we are examining improvements to this.

References

[Far88] C. Farhat. A simple and efficient automatic FEM domain decomposer. *Computation and Structures*, 28(5):579–602, 1988.

[GBD+94] A. Geist, A. Beguelin, J. Dongarra, J. Weicheng, R. Manchek, and V. Sunderam. *PVM: A Users' Guide and Tutorial for Networked Parallel Computing*. MIT Press, 1994.

[GS96] A. V. Gerbessiotis and C. J. Siniolakis. Deterministic sorting and randomized median finding on the BSP model. In *8th ACM Symposium on Parallel Algorithms and Architectures (SPAA '96)*. ACM Press, June 1996.

[Hen98] B. Hendrickson. Graph partitioning and parallel solvers: Has the Emperor no clothes. In *Irregular '98*, 1998.

[JáJ92] J. JáJá. *An Introduction to Parallel Algorithms*. Addison-Wesley, 1992.

[KK96] G. Karypis and V. Kumar. Parallel multilevel k-way partitioning scheme for irregular graphs. In *Supercomputing 1996*, 1996.

[KL70] B. W. Kernighan and S. Lin. An efficient heuristic for partitioning graphs. *Bell System Technical Journal*, 49(2):291–307, 1970.

[RLS98] M. Rudgyard, D. Lecomber, and T. Schönfeld. *The COUPL+ User Manual*, 1998.

[RSD96] M. Rudgyard, T. Schönfeld, and I. D'Ast. A parallel library for CFD and other grid-based applications. *Lecture Notes in Computer Science*, 1067, 1996.

[Sim91] H. D. Simon. Partitioning of unstructured problems for parallel processing. *Computational Systems Engineering*, 36(5):745–764, 1991.

[SOHL+95] M. Snir, S. W. Otto, S. Huss-Lederman, D. W. Walker, and J. Dongarra. *MPI: The Complete Reference*. MIT Press, 1995.

[Val90] L. Valiant. A bridging model for parallel computation. *Communications of the ACM*, 33(8):103–111, 1990.

Dynamic Grid Adaptation for Computational Magnetohydrodynamics

R. Keppens[1], M. Nool[2], P.A. Zegeling[3], and J.P. Goedbloed[1]

[1] FOM-Instituut voor Plasma-Fysica Rijnhuizen, P.O.Box 1207
3430 BE Nieuwegein, The Netherlands
[2] Centre for Mathematics and Computer Science, P.O.Box 94079
1090 GB Amsterdam, The Netherlands
[3] Department of Mathematics, University of Utrecht, P.O.Box 80010
3508 TA Utrecht, The Netherlands

Abstract. In many plasma physical and astrophysical problems, both linear and nonlinear effects can lead to *global* dynamics that induce, or occur simultaneously with, *local* phenomena. For example, a magnetically confined plasma column can potentially posses global magnetohydrodynamic (MHD) eigenmodes with an oscillation frequency that matches a local eigenfrequency at some specific internal radius. The corresponding linear eigenfunctions then demonstrate large-scale perturbations together with fine-scale resonant behaviour. A well-known nonlinear effect is the steepening of waves into shocks where the discontinuities that then develop can be viewed as extreme cases of 'short wavelength' features. Numerical simulations of these types of physics problems can benefit greatly from *dynamically controlled grid adaptation schemes.*
Here, we present a progress report on two different approaches that we envisage to evaluate against each other and use in multi-dimensional hydro- and magnetohydrodynamic computations. In *r-refinement*, the number of grid points stays fixed, but the grid 'moves' in response to persistent or developing steep gradients. First results on 1D and 2D MHD model problems are presented. In *h-refinement*, the resolution is raised locally without moving individual mesh points. We show 2D hydrodynamic 'shock tube' evolutions where hierarchically nested patches of subsequently finer grid spacing are created and destroyed when needed. This *adaptive mesh refinement* technique will be further implemented in the Versatile Advection Code, so that its functionality carries over to any set of near conservation laws in one, two, or three space dimensions.

1 Introduction

Computational magnetohydrodynamics is rapidly developing into a standard tool for investigating the behaviour of a plasma (a charge-neutral 'soup' of ions and electrons). The numerical algorithms used in state-of-the-art software packages for multi-dimensional MHD studies heavily borrow on well-established techniques employed in computational fluid dynamics. However, significant complications arise due to the presence of a magnetic field, together with its dynamical influence. E.g., Riemann solver based methods must allow for the presence of three basic wave modes in the plasma: a fast magnetosonic, an Alfvén, and a slow magnetosonic signal travel outwards to communicate localized, isolated perturbations to further out regions. In addition, the magnetic field itself satisfies a basic law that constrains possible solutions of MHD problems: the absence of magnetic charges or monopoles cause the field to be solenoidal $\nabla \cdot \mathbf{B} = 0$.

In spite of these complications, various methods have been developed and applied successfully for simulating magnetically controlled fluid dynamics. The resulting richness in physics phenomena often involve both long and very short lengthscales. To name but a few recently investigated topics:

- **Sunspot eigenoscillations:** the natural vibration modes of sunspots, when modeled as magnetic flux tubes embedded in unmagnetized surroundings, include so-called *leaky, resonantly damped modes* [11]. These modes correspond to global oscillations of the sunspot that affect the entire surroundings through outwards travelling sound waves, but also have internal narrow boundary layers where energy is dissipated Ohmically.
- **Secondary, induced plasma instabilities:** resistive MHD studies of velocity shear layers, susceptible to the Kelvin-Helmholtz instability (known from wind-induced ripples on a pond), demonstrated how small-scale reconnection events can occur by secondary tearing instabilities [8].
- **Complex interacting bow shock patterns:** numerical simulations of idealized plasma flow problems around perfectly conducting, rigid objects, revealed how under certain inflow conditions, the resulting bow shock consists of several small-scale and large-scale features with interconnecting weak and strong discontinuities [3].

It should be clear that numerical simulations of these plasma physical processes need to employ a sufficiently high resolution to capture both the fine-scale structure and the overall dynamics. For steady problems, a priori knowledge of the regions where a high spatial resolution is needed can be incorporated by using a static, stretched grid. However, for unsteady problems where typically long-term interactions are of interest, a dynamically adjusted grid resolution is needed. Therefore, we are currently assessing the use of two different grid adaptation schemes for multi-dimensional hydro- and magnetohydrodynamic problems. We demonstrate the workings of a moving grid method on some MHD problems in Sect. 2, and explain in some more detail the patch-based adaptive mesh refinement (AMR) scheme [1] in Sect. 3. We are implementing the latter approach in the Versatile Advection Code [12] [VAC, see http://www.phys.uu.nl/~toth]. The VAC software has already demonstrated its capacities for doing multi-dimensional magneto-fluid-dynamical simulations in a wealth of astrophysical and fundamental plasma physical applications. The lack of adaptivity in the mesh geometry has so far been compensated by the fact that we can run on massively parallel platforms [7, 14]. Still, to make further quantitative parametric studies into fully nonlinear regimes, an efficient grid adaptivity is desirable.

2 Adaptive Method Of Lines

In the Method Of Lines (MOL) approach, the grid points reposition themselves dynamically in accord with the local resolution requirements [17]. This means that the new grid point positions must also be calculated simultaneously with the physical variables (like density, momenta, energy, etc.) at these new locations. The governing physics equations are therefore first transformed to new coordinates, i.e. $\xi \equiv \xi(\mathbf{x}, t)$ and $\theta = t$ where $\mathbf{x}$ and t denote the original cartesian coordinates and time, respectively. In this coordinate transformation, the $\xi(\mathbf{x}, t)$ itself obeys a suitably constructed partial differential equation that controls the mesh movements. To obtain an efficient and gradually adjusting adaptive grid, a so-called equidistribution principle is being used, enhanced with smoothing procedures in the spatial and the time direction. This adaptive grid PDE, together with the transformed governing PDE model, is then semi-discretized and one obtains a large system of ordinary differential equations. This system can be solved with an appropriate stiff time-integrator.

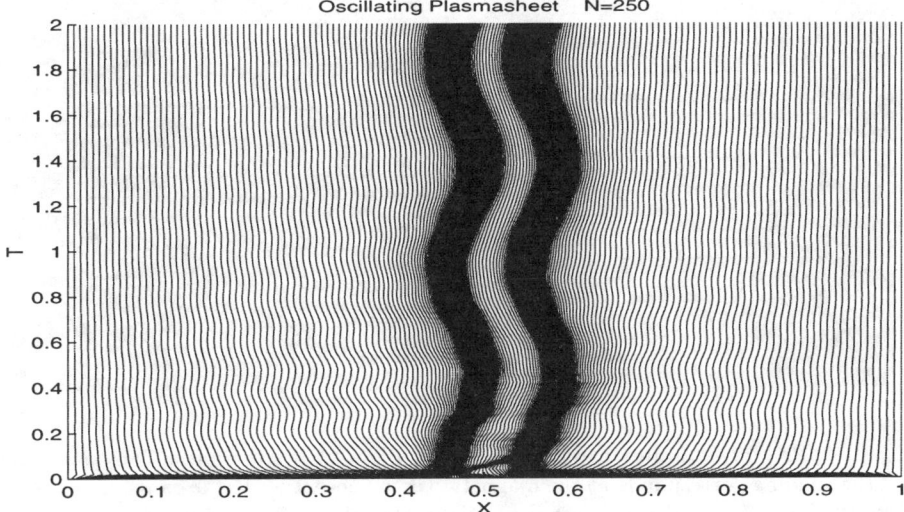

Fig. 1. The grid history in a 1D MHD simulation of an oscillating plasma sheet embedded in a vacuum. Starting with an equidistant grid of 250 grid points, the sheet boundaries are automatically recognized as regions where grid points need to be clustered. After this rapid initial adjustment (prior to times $T < 0.05$), the mesh clearly follows the oscillation.

We have implemented and applied this MOL-approach to various MHD model problems [18]. The earliest study by Dorfi and Drury [4] already demonstrated this method on one dimensional Euler shock tubes. We have successfully tested the method to similar MHD shock problems and are evaluating different means to generalize the method to two and three space dimensions. The difficulty in generalizing MOL methods to more than one space coordinate is that the corners of 2D or 3D cells should be prevented from folding over onto each other. Also, in multidimensional MHD applications, we must pay particular attention to the solenoidal condition on the magnetic field.

Two results are shown here: Fig. 1 shows the time history of the grid point locations for a 1.5D MHD problem first introduced by Tóth, Keppens, and Botchev [15]. A high density plasma sheet, embedded in a 'vacuum' bounded by rigid walls, is set into motion by an initial imbalance in the magnetic pressure between the left and right vacuum region. The resulting dynamics is a linear oscillation of the plasma sheet which retains its identity. The oscillation is governed by alternating compressions and expansions of the magnetic field trapped in the vacuum regions. In the grid line history, the originally uniformly spaced mesh rapidly concentrates around the discontinuities that form the plasma sheet edges, and are seen to follow the slow waving motion of the sheet.

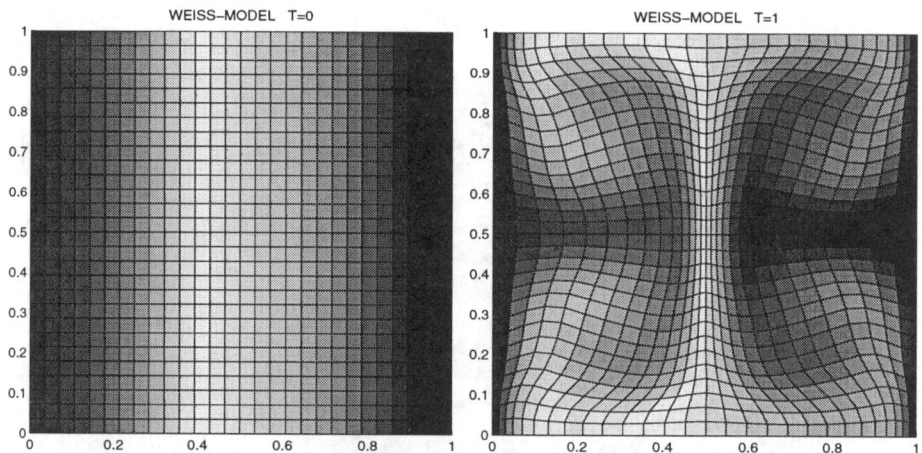

Fig. 2. A 2D kinematic flux expulsion. The left panel shows the initial cartesian mesh and the shading corresponds to the magnetic vector potential. Right panel: an imposed four-cell convection pattern causes the initially straight, uniform field to distort, which is recognized and followed by the 2D grid cell movements.

In Fig. 2, a 2D kinematic flux expulsion is simulated. As in the original work by Weiss [16], a prescribed convection pattern [velocity distribution $\mathbf{V}(\mathbf{x})$] is used in the induction equation for the magnetic field $\mathbf{B}$, namely

$$\frac{\partial \mathbf{B}}{\partial t} = \nabla \times (\mathbf{V} \times \mathbf{B}) - \nabla \times (\eta \nabla \times \mathbf{B}), \tag{1}$$

where η indicates a magnetic diffusivity. Starting with an initially uniform field and mesh, the field lines deform in response to perpendicular flow, while parallel flow simply follows the field lines. The diffusion becomes important in regions of strong gradients only. In the figure, the shading corresponds to the magnetic field potential: field lines would be isolines in this plot. One can see that at later times, some field lines are curled up and the grid is distorted to capture the localized strong variations. In a forthcoming publication [18], we plan to discuss these and other problems in detail and compare them with high resolution solutions on static grids, obtained with the Versatile Advection Code [12]. To get similarly accurate solutions on a non-adaptive grid, many more grid points must be used in each space direction.

3 Adaptive Mesh Refinement

One of the best known Adaptive Mesh Refinement (AMR) methods is the one originally developed by Berger [1]. The AMR philosophy is to allow for a user-defined number of grid levels (indexed by l), that have fixed refinement ratios r_l between their spatial step sizes Δx_l (time steps Δt_l), so that

$$r_l \equiv \frac{\Delta x_{l-1}}{\Delta x_l} \equiv \frac{\Delta t_{l-1}}{\Delta t_l}. \tag{2}$$

In a patch-based approach, a refinement criterium applied to all grid patches on level l yields a collection of (scattered) points where a higher resolution is needed. Such a refinement criterium can be based on physical quantities – like a flow divergence or a current – exceeding user specified threshold values. For efficiency, these quantities can be estimated from a low order solution on the grids while the solution method used in the actual time integration can be of higher order. Another refinement criterium often used in AMR implementations is a point-to-point comparison between the conservative variables (e.g. density) obtained by a normal, 'fine' step on grid patch $G_{i,l}$ of resolution $[\Delta x_l, \Delta t_l]$ and by a 'coarse' step of resolution $[2\Delta x_l, 2\Delta t_l]$. By saving previous time steps of the solution on patch $G_{i,l}$, this only involves one coarse and one fine time step advance, which again can be of low order.

In all cases, the points thus flagged for refinement are clustered in groups and surrounded by clouds of 'buffer' points to anticipate the expected spreading of the dynamics over a larger area. All the resulting points are then grouped in rectangles (in 2D), which by subsequent bisections form the most efficient next candidate grid patches on level $l+1$. Extra measures can be taken to ensure a proper nesting of grids: each level $l+1$ grid must be entirely contained in level l grids with at least one grid cell of a level l grid neighbouring its sides. Exceptions are possible near the computational domain edge.

The time integration must proceed in a well-defined order, such that all grids at all levels agree in the physical solution after each time step $t_1^n \rightarrow t_1^n + \Delta t_1$ on the coarsest grid(s) present. A hypothetical sequence of three subsequent timesteps is schematically represented in Fig. 3, showing the possibility for grid level creation and destruction. Starting from time t_1^n, the n-th time step as judged from the level 1 grids, the scheme is traversed from left to right, bottom

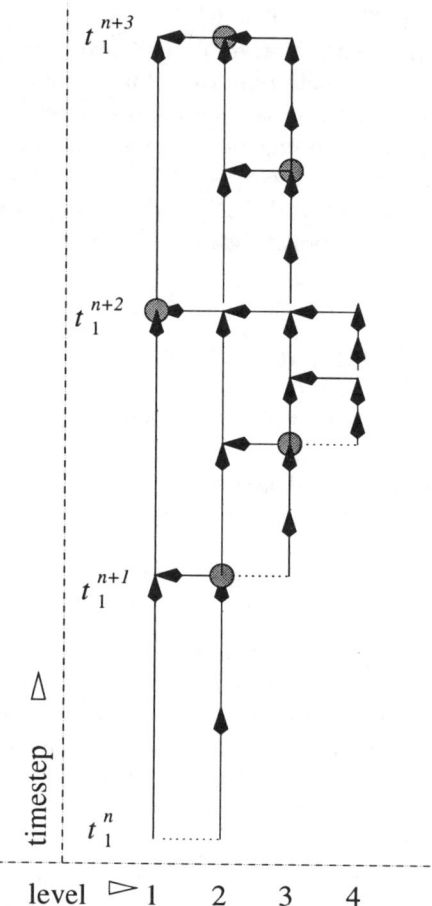

Fig. 3. A hypothetical sequence of three time steps, in an AMR simulation with a maximal allowed nesting level of 4. Vertical 'advance' arrows time-integrate all grids at that particular level; horizontal 'update' arrows pass the more accurate solutions down the level tree; and 'refine' actions (grey circles) may lead to higher level creation or destruction.

to top, and with horizontal 'update' arrows preceeding vertical time 'advance' steps. In the first time step shown, the 'advance' of level 1 is followed by two 'advance' steps on level 2, which are the only levels present at time t_1^n. When level 2 has caught up in time with level 1 (both arrived at time t_1^{n+1}), the coarse solution is 'updated' – indicated by a horizontal arrow – with the finer level 2 solution, where available. This process continues in the second and third time step. However, the sequence is complicated by allowing for newly created (or destroyed) grids on levels $l+1$ up to a maximally allowed level (taken as 4 in Fig. 3). This happens at the locations marked by the grey circles: the grids at that level are unchanged, but all higher level grids can suddenly appear (after the first and halfway in the second time step), disappear (after the second timestep), or simply get rearranged or be left unchanged (halfway in the third time step). The criterion for when a specific 'refine' action (grey circle) takes place is simple: when k timesteps are taken on a certain level, it is evaluated for refinement ($k = 2$ in Fig. 3). However, the maximally allowed finest level 4 is never refined, and a downward cycle of update steps should not lead to duplicate refinements.

We have a Fortran 90 implementation of an AMR scheme, usable for the Euler equations in two space dimensions. The integrator is a finite volume, conservative Flux Corrected Transport [2] algorithm. It should be clear that the update steps mentioned above also involve 'fix' operations at boundaries between level l and level $l + 1$ grids: to ensure global conservation, the fluxes as obtained by the addition of the fine cell fluxes that make up one coarse level l cell replace the fluxes obtained from the level l cells that are covered by a finer mesh.

As an example calculation, we show in Fig. 4 a two dimensional generalization of Harten's [6] shock tube problem, where the bottom right hand corner of a rectangular domain has different constant state variables than those in the rest of the domain. The simulation allows for four grid levels, which are automatically created at time $t = 0$ and nicely follow the discontinuities present. At a slightly later time, the discontinuities in each direction develop locally in combinations of rarefactions, shocks and contact discontinuities. Note in particular how the hierarchically nested grid structure rearranges itself to capture the evolving flow features. Thereby, grids can merge, disappear, shrink or grow in size as imposed by the physics.

We will further translate the Fortran 90 code into LASY syntax [13], so that both 1D, 2D, and 3D applications can be run with the same source code. The coupling with the Versatile Advection Code will open up the possibility to apply the AMR technique to any set of (near) conservation laws, like the (resistive) MHD equations.

4 Conclusions and Outlook

This progress report summarizes our continuing efforts to evaluate and exploit grid adaptation schemes in challenging magnetohydrodynamic computations. We demonstrated the workings of two different approaches, r-refinement and h-refinement, for some idealized model problems. The application of MOL-techniques to MHD problems, in particular in 2D and 3D versions, is a novel research area which should be pursued further along the lines indicated in this manuscript. The more established AMR technique has been applied in MHD problems recently, e.g. see [10, 9, 5], but a dimension independent implementation in combination with a choice in the actual set of conservation laws to solve, will become feasible for the first time when we finish our efforts.

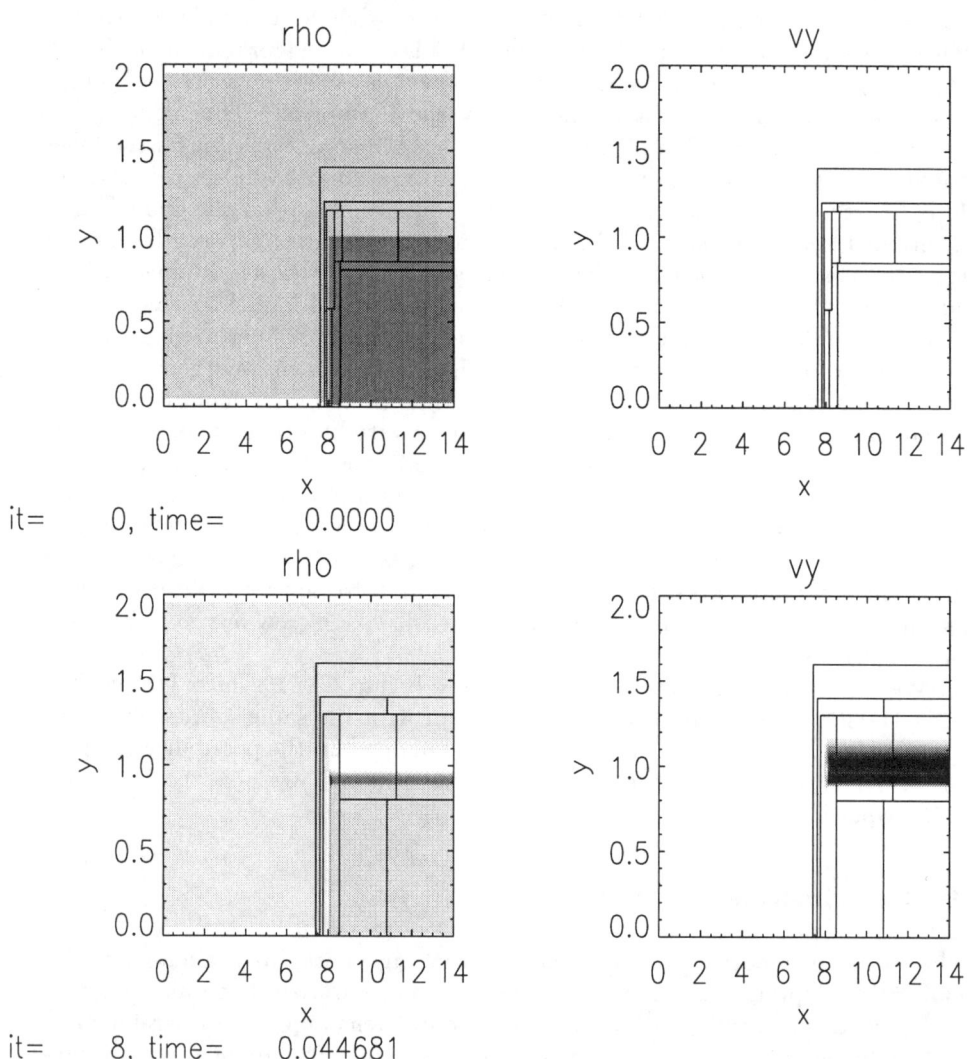

Fig. 4. A 2D hydrodynamic shock tube problem. We show density ρ (left) and y-velocity (right) at times $t = 0.0$ (top) and eight CFL-limited timesteps later (bottom). Four refinement levels, with $r_l = 2, l = 2, 3, 4$, automatically form a nested structure that follows the initial discontinuity. Level 1 is the full square, and the thin lines are the boundaries of the grid patches, which are nested into that. As time evolves, the grids adjust dynamically: note how at $t = 0.0$, five grids on level 4 were formed, which have merged and broadened into three level 4 grids at the last time shown. The underlying level 2 and 3 grids also changed, always ensuring a proper nesting.

The number of applications that become amenable to large-scale numerical simulations promise to keep us and other physicists alike busy for the years to come.

Acknowledgements. This work was performed as part of the research programme of the 'Stichting voor Fundamenteel Onderzoek der Materie' (FOM) with financial support from the 'Nederlandse Organisatie voor Wetenschappelijk Onderzoek' (NWO). Part of this work is done in the project on 'Parallel Computational Magneto-Fluid Dynamics', funded by the NWO Priority Program on Massively Parallel Computing. Another part is sponsored by the 'Stichting Nationale Computerfaciliteiten' (NCF, Grant # NRG 98.10), also for the use of supercomputer facilities.

References

1. Berger, M.J.: Data structures for adaptive grid generation, SIAM J. Sci. Stat. Comput. **7**(3), 904 (1986)
2. Boris, J.P., Book, D.L.: Flux-corrected transport. I.SHASTA, A fluid transport algorithm that works, J. Comput. Phys. **11**, 38 (1973)
3. De Sterck, H., Low, B.C., Poedts, S.: Characteristic analysis of a complex two-dimensional magnetohydrodynamic bow shock flow with steady compound shocks, Phys. of Plasmas **6**, 954 (1999)
4. Dorfi, E.A., Drury, L. O'C.: Simple adaptive grids for 1-D initial value problems, J. Comput. Phys. **69**, 175 (1987)
5. Friedel, H., Grauer, R., Marliane, C.: Adaptive mesh Refinement for Singular Current Sheets in Incompressible Magnetohydrodynamic Flows, J. Comput. Phys. **134**, 190-198 (1997)
6. Harten, A.: High resolution schemes for hyperbolic conservation laws, J. Comput. Phys. **49**, 357 (1983)
7. Keppens, R., Tóth, G.: Simulating Magnetized Plasmas with the Versatile Advection Code, in *VECPAR'98 - Third International Conference for Vector and Parallel Processing*, Lecture Notes in Computer Science, **1573**, edited by J. M. L. M. Palma, J. Dongarra and V. Hernandez p. 680-690 (Springer-Verlag, 1999)
8. Keppens, R., Tóth, G., Westermann, R.H.J., Goedbloed, J.P.: Growth and saturation of the Kelvin-Helmholtz instability with parallel and anti-parallel magnetic fields, J. Plasma Phys. **61**, 1 (1999)
9. Powell, K.G., Roe, P.L., Linde, T.J., Gombosi,T. I., De Zeeuw, D.L.: A Solution-Adaptive Upwind Scheme for Ideal Magnetohydrodynamics, J. Comput. Phys. **154**, 284-309 (1999)
10. Steiner, O., Knölker, M., Schüssler, M.: Dynamic interaction of convection with magnetic flux sheets: first results of a new MHD code, in *Proc. NATO advanced research workshop ASI Series C-433*, Solar Surface Magnetism, edited by R.J. Rutten and C.J. Schrijver, p. 441-470 (Kluwer Dordrecht, 1994)
11. Stenuit, H., Keppens, R., Goossens, M.: Eigenfrequencies and optimal driving frequencies of 1D non-uniform magnetic flux tubes, Astron. & Astrophys. **331**, 392 (1998)
12. Tóth, G.: Versatile Advection Code, in *Proceedings of High Performance Computing and Networking Europe 1997*, Lecture Notes in Computer Science, **1225**, edited by B. Hertzberger and P. Sloot, p. 253–262 (Springer-Verlag, 1997)

13. Tóth, G.: The LASY Preprocessor and its Application to General Multi-Dimensional Codes, J. Comput. Phys. **138**, 981 (1997)
14. Tóth, G., Keppens, R.: Comparison of Different Computer Platforms for Running the Versatile Advection Code, in *Proceedings of High Performance Computing and Networking Europe 1998*, Lecture Notes in Computer Science, **1401**, edited by P. Sloot, M. Bubak and B. Hertzberger p. 368-376 (Springer-Verlag, 1998)
15. Tóth, G., Keppens, R., Botchev, M. A.: Implicit and semi-implicit schemes in the Versatile Advection Code: numerical tests, Astron. & Astrophys. **332**, 1159 (1998)
16. Weiss, N.O.: The expulsion of magnetic flux by eddies, Proc. Roy. Soc. A **293**, 310 (1966)
17. Zegeling, P.A.: r-refinement for evolutionary PDEs with finite elements or finite differences, Applied Numer. Math. **26**, 97 (1998)
18. Zegeling, P.A., Keppens, R.: Adaptive Method of Lines for Magneto-Hydrodynamic PDE Models, in preparation.

Parallelization of Irregular Problems Based on Hierarchical Domain Representation

Fabrizio Baiardi, Sarah Chiti, Paolo Mori, and Laura Ricci

Dipartimento di Informatica, Universitá di Pisa
Corso Italia 40, 56125 - Pisa
{baiardi,chiti,mori,ricci}@di.unipi.it

Abstract. Irregular problems require the computation of some properties for a set of elements that are irregularly distributed in a domain. The distribution may change at run time in a way that cannot be foreseen in advance. Most irregular problems satisfy a locality property because the properties of an element e depend on the elements that are "close" to e. We propose a methodology to develop a highly parallel solution based upon a load balancing strategy that respects locality because e and most of the elements close to e are mapped onto the same processing node. We also discuss the update of the mapping at run time to recover an unbalancing, together with strategies to acquire data on elements mapped onto other processing node. The proposed methodology is applied to the multigrid adaptive problem and some experimental results are discussed.

1 Introduction

Several problems in computer science require the computation of some properties, i.e. speed, position, temperature, illumination etc., for each element in a domain of interest. The computation is iterated either to simulate the system evolution in an interval of time or to improve the accuracy of the results. A problem is irregular if the elements are distributed in the domain in a non homogeneous and dynamic way that cannot be foreseen in advance.

Some important examples of irregular problems are: the Barnes-Hut method for n-body problems [3], the adaptive multigrid method for the solution of partial differential equations [6] and the hierarchical radiosity methods to determine the global illumination of a scene [9].

In all these problems, the properties of an element e_i depend upon those of some other elements, the neighbors of e_i. A problem dependent neighborhood relation determines which elements affect the properties of e_i, but the probability that e_j affects the properties of the e_i is inversely related to the distance between e_i and e_j. In the following, this property will be denoted as *locality*.

A key point in the development of a highly parallel solution of an irregular problem is a load balancing strategy that maps the elements onto processing nodes (p-nodes) while respecting locality, i.e. that maps most of the neighbors of e_i onto the same p-node of e_i . Furthermore, the strategy should also define how the mapping is updated when and if the distribution changes.

Several techniques have been developed to parallelize irregular problems on parallel architectures with distributed memory. Two different parallelization strategy called *Costzone* and *Orthogonal Recursive Bisection* have been described in [14], [16] and [17]. These techniques are based on two different kind of hierarchical decomposition of the domain. The ordering of domain elements throught *space-filling curves* has been adopted in [8], [11] and [19]. Another parallelization approach for irregular problems, called CHAOS, is described in [10] and [15]. This approach requires that the program consists of a sequence of clearly demarcated concurrent loop-nests.

This paper proposes a parallelization methodology for irregular problems that integrates two strategies: a load balancing strategy that respects locality and one to collect remote data, i.e. data of elements mapped onto a distinct p-node. The methodology is independent of the distributed memory parallel architecture, and it only assumes that the p-nodes are connected by a sparse interconnection network. The rest of the paper is organized as follows: sect. 2 describes a hierarchical representation of the domain, sect. 3 discusses the data mapping and the load balancing technique, sect. 4 presents the strategy to collect remote data. The application of our methodology to the adaptive multigrid method and some experimental results are presented in Sect. 5. The application to the Barnes-Hut method has already been described in [2]

2   A Hierarchical Representation of the Domain

In the following, we assume that the problem domain belongs to a space with n dimensions. The proposed methodology exploits a hierarchical representation of the problem domain. At each level the domain is partitioned into a set of n-dimensional *spaces*. The procedure that partitions a space S is recursive and it starts from the space that represents the whole domain. If a problem dependent condition is satisfied, S is partitioned by halving each side to produce 2^n equal subspaces, and the procedure is applied to each resulting space. The spaces including a large number of elements are partitioned into finer spaces than the other ones. The whole decomposition is represented through the *Hierarchical Tree*, H-Tree. Each node of the H-Tree, hnode, represents a space and the root represents the whole problem domain. Each space S considered in the decomposition is represented by one hnode and this hnode records information on the elements in S. Larger spaces are paired with abstract information, while smaller ones are paired with a more detailed information. In the following, *space(N)* denotes the space represented by the hnode N, while *node(S)* denotes the hnode representing the space S. We notice that each hnode either is a leaf or has 2^n sons. If N is a leaf, *space(N)* is not decomposed.

Because of the non uniform distribution of the elements, two distinct subtrees rooted in the same hnode may have different heights. If the number of elements and/or their distribution change during the computation, the partition of the domain and the H-Tree are to be updated according to the current distribution. As soon as *space(N)* is partitioned, 2^n sons of N are inserted, while as soon as *space(N)* is no longer partitioned, the sons of N are pruned.

Our methodology assumes that the H-Tree cannot be replicated in each p-node, because of its memory requirement. Instead, we consider $np + 1$ subset of the H-Tree, where np is the number of the p-nodes. One subset is the replicated H-Tree, replicated in all the p-nodes. Each of the other subsets is stored in one p-node only and it is the private H-Tree of the p-node.

3 Data Mapping and Runtime Load Balancing

To take locality into account, we propose a three step mapping: *i)* spaces ordering; *ii)* determination of the computational load of every hnode and *iii)* order preserving mapping of the spaces onto the p-nodes.

The spaces are ordered through a space filling curve built on the spaces hierarchy representing the domain decomposition. Space filling curves are a family of curves that visit any point of a given space [12]. The curve is built starting from the lowest level spaces, i.e. from the first partition of the problem domain. These spaces are visited in the order stated by the characteristic figure of the adopted curve. If a space has been partitioned, then all its subspaces are visited in a recursive way, before the next space at the same level. Any space filling curve *sf* also defines a visit *v(sf)* of the H-Tree that returns a sequence $S(v(f)) = [N_0, N_1, ..., N_m]$ of hnodes. Alternative space filling curves may be adopted because the curve dependent aspects of *v(sf)* may be encapsulated into a function *next_son(N)*, that determines the next son of N to be visited. If implemented through a table look-up, *next_son(N)* is computed in a constant time.

The *computational load* of a hnode N is a problem dependent metric that evaluates the amount of computations on the elements in *space(N)*. According to the considered problem, the load can be *i)* constant during the computation and equal for all the hnodes, *ii)* constant during the computation but distinct for each hnode, or *iii)* variable during the computation and distinct for each hnode.

The np p-nodes of the distributed memory architecture are ordered too. A p-node P_j immediatly precedes P_k in the ordered sequence SP, if the cost of an interaction between P_j and P_k is not larger than the cost of the same interaction between P_j and any other p-node. Since each p-node executes one process, P_k also denotes the process executed on the k-th p-node of SP.

To preserve the ordering among the spaces, they are mapped by partitioning the hnodes through a blocking strategy. The sequence $S(v(sf))$ is partitioned into np segments, i.e. into np subsequences of consecutive hnodes. The overall load of a segment should be as close as possible to *average_load*, i.e. to the ratio between the overall computational load and the number of p-nodes. We cannot require that the load of each segment is equal to *average_load*, because each hnode, and its load, is assigned to one process only. In the following, $= (S, C)$, where S is a segment and C is a constant, denotes that the load of S is as close as possible to C. Due to the large number of elements, the difference between *average_load* and the assigned workload is negligible. Then, the first segment is mapped onto the p-node P_0, the second onto P_1 and so on. The resulting mapping satisfies the *range property*: if the hnodes N_i and N_{i+j} are assigned to process P_h, then all the hnodes in-between N_i and N_{i+j} in $S(v(sf))$, are assigned to P_h as well.

After the data distribution, each process P_h builds the replicated H-Tree and its private H-Tree. The private H-Tree of P_h includes a hnode N if $space(N)$ is assigned to P_h. The replicated H-Tree is the union of the paths from the H-Tree root to the root of each private H-Tree. Each hnode N of the replicated H-Tree records the position of $space(N)$ in the domain and the identifier of the owner process. In some problems, the intersection among the private H-Tree and the replicated H-Tree includes the roots of the private H-Tree only. In other problems, the private H-Trees and the replicated H-Tree are partially overlapped.

In all the problems that either emulates the evolution of a system or achieves the required accuracy of the results through iteration, the data mapping chosen at the i-th iteration could result in an unbalanced load at a later iteration. We define a procedure to correct an unbalance while minimizing the correspondly overhead. To detect when the procedure has to be applied, each process periodically broadcasts its workload, and it computes *max_unbalance*, the current maximum load unbalance, that is the largest difference between *average_load* and the workload of each process. If *max_unbalance* is larger than a tolerance threshold T, then each process executes the procedure. The threshold avoids that the procedure is executed to correct a very low unbalance. Let us suppose that the workload of P_i is *average_load* + C, $C > T$, while that of P_j, $i < j$, is *average_load* - C. To solve the unbalance P_i cannot send to P_j a set S of hnodes where $= (S, C)$ because the resulting allocation violates the range property. The correct procedure can be sketched as a shift of the spaces that involves all the processes in-between P_i and P_j. Let us define $Prec_i$ as the set of processes $[P_0...P_{i-1}]$ that precede P_i in the sequence SP, and $Succ_i$ as the set of processes $[P_{i+1}...P_{np}]$ that follow P_i in SP. Furthermore, $Sbil(Prec_i)$ and $Sbil(Succ_i)$ are, respectively, the global load unbalances of the sets $Prec_i$ and $Succ_i$. If $Sbil(Prec_i) = C > T$, i.e. processes in $Prec_i$ are overloaded, P_i receives from P_{i-1} a segment S where $= (S, C)$. If, instead, $Sbil(Prec_i) = C < -T$, P_i sends to P_{i-1} a segment S where $= (S, C)$. The same procedure is applied to $Sbil(Succ_i)$ but, in this case, the hnodes are either sent to or received from P_{i+1}.

To respect the range property, if $[N_q....N_r]$ is the segment of hnodes it has been assigned, P_i sends to P_{i-1} a segment $[N_q....N_s]$, with $q \leq s \leq r$, while it sends to P_{i+1} a segment $[N_t....N_r]$, with $q \leq t \leq r$.

4 Fault Prevention

Each process computes the properties of all the elements in the spaces it has been assigned. To compute the properties of e, the process needs those of the neighbors of e that may have been allocated onto other p-nodes. A simple and general strategy to collect such remote data is *request/answer*. During the computation, as soon as P_h needs data of a space S assigned to P_k, it suspends the computation and sends a request to P_k. This strategy requires two communications for each remote data, one from P_h to P_k, and one from P_k to P_h.

To reduce this overhead, we introduce the *fault prevention* strategy. P_k, the owner of the space S, determines, through the neighborhood stencil, which processes require the data of S, and it sends to these processes the data, without any explicit request. To determine all the data to be sent to P_h, P_k exploits the information in the replicated H-Tree on the subspaces assigned to P_h. In general, P_k approximates the data that P_h requires, because the replicated H-Tree records a partial information only. The approximation is always safe, i.e. it includes any data P_h needs.

Fault prevention requires at most one communication for each other p-node to collect the remote data but, if the accuracy of the approximation is low, most of the data sent are useless. In some problems, the information in the replicated H-Tree enables P_k to determine a set of data that is not much larger than the one required by P_h. In other problems, instead, to improve the accuracy of the approximation, processes exchange some information about their private H-Trees before the fault prevention phase. Also the time to compute the data to be sent is rather important, because it cannot be larger than the one to explicitly request the data to the other p-nodes.

5 Adaptive Multigrid Methods

Adaptive multigrid methods are fast iterative methods based on multi level paradigms to solve partial differential equations in two or more dimensions [6], [7]. They may be applied to compute the turbulence of incompressible fluids [5], for macromolecular electrostatic calculation in a solvent [18], to solve plane linear elasticity problems [4] and so on.

The adaptive method builds the grid hierarchy during the computation, accordingly to the considered partial differential equation. Each grid of the hierarchy partitions the domain, or some parts of it, into a set of square spaces; the value of the equation is computed in the corners of each square. The hierarchy is built during the computation starting from an uniform grid, the level 0 of the hierarchy. Let us suppose that, at a level l, a square A has been discretized though the grid g. To improve the accuracy of the values in A, a grid finer than g is added at level $l+1$. Also the new grid represents A, but it doubles the points of g on each dimension. This doubles the accuracy of the discretization in A. As the computation goes on, finer and finer grids may be added to the hierarchy until the desidered accuracy has been reached in each square.

5.1 Multigrid Operators and V-Cycle

In the following, we consider the solution of the second order *Poisson differential equation* in two dimensions, subject to the *Dirichlet boundary conditions*:

$$-\frac{d^2u}{dx^2} - \frac{d^2u}{dy^2} = f(x,y) \qquad \Omega = 0 < x < 1, \ 0 < y < 1 \qquad (1)$$

$$u(x,y) = h(x,y) \qquad (x,y) \in \delta\Omega \qquad (2)$$

To solve the considered equation, multigrid operators are applied on each grid of the hierarchy in a predefined order, the V-cycle. We briefly describe the main multigrid operators and the V-cycle; for a complete description see [7].

The *smoothing (or relaxation) operator* usually consists of some iterations of the Gauss-Seidel method or the Jacobi one and it is applied on each grid g to improve the approximation of the current solution on g.

The *restriction operator* maps the current solution on the grid at level l onto the grid at level l-1. The value of each point p on the grid at level l-1 is a weighted average of the value of the neighbors of p in the grid at level l.

The *prolongation operator* maps the current solution on the grid at level l-1 onto the grid at level l. If a point exists on both grids, its value is copied. The value of any other points at level l is an interpolation of the value of the neighbor of p on the grid at level l-1.

The *refinement operator*, if applied to a grid, or to a part of it, at level l, adds a new grid to the hierarchy at level l+1. The new grid represents the same square but it doubles the number of points on each dimension.

The V-cycle includes a downward phase and an upward one. The downward phase applies the smoothing operator to each grid, from the highest level to the one at level 0. Before applying this operator to the grid at level l, the restriction operator maps the values on the grid at level l+1 onto the one at level l. The upward phase is symmetric to the downward one; the smoothing operator is applied to each grid, from that at level 0 to the highest level ones. Before applying the smoothing operator to a grid at level l, the prolongation operator maps the values of the grid at level l-1 to the one at level l.

At the end of the V-cycle, the results are evaluated through an error estimation criteria. The refinement operator is applied to all the squares violating the criteria before starting another V-cycle.

5.2 Data Mapping

The resolution of partial differential equations through the adaptive method is an irregular problem, because the discretization of the domain and, consequently, the distribution of the points, are not uniform and not foreseeable. Moreover, adaptive methods are highly dynamic because the grid hierarchy is a function of the considered domain.

The load balancing procedure should take into account two aspects of locality because the value of a point p on the grid g at level l is function of the values of the neighbors of p *i)* on the same grid g for the smoothing operator (intra-grid or horizontal locality) *ii)* on the grids at level l+1 (if it exists) and l-1 for the prolongation and restriction operators (inter-grid or vertical locality).

To apply the proposed methodology to this problem, the square spaces of all the grids of the hierarchy are ordered by visiting the domain through a space filling curve, and they are mapped as shown in sect. 3.

In the adaptive method, the hnodes at level l represent all the squares of the grids at level l, each hnode has either 4 sons or none and the squares associated to the sons of the hnode N represent the same square of N, but the number of points on each dimension is doubled.

To determine the computational load of an hnode we notice that the number of operations is the same for each point of a grid and does not change during the computation. Hence, the same computational load is assigned to each point, i.e. to each hnode, and we assign to each p-node the same number of squares.

In general, the domain subset assigned to each process is a set of squares that belong to grids at distinct levels. We denote by $Do(P_h)$ the subdomain assigned to process P_h. For each square it has been assigned, a process has to compute one point, the rightmost downward corner of the square.

The private H-Tree of process P_h includes all the hnodes representing the squares assigned to P_h, while the replicated H-Tree includes all the hnodes on the paths from the root of the H-Tree to the roots of the private H-Trees.

A hnode can belong to more than one tree, because the computation is executed both on intermediate hnodes and on the leaves. To show that the replicated H-Tree and the private H-Tree are not disjoint, consider a hnode N assigned to the process P_h. If one of the descendant of N has been assigned to P_k, $h \neq k$, N belongs to the private H-Tree of P_h, because P_h computes the value of the points in $space(N)$, and to the replicated H-Tree because it belongs to the path from the H-Tree root to the root of the private H-Tree of P_k.

5.3 Fault Prevention

Each process P_h applies the multigrid operators, in the order stated by the V-cycle, to the points of the squares in $Do(P_h)$.

Let us define $Pe(P_h)$, the boundary of $Do(P_h)$, as the sets of the squares in $Do(P_h)$ such that one of the neighbors does not belong to $Do(P_h)$. $Pe(P_h)$ depends upon the neighborhood relation of the operator op that is considered. To apply op to the points in $Pe(P_h)$, P_h has to collect the values of points in squares assigned to other processes. Let us define $I_{h,op,liv}$ as the set of squares of the domain not belonging to $Do(P_h)$ and including the points whose values are required by P_h to apply op to the points in the subgrid at level liv of $Do(P_h)$. Each of these squares belongs to $Pe(P_z)$, for some $z \neq h$. The values of points in $I_{h,op,liv}$ are exchanged among the processes just before the application of op, because they are updated by the operators applied before op in the V-cycle.

If fault prevention is adopted, P_h does not compute $I_{h,op,liv}$; instead, each process P_k determines the squares in $Pe(P_k)$ belonging to $I_{h,op,liv}$, $\forall h \neq k$. P_k exploits the information in the replicated H-Tree about $Do(P_h)$ to determine $I_{h,op,liv}$. Since this information could be not detailed enough, P_k computes an approximation AI of $I_{h,op,liv}$; in order to make a safe approximation, P_k includes in AI all the points that could have a neighbor at level liv in $Do(P_h)$ according to the neighborhood stencil of op. Then P_k sends to P_h, without any explicit request, the values of the points in AI. To show that, due to the approximation, some of these values may be useless for P_h, suppose that $Do(P_k)$ and $Do(P_h)$

share a side, that $Do(P_k)$ and $Do(P_h)$ have been uniformly partitioned until, respectively level l and level l-m, and that the neighborhood stencil of op for the point p involves only the points on the same level of p. Since P_k does not know l-m, it could send to P_h some of its square on $Pe(P_k)$ at level higher than l-m that are useless for P_h, because it has no point on these levels.

To reduce the amount of useless data, we introduce *informed fault prevention*. If $Do(P_h)$ and $Do(P_k)$ share a side, P_h sends to P_k, before the fault prevention phase, the depth of each square in $Pe(P_h)$ that could have a neighbor in $Pe(P_k)$, l-m in the previous example. This information allows P_k to improve the approximation of the set of points to be sent to P_h. The information on the depth of the squares in $Pe(P_k)$ is sent by P_k at the beginning of each V-cycle and it is correct until the end of the V-cycle, when the refinement operator may add a new grid. If the load balance procedure, that updates $Pe(P_k)$, is applied, then at the beginning of the V-cycle P_k has to send the depth of all the squares on $Pe(P_k)$. Otherwise, since the refinement operator cannot remove a grid, each process has to send information on the squares of the new grids only.

The informed fault prevention technique is applied to the refinement operator too, but in this case the set of data to be sent to each process is always approximated. In fact, whether the process P_h, that owns the square of a point p, needs or not the square of the point q, owned by P_k, depends upon the value of p. Since P_h, at the beginning of the V-cycle, sends to P_k the depth of squares in $Pe(P_h)$, but not the values of the points, it could receive some useless values.

5.4 Experimental Results

We present some experimental results of the parallel version of the adaptive multigrid algorithm resulting from our methodology. The parallel architecture we consider is a Cray T3E [1]; each p-node has a DEC 21164 EV5 processor and 128Mb of memory. The interconnection network is a torus. The programming language is C extended with the Message Passing Interface primitives [13].

The simulations solve two problems derived from the equation (1), with $f(x, y) = 0$ and two different boundary conditions (2), denoted by $h1$ and $h2$:

$$h1(x, y) = 10 \qquad h2(x, y) = 10 \cos(2\pi(x - y)) \frac{\sinh(2\pi(x + y + 2))}{\sinh(8\pi)}$$

The solution of the Poisson equation is simpler than other equations such as the Navier-Stokes one. Hence, the ratio between computational work and parallel overhead is low and this is a significant test for a parallel implementation.

Figure 1 compares the remote data collecting techniques. We plot the total number of communications for request/answer (req/ans), for fault prevention (fp) and for informed fault prevention (ifp). In both problems, the number of communications of fault prevention and of informed fault prevention are, respectively, less than 61% and 52% than those of request/answer. As previously explained, because of the refinement operator, the number of communications of informed fault prevention is larger than 50% of the request/answer one, but the amount of useless data is less than 2%.

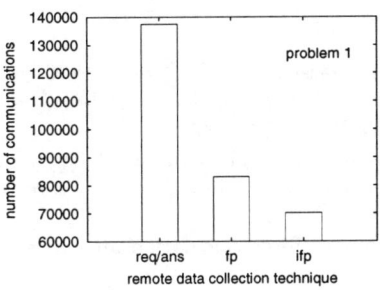
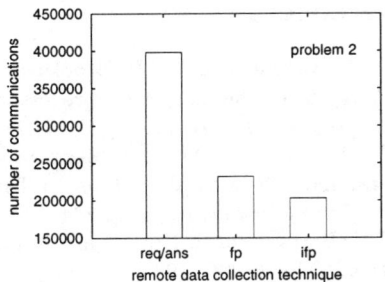

Fig. 1. A Comparison of remote data collection techniques

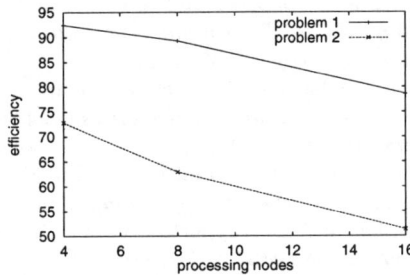

Fig. 2. Efficiency for problems with fixed data dimension

Figure 2 shows the efficiency of the parallel multigrid algorithm for the two problems, for a fixed number of initial points, 2^{14}, the same maximum grid level, 12, and a variable number of p-nodes. These simulations exploit informed fault prevention. The low efficiency resulting in the second problem is due to an highly irregular grid hierarchy. However, even in the worst case, our solution achieves an efficiency larger than 50% even on 16 p-nodes.

6 Conclusions

This paper has presented a methodology for the parallelization of irregular problems based upon the hierarchical structuring of the domain, a mapping strategy based upon space-filling curves and a technique, fault prevention, that reduces the communications overhead by preventing the data faults. This methodology has been previously applied to parallelize the n-body problem [2]. The results of our numerical experiments show that this approach achieves good performances on high parallel distributed memory architectures.

We plan to extend the approach by considering other irregular problems, such as hierarchical radiosity, in order to define a package that simplify the developement of parallel solutions to irregular problems. A further development concerns the evaluation of our approach in the case of networks of workstations.

References

[1] E.C. Anderson and J.P. Brooks and C.M. Gassi and S.L. Scott. Performance Analysis of the T3E Multiprocessor *SC'97: High Performance Networking and Computing: Proceedings of the 1997 ACM/IEEE SC97*, 1997.

[2] F. Baiardi, P. Becuzzi, P. Mori, and M. Paoli. Load balancing and locality in hierarchical N-body algorithms on distributed memory architectures. *Lecture Notes in Computer Science*, 1401:284–293, 1998.

[3] J.E. Barnes and P. Hut. A hierarchical O(nlogn) force calculation algorithm. *Nature*, 324:446–449, 1986.

[4] P. Bastian, S. Lang, and K. Eckstein. Parallel adaptive multigrid methods in plane linear elasticity problems. *Numerical linear algebra with applications*, 4(3):153–176, 1997.

[5] P. Bastian and G. Wittum. Adaptive multigrid methods: The UG concept. In *Adaptive Methods – Algorithms, Theory and Applications*, volume 46 of *Notes on Numerical Fluid Mechanics*, pages 17–37, 1994.

[6] M. Berger and J. Oliger. Adaptive mesh refinement for hyperbolic partial differential equations. *Journal of Computational Physics*, 53:484–512, 1984.

[7] W. Briggs. *A multigrid tutorial*. SIAM, 1987.

[8] M. Griebel and G. Zumbusch. Parallel multigrid in an adaptive PDE solver based on hashing and space-filling curves. *Parallel Computing*, 25(7):827–843, 1999.

[9] P. Hanrahan, D. Salzman, and L. Aupperle. A rapid hierarchical radiosity algorithm. *Computer Graphics*, 25(4):197–206, 1991.

[10] Y.S. Hwang, R. Das, J.H. Saltz, M. Hodoscek, and B.R. Brooks. Parallelizing molecular dynamics programs for distributed-memory machines. *IEEE Computational Science & Engineering*, 2(2):18–29, 1995.

[11] M. Parashar and J.C. Browne. On partitioning dynamic adaptive grid hierarchies. In *Proceeding of the 29th annual Hawaii international conference on system sciences*, 1996.

[12] J.R. Pilkington and S.B. Baden. Dynamic partitioning of non–uniform structured workloads with space filling curves. *IEEE Transaction on parallel and distributed systems*, 7(3):288–299, 1996.

[13] M. Prieto, D. Espadas, I.M. Llorente, and F. Tirado. Message passing evaluation and analysis on Cray T3E and SGI Origin 2000 systems. *Lecture Notes in Computer Science*, 1685:173–182, 1999.

[14] J.K. Salmon. *Parallel hierarchical N-body methods*. PhD thesis, California Institute of Technology, 1990.

[15] S. Sharma, R. Ponnusamy, B. Moon, Y. Hwang, R. Das, and J. Saltz. Runtime and compile-time support for adaptive irregular problems. In *Proceedings of Supercomputing*, pages 97–106, 1994.

[16] J.P. Singh. *Parallel hierarchical N-body methods and their implications for multiprocessors*. PhD thesis, Stanford University, 1993.

[17] J.P Singh, C. Holt, T. Totsuka, A. Gupta and J.L. Hennessy. Load balancing and data locality in adaptive hierarchical n-body methods: Barnes-Hut, Fast Multipole and Radiosity *Journal of Parallel and Distributed Computing*, 27(2):118–141, 1995.

[18] Y.N. Vorobjev and H.A. Scheraga. A fast adaptive multigrid boundary element method for macromolecular electrostatic computations in a solvent. *Journal of Computational Chemistry*, 18(4):569–583, 1997.

[19] M.S. Warren and J.K. Salmon. A parallel hashed oct-tree N-body algorithm. In *Proceedings of Supercomputing '93*, pages 12–21, 1993.

Dynamic Iterative Method for Fast Network Partitioning

Chang-Sung Jeong*, Young-Min Song, and Sung-Up Jo

Department of Electronics Engineering, Korea University
1-5Ka, Anam-dong, Sungbuk-ku, 136-701, Korea
Tel: (02) 3290-3229
csjeong@charlie.korea.ac.kr

Abstract. In this paper, we address multiway network partitioning problem of dividing the cells of network into multiple blocks so as to minimize the number of nets interconnecting cells in different blocks while balancing the blocks' sizes. The sequential iterative improvement algorithm for the problem consists of several passes each of which is performed by repeatedly iterating the move operation. Therefore, the whole execution time taken by the algorithm is greatly affected by the number of the move operations and the execution time for each move operation. We present a fast parallel algorithm for solving the multiway network partiotioning problem by reducing both of them. We propose a new dynamic iterative method which reduces the number of move operations executed at each pass dynamically, and hence speed up the whole algorithm sharply. Moreover, we reduced the execution time of each move by its parallelization using the proper cell distribution.

1 Introduction

In this paper, we address *b-way network partitioning problem* of dividing the cells of the network into b blocks so as to minimize the number of nets interconnecting cells in different blocks while balancing the blocks' sizes. It has numerous applications such as VLSI layout, task assignment, placement of components onto printed circuit boards, etc[1-5]. Since the network partitioning problem is NP-hard, attempts to solve the problem have been concentrated on finding heuristics which yield approximate solutions in polynomial time.

Kernighan and Lin suggested a heuristic procedure for 2-way graph partitioning problem, which became the basis for most of the iterative improvement method [1]. They solve the problem by repeatedly performing the pairwise exchange of cells residing in different two blocks. Several variations on their algorithm have been reported and used for VLSI applications practically [2,3,5]. Instead of making repeated use of 2-way pairwise exchange as in the previous algorithms, Krishnamurthy and Sanchis proposed a multi-way network partitioning algorithm which adopted the concept of cutset gain, i.e. the cell with the largest gain in cutset cost is selected as the candidate for the one move operation with the objective of minimizing cutset cost[3,5]. Also the candidate cell

---

* This work has been supported by KRF

must satisfy some size constraint so as to achieve the final size balancing. They treat the two objectives in the network partitioning problem, i.e., minimizing cutset cost and balancing the blocks' sizes in not a coherent but respective ways, having some difficulties in meeting both objectives. Kim and et. al. presented an iterative improvement algorithm which can achieve the two objectives in a coherent way by combining the cutset cost and balance cost into a single cost function of weighted sum, and showed that their method outperforms the Sanchis[4]. (From now on, we shall refer to the algorithm proposed by Kim and et. al. as KL algorithm for simplicity.) Our parallel algorithm is based on KL algorithm.

The KL algorithm for the network partitioning problem consists of several passes each of which is performed by repeatedly iterating the move operation. Therefore, the whole execution time taken by the algorithm is greatly affected by the number of the move operations and the execution time for each move operation. We present a fast parallel algorithm for solving the multiway network partitioning problem by reducing both of them. We propose a new dynamic iterative method which reduces the number of move operations executed at each pass dynamically, and hence speed up the whole algorithm sharply. Moreover, we reduce the execution time of each move by its parallelization using the proper cell distribution and load balancing schemes. Our algorithm is implemented on a hypercube network of transputers, and we shall show that ours outperforms the sequential one in that it will decrease the overall execution time significantly compared to the original sequential one without degrading the whole performance of the problem.

The outline of our paper is as follows: In section 2 we explain some definitions and notations used in the description of our algorithm. In section 3, we describe KL algorithm, and in section 4 present a parallel algorithm for solving the network partitioning problem. In section 5, we describe the experimental result. In section 6, we give a conclusion.

2 Network Partitioning Problem

A network consists of a set of cells $C = \{C_1, C_2, .., C_c\}$ connected by a set of nets $N = \{N_1, N_2, .., N_n\}$. Each net connects two or more cells and each cell is on at least one net. We denote by c the number of cells in the network and by n the number of nets. A b-way partition of a network is a partition of the cells of the network into b blocks, represented by a b tuple $B = \{B_1, B_2, ..., B_b\}$. Each block B_i is a set of cells whose union is the set of the entire cells in the network. The *cutset* of a partition is the set of nets with cells in two or more blocks of the partition. The *cutset cost* W_c of a partition is equal to the sum of all nets' costs, each of which is obtained by assigning cost $k-1$ to each net with cells in k blocks. Similarly, the *balance cost* W_b is defined as the sum of the balance costs of all cells such that a partition with a lower balance cost is a better balanced partition. The total cost of a partition is defined as $W_c + f \cdot W_b$, where the balancing factor f is some nonnegative value which may be set by the system designer. The *gain* $G(C_i, B_d)$ of each cell C_i in block B_s is defined as the value by which the total cost decreases when C_i is moved from B_s to B_d. We shall not describe how the gain of each cell is obtained, but refer to [4] for more detail.

3 KL Algorithm

In this section, we give a brief overview of KL algorithm. Given a network NT, it finds a b way partition $PART[1..c]$, where $PART[i]$ represents a block where cell C_i belongs to. The basic approach is to start with an arbitrary initial partition and to improve it by iteratively choosing a set of cells to be moved. The description of the algorithm is shown in Algorithm KL below. It consists of a series of passes. At step 1) of each pass, we repeatedly iterate the move operation as many times as that of cells, and at step 2), find, from the information produced from the move operations, the partition with the smallest total cost which shall be used as the starting partition for the next pass. Passes are performed until no improvement in the total cost can be obtained. Initially, all the cells are free cells with $Free$ array values set to 0. The mth move operation executes the followings: 1) updating the gains of all the free cells, 2) find the free cell C_l with the largest gain $G(C_l, B_j)$, 3) locking the cell C_l so that it may not be chosen in the next move operation, and 4) storing the information about cell C_l including its gain $G(C_l, B_j)$ and destination block B_j to be moved. After step 1), the cells are sorted in Tmp_gain array according to their gains in nonincreasing order, and the information about the cells, their gains and destination blocks are stored in Tmp_C, Tmp_gain and Tmp_B arrays respectively. At step 2.1), we find a set M_p of the cells to be moved by choosing k_p which maximizes the partial sum of gains in Tmp_gain, and at step 2.2), those cells in M_p are moved to their destination blocks by changing their corresponding contents of $PART$ array. If $T \leq 0$, we have arrived at a locally optimum partition, and the algorithm stops.

Algorithm KL

Input: A network NT, b and an initial partition $PART[1..c]$, where b is a number of blocks and $PART[i]$ is a block where the cell C_i belongs to.
Output: A final partition $PART[1..c]$.

For each pass $p = 1, 2, ...$ do the following:
1). For each move $m = 1, 2, 3, .., c$ do the following:
1.1). Calculate, for each free cell C_i, its gain $G(C_i, B_j)$, for every block B_j.
1.2). Find the cell C_l with the largest gain $G(C_l, B_j)$.
1.3). Lock the cell C_l by setting $Free[l]$ to 1.
1.4). Store the largest gain $G(C_l, B_j)$ into $Tmp\_gain[m]$, and l and B_j into $Tmp\_C[m]$ and $Tmp\_B[m]$ respectively.

2). Find the best partition for pass p as follows:
2.1). Calculate k_p which maximize $T = \sum_{i=1}^{k_p} Tmp\_gain[i]$. If $T \leq 0$, then stop
2.2). For each $i = 1, .., k_p$, set $PART[Tmp\_C[i]]$ to $Tmp\_B[i]$.

End Algorithm.

4 Parallel Algorithm

As we see in the previous section, KL algorithm consists of a series of passes, each of which in turn iterates the move operations. Therefore, the whole execution time taken by KL algorithm is greatly affected by the number of the move operations and the execution time of each move operation. The basic approach of our parallel algorithm is to reduce both of the number of the move operations and the execution time of each move operation in order to speed up the whole algorithm. The former is achieved by using the dynamic iterative method, and the latter by the parallelization of each move, which shall be described in detail in the subsequent sections.

4.1 Dynamic Iterative Method

In this subsection, we describe a dynamic iterative method which decreases the number of move operations at each pass dynamically and hence reduces the overall number of move operations in the whole algorithm. We define the *acceptance ratio* AR_p of pass p as $\frac{k_p}{n_p}$, and the *rejection ratio* RR_p as $1 - AR_p$ respectively, where n_p is the number of move operations at pass p. The *average acceptance ratio* AAR is defined as the average of the acceptance ratios for all the passes. For KL algorithm, n_p is fixed to c. Note that k_p represents the number of move operations which contribute to the partition of pass p. That is, only the cells C_l obtained in the first k_p move operations at step 1) of KL algorithm are moved to their destination blocks at step 2). Therefore, if we can decrease the number of move operations down to the point that is greater than, but close to k_p at each pass p dynamically, we can reduce the time taken for the pass sharply without degrading the overall total cost of the network partition.

Figure 1 shows an example of acceptance ratio graph for each pass of KL algorithm, where $b = 10$, $c = 400$, and $n_p = c$. As we see in the figure, the acceptance ratio of each pass tends to decrease rapidly as the number of pass increases, that is, $AR_1 = 0.73, AR_2 = 0.40, AR_3 = 0.13, ..$, and so on. The average acceptance ratio of the graph is 0.18, which tells that the algorithm spends too much time on the move operations which shall be rejected. This is due to the fact that the number of cells to be moved usually becomes smaller rapidly as the algorithm approaches its optimal state. Therefore, we can speed up each pass without degrading the performance of the algorithm by reducing n_p, the number of move operations in each pass p dynamically while satisfying the condition $n_p \geq k_p$. We define n_p as $\frac{c}{f(i)}$ in our parallel algorithm, where $f(i) = i * p$ is a dynamic function proportional to the pass number p with

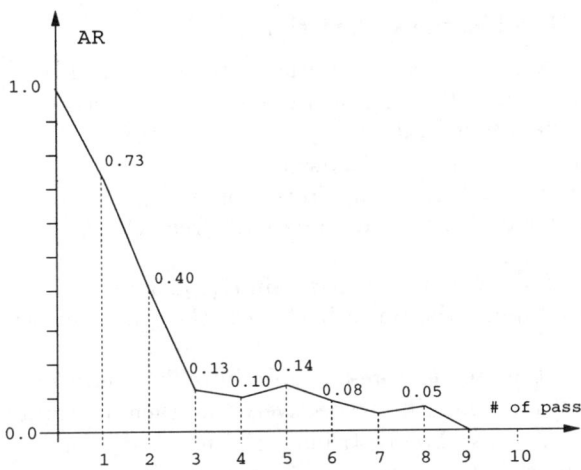

Fig. 1. Acceptance ratio graph for a network: $b = 10$, $c = 400$, and $n_p = c$

constant i. Our experimental result for various sample networks shows that it always satisfies the condition $n_p \geq k_p$ when $i = 1$. As i increases, each pass can be executed faster, but the condition may be violated with some loss in the cost of the network partition at some passes, which may result in the overall delay due to the increase in the number of passes.

4.2 Algorithm

The description of the parallel algorithm is shown in Parallel Algorithm Network_Partition below. Initially, the cells are distributed among PEs(Processing Elements) so that each PE stores a subset of cells together with the information about their network configuration. At step 1) of each pass, we repeatedly iterate the move operation as many times as that inversely proportional to the pass number. At each mth move operation of step 1), every PE storing the cells updates their gains, finds the cell with the maximum gain in parallel, and distributes the maximum gain to the other PE's for the calculation of the best partition which maximizes the partial sum of gains. Let CS_i be a set of cells distributed in PE i. After step 1), each PE i stores, for all cells, their gains in TMP_gain array, and for each cell C_l in CS_i, its associated move number m and destination block B_j in $TMP\_M[l]$ and $TMP\_B[l]$ respectively if C_l is locked at step 1.3). Finally, the maximum partial sum is obtained by prefix operation at step 2.1), and the cells with gains in the maximum partial sum are moved to their new destination blocks at step 2.2). The distribution at step 2.2 allows load balancing among PEs such that each PE works on $\frac{n_p}{m}$ elements in Tmp_gain array for the calculation of partial sum at step 2).

Parallel Algorithm Network_Partition

Input: A network NT, b and an initial partition $PART[1..c]$, where b is a number of blocks and $PART[i]$ is a block where the cell C_i belongs to.
Output: A final partition $PART[1..c]$.
For each pass $p = 1, 2, ...$ do the following:
1). For each move $m = 1, 2, 3, ..., n_p$ do the following:
1.1). Calculate, for each PE containing each free cell C_i, its gain $G(C_i, B_j)$, for every block B_j.
1.2). Find the cell C_l with the largest gain $G(C_l, B_j)$.
1.3). The PE containing the free cell C_l locks the cell C_l by setting $Free[l]$ to 1.
1.4). The PE containing C_l stores its gain into $Tmp\_gain[m]$, and m and B_j into $Tmp\_M[l]$ and $Tmp\_B[l]$ respectively, and then distribute $Tmp\_gain[m]$ to all the other PE's. The distribution is necessary for load balancing of PEs such that each PE works on $\frac{n_p}{m}$ elements in Tmp_gain array for the calculation of partial sum at step 2)).
2). Find the best partition for pass p as follows:
2.1). Calculate k_p which maximize $T = \sum_{i=1}^{k_p} Tmp\_gain[i]$. Each PE stores $\frac{n_p}{m}$ elements in $Tmp\_gain[m]$. If $T \le 0$, then stop
2.2). For each PE with C_l such that $Tmp\_M[l] \le k_p$, set $PART[l]$ to $Tmp\_B[l]$.
End Algorithm.

4.3 Time Complexity Analysis

According to the scheme of communication between processors, parallel machine can be divided into shared memory model(SMM) and distributed memory model(DMM) [6]. In DMM each processor communicates with other processors through a fixed interconnection network[7]. Cube-connected computer(CCC) and Mesh-connected computer(MCC) are the well known examples of DMM, and have been long used in a wide variety of applications [7–14]. We shall use a cube-connected computer(CCC) as our model of parallel computation. In CCC each processor is connected to the other processors whose indices are different in only one bit position from its index, and it takes $O(logm)$ for distribution and prefix operations on CCC with m processors which is faster than $O(\sqrt{m})$ taken on MCC or torus.

In our parallel algorithm on CCC with m processors, each move operation takes $O(\frac{c}{m} \cdot b + logm)$, and hence step 1) at each pass p takes $O(n_p \cdot (\frac{c}{m} \cdot b + logm))$. Step 2) at each pass p takes $O(\frac{n_p}{m} + logm)$, since each PE works on $\frac{n_p}{m}$ elements of Tmp_gain array. Therefore, each pass p takes $O(n_p \cdot (\frac{c}{m} \cdot b + logm))$. For KL algorithm, each move takes $O(bc)$, and hence each pass takes $O(bc^2)$. Since n_p becomes much smaller than c as the pass p increases in the dynamic iterative method, the total number of move operations is sharply reduced compared to KL algorithm. Moreover, the time complexity of each move operation is reduced significantly through the parallelization of the move operations compared to that of KL algorithm. Therefore, our parallel algorithm becomes much faster than KL algorithm in terms of the total time complexity.

Table 1. Average acceptance ratio for KL algorithm and our parallel algorithm

s_{max}	KL algorithm	Ours
1	0.169	0.513
5	0.177	0.483
10	0.149	0.466
20	0.136	0.485

Table 2. Performance comparison between KL algorithm and our parallel algorithm

b	N_c	c	s_{max}	KL algorithm	Ours(PE #)				
					1	2	4	8	16
10	5	400	1	96.1	44.1	25.5	19.1	10.1	5.8
			5	68.5	38.5	23.6	18.2	9.0	4.1
			10	93.8	44.2	24.6	19.8	10.1	5.7
			20	105.3	47.8	26.4	21.2	10.8	6.2
		800	1	365.5	167.8	96.0	74.2	32.2	19.5
			5	427.0	182.8	99.4	78.5	35.4	23.4
			10	409.4	180.0	97.9	76.1	34.8	21.6
			20	506.4	200.4	108.7	88.5	38.7	25.3

5 Experimental Result

We implemented our parallel algorithm on a network of sixteen transputers with hypercube connection as CCC. To evaluate the performance of the proposed algorithm, experiments with various characteristics such as number of cells(c), number of nets(n), maximum size among all cells(s_{max}), and averge number of nets on a network(N_c) were performed and compared with KL algorithm. Table 1 shows the average acceptance ratio in KL and our parallel algorithm respectively for a network with $b = 10, c = 400, n = 400, N_c = 5$. The acceptance ratio of ours is larger than that of KL algorithm by 3 ~ 4 times, which tells that ours does not waste much time on the unnessary move operations for the best partition of each pass, speeding up the whole algorithm. Table 2 shows the comparison of total execution time taken for KL algorithm and our parallel algorithm for different c and s_{max}. The execution time was measured in seconds. Note that our parallel algorithm outperforms KL algorithm by more than two times when using one processing element, and faster than KL algorithm by more than the number of processing elements when using two or more processing elements while producing the same result.

6 Conclusion

The execution time of the KL algorithm depends largely on the total number of move operations and the execution time of each move operation, since the algorithm consists of a series of move operations. In this paper, we have presented a fast parallel algorithm for solving multiway network partioning problem by reducing both of them. Using the fact that a large portion of move operations does not contribute to the partition of each pass, we have proposed a new dynamic iterative method which decreases the number of move operations dynamically as the passes goes on, and hence sharply speeds up the whole algorithm. Moreover, we reduced the execution time of each move by its parallelization using the proper cell distribution. We have implemented our parallel algorithm on a hypercube network of transputers, and have shown that ours is superor to KL algorithm for single processor and the execution time of ours is decreased significantly compared to KL algorithm without degrading the whole performance of the problem. For future work, we are going to work for further refinement in selecting the dynamic function.

References

1. B. W. Kernighan and S. Lin, An Efficient Heuristic Procedure for Partitioning Graphs, Bell Syst. Tech. J., Vol. 49, Feb. 1970, pp. 291-307.
2. C. M. Fiduccia and R. M. Mattheyses, A Linear Time Heuristic for Improving Network Partitions, Proc. 19th Design Automation Conf., 1982, pp. 175-181.
3. B. Krishnamurthy, An Improved Min-cut Algorithm for Partitioning VLSI Networks, IEEE Trans. on Computers, Vol. C-33, May 1984, pp. 438-446.
4. J. U. Kim, C. H. Lee and M. H. Kim, An Efficient Multiple-way Network Partitioning Algorithm, Computer Aided Design, Vol. 25, Oct. 1993, pp. 269-280..
5. L. A. Sanchis, Multiple-way Network Partitioning, IEEE Trans. on Computers, Vol. C-38, Jan. 1989, pp. 62-81.
6. H. J. Siegel, A Model of SIMD Machines and a Comparison of Various Interconnection Networks, IEEE Trans. on Computers, Vol. 28, 1979 907-917.
7. C. K. Koc, A. Guvenc and B. Bakkaloglu, Exact Solution of Linear Equation on Distributed Memory Multiprocessors, Parallel Algorithms and Applications, Vol. 3, pp. 135-143.
8. B. Murthy, K. Bhuvaneswari, and C. S. Murthy, A New Algorithm based on Givens Rotations Solving Linear Equations on Fault-Tolerant Mehs-Connected Processors, IEEE Trans. on Parallel and Distributed Systems, 1998, pp. 825-832.
9. V. Kumar and et. al., Introduction to Parallel Computing: Design and Analysis of Algorithms, Benjamin/Cummings, 1994.
10. Q. M. Malluhi, M. A. Bayoumi and T. R. Rao, An Efficient Mapping of Multilayer Perception with Backpropataion Anns on Hypercubes, Proceedings of 5th IEEE symposium on Parallel and Distributed Processing, Dallas, Texas, 1993, pp. 368-375.
11. S. S. Gupta and B. P. Sinha, A Simple O(logN) Time Parallel Algorithm for Testing Isomorphism of Maximal Outerplanar Graphs, Journal of Parallel and Distributed Computing 56, 1999, pp. 144-155.
12. K. D. Sajal, Parallel Graph Algorithms for Hypercube, Parallel Computing, Vol. 13, 1990, pp. 177-184.
13. R. A. Ayoubi and M. A. Bayoumi, Bitonic Sort on the Connection Machine, Parallel Algorithms and Applications, Vol. 3, 1994, pp. 151-161.
14. Y. Shih and J. Fier, Hypercube Systems and Key Applications, Parallel Processing for Supercomputing and AI, McGraw-Hill, 1990.

ParIC: A Family of Parallel Incomplete Cholesky Preconditioners

Mardochée Magolu monga Made[1]* and Henk A. van der Vorst[2]

[1] Université Libre de Bruxelles
Service des Milieux Continus, CP 194/5
50, avenue F.D. Roosevelt, B-1050 Brussels, Belgium
magolu@ulb.ac.be
http://homepages.ulb/~magolu/
[2] Utrecht University, Mathematical Institute, Mailbox 80.010,
3508 Utrecht, The Netherlands
vorst@math.uu.nl
http://www.math.uu.nl/people/vorst/

Abstract. A class of parallel incomplete factorization preconditionings for the solution of large linear systems is investigated. The approach may be regarded as a generalized domain decomposition method. Adjacent subdomains have to communicate during the setting up of the preconditioner, and during the application of the preconditioner. Overlap is not necessary to achieve high performance. Fill-in levels are considered in a global way. If necessary, the technique may be implemented as a global re-ordering of the unknowns. Experimental results are reported for two-dimensional problems.

1 Introduction

Krylov subspace based iterative methods are quite popular for solving large sparse preconditioned linear systems

$$B^{-1}Au = B^{-1}b , \qquad (1)$$

where $Au = b$ denotes the original system, and B denotes a given preconditioning matrix (see, e.g., [1,8]). The main operations within Krylov subspace methods are following: sparse matrix–vector multiplication(s); vector updates; dot products; setting up of the preconditioner; application of the preconditioner, that is, solve w from $Bw = r$, for given r. As is well-known, the handling of steps involving the preconditioner may be problematic on parallel platforms. In general, there is a trade-off between high level parallelism and fast convergence [6], especially when B is taken as an incomplete Choleky (IC) factorization of A [14, 15]. We refer to [5,7] for a review of all commonly used techniques for achieving parallelism. Till recently, most of works (if not all) on parallel *global* IC type methods concentrate on fill level zero preconditionings, for which the sparsity

---

* Research supported by the Commission of the European Community, under contract nr. 25009, within the ESPRIT IV project.

structure of A is preserved (see, e.g., [9, 10, 16]). We propose two techniques that allow high fill levels in a global preconditioner. A key feature of our approach is that adjacent subdomains have to exchange data during the computation of the preconditioning matrix factors, and during the application of the preconditioner. In contrast to classical domain decomposition methods, there is no overlap. A special treatment of interior boundary layers (interfaces) allows to alleviate the degradation of the convergence rate. In each situation, there exists an implicit global (re-)ordering of the unknowns. Experimental results are reported for two-dimensional problems, on a 16-processor SGI Origin 2000, showing that our methods compare favourably with classical overlapping domain decomposition methods.

2 Background

We consider the following self-adjoint second order two-dimensional elliptic PDE

$$
\begin{aligned}
-(p\,u_x)_x - (q\,u_y)_y &= f(x,y) && \text{in } \Omega = (0,1) \times (0,1) \\
u &= 0 && \text{on } \Gamma \\
u_n &= 0 && \text{on } \partial\Omega \backslash \Gamma \ .
\end{aligned}
\qquad (2)
$$

Γ denotes a nonempty part of the boundary $\partial\Omega$ of Ω. The coefficients p and q are positive, bounded and piecewise constant. We discretize (2) over a uniform rectangular grid of mesh size h in both directions with the five-point box integration scheme. The lexicographical ordering in the (x,y)-plane is used to number the unknowns. The resulting system matrix A is a nonsingular block-tridiagonal, irreducibly diagonally dominant, Stieltjes (that is, symmetric positive definite and none of its offdiagonal entries is positive) matrix. A popular method for solving such a system is the preconditioned conjugate gradient (PCG) method, combined with an incomplete factorization as preconditioning technique (see, e.g., [1, 8]). Fig. 1 shows an incomplete LPL^t factorization algorithm, where

$$\mathcal{D} = \{\, (k,i) \mid lev(l_{k,i}) > \ell \,\}$$

stands for the set of discarded fill-in entries. The integer ℓ denotes a user specified maximal fill level. With respect to the notation of Fig. 1, $lev(l_{k,i})$ is defined as following:

<u>Initialization:</u> $lev(l_{k,i}) := \begin{cases} 0 & \text{if } l_{k,i} \neq 0 \text{ or } k = i \\ \infty & \text{otherwise} \end{cases}$

<u>Factorization:</u> $lev(l_{k,i}) := \min\{\, lev(l_{k,i}),\, lev(l_{i,j}) + lev(l_{k,j}) + 1 \,\}$.

Any gridpoint j that is connected, with respect to the graph of L, with two gridpoints i and k such that $j < i < k$ (say, $l_{i,j} \neq 0$ and $l_{k,j} \neq 0$) gives rise to a fill-in entry (or a correction) in position (k,i) of L.

Compute P and L $(B = LPL^t$ with $diag(L) = I)$

Initialization phase

$p_{i,i} := a_{i,i}$, $\quad i = 1, 2, \cdots, n$

$l_{i,j} := a_{i,j}$, $\quad i = 2, 3, \cdots, n$, $\quad j = 1, 2, \cdots, i-1$

Incomplete factorization process

do $j = 1, 2, \cdots, n-1$

    compute parameter ρ_j

    do $i = j+1, j+2, \cdots, n$

$$p_{i,i} := p_{i,i} - \frac{l_{i,j}^2}{p_{j,j}}$$

$$l_{i,j} := \frac{l_{i,j}}{p_{j,j}}$$

        do $k = i+1, i+2, \cdots, n$

            if $(k, i) \notin \mathcal{D}$ $l_{k,i} := l_{k,i} - l_{i,j} l_{k,j}$

        end do

    end do

end do

Fig. 1. Standard incomplete factorization (IC).

3 Parallel Incomplete Cholesky Preconditioners

3.1 Explicit Pseudo-overlap

For simplicity we assume that the grid is divided into p stripes, as illustrated on Fig. 2. We impose the following conditions:

(c1) for each subdomain the computation starts at the bottom layer gridpoints, and finishes at the top layer gridpoints. The actual computations start from two sides: for the subdomains in the upper side of the physical domain, the flow of computations is downward;

(c2) immediately after the computations at the bottom layer gridpoints of a subdomain ($\mathcal{P}_j$, $j \notin \{0, p-1\}$) have been completed, the relevant corrections for the top layer gridpoints of the (appropriate) adjacent subdomain are packed and sent (preferably by means of some nonblocking communication);

(c3) the numbering decreases or increases in the same way for neighbouring points, for the bottom layer gridpoints and the top layer gridpoints of adjacent subdomains. This facilitates the implementation (communication). Each top layer gridpoint has "to know" where corrections come from.

Definition 1. *Given that communication only involves the gridpoints in the bottom and top layer, the union of adjacent layers is referred to as the pseudo-overlapping region (or simply the pseudo-overlap). Equivalently, if $\mathcal{P}_r$ has to*

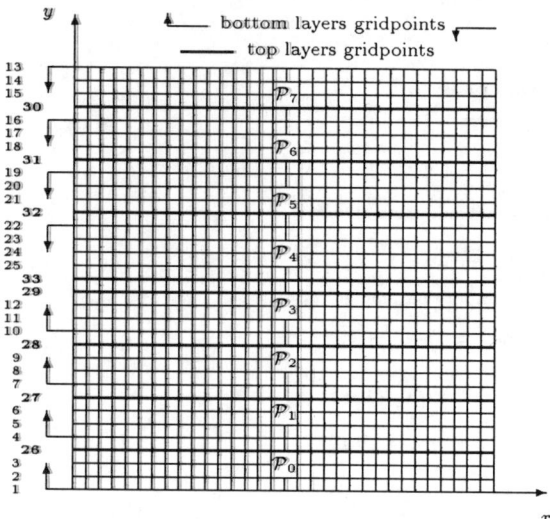

Fig. 2. Decomposition of the grid into stripes, and assignment of subdomains to processors, for $n_x = 32$, $n_y = 33$ and $p = 8$. Arrows indicate the progressing direction of the line numbering per subdomain. Numbers along the y-axis give an example of global (line) ordering, which satisfy all the required conditions. Within each horizontal line, gridpoints are ordered lexicographically.

send data to $\mathcal{P}_s$ during the incomplete factorization process, we will say that $\mathcal{P}_s$ is *pseudo-overlapped* by $\mathcal{P}_r$.

The rate of convergence of a parallel IC(0), executed under the conditions as described above, will degrade as the number of subdomains increases ($p > 2$ for stripes type partitionings [2, 13], and $p > 4$ for more general partitionings [6, 19]; see also [9, 10, 16]). In order to explain why this occurs, we make, in Fig. 3, a zoom of an interface between two adjacent subdomains. We use a stencil graph notation [11]: a diagonal entry $a_{i,i}$ is represented by circle number i; the edge $\{i, j\}$ (here horizontal and vertical lines) corresponds to a nonzero offdiagonal entry $a_{i,j}$. Oblique line represent (rejected) level-1 fill-in entries that are not significantly different from the case of $p = 1$. Thick lines (the arcs) are the (neglected) level-1 fill-in entries that would not arise if a global *natural* ordering (or some other equivalent ordering) were used. Such level-1 fill-in entries are responsible for the deterioration of the convergence. For IC(0), rejected level-1 fill entries determine the remainder matrix $R = B - A$. In the terminology of Doi and Lichnewsky [3, 4], the bottom layer gridpoints of $\mathcal{P}_{s+1}$ (see gridpoints marked with $\star$ in Fig. 3), which induce the additional level-1 fill entries, are called *incompatible nodes*.

A way to improve the performance consists in accepting *enough* fill-in entries generated by the parallelization strategy: increase both the pseudo-overlap width

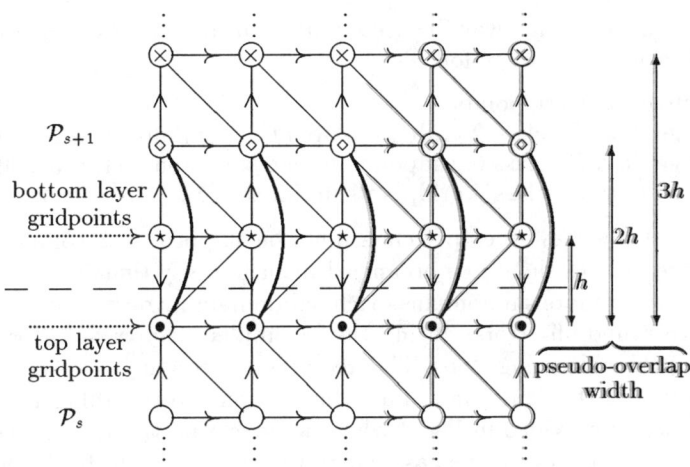

Fig. 3. Part of graph of matrix A assigned to two adjacent subdomains ($\mathcal{P}_s$ and $\mathcal{P}_{s+1}$). Oblique lines and thick lines are (rejected) level 1 fill-in entries.

($\varpi = h, 2h, 3h, \ldots$) and the fill level ($\ell_\varpi \geq 0, 1, 2, \ldots$) in the pseudo-overlapping region(s).

Definition 2. *We denote by* $\text{ParIC}(\ell; \varpi, \ell_\varpi)$ *any standard* $\text{IC}(\ell)$ *combined with the parallelization strategy as described above. This reads parallel IC with interior fill level ℓ, pseudo-overlap width ϖ, and pseudo-overlap fill level ℓ_ϖ. In the specification of ϖ, k will stand for kh, in order to include variable mesh size problems and (graphs of) matrices that do not stem from discretized PDEs.*

3.2 Implicit Pseudo-overlap

The parallelization technique discussed so far may be rather tedious to apply, when some subdomains have a high number of neighbours and the grid is not well structured. The alternative method, with an ordering induced pseudo-overlapping strategy, that we shall describe now, may be easily used to tackle intricate geometries and partitionings. For the purposes of our exposition, let us think of each small (grid) square of Fig. 2 as a finite element, and assume that the finite elements have been partitioned into p subdomains by means of some automatic mesh partitioning algorithm. Then, *in each subdomain*, the (local) unknowns are re-numbered class by class, consecutively, as following:

1. class 1: all interior gridpoints are numbered (firstly);
2. class 2: next follow all gridpoints, if any, that belong to two subdomains;
3. class 3: next follow all gridpoints, if any, that belong to three subdomains;
4. etc ...

In doing so, we obtain a (generalization of a) reverse variant of an ordering discussed in [9] (see also [16]). A global renumbering of the gridpoints may be achieved in a similar way. The computation and the exchange of data is performed, class by class, as follows:

1. compute class 1 gridpoints;
2. exchange data for class 2 gridpoints updates; compute class 2 gridpoints;
3. exchange data for class 3 gridpoints updates; compute class 3 gridpoints;
4. exchange data for class 4 gridpoints updates; etc ...

Any step that involves an empty class must be skipped. The computation and the exchange of data should be organized in such a way that, at each gridpoint shared by two or more subdomains, each subdomain involved obtains the same value, up to round-off errors, during the incomplete factorization process, and during the preconditioning steps. This requires to drop any connection between two gridpoints of the same class, but which belong to two different interfaces. An illustration is provided in Fig. 4 where we give a partitioning of the physical domain into 2×4 boxes. In the case where the connection to be dropped corresponds to some entry $a_{i,j}$ of the original system matrix, the dropped value may be added to the diagonal entries $a_{i,i}$ and $a_{j,j}$. The rowsum of the system matrix is then preserved.

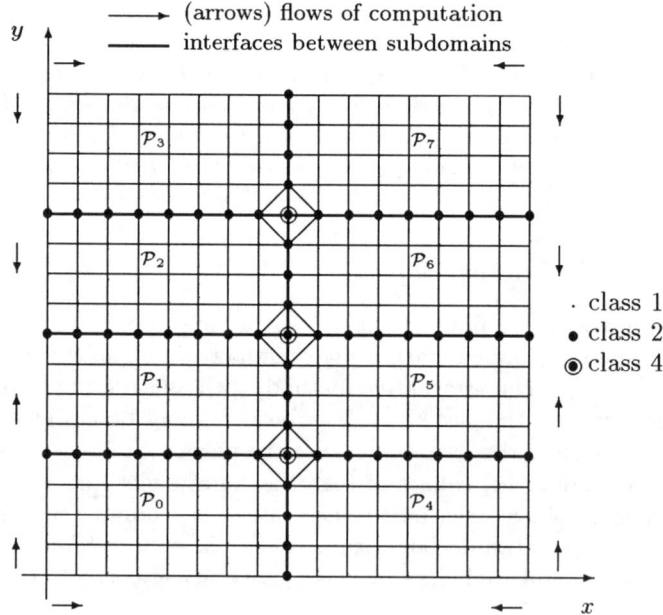

Fig. 4. Decomposition of the grid into 2×4 subdomains, assignment of subdomains to processors, and partitioning into classes. Oblique lines correspond to forbidden level-1 fill-in entries.

Definition 3. *Now the pseudo-overlap will be implicitly determined by the local numberings of unknowns, and the fill level taken inside each subdomain. We shall* denote this more general parallel IC simply by ParIC(ℓ). It can be easily extended to include the case of subdomains with different fill levels.

4 Numerical Experiments

The zero vector is used as initial guess, and the PCG is stopped as soon as the residual vector r satisfies $||r||_2 / ||b||_2 \leq 10^{-6}$. The test is performed only once $||B^{-1}r||_2 / ||B^{-1}b||_2 \leq 10^{-6}$ is satisfied. The computations are carried out in double precision Fortran on a 16-processor SGI Origin 2000 (195 MHz), using the MPI library for communications. The preconditionings include: ParIC($\ell; \varpi, \ell_\varpi$), $\ell_\varpi \geq \varpi - 1$, with the stripes (or $1 \times p$) partitionings; ParIC(ℓ) with $2 \times p$ partitionings; and the additive Schwarz with overlap AS($\ell; \varpi$), each local problem is handled with one IC(ℓ) solve, ϖ stands here for the actual overlap width. We use $\varpi = h_0$, h, $2h$, where h_0 means that only one line of nodes is shared by adjacent subdomains. No global coarse grid correction has been added, as suggested in [17, 18]), to improve the performance of the preconditionings involved.

Problem 1. $p = q = 1$, $\Gamma = \Omega$, $u(x,y) = x(x-1)y(y-1)e^{xy}$, and $h = 1/(n_y+1)$.

Problem 2. $\Gamma = \{(x,y); 0 \leq x \leq 1, y = 0\}$, $h = 1/n_y$,

$$p = q = \begin{cases} 100 \text{ in } (1/4, 3/4) \times (1/4, 3/4) \\ 1 \text{ elsewhere }, \end{cases} \quad f(x,y) = \begin{cases} 100 \text{ in } (1/4, 3/4) \times (1/4, 3/4) \\ 0 \text{ elsewhere } . \end{cases}$$

We first collect in Figs. 5 and 6, and in Table 1, the results of our numerical experiments performed with the stripes partitionings. We use the parallel speed-up, which is the ratio between the execution time of the parallel algorithm on one processor and the time on p processors.

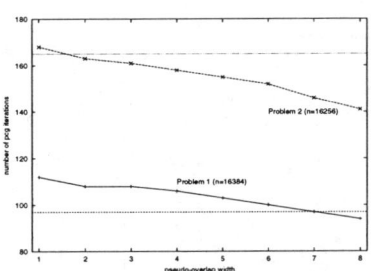

Fig. 5. Effects of pseudo-overlap width ϖ on the number of pcg iterations, for 1×8 processors and ParIC($0; \varpi, \varpi - 1$). Horizontal lines display the number of pcg iterations for sequential IC(0).

From all the observed results, the following trends are evident. It is advantageous to use increased pseudo-overlap. In particular, for difficult (large size) problems (see Fig. 5), the degradation of the performance is (more than) mastered when one accepts some fill-in entries generated by the parallelization strategy. ParIC($4; 5, 4$) is in general twice as fast as ParIC($0; 1, 0$). For both methods,

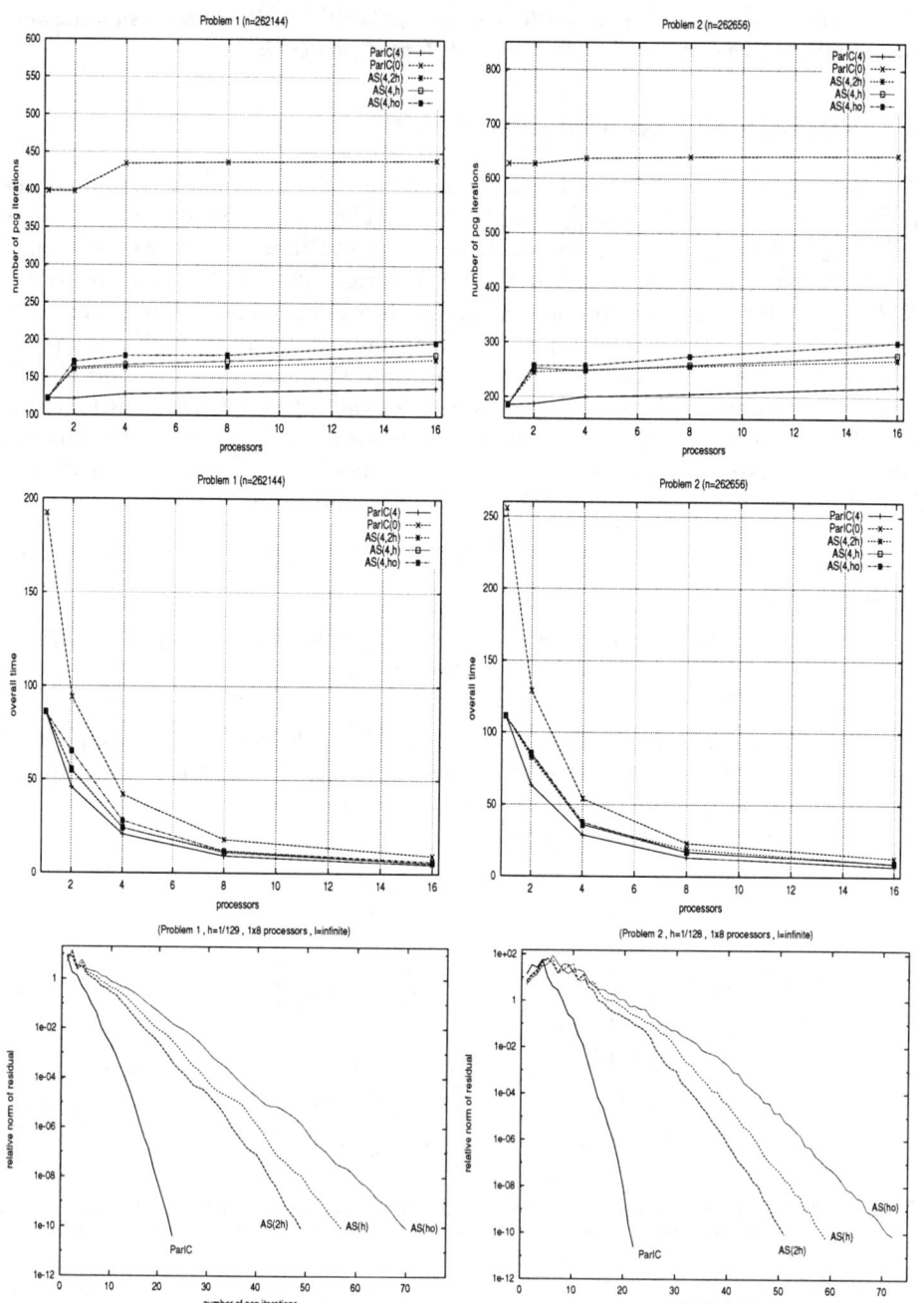

Fig. 6. Number of PCG iterations and overall computational time for ParIC(0;1,0), ParIC(4;5,4), AS(4,h_0), AS(4,h), AS(4,$2h$), and $1 \times p$ partitionins (stripes). Evolution of the relative residual error for 1×8 processors; the fill-in level $\ell = \infty$ (locally) for each preconditioner involved

Table 1. Speed-up for stripes partitionings ($1 \times p$).

		\multicolumn{4}{c	}{Problem 1}	\multicolumn{4}{c}{Problem 2}
Precond.	$p =$	2	4	8	16	2	4	8	16
ParIC(0;1,0)		2.04	4.57	10.66	19.94	1.98	4.73	10.99	19.66
ParIC(4;5,4)		1.88	4.15	9.24	17.96	1.76	3.86	8.37	15.92
AS(4,h_0)		1.32	3.08	7.19	13.12	1.30	2.96	6.46	12.21
AS(4,h)		1.56	3.56	7.38	14.99	1.32	3.11	6.53	11.96
AS(4,2h)		1.58	3.56	7.29	15.46	1.35	3.08	5.78	12.26

the speed-up is high, and in general better than for AS methods. AS methods must be applied with a sufficiently large overlap width, in order to achieve performance comparable to ParIC methods, which dramatically increases the computational complexity. This holds even if each local problem is solved exactly. Observe that, for $p = 2$, ParIC($\infty; \varpi_{max}, \infty$), which is equivalent to ParIC(∞), becomes a direct solver, whereas AS remains an iterative one.

Table 2 shows the performance of ParIC(4) combined with various partitionings. For stripes (or $1 \times p$) partitionings, ParIC(4) is mathematically equivalent to ParIC(4;5,4). It appears that, for difficult problems, it would be interesting to use partitionings better than the simple stripes ones.

Table 2. Problem 1, $h^{-1} = 513$, $n = 262144$. Problem 2, $h^{-1} = 512$, $n = 262656$. Comparison of various partitionings (part). Number of PCG iterations (iter); elapsed time in seconds for: the computation of the preconditioning matrix (fact), the solver, and overall time; for ParIC(4).

	Problem 1		Time		Problem 2		Time	
part	iter	fact	pcg	overall	iter	fact	pcg	overall
1	122	4.76	80.59	86.20	185	4.84	106.17	111.61
1×2	122	2.48	42.84	45.82	187	2.48	60.60	63.45
2×1	122	2.43	42.35	45.25	150	2.34	47.54	50.14
1×4	128	1.24	19.30	20.78	200	1.22	27.48	28.88
2×2	127	1.42	19.46	21.18	159	1.44	22.76	24.37
1×8	131	0.65	8.50	9.32	205	0.62	12.60	13.33
2×4	135	0.74	8.45	9.30	167	0.74	9.66	10.51
1×16	137	0.37	4.26	4.80	219	0.33	6.49	7.01
2×8	137	0.41	4.54	5.12	172	0.40	5.11	5.61

References

1. R.F. Barret, M. Berry, T.F. Chan, J. Demmel, J. Donato, J.J. Dongarra, V. Eijkhout, R. Pozo, C. Romine, and H. van der Vorst: *Templates for the Solution of Linear Systems : Building Blocks for Iterative Methods* (SIAM, Philadelphia, 1994).
2. R. Beauwens, L. Dujacquier, S. Hitimana and M. Magolu monga Made: MILU factorizations for 2-processor orderings. In: I.T. Dimov, Bl. Sendov and P.S. Vassilevski (eds.), Advances in Numerical Methods and Applications $\mathcal{O}(h^3)$ (World Scientific, Singapore, 1994) 26–34.
3. S. Doi and A. Lichnewsky: A graph-theory approach for analyzing the effects of ordering on ILU preconditioning. INRIA report 1452, INRIA-Rocquencourt, France, 1991.
4. S. Doi and T. Washio: Ordering strategies and related techniques to overcome the trade-off between parallelism and convergence in incomplete factorizations. Parallel Comput. **25**, (1999) 1995–2014.
5. J.J. Dongarra, I.S. Duff, D.C. Sorensen and H.A. van der Vorst: *Numerical Linear Algebra for High-Performance Computers* (SIAM, Philadelphia, 1998).
6. I.S. Duff and G.A. Meurant: The effect of ordering on preconditioned conjugate gradients. BIT **29**, (1989) 635–657.
7. I.S. Duff and H.A. van der Vorst: Developments and Trends in the Parallel Solution of Linear Systems. Parallel Comput. **25**, (1999) 1931–1970.
8. G.H. Golub and C.F. van Loan: *Matrix Computations* (third ed.) (The John Hopkins University Press, Baltimore, Maryland, 1996).
9. G. Haase: Parallel incomplete Cholesky preconditioners based on the nonoverlapping data distribution. Parallel Comput. **24**, (1998) 1685–1703.
10. M. Magolu monga Made: Implementation of parallel block preconditionings on a transputer-based multiprocessor. Future Generation Computer Systems **11**, (1995) 167–173.
11. M. Magolu monga Made: Taking advantage of the potentialities of dynamically modified block incomplete factorizations. SIAM J. Sci. Comput. **19**, (1998) 1083–1108.
12. M. Magolu monga Made and B. Polman: Efficient planewise like preconditioners to cope with 3D problems. Numer. Linear Algebra Appl. **6**, (1999) 379–406.
13. M. Magolu monga Made and H.A. van der Vorst: Parallel incomplete factorizations with pseudo-overlapped subdomains. Technical Report, Service des Milieux Continus, Université Libre de Bruxelles, February 2000.
14. J.A. Meijerink and H.A. van der Vorst: An iterative solution method for linear systems of which the coefficient matrix is a symmetric M-matrix. Math. Comp. **31**, (1977) 148–162.
15. J.A. Meijerink and H.A. van der Vorst: Guidelines for the usage of incomplete decompositions in solving sets of linear equations as they occur in practical problems. J. Comp. Physics **44**, (1981) 134–155.
16. Y. Notay: An efficient Parallel Discrete PDE Solver. Parallel Comput. **21**, (1995) 1725–1748.
17. Y. Notay and A. Van de Velde: Coarse grid acceleration of parallel incomplete preconditioners. In: S. Margenov and P. Vassilevski, (eds.), Iterative Methods in Linear Algebra II. IMACS series in Computational and Applied Mathematics **3**, (1996) 106–130.
18. B.F. Smith, P.E. Bjorstad and D. Gropp: *Domain Decomposition : Parallel Multilevel Methods for Elliptic Partial Differential Equations* (Cambridge University Press, Cambridge, 1996).
19. H.A. van der Vorst: High performance preconditioning. SIAM J. Sci. Statist. Comput. **10**, (1989) 1174–1185.

A Parallel Block Preconditioner Accelerated by Coarse Grid Correction

C. Vuik[1] and J. Frank[1,2]

[1] Delft University of Technology, Faculty of Information Technology and Systems
Department of Applied Mathematical Analysis
P.O. Box 5031, 2600 GA Delft, The Netherlands
c.vuik@math.tudelft.nl
[2] Center for Mathematics and Computer Science (CWI)
P.O. Box 94079, 1090 GB Amsterdam, The Netherlands
jason@cwi.nl

Abstract. A block-preconditioner is considered in a parallel computing environment. This preconditioner has good parallel properties, however the convergence deteriorates when the number of blocks increases. Two different techniques are studied to accelerate the convergence: overlapping at the interfaces and using a coarse grid correction. It appears that the latter technique is indeed scalable, so the wall clock time is constant when the number of blocks increases. Furthermore the method is easily added to an existing solution code.
Keywords: Parallel Krylov subspace methods; block preconditioner; overlapping subdomains; coarse grid correction.

1 Introduction

Domain decomposition arises naturally in computational fluid dynamics applications on structured grids: complicated geometries are broken down into (topologically) rectangular regions and discretized in general coordinates, see e.g. [25], applying domain decomposition to iteratively arrive at the solution on the global domain. This approach provides easy exploitation of parallel computing resources, and additionally offers a solution to memory limitation problems.

In this paper we present a parallel implementation of a Krylov accelerated block Gauss-Jacobi method for the DeFT Navier-Stokes solver. This research is a continuation of our work presented in [2, 3, 24, 9, 22]. For an overview of the literature of related parallel methods we refer to [24, 9]. We report results for a Poisson problem on a square domain, which is representative of the pressure system which must be solved for the pressure correction method used in DeFT.

The main parallel operations required in Krylov subspace methods are distributed matrix-vector multiplications, vector updates, inner products, and precon-ditioner-vector multiplications. For many problems, the matrix-vector multiplications require only nearest neighbor communications, and are very efficient. Vector updates are also easy to parallelize. Inner products require global communication so one has to be careful in their parallel implementation. This aspect of the Krylov subspace solver has been addressed in [9]. The parallelization of non-overlapping block preconditioner operations is also trivial. However, the convergence behavior of the preconditioner deteriorates considerably, when the number of blocks increases (compare [9]).

In this paper we consider the acceleration of parallel block preconditioned Krylov subspace methods by overlapping and deflation. The parallel implementation is based on MPI subroutines [11]. The details of the block preconditioner and its convergence are given in Section 2. In Section 3, overlapping of the subdomains is defined, which can be used to enhance the convergence of the block preconditioner. Coarse grid correction, presented in Section 4, is another promising technique to accelerate the block preconditioned method. Section 5 contains numerical experiments illustrating the convergence of the various parallel iterative methods.

2 Block Preconditioning and Krylov Subspace Methods

2.1 The Block Jacobi Preconditioner

We consider an elliptic partial differential equation discretized using a cell-centered finite difference method on a computational domain Ω. Let the domain be the union of M nonoverlapping subdomains Ω_m, $m = 1, \ldots, M$. Discretization results in a sparse linear system $Ax = b$, with $x, b \in \mathrm{R}^N$. When the unknowns in a subdomain are grouped together one gets the block system:

$$\begin{bmatrix} A_{11} & \cdots & A_{1M} \\ \vdots & \ddots & \vdots \\ A_{M1} & \cdots & A_{MM} \end{bmatrix} \begin{pmatrix} x_1 \\ \vdots \\ x_M \end{pmatrix} = \begin{pmatrix} b_1 \\ \vdots \\ b_M \end{pmatrix}. \tag{1}$$

In this system, the diagonal blocks A_{mm} express coupling among the unknowns defined on Ω_m, whereas the off-diagonal blocks A_{mn}, $m \neq n$ represent coupling across subdomain boundaries. The only nonzero off-diagonal blocks are those corresponding to neighboring subdomains.

In order to solve system (1) with a Krylov subspace method we use the block Jacobi preconditioner:

$$K = \begin{bmatrix} A_{11} & & \\ & \ddots & \\ & & A_{MM} \end{bmatrix}.$$

When this preconditioner is used, systems of the form $Kv = r$ have to be solved. Since there is no overlap the diagonal blocks $A_{mm} v_m = r_m$, $m = 1, \ldots, M$ can be solved in parallel. In our method these systems are solved by an iterative method. Since the number of inner iterations may vary in each outer iteration, the effective preconditioner is nonlinear and varies in each outer iteration.

We use RILU preconditioned GMRES [1, 21] to solve the subdomain problems to within a fixed tolerance. Additionally, a blockwise application of the RILU preconditioner is used.

2.2 The Krylov Subspace Methods

In this paper we consider linear systems where the coefficient matrix is symmetric or non-symmetric. For a symmetric matrix we use a parallel version of the preconditioned Conjugate Gradient method [17, 6]. For this method the preconditioner should be the same in every outer iteration. This means that only two choices for the preconditioner can be used: a block RILU preconditioner or solving the subdomain problems with a small tolerance. In the latter choice the preconditioner is close to K^{-1}.

In our application, pressure correction, the pressure system resembles a discretized Poisson equation; however, the coefficient matrix may be non-symmetric. For the non-symmetric case we use the GCR method [8, 20]. This is a Krylov subspace methods which allows a variable preconditioner.

Algorithm: GCR
Given: initial guess x_0
$r_0 = b - Ax_0$
for $k = 1, \ldots,$ convergence
 Solve $K\tilde{v} = r_{k-1}$ (approximately)
 $\tilde{q} = A\tilde{v}$
 $[q_k, v_k] = $ **orthonorm** $(\tilde{q}, \tilde{v}, q_i, v_i, i < k)$
 $\gamma = q_k^T r_{k-1}$
 Update: $x_k = x_{k-1} + \gamma v_k$
 Update: $r_k = r_{k-1} - \gamma q_k$
end

The function **orthonorm()** takes input vectors $\tilde{q}$ and $\tilde{v}$, orthonormalizes $\tilde{q}$ with respect to the q_i, $i < k$, updating $\tilde{v}$ as necessary to preserve the relation $\tilde{q} = A\tilde{v}$, and returns the modified vectors q_k and v_k. Using the conclusions of [9] we choose the following orthogonalization methods: Reorthogonalized Classical Gram-Schmidt when the subdomain size is small, otherwise we take the Modified Gram-Schmidt method.

2.3 The Convergence Behavior of the Block Preconditioned GCR Method

As a test example, we consider a Poisson problem, discretized with the finite volume method on a square domain. We do not exploit the symmetry of the Poisson matrix in these experiments. The domain is composed of a $\sqrt{p} \times \sqrt{p}$ array of subdomains, each with an $n \times n$ grid. With $h = \Delta x = \Delta y = 1.0/(n\sqrt{p})$ the discretization is

$$4u_{i,j} - u_{i+1,j} - u_{i-1,j} - u_{i,j-1} - u_{i,j+1} = h^2 f_{i,j}.$$

The right hand side function is $f_{i,j} = f(ih, jh)$, where $f(x,y) = -32(x(1-x) + y(1-y))$. Homogeneous Dirichlet boundary conditions $u = 0$ are defined on $\partial\Omega$, implemented by adding a row of ghost cells around the domain, and enforcing the condition, for example, $u_{0,j} = -u_{1,j}$ on boundaries. This ghost cell scheme allows natural implementation of the block preconditioner as well.

For the tests of this section, GCR is restarted after 30 iterations, and modified Gram-Schmidt was used as the orthogonalization method for all computations. The solution was computed to a fixed tolerance of 10^{-6}. The subdomain approximations will be denoted as follows:

- GMR6 = GMRES with a tolerance of 10^{-6}, (preconditioned with RILU)
- GMR2 = GMRES with a tolerance of 10^{-2}, (preconditioned with RILU)
- GMR1 = GMRES with a tolerance of 10^{-1}, (preconditioned with RILU)
- RILU = one application of an RILU preconditioner.

We compare results for a fixed problem size on the 300×300 grid using 4, 9, 16 and 25 blocks. In Table 1 the iteration counts are given: both the number of outer iterations and the average number of inner iterations (in parentheses). Note that for all preconditioners the number of outer iterations increases when the number of blocks grows. This implies that the parallel efficiency decreases when one uses more processors. In the next sections we present two different approaches to diminish this drawback.

	$p = 4$	$p = 9$	$p = 16$	$p = 25$
GMR6	78(68.4)	83(38.7)	145(31.4)	168(26.4)
GMR2	86(15.7)	118(15.7)	168(13.7)	192(10.9)
GMR1	139(13.6)	225(9.3)	287(7.1)	303(5.9)
RILU	341(1)	291(1)	439(1)	437(1)

Table 1. Number of iterations for various number of blocks

3 Overlapping of the Subdomains

It is well known that the convergence of an overlapping block preconditioner is nearly independent of the subdomain grid size when the physical overlap region is constant (see [16, 19, 7, 18]).

To describe the overlapping block preconditioner we define the subdomains $\Omega_m^* \subset \Omega$. The domain Ω_m^* consists of Ω_m and n_{over} neighboring grid points (see Figure 1).

The matrix corresponding to this subdomain is denoted by A_{mm}^*. Application of the preconditioner goes as follows: given r compute v using the steps

1. r_m^* is the restriction of r to Ω_m^*,
2. solve $A_{mm}^* v_m^* = r_m^*$ in parallel,
3. form v_m, which is the restriction of v_m^* to Ω_m.

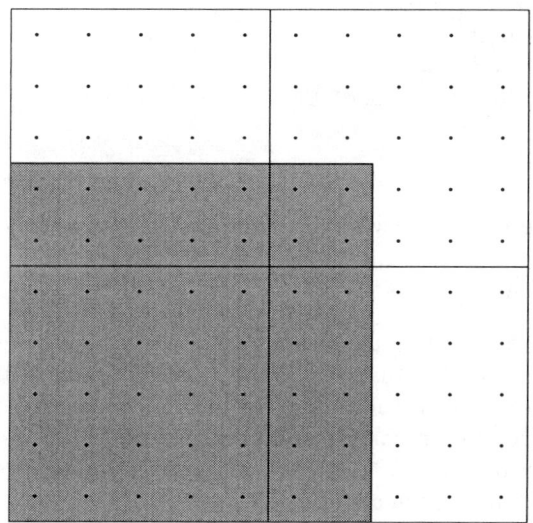

Fig. 1. The shaded region is subdomain Ω_1^* for $n_{over} = 2$

A related method is presented by Cai, Farhat and Sarkis [5, 4]. A drawback of overlapping subdomains is that the amount of work increases proportional to n_{over}. Furthermore it is not so easy to implement this approach on top of an existing software package.

4 Coarse Grid Correction

We present the coarse grid correction only for the symmetric case. In our implementation we use the Deflated ICCG method as defined in [23] (see also [15, 13, 14, 12]). To define the Deflated ICCG method we need a set of projection vectors $v_1, ..., v_M$ that form an independent set. The projection on the space A-perpendicular to span$\{v_1, ..., v_M\}$ is defined as

$$P = I - VE^{-1}(AV)^T \text{ with } E = (AV)^T V \text{ and } V = [v_1...v_M].$$

The solution vector x can be split into two parts $x = (I - P)x + Px$. The first part can be calculated as follows $(I - P)x = VE^{-1}VAx = VE^{-1}V^Tb$. For the second part we project the solution x_j obtained from DICCG to Px_j. DICCG consists of applying CG to $L^{-T}L^{-1}P^TAx = L^{-T}L^{-1}P^Tb$.

The Deflated ICCG algorithm reads (see Reference [23]):

DICCG
$j = 0$, $\hat{r}_0 = P^T r_0$, $p_1 = z_1 = L^{-T} L^{-1} \hat{r}_0$;
while $\|\hat{r}_j\|_2 >$ accuracy **do**
$\quad j = j + 1; \ \alpha_j = \frac{(\hat{r}_{j-1}, z_{j-1})}{(p_j, P^T A p_j)}$;
$\quad x_j = x_{j-1} + \alpha_j p_j$;
$\quad \hat{r}_j = \hat{r}_{j-1} - \alpha_j P^T A p_j$;
$\quad z_j = L^{-T} L^{-1} \hat{r}_j; \ \beta_j = \frac{(\hat{r}_j, z_j)}{(\hat{r}_{j-1}, z_{j-1})}$;
$\quad p_{j+1} = z_j + \beta_j p_j$;
end while

For the coarse grid correction we choose the vectors v_m as follows:

$$v_m(i) = 1, \ i \in \Omega_m, \text{ and } v_m(i) = 0, \ i \notin \Omega_m. \tag{2}$$

We are able to give a sharp upperbound for the effective condition number of the deflated matrix, used with and without classical preconditioning [10]. This bound provides direction in choosing a proper decomposition into subdomains and a proper choice of classical preconditioner. If grid refinement is done keeping the subdomain resolutions fixed, the condition number can be shown to be independent of the number of subdomains.

In parallel, we first compute and store $((AV)^T V)^{-1}$ in factored form on each processor. Then to compute $P^T A p$ we first perform the matrix-vector multiplication $w = Ap$, requiring nearest neighbor communications. Then we compute the local contribution to the restriction $\tilde{w} = V^T w$ and distribute this to all processors. With this done, we can solve $\tilde{e} = ((AV)^T V)^{-1} \tilde{w}$ and compute $(AV)^T \tilde{e}$ locally. The total communications involved in the matrix-vector multiplication and deflation are a nearest neighbor communication of the length of the interface and a global gather-broadcast of dimension M.

5 Numerical Experiments

5.1 Block Preconditioner Results

In Table 1 we present the iteration counts for a problem on the 300×300 grid. Table 2 contains the corresponding timing results in seconds on the Cray T3E. The fastest solutions in each case are obtained with the least accurate subdomain approximation - namely, the block RILU preconditioner. Therefore, we use this choice in our timing measurements in Section 5.4.

5.2 Block Preconditioner and Overlap

We consider a Poisson problem on a square domain with Dirichlet boundary conditions and a constant right-hand-side function. The problem is discretized

	$p=4$	$p=9$	$p=16$	$p=25$
GMR6	685	178	143	79
GMR2	167	102	63	37
GMR1	222	118	66	39
RILU	65	26	22	15

Table 2. Wall clock times in seconds on the Cray T3E

by cell-centered finite differences. We consider overlap of 0, 1 and 2 grid points and use A_{mm}^{-1} in the block preconditioner. Table 3 gives the number of iterations necessary to reduce the initial residual by a factor 10^6 using a decomposition into 3×3 blocks with subgrid dimensions given in the table. Note that the number of iterations is constant along the diagonals. This agrees with domain decomposition theory that the number of iterations is independent of the subdomain grid size when the physical overlap remains the same (see [16, 19, 7, 18]).

	overlap		
grid size	0	1	2
5 × 5	10	8	7
10 × 10	14	9	8
20 × 20	19	13	10
40 × 40	26	18	14

Table 3. Iterations for various grid sizes

In the second experiment we take a 5×5 grid per subdomain. The results for various number of blocks are given in Table 4. Note that without overlap the number of iterations increases considerably, whereas the increase is much smaller when 2 grid points are overlapped. The large overlap (2 grid points on a 5×5 grid) that has been used in this test, is not affordable for real problems. In Section 5.4 we present results with large subdomain grids without overlap.

5.3 Coarse Grid Correction

We do the same experiments using the coarse grid correction (see Table 4). Initially we see some increase of the number of iterations, however, for more than 16 blocks the increase levels off. This phenomenon is independent of the amount of overlap. The same conclusion holds when block RILU is used instead of fully solving the subdomain problems.

5.4 Timing Results of Coarse Grid Correction

Finally we present some timing results on the Cray T3E for a problem on a 480×480 grid. The results are given in Table 5. In this experiment we use GCR

	overlap			overlap+cgc		
decomposition	0	1	2	0	1	2
2 × 2	6	5	4	6	4	4
3 × 3	10	8	7	11	6	6
4 × 4	15	9	7	14	9	6
5 × 5	18	12	9	16	10	8
6 × 6	23	13	10	17	11	9
7 × 7	25	16	12	17	12	10
8 × 8	29	17	12	18	13	10
9 × 9	33	19	14	18	14	11

Table 4. Iterations for various block decompositions with and without coarse grid correction (subdomain grid size 5 × 5)

with the block RILU preconditioner combined with coarse grid correction. Note that the number of iterations decreases when the number of blocks increases. This leads to an efficiency larger than 1. The decrease in iterations is partly due to the improved approximation of the RILU preconditioner for smaller subdomains. On the other hand when the number of blocks increases, more small eigenvalues are projected to zero which also accelerates the convergence (see [10]). We expect that there is some optimal value for the number of subdomains, because at the extreme limit there is only one point per subdomain and the coarse grid problem is identical to the original problem so there is no speedup at all.

p	iterations	time	speedup	efficiency
1	485	710	-	-
4	322	120	5	1.2
9	352	59	12	1.3
16	379	36	20	1.2
25	317	20	36	1.4
36	410	18	39	1.1
64	318	8	89	1.4

Table 5. Speedup of the iterative method using a 480 × 480 grid

6 Conclusions

From the experiments presented in this paper we conclude that overlapping of the subdomains makes the parallel iterative method more or less independent of the subdomain grid size. A drawback is that overlapping is not so easy to implement on top of an existing software package

Coarse grid correction implemented by deflation can easily be used in combination with existing software. It appears that the number of iterations decreases when the number of processors increases. This leads to efficiencies larger than 1. So we conclude that coarse grid correction is a very efficient technique to accelerate parallel block preconditioners.

Acknowledgment. The authors thank HPαC for providing computing facilities on the Cray T3E.

References

1. O. Axelsson and G. Lindskog. On the eigenvalue distribution of a class of preconditioning methods. *Num. Math.*, 48:479–498, 1986.
2. E. Brakkee, A. Segal, and C.G.M. Kassels. A parallel domain decomposition algorithm for the incompressible Navier-Stokes equations. *Simulation Practice and Theory*, 3:185–205, 1995.
3. E. Brakkee, C. Vuik, and P. Wesseling. Domain decomposition for the incompressible Navier-Stokes equations: solving subdomain problems accurately and inaccurately. *Int. J. for Num. Meth. Fluids*, 26:1217–1237, 1998.
4. X.-C. Cai, C. Farhat, and M. Sarkis. A minimum overlap restricted additive Schwarz preconditioner and applications in 3D flow simulations. In J. Mandel, C. Farhat, and X.-C. Cai, editors, *The Tenth International Conference on Domain Decomposition Methods for Partial Differential Equations*, pages 479–485, Providence, 1998. AMS.
5. X.-C. Cai and M. Sarkis. A restricted additive Schwarz preconditioner for general sparse linear systems. *SIAM J. Sci. Comput.*, 21:792–797, 1999.
6. E. de Sturler and H.A. van der Vorst. Reducing the effect of global communication in GMRES(m) and CG on parallel distributed memory computers. *Appl. Num. Math.*, 18:441–459, 1995.
7. Eric de Sturler. Incomplete block LU preconditioners on slightly overlapping subdomains for a massively parallel computer. *Applied Numerical Mathematics*, 19:129–146, 1995.
8. S.C. Eisenstat, H.C. Elman, and M.H. Schultz. Variational iterative methods for nonsymmetric systems of linear equations. *SIAM J. Num. Anal.*, 20:345–357, 1983.
9. J. Frank and C. Vuik. Parallel implementation of a multiblock method with approximate subdomain solution. *Appl. Num. Math.*, 30:403–423, 1999.
10. J. Frank and C. Vuik. On the construction of deflation-based preconditioners. MAS-R 0009, CWI, Amsterdam, 2000.
11. W. Gropp, E. Lusk, and A. Skjellum. *Using MPI, portable programming with the Message-Passing Interface*. Scientific and Engineering Computation Series. The MIT Press, Cambridge, 1994.
12. C. B. Jenssen and P. Å. Weinerfelt. Coarse grid correction scheme for implicit multiblock Euler calculations. *AIAA Journal*, 33(10):1816–1821, 1995.
13. L. Mansfield. On the conjugate gradient solution of the Schur complement system obtained from domain decomposition. *SIAM J. Numer. Anal.*, 27(6):1612–1620, 1990.
14. L. Mansfield. Damped Jacobi preconditioning and coarse grid deflation for conjugate gradient iteration on parallel computers. *SIAM J. Sci. Stat. Comput.*, 12(6):1314–1323, 1991.

15. R. A. Nicolaides. Deflation of conjugate gradients with applications to boundary value problems. *SIAM J. Numer. Anal.*, 24(2):355–365, 1987.
16. G. Radicati and Y. Robert. Parallel conjugate gradient-like algorithms for solving sparse nonsymmetric linear systems on a vector multiprocessor. *Parallel Computing*, 11:223–239, 1989.
17. M.K. Seager. Parallelizing conjugate gradient for the Cray X-MP. *Paral. Comp.*, 3:35–47, 1986.
18. K.H. Tan. *Local coupling in domain decomposition*. PhD thesis, University Utrecht, Utrecht, 1995.
19. W.P. Tang. Generalized Schwarz splittings. *SIAM Journal on Scientific and Statistical Computing*, 13:573–595, 1992.
20. H.A. van der Vorst and C. Vuik. GMRESR: a family of nested GMRES methods. *Num. Lin. Alg. Appl.*, 1:369–386, 1994.
21. C. Vuik. Fast iterative solvers for the discretized incompressible Navier-Stokes equations. *Int. J. for Num. Meth. Fluids*, 22:195–210, 1996.
22. C. Vuik and J. Frank. A parallel implementation of the block preconditioned GCR method. In P. Sloot, M. Bubak, A. Hoekstra, and B. Hertzberger, editors, *High-Performance Computing and Networking, Proceeding of the 7th International Conference, HPCN Europe 1999, Amsterdam, The Netherlands, April 12-14, 1999*, Lecture Notes in Computer Science 1593, pages 1052–1060, Berlin, 1999. Springer.
23. C. Vuik, A. Segal, and J.A. Meijerink. An efficient preconditioned CG method for the solution of a class of layered problems with extreme contrasts in the coefficients. *J. Comp. Phys.*, 152:385–403, 1999.
24. C. Vuik, R.R.P. van Nooyen, and P. Wesseling. Parallelism in ILU-preconditioned GMRES. *Paral. Comp.*, 24:1927–1946, 1998.
25. P. Wesseling, A. Segal, and C.G.M. Kassels. Computing flows on general three-dimensional nonsmooth staggered grids. *J. Comp. Phys.*, 149:333–362, 1999.

Towards an Implementation of a Multilevel ILU Preconditioner on Shared-Memory Computers[*]

Arnold Meijster[1] and Fred Wubs[2]

[1] Computing Centre of the University of Groningen
A.Meijster@rc.rug.nl, http://www.rug.nl/hpc/people/arnold
[2] Research Institute of Mathematics and Computer Science
P.O. Box 800, 9700 AV GRONINGEN, The Netherlands
F.W.Wubs@math.rug.nl, http://www.math.rug.nl/~wubs

Abstract. Recently, substantial progress has been made in the development of multilevel ILU-factorizations. These methods are attractive for very large problems due to their good convergence properties. We consider the parallelization of the instance MRILU, where we restrict to a version intended for scalar problems. The most time consuming parts in using MRILU are repeated multiplication of two sparse matrices in the construction phase and the multiplication of a sparse matrix and a full vector in the solution phase. Algorithms for these operations, as well as matrix transposition, are presented and have been tested on a Cray J90.

1 Introduction

Many physical phenomena can be described by partial differential equations (PDEs). Irrespective whether one likes to compute eigenvalues, to use implicit time-integration methods or to do continuation on a steady solution one ends up with a linear system to be solved, which is often very large and sparse. In almost all cases, the solution thereof forms the bottleneck with respect to computation time. For such problems, users like to have the availability of a black-box linear-system solver, which can handle complicated systems of PDEs reasonably efficient. Developing a special purpose solver may take too much time and a direct solver is far too expensive.

An important class of iteration methods for linear systems form preconditioned CG methods and among the preconditioners Incomplete LU factorizations play a dominant role. If we consider the problem $Ax = b$ then for the ILU-factorization holds $A = LU + R$ and the CG-type method is applied to the preconditioned system

$$L^{-1}AU^{-1}\tilde{x} = L^{-1}b, \qquad \tilde{x} = Ux$$

For the classical ILU and modified ILU factorization using the same fill as the original matrix, the number of flops needed to gain a fixed amount of digits increases with the grid size which is unfavorable for very large problems.

---
[*] This research has been supported by the Stichting Nationale Computerfaciliteiten (National Computing Facilities Foundation, NCF).

The ideal case would be if the preconditioned matrix $L^{-1}AU^{-1}$ is close to identity, which is obviously the case if R is small or in other words if the factorization is nearly exact. This is accomplished in the multi-level ILU factorization. As accelerator, the classical CG method can be used when the preconditioned matrix is symmetric positive definite and e.g. Bi-CGSTAB or GMRES when it is not. However, it is our experience, that the choice of accelerator is far less critical than the preconditioning.

Our multilevel ILU method MRILU (for details see [3]) has already successfully been applied to the incompressible Navier-Stokes equations (in some cases extended with a k-ε turbulence model), to Rayleigh-Bénard flow, and to convection-diffusion problems in two and three dimensions. For a comparison of a variety of solvers including MRILU on Laplace-like equations see [2]. Similar methods like MRILU are described in [1, 8, 10]. Furthermore, there is also a link to algebraic multigrid, e.g. [7].

As was recognized also by others, e.g. [8], multilevel ILU methods can be parallelized. The basic steps in the factorization phase form the search of an independent set, multiplication of two sparse matrices, transposition, and for the solution phase the multiplication of a sparse matrix with a full vector. It appeared that the construction of the independent set is least critical, hence we confined our attention to the other parts. We parallelized already existing code, in which sparse matrices are stored in CSR format (section 3), and did not want to rewrite the whole code. We thus decided to stick to the CSR format[1].

The experiments in this paper have all been carried out on a Cray J90. Loops which can be parallelized are coded such that the compiler automatically generates parallel tasks for these loops. During run time execution the creation of these tasks is handled by the operating system and does involve some overhead. At the end of the loops all tasks have to synchronize (i.e. wait until all tasks have finished). This introduces some overhead as well.

In the remainder of the paper we will describe MRILU and indicate the parts to be parallelized (section 2). In section 3 we introduce data structures for storing sparse matrices, and an extension of this format for concurrent computation. Section 4 discusses parallel multiplications using these formats. In section 5 a parallel transposition of a matrix is presented. The latter two sections contain practical results obtained on a Cray J 90. Conclusions are drawn in section 6.

2 MRILU

In this section the multi-level ILU method MRILU (MR stands for matrix renumbering) will be described in short and we will comment on the parallelization aspects. A more detailed description can be found in [3]. In Algorithm 1 the basic steps in the factorization process are given. For a sparse matrix the partitioning of step 1 can always be made by extracting a set of unknowns that

---

[1] Though the matrix-vector multiplication using CSR format is becoming part of sparse BLAS, it is currently not implemented for the Cray J90.

> Set $A^{(0)} = A$
> **for** i=1..M
> 1. Reorder and partition $A^{(i-1)}$, to obtain $\begin{bmatrix} A_{11} & A_{12} \\ A_{21} & A_{22} \end{bmatrix}$,
>   such that the matrix A_{11} is sufficiently diagonal dominant.
> 2. Approximate A_{11} by a diagonal matrix $\tilde{A}_{11}$.
> 3. Drop small elements in A_{12} and A_{21}.
> 4. Make an incomplete LU factorization, $\begin{bmatrix} I & 0 \\ \tilde{A}_{21}\tilde{A}_{11}^{-1} & I \end{bmatrix} \begin{bmatrix} \tilde{A}_{11} & \tilde{A}_{12} \\ 0 & A^{(i)} \end{bmatrix}$,
>   where $A^{(i)} = A_{22} - \tilde{A}_{21}\tilde{A}_{11}^{-1}\tilde{A}_{12}$ (Schur complement of $\tilde{A}_{11}$).
> **endfor**
> Make an exact (or accurate incomplete) factorization of $A^{(M)}$.

Algorithm 1. The basic steps of MRILU.

are not directly connected, the so-called independent set. By allowing also weak connections, which are deleted in step 2, this set can be enlarged.

The dropping strategy used in steps 2 and 3 is based on the ratio of the element at hand and the diagonal element, and on the amount dropped so far in the corresponding row and column.

Since $\tilde{A}_{11}$ is diagonal, also its inverse and consequently the new Schur complement constructed in step 4 will be sparse, which makes it possible to repeat the process. However, the fill slowly increases in subsequent steps.

Let us discuss briefly the parallelization aspects of the respective steps. The independent set needed to create the ordering in step 1 is not unique, and finding the largest one is even an NP-complete problem. Hence, one usually uses some greedy algorithm to find an independent set. This process is sequential in nature and hence hard to parallelize, however some attempts have been made [5,6] but we expect that this leads to smaller independent sets. Since in our case the independent set selection is not the most critical part we did not parallelize it.

The parallelization of the dropping in steps 2 and 3 is straight-forward. The construction of the Schur-Complement is the more difficult one, especially the multiplication of the two sparse matrices $\tilde{A}_{21}$ and $\tilde{A}_{11}^{-1}\tilde{A}_{12}$ (the multiplication with the diagonal matrix is easy). So in general we are interested in speeding up the multiplication of two sparse matrices.

In the solution phase the L and U factor have diagonal blocks. So solving a system with these matrices amounts to multiplication of a sparse matrix and a full vector (see also the level-scheduling idea in [9]). We studied this step already for scalar equations in [4] but we will reconsider it here.

3 Data Structure

We have chosen to adopt two storage schemes. The *CSR format* (Compressed Sparse Row), which stores matrices row-wise, and the *CSC format* (Compressed

Sparse Column), which stores matrices column-wise. The reason for using two formats is the difficulty one will encounter in creating the L matrix in CSR format. In this matrix one wants to store A_{21} in each reduction step. If L were to be in CSR format then in each step A_{21} has to be merged with the hitherto formed L, which can be avoided by transposing A_{21} into CSC format and storing it thus in L. Efficient transposition is discussed in section 5.

3.1 CSR/CSC Format

The CSR data structure consists of a 5-tuple (nr, nc, cf, col, beg), where nr and nc are integers representing the number of rows and columns, respectively. The elements cf, col, and beg are arrays. The array cf contains the non-zero entries of the matrix, stored row-wise. These values are floating-point numbers. The array col is of the same length as cf, and beg has length $nr + 1$. Both arrays are integer valued. For entry $cf(i)$, its corresponding column number is found in $col(i)$. Since entries of the same row are stored consecutively in cf, we only need to know the index of the first entry of this row, and the index of the last entry. Therefore, rows are not stored explicitly (as in the case of columns), but only the index of the first element of each row is stored in the array beg. The index of the first element of row i is stored in $beg(i)$, while the index of its last element is $beg(i + 1) - 1$. The array beg has length $nr + 1$, since we need to know the index of the last element of row nr. In the next section this format is extended for use on shared memory architectures. An example of the format is given at the end of that section. The CSC format is a sort of natural dual of the CSR format. Instead of storing entries row-wise, they are stored column-wise.

3.2 Extension for Parallelism

If we try to implement algorithms on these data structures using multiple processors (cpu's), it is natural to split the data representation in as many (preferably) equal sized chunks as there are cpu's. Each cpu is assigned a chunk, from now on called its *private chunk*. On a shared memory computer, each cpu can access each memory location. In view of this machine model, it might appear a bit strange to split these data structures in chunks, since each processor can access the entire structure. Still it is useful to do this, for two reasons. The first is to reduce the amount of synchronization as much as possible. If each cpu may only modify a 'private' part of a shared resource, there is no need to protect this shared resource against simultaneous updates of this resource by more than one cpu. Such a protection is always expensive, regardless whether we program these protections explicitly ourselves (using semaphores), or let a compiler generate a data-parallel executable by using loop-parallelism. A second reason is that this data-layout mimics to some extent the distributed memory model, and thus the resulting algorithms are, with some effort, likely to be portable to distributed memory machines using message passing.

Assuming that the most frequently used operations are multiplications of matrices with vectors, it is natural to distribute the CSR data structure row-wise.

We assume that the distribution of non-zero entries in the matrix is reasonably uniform, i.e. the number of non-zero elements per row does not vary very much. Thus, if we assign to each processor (almost) the same number of rows in its chunk, we probably get a reasonable load-balance.

Let us assume that we deal with a matrix consisting of nr rows, and have the availability of P cpu's, numbered $p_1, \ldots, p_P$. Then, we distribute the matrix as follows. We compute $c = \lfloor nr/P \rfloor$. If P is a divisor of nr each processor is assigned a chunk with exactly c rows. If P is not a divisor of nr, we compute the remainder $r = nr - c * P$. It is possible to assign a single processor to deal with these r extra rows, but this might introduce significant load-imbalance (worst case $r = P - 1$). Therefore, we decide that each cpu which has a processor identification number less or equal to r is assigned $c + 1$ rows, resulting in an imbalance of only a single row. This results in the following simple algorithm to compute the lower bound (lwb) and the upper bound (upb) of the chunk assigned to processor p if the number of rows is n:

```
procedure chunk (n, p, nprocs : integer; var lwb, upb : integer);
    c := n / nprocs;  r := n - c * nprocs;
    if p ≤ r
    then lwb := (p − 1) * c + p;  upb := lwb + c
    else lwb := (p − 1) * c + r + 1;  upb := lwb + c - 1
end
```

Thus, we augment the CSR format with an integer array par which has length $P + 1$, which has basically the same structure as the array beg. For processor p, the starting row of its chunk can be found in $par[p]$, and the last row of its chunk can be found in $par[p + 1] - 1$. The augmented CSR structure will from now on be called the *PCSR* (parallel CSR) structure.

As an example, using $P = 3$, we would find for the 5×5 matrix A the following PCSR representation[2].

$$A = \begin{pmatrix} a & 0 & b & 0 & 0 \\ c & d & e & f & 0 \\ g & 0 & h & 0 & 0 \\ i & 0 & 0 & j & k \\ l & 0 & 0 & m & n \end{pmatrix}$$

nr	5											
nc	5											
par	1	3	5	6								
beg	1	3	7	9	12	15						
cf	a	b	c	d	e	f	g	h	i	j	k	l	m	n
col	1	3	1	2	3	4	1	3	1	4	5	1	4	5

For the CSC format, the same distribution technique is used on the columns of the matrix. This format will from now on be called the *PCSC* format.

4 Matrix Multiplication Algorithms

One of the most common operations on (sparse) matrices is multiplication. Two variants are needed. Let A, B, and C be sparse matrices, and x, y be full vectors. We have to deal with the following two cases.

---
[2] The entries are deliberately chosen to be symbolic, instead of actual numbers, in order to avoid confusion between entries and indices.

- $y := A \times x$, i.e. a sparse matrix times a full vector, and the result is stored in a full vector.
- $C := A \times B$, i.e. a sparse matrix times a sparse matrix, and the result is stored in a new sparse matrix.

4.1 Multiplication of a Sparse Matrix with a Full Vector

We have to perform nr inner products between the rows of A, and the vector x. We choose therefore to represent A using the PCSR format. In this format the entries are stored consecutively, and the first and last element of a row is directly accessible, which makes the algorithm for performing this multiplication relatively simple. The sequential algorithm is given below(left). A direct parallelization of this algorithm is to distribute the outer loop using the *par* field of the PCSR structure, and use private variables for r, and c resulting for cpu p in the algorithm on the right.

for $r := 1$ to nr → $\quad y(r) := 0$ $\quad$ for $c := beg(r)$ to $beg(r+1) - 1$ → $\quad\quad y(r) := y(r) + cf(c) * x(col(c))$	for $r := par(p)$ to $par(p+1) - 1$ → $\quad y(r) := 0$ $\quad$ for $c := beg(r)$ to $beg(r+1) - 1$ → $\quad\quad y(r) := y(r) + cf(c) * x(col(c))$

Assuming reasonable load-balancing, we expect a speedup linear in the number of processors on shared memory machines. If the architecture employs vector-pipes to speed up vector operations the algorithm ought to be modified to gain performance from parallelization as well as vectorization. The inner loop is vectorizable, however the length of this loop is of size $beg(r+1) - beg(r)$, which is in practical cases smaller than the length of the vector pipes. Therefore we try to lengthen the inner loop as much as possible. This is unfortunately only partly realizable, since row-wise addition is unavoidable. We introduce on each cpu a private auxiliary array tmp, and initialize it such that $tmp(i) = x(col(i))$, and distribute its initialization out of the inner loop, such that we can unroll it completely. When this array has been constructed an element-wise product $tmp := cf * tmp$ can be computed in a fully vectorizable loop. From this new result, it is easy to compute the final result using simple summation, in smaller vectorizable loops. This leads to the following algorithm for processor p:

$lwb := beg(par(p))$; $upb := beg(par(p+1))$; for $r := lwb$ to $upb - 1$ → $\quad tmp(r) := x(col(r))$; for $r := lwb$ to $upb - 1$ → $\quad tmp(r) := tmp(r) * cf(r)$; for $r := par(p)$ to $par(p+1) - 1$ → $\quad y(r) := 0$; $\quad$ for $c := beg(r)$ to $beg(r+1) - 1$ → $\quad\quad y(r) := y(r) + tmp(c)$	Although the length of this algorithm is longer than the trivial parallel solution, the amount of computational work is the same. The compiler, however, can vectorize the last two loops, and thus performance increase might be expected. The amount of extra memory needed is linear in the total number of non-zero entries, i.e. the size of the array cf.

We performed a performance test, the results of which are shown in the following table. Since absolute timings give only a measure of the performance of the cpu's, and not of the algorithm itself, we only present speed-up factors (i.e. T_1/T_p, where T_p is the absolute time measured using p cpu's). We computed $y = Ax$, where A is a $10^4 \times 10^4$ matrix, with on average 10 non-zeroes per row.

NCPUS	1	2	4	6	8	10	12	14	16
Speed-up	1.0	1.9	3.7	5.0	6.1	7.0	7.7	8.2	8.8

On a small number of cpu's the algorithm scales almost linearly in the number of processors. If we use more than 4 cpu's some degradation is observed, which can mainly be accounted for by the fact that the data structure uses several indirections, which easily results in slightly different running times per cpu. The total execution time however, is equal to the running time of the task which ran longest. Also the creation of processes results in some operating system overhead. However, even on 16 cpu's an efficiency of more than 50% is achieved.

4.2 Multiplication of Two Sparse Matrices

We consider the 'assignment' $C := A \times B$, where A and B are sparse matrices. We start with explaining how the parallelization can be done by means of a multiplication of two full matrices in order to not obscure the approach by details on the handling of the sparsity. The sequential algorithm for multiplying an $M \times N$ matrix A, with an $N \times R$ matrix B consists of three nested loops.

```
for r := 1 to M →   (* rows of A *)
  for c := 1 to R →   (* columns of B *)
    c(r, c) := 0
    for i := 1 to N →   (* dot product *)
      c(r, c) := c(r, c) + a(r, i) * b(i, c)
```

Clearly, each iteration of the outer loop can be performed independent of all other iterations, which allows a direct parallelization. We simply use the partitioning of the data using the PCSR format. It is useless to parallelize the second loop as well, since it would only interfere with the parallelization of the outer loop.

```
lwb := parA(p);
upb := parA(p + 1);
for r := lwb to upb − 1 →
  for c := 1 to ncB →   (* initialize row r of C *)
    c(r, c) := 0
  for i := 1 to ncA →   (* adapt row r of C *)
    for c := 1 to ncB →   (* all columns of B *)
      c(r, c) := c(r, c) + a(r, i) * b(i, c)
```

In the actual implementation we have to deal with the sparsity. On each processor we introduce a temporary full array d in which we will store all the intermediate contributions. Using reference arrays we keep precisely track of where elements in d are stored. To limit the length of the presentation we have not included these operations in the following code.

```
lwb := parA(p); upb := parA(p + 1);
for c := 1 to ncB → (* initialize temporary full row of C *)
    d(c) := 0;
initialize reference arrays for d
for r := lwb to upb − 1 →
    for i := begA(r) to begA(r + 1) − 1 → (* adapt row r of C *)
        for c := begB(colA(i)) to begB(colA(i) + 1) − 1 →
            d(colB(c)) := d(colB(c)) + cfA(i) ∗ cfB(c);
        adapt referencing d;
    copy d to begC(p), colC(p), cfC(p) using the reference arrays;
    reset d and reference arrays;
```

The resetting in the last step can be done using the same reference arrays in a time proportional to the fill. In the actual implementation this is performed together with the copying as shown in the previous program line. Indeed in this line we see that every processor has its own $begC$, $colC$ and cfC array. This is necessary since we do not know in advance the number of nonzero entries of C on a certain processor. Hence, afterwards we assemble the parts of the sparse matrix into the global $begC$, $colC$ and cfC array, which can be executed in parallel apart from a loop with length of the number of tasks.

We note that due to the fact that the reference arrays are constructed in order of occurrence of a new fill in d that the column numbers on a row are not necessarily ordered, which does not affect the remainder of the program.

We studied the speed-up of this algorithm on the product of two sparse matrices of order ($10^3 \times 10^3$) with an average fill of 10 per row.

NCPUS	1	2	4	6	8	10	12	14	16
Speed-up	1.0	1.6	3.3	4.8	6.3	8.2	8.8	9.9	11

The speed-up is reasonable. This time, performance degradation is less severe on a larger number of cpu's. This can be explained by the fact that the amount of computation per task is a lot larger. The waiting time at the synchronization point at the end of the tasks is about the same as in the previous case (matrix times vector), but it has become relatively small compared to computation time.

5 Transpose of a PCSR Matrix

In this section we consider the transposition of a sparse matrix. The input and the output of the algorithm is a matrix stored in PCSR format. The process of transposing a PCSR matrix consists of three stages. The first, and the last being highly parallel, while the middle is purely sequential. The sequential part, however, is negligible in computation time.

Let us assume we deal with an $m \times n$ matrix (m rows) A and we want to implement $B := A^T$ using P processors. A problem is to determine the beg and par array corresponding with the destination array B. The computation of these arrays is done in two stages. In the first stage, each processor computes a private histogram of the number of elements per column in its private part of the source

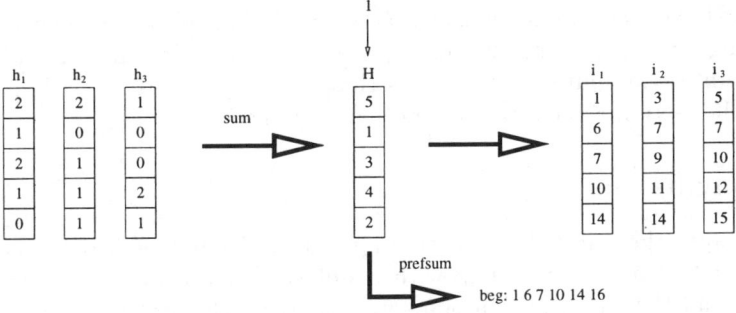

Fig. 1. Parallel transpose of a matrix using PCSR format.

matrix. This is a trivial linear time loop. The size of a histogram is n integers, and thus the total amount of extra memory is $P \times n$. Clearly, each processor can compute its private histogram, without any communication with any other process. Since the amount of work per processor is almost the same, we expect each processor to finish the first stage at approximately the same time. In Fig. 1 at the left side three histograms (h_1, h_2, h_3) are given. In the second stage, these histograms are summed resulting in the histogram H which shows the number of elements per column for the whole source matrix, and therefore the number of elements per row of the destination matrix. The summing of these histograms can only be performed when all 'private' histograms have been computed, i.e. each cpu has finished the first stage. This summing is performed in the second stage, on only one cpu. From the summed histogram H the beg array for B is easily obtained as follows. Shift the histogram H by one index, and insert 1 at H(1). Then we have $beg(i+1) = \sum_{k=1}^{i} H(i)$ (a so-called prefix-sum), and $beg(1) = 1$. This beg array gives also the begin locations of the columns to be built by the matrix part on the first processor (i_1). For the second processor we simply have to add i_1 and h_1 to find the begin locations for the columns of the matrix on that processor. Similarly i_3 is the sum of i_2 and h_2. In the last stage, all the elements of the source matrix are visited and copied into the destination matrix using the data structure obtained in the previous stage. This stage, just like the first one, scales linearly in the number of cpu's.

We computed $B = A^T$, where A is a $10^4 \times 10^4$ matrix, with on average 10 non-zeroes per row. The speed-up results are in the table below.

NCPUS	1	2	4	6	8	10	12	14	16
Speed-up	1.0	1.7	3.6	3.9	4.4	5.1	5.1	4.9	4.6

Two stages are highly parallel, while the middle stage is sequential. Hence, the processes have to wait for each-other when all processes have computed their private histograms and when the global histogram H has been computed. The merging of the private histograms into the global histogram is performed by process 1, which also creates the other processes at startup. The speed-ups are similar to the expected ones considering Amdahl's law. However, they are slightly decreasing for a large number of cpu's, due to process creation overhead. Besides the amount of computation is far less than in the previous algorithms, making the middle section relatively even more a bottleneck.

6 Conclusions

In this paper the parallelization of three essential parts of MRILU is studied: the product of a sparse matrix with a full vector, the product of two sparse matrices and the transposition of a sparse matrix. The matrices are all in CSR-format which we extended to a parallel format PCSR by giving each parallel task a number of rows of the matrix. We found that the smaller the parallel tasks the sooner the speed-up drops as the number of processors grows. For the transposition the speed-up drops if more than 4 processors are used and even the runtime becomes constant due to a small sequential part. A maximum speed-up of about 5 was observed. The matrix-vector product scales linearly for up to 4 processors and after that the speed-up increases more slowly. On 16 processors a speed-up of about 9 was observed. The matrix-matrix multiplication has the largest parallel task and there a speed-up of 11 on 16 processors is observed. Thus, on average we found a speed-up of about an order of magnitude, which will, once implemented seriously improve the performance of MRILU.

References

1. R.E. Bank and C. Wagner. Multilevel ILU decomposition. *Numer. Math.*, 82:543–576, 1999.
2. E.F.F. Botta, K. Dekker, Y. Notay, A. van der Ploeg, C. Vuik, F.W. Wubs, and P.M. de Zeeuw. How fast the Laplace equation was solved in 1995. *Appl. Numer. Math.*, 24:439–455, 1997.
3. E.F.F. Botta and F.W. Wubs. Matrix Renumbering ILU: An effective algebraic multilevel ILU-preconditioner for sparse matrices. *SIAM J. Matrix Anal. Appl.*, 20(4):1007–1026, 1999.
4. E.F.F. Botta, F.W. Wubs, and A. van der Ploeg. A fast linear-system solver for large unstructured problems on a shared-memory computer. In O. Axelsson and B. Polman, editors, *Proceedings of the Conference on Algebraic Multilevel Iteration Methods with Applications*, pages 105–116, Nijmegen, The Netherlands, 1996. University of Nijmegen.
5. M.T. Jones and P.E. Plassman. A parallel coloring heuristic. *SIAM J. Sci. Comput.*, 14(3):654–669, 1993.
6. M. Luby. A simple parallel algorithm for the maximal independent set problem. *SIAM J. Comput.*, 4:1036–1053, 1986.
7. A. Reusken. On a robust multigrid solver. *Computing*, 56(3):303, 1996.
8. Y. Saad. ILUM: A multi-elimination ILU preconditioner for general sparse matrices. *SIAM J. Sci. Comput.*, 17(4):830–847, 1996.
9. Y. Saad. *Iterative Methods for Sparse Linear Systems*. PWS, 1996.
10. Y. Saad and J. Zhang. BILUM: Block versions of multi-elimination and multilevel ILU preconditioner for general sparse linear systems. Technical Report UMSI 97-126, University of Minnesota, Minneapolis, 1997.

Application of the Jacobi–Davidson Method to Spectral Calculations in Magnetohydrodynamics

A.J.C. Beliën[1], B. van der Holst[1], M. Nool[2],
A. van der Ploeg[3], and J.P. Goedbloed[1]

[1] FOM-Instituut for Plasma Physics 'Rijnhuizen', P.O. Box 1207
3430 BE Nieuwegein, The Netherlands
[2] Centre for Mathematics and Computer Science, P.O. Box 94079
1090 GB Amsterdam, The Netherlands
[3] MARIN, P.O. Box 28, 6700 AA Wageningen, The Netherlands

Abstract. For the solution of the generalized complex non-Hermitian eigenvalue problems $Ax = \lambda Bx$ occurring in the spectral study of linearized resistive magnetohydrodynamics (MHD) a new parallel solver based on the recently developed Jacobi–Davidson [18] method has been developed. A brief presentation of the implementation of the solver is given here. The new solver is very well suited for the computation of some selected interior eigenvalues related to the resistive Alfvén wave spectrum and is well parallelizable. All features of the spectrum are easily and accurately computed with only a few target shifts.

1 Introduction

Plasma is the single most occurring state of matter in the universe. It is characterized by such high temperatures that almost all atoms are ionized completely. In most situations where plasmas occur a magnetic field is present which will interact with the charged plasma particles. On a global scale, i.e., on time and length scales much larger than typical kinetic time and length scales, the interaction of plasma with magnetic fields can be described by magnetohydrodynamics (MHD). This theory is scale invariant so that it is applicable to such diverse and utterly different objects as stellar winds, coronal loops, and thermonuclear fusion plasmas in tokamaks, to name a few.

A key aspect of the MHD analysis of plasmas is the study of waves and instabilities. In thermonuclear fusion plasmas the goal is to confine a dense hot plasma for as long as is necessary to reach ignition. MHD instabilities limit the densities that can be obtained and, hence, have a negative impact on fusion operation. In fact, the main activity of the MHD fusion theory in the last thirty years has been to increase this limit by optimizing the plasma profiles and plasma cross-section with respect to stability, see for example Ref. [5]. From a plasma physics point of view this emphasis on (in)stability does not do justice to the importance of the stable part of the spectrum. Besides the fact that a multitude of observed phenomena in astrophysical and laboratory plasmas are wave-like, detailed knowledge of the spectrum of waves allows for MHD spectroscopy (see, e.g., Refs. [10, 1, 9, 2, 4]). Free oscillations (waves) play a dominant role in MHD spectroscopy. It is the computation of these free eigen-oscillations that we are interested in in this paper.

Mathematically, free oscillations and instabilities of a resistive MHD plasma are described by a large complex non-Hermitian generalized eigenvalue problem. Until recently, the large scale spectral codes CASTOR [16, 12] and POLLUX [6] that we employ for the computation of tokamak and coronal loop spectra calculated them either by solving for the whole spectrum with a direct dense matrix method like QR or using inverse vector iteration for a selection of eigenpair solutions in the neighborhood of a target value [11]. The QR method is limited to very coarse meshes (due to its storage and computation requirements) and large values of the resistivity (to get reasonably converged results). Therefore, the eigenvalues obtained with QR are normally used as initial guesses for the inverse vector iteration only. For the application of interest, viz., the computation of the resistive Alfvén wave spectrum for realistic small values of the resistivity ($\eta < 10^{-7}$), the QR eigenvalues are very poor initial guesses. The original versions of the CASTOR and POLLUX codes contained a two-sided Lanczos solver with no orthogonalization [3, 11]. Since no orthogonalization is used criteria are implemented that do away with occurring spurious eigenvalues. However, these criteria are not bullet-proof. Not all spurious eigenvalues are removed and occasionally proper eigenvalues are labeled as spurious and removed. Furthermore, the accuracy of the computed eigenvalues is generally poor and the inverse vector iteration has to be used to get good converged eigenvalues. We needed a more robust iterative solver to obtain several eigenvalues at once within a specified region of the complex plane. We have opted for the Jacobi–Davidson method.

The Jacobi–Davidson (JD) subspace iteration method is a new and powerful technique for solving non-symmetric eigenvalue problems in a sequence of steps [18–20]. It is extremely suitable for solving the resistive Alfvén spectrum since such a spectrum consists of many complex branches. To obtain converged results on the interior Alfvén modes, an effective search mechanism is highly desirable. Such a search mechanism is provided by the JD algorithm. Compared with the Arnoldi method, where the projection matrices are always upper Hessenberg, these matrices in JD are transformed to upper Hessenberg each step. This allows the restart technique of JD to be simpler. Another difference is the fact that by selecting the Ritz value that has the maximum absolute value the JD method can be accelerated (see also [14]).

In this paper we briefly describe the JD method and its parallel implementation for use in MHD spectral computations and apply it to the calculations of some resistive MHD spectra in tokamaks.

2 Equilibrium, Spectral Equations, and Core Numerical Problem

For the description of small amplitude waves and instabilities we exploit the linearized resistive MHD equations for the evolution of the density ρ, velocity **v**, pressure p, and magnetic field $\mathbf{B}$, where resistivity is denoted by η:

$$\frac{\partial \rho_1}{\partial t} = -\nabla \cdot (\rho_0 \mathbf{v}_1), \tag{1}$$

$$\rho_0 \frac{\partial \mathbf{v}_1}{\partial t} = -\nabla p_1 + (\nabla \times \mathbf{B}_0) \times \mathbf{B_1} + (\nabla \times \mathbf{B}_1) \times \mathbf{B_0}, \tag{2}$$

$$\frac{\partial p_1}{\partial t} = -\mathbf{v}_1 \cdot \nabla p_0 - \gamma p_0 \nabla \cdot \mathbf{v}_1 + (\gamma - 1) \eta (\nabla \times \mathbf{B}_1)^2, \tag{3}$$

$$\frac{\partial \mathbf{B}_1}{\partial t} = \nabla \times (\mathbf{v}_1 \times \mathbf{B}_0) - \nabla \times (\eta \nabla \times B_1). \tag{4}$$

The subscripts 0 and 1 indicate equilibrium and perturbation quantities, respectively. Together with boundary conditions this constitutes the spectral MHD formulation. The equilibrium quantities obey the ideal static force balance:

$$\nabla p_0 = (\nabla \times \mathbf{B}_0) \times \mathbf{B}_0, \qquad \nabla \cdot \mathbf{B}_0 = 0. \tag{5}$$

Discretization of Eqs.(1-4) using finite elements in the direction across the magnetic field and Fourier harmonics in the poloidal and toroidal directions results in a generalized eigenvalue problem for the case of free plasma oscillations and instabilities

$$\mathbf{A}\mathbf{x} = \lambda \mathbf{B}\mathbf{x} \tag{6}$$

where λ is an eigenvalue. Its imaginary part describes oscillations, its real part instabilities and damping. Matrix $\mathbf{B}$ is a complex Hermitian positive definite block tridiagonal matrix while matrix $\mathbf{A}$ is a complex non-Hermitian block tridiagonal matrix. Both $\mathbf{A}$ and $\mathbf{B}$ are of size $M \times M$ with $M = nN$. The block size n is a always a multiple of 16. The number of diagonal blocks is denoted by N.

3 An Eigenvalue Solver Based on the Jacobi–Davidson Algorithm

The Jacobi–Davidson method [18–20] is based on two concepts: the application of a Ritz-Galerkin approach to the eigenvalue problem (6) with respect to a subspace spanned by an orthonormal basis of low dimensionality, and the construction of a new search vector orthogonal to the eigenvector approximations that have been obtained so far.

To obtain a new search direction the JD algorithm solves a system of linear equations called the correction equation. An appropriate preconditioner has to be applied to obtain fast convergence such that the correction equation can be solved to some modest accuracy using only a few steps of, e.g., GMRES [17]. Because $\mathbf{A} - \sigma \mathbf{B}$, where σ is the target value in the vicinity of which eigenvalues are sought, has relatively full blocks, due to the structure of the MHD equations, a parallel complete block LU-decomposition has been shown to be a fast and robust method to solve the correction equation [14].

Both the construction and the application of L and U can be parallelized efficiently by, prior to decomposition, reordering block rows and columns based on Domain Decomposition and block Cyclic Reduction (DDCR). Due to the special block tridiagonal form of $\mathbf{A} - \sigma\mathbf{B}$ its LU-decomposition can be constructed without excessive memory requirements or a large number of operations. On account of the cyclic reduction part of the decomposition, which starts on all processors, while half of the active processors becomes idle after each step, we may not expect linear speed-up. However, the overall performance of DDCR is quite good due to the domain decomposition part (see [15]). Since the construction and application of the LU-decomposition form the most time consuming part this guarantees good overall parallel performance as well. In Fig. 1, we show the Gflop/s rate for the DDCR-decomposition and the corresponding solution process SOLDDCR on a CRAY-T3E for different block sizes and number of diagonal blocks. From this figure it is clear that the total number of diagonal blocks, N, has a far greater influence on the scalability than the block size n.

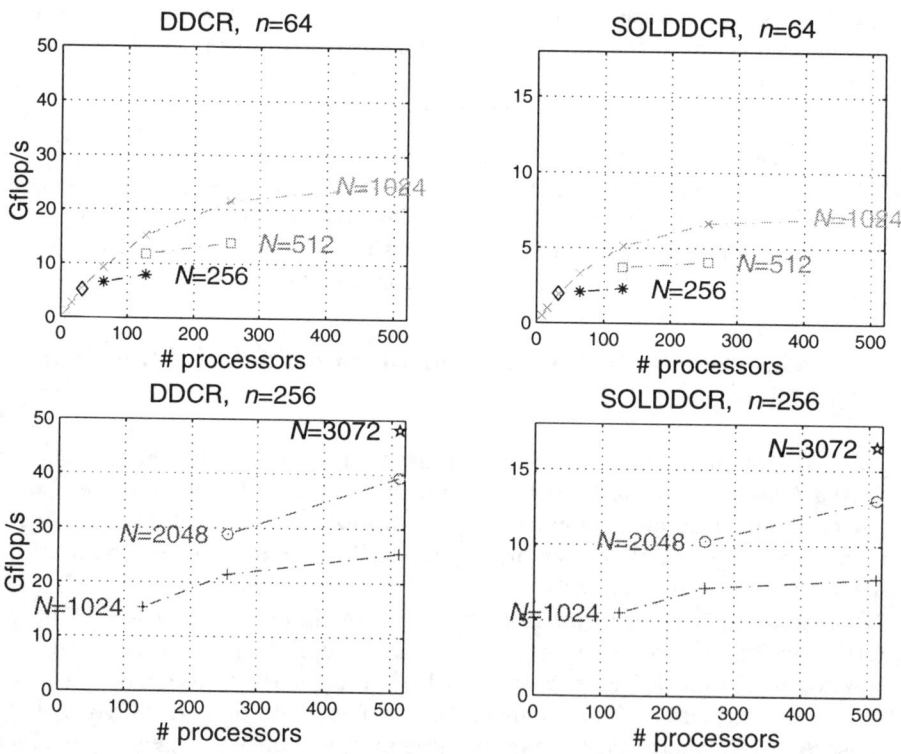

Fig. 1. Gflop/s-rate on a CRAY-T3E, with a single-node peak performance of 600 Mflop/s, for the construction (left) and application (right) of the LU-decomposition based on a domain decomposition and block cyclic reduction method (DDCR). The results are obtained for block sizes $n = 64$ and $n = 256$. The number of diagonal blocks varies from $N = 256$ till $N = 3072$.

With the definition of $Q \equiv (LU)^{-1}B$ we apply JD to the standard eigenvalue problem

$$\mathbf{Qx} = \mu \mathbf{x}, \qquad (7)$$

with the inverse shifted eigenvalues $\mu = 1/(\lambda - \sigma)$ which automatically make the desired eigenvalues λ, close to the target σ, the dominant ones and, hence, 'easy' to compute. When the target σ is close to an eigenvalue, small pivot elements can be generated in the LU-decomposition of $\mathbf{A} - \sigma \mathbf{B}$ and the application of Q may influence the computed spectrum. In such cases using smaller tolerances or exploiting different target values can give information on the accuracy. It should be noted that numerical experiments with a similar algorithm using harmonic Ritz values applied to the generalized eigenvalue problem is a more promising approach [13, 18]. The main advantage of that approach is that the LU-decomposition is only used as a preconditioner and not as a shift and invert operator. However, the method exploiting harmonic Ritz values consumes more memory and costs approximately 20% more execution time per Jacobi–Davidson iteration step.

At the k-th step of the JD iteration algorithm the approximation of the eigenvector can be written as $\mathbf{V}_k \cdot \mathbf{s}$ where $\mathbf{V}_k$ is the $M \times k$ matrix whose columns are the k search vectors. The search directions have been made orthonormal to each other using the Modified Gram-Schmidt procedure. The vector $\mathbf{s}$ and the approximation θ of an eigenvalue of $\mathbf{V}_k \cdot \mathbf{s}$ are constructed such that the residual $r = (\mathbf{Q} \cdot \mathbf{V}_k - \theta \mathbf{V}_k) \cdot \mathbf{s}$ is orthogonal to the k search directions. The Rayleigh-Ritz requirement then leads to

$$\mathbf{V}_k^* \cdot \mathbf{Q} \cdot \mathbf{V}_k \cdot \mathbf{s} = \theta \mathbf{s}, \qquad (8)$$

where θ and $\mathbf{s}$ are an eigenpair of the small matrix $\mathbf{V}_k^* \cdot \mathbf{Q} \cdot \mathbf{V}_k$ of size k. A proper restart technique keeps the size of the matrix on the left-hand side very small compared to M enabling the solution with a direct method, for example QR.

The calculation of the eigenvalues and eigenvectors of the small projected system has not been parallelized. Actually, it is performed by all processors, because then after the calculation on each processor all information is available without communication. The projected system (8) of the Jacobi–Davidson process contains a lot of information about the eigenvalues in the neighborhood of the target σ. The question arises how much information may be thrown away when a restart is performed, necessarily to keep the projected system small, without slowing down the convergence behavior too much. Obviously, in the parallel case the size of the projected system plays even a more important role in wall clock time than in the sequential case. In [13], it is shown that it depends on the value M/p, where p is the number of processors, whether restarts lead to a reduction in the wall clock time for the parallel JD process.

The maximum problem size depends on the number of processors p and the total amount of memory per processor. Our parallel implementation requires per processor

$$128 N_p n^2 + 16 N_p (3m + N_{\mathrm{ev}} + 7) + 96 n^2 + 64 m (m + 1) \text{ bytes}, \qquad (9)$$

as is shown in [13], where N_p denotes the maximum number of diagonal blocks on one processor, m the maximum allowed size of the projected system and N_{ev} stands for the desired number of eigenvalues.

4 Application to Resistive Toroidal Spectra

As an example, we have used the JD solver to solve the stable resistive spectrum of a tokamak with circular cross-section and an inverse aspect ratio $\epsilon = 0.2$. The results are shown in Fig. 2.

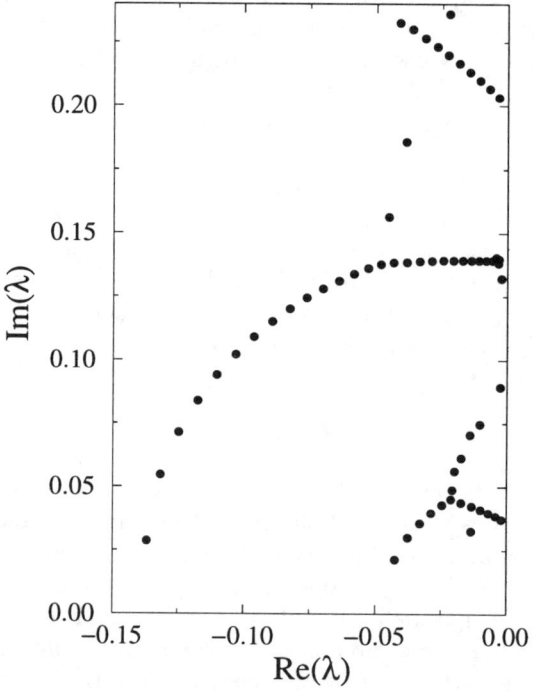

Fig. 2. The resistive spectrum for unit toroidal mode number, inverse aspect ratio $\epsilon = 0.2$, and resistivity $\eta = 2.5 \times 10^{-5}$. Purely damped eigenmodes have been left out.

The resistive spectrum was calculated with 1000 radial points and 4 poloidal harmonics, corresponding to matrices **A** and **B** with a block size of $n = 64$ and $N = 1000$ diagonal blocks. The acceptance criterium of a Ritz eigenpair leading to a eigenpair of the original problem (6) is that the 2-norm of the residual is smaller than 10^{-6} times the eigenvalue. The implementation of the JD solver is such that it terminates when N_{ev} are found or the maximum number of iterations is surpassed. The computations were done with $N_{ev} = 10$ to keep the size of the matrix occurring in Eq. (8) as small as possible. Several target shifts are necessary to map out the spectrum and many eigenvalues were found by more than one shift. If the target is not well-chosen, it may happen that less than N_{ev} eigenvalues will be found. In that case a new target based on the obtained spectrum must be taken rather than to increase the number of iterations.

Inspection of the number of radial nodes of the eigenfunctions can be used to ensure that all eigenvalues have been found. For smaller resistivity the topology of the spectrum remains the same but the number of eigenvalues situated on the curves shown in Fig. 2 increases (the number is proportional to $\eta^{-1/2}$). One then has the choice of enlarging N_{ev} so that more eigenvalues can be found for one specified target shift at the expense of larger matrices in the Rayleigh-Ritz part, or using more target shifts. The fact that new target shifts require new LU-decompositions, which are the computationally most intensive operations, (though well parallelizable) the first option seems the best.

Physically, the eigenvalues are converged as well: taking more poloidal harmonics or more radial grid points into account results in changes that are smaller than the acceptance criterium for the Ritz pairs. The physical interpretation of resistive spectra, like the one shown in Fig. 2, can be found in Ref. [7].

5 Conclusions and Outlook

The Jacobi–Davidson method appears to be an excellent method for parallel computation of resistive Alfvén spectra that occur in studies of the linearized interaction between plasmas and magnetic fields. The method discussed in this paper is based on solving projected eigenvalue problems of an order typically less than 30.

Within Jacobi–Davidson a parallel method to compute the action of the inverse of the block-tridiagonal matrix $A - \sigma B$ is used. In this approach, called DDCR, a block-reordering based on a combination of Domain Decomposition and Cyclic Reduction is combined with a complete block-tridiagonal LU decomposition of $A - \sigma B$, so $LU = A - \sigma B$. Both the construction of L and U and the triangular solves parallelize well.

We have successfully applied the JD-solver to the problem of computing branches of the stable resistive Alfvén spectrum with only a few target values σ (smaller resistivity requires more targets) and with an accuracy of 10^{-6}.

This has given us enough confidence to endeavor into the much more complex field of computing the spectra for plasmas with background flows, which is presently under taken [8].

Acknowledgments. This work was performed as part of the research programme of the 'Stichting voor Fundamenteel Onderzoek der Materie' (FOM) with financial support from the 'Nederlandse Organisatie voor Wetenschappelijk Onderzoek' (NWO). Part of this work is done in the project on 'Parallel Computational Magneto-Fluid Dynamics', funded by the NWO Priority Program on Massively Parallel Computing. Another part is sponsored by the 'Stichting Nationale Computerfaciliteiten' (NCF), also for the use of supercomputer facilities.

References

1. Beliën, A.J.C., Poedts, S., and Goedbloed, J.P., Phys. Rev. Lett. **76**, 567 (1996).
2. Beliën, A.J.C., Poedts, S., and Goedbloed, J.P., Astron. Astrophys. **322**, 995 (1997).
3. Cullum, J., Kerner, W., and Willoughby, R.A., IBM RC14190, February 1988, IBM T.J. Watson Research Center, Yorktown Heights, in Computer Physics Communications, **53** (North-Holland, Amsterdam, 1988).
4. De Ploey, A., van der Linden, R.A.M, and Beliën, A.J.C., to appear in Physics of Plasmas (2000).
5. Freidberg, J.P., "Ideal magnetohydrodynamics", Plenum Press, New York, 1987.
6. Halberstadt, G., and Goedbloed, J.P., Astron. Astrophys. **301**, 559 & 577 (1995).
7. van der Holst, B., Beliën, A.J.C., Goedbloed, J.P., Nool, M., and van der Ploeg, A., Physics of Plasmas **6**, 1554 (1999).
8. van der Holst, B., Beliën, A.J.C., Goedbloed, J.P., "New Alfvén Continuum Gaps and Global Modes Induced by Toroidal Flow", Phys. Rev. Lett., to appear (March 2000).
9. Holties, H.A., Fasoli, A., Goedbloed, J.P., Huijsmans, G.T.A., and Kerner, W., Physics of Plasmas **4**, 709 (1997).
10. Huijsmans, G.T.A., Kerner, W., Borba, D., Holties, H.A., and Goedbloed, J.P., Physics of Plasmas **2**, 1605 (1995).
11. Kerner, W., J. Comput. Phys. **85**, 1 (1989).
12. Kerner, W., Goedbloed, J.P., Huijsmans, G.T.A., Poedts, S., and Schwartz, E., J. Comput. Phys. **142**, 271 (1998).
13. Nool, M., and van der Ploeg, A., "Parallel Jacobi–Davidson for Solving Generalized Eigenvalue Problems", in Palma, J.M.L.M., Dongarra, J., and Hernandez, V., editors, Proceedings of the Third International Meeting on Vector and Parallel Processing, volume 1573 of Lecture Notes in Computer Science, p. 58–71, Springer-Verlag, Berlin, 1999.
14. Nool, M., and van der Ploeg, A., "A Parallel Jacobi–Davidson-type Method for Solving Generalized Eigenvalue Problems in Magnetohydrodynamics" to appear in SIAM J. on Scientific Computing.
15. van der Ploeg, A., "Reordering Strategies and LU-decomposition of Block Tridiagonal Matrices for Parallel Processing", Technical Report NM-R9618, CWI, Amsterdam, October 1996.
16. Poedts, S., Kerner, W., Goedbloed, J.P., Keegan, B., Huijsmans, G.T.A., and Schwartz, E., Plasma Phys. Controlled Fusion **34**, 1397 (1992).
17. Saad, Y., Schultz, M.H., SIAM Journal of Scientific and Statistical Computing, **17**, 856, 1986.
18. Sleijpen, G.L.G., and van der Vorst, H.A., SIAM J. Matrix Anal. Appl. **17**, 401 (1996).
19. Sleijpen, G.L.G., Booten, G.L., Fokkema, D.R., and van der Vorst, H.A., BIT, **36**, 595 (1996).
20. Sleijpen, G.L.G., van der Vorst, H.A., and Bai, Z, Chapters 4, 5, 7, and 8 in "Templates for the Solution of Algebraic Eigenvalue Problems: A Practical Guide", eds. Bai, Z., Demmel, J., Dongarra, J., Ruhe, A., and van der Vorst, H.A, SIAM, to appear in 2000.

PLFG: A Highly Scalable Parallel Pseudo-random Number Generator for Monte Carlo Simulations

Chih Jeng Kenneth Tan[1]* and J. A. Rod Blais[2]

[1] High Performance Computing Center
The University of Reading
Reading, United Kingdom
cjtan@acm.org

[2] Pacific Institute for the Mathematical Sciences
Department of Geomatics Engineering
University of Calgary
Calgary, Alberta, Canada
blais@ucalgary.ca

Abstract. In this paper, a parallel pseudo-random generator, named PLFG, is presented. PLFG was designed specifically for MIMD parallel programming, implemented using Message Passing Interface (MPI) in C. It is highly scalable and with the default parameters chosen, it provides an astronomical period of at least $2^{29} \left(2^{23209} - 1\right)$. Its scalability and period is essentially limited only by the hardware architecture on which it is running on. An implementation in MPI guarantees portability across the large number of high-performance parallel computers, ranging from clusters of workstations to massively parallel processor machines, supported by MPI. PLFG has been subjected to the 2D Ising model Monte Carlo simulation test with the Wolff algorithm. Results from the test show that the quality of the pseudo-random numbers generated are comparable to that of other more commonly used parallel pseudo-random generator. Timing results show that PLFG is faster than some PPRNGs, and on par with others.

Keywords: Pseudo-random number generator, Lagged Fibonnaci Generator, Parallel computation, Monte Carlo method, Randomized computation

1 Introduction

Even in this day and age, when parallel computing is increasingly being used across various scientific computations, the majority of pseudo-random number generators (PRNGs) used and continue to be developed are still sequential PRNGs. Only a small fraction of PRNGs belong to the parallel PRNG (PPRNG) group.

---

* Previously at the Department of Computer Science, The University of Liverpool, Liverpool, United Kingdom, and the Remote Sensing Laboratory, Department of Geomatics Engineering, University of Calgary, Calgary, Alberta, Canada.

Of the computations done on parallel machines, Monte Carlo method, or Markov Chain Monte Carlo method in general, require the highest quality of pseudo-random numbers [12]. Monte Carlo simulations are parallel by nature, and as noted by Metropolis and Ulam in 1949 [10], the simulations may (theoretically) be partitioned into n steps and executed on n number of processors with minimal inter-processor communication. However, this will also require PRNG of high quality, high speed and long period. The pseudo-random numbers used on the each of the processors and between the processors have to be statistically independent. Also, the period of the PRNG has to be at least the same length as the number of pseudo-random numbers required in the simulation. If there are n steps in the simulation but the PRNG used has a period $\frac{n}{m} < n$, then the same $\frac{n}{m}$ points are being sampled m times, rather than sampling on all the n points. Simulation results obtained under such conditions will be erroneous.

In this paper, PLFG, a PPRNG based on the lagged Fibonacci algorithm developed by the authors, is shown. First, a brief outline of the lagged Fibonacci algorithm is given. Several methods of parallelizing PRNGs are then given. The algorithm for PLFG follow suit. Results of tests conducted are also presented.

2 Lagged Fibonacci Generators

In general, lagged Fibonacci generators (LFGs) are of the form

$$x_i = (x_{i-p} \odot x_{i-q}) \bmod M$$

where x_i is the pseudo-random number to be output, and $\odot$ is a binary operation performed with the operands x_{i-p} and x_{1-q}. The values p and q are called the lag values, and $p > q$. The operand $\odot$ commonly used are addition (or subtraction), multiplication or bitwise exclusive OR (XOR). The value of M is typically a large integer value or 1 if x_i is a floating point number. When XOR operation is used, mod M is dropped. An array of length p is used to store the previous p values in the sequence.

XOR operations give the worst pseudo-random numbers, in terms of their randomness properties [4, 1, 14]. Additive LFGs were more popular than multiplicative LFGs because multiplication operations were considered slower than addition or subtraction operations, despite the superior properties of multiplicative LFGs noted by Marsaglia in 1984 [4]! Tests conducted by comparing operation execution times have shown that, with current processors and compilers, multiplication, addition and subtraction operations are of similar speeds. Thus, the argument preferring additive operations over multiplicative operations is nulled and multiplicative operations should be used.

The parameters p, q and M should be chosen with care, to obtain a long period and good randomness properties. [2] suggested lag values be greater than 1279. Having large lag values also improves randomness since smaller lags lead to higher correlation between numbers in the sequence [4, 1, 2]. In LFGs, the key purpose of M is to ensure that the output does not exceed the range of the data type.

Initializing the lag table before any pseudo-random numbers are generated with the LFG is also of critical importance. The initial values have to be statistically independent. To obtain these values, another PRNG is often used.

With $M = 2^e$, where e is the total number of bits, additive LFG have a period $2^{e-1}(2^p - 1)$. The period of multiplicative LFG, however, is shorter than that of additive LFG: $2^{e-3}(2^p - 1)$. This shorter period should not pose a problem for multiplicative LFGs, if the value of p is large.

Since the LFG shown above uses two past random numbers to generate the next, it is said to be a "two-tap LFG". Empirical tests have shown that LFGs with more taps, three-tap LFG or four-tap LFG for example, give better results in statistical tests, at the minimal extra time cost of accessing the lag table and performing the $\odot$ operation.

If a multi-tap additive LFG is used, M can be chosen to be the largest prime number that fits in the data type [3]. It can be proven using the theory of finite fields that random numbers generated in such a manner will be a good source of random numbers. For a complete discussion, see [3].

3 The Design of PLFG

3.1 Parallelizing the Pseudo-random Number Generator

There are several techniques for parallelizing pseudo-random number generators. The well known ones being leap frog, sequence splitting, independent sequences and shuffling leap frog [2, 15]. All these techniques, except the method of independent sequence, require arbitrary elements of the sequence to be generated efficiently [2]. While it is technically possible to parallelize LFGs with techniques like leap frog, sequence splitting and shuffling leap frog, the amount of inter-processor communication that would be required makes it impractical to parallelize LFGs in such a manner. But LFGs can be parallelized very easily with the independent sequence method, and may be very efficient. PRNG parallelization by independent sequences is also a recommended technique [11]. PLFG was parallelized with this technique.

Independent sequences are obtained by having multiple generators running on multiple processors, but seeded independently. When a PPRNG with independent sequences is being used, it is analogous to running the simulation multiple times, each time with a different PRNG. This is highly desirable in Monte Carlo simulations since the variance can be reduced by $\sqrt{n}$ if n independent trials are being carried out.

3.2 Seeding the Pseudo-random Number Generator

Some pseudo-random number generators like the Linear Congruential Generator (LCG), seen in most implementations of the standard rand() call in ANSI Standard C, need only one parameter for initialization. However, this is not the case with LFGs. Initialization of an LFG is by initializing the lag table, which is an array of length $\geq p$.

In the case of a PPRNG, for a sequence to be independent from another, all the bits in the whole bit table of one generator has to be independent from that of another generator at any point in time. That is to say, after n states (i.e.: after n pseudo-random numbers have been generated), the bit in one bit table has to be independent from that of another bit table. But for that to be possible, the initial bits in all the bit tables have to be independent since the bits of subsequent elements that are generated are pushed onto the bit table. For an LCG which requires only one seed and one previous element to generate the next element, the whole bit table is made up of the bits of that one element. But for an LFG, the bit table is the collection of bits in all p elements of the lag table. The stringent requirement for statistical independence between the seed elements of the lag table on each processor cannot be stressed any further, to avoid long range correlation [3].

Where most existing LFGs, sequential or parallel, use an LCG to initialize the lag table, PLFG differs. In the case of PLFG, the lag tables of the sequences are initialized with pseudo-random numbers generated by a sequential PRNG, the Mersenne Twister [9]. The Mersenne Twister used has the Mersenne prime period, $2^{19937}-1$, thus known as the MT19937. This generator has passed several statistical tests for randomness, including DIEHARD [9].

3.3 The PLFG Algorithm

Mascagni et. al. in [6, 7, 8, 5] proposed that the PPRNG be parallelized by splitting the sequence into its maximal periods, which each processor will generate sequences within its designated maximal period. Parallelization by dividing the state space of the sub-maximal-period generators were also proposed. These techniques have proven to be very useful. However, in PLFG, these steps are ignored. Rather, the lag tables for the all processors are simply initialized using the MT19937.

For a parallel program to use PLFG, the processors on the machine will have to be divided into two major groups: processors with a PRNG (PLFG processors) and processors without a PRNG. In addition, the master node is set to node 0. Such a division scheme allows the program running on p processors to have one pseudo-random number stream per processor.

Before PLFG can be used, it must be initialized. The initialization algorithm is as follows:

A-1 Processor 0 sets m processors to be used as PLFG processors.
A-2 List of PLFG processor IDs broadcasted to all processors by processor 0.
A-3 Processor 0 initializes all m lag tables using the MT19937 PRNG.
A-4 Lag tables broadcasted by processor 0 to all PLFG processors.

For a PLFG processor, $\alpha_i, 0 \leq i < m$, generating a vector of r pseudo-random numbers is by simply calling the generating function. For one without a PRNG, β, the following message passing scheme will have to be used:

B-1 Processor β requiring r pseudo-random numbers sends request to processor α_i.

B-2 Processor α_i receives requesting message, generates pseudo-random number vector of length r, then sends result back to processor β.

PLFG was designed to be a two-tap LFG-based generator, and the lag table parameters were chosen to be $p = 23209$, $q = 9739$. These lag values were recommended by Knuth, in [3]. There is nothing in the design of PLFG that prohibits the extension of the PPRNG, to have more than two taps. The round-robin algorithm used in PLFG ensures that the actual size of the lag table is of length p only, minimizing the concern for a large memory footprint.

Each pseudo-random number generated by PLFG is of type unsigned long in ANSI Standard C. On 32-bit architecture machines, unsigned long will typically be a 32-bit data type. But on some 64-bit architecture machines, unsigned long is a 64-bit data type. This dependence on machine architecture results in the variability in the period of the pseudo-random number sequence generated by PLFG. Therefore, on 32-bit architecture machines, the period remains to be $2^{29} \left(2^{23209} - 1\right)$, but the period is more than double this value on 64-bit architecture machines, being $2^{61} \left(2^{23209} - 1\right)$. This apparent weakness due to the dependency on machine architecture is indeed a feature: when machine with wider word size becomes available, the period of PLFG automatically expands! The issue of using PLFG for simulations running on heterogeneous workstation clusters with workstations having different unsigned long data type sizes may be a concern, however.

The number of processors designated as PLFG processors, m, is limited by the period of MT19937. However, since the period of MT19937, used for initializing the lag table, is $2^{19937} - 1$, the maximum number of PLFG processors is $\frac{2^{19937}-1}{23209}$. This limit is not really a limit at all, in practical applications. Thus, PLFG is in practice scalable to as many processors as may be required for a simulation. Coupling the long period of MT19937 with the long period of each of the independent sequences in PLFG, the probability that the sequences will overlap is minimal.

When each processor is given its own generator, it is obvious that independent sequences can be generated easily with minimal communication. The only communication involved is in the initialization stage when the lag tables are sent from the master node to all the other nodes. After initialization, there is zero communication until shutdown!

The message passing scheme used to generate the pseudo-random numbers for processors without a PRNG may present a drawback, especially when it is used on a cluster interconnected with a high latency and low bandwidth network. Under such conditions, the capability of generating a vector of r pseudo-random numbers may be used to mitigate the network communication overhead, by requesting for a long vector of pseudo-random numbers with each message to the PLFG processor. However, when used on a high performance network, the message passing scheme should not pose as a problem at all. Such a design has the advantage that a small number of processors can be designated as PLFG processors to generate pseudo-random numbers only, and this small pool of processors can be shared among all the other compute processors. Alternatively, each of the processors may have its own PRNG.

4 Tests Conducted and Current Use

PLFG has been subjected to the test using 2D Ising model Monte Carlo simulation with Wolff algorithm.[1] The results obtained (see Table 1) using identical test parameters show that the quality of the pseudo-random number generator is comparable to those provided in the Scalable Pseudo-random Number Generator (SPRNG) package developed at the National Center for Supercomputing Applications, University of Illinois at Urbana-Champaign.[2]

Generator	Energy	Energy error
PLFG	-1.442187	0.010877
SPRNG Multiplicative LFG	-1.439941	0.013123
SPRNG Additive LFG	-1.466406	0.013341
SPRNG Combined Multiple Recursive Generator	-1.480273	0.027208

Table 1. Comparison of results of 2D Ising model Monte Carlo simulation test with Wolff algorithm.

As this paper is being written, work are being done to subject PLFG to further statistical tests. Researchers are invited to test PLFG.

Timing tests for generating 10^6 random numbers have also been conducted. Table 2 shows results for tests conducted on both a cluster of DEC Alpha machines with Alpha 21164 500 MHz processors[3], connected via Myrinet, and a dual processor Intel x86 machine with Pentium Pro 200 MHz processors[4]. For tests on the DEC Alpha cluster, 20 processors were used. From Table 2, it can be seen that the speed of PLFG is on par with that of other PPRNGs.

A number of research projects on Monte Carlo solutions for both sparse and dense matrix computations [13], Monte Carlo solutions for computational finance problems, are now being conducted using PLFG as the PRNG. These applications will further verify the randomness of the pseudo-random numbers. Results of these research projects will be reported at a later time, as they become available.

---

[1] The source code for the test was ported from the Scalable Pseudo-random Number Generator (SPRNG) package.
[2] At the time of writing, SPRNG Version 2.0 has just been announced. The version of SPRNG considered here is Version 1.0.
[3] The Alpha 21164 processors have on-chip instruction and data L1 caches of 8Kb each and 96Kb L2 cache.
[4] The Pentium Pro processors have on-chip instruction and data L1 caches of 8Kb each, and 256 Kb L2 cache

Generator	Average time (sec.)	
	Intel Pentium Pro	DEC Alpha
PLFG	0.614	0.251
SPRNG Multiplicative LFG	0.720	0.187
SPRNG 64-bit LCG	1.510	0.078
SPRNG Additive LFG	0.270	0.260
SPRNG Combined Multiple Recursive Generator	2.910	0.238

Table 2. Average time taken for generating 10^6 pseudo-random numbers.

5 PLFG Code in C

Platforms which PLFG has been tested on are shown in Table 3. Since the code does not rely on any MPI implementation-specific functions nor any system-specific calls, it is expected to work with any implementations of MPI, on any ANSI C Standard-compliant platforms. The code may be used in any MPI programs written in C++, just like any other ANSI Standard C-based libraries. PLFG does rely on the GNU Multi Precision library, which is available for a wide variety of platforms, for large integer operations. The rationale for using the GNU Multi Precision library is again to maintain efficiency and portability.

Architecture	Operating System	Compiler(s)	MPI Implementation
DEC Alpha cluster	DEC UNIX 4.0	DEC C 5.6-071	MPICH 1.1.2
HP PA-RISC cluster	HP-UX 10.20	HP C 10.32.03	MPICH 1.1.2
Sun SPARC cluster	Solaris 2.6	Sun C 4.2	MPICH 1.0.11
Intel x86 cluster	Linux 2.2.14	EGCS 1.1.2	MPICH 1.2.0

Table 3. Platforms on which PLFG has been tested to work.

The code can be downloaded via anonymous FTP from ftp.csc.liv.ac.uk:/pub/cjtan/plfg/plfg.current.tar.gz.

6 Conclusion

In PLFG, it is shown that splitting maximal periods of a sequence or splitting equivalence classes onto multiple processors in order to parallelize the PRNG may be neglected. The lag tables can be initialized with another PRNG with good statistical qualities, speeding up the initialization of the multiplicative LFG. From the results of the Monte Carlo solution for 2D Ising model tests conducted, it can be seen that PLFG is comparable to other commonly used PPRNGs. Coarse-grained parallelism employed in parallelizing PLFG and its scalability makes it extremely suitable for Monte Carlo simulations.

Further work can be done to increase the number of taps used in `PLFG`, and test for statistical qualities. Work on finding larger lag table values are ongoing at the moment.

From the programming perspective, the code can be made more flexible to allow user specification of the number of taps to be used at run-time.

Acknowledgment

Work on this research was started in the Remote Sensing Laboratory, Department of Geomatics Engineering, University of Calgary, Calgary, Alberta, Canada. Part of the work on this research was done at the Department of Computer Science, The University of Liverpool, Liverpool, United Kingdom. The University of Calgary MACI Project provided part of the computational resources of this project.

The first author is indebted to Maria Isabel Casas Villalba, now at Norkom Technologies, Dublin, Ireland, for her initial prompt to study pseudo- and quasi-randomness, and would like to thank Vassil N. Alexandrov, from the Department of Computer Science, The University of Reading, United Kingdom, for the fruitful discussions.

References

[1] CODDINGTON, P. D. Analysis of Random Number Generators Using Monte Carlo Simulation. *International Journal of Modern Physics C5* (1994).

[2] CODDINGTON, P. D. Random Number Generators for Parallel Computers. *National HPCC Software Exchange Review*, 1.1 (1997).

[3] KNUTH, D. E. *The Art of Computer Programming, Volume II: Seminumerical Algorithms*, 3 ed. Addison Wesley Longman Higher Education, 1998.

[4] MARSAGLIA, G. A Current View of Random Number Generators. In *Computing Science and Statistics: Proceedings of the XVI Symposium on the Interface* (1984).

[5] MASCAGNI, M., CEPERLEY, D., AND SRINIVASAN, A. SPRNG: A Scalable Library for Pseudorandom Number Generation. In *Proceedings of the Third International Conference on Monte Carlo and Quasi Monte Carlo Methods in Scientific Computing* (1999), J. Spanier, Ed., Springer Verlag.

[6] MASCAGNI, M., CUCCARO, S. A., PRYVOR, D. V., AND ROBINSON, M. L. A Fast, High Quality, and Reproducible Parallel Lagged-Fibonacci Pseudorandom Number Generator. *Journal of Computational Physics 119* (1995), 211 – 219.

[7] MASCAGNI, M., CUCCARO, S. A., PRYVOR, D. V., AND ROBINSON, M. L. Parallel Pseudorandom Number Generation Using Additive Lagged-Fibonacci Recursions. In *Lecture Notes in Statistics*, vol. 106. Springer Verlag, 1995, pp. 263 – 277.

[8] MASCAGNI, M., CUCCARO, S. A., PRYVOR, D. V., AND ROBINSON, M. L. Recent Developments in Parallel Pseudorandom Number Generation. In *Proceedings of the Sixth SIAM Conference on Parallel Processing for Scientific Computing* (1995).

[9] MATSUMOTO, M., AND NISHIMURA, T. Mersenne Twister: A 623-Dimensionally Equidistributed Uniform Pseudo-Random Number Generator. *ACM Transactions on Modeling and Computer Simulation 8*, 1 (January 1998).

[10] METROPOLIS, N., AND ULAM, S. The Monte Carlo Method. *Journal of the American Statistical Association*, 44 (1949), 335 – 341.
[11] SRINIVASAN, A., CEPERLEY, D., AND MASCAGNI, M. Testing Parallel Random Number Generators. In *Proceedings of the Third International Conference on Monte Carlo and Quasi-Monte Carlo Methods in Scientific Computing* (1998).
[12] SRINIVASAN, A., CEPERLEY, D. M., AND MASCAGNI, M. Random Number Generators for Parallel Applications. Tech. rep., National Center for Supercomputing Applications, 1995.
[13] TAN, C. J. K., CASAS VILLALBA, M. I., AND ALEXANDROV, V. N. Accuracy of Monte Carlo Solution for Systems of Linear Algebraic Equations with PLFG and rand(). Tech. rep., Parallel, Emergent and Distributed Architectures Laboratory, Department of Computer Science, University of Reading, United Kingdom, 2000.
[14] VATTULAINEN, I., ALA-NISSILA, T., AND KANKAALA, K. Physical Models as Tests of Randomness. *Physics Review E52* (1995).
[15] WILLIAMS, K. P., AND WILLIAMS, S. A. Implementation of an Efficient and Powerful Parallel Pseudo-random Number Generator. In *Proceedings of the Second European PVM Users' Group Meeting* (1995).

$_{par}$SOM: Using Parallelism to Overcome Memory Latency in Self-Organizing Neural Networks

Philipp Tomsich, Andreas Rauber, and Dieter Merkl

Institute of Software Technology
Vienna University of Technology
Favoritenstraße 9-11/188
A-1040 Wien, Austria
{phil,andi,dieter}@ifs.tuwien.ac.at

Abstract. The self-organizing map is a prominent unsupervised neural network model which lends itself to the analysis of high-dimensional input data. However, the high execution times required to train the map put a limit to its application in many high-performance data analysis application domains, where either very large datasets are encountered and/or interactive response times are required.
In this paper we present the $_{par}$SOM, a software-based parallel implementation of the self-organizing map, which is particularly optimized for the analysis of high-dimensional input data. This model scales well in a parallel execution environment, and, by coping with memory latencies, a better than linear speed-up can be achieved using a simple, asymmetric model of parallelization. We demonstrate the benefits of the proposed implementation in the field of text classification, which due to the high dimensionalities of the data spaces encountered, forms a prominent application domain for high-performance computing.

1 Introduction

The self-organizing map (SOM) [4] is a prominent unsupervised neural network model providing a topology-preserving mapping from a high-dimensional input space onto a two-dimensional map space. This capability allows an intuitive analysis and exploration of unknown data spaces, with its applications ranging from the analysis of financial data, medical data, to time series data [4, 2, 11]. Another rather prominent application arena is text classification, where documents in a collection are clustered according to their content. However, in many applications the dimensionalities of the feature spaces to be analyzed as well as the amount of data encountered are too large to allow a fast, interactive training of the neural network. Still, interactive training and mapping are highly desirable, as they open a huge field of applications in the context of data analysis in Data Warehouses or in the field of Digital Libraries, to name just two examples. In both cases, interactive response times are a must.

Existing hardware implementations [4, 3, 5] have not gained wide-spread acceptance because of the limitations they put both on the dimensionality of the data to be analyzed or the size of the map. On the other hand, high-performance computing provides the technology necessary for the desired speedup.

In this paper we present the *par*SOM, a software-based parallel implementation of the SOM, using a simple asymmetric model of parallelization. Apart from the obvious speedup gained by having the training process executed in parallel for the various neural processing units in a SOM, we show that the use of cache effects allows a better-than-linear speedup for the resulting implementation. We have implemented and evaluated the *par*SOM on a number of platforms, ranging from off-the-shelf dual-processor Celeron systems, the 16-processor SGI Power Challenge, to the 64-processor SGI Cray Origin 2000 using an application scenario from the digital library domain as a test-bed for our performance analysis.

The remainder of the paper is organized as follows. Section 2.1 provides an introduction to the architecture and the training algorithm of a self-organizing map, with a serial implementation discussed in Section 2.2. The actual algorithm implemented in the *par*SOM system is presented in Section 2.3. Our application scenario from the field of Digital Libraries is introduced in Section 3, followed by an analysis of experimental results in Section 4. Final conclusions and an outlook are presented in Section 5.

2 Self-Organizing Maps

2.1 Architecture and Training

The *self-organizing map* [4] basically provides a form of cluster analysis by producing a mapping of high-dimensional input data $x, x \in \Re^n$ onto a usually 2-dimensional output space while preserving the topological relationships between the input data items as faithfully as possible. This model consists of a set of units, which are arranged in some topology where the most common choice is a two-dimensional grid. Each of the units i is assigned a weight vector m_i of the same dimension as the input data, $m_i \in \Re^n$. In the initial setup of the model prior to training, the weight vectors are filled with random values.

During each learning step, the unit c with the highest activity level, i.e. the *winner* c with respect to a randomly selected input pattern x, is adapted in a way that it will exhibit an even higher activity level at future presentations of that specific input pattern. Commonly, the activity level of a unit is based on the Euclidean distance between the input pattern and that unit's weight vector, i.e. the unit showing the lowest Euclidean distance between it's weight vector and the distance between the input pattern and that unit's weight vector, i.e. the unit showing the lowest Euclidean distance between it's weight vector and the presented input vector is selected as the winner. Hence, the selection of the winner c may be written as given in Expression (1).

$$c : ||x - m_c|| = \min_i \{||x - m_i||\} \tag{1}$$

Adaptation takes place at each learning iteration and is performed as a gradual reduction of the difference between the respective components of the input vector and the weight vector. The amount of adaptation is guided by a learning-rate α that is gradually decreasing in the course of time. This decreasing nature of adaptation strength ensures large adaptation steps in the beginning of the learning process where the weight vectors have to be tuned from their random initialization towards the actual requirements of the input space. The ever smaller adaptation steps towards the end of the learning process enable a fine-tuned input space representation.

As an extension to standard competitive learning, units in a time-varying and gradually decreasing neighborhood around the winner are adapted, too. According to [9] we may thus refer to the self-organizing map as a neural network model performing a spatially smooth version of k-means clustering.

The neighborhood of units around the winner may be described implicitly by means of a neighborhood-kernel h_{ci} taking into account the distance—in terms of the output space—between unit i under consideration and unit c, the winner of the current learning iteration. This neighborhood-kernel assigns scalars in the range of $[0, 1]$ that are used to determine the amount of adaptation ensuring that nearby units are adapted more strongly than units further away from the winner. A Gaussian may be used to define the neighborhood-kernel as given in Expression (2) where $||r_c - r_i||$ denotes the distance between units c and i within the output space, with r_i representing the two-dimensional vector pointing to the location of unit i within the grid.

$$h_{ci}(t) = e^{-\frac{||r_c - r_i||}{2 \cdot \delta(t)^2}} \qquad (2)$$

It is common practice that in the beginning of the learning process the neighborhood-kernel is selected large enough to cover a wide area of the output space. The spatial width of the neighborhood-kernel is reduced gradually during the learning process such that towards the end of the learning process just the winner itself is adapted. Such a reduction is done by means of the time-varying parameter δ in Expression (2). This strategy enables the formation of large clusters in the beginning and fine-grained input discrimination towards the end of the learning process.

In combining these principles of self-organizing map training, we may write the learning rule as given in Expression (3). Please note that we make use of a discrete time notation with t denoting the current learning iteration. The other parts of this expression are α representing the time-varying learning-rate, h_{ci} representing the time-varying neighborhood-kernel, x representing the currently presented input pattern, and m_i denoting the weight vector assigned to unit i.

$$m_i(t+1) = m_i(t) + \alpha(t) \cdot h_{ci}(t) \cdot [x(t) - m_i(t)] \qquad (3)$$

A simple graphical representation of a self-organizing map's architecture and its learning process is provided in Figure 1. In this figure the output space consists of a square of 36 units, depicted as circles. One input vector $x(t)$ is randomly chosen and mapped onto the grid of output units. In the second step of the learning process, the winner c showing the highest activation is selected. Consider the

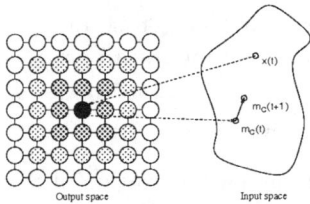

Fig. 1. Architecture and training process of a self-organizing map

winner being the unit depicted as the black unit in the figure. The weight vector of the winner, $m_c(t)$, is now moved towards the current input vector. This movement is symbolized in the input space in Figure 1. As a consequence of the adaptation, unit c will produce an even higher activation with respect to input pattern x at the next learning iteration, $t+1$, because the unit's weight vector, $m_c(t+1)$, is now nearer to the input pattern x in terms of the input space. Apart from the winner, adaptation is performed with neighboring units, too. Units that are subject to adaptation are depicted as shaded units in the figure. The shading of the various units corresponds to the amount of adaptation and thus, to the spatial width of the neighborhood-kernel. Generally, units in close vicinity of the winner are adapted more strongly and consequently, they are depicted with a darker shade in the figure.

2.2 Implementing SOMs in Software

A serial implementation of the SOM iterates through the training algorithm. This involves scanning all units to find the winner and then modifying all units to reflect the change. As the map will usually be implemented as an array of weight-vectors, this array is traversed twice. From this, it should be clear that a major bottleneck in a software-based implementation of a self-organizing map consists in the memory bandwidth. During every iteration of the training algorithm every unit within the map is touched twice. First, during the search for the best matching unit within the map and then again, when the weights are adjusted. Due to the fact that we are using high-dimensional inputs with multiple thousands of inputs, it is not uncommon that a rather small map requires in excess of 5 MBs. A medium-sized map may easily occupy tens of megabytes. This is beyond the capacity of most cache memories and therefore incurs large overheads in its memory accesses. The cache thrashing can be lessened a little by using a technique known as *loop inversion*. Instead of traversing the unit-array from its beginning to its end when traversing it for the second time (i.e., when adjusting the weights), the array is traversed backwards to exploit the cache.

Another major bottleneck results from the huge number of floating point operations required. Training a 10x15 map of 4000 features in 20000 iterations requires in excess of 72 billion arithmetic floating point instructions and 6 million floating point comparisons. In order to provide high performance in such applications, multiple processors can be used in parallel. This becomes a viable alternative, as the distance calculation and the adaption of the weights can be executed independently for each unit. These independent operations constitute the major part of the runtime, calling for a parallel implementation.

2.3 The *parSOM* Implementation

Two different strategies for the parallelization of SOMs are available:

- **Vectorization.** It is possible to use multiple processors or SIMD techniques to operate on the vector-components in parallel. Modern microprocessors are sometimes fitted with SIMD nodes for the benefit of multimedia applications. For example, the Altivec unit used in the MPC7400 PowerPC RISC processor features a flexible vector unit with 32 128bit registers. It is capable of operating in parallel on 4 single precision floating-point numbers. Although SIMD units cut down the pure processing time, the load on the memory bus is increased. Special compiler support is necessary for this optimization.
- **Partitioning.** The map is partitioned into multiple sections, which are maintained by different processors. One of the major benefit of this strategy is the distribution of required memory across multiple units. This allows both for the handling of far larger maps than serial implementations and for a faster execution, as multiple processors are used.

For our current implementation, we use partitioning, with one master controls multiple slaves. As depicted in Figure 2 three major synchronization points occur in this implementation. The data transfers are very small, as only the offset of the input vector (within a replicated file) is broadcast; the data packets for computing the minima consist of a Cartesian coordinate and a floating point number signifying the error at that coordinate. The approach chosen in our work is based on a communication-oriented computation and synchronization model. It uses a number of information and condition broadcasts to synchronize its operations. In addition, we view a multi-dimensional self-organizing map as a collection of units with associated coordinates. This allows us to distribute the units freely between the threads.

Any subsets of the units can be easily represented in a linear array. By segmenting the SOM into multiple regions, we can effectively distribute the memory usage across multiple processing nodes. Since training the SOM requires the entire map segment to be operated on, the distribution of the map across multiple processing nodes with independent memories (e.g., on ccNUMA architectures) reduces the pressure on the memory hierarchy and makes it possible to fit the map segments entirely into L2 caches on the respective nodes.

A master thread selects an input vector and broadcasts it to all its slave threads. All threads search for their local minima, i.e. best-matching units concurrently. These best-matches are sent back to the master, which determines a global match. All slaves are notified of the global match and they modify their units relative to the location of it independently. The data packets exchanged at the synchronization points are very small, consisting of a spatial coordinate and an error value at most. This algorithm should scale very well to any number of processors in theory, as the search for the local minima and the training

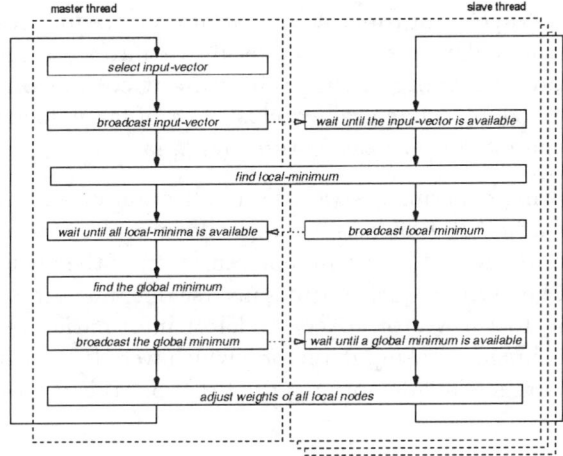

Fig. 2. A flowchart of the computation using an asymmetric model of parallelization. Three major synchronization points occur in this implementation.

process account for almost the entire execution time of the program. Remember, that both finding the local minima and training involves operations on arrays of large vectors of floating point numbers. For this reason, the additional work of the master thread is negligible and does not affect the overall performance. However, for small maps and very fast processors, scalability problems may arise due to the synchronization overheads.

3 SOMs and Digital Libraries

Information retrieval in digital libraries is often based on similarity search [1], such that a user queries the document repository by example. A typical implementation for this is based on the use of self-organizing maps (sometimes referred to as Kohonen maps). These are a type of unsupervised-learning neural networks, that provide the capability to the high-dimensional feature vectors associated with documents into a low-dimensional Cartesian space. Once this mapping in concluded, it can be used to determine a clustering of documents. Similar documents should belong to the same or neighboring clusters. This provides a unique opportunity to visualize the clustered documents.

The quality of these clusters depends largely on the use of appropriate learning functions and distance-based feedback. Unfortunately these functions need to be applied in an iterative manner for multiple thousands of iterations. This is leads to execution times for every training run that fall far outside of what can be considered appropriate for interactive operation. This implies that documents can not be clustered on the fly (e.g., for web-searches or for searches using iterative refinement). The rapid growth of digital content outpaces the increase of the computing power of traditional single-processor systems. However, the most important upcoming applications of digital libraries require the just this interactive response:

- **Web-based search-engines.** A visualization of search results is possible by mapping the documents from the result set into a 2-dimensional space.
- **Fine-grained clustering of large document collections.** Various types of documents require a fine-grained clustering of large document collections with a high number of relevant features (e.g., laws).

For the following experiments we use a collection of 420 articles from the *TIME Magazine* from the 1960's as a sample document repository. The articles are parsed to create a word histogram representation of the documents using the classical $tf \times idf$, i.e. term frequency times inverse document frequency weighting scheme according to the vector space model of information retrieval [10]. We thus obtain 4012-dimensional input vectors, which were further used for training SOMs of 10×15 units. The memory required to store this map exceeds 4.5MBs.

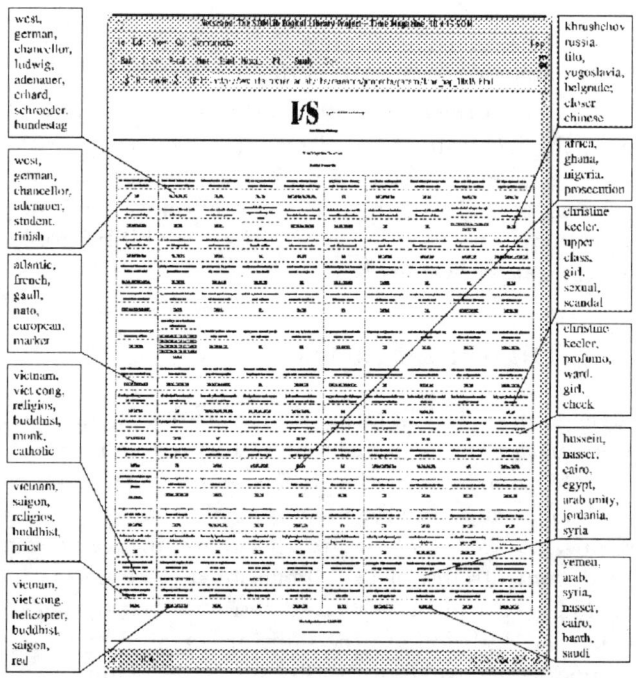

Fig. 3. A map trained with 420 articles from the *TIME Magazine*.

The resulting SOM is depicted in Figure 3. As a closer look at the resulting mapping reveals, the SOM succeeded in producing a topology preserving mapping from the 4012-dimensional input space onto the 2-dimensional map space, clustering the articles by topic. This can be easily demonstrated using the labels automatically extracted from the trained SOM by applying a variation of the *labelSOM* technique [6].

For example, we find a number of articles covering the war in Vietnam in the lower left corner of the map. Another rather large cluster in the lower right corner of the map is devoted to articles on the situation in the Middle East, covering Nasser's attempts to create an arab union. Due to space considerations, we cannot present the complete map in full detail. However, the map is available at http://www.ifs.tuwien.ac.at/ifs/research/projects/parsom/ for interactive exploration. For a closer analysis of the application of SOMs in the field of information retrieval, refer to [8, 7].

The resulting representation of documents due to its topical organization offers itself to interactive exploration. However, to allow its application in an information retrieval or data mining setting, providing competitive response times is crucial, calling for a parallel processing of the SOM training process.

4 Analysis of Experimental Results

Table 1 summarizes the execution time for training the map using $_{par}$SOM. We used POSIX threads, a shared address space and user-level spinlocks in our implementation. The test were conducted on a number of multi-processor machines with dissimilar memory architectures:

- **Dual Celeron System.** Two Intel Celeron processors using a shared memory bus with a total throughput of 264MB/s. Each processor uses a 256KB level 2 cache.
- **SGI PowerChallenge XL.** A 16-processor system with a shared-bus. Each processor has a 2MB level 2 cache available and the total throughput to the main memory is 1.2GB/s.
- **SGI Cray Origin 2000.** A 64-processor system which adheres to a cc-NUMA architecture. The system uses 32 SMP node boards, where each processor has 4MB of cache memory;

threads	SGI PowerChallenge time elapsed	relative	speedup	SGI Origin 2000 time elapsed	relative	speedup	Dual Celeron PC time elapsed	relative	speedup
1	1555.591	1.0000	1.0000	461.940	1.0000	1.0000	1521.900	1.0000	1.0000
2	533.265	0.3428	2.9171	209.790	0.4541	2.2021	1037.850	0.6819	1.4664
3	231.964	0.1491	6.7069	150.223	0.3252	3.0750			
4	180.035	0.1157	8.6430	117.100	0.2534	3.9448			
5	128.025	0.0886	11.3122	114.930	0.2488	4.0178			
6	117.228	0.0753	13.2802	102.134	0.2211	4.5216			
7	96.632	0.0621	16.1030	90.817	0.1966	5.0854			
8	91.481	0.0588	17.0068	83.240	0.1802	5.5493			
9	83.333	0.0535	18.6915						
10	80.993	0.0520	19.2307						
11	76.701	0.0493	20.2839						
12	70.390	0.0452	22.1238						

Table 1. Execution times and speedups on multiprocessor systems

The dual-processor Celeron system is a commodity personal computer. The small cache sizes of the processors require large amounts of memory transfers.

Due to the limited memory bandwidth, not even a linear speedup is achievable in this configuration. The only apparent cause for the bad scalability can be attributed to memory contention. On this system, inverting the training loop does not cause a significant performance gain, which is further evidence that the scalability is limited primarily by the memory bandwidth.

An entirely different performance characteristic was obtained on our principal development platform. On the SGI PowerChallenge, which uses a shared memory architecture, a 22-fold increase in performance could be achieved with only 12 processors. Although this may seem paradox, it should be expected. The performance increases very quickly for the first few threads, until the amount of cache memory exceeds the size of the map (compare Figure 4). After that, the performance increase continues in a linear fashion. We provided time measurements up to 12 processors only, as the machine was not available for dedicated testing. The precision of these measurements for more than 10 processors was also affected by the background load. The scalability on this system remains unaffected by the available memory bandwidth: the entire SOM is distributed through the cache memories of the various processor, effectively avoiding bus contention. Interestingly, loop inversion leads to a large increase in performance for the uniprocessor version on this machine. This optimization resulted in a 22% increase in performance.

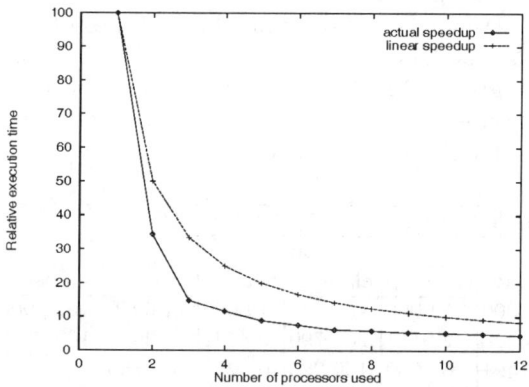

Fig. 4. Performance on a SGI PowerChallenge: Cache effects on a shared-memory system with a high memory bandwidth can lead to a better than linear speedup.

The performance gains on the SGI Origin were less impressive than those on the PowerChallenge. Although the current implementation scales to 8 processors and achieves a 5.5-fold performance increase, it still offers room for improvement. This is due to our dependence on POSIX threads and how these are implemented in IRIX 6.5. Instead of allocating a separate heap for each thread, a shared heap is used, which has its affinity hard-wired to the processor of the main thread. Understandably, performance degrades in such an environment, if many threads are used, as the traffic on the memory backbone increases. Nonetheless, the speedup between the single-processor and dual-processor implementation shows again how the availability of sufficient cache memory to hold the entire SOM has a significant performance impact. We are currently working on a MPI-based implementation to create an optimal version for this architecture. We expect MPI to solve the performance problem on this architecture, as it provides distinct heaps to the different processes.

5 Conclusions and Future Work

Our experiments with parallel, software-based implementations of self-organizing maps for Digital Libraries have shown that interactive response times—even for high dimensional feature spaces—can be achieved using supercomputing resources. The used algorithms scale well and gain considerable performance due to cache effects. Parallelism effectively becomes a method to overcome the memory latencies in self-organizing neural networks.

We are currently adding MPI support. Dynamic load-balancing becomes a necessity, if non-dedicated computing resources are used, as the background load may severely affect the performance of the system. In such environments, a dynamic redistribution of the map between nodes becomes necessary to provide peak performance. Support for SIMD extensions to common microprocessors and vector computers will be also be added in future versions.

References

1. P. Adam, H. Essafi, M.-P. Gayrad, and M. Pic. High-performance programming support for multimedia document database management. In *High Performance Computing and Networking*, 1999.
2. G. DeBoeck and T. Kohonen. *Visual Explorations in Finance*. Springer Verlag, Berlin, Germany, 1998.
3. D. Hammerstrom. A VLSI architecture for high-performance, low-cost, on-chip learning. In *International Joint Conference on Neural Networks*, 1990.
4. T. Kohonen. *Self-Organizing Maps*. Springer Verlag, Berlin, Germany, 1995.
5. J. Liu and M. Brooke. A fully parallel learning neural network chip for real-time control. In *International Joint Conference on Neural Networks*, 1999.
6. A. Rauber. LabelSOM: On the labeling of self-organizing maps. In *Proc. International Joint Conference on Neural Networks*, Washington, DC, 1999.
7. A. Rauber and D. Merkl. Finding structure in text archives. In *Proc. European Symp. on Artificial Neural Networks (ESANN98)*, Bruges, Belgium, 1998.
8. A. Rauber and D. Merkl. Using self-organizing maps to organize document collections and to characterize subject matters. In *Proc. 10th Conf. on Database and Expert Systems Applications*, Florence, Italy, 1999.
9. B. D. Ripley. *Pattern Recognition and Neural Networks*. Cambridge University Press, Cambridge, UK, 1996.
10. G. Salton. *Automatic Text Processing: The Transformation, Analysis, and Retrieval of Information by Computer*. Addison-Wesley, Reading, MA, 1989.
11. O. Simula, P. Vasara, J. Vesanto, and R. R. Helminen. The self-organizing map in industry analysis. In L.C. Jain and V.R. Vemuri, editors, *Industrial Applications of Neural Networks*, Washington, DC., 1999. CRC Press.

Track II

Web-Based Cooperative Applications Track

Towards an Execution System for Distributed Business Processes in a Virtual Enterprise

L.M. Camarinha-Matos and C. Pantoja-Lima

New University of Lisbon, Faculty of Sciences and Technology
Quinta da Torre, 2825 Monte Caparica, Portugal
cam@uninova.pt

Abstract. The implantation of the virtual enterprise paradigm requires the design and development of a flexible execution environment to support the distributed business processes that materialize the cooperation in a network of enterprises. A configurable architecture for such execution system is proposed focusing the support for multi-level process coordination. A set of examples to illustrate the adopted concepts and developed tools are discussed.

1 Introduction

The paradigm of virtual enterprise (VE) represents a major area of research and technological development for industrial enterprises and an important application area for web-based cooperative environments. A virtual enterprise is usually defined as a temporary alliance of enterprises that come together to share their skills, core competencies, and resources in order to better respond to business opportunities, and whose cooperation is supported by computer networks [4], [7].

Two keyword elements in this definition are the networking and the cooperation. As a result of the globalization of the economy and new market characteristics, there is a clear trend for the manufacturing business processes not to be carried out by a single enterprise any more. In order to reach world class, companies are forced to embark in cooperation networks where every enterprise is just one node adding some value (a step in the manufacturing chain) to the entire production cycle. As such, in virtual enterprises, manufacturers do not produce complete products in isolated facilities. Rather, they operate as nodes in a network of suppliers, customers, engineers, and other providers of specialized service functions. Business processes (BP) take place, therefore, in a distributed environment of heterogeneous and autonomous nodes.

The materialization of this paradigm, although enabled by recent advances in communication technologies, computer networks, and logistics, requires an appropriate architectural framework and support tools. Namely, it requires the definition of a suitable open reference architecture for cooperation, the development of a flexible and configurable BP supporting platform, and the development of appropriate protocols and mechanisms.

2 Coordination Issues

2.1 The Need for Coordination

Coordination of processes and activities has been identified as a key element in the operation of virtual enterprises [2], [4], [5], [9]. Coordination has to do with the proper management of dependencies among activities [10] and correspond, in accordance with the view of [6], to the process of "gluing" together active pieces (enterprise activities) with a purpose.

From a *bottom up* perspective, activities carried out by a company are usually organized in "clusters" of inter-related activities called *processes* (*business processes*). The composition of each process is designed in order to achieve a (partial) specific goal. When properly "orchestrated", a combination of various processes taking place in different members of the VE, i.e. a distributed business process (DBP), will lead to the achievement of the global goal of the VE.

Alternatively, a CIM-OSA-like *top down* view can be considered according to which a VE business process can be decomposed into a hierarchy of sub-business processes and enterprise activities. The enterprise activities, that are supported by the CIM-OSA's Implemented Functional Operations, or the PRODNET's services, represent the basic building blocks (libraries of services) each enterprise per se, and the VE as a whole, have to actually realize the assigned processes.

2.2 Need for Multi-level Coordination

As it is implicit in the notion that a VE business process can be decomposed into a hierarchy of sub-business processes, the coordination issues must be analyzed at different levels of abstraction. This need is clearly illustrated by the PRODNET pilot demonstration [4] of manufacturing a bicycle in a VE environment.

According to this example, different parts of the full manufacturing process may be assigned to different members of the VE according to their core skills, resources and roles in the VE. For instance, one node A becomes responsible for assembling all components in a final product, another node B is responsible for manufacturing the metallic parts, a node C supplies the plastic components, a node D produces moulds for plastic components, etc. Once a global business process is defined and scheduled, and responsibilities are assigned to each individual partner, the successful achievement of the common goal – delivery of the final product to the client –depends on the proper and timely operation of each VE member (and each supporting service in each VE member). A delay in one node, if not taken care of in time, may jeopardize the common goal. Therefore, one enterprise may assume the role of VE coordinator and manage (supervise) the interdependencies among the various (distributed) BPs. On the other hand, inside each company, the (local) assigned business process(es) may, on its (their) turn, be decomposed into several sub-processes whose activities are supported (performed) by various service functions available in the enterprises.

For instance, a "produce bike frame" process may involve:

"*Receive order from the coordinator (via EDI)*", "*design frame*" *(using a CAD system)*; "*generate process plan*" *(using a CAPP system)*, "*plan production*" *(using a PPC system)*, "*supervise production*" *(using a shop-floor control system)*; "*send periodic progress reports to the VE coordinator*" *(using the PPC system), etc.*

The interdependencies (sequence/parallelism, synchronization, data flow, precedence conditions) among these activities need to be properly managed.

This "decomposition" process may continue and some activities may imply a sequence of fine-grained activities. For instance, the "receive order" process can be decomposed into:

"*Identify arriving message*", "*parse order coded in EDIFACT format*", "*temporarily store order data*", "*ask advise from human operator on what to do with the order*", "*notify ERP/PPC of the order arrival*", *etc.*

Some projects, such as VIVE [15], have devoted some effort to the identification of the major BPs and their decomposition for some key areas of manufacturing. However, although important, modeling BPs brings little benefit to the actual operation of the VE if not supported by an *execution infrastructure* – the **VE execution system**.

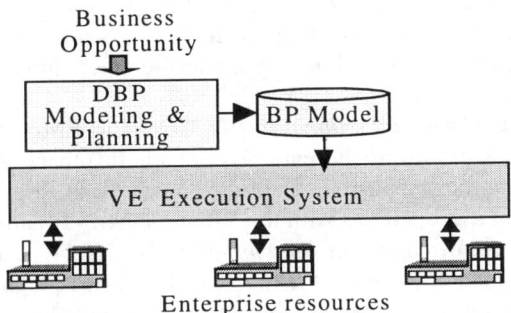

Fig. 1. VE execution system

Considering the distributed nature of a VE, the structure of a minimal VE execution system should comprise: a process execution and coordination system, a distributed information management system, and a safe communications infrastructure.

The enterprise resources (enterprise applications, human resources, machines, etc.) will carry out the actual execution of enterprise activities under the control of the process execution and coordination system. By analogy with the computer systems, this infrastructure can be seen as the "*operating system*" of the VE.

One aspect to retain here is the heterogeneity of the processing nodes / VE members. The enterprises that combine efforts to form a VE pre-exist and have their own legacy systems and distinct enterprise culture. Therefore, in addition to requirements of portability, it is also necessary to support a high degree of configurability in order to cope with the particular context of each enterprise.

Portability is reflected as a requirement to adopt industry standards such as WPDL (or PIF) for process definition, EDIFACT for the exchange of business messages, STEP for the exchange of technical product data, etc. The configurability characteristic shall provide a flexible way to adapt the behavior of the execution system to the specificity of each local enterprise and to hide, to some extent, this specificity from the other members of the network.

Another characteristic of this environment is the hybrid nature of the execution system. Some processes may be carried out automatically by services offered by enterprise applications (ERP, PDM, etc.), while other processes are performed by humans. It may also happen that a given process is executed differently (automatically or human-based) if assigned to different enterprises.

3 A Coordination Kernel

Taking these requirements into account, the PRODNET project [4] designed and developed a prototype of an execution infrastructure for distributed BPs. From the activity/process coordination point of view, the PRODNET coordination subsystem considers three levels of abstraction:
- *Core Cooperation Layer (CCL)*. The CCL is responsible for the basic interactions among VE members offering support for safe communications, exchange of business messages, sharing and management of cooperation information, federated information queries, etc.
- *Enterprise Management Functionalities (EMF)*. The EMF is responsible for the coordination of activities at the enterprise level. In other words, the EMF deals with coordinating the responsibilities of the enterprise towards the accomplishment of its assigned BPs or contracts with the VE and other VE-partners.
- *Virtual Enterprise Management Functionalities (VMF)*. The coordination aspects at the VE level are considered in the third level. In principle, only the node playing the VE coordinator role will use this layer to monitor, assist, and modify the necessary activities related to the VE goal achievement. The VE Management Functionalities (VMF) resort to the services provided by the CCL and EMF of its node to communicate with the other nodes of the VE.

Although in its current implementation only three levels are considered, the model can be easily generalized to any number of levels in order to cope with any BP tree. In fact, in addition to the VE coordination role, responsible for the global BP, other enterprises may assume the role of coordinators of sub-business processes that might be decomposed and performed by a sub-consortium of enterprises. This idea of dynamic formation of sub-consortia inside a community of cooperating agents (a VE) in order to coordinate sub-business processes is being exploited by the MASSYVE project [14].

These sub-consortia are formed for the sole purpose of facilitating the coordination of activities involved in the related sub-business processes. Once a sub-business process ends, the sub-consortium "dissolves" and its members may become involved in other sub-consortia dynamically formed as the execution of the VE BPs evolves. For instance, enterprise C (in Fig. 2) coordinates sub_VE1 and is a member of sub_VE2. Under this model, the formation of a sub-consortium inside a VE follows a similar process as the formation of the VE itself. In other words, a VE may recursively give raise to internal sub-VEs.

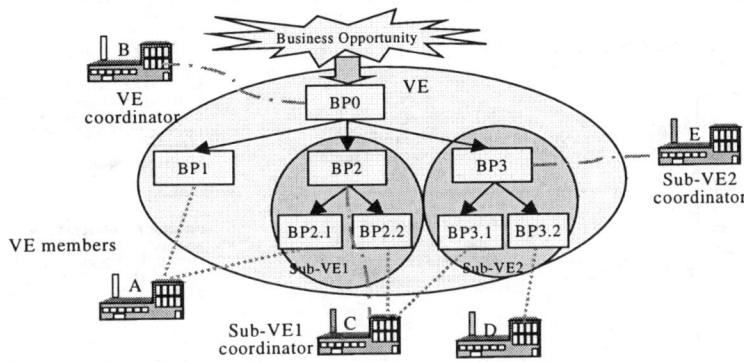

Fig. 2. Coordination and hierarchies of BPs

3.1 Structure of the PRODNET Coordination Kernel

The two main components of the PRODNET coordination kernel are the Local Coordination Module (LCM) [4] and the federated Distributed Information Management System (DIMS) [1].

Similarly to some other projects (e.g. NIIIP, VEGA, WISE), PRODNET adopted a workflow-based approach for coordination of VE-related activities. Combined with a graphical specification language and a graphical editor, this approach proved to be an efficient way to implement flexible and easily re-configurable, by non-programmers, activity coordination strategies. In this way, the needs of different companies and different business processes are properly accommodated.

The integration and management of the VE information is supported in PRODNET by an innovative distributed / federated information management approach [1]. The federated database approach has proven to provide particularly attractive features addressing some challenging information management criteria such as the openness, autonomy, heterogeneity, and information privacy, what are the main pre-conditions for trust building among SMEs involved in a virtual organization domain.

The very tight integration of activity coordination and federated information management provides a sound basis to support the advanced VE coordination functionalities.

Another major component of the PRODNET execution system is the PRODNET Communications Infrastructure (PCI) which is responsible for providing communication channels between VE members and a set of functionalities to guarantee safe communications, authentication, auditing services, heuristics for communication channel selection, etc.

The execution environment is complemented by a library of service functions that support the activities at each level of abstraction (Fig. 3). For instance, at the CCL level, services for processing EDIFACT and STEP messages are offered. At the EMF level services such as product negotiation or issuing product purchase order can be found. At the VMF level services related with the VE global coordination, such as getting production status from VE members or search for new partners, are included.

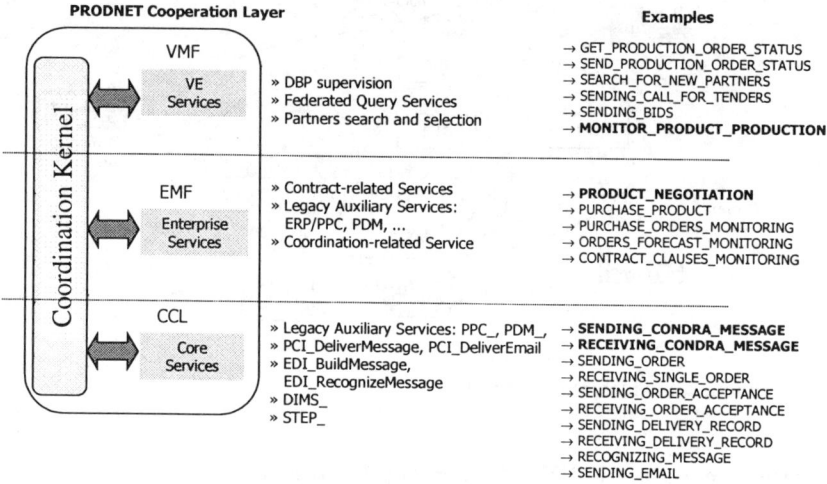

Fig. 3. Examples of services at different abstraction levels

A network of cooperating nodes (network of enterprises) therefore composes the execution environment for DBPs in PRODNET and each node includes the coordination kernel, the safe communications infrastructure, and the support service functions.

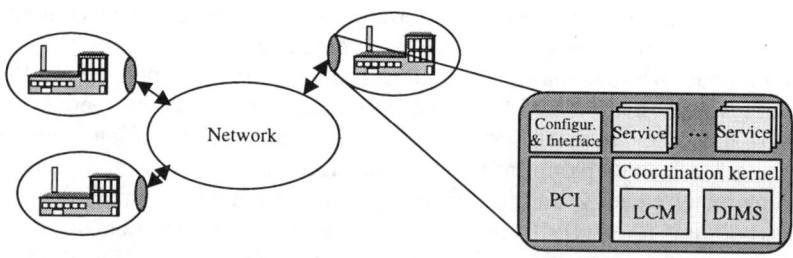

Fig. 4. The PRODNET execution environment

In terms of implementation it is necessary to take into account that enterprises that come together to form a VE pre-exist and have their own legacy systems (ERP / PPC, SCM, CAD, CAPP, PDM, etc.). Therefore, the implantation of the BP execution environment has to take into account the migration of these legacy systems as they will support many of the necessary services.

Following a common procedure in systems integration, PRODNET "recovers" the existing legacy applications by adding to them a new layer (the PRODNET

Cooperation Layer) comprising the coordination kernel, the communications subsystem, and other basic services, and by establishing the necessary mappings between this layer and the enterprise applications. The communication between PCL and legacy applications is supported by a client-server interaction. PCL provides a "client library" which guarantees the communication from the legacy applications to PCL. Each legacy application provides a similar library to PCL.

Future trends, such as the OAG initiative [11], towards the "componentization" of enterprise applications and definition of standard interoperation mechanisms among these functions, raise the expectation that future enterprise applications can be seen as a library of service functions that can be flexibly used by the coordination system.

3.2 Graphical Modeling of BPs

For the modeling of BPs a graphical language and an associated editor that borrows many ideas from the workflow management systems, trying to be compliant with the WfMC Reference Model, but including some adaptations to the VE environment is developed (Fig. 5).

The following modeling primitives are supported:
- Sequences of activities that might invoke external services or other sub-activities.
- Sub-workflow definition, a mechanism to support nested BPs definition.
- Data flow management for parameter passing when activating services / sub-activities, i.e. data that is essential for the process execution control flow. This is the explicit data exchange. A form of implicit data exchange is also supported be the DIMS module.
- Splits and joins that can be subjected to the logical operators AND / XOR.
- Simple and conditional transitions.
- Temporized and cyclic activities, providing the high-level coordination facility that is needed for instance in the case of monitoring contract clauses among VE members.
- Flexible configuration of catalogs of external services and relevant data.
- Workflow instances and memory spaces. For each execution of a workflow model an instance is created with its memory space. The explicit data flow associated to an instance (relevant data) is only valid inside the memory space of that instance.
- As an enterprise may be involved, at the same time, in various VEs, there is also a memory space separation per VE participation.
- Management of waiting lists. Each time an instance of a workflow model needs to wait for the conclusion of an external service it is put in a waiting list. Waiting lists are also used for instances waiting for temporized activities. Signals can be sent to the waiting lists manager to provoke changes in the status of workflow model instances.

Hierarchical BPs can be represented using the concept of sub-workflow as illustrated in Fig. 6.

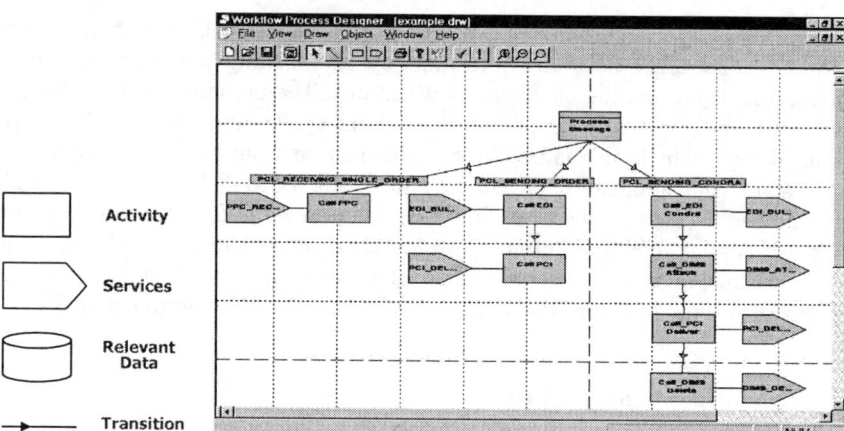

Fig. 5a. Graphical process modeling primitives   **Fig 5b.** Graphical process model editor

Sub-workflows also provide a basic degree of *reusability* in the model, since a sub-workflow can be used several times in a workflow model, like a sub-routine. The more frequent tasks can be modeled as sub-workflows and used as many times as necessary. This feature allows the creation of a library of sub-workflows (templates) that model the processes frequently performed.

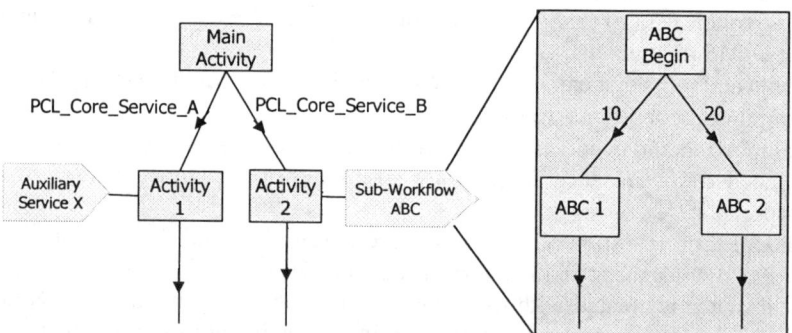

Fig. 6. Activity implemented as sub-workflow

The graphical process editor associated to LCM supports a fundamental characteristic in a VE environment that is the ability to support a gradual migration of the company's legacy business practices towards a VE environment operation. In a first step, it is necessary to guarantee that the company's privacy, autonomy as well as its way of making business will be preserved, as far as possible. When the level of trust between the enterprise and the VE partners increases, or a better understanding of the VE operation is achieved, a new behavior of the cooperation layer can be defined by a BP expert without the need of programming. One of the reasons to adopt a workflow-based modeling approach instead of a more specialized language for coordination such as LINDA or MANIFOLD [3] is because it is easy to use by BP experts with limited programming skills. The output of the graphical editor is stored as a WPDL file, following the WfMC proposed syntax [16].

Fig. 7. BP edition and execution environment

The LCM includes the process executor (enactment engine) and a monitoring and

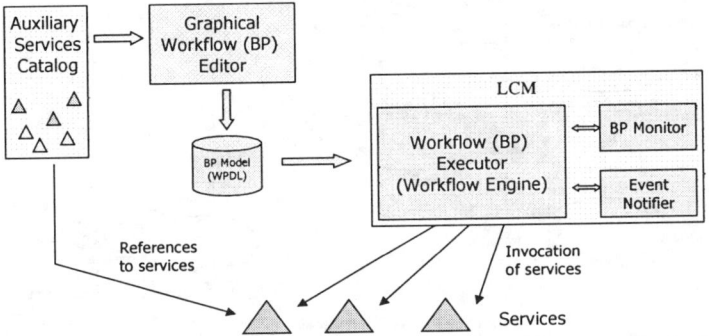

event notification module that supports a user interface with a graphical visualization of (and interaction with) the processes under execution (Fig. 7).

4 An Example

In order to illustrate the operation of the described execution system, let us consider an example of two enterprises involved in the business process of developing a new bicycle and that have to cooperate in the design of a component (a pedal, for instance). In this scenario, enterprise A prepares a first model of the desired component using a CAD system and defines some additional characteristics for this part using a PDM (Product Data Management) system. This technical specification is then sent to enterprise B. Enterprise B might accept the proposed design or suggest changes that have to be negotiated with enterprise A until an agreement is reached.

In order to support this process, there is a need for a core service (i.e. a service at the CCL level) that takes care of sending out a message with the technical specifications of the required part. A solution for that can be the encapsulation of a STEP model in a EDIFACT CONDRA message. Fig. 8 shows a simplified process model (workflow model) for this task. As illustrated in this figure, the implementation of a core service resorts to other more basic services (the auxiliary services) offered by the PRODNET infrastructure such as *EDI_BuildMessage*, *DIMS_AttachReferences*, etc.

Different implementation solutions can be found for this core service. For instance, the product model may be sent using a proprietary format (agreed between the two companies) without resorting to EDIFACT. On a similar way, specific processes have to be modeled, at the CCL level, for each of the core services supported by the enterprise cooperation layer. Examples are: *Send_order*, *Receive_order*, *Order_acceptance*, *Send_delivery_record*, etc. The graphical process editor provides a flexible tool for a system manager (BP expert) to define the set of core services for each enterprise as well as the set of available auxiliary services in the installed infrastructure. In the same way, but not necessarily following the same models, it is necessary to define the core services of enterprise B.

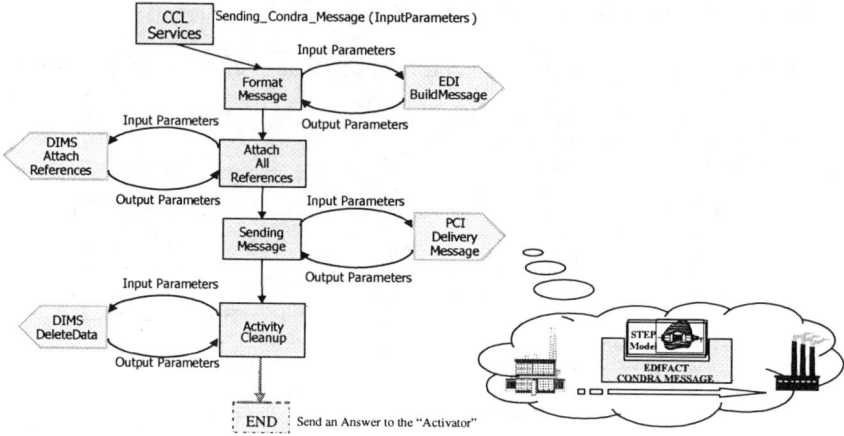

Fig. 8. Example of core service (CCL-level process)

Using this basic level of the cooperation layer it is possible to implement cooperation processes between enterprises. If no other higher level services are available, then the coordination of the higher level processes has to be done by the human operators or to be supported by some hard-coded fixed procedures implemented in the interface of the enterprise applications. For instance, the core service presented in Fig. 8 could be invoked by a PDM application or explicitly launched by the PDM user. This was the situation implemented in the PRODNET demonstrator that was sufficient to demonstrate how complex cooperation scenarios can be implemented on top of a basic infrastructure [4].

It is important to note that it is not the sender who decides which applications / services will be used on the receiver's side. A message is sent independently of the type of processing that the destination will apply. A lookup table at each site establishes this correspondence (Fig. 9). It is, however, necessary that a *common taxonomy* of messages types is shared by all members of the VE. If a standard protocol is used (like EDIFACT), the common set of types is guaranteed. For proprietary types it is necessary to reach an agreement among all VE members. It is at configuration time, when the behavior of the coordination module is defined, that a specification of which local processes are triggered by each type of arriving message is made. In fact this is necessary even when the messages are EDIFACT-based as only a subset of EDIFACT messages are used for each two enterprises.

In this way, the enterprise keeps its autonomy not only in terms of deciding the specific structure of its processes (which can evolve), but also in terms of deciding which processes to apply to each type of event. In our opinion this is a more pragmatic solution when compared to other approaches as the one proposed by the WISE project [2]. In WISE, companies declare (register) their services in a global catalog that is then used by the BP modeler to build BP models.

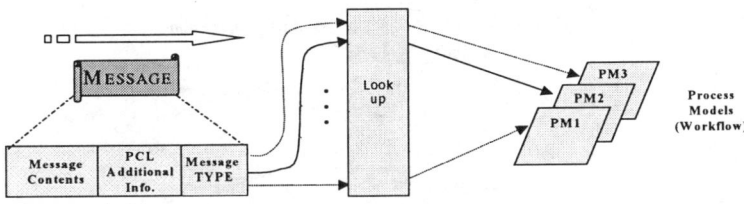

Fig. 9. Association of message types to processing services

If the second coordination level (EMF) is available, than it is possible to implement a flexible, automatic or computer-assisted, coordination mechanism for higher-level processes. For instance, Fig. 10 illustrates a simple enterprise-level process describing the steps involved, from the enterprise A side, in the negotiation of a part with company B.

In this example, the activities "Design model" and "Prepare PDM data" are performed by services offered by the enterprise applications CAD and PDM respectively. An alternative to a direct invocation of a CAD (or PDM) application can be the activation of a user interface program that *notifies* the human operator about the tasks to be done (see the *Event notifier* component in Fig. 7). In this way, the same basic modeling and control mechanisms can be used for automatic or manual (computer-assisted) systems. This example also shows that the same workflow-based modeling mechanisms can be used to model this type of BPs.

The activity "Send product model" is performed by the core service described above (Fig. 8). This example illustrates how lower level processes can support a higher level process. The instantiation of a lower level workflow model is done following the same invocation mechanism used to call application services, i.e. some input and output parameters are used to establish the interaction between the two processes in a similar way as the invocation of a procedure in a traditional programming language. The LCM engine interprets the process description language, assuring the proper parameter passing in the right memory space and right VE space.

In the case of processes mainly executed by humans, the workflow type of control is however too rigid. People like to keep their freedom regarding the way they work. That is the case, for instance, in *concurrent engineering* where a team of designers, possibly located in different enterprises, cooperate in a joint project. Product design, like any other creative process evolves in a kind of non-deterministic flow. Several approaches to develop *flexible workflow* systems have been proposed [8]. One solution for the coordination flexibility was first introduced in the CIM-FACE system [13]. In this system, instead of rigid precedence rules, other types of relationships, inspired in the Allen's temporal primitives, are possible: start_before, finish_during, start_after, finish_after, do_during, etc. Other constraints such as pre- and post-conditions can be specified. These mechanisms, previously developed in CIM-FACE, are not included, however, in the current implementation of LCM.

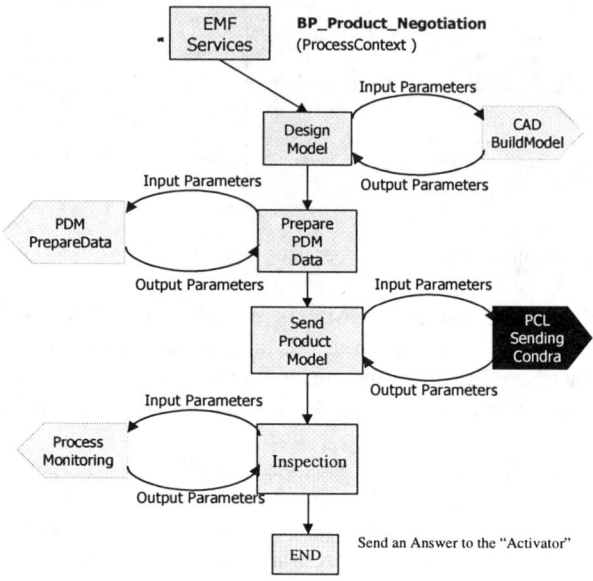

Fig. 10. Example of an EMF-level process

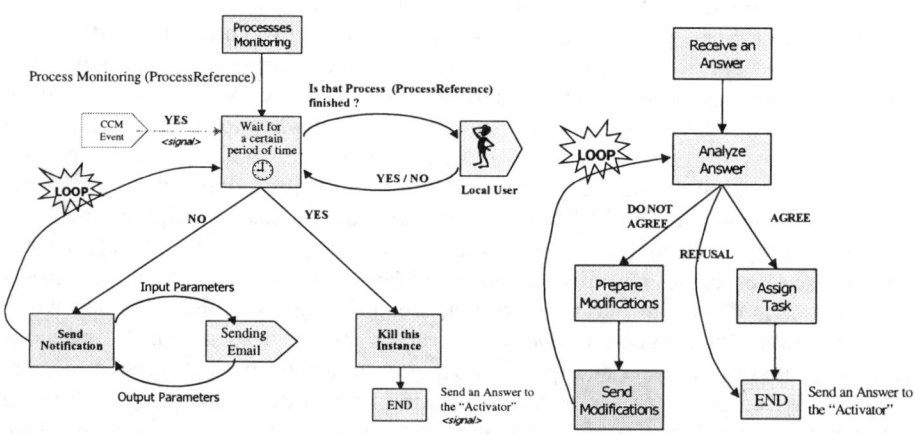

Fig. 11. Generic "wait answer" process iteration   **Fig. 12.** Example of negotiation

The "inspection" step can be performed by another process, the "monitoring process". This can be a generic process used whenever some acceptance (or cancellation) is required from a VE partner, as illustrated by Fig. 11. This process periodically checks for a termination event that lets the calling process finish. Examples of termination events are a signal coming from a lower level process (CCL event) or information provided by a user interface (for the cases that a confirmation arrived by fax or phone). If no termination event arrives during a specified time period, a notification is sent out (by email, for instance).

For the particular example of product negotiation, a specific process considering several iterations could be used (Fig. 12). In this case the answer from enterprise B can be a simple acceptance, a proposal for a design modification in the part (that enterprise A may accept or propose another alternative, initiating a new cycle), or a simple refusal (unable to perform the required task).

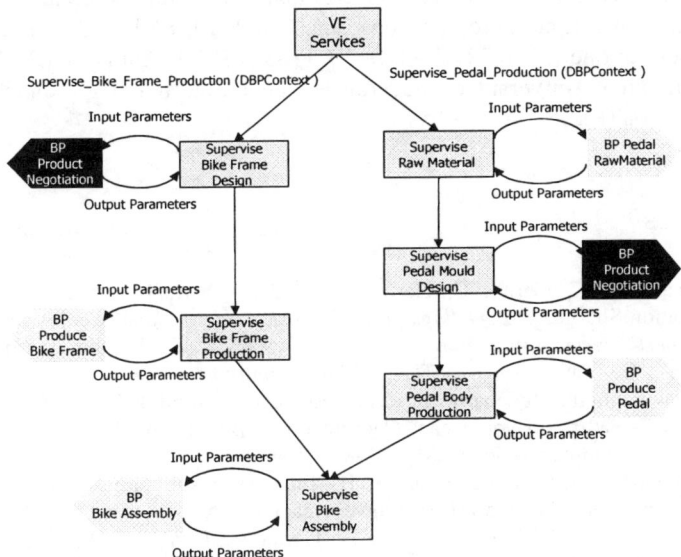

Fig. 13. Examples of BPs at the VEF level

Examples of coordination at the upper level (VE level) are shown in Fig. 13. Two supervision processes are illustrated here. These processes use lower level processes (such as Product Negotiation).

In the implemented coordination system the *process execution history* can be maintained, resorting to services of DIMS, to be used for auditing purposes or data mining towards process performance improvement.

5 Conclusions and Further Work

A flexible and open execution environment is a fundamental requirement for the implementation of virtual enterprises. The PRODNET II project developed an example of such infrastructure that combines diverse support technologies and addresses a wide range of requirements set by the VE paradigm with particular focus on the characteristics of SMEs. In particular, the workflow-based multi-level coordination approach developed by PRODNET II seems particularly adapted to the configurability and flexibility requirements.

The essential features that validate the model are implemented and tested in the PRODNET demonstration scenario, based on a Windows-NT environment, however further work is required namely in terms of flexible planning and scheduling of distributed business processes. The MASSYVE project is addressing these issues and an effort towards an integration of results of both projects is being pursued.

Acknowledgements. This work was funded in part by the projects Esprit PRODNET II and INCO MASSYVE. The authors also thank the valuable contributions of their partners in the mentioned projects consortia: CSIN (P), ESTEC (P), HERTEN (BR), Lichen Informatique (F), MIRALAGO (P), ProSTEP (D), Uninova (P), University of Amsterdam (NL), Universidade Federal de Santa Catarina (BR), and Universidade Nova de Lisboa (P).

References

1. Afsarmanesh, H.; Garita, C; Hertzberger, L.O. - Virtual Enterprises and Federated Information Sharing. *Proceedings of the 9th IEEE International Conference on Database and Expert Systems Applications", DEXA '98*, pages 374-383, Vienna, Austria, Aug 1998.
2. Alonso, G. at al, - *The WISE approach to Electronic Commerce*, http://www.inf.ethz.ch/department/IS/iks/research/wise.html, Feb 15, 1999.
3. Arbab, F. – What do you mean, Coordination?, Bulletin of the Dutch Association for Theoretical Computer Science, Mar 1998.
4. Camarinha-Matos, L.M; Afsarmanesh, H. (Ed.s)– Infrastructures for virtual enterprises – Networking industrial enterprises, Kluwer Academic Publishers, Oct 1999.
5. Camarinha-Matos, L.M; Afsarmanesh, H.; Garita, C.; Lima, C. - Towards an architecture for virtual enterprises, *J. of Intelligent Manufacturing*, Vol. 9, Issue 2, Apr 98, pp 189-199.
6. Carriero, N.; Gelertner, D. – Coordination languages and their significance, Communications of the ACM, Vol. 35 (2), 1992, pp. 97-107.
7. Goranson, H.T. – Agile virtual enterprises: cases, metrics, tools, Quorum books, 1999.
8. Heinl, P.; Horn, S.; Jablonski, S.; Neeb, J.; Stein, K.; Teschke, M. – A comprehensive approach to flexibility in workflow management systems, Joint Conf. On Work Activities Coordination and Collaboration, San Francisco, USA, 1999.
9. Klen, A.; Rabelo, R.; Spinosa, L.M.; Ferreira, A.C. - Integrated Logistics in the Virtual Enterprise: the PRODNET-II Approach, Proceedings of IMS'98 - 5th IFAC Workshop on Intelligent Manufacturing Systems, Gramado, Brazil, 9-11 Nov 1998.
10. Malone, T. W.; Crowston, K. – The interdisciplinary study of coordination, ACM Computing Surveys, Vol. 26 (1), 87-119, Mar 1994.
11. OAG – Open Applications Integration White Paper, Open applications Group, 1997.
12. Osorio, A.L.; Antunes, C.; Barata, M. – The PRODNET Communication Infrastructure, in Infrastructures for virtual enterprises, Kluwer Academic Publishers, Oct 1999.
13. Osorio, A.L.; Camarinha-Matos, L.M. – Support for Concurrent Engineering in CIM-FACE, in Balanced Automation Systems, Chapman and Hall, pp.275-286, 1995.
14. Rabelo, R.; Camarinha-Matos, L.M.; Afsarmanesh, H. - Multi-agent-based agile scheduling, *Journal of Robotics and Autonomous Systems* (Elsevier), Vol. 27, N. 1-2, April 1999, ISSN 0921-8890, pp. 15-28.
15. VIVE – VIVE Reference model, 1999, www.ceconsulting.it/VIVE/Results/default.html.
16. WfMC - Workflow Management Coalition (1994) - The Workflow Reference Model - Document Nr. TC00 - 1003, Issue 1.1, Brussels Nov 29, 1994.

Towards a Multi-layer Architecture for Scientific Virtual Laboratories

H. Afsarmanesh, A. Benabdelkader, E. C. Kaletas, C. Garita, and L.O. Hertzberger

University of Amsterdam, Kruislaan 403, The Netherlands
{hamideh,ammar,kaletas,cesar,bob}@wins.uva.nl

Abstract: In order to assist researchers with conducting their complex scientific experimentation and to support their collaboration with other scientists, modern advances in the IT area can be properly applied to the domain of experimental sciences. The main requirements identified in this domain include management of large data sets, distributed collaboration support, and high-performance issues, among others. The Virtual Laboratory project initiated at the University of Amsterdam aims at the development of a hardware and software reference architecture, and an open, flexible and configurable laboratory framework to enable scientists and engineers to work on their experimentation problems, while making optimum use of modern information technology approaches. This paper first describes the current design of a multi-layer architecture for this Scientific Virtual Laboratory, and then focuses further on the cooperative information management layer of this architecture, and exemplifying its application to experimentation domain of biology.

1. Introduction

The rapid improvement in the networking technology, computing systems, and information management methodologies allow application developers to solve some of the problems they face when designing and developing their own complex applications. The problems in the engineering and scientific domains are quite diverse and complex in nature. For instance, the complexity is faced when remotely controlling/monitoring a physical apparatus, running activities that require excessive computational resources or parallel/distributed machines, requesting collaboration within a distributed community that involves scientists with different interests, looking for the necessary information requested by the application from various information resources, or when trying to share the local results with external applications from the outside world.

In most ongoing design and development work in the area of Virtual Laboratory, researchers focus on certain specific aspects, for instance mostly related to the distance problem. Here for instance, they introduce mechanisms to remotely control devices, for the video conferencing, and file sharing mechanisms to enable the co-working among scientists [1, 2, 3].

Depending on the application interest (e.g. education, games, experiments, chemistry, aerospace, and government) and due to the variety of the possible application fields, the concept of the "virtual laboratory" has been associated to many meanings and refers to many spectrums. In the area of education for example, virtual laboratories can be applied to start a chemical reaction, or to see the basic mechanics rules at work while sitting in front of the computer screen [4]. In other areas, virtual laboratories are used as a simulator to study dangerous situations, for instance in the fields of aerospace [5] and nuclear engineering [6]. Another common usage of the "virtual laboratory" concept is its consideration as a replacement for specific task-oriented physical facilities in experimental computational environment.

However, most of the interpretations of the virtual laboratory concept are in fact limited in the sense that they provide a solution to some specific problem in certain application domains. Very seldom there are cases where the virtual laboratory concept is not restricted to one particular case, and offers means to support multidisciplinary users with uniform high-level interfaces to a general distributed and collaborative environment. Although in [7] the need for such an open environment is explained and the main requirements for such virtual laboratories are identified, we observe that this environment is not yet developed and the scientific community still lacks a reference framework that covers many aspects of a real collaborative multi-disciplinary experimental environment.

The Virtual Laboratory (VL) is a four-year project (1999-2003), initiated at the University of Amsterdam. This paper describes our ideas on the design of a multi-layer architecture, representing the fundamental functionalities to be supported by the laboratory. This project aims at the design and development of an open, flexible and configurable laboratory framework providing the necessary hardware and software to enable scientists and engineers to work on their problems via experimentation, and overcome many obstacles while making optimum use of the modern Information Technology. The realization and development plan for the VL is outside the scope of this paper and is the subject discussed in the forthcoming papers. The VL will develop the necessary technical and scientific computing framework to fulfil the requirements in different scientific application domains. The domains of experimentation such as physics, biology, and systems engineering in specific are considered for which the project will develop certain application cases during the first phase of the project prototyping and evaluation. The project plans, on one hand, to fulfill the main requirements of these domains as the VL generic functionalities, e.g. the support for the collaboration among scientists and information management, among many others. On the other hand, the framework will be flexible and configurable, in order to be extended and support specific application oriented requirements. The application cases mentioned above will be used to test this flexibility of the framework.

The remainder of the paper is organized as follows: Section 2 describes the VL requirements and important considerations taken into account, when designing such a framework. A multi-layer architecture for the Virtual Laboratory is described in Section 3. The Virtual-lab Information Management for Cooperation (VIMCO) layer of VL is described in Section 4, while an example application case for the VIMCO layer addressing the DNA-micro array is described in Section 5. Finally, Section 6 concludes the paper.

2. Main Considerations

When designing a generic multi-disciplinary VL framework, there are certain requirements and considerations to take into account which are inherent in the experimental science domains. The VL must adhere to the important characteristics and requirements stated by these domains. Therefore, the architecture of the VL must be able to satisfy the requirements imposed by these characteristics.

One of the most important characteristics of the experimental science domain is that researchers need to manipulate large data sets produced by physical devices. In order to convert this data into valuable information, different processes need to be applied on these data sets. These processes typically demand high-performance computing support and large data storage facility. Also, since everyday more and more scientists are involved in the experimental science domain, the efficient utilization of the data sets is becoming mandatory to support the collaboration among these experts. Furthermore, due to the enormous amount of generated data there is a need that the data sets resulting from different sorts of measurements are combined and inter-linked together to provide a better insight on the problem under study. These characteristics are fundamental to the experimental science, and hence indispensable to the VL.

Considering the several application domain characteristics presented above, there are many design issues that need to be incorporated in the VL architecture. Below, we address three of these key issues for consideration, namely, the proper management of large data sets, information sharing for collaboration, and distributed resource management.

The first issue addresses the management/handling of large data sets for which the size of data generated by the devices connected to the VL can be very large. The data size ranges for instance between 1.2 MB for the data generated every second by a Micro-beam device, to 60 GB for every slice scanned from a human brain tissue, generated by a CT scanner. This excessive amount of data must be stored, filtered, classified, summarized, merged, inter-linked, and made available to the programs using it with the required performance [9].

Advanced collaboration facility is the second characteristic to take into account for the VL architecture design. Not only the efficient storage and retrieval of such data is challenging, a more challenging issue is to provide means to efficiently exchange such large amounts of data among collaborating scientists, organizations, or research centers. Moreover, the data is inherently of a heterogeneous nature and there may not even exist a standard representation for such a data. One of the primary goals of the VL is to support the increase of collaboration among scientists and scientific centers. This brings up the need for a comprehensive, advanced interoperation/collaboration facility, which must be supported by the Internet-tools, Web-tools and other support-tools.

Furthermore, the distributed nature of the VL together with the high performance and massive computation and storage requirements brings up the third important characteristic, the distributed resource management issue. Therefore, utilization of an adequate hardware resource manager within the Virtual Laboratory must also be considered.

3. VL Multi-layer Architecture Design

The general design of the VL architecture is based on multiple functionality layers, so that the application-specific and domain-specific computational and engineering issues can be separated from the generic computing aspects. The generic computing aspects serve a broad range of domains, and include issues related to distributed computing, networking, basic information management, communication/collaboration and visualization. Further, application-specific or domain-specific issues, on the other hand, can be defined on top of these generic issues.

As such, the VL reference architecture is primarily composed of five functional layers, as illustrated in Figure 1. These layers are briefly described below.

Layer 1
The computing and networking layer provides the high-bandwidth low-latency communication platform, which is necessary both for making the large data sets available to the collaborating experts and for the physical or logical distribution of the connected external devices and the client community that uses the laboratory facilities.

The gigabit networking technology being set at the University of Amsterdam, and the Globus distributed resource management system are considered for the development of VL environment. The Globus [8] system addresses the needs of the high performance applications that require the ability to exploit diverse, geographically distributed resources. A low-level toolkit provides basic mechanisms for the communication, authentication, network information, and data access. These mechanisms are used to construct various higher level services, such as parallel programming tools and schedulers for multidisciplinary VL applications. Since the Globus system offers the resource management required for distributed computing, it can be well used for the development of some VL internal components.

Layer 2
The VIMCO co-operative information management layer provides archiving services as well as the information handling and data manipulation within the virtual laboratory. This layer supports a wide range of functionality ranging from the basic storage and retrieval of information (e.g. for the raw data and processed results) to advanced requirements for intelligent information integration and federated database facilities to support the collaboration and information sharing among remote centers.

Layer 3
The COMCOL layer enables the communication with external devices connected to the laboratory, as well as the secure communication and collaboration between users within and outside the laboratory. The Web will play the major role for the communication.

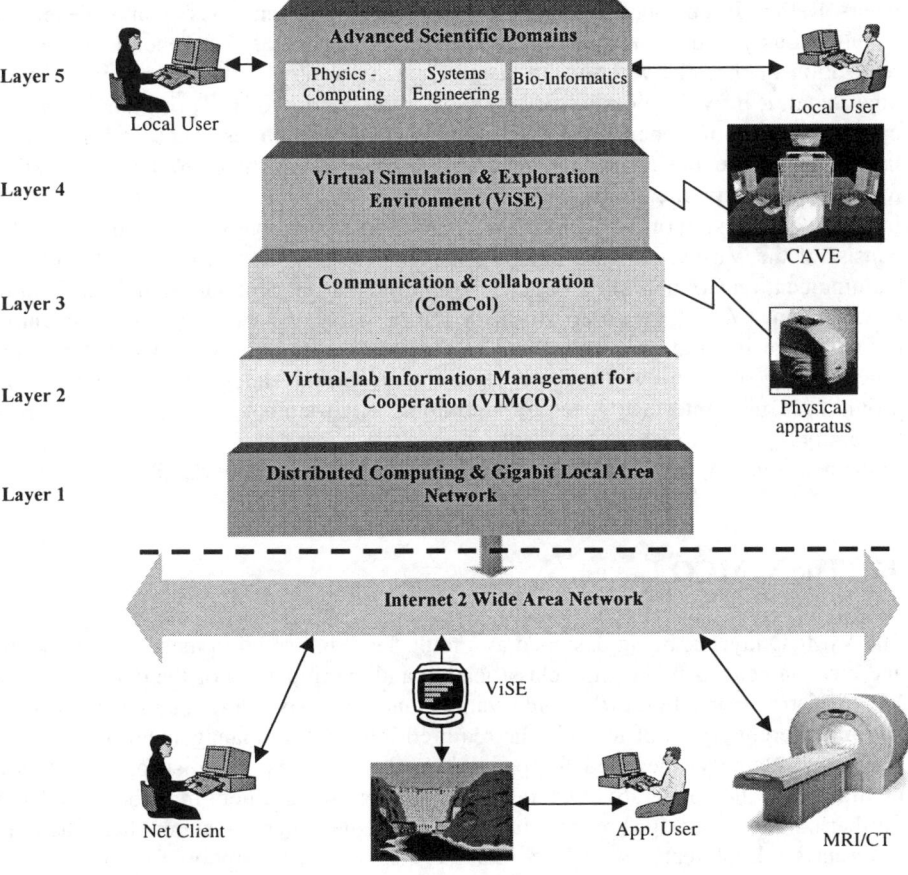

Fig. 1. Functional layers within the Virtual Laboratory

Layer 4
The VISE layer presents a generic environment in which scientific visualization, interactive calculation, geometric probing and context-sensitive simulations are supported.

Layer 5
The fifth layer is the application dependent part of the VL. At this layer, interfaces will be present and application-specific and domain-specific tools will be provided in order to enable users to make their experiments, using the functionality provided by the other layers in the VL.

This primary multi-layered design represents the functionalities supported by the VL, and does not imply that the layers at the higher levels are dependent on the lower layers. Rather, it represents the fact that on one hand different layers can be developed simultaneously and somewhat independently, as a part of the general design and realization of the VL, without the need of extensive day to day interaction. On the other hand, it provides possibilities for a clear description of the primitive laboratory operations and components and their individual functionalities. These primitives are later used in the design of the general VL abstract machine, to be described in forthcoming VL publications.

In the layered design of the VL, the three main functional layers that form its skeleton consist of the Virtual-lab Information Management for Cooperation (VIMCO), the Communication Collaboration (ComCol), and the Virtual Simulation and Exploration Environment (ViSE). In order to satisfy the needs of the users of VL in scientific domains, the integration of elements of these layers is necessary. Considering the functionality offered by the three layers, their proper integration, together with providing a user interface access/programming environment is vital for the Virtual Laboratory.

In the next section, the VIMCO layer is further analyzed in more details.

4. The VIMCO Layer

The VIMCO layer is being designed as a multi-level information management system and environment to support the classification and manipulation of the data within the VL platform. Considering the wide variety and excessive amount of data handled within different layers of the VL, the required information management mechanisms may vary. Namely, the need for parallel database extensions, distributed database facilities, and the cooperative/federated information management must be considered. Furthermore, the design of the information management system shall facilitate different kinds of features on large data sets, for instance support for structured as well as binary data access, data integration of data from several sources location transparency for remote data access, data mining, secure and authorized access to shared data among different nodes, and the intelligent data handling.

The general design objectives of the VIMCO layer within the VL cover the areas of fundamental database research and development to support complex domains. For simplicity reasons, only three main focus areas of development are addressed in this paper. The first area focuses on the *Data Archive*; archiving the wide variety of data necessary to be handled within the VL, supporting their categorization and storage. The second area concentrates on the development of *ARCHIPEL*; a generic co-operative information management framework supporting the node autonomy, and the import/export of data based on information visibility and access rigths among nodes. The last area focuses on the analysis, modeling, and provision of support mechanisms for the information management requirements of a specific application, the DNA micro-array application from the domain of bio-informatics.

4.1. Data Archive

The work on data archiving focuses on the design and development of an information brokerage system to archive the wide variety of data with different modalities and from different sources. This includes all the data generated through specific research and application domains supported by the VL. For instance, a database schema for the information handled by the ViSE and ComCol layers of the VL, and a catalogue/archive schema has been developed using the Dublin Core MetaData standard. This catalogue/archive schema has been refined to achieve a more scalable and extendable archive meta-schema, able to capture the raw/processed data, the experiment and scientist related information, and hardware (devices) and software characteristics. The designed schema is easily extendable to cope with the future modifications with the flexible addition of new experiment types.

4.2. Archipel

The research and development performed within ARCHIPEL, a generic co-operative information management framework, covers the fundamental data management infrastructures and mechanisms necessary to support the forthcoming advanced applications for the VL. ARCHIPEL is an object-oriented database management system infrastructure, supporting the storage and manipulation of inter-related data objects distributed over multiple sites. ARCHIPEL supports a dynamic and wide variety of database configurations (e.g. distributed or federated) and unites different object distribution strategies in a single architectural model. The framework defined for ARCHIPEL improves the accessibility to large databases in data intensive applications and provides access to a variety of distributed sources of information. The architecture of ARCHIPEL has its roots in the PEER federated/distributed database system [15], which has been previously developed within the University of Amsterdam. PEER federated database systems and its second and third generation developments the DIMS, and the FIMS systems are already applied to several ESPRIT projects [10, 11, 12, 13, 14]. In the ARCHIPEL system, the nodes store and manage their data independently, while in the case of common interest they can work together to form a co-operation network at various ARCHIPEL levels. This nesting of nodes at different ARCHIPEL levels allows for a variety of configurations, where, for instance, certain kinds of co-operations are formed to enhance the performance of data sharing, while others are formed to enhance the complex data sharing in a distributed or federated fashion.

4.3. DNA Micro-array Application

In this focus area, information management requirements of the bio-informatics domain are studied and the necessary functionality to support this domain is designed. Special emphasis is paid to the nature, structure and size of the local databases as well as the data integration with other information sources, and the nature of the user interface in order to form a base for the effective integration with the developments in the other layers. The DNA micro-array application [16] is described in more details in the next section.

5. An Example Application Case for the VIMCO Layer: The DNA Micro-array

In general, DNA micro-arrays allow genome-wide monitoring of changes in gene expressions; induced in response to physiological stimuli, toxic compounds and disease or infection [16]. Depending on the organism and the exact design of the experiment, the number of useful data points produced *per experiment* will range from around 12000 to 20000 for a simple organism like yeast upwards to 200000 to 300000 for man.

Today, the micro-array technology makes it possible to study the characteristics of thousands of genes in a single experiment. Thus, it supports the following tasks:
1. Identify genes responsible for a given physiological response.
2. Monitor physiological changes occurring during disease progression or in industrial organisms to identify cellular responses at specific stages of the production process.
3. Better understand the mechanisms of gene regulation and identify transcription factors responsible for coordinate expression of genes displaying similar responses.
4. Assign functions to novel genes.

5.1. The Experiment

Within each DNA micro-array experiment, one or more experimental conditions (experimental parameters) are different and will be adjusted. These parameters may represent different target RNA extractions or various environmental conditions (e.g., temperature). As shown in Figure 2, every micro-array experiment is mainly accomplished in three steps: pre-experiment, experiment, and results analysis.

i. Pre-experiment

This is the stage where the necessary information about the genes, properties of the RNA extractions, results of current research, and images are gathered. This information may be locally available (in the local database), or can be accessed from remote information servers.

After gathering the necessary information, the DNA pieces (genes) to be analyzed are prepared and placed into the array (by an array preparation device called arrayer), and the target RNAs are extracted and labeled for hybridization.

ii. Experiment

During the experiment stage the prepared array (at the previous stage) is hybridized (put into reaction) with the target RNA solution(s). A scanner then scans the hybridized array. The information about the array, array spots, the target, the source of the target, extraction and hybridization protocols used, and the information about

the scientist submitting the experiment are stored in the database. This information consists of:
- A table showing which gene is on which spot (including the information about the gene on that spot)
- Information about the target RNAs
- Experimental conditions (parameters for the experiment)
- Information about the experiment (samples used, date, protocols used, etc.)
- Information about the scientist.

iii. Analysis

In this stage, the results of the experiment are received from the scanner. Currently, this information is stored as images. The images of the experiment are fed into a software program and compared to see the effect of a change in those parameters. The generated data from the comparison program is stored in the database.
The locally generated data is later retrieved and analyzed. This step may also involve retrieving data from remote information sources, and analyzing the whole set of data.

5.2. Data Management Approach

The DNA micro-array application requirements are going to be studied and addressed as a case for the Virtual Laboratory project. In this paper, specific attention is dedicated to the functionality required to be provided by the VIMCO layer. From the scenario described above, two main basic requirements are identified for the proper VIMCO information management of such a large-inter-linked scientific data:
1. Ability to compare experiment results with both local data sets and with other results made available by external scientific centers, and ability to easily visualize the response patterns.
2. Ability to link experiment results for each gene with other local and external information on that gene. This information is extremely heterogeneous and may involve a wide range of elements, being for instance a sequence (sequence variants, mutations, possible association with disease), the predicted protein sequence, 3D-structure, cellular location, biochemical function, interactions with other molecules, or regulation pattern.

Besides these basic requirements, many advanced data management functions are necessary to organize and analyze the large-scale expression data that is generated by the micro-array experiments. Therefore, various tools need to be designed and developed for data retrieval and analysis in order to facilitate other more specific tasks such as:
- Management of large quantities and wide variety of data.
- Identification of patterns and relationships on individual experiment data and also across multiple experiments
- Interactive or real time queries.
- Data loading
- Input of data from different sources
- Interpretation of the complex hybridization results

- Integration of external information resources
- Data mining

Furthermore, the stored results of the experiments can be made available via Internet by means of different web applications. These applications include: results/summaries browser, SQL query evaluators, and image analysis.

Related to this web information access, there is also a need for remote data gathering support. This process corresponds to the use, interpretation, and storage of the information that is offered by other remote systems by means of web interface tools. At this point, the information from the other remote sources is browsed, selected, and may be partially stored in the local database or separately in another repository.

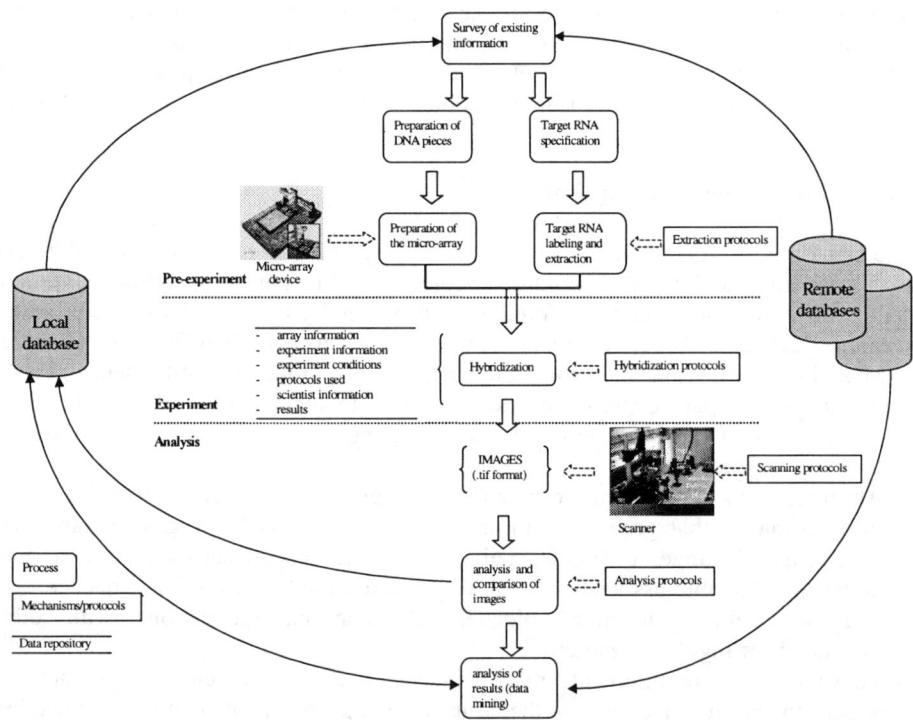

Fig. 2. General description of the micro-array experiment case

5.3. Functionality Offered by VIMCO

After studying the micro-array scenario described above, some general functionality is identified that the VIMCO layer should support for the DNA micro-array application. In this section, these identified requirements are briefly addressed (see Figure 3).

- **VIMCO Database Server**

The VIMCO database server back-end is represented so far by the Oracle and the Matisse database servers [17].

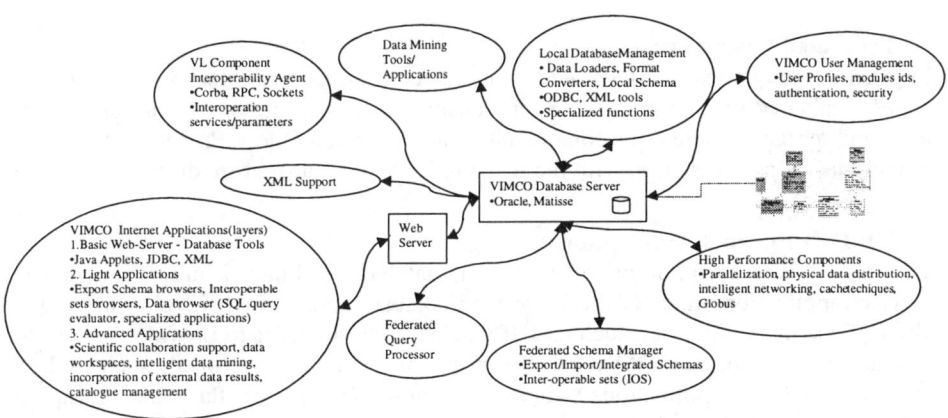

Fig. 3. General VIMCO functionality diagram.

- **Local Database Management**

The local database management functionality basically encompasses the "traditional" database application services. Most of these services are already provided by the database (DB) vendors and could be directly used by some applications. The services include data loaders, for instance, for direct loading of ASCII files into the database, format converters to load specific data into the database, ODBC access, XML functions to store/get XML documents into/from the database, and specialized functions for VIMCO to develop higher-level functions to support specific operations on the database.

- **VIMCO User Management**

This functionality includes user profiles management, id-management of the modules in the VL, user authentication, and security issues.

- **High Performance Components**

The use of high performance techniques in the VIMCO layer needs to be applied at different layers. The techniques include parallelization, physical data distribution, intelligent networking and caching techniques, and functionality provided by Globus.

- **Federated Schema Manager**

The federated database architecture applied in the VIMCO layer is in charge of the management of data sharing, exchange and integration from many autonomous data sources.

- **Federated Query Processor**

In general, the queries in a federated system need to be decomposed, sent to different nodes, evaluated on their proper export schemas, results sent back to the originating node and merged to make the final result. The Federated Query Processor takes care of these tasks in conjunction with the federated schema manager module.

- **VIMCO Internet Applications**

The use of web technology together with database techniques are required. The services/applications include the basic web-database tools which involve the use of already existing technology such as JDBC, XML, etc. in order to provide web access to database information. Also other applications will be developed, such as light internet database applications which are built on top of the first group of applications/services, and other advanced internet database collaborative applications for specific scenarios.

- **VL Components Interoperability Agent**

Depending on the interoperability approach among the VL modules (e.g. based on CORBA, RPC, sockets, etc.), VIMCO will develop specific functionalities complying with the given "VL interoperation standard", in order to make its services available to other modules. The idea is to encapsulate all these interoperation tasks in one module (agent) if possible.

- **Data Mining Tools/ Applications**

Once the information about the experiments is collected it must be processed, analyzed, and the results presented in such a way that valuable knowledge could be gained from the identification of patterns within the data. It is important to present all the information together in a summarized and concise way, that allows very efficient visual evaluation and cross-comparison.

6. Conclusions

From the current trends in emerging complex application domains and their requirements, it is foreseen that increasing computational power and using high-bandwidth network structures in the information and communication technology will play a major role in near future. Collaborative scientific application domains certainly are in need of these technology advances. The Virtual Laboratory framework described in this paper is a step in this direction providing an open, flexible environment to support both current and future applications and their emerging requirements, for the challenging field of scientific collaboration. In this framework, the paper focuses on the design of VIMCO, the cooperative information management layer of the VL project at the University of Amsterdam. VIMCO aims at supporting the requirements set for the information sharing and exchange among a wide variety of collaborating scientists in the Virtual Laboratory. The paper addresses the multi-layer architecture of the VL development, and some details about the functionality of the VIMCO. Furthermore one specific application case for VL, the DNA micro-array, is described and the VIMCO functionality required to support this application is addressed.

References

1. J. Fisher- Wilson, Working in a Virtual Laboratory - Advanced technology substitutes for travel for AIDS researchers, The Scientist, v. 12, n. 24, p. 1, December 7, 1998.
2. K. Nemire, Virtual Laboratory for Disabled Students: Interactive Metaphors and Methods, California State University, Northridge Center on Disabilities, Proceedings of the Virtual Reality Conference, 1994.
3. Caterpillar Inc. Distributed Virtual Reality (DVR) Project, http://www.ncsa.uiuc.edu/VEG/DVR/
4. Virtual Laboratory, University of Oregon, Department of Physics, http://jersey.uoregon.edu/vlab/
5. Ross, M. Virtual Laboratory Expands NASA Research – Aerospace Technology Innovation, Volume 5, Number 6, November/December 1997
6. Texas A&M University/EPF Ecole d'Ingenieurs Nuclear Engineering Design Virtual Laboratory, http://trinity.tamu.edu/COURSES/NU610/
7. W. Johnson et al. The Virtual Laboratory: Using networks to Enable Widely Distributed Collaboratory Science, Ernest Orlando Lawrence Berkeley National Laboratory, Formal Report, LBL-37466, 1997.
8. I. Foster, C. Kesselman, The Globus Project: A Status Report. Proc. IPPS/SPDP '98 Heterogeneous Computing Workshop, pg. 4-18, 1998.
9. A. Benabdelkader, H. Afsarmanesh, E. C. Kaletas, and L.O. Hertzberger. Managing Large Scientific multi-media Data Sets. To be presented in the Workshop on advanced Data Storage / Management Techniques for High Performance Computing, February 23-25, 2000, Warrington, UK.
10. H. Afsarmanesh, A. Benabdelkader, and L.O. Hertzberger. «*Cooperative Information Management for Distributed Production Nodes*», *In Proceedings of the 10th IFIP International Conference PROLAMAT '98*, Trento, Italy: Chapman & Hall Press.
11. H. Afsarmanesh, A. Benabdelkader, and L.O. Hertzberger. A Flexible Approach to Information Sharing in Water Industries. *In Proceedings of the* International Conference on Information Technology CIT'98, December 21-23, 1998. Bhubaneswar, India.
12. L.M. Camarinha-Matos and H. Afsarmanesh. Flexible Coordination in Virtual Enterprises. In *Proceedings of the 5th IFAC Workshop on Intelligent Manufacturing Systems, IMS'98*, pages 43-48 Gramado, Brazil, November 98.

13 H. Afsarmanesh, C. Garita, L.M. Camarinha-Matos, and C. Pantoja-Lima. Workflow Support for Management of Information in PRODNET II. In *Proceedings of the 5th IFAC Workshop on Intelligent Manufacturing Systems, IMS'98*, pages 49-54, Gramado, Brazil, November 98.
14 L.M. Camarinha-Matos, H. Afsarmanesh, C. Garita, and C. Lima. Towards an Architecture for Virtual Enterprises. *Journal of Intelligent Manufacturing*, Volume 9, Issue 2, pages 189-199, Chapman and Hall publication, April 1998.
15 F. Tuijnman and H. Afsarmanesh. Sharing Complex Objects in a Distributed PEER Environment. In *13th International Conference on Distributed Computing Systems*, pages 186-193. IEEE, May 1993.
16 Alan Robinson. Life Sciences Research / Gene Expression, Request for Information (LSR RFI3) European Bioinformatics Institute EMBL Outstation - Hinxton, Cambridge, UK.
17 Matisse Database System Manuals, Copyright © 1996 ADB S.A. 3rd Edition - June 98, France.

Modelling Control Systems in an Event-Driven Coordination Language

Theophilos A. Limniotes and George A. Papadopoulos

Department of Computer Science, University of Cyprus
75 Kallipoleos Str, P.O.Box 20537
CY-1678 Nicosia, Cyprus
{theo,george}@cs.ucy.ac.cy

Abstract. The paper presents the implementation of a railway control system, as a means of assessing the potential of coordination languages to be used for modelling software architectures for complex control systems using a components-based approach. Moreover, with this case study we assess and understand the issues of real time, fault tolerance, scalability, extensibility, distributed execution and adaptive behaviour, while modelling software architectures. We concentrate our study on the so-called control- or event-driven coordination languages, and more to the point we use the language Manifold. In the process, we develop a methodology for modelling software architectures within the framework of control-oriented coordination languages.
Keywords. Concepts and languages for high-level parallel programming; Distributed component-based systems; Software Engineering principles; High-level programming environments for Distributed Systems.

1 Introduction

A number of programming models and associated software development environments for parallel and distributed systems have been proposed, ranging from ones providing elementary parallel constructs (such as PVM and MPI) to ones offering higher level logical abstractions such as skeletons, virtual shared memory metaphors (such as Linda), coordination models ([4]), software architectures ([5]), etc. It would be interested to examine the potential of those models in modelling real-life non-trivial applications and in the process develop a software engineering methodology for their use.

In this paper we present some of the main components of implementing a railway control system using the coordination language Manifold ([1]). It is a typical real world problem which, apart from its operational aspects, it necessitates the addressing of several other requirements that relate to *real-time*, *fault tolerance* and *adaptive behaviour*. In the process, we present a methodology for developing such real-life applications using the control-driven coordination metaphor.

The rest of the paper is organised as follows: the next section describes briefly the case study while the following one is a brief introduction to the coordination language

Manifold. The description of the implementation of the case study in Manifold is then presented with some concluding remarks at the end of the paper.

2 Description of the Case Study

The scope of designing a control system is to process raw data obtained from the environment through sensing devices and gauges, determine the model parameters that describe the environment, decide interdependencies of change of states, adapt problem solving routines, and provide control information to the users. Furthermore, other requirements to be met include *functionality, timeliness, fault-tolerance, degraded modes, extensibility,* and *distribution.*

In particular, we study the modelling of a railway system ([2]) consisting of railway tracks, junctions, and platforms. A number of trains is expected to be travelling across the network. The system should be able to monitor the position of each of the trains, access the current situation of the trains, be able to predict and cope with future developments, and take into account timetables (and follow the specified schedules as much as possible). Moreover, speed variation should be avoided whenever possible (speed adjustments) and direction adjustments should be supported. Other desirable features should include *scalability* and *extensibility* (e.g. modifying the network topology) and *fault tolerance* (e.g. coping with train failures).

3 Manifold

Manifold ([1]) is a control-driven coordination language. In Manifold there are two different types of processes: *managers* (or *coordinators*) and *workers*. A manager is responsible for setting up and taking care of the communication needs of the group of worker processes it controls (non-exclusively). A worker on the other hand is completely unaware of who (if anyone) needs the results it computes or from where it itself receives the data to process. Manifold possess the following characteristics:
- *Processes.* A process is a *black box* with well defined *ports* of connection through which it exchanges *units* of information with the rest of the world.
- *Ports.* These are named openings in the boundary walls of a process through which units of information are exchanged using standard I/O type primitives analogous to read and write. Without loss of generality, we assume that each port is used for the exchange of information in only one direction: either into (*input* port) or out of (*output* port) a process. We use the notation p.i to refer to the port i of a process instance p.
- *Streams.* These are the means by which interconnections between the ports of processes are realised. A stream connects a (port of a) producer (process) to a (port of a) consumer (process). We write p.o -> q.i to denote a stream connecting the port o of a producer process p to the port i of a consumer process q.
- *Events.* Independent of streams, there is also an event mechanism for information exchange. Events are broadcast by their sources in the environment, yielding *event*

occurrences. In principle, any process in the environment can pick up a broadcast event; in practice though, usually only a subset of the potential receivers is interested in an event occurrence. We say that these processes are *tuned in* to the sources of the events they receive. We write e.p to refer to the event e raised by a source p.

Activity in a Manifold configuration is *event driven*. A coordinator process waits to observe an occurrence of some specific event (usually raised by a worker process it coordinates) which triggers it to enter a certain *state* and perform some actions. These actions typically consist of setting up or breaking off connections of ports and channels. It then remains in that state until it observes the occurrence of some other event which causes the *preemption* of the current state in favour of a new one corresponding to that event. Once an event has been raised, its source generally continues with its activities, while the event occurrence propagates through the environment independently and is observed (if at all) by the other processes according to each observer's own sense of priorities. Figure 1 below shows diagrammatically the infrastructure of a Manifold process.

The process p has two input ports (in1, in2) and an output one (out). Two input streams (s1, s2) are connected to in1 and another one (s3) to in2 delivering input data to p. Furthermore, p itself produces data which via the out port are replicated to all outgoing streams (s4, s5). Finally, p observes the occurrence of the events e1 and e2 while it can itself raise the events e3 and e4. Note that p need not know anything else about the environment within which it functions (i.e. who is sending it data, to whom it itself sends data, etc.).

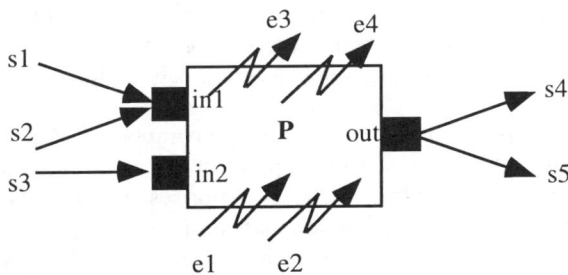

Fig. 1.

Note that Manifold has already been implemented on top of PVM and has been successfully ported to a number of platforms including Sun, Silicon Graphics, Linux, and IBM AIX, SP1 and SP2. For more information on the language and its potential uses we refer the interested reader to [1] and [5].

4 Implementing the Railway System as Components

Our aim is to construct a system that consists of static and dynamic components. The static ones are those that comprise a railway network, namely platforms, tracks, and junctions; we refer to them as rail components. These can be altered in the sense that entities of this kind can be added or removed, but these operations are not continual and time dependent as it happens with the dynamic components. The number and variation of both types of components directly relates to the scalability of the system.

The dynamic components are the trains themselves. There is a potential number of trains remaining dormant, that can be activated at any moment. Their 'course' or 'route' can be seen as a collection of static objects that the train's trip comprises. Other dimensions of continual change are the location of the train that is time dependent and the status of the train that is indirectly related to the status of the 'next one to visit' rail component. The two entities above can trigger at any time the redefinition of part of the route, including the destination platform of the train.

The railway system model built in Manifold consists of coordinating modules at a higher level and computational components at a lower level. In this particular application the computational components are written in C. Due to lack of space, we refrain from describing these modules in this paper. Irrespectively of its functionality, each module represents a building block in the architecture of the system. At one level the coordination components specify a number of transition pipelines that constitute the deportment of a process. At a higher level another transition system defines the interaction among such processes. Thus the pattern of the system formed can be described as transitions (at a higher level) to predefined algorithms (at a lower level). It should be made clear at this stage that the first level can have processes written in both computational languages which have their external behaviour observable, and processes written in Manifold defining different sets of transitions into the process structure.

Here follows a description of the coordination component. We should emphasize at this point that only the most important parts of the code are shown below in the presented manifolds. The whole application comprises around 2,000 lines of Manifold and C code. Thus, the presented functionality of the major components described below is somewhat simplified. Note that the names within the boxes refer to atomic (C code) processes, which are called by the respective manifolds and perform some lower level (namely irrelevant to the coordination protocols used) function.

Train Coordinator Code. An instance of this Manifold receives at a port the identification number of a train that is about to depart. For the current model this instance has to be predefined in Main, so that the particular train's id can be associated with a particular number of already activated Train instances that can run in parallel.

Let such a Train instance be train1. This train1 will have to observe events that are only broadcast by particular Platforms and Junctions (which the train in question will go through during its trip) to which train1 sends requests for passing. A particular input port (say train1.rail) is assigned for this job. This port then is passed as a parameter to the manner CheckInput whose function is to make any process dereferenced from this input port an observable source of events.

Similarly, provided that we have such a reference of a `Train` instance (`&train1` that is), connected to the input port of another process instance which in turn is also a parameter for `CheckInput`'s input port (say `platfrom1.train`), `train1` will be able to have its events *observable* by that other process instance. Moreover it has to be sorted out through which port each event of another process will be made *available* to `train1` This can be done by dereferencing a local event to another input port (say `train1.get`) which has to be connected to the output port of the event it expects (say `platform1.send`). An example of such code follows:

```
Train ()
port in rail
port in get1, get2, get3./* get three different events */
port in depart    /* get train id from Actual Rail info*/
{
 event get_thru deref get1.
 event get_thru_next deref get2.
 event gone deref get3.
 process i is variable(1).
 begin:
        CheckInput(rail);
        while true do {
        begin:
            /* steps of going through a rail component */
            if (i==1) then (raise(get_thru),WAIT);
            if (i==2) then (raise(get_thru_next),WAIT);
            if (i==3) then (raise(gone),WAIT);
            if (i==4) then (post(end)).
            /*reply to events above done */
            get_thru | get_thru_next | gone:i=i+1.
        end:. }
}
```

Platform Coordination Code. The instances of this Manifold are activated at the beginning of `Main()` as 'auto' instances. Each platform instance is given its id as a parameter at the declaration statement. In that way, each platform instance is associated with a particular platform right from the beginning.

Each `Platform` instance has to observe events which, in fact, are requests from `Train` instances. A particular input port is assigned for this purpose. This port then is passed as a parameter to an input port (say `platform1.train`) of the manner `CheckInput`, so that any instance of a process dereferenced from that input port, is an observable source of events.

A `Platform` instance can 'reply' to events of a `Train` instance, which in this case can be consider as requests, when its reference, i.e. `&platform1` is connected to the input port of that train which in turn is also a parameter for `CheckInput`'s port. The platform code follows below.

```
Platform(Manifold Get_Block_Status(event),
         Manifold Get_Block_Free, port in id)
```

```
port in   train, next_id.
port out  send1, send2, send3.
{
 event OK_free, OK_free_next, gone_OK, replies.
 event replies.
 auto process ID is variable.
 auto process next_ID is variable.
 begin: heckInput(train);
        (id->ID,next_id->next_ID,post(replies)).
 replies: while true do {
          begin:(&OK_free->send1,
                 &OK_free_next->send2,&gone_OK->send3).
          OK_free.*train:{ event im_free.
                           auto process gs is
                             Get_Block_Status(im_free).
                           begin:Trip(train);
                                 ID->(->gs,->ID).
                           im_free.gs:raise(OK_free).
                  end:.}.
          OK_free_next.*train:{ event im_free.
                           auto process gs is
                           Get_Block_Status(im_free).
                             begin:Trip(train);
                             next_ID->(->gs,->next_ID).
                    im_free.gs:{ auto process gf is
                                         Get_Block_Free.
                                 ID->gf;
                       (raise(OK_free_next),raise(OK_free)).
                                     }. end:.}.
          gone_OK.*train:{ auto process gf is
                                         Get_Block_Free.
                             begin: next_ID-> gf;
                                    Trip(train);

 (raise(gone),raise(OK_free_next),raise(OK_free)).
 } } }.
```

Junction Coordination Code. The operation of a Junction instance is basically the same as that of a platform, and the id of the next track that the train will pass is provided by

```
train.next_track->junction.next_id.
```

as well as

```
train.next_track2->junction.next_id2
```

which is an alternative choice for a diversion, in case that the first track is occupied. The availability of a second choice in its route is decided by the train instance itself. The junctions instance role in this is that it checks serially for this second choice (if it is free) if and only if checking the first choice has not preempted the control to the 'im_free' state.

Assembling All Components Together. The above described major components are assembled together as indicated by the figure below. The following description is concerned with the externally observable behaviour, and as all but one atomic are activated from within the coordinator manifolds, we are not concerned with these atomics at this level.

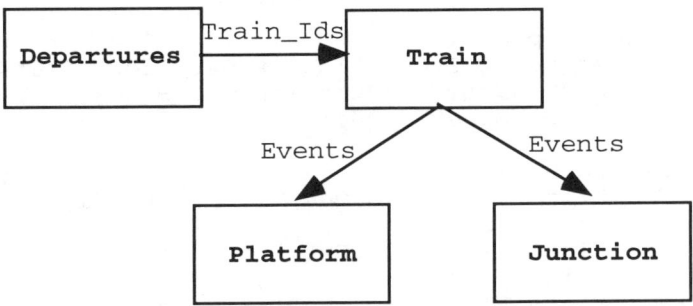

Fig. 2.

To show a particular scenario, we consider a simple case (effectively a snapshot of a more general scenario) of 3 trains travelling on a railway configuration involving a junction and a platform. There are two tracks connected to the junction and one tracj leaving the junction and ending at the platform. Two of the trains are approaching the junction travelling on either track, whereas the third one is stationed at the platform. In Main there are two streams created that connect two output ports of a Departure atomic with two input ports of two other instances of the Train coordinator. All instances of Platform and Junction controllers can accept requests. So, there are two Train instances running in parallel with this model, sending out requests to the already activated Platform and Junction instances. For the requests and replies there should be streams hardwired between pairs of 'send' and 'get' ports of particular Platform/Junction and Train instances respectively. Also, a guard is required to be placed at Train's 'end_port' in order to be able to preempt to a state that defines a new transition configuration. Finally, we have to hardwire the reference of each instance of a process with the dereferenced port of the instance that it has to send an event to.

A number of Train instances have to be activated at the begin state of Main, together with an instance Departures atomic. The Departures is a computational process that reads records from the Time_Scheduler and produces at its output ports train ids. These are the values to a transition between the Departure and the Train instances. The rest of the behaviour of the system is then decided by the hardwired connectivity between Train instances on the one hand and Platform/Junction instances on the other. This connectivity is entered as sets of transitions in separate states. Control is set from one state to another by guards set on Train instances in such a way as to indicate that the procedure of going through a component has ended. That is

guard(train1.end_port,a_connected,new_state).

The action of this non-transitory guard is that it preempts control to new_state, whenever there is one connection at least at the arrival side. The Main manifold for this scenario is shown below.

```
Main()
{
 event step1,step2, step11, step21, step22.

 auto process platf1 is
Platform(Get_Block_F,Get_Block_P,Get_Block_S,104).
 auto process junct1 is Junction(Get_Block_F,
Get_Block_J,Get_Block_S,201).

process train1 is
Train(TrackInfo,PlatfInfo,Get_Track,Get_ATI_R,1).
 process train2 is
Train(TrackInfo,PlatfInfo,Get_Track,Get_ATI_R,1).
 process train3 is
Train(TrackInfo,PlatfInfo,Get_Track,Get_ATI_R,3).

 begin: "main starts"->stdout;
          (activate(train1), activate(train2),
           activate(departures),
           departures.train1->train1.depart,
           departures.train2->train2.depart,
           post(step21)).
 step11: (train1.next_track->platf1.next_track,
           guard(train1.l3,a_connected,step22),

           (platf1.send->)  -> train1.go,
           (platf1.send->)  -> train2.go,
           (platf1.send->)  -> train3.go,

           (platf1.send2->) -> train1.go2,
           (platf1.send2->) -> train2.go2,
           (platf1.send2->) -> train3.go2,

           (platf1.send3->) -> train1.go3,
           (platf1.send3->) -> train2.go3,
           (platf1.send3->) -> train3.go3,

           (&platf1->) -> train1.rl,
           (&train1->) -> platf1.tr,
           (&platf1->) -> train2.rl,
           (&train2->) -> platf1.tr).
  step21: (train1.next_track->junct1.next_track,
           train1.next2->junct1.next_track2,
           guard(train1.end_comp,a_connected,step11),

           (junct1.send->)  ->  train1.go,
           (junct1.send2->) ->  train1.go2,
           (junct1.send3->) ->  train1.go3,
```

```
            (&junct1->) -> train1.rl,
            (&train1->) -> junct1.tr).

    step22:"step22 activated"->stdout.
}
```

5 Conclusions

This case study is concerned with the implementation of a railway system using a control-driven coordination language. There are already two implementations of a similar system; one is based on a data-flow architecture ([7]) and the other uses another control-driven language, namely ConCoord ([3]). In the data-flow model all processes act through decisions based on a Global Data Store. In that particular application, the behaviour of control processes is based upon the event action model, which in turn is based on the values of the above data structure. There is no direct reflection of the system's architecture in this, as the controller processes communicate only with the data store.

A characteristic feature of control-driven systems is that coordination is achieved through changes of states in processes or through broadcasting of events. Particularly in Manifold the Global Data Store manipulation, is left entirely to the computational processes, which reflect changes with the raising of events. To the coordinating building blocks, data and their values mean nothing. All they cope with is the handling of event occurrences with the preemption to new states, and the connection and definition of communication in streams between such components.

Regarding the issues of precision, synchronization and interaction with the environment, the entities requiring timing for changing their state, rely on coordination processes, with a request-reply event system. For every route component that the train has to cross, there is a check on the components' status before the train is allowed to proceed. The event-triggering mechanism of Manifold provides a natural synchronization mechanism for these types of control applications. The timing constraints imposed can exhibit (soft) real-time behaviour (albeit for lack of space we have not elaborated in detail on this issue in this paper).

Finally, Manifold's philosophy on coordinators being treated as black boxes, aware only about what is happening in their immediate vicinity • namely their input and output ports • provides a natural way for achieving extensibility and adaptability (dynamic reconfiguration of the railway topology). Each component in the system (trains, junctions and platforms) operates in a quite autonomous manner, continuously consulting and updating the global state and communicating with the other components by means of events. Thus, a train only has to deal with its immediate challenge (crossing a junction or approaching a platform), while asynchronously the global state may be changing, i.e. the railway topology may be altered, the route of the train may be modified, etc.

Acknowledgments

This work has been partially supported by the INCO-DC KIT (Keep-in-Touch) program 962144 „Developing Software Engineering Environments for Distributed Information Systems" financed by the Commission of the European Union.

References

1. F. Arbab, 'The IWIM Model for Coordination of Concurent Activities', *First International Conference on Coordination Models, Languages and Applications (Coordination'96)*, Cesena, Italy, 15-17 April, 1996, LNCS 1061, Springer Verlag, pp. 34-56.
2. E. de Jong, 'Software Architectures for Large Control System: A Case Study Description', *Second International Conference on Coordination Models, Languages and Applications (Coordination'97)*, Berlin, Germany, 1-3 Sept., 1997, LNCS 1282, Springer Verlag, 1997, pp. 150-156.
3. A. A. Holzbacher, M. Perin, M. Suhold, 'Modelling Railway Control Systems Using Graph Grammars: A Case Study', *Second International Conference on Coordination Models, Languages and Applications (Coordination'97)*, Berlin, Germany, 1-3 Sept., 1997, LNCS 1282, Springer Verlag, 1997, pp. 172-186.
4. G. A. Papadopoulos and F. Arbab, 'Coordination Models and Languages', *Advances in Computers*, Marvin V. Zelkowitz (ed), Academic Press, Vol. 46, August, 1998, 329-400.
5. G. A. Papadopoulos, 'Distributed and Parallel Systems Engineering in Manifold', *Parallel Computing*, Elsevier Science, special issue on Coordination, 1998, Vol. 24 (7), pp. 1107-1135.
6. M. Shaw, R. DeLine, D. V. Klein, T. L. Ross, D. M. Young and G. Zelesnik, 'Abstractions for Software Architecture and Tools to Support Them', *IEEE Transactions on Software Engineering* **21 (4)**, 1995, pp. 314-335.
7. S. Stuurman and J. van Katwijk, 'Evaluation of Software Architectures for a Control System: A Case Study', *Second International Conference on Coordination Models, Languages and Applications (Coordination'97)*, Berlin, Germany, 1-3 Sept., 1997, LNCS 1282, Springer Verlag, 1997, pp. 157-171.

Ruling Agent Motion in Structured Environments

Marco Cremonini[1], Andrea Omicini[1], and Franco Zambonelli[2]

[1] LIA – DIES, Università di Bologna
Viale Risorgimento 2, 40126 Bologna, Italy
{mcremonini,aomicini}@deis.unibo.it
[2] DSI -Università di Modena e Reggio Emilia
Via Campi 213/b, 41100 Modena, Italy
franco.zambonelli@unimo.it

Abstract. The design and development of cooperative Internet applications based on mobile agents require appropriate modelling of both the physical space where agents roam and the conceptual space of mobile agent interaction. The paper discusses how an open, Internet-based, organisation network can be modelled as a hierarchical collection of locality domains, where agents can dynamically acquire information about resource location and availability according to their permissions. It also analyses the issue of how agent motion can be ruled and constrained within a structured environment by means of an appropriate coordination infrastructure.

1 Introduction

Mobile agents are a promising technology for the design and development of cooperative applications on the Internet [3, 5, 12, 13]. Due to their capability of autonomously roaming the Internet, mobile agents can move locally to the resources they need – let them be users, data, or services – and there interact with them. This can provide for saving bandwidth and, by embedding some sort of intelligence into the agents, for enabling complex and dynamic interaction protocols to be defined, as needed in an open and unpredictable environment as the Internet.

Nowadays, researches in the area of mobile agents have greatly emphasised the dichotomy between mobile, possibly intelligent agents, and static ones. On the one hand, the community of distributed artificial intelligence has mostly focussed the aspects of agent intelligence and of autonomous planning capability [11]. On the other hand, the community of distributed systems has widely discussed the effectiveness of mobility with respect to traditional client/server solutions [8] and developed systems to make it practical. This way, the problems mostly tackled by designers of mobile agent applications have been how mobility and intelligence could be exploited to achieve complex cooperative tasks (for example in the e-commerce, or in the information retrieval area) and which technological solutions are better suited to face issues that mobility makes more critical, such as reliable message delivery and security.

However, in order to fully exploit the potentiality of mobile agents, we argue that also structural models for the environment where agents move and interact are needed and have to be supported by appropriate infrastructures.

In this paper, we suggest to take more deeply into account the need of a supporting infrastructure, and we focus on two specific issues of agent mobility:

- Most of the resources available in the Internet may be *a-priori* unknown to agent systems, and they are likely to be subject to different access control policies. In this context, we argue that an infrastructure should be in charge of supporting agents in the process of acquiring knowledge about the environment and the resources they need to handle.
- The Internet is not merely a flat collection of nodes, but mostly a dynamically structured network of domains, belonging to different organisations and ruled in different ways. In this context, we argue that the agent motion should be governed by an infrastructure that models such a scenario.

With regard to the former issue, we believe that an infrastructure should help improving agent efficiency both by driving them along specific itineraries, and by providing them with the ability of acquiring knowledge about the structure of the network in a dynamic way.

In our opinion, the two issues are strictly related. In this paper, we argue that the same abstractions and mechanisms exploited by an infrastructure to make agents aware of the network topology and of the organisation structure can be used to control and govern their motion. Agents should have the possibility to move and interact only with those part of the infrastructure and with those resources strictly needed to complete their tasks. Access control techniques applied to agent interaction and mobility should be adopted for this purpose. This topic is discussed in Section 2.

Then, in Section 3, we describe an infrastructure for mobile agent applications in structured environments, which is taken as a case study. The proposed infrastructure, by exploiting the TuCSoN coordination model based on programmable tuple space [13], provides mobile agent applications with:

- *dynamic knowledge acquisition*, which let agents free from the charge of an a-priori knowledge of the environment;
- *dynamic exploration constraining*, which forces agents to follow only specifically authorised paths within an organisation's network;
- *uniform information-based interaction pattern*, for both agent-to-agent communication and resource access.

Section 4 concludes the paper and briefly discusses some related works.

2 Ruling Agent Motion

In this section, we assume virtual enterprises as a case study to discuss some problems related to agent mobility in the Internet. Furthermore, we show how an appropriate infrastructure should support agents in the dynamic acquisition of knowledge about the network structure and in their motion.

2.1 The Case of Virtual Enterprises

More and more, business activities cross the boundaries of single companies and encompass federations – possibly of a world-wide size – of independent partners, working together on common projects. These federations, of either short or long duration, require support for the electronic management of inter-organisational business processes. This is likely to be a complex task, due to the need of adapting the processes of single companies to the emerging ones of the federation, as well as to the inescapable heterogeneity of data and services that the overall electronic infrastructure of the virtual enterprise is likely to exhibit.

For these reasons, and due to the possibly world-wide area of the virtual enterprise, mobile agents are likely to be fruitfully exploited in this context. In fact, agents can move across the nodes of the virtual enterprise and effectively handle problems such as the dynamic adaptation of processes or the heterogeneity of data and services, by exploiting their embedded intelligence and the capability of locally accessing data and resources.

However, in this context, it is unrealistic to assume that agents have a complete knowledge of the whole infrastructure, of the enterprise's network organisation, as well as of its resources. In fact, virtual enterprises are intrinsically dynamic, may be highly distributed, and have no centralised authority, being composed by independent partners. Consequently, it is very complex and often not convenient to provide agents with the detailed knowledge about the structure of all the federation's nodes and resources each time a new federation is formed. Also, basic security reasons suggest that a company is not likely to disclose all its data and resources to the virtual enterprise applications, but only a sub-set of them (i.e., the ones required for the processes in the virtual enterprise to proceed correctly). Thus, agents must be able to deal with the fact that the access to several of the resources and/or to several of the nodes in the network might be denied.

From the virtual enterprise example, it results that a suitable infrastructure for mobile agents in the Internet must provide two facilities:

- *dynamic knowledge acquisition* – during the agent motion both the resource availability and the organisation topology should be considered, in general, *a-priori unknown* or only *partially known*. Therefore, agents should be supported by the organisation infrastructure in the acquisition of knowledge about resource availability and location in order to plan their actions and achieve their goals.
- *dynamic exploration constraining* – both the desired destinations of a mobile agent and the ones it is effectively permitted to reach within an organisation network should be considered *a-priori unpredictable*. Therefore, agents should not be allowed to freely roam an organisation sub-network. On the contrary, their motion and interaction should be ruled and constrained by the infrastructure according with access control policies.

We call *exploration* the interaction of an agent with an infrastructure that provides for the above two facilities, and *explorable topology* the resulting interaction space, whose knowledge an agent acquires dynamically, incrementally, and accordingly to access control policies, by querying and moving around the infrastructure.

2.2 Dynamical Knowledge Acquisition

In our everyday experience, if we want to search and retrieve some information from the Web, we rarely get it directly from the first page of a site. Since we might not know the exact location of the page we need, we usually exploit search engines and then navigate through a set of links until some interesting information is found.

When mobile agents are involved, the scenario is likely to be the same. For example, consider a mobile agent application in charge of retrieving information from Internet sites by using agents that roam between Internet sites and query information sources. Agents, in general, cannot retrieve the required information from an a-priori known Internet host, but they are forced to explore complex structures to reach the location where information is stored. For instance, when information about the product of a specific company is desired, the agent could start the exploration by migrating to the company's main location (correspondingly to the home page on the Web). There the support for agent applications should allow it to produce its query and reply with information useful to refine the exploration. For example, in the case of a virtual enterprise composed by a main company and several subsidiaries, the main company's site, when required to provide information about a product, could reply with the location of the subsidiary company responsible for the product itself. Then, the agent could migrate to the node of the subsidiary company and repeat its query. There, the subsidiary's node could provide the agent with the commercial office location, thus incrementally augmenting the agent knowledge about the organisation' topology with this additional permitted destination. This way, the agent is allowed to migrates within an infrastructure and the exploration goes on.

The point here is that information is seldom centralised and often its management is delegated to local authorities (e.g. each of the federated subsidiary). This is true for the Web as well as for virtual enterprises. Thus, mobile agent applications have necessarily to deal with these situations where information are incrementally acquired and organised within a complex network infrastructure.

Supporting agents in dynamically acquiring knowledge is also likely to reduce the number of exceptions that agents must handle. The exploration plan could be refined or re-built as needed at each step by exploiting the information gathered, so as to prevent agents to move to nodes that will deny them the access to their resources, thus forcing them to handle a number of exceptions.

In order to support the dynamic exploration of agents, the infrastructure has to model the structure of the network, to keep into account the locality relationship (either logical or physical) between the nodes of the network. This requires different roles to be assigned to the nodes of the network. The nodes where agents execute to access specific data and resources play the role of *places*. The nodes exploited by agents to acquire information about the network structure and resources play the role of *gateways*. Since a gateway provides information for a limited set of places (a single centralised repository is unfeasible in complex and large environments), we call *domain* the set of nodes composed by the gateway and the places for which it provides information.

Moreover, a hierarchical topology of gateways seems the most manageable model for organisational structures (see Figure 1). A hierarchical structure naturally maps into the typical physical structure of the Internet, where the nodes of an administrative domain are often protected by a firewall and possibly clustered in higher level structures.

It is worth noting that the concepts of gateway and place do not automatically imply the definition of a unique hierarchical structure: a place can be part of different domains (there can be more than one gateway to provide information about that node), and a gateway can be a place in its turn (it can have local resources, as well). This can be very useful to model complex and dynamic network topologies, as those deriving from virtual organisations. As a result, a virtual enterprise, formed by several federated organisations, could be modelled as a composition of tree-like infrastructures of gateways and domains.

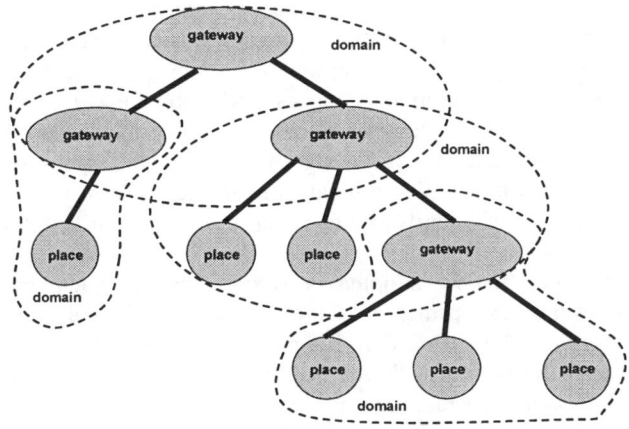

Fig. 1. The supporting infrastructure

The abstractions introduced above are used both to model the physical topology of a company and to rule the motion of mobile agents [6]. Gateways are the key elements for the support of the agent dynamic acquisition of knowledge. They are exploited to provide mobile agents with a multi–layered description of the network topology, where each gateway only describes a single level (the structure of its associated domain). By allowing information about the network topology and the organisation to be acquired incrementally by need, whenever crossing gateways and entering new domains, the task of mobile agents result simplified, because the complexity of knowledge management is shifted from the agents to the infrastructure.

2.3   Dynamic Exploration Constraining

In the above subsection, we have described how an infrastructure can model the physical and logical topology of a network and help agents in the exploration phase.

From the system viewpoint, further considerations have to be done: an agent should not be allowed to move to those nodes which do not hold accessible resources. On the one hand, computational resources would be wasted if agents were free to reach nodes where no resources can be accessed. In this case, access control should be enforced by limiting the number of unauthorised attempts of accessing local resources. On the other hand, agents may be required to dynamically handle exceptions when they are denied access to a resource. Since exception handling negatively impact on application efficiency, also from the agent viewpoint unauthorised accesses to resources should prevented.

For the aforementioned reasons, when agents have to interact with the gateway infrastructure to acquire knowledge, gateways could also be exploited to guide and constrain agent movements and their access to resources. Information provided to agents by gateways should depend from agents' identities and from privileges that an agent is enabled to acquire. In this acceptation, we elect knowledge about the network structure as a first-class information resource, to be managed accordingly to specific access control policies, as any other network resource. Only agents which are authorised to access to specific resources on a node will be given information about that node and about the resources allocated there.

However, the pre-condition for applying an access control technique is that entities whose access might be authorised (mobile agents, in our case) must be authenticated. As a result, gateways could be also exploited for the authentication of incoming agents on the behalf of all domains' places and sub-gateways. Places and sub-gateways of a given domain should reject each agent that was not previously authenticated by a gateway. By enforcing this policy, the agent motion can be controlled and limited within a hierarchical infrastructure. From the implementation viewpoint, this can be efficiently built upon current security frameworks and cryptographic tools [2].

In summary, the approach we adopted is to state that the agent motion must be governed following the basic principle of the *least privilege*, as usually done for the resource access control. This means that, if a place holds resources that an agent is not allowed to interact with, then also the migration of the agent to that place should be not authorised. The same consideration could be extended to a sub-tree of the whole structure: each sub-tree of the infrastructure could be explored by an agent only if it logically contains some resources needed by the agent to complete its task.

3 The TuCSoN Infrastructure

In this section we describe the infrastructure that we have defined to rule agent motion in a structured environment. The infrastructure is based on the TuCSoN coordination model, which enables the interaction between agents and resources to be uniformly managed in an information-oriented fashion [13].

3.1 The TuCSoN Coordination Model

The TuCSoN coordination model defines an interaction space spread over a collection of Internet nodes and built upon a multiplicity of independent tuple-based communication abstractions called *tuple centres* [13]. Each tuple centre is associated to a node and is denoted by a locally unique identifier. Each node possibly hosts a multiplicity of tuple centres, providing its own local TuCSoN name space (the set of the tuple centre identifiers), and virtually implements each tuple centre as an Internet service. Any tuple centre can then be identified via either its full Internet (absolute) name or its local (relative) name. This supports the twofold role of agents as network-aware entities explicitly accessing to a remote tuple centre, and as local entities of their current execution node. An operation on a remote tuple centre must be invoked specifying its full Internet name, as in tcID@some.node?op(tuple), while a local tuple centre can be accessed by specifying its local identifier only, as in tcID?op(tuple).

To overcome the limits of the original Linda model [9], TuCSoN tuple centres enhance tuple spaces with the notion of *behaviour specification*: each tuple centre can be programmed so as to implement its own observable behaviour in response to communication events. Instead of simply triggering the basic pattern–matching mechanism of the Linda model, the invocation of any of the TuCSoN basic communication primitives performed by a given agent can be associated to specific computational activities, called *reactions* [7]. The result of the invocation of a communication primitive is perceived by agents as a single-step transition of the tuple centre state, which combines altogether the effects of the primitive itself and of all the reactions it has triggered. Thus, a new observable behaviour can be defined for a tuple centre, where global coordination laws can be embedded.

Due to their programmability, different tuple centres can exhibit different behaviours in response to the same access event. In addition, since an access event is also characterised by the identity of the agent that performs it, the same tuple centre can behave with different behaviours in response to the same access event performed by two different agents. This can be fruitfully exploited to make tuple centres act both as the media used by agent to communicate with each other (by integrating in the form of behaviour specification any needed communication protocol) and as the media used for accessing any resource provided by a node (by representing data in the form of tuples and services in the form of behaviour specification).

3.2 TuCSoN as the Engine for Agent Motion and Access Control

TuCSoN tuple centres can be exploited for ruling agent motion in a structured environment and to handle the access control policies. To support the agent motion, and by considering the agent/gateway interaction, we have mediated both knowledge acquisition and constraints on agents' movements via tuple centres. Differently from nodes holding resources, for which more application-oriented tuple centres could be defined, each gateway is supposed to implement a *default tuple centre*, which represents the standard communication media that all the agents know and are able to interact with (see Figure 2). Hence, when an agent arrives on a gateway, it can query the default tuple centre to get the structure of the domain, in order to be enabled to proceed in the exploration. The default tuple centre provides the agent with:

- information about those nodes that authorise the execution of the agent;
- information about the available resources located on the nodes, i.e. those tuple centres that the agent can access;
- information about the privileges granted by tuple centres to the agent.

This way, the agent is dynamically informed about the structure of the network, and bounded in the motion towards the resources it can actually access. Avoiding useless migrations improves both agents and supporting infrastructure efficiency. Even more, since a tuple centre can be programmed to behave differently to accesses performed by different agents, it is possible to provide different agents with different information about the structure of the domain, depending on the permissions these agents have.

Similarly, when an agent arrives on a place, it can access the resources via local tuple centres. Again, a suitable programming of the tuple centres permits to develop the required access control policies to protect local resources. This leads to a very uniform model, where inter-agent communication, agent's access to the resources on a node, and agent's access to the information about the network structure, all exploit the same tuple-based interaction model.

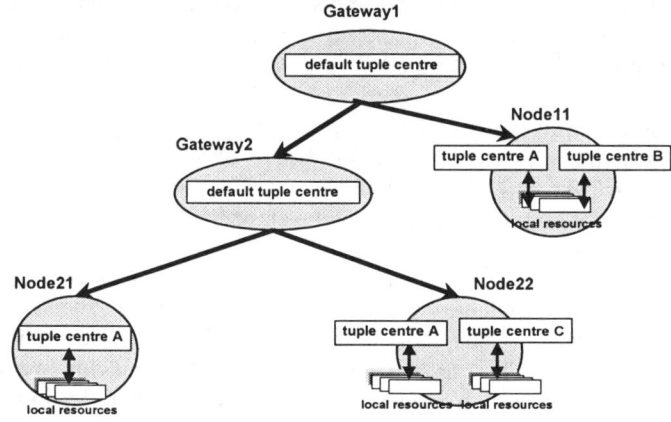

Fig. 2. Tuple centres and hierarchical topology

Table 1 reports the typical pseudo-code that an agent executes while moving in a structured network ruled by the TuCSoN infrastructure.

Table 1. Pseudo-code for the agent exploration protocol in TuCSoN

Exploration	
`<goto d>`	*migration to gateway* `d`
`<identify>`	`d` *authenticates the agent on behalf of all the places of its domain*
`?read(subdomlist)` `?read(placelist)` `?read(commspace)`	*access the default tuple centre of the gateway and obtain information about accessible sub-domains* (`subdomlist`), *places* (`placelist`), *and tuple centres* (`commspace`).
`<for pl in placelist do>`	*exploration of the accessible places of the domain*
`<goto pl>`	*migration to place* `pl`
Local interaction	
`<for tc in commspace do>`	*for all the visible tuple centres of place* `pl`
`tc?op(Tuple)`	*ask tuple centre* `tc` *of place* `p` *to execute* `op` *on* `Tuple`, *if authorised by* `pl`

4   Conclusions and Related Work

The novelty of this paper is to discuss how a homogeneous, well-balanced infrastructure for agent systems could be arranged to deal with the topological structure of a dynamic and unpredictable environment, the coordination of interactions among autonomous, mobile agents, and the access control applied to the exploration of an infrastructure. These issues have been often recognised as being relevant in agent-based systems, but not yet fully addressed in their complexity and in their reciprocal dependencies by the agent researches. Hence, it has been described how the TuCSoN framework allows to support that infrastructure in a coherent and sound fashion.

Considering systems related with TuCSoN, *ActorSpace* [10] is one of the most relevant proposals in the area of agent-oriented infrastructures and provides an underlying platform for agent systems that enables to control the access to, and the management of resources. ActorSpace explicitly models the location of agents on particular hosts and the amount of computational resources that an agent is allowed to consume. Resource interaction are mediated by means of proxies, which are components of the agent's behaviour specifically customised to interact with a certain type of resource. Differently from ActorSpace, in TuCSoN agents interact in a standard way with resources and have no need to embed components (i.e. proxies) specifically tailored for the different resource types. This better management of the heterogeneity of the interaction space is one of the main benefits derived from the adoption of a tuple-based coordination model. Moreover, issues derived from the structure of the environment are not explicitly addressed in ActorSpace.

LIME [14] proposes an interesting way for coordinating mobile agents and hosts (possibly mobile, too). In LIME, each mobile agent carries a tuple space. Then, when two or more agents are co-located in a host, their tuple spaces are dynamically recomputed in such a way that, for each mobile agent, the content of a tuple space is logically merged with those of the other agents and the one of the host. The resulting global tuple space is shared among all the agents and the host. Differently from TuCSoN, LIME principally investigates problems related to inter-agent communication, while aspects concerning the need of an infrastructure seems quite ignored. Consequently, also problems concerning access control and topology are not considered.

Secure Spaces [15] is a relevant proposal in the area of coordination models. It modifies Linda in order to be able to offer security guarantees to mutually untrusting programs. Some locking mechanisms to protect data stored in the tuple set have been developed by means of cryptographic techniques. If compared with TuCSoN's goal of controlling the agent interaction with an organisational infrastructure, it addresses the issue of controlling the agent access to available tuples in a shared blackboard.

Ambit [4] is a formal model that recognises the hierarchical structure of the Internet and explicitly models the migration of active entities across protected domains. It is one of the few works that, like TuCSoN, considers together the structure of a network and security problems in an agent-based scenario. Differently from TuCSoN, it is focused on the formal representation of the model, while a real infrastructure defined in terms of effective technological solutions is not provided.

Further work will be devoted to the identification of suitable high-level design patterns for mobile agent [1] and to the integration with the Jini technology [16], whose JavaSpaces coordination model is suitable to be enhanced according to the TuCSoN's model.

References

1. Aridor, Y., Lange, D.B., Agent Design Patterns: Elements of Agent Application Design, in Proceedings of Autonomous Agents '98, ACM Press, 1998.
2. Arsenault, A. and Turner, S., Internet X.509 Public Key Infrastructure PKIX Roadmap. IETF Internet Draft, PKIX Working Group, March 1999.
3. Cabri, G., Leonardi, L., Zambonelli, F., Coordination Models for Internet Applications based on Mobile Agents, IEEE Computer, Feb. 2000.
4. Cardelli, L., Gordon, A., Mobile Ambients, Foundations of Software Science and Computational Structures, LNCS no. 1478, Springer-Verlag, 1998.
5. Chess, D., Harrison, C., Kershenbaum, A., Mobile Agents: Are They a Good Idea?, RC 19887, IBM Research Division, 1994.
6. Cremonini, M., Omicini, A., Zambonelli, F., Multi-agent systems on the Internet: Extending the scope of coordination towards security and topology. In F.J. Garijo and M.Boman, (ed.), Multi-Agent Systems Engineering - Proc. of the 9th European Workshop on Modelling Autonomous Agents in a Multi-Agent World (MAMAAW'99), LNAI no. 1647, pages 77-88, Springer-Verlag, 1999.
7. Denti, E., Natali, A., Omicini, A., On the expressive power of a language for programming coordination media. In Proceedings of the 1998 ACM Symposium on Applied Computing (SAC'98), pages 169-177, Atlanta (GA), 1998.
8. Fuggetta, G., Picco, G.P., Vigna, G., Understanding Code Mobility, IEEE Transactions on Software Engineering, May 1998.
9. Gelernter, D., Carriero, N., Coordination languages and their significance, Communications of the ACM, 35(2) 97-107, Feb. 1992.
10. Jamali, N., Thati, P., Agha, G. A., An Actor-based Architecture for Customising and Controlling Agent Ensembles. IEEE Intelligent Systems, Special Issue on Intelligent Agents, 38-44, March-April 1999.
11. Jennings, N. R., Sycara, K., and Wooldridge, M., A Roadmap of Agent Research and Development International Journal of Autonomous Agents and Multi-Agent Systems 1(1) 7-38, 1998.
12. Karnik, N. M., Tripathi, A. R., Design Issues in Mobile-Agent Programming Systems, IEEE Concurrency, 6(3), July-Sept. 1998, pp. 52-61.
13. Omicini, A., Zambonelli, F., Coordination for Internet Application Development, Journal of Autonomous Agents and Multi-Agent Systems, 2(3), Sept. 1999.
14. Picco, G.P., Murphy, A., Roman, G.C., LIME: Linda Meets Mobility. In Proc. of the 21st International Conference on Software Engineering (ICSE'99), Los Angeles, California, May 1999.
15. Vitek, J., Bryce, C., Oriol, M., Coordinating Agents with Secure Spaces, In Proceedings of Coordination '99, Amsterdam, The Nederlands, May 1999.
16. Waldo, J., The Jini Architecture for Network-centric Computing. Communications of the ACM, 42(7), July 1999.

Dynamic Reconfiguration in Coordination Languages

George A. Papadopoulos[1] and Farhad Arbab[2]

[1] Department of Computer Science, University of Cyprus, 75 Kallipoleos Str.
P.O.B. 20537, CY-1678 Nicosia, Cyprus
george@cs.ucy.ac.cy

[2] Department of Software Engineering, Centre for Mathematics and Computer Science (CWI), Kruislaan 413, 1098 SJ Amsterdam, The Netherlands
farhad@cwi.nl

Abstract. A rather recent approach in programming parallel and distributed systems is that of coordination models and languages. Coordination programming enjoys a number of advantages such as the ability to express different software architectures and abstract interaction protocols, supporting multilinguality, reusability and programming-in-the-large, etc. In this paper we show the potential of control- or event-driven coordination languages to be used as languages for expressing dynamically reconfigurable software architectures. We argue that control-driven coordination has similar goals and aims with reconfigurable environments and we illustrate how the former can achieve the functionality required by the latter.
Keywords: Coordination Languages and Models; Software Engineering for Distributed and Parallel Systems; Modelling Software Architectures; Dynamic Reconfiguration; Component-Based Systems.

1 Introduction

It has recently been recognized within the Software Engineering community, that when systems are constructed of many components, the organization or architecture of the overall system presents a new set of design problems. It is now widely accepted that an architecture comprises, mainly, two entities: *components* (which act as the primary units of computation in a system) and *connectors* (which specify interactions and communication patterns between the components).

Exploiting the full potential of massively parallel systems requires programming models that explicitly deal with the concurrency of cooperation among very large numbers of active entities that comprise a single application. Furthermore, these models should make a clear distinction between individual components and their interaction in the overall software organization. In practice, the concurrent applications of today essentially use a set of ad hoc templates to coordinate the cooperation of active components. This shows the need for proper *coordination languages* ([2,15]) or *software architecture languages* ([18]) that can be used to explicitly describe complex coordination protocols in terms of simple primitives and structuring constructs.

Traditionally, coordination models and languages have evolved around the notion of a *Shared Dataspace*; this is a common area accessible to a number of processes cooperating together towards the achievement of a certain goal, for exchanging data. The first language to introduce such a notion in the Coordination community was Linda with its Tuple Space ([1]), and many related models evolved around similar notions ([2]). We call these models *data-driven*, in the sense that the involved processes can actually examine the nature of the exchanged data and act accordingly.

However, many applications are by nature event-driven (rather than data-driven) where software components interact with each other by posting and receiving events, the presence of which triggers some activity (e.g. the invocation of a procedure). Events provide a natural mechanism for system integration and enjoy a number of advantages such as: (i) waiving the need to explicitly name components, (ii) making easier the dynamic addition of components (where the latter simply register their interest in observing some event(s)), (iii) encouraging the complete separation of computation from communication concerns by enforcing a distinction of event-based interaction properties from the implementation of computation components. Event-driven paradigms are natural candidates for designing coordination rather than programming languages; a „programming language based" approach does not scale up to systems of event-driven components, where interaction between components is complex and computation parts may be written in different programming languages.

Thus, there exists a second class of coordination models and languages, which is control-driven and state transitions are triggered by raising events and observing their presence. A prominent member of this family (and a pioneer model in the area of control-driven coordination) is Manifold ([4]), which will be the primary focus of this paper, Contrary to the case of the data-driven family where coordinators directly handle and examine data values, here processes are treated as black boxes; data handled within a process is of no concern to the environment of the process. Processes communicate with their environment by means of clearly defined interfaces, usually referred to as *input* or *output ports*. Producer-consumer relationships are formed by means of setting up *stream* or *channel* connections between output ports of producers and input ports of consumers. By nature, these connections are *point-to-point*, although *limited broadcasting* functionality is usually allowed by forming 1-n relationships between a producer and n consumers and vice versa. Certainly though, this scheme contrasts with the Shared Dataspace approach usually advocated by the coordination languages of the data-driven family. A more detailed description and comparison of these two main families of coordination models and languages can be found in [15].

It has become clear over the last few years that the above mentioned principles and characteristics are directly related to the needs of other similar abstraction models, notably *software architectures* and *configuration* languages such as Conic/Durra ([5]), Darwin/Regis ([9]), PCL ([19]), POLYLITH ([17]), Rapide ([7,10]) and TOOLBUS ([6]). The configuration paradigm also leads naturally to the separation of the component specifying initial and evolving configuration from the actual computational component. Furthermore, there is the need to support reusable (re-) configuration patterns, allow seamless integration of computational components but also substitution of them with others with additional functionality, etc.

In this paper we use the *control-* or *event-driven* coordination language Manifold ([4]) to show how it can be used for developing dynamically evolving configurations of components. The important characteristics of Manifold include compositionality, inherited from the data-flow model, anonymous communication, evolution of coordination frameworks by observing and reacting to events and complete separation of computation from communication/configuration and other concerns. These characteristics lead to clear advantages in large distributed applications.

The rest of the paper is organised as follows: The next section is a brief presentation of the language. The following one illustrates the usefulness of the framework for developing dynamic reconfiguration abstractions. The paper ends with some conclusions, comparison with related work and reference to future activities.

2 The Coordination Language Manifold

Manifold ([4]) is a coordination language which is control- (rather than data-) driven, and is a realisation of a new type of coordination models, namely the Ideal Worker Ideal Manager (IWIM) one ([3]). In Manifold there exist two different types of processes: *managers* (or *coordinators*) and *workers*. A manager is responsible for setting up and taking care of the communication needs of the group of worker processes it controls (non-exclusively). A worker on the other hand is completely unaware of who (if anyone) needs the results it computes or from where it itself receives the data to process. More information on Manifold can be found in [11] and in the paper „Modelling Control Systems in an Event-Driven Coordination Language" in these proceedings. Note that Manifold has already been implemented on top of PVM and has been successfully ported to a number of platforms including Sun, Silicon Graphics, Linux, and IBM AIX, SP1 and SP2. The language has been used successfully for coordinating parallel and distributed applications ([14,16]), modelling activities in information systems ([13]) and expressing real-time behaviour ([12]).

3 A Case Study in Dynamic Reconfiguration: The Patient Monitoring System

In this section we apply control- or event-driven coordination techniques to model a classical case in dynamic reconfiguration, namely coordinating activities in a patient monitoring system ([19]). In the process, we take the opportunity to introduce further features of Manifold. Due to lack of space, we consider a somewhat simplified version of a real patient monitoring system, but we will hopefully be able to persuade the reader that any additional functionality can be handled equally well by our model.

The basic scenario involves a number of monitors - one for every patient - recording readings of the patient's health state, and managed by a number of nurses. A nurse can concurrently manage a number of monitors; furthermore, nurses can come and go and thus an original configuration between available nurses and monitors can subsequently change dynamically. A monitor is periodically asked by its supervising nurse to send available readings for the corresponding patient, and it does so. However, a monitor can also on its own send data to the nurse if it notices an abnormal situation. A nurse is responsible for periodically checking the patient's state by asking the corresponding monitor for readings; furthermore, a nurse should also respond to receiving abnormal data readings. Finally, if a nurse wants to leave, s/he notifies a supervisor and waits to receive permission to go; the supervisor will send to the nurse such a permission to leave once all monitors managed by this nurse have been re-allocated to other available nurses. We start with the code for a monitor:

```
Manifold Monitor
{
  port output normal, abnormal.

  stream reconnect BK input    -> *.
  stream reconnect KB normal   -> *.
  stream reconnect KB abnormal -> *.

  event abnormal_readings, normal_readings.
  priority abnormal_readings > normal_readings.

  auto process check_readings is
            AtomicMonitor(abnormal_readings) atomic.

  begin: (guard(input,full, normal_readings),
          terminated(self)).

  abnormal_readings: &self -> abnormal;
                     check_readings -> abnormal;
                     post(begin).
     normal_readings: &self -> normal;
                      check_readings -> normal;
                      post(begin).
}
```

A `Monitor` Manifold comprises three ports: the default `input` port and two output ports, one for sending normal readings and another one for sending abnormal readings. The streams connected to these ports from the responsible nurses have been declared to be `BK` (break-keep) for the input port and `KB` (keep-break) for the two output ports, signifying the fact that in the case of a nurse substitution the data already within the streams to be transmitted to or from the monitor will remain in the corresponding sttream until some other nurse has re-established connection with the monitor. Two local events have been declared, `normal_readings` for the case of handling periodical data readings and `abnormal_readings` for handling the exception of detecting abnormal readings. Note that, for obvious reasons, the priority of the latter has been declared to be higher than that of the former. In the case that both events have been raised (e.g. immediately after a periodic reading an abnormal situation has been detected), the monitor will serve first `abnormal_readings` (if priority had not been specified, the language would have made a non-deterministic choice). Finally, note that `Monitor` collaborates closely with the process `check_readings`, an instance of the predefined *atomic* Manifold `AtomicMonitor`. Atomic Manifolds (and associated processes) are ones written in some other language (typically C or Fortran for the case of the Manifold system). In this case, `AtomicMonitor` can be seen as the device driver for the monitor device.

Initially, `Monitor` sets a *guard* to its input port, which will post the event `normal_readings` upon detecting a piece of data in the port. This piece of data is interpreted by `monitor` as being a periodic request by the responsible for this monitor nurse to get the data readings. It then suspends, by means of calling the predefined primitive `terminated(self)`, and waits for a notification (by means of the corresponding event being posted) by either the guard to send periodic data readings or the process `check_readings` that some abnormal situation has been detected. Upon detecting the presence of either of the two involved events, `Monitor` changes to the corresponding state, and sends to the respective output port first its own id (`&self`) followed by the actual data readings as provided by `check_readings`. It then loops back to the first (waiting) state, by posting the event `begin`. It is important to note that `Monitor` works quite independently from its environment. For instance, it has no knowledge or concern about which nurse (if anyone at all!) is receiving the data it sends. Nor is it affected by any changes in the configuration of the nurses set up. The code for a nurse is shown below:

```
Manifold Nurse
{
 port in normal, abnormal.

 stream reconnect KB normal    -> *.
 stream reconnect KB abnormal  -> *.
 stream reconnect BK output    -> *.

 event got_abnormal, got_normal,
       read_data, leave, ok_go.
 priority got_abnormal > got_normal.

 auto process wakeup is WakeUp(read_data,leave).
 auto process process_data is ProcessData.

 begin: (guard(abnormal,full,got_abnormal),
         guard(normal,full,got_normal),
         guard(abnormal,a_disconnected,ok_go),
         ternimated(self)).

 read_data: („SEND_DATA" -> output, post(begin)).

 got_abnormal: process monitor deref abnormal.
               (monitor.abnormal -> process_data,
                post(begin)).

 got_normal: process monitor deref normal.
             (monitor.normal -> process_data,
              post(begin)).

 leave: (raise(go), post(begin)).

 ok_go: .
}
```

A nurse has two input ports and one output port, a mirror image of how a monitor is defined. It collaborates with two atomic processes: `wakeup` is responsible for periodically asking the nurse to order a monitor to send data and `process_data` does the actual processing of data. Furthermore, `wakeup` also monitors how long a nurse can be on duty. After setting guards into the two input ports (the second guard for `abnormal` is explained below), `nurse` suspends waiting for either `wakeup` to ask her for periodic readings or some monitor to send her abnormal data readings. In the former case, `nurse` sends a notification to all the monitors that it controls and upon receiving back data it forwards them to `process_data`. In either of the two cases, a monitor will first send its own id which is then being dereferenced (by means of the `deref` primitive) to yield a process reference for the monitor in question. This id is then used to connect the monitor's appropriate output port to the input port of `process_data` so that the readings can be transmitted. This process is being repeated until `wakeup` lets the nurse know that it can now ask permission to leave. The nurse raises the event `go` (note here that an event can either be „posted" in which case its presence is known only to the Manifold within which it was posted, or „raised" in which case its presence is known only outside the Manifold within which it was raised), and waits for a notification that it is allowed to leave. Note also that until such a notification has been provided, the nurse is still responsible for its monitors. The requested notification will be provided implicitly by noting that one of the nurse's input ports has now no connections to a monitor. This will be detected because of the presence of the second guard in the `abnormal` input port with the directive `a_disconnected`. The nurse can then leave the system.

A number of abstractions have been introduced in the modelling of the nurse, which are of importance to a dynamic configuration paradigm. A nurse is unaware of the number of monitors it supervises. Thus, monitors can be added or deleted from its list of responsibility without affecting the pattern of the nurse's behaviour. Also, the decision on whether a nurse should leave (which can be as simple as noting when a specified time interval has passed or as complicated as taking into consideration additional parameters such as specialization of work, priorities in types of duty, redistribution of workload, etc.), is encapsulated into different components which, as in the case of the number of monitors, they do not affect the basic pattern of behaviour for the nurse. Furthermore, the actual processing of data, which can vary depending on the type of monitor or the patient's case, is also abstracted away. All in all, the *policies* that define the nurses' behaviour have been separated from the actual work to be done, and they can therefore be changed easily, dynamically, and without affecting the interaction patterns between the involved components. The code for the supervisor is as follows:

```
Manifold Supervisor (process setup)
{
  event get_modify.

  begin: (guard(input,full,get_modify),
          terminated(self)).

  go.*nurse: (&nurse -> setup, post(begin)).

  get_modify:process new_nurse deref tuplepick(input,1).
             process mon1 deref tuplepick(input,2).
             process mon2 deref tuplepick(input,3).
             process mon3 deref tuplepick(input,4).
             new_nurse -> (-> mon1, -> mon2, -> mon3),
```

```
                mon1.abnormal -> new_nurse.abnormal,
                mon1.normal -> new_nurse.normal
                mon2.abnormal -> new_nurse.abnormal,
                mon2.normal -> new_nurse.normal
                mon3.abnormal -> new_nurse.abnormal,
                mon3.normal -> new_nurse.normal,
                activate(new_nurse), post(begin).
}
```

The Supervisor Manifold is responsible for monitoring a number of nurses. It collaborates closely with the atomic process setup, which maintains and enforces the policy of the environment with respect to issues such as how many nurses should be active concurrently, how many monitors each nurse should be responsible for, how can the workload of monitor responsibility be distributed evenly to the available nurses, the minimum and maximum amounts of time each nurse should be working before asking to be relieved from duty, etc. Upon receiving a request by some nurse to be allowed to leave, Supervisor passes the id of that nurse to setup which, among other things, keeps record of which monitors each nurse is responsible for. A new nurse to take over is found and its id along with the ids of the monitors to handle is passed back to Supervisor; the latter then sets up the stream connections between the new nurse and the monitors and activates the new nurse (for the sake of brevity here we have assumed a simple scenario where the old nurse is responsible for three monitors, all of which are passed to the new nurse; this of course does not have to be the case). The transferring of the streams connecting the monitors to the new nurse causes the disconnection of the input ports of the old nurse from the whole apparatus. We recall that the nurse has set up a guard process at its input ports which will get activated when it detects such a disconnection; upon the disconnection of its input port abnormal from all monitors involved, the old nurse leaves the system and its associated process terminates execution gracefully.

An initial setup with three monitors and two nurses is shown below within the Main Manifold which is the first one to commence execution in a Manifold program:

```
Manifold Main()
{
  event go, ok_go.

  auto process n1 is Nurse.
  auto process n2 is Nurse.
  auto process m1 is Monitor.
  auto process m2 is Monitor.
  auto process m3 is Monitor.
  auto process setup is Setup (n1,n2,m1,m2,m3) atomic.
  auto process supervisor is Supervisor(setup).

  begin: (n1-> m1, n2 -> ( -> m2, -> m3),
          m1.normal->n1.normal,m1.abnormal->n1.abnormal,
          m2.normal->n2.normal,m2.abnormal->n2.abnormal,
          m3.normal->n2.normal,m3.abnormal->n2.abnormal).
}
```

In the above piece of code, the instances of the Manifolds forming the initial configuration are first created and activated. In particular, the initial configuration comprises two monitors and three nurses, where the first nurse is responsible for the first monitor and the second one for the rest. A setup process (an instance of Setup) is also activated and given the previously mentioned five processes as parameters. Finally, a supervisor process is also created and given setup as parameter. The begin state of Main sets up the configuration by creating stream connections between the appropriate input-output ports. After that, the apparatus is left to its own devices, evolving dynamically as was described previously during the presentation of the code for Monitor, Nurse and Supervisor.

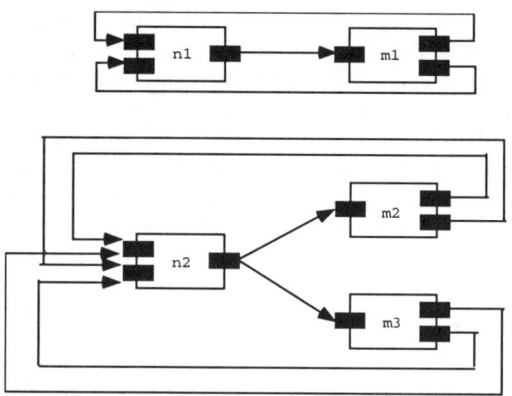

Fig. 1.

We should probably clarify at this point that the actual evolution of the configuration (i.e. the creation of new instances of Nurse to substitute other instances) is performed within the Setup Manifold, which effectively administers the whole environment, keeps track of changes, etc. We would expect that Manifolds with such a complicated behaviour are atomic processes written, say, in C for convenience and ease of expressiveness. The initial configuration is shown diagramatically below:

4 Conclusions - Related and Further Work

In [8], while describing another configuration mechanism based on I/O abstractions, a number of desirable properties that configuration models should possess are listed. These properties are active and reactive communication, connection-oriented and user-specifiable configuration and support for a variety of communication schemes such as implicit, direct, multi-way, and use of continuous streams. It is worth mentioning here that Manifold supports all these schemes as first class citizens. In addition, Manifold supports complete separation of computation from coordination concerns and a control-driven specification of system transformations, which unlike I/O abstractions, is, in our opinion, more appropriate for configuration programming.

Note that in Manifold, unlike in many other coordination models and languages, a component is oblivious not only to bindings produced by other components but also to whether or not communication is taking place at all or what type of communication this is. This frees the programmer from having to establish when it is the best moment to send and/or receive messages. And of course, the language enjoys the ability for dynamic system reconfiguration without the need to disrupt services or the components having mutual knowledge of structure or location — point-to-point or multicast communications can be configured independently of the computation activity and mapped appropriately onto the underlying architecture.

Furthermore, the stream (or channel) connections that Manifold supports as the basic mechanism for communication between computation components, provide a natural abstraction for supporting continuous datatypes such as audio or video and make this coordination model and its associated language ideal for configuring the activities in, say, distributed multimedia environments. We are currently exploiting this characteristic of Manifold in a recently commenced research project where the language will be used to manage and coordinate, among other activities, the data produced or consumed by media servers.

Finally, Manifold advocates a liberal view of dynamic reconfiguration and system consistency. Consistency in Manifold involves the integrity of the topology of the communication links among the processes in an application, and is independent of the states of the processes themselves. Other languages, such as Conic, limit the dynamic reconfiguration capability of the system by allowing evolution to take place only when the processes involved have reached some sort of a safe state (e.g. quiescence). Manifold does not impose such constraints; rather, by means of a plethora of suitable primitives, it provides programmers the tools to establish their own safety criteria to avoid reaching logically inconsistent states. Furthermore, primitives such as guards, installed on the input and/or output ports of processes, inherently encourage programmers to express their criteria in terms of the externally observable (i.e., input/output) behavior of (computation as well as coordination) processes.

Acknowledgments

This work has been partially supported by the INCO-DC KIT (Keep-in-Touch) program 962144 „Developing Software Engineering Environments for Distributed Information Systems" financed by the Commission of the European Union.

References

1. S. Ahuja, N. Carriero and D. Gelernter, „Linda and Friends", *IEEE Computer* **19 (8)**, 1986, pp. 26-34.
2. J. - M. Andreoli, C. Hankin and D. Le Métayer, *Coordination Programming: Mechanisms, Models and Semantics*, World Scientific, 1996.
3. F. Arbab, „The IWIM Model for Coordination of Concurrent Activities", *Coordination'96*, Cesena, Italy, 15-17 April, 1996, LNCS 1061, Springer Verlag, pp. 34-56.
4. F. Arbab, I. Herman and P. Spilling, „An Overview of Manifold and its Implementation", *Concurrency: Practice and Experience* **5 (1)**, 1993, pp. 23-70.
5. M. R. Barbacci, C. B. Weinstock, D. L. Doubleday, M. J. Gardner and R. W. Lichota, „Durra: A Structure Description Language for Developing Distributed Applications", *Software Engineering Journal*, IEE, March 1996, pp. 83-94.

6. J. A. Bergstra and P. Klint, „The TOOLBUS Coordination Architecture", *Coordination'96*, Cesena, Italy, 15-17 April, 1996, LNCS 1061, Springer Verlag, pp. 75-88.
7. C. Chen and J. M. Purtilo, „Configuration-Level Programming of Distributed Applications Using Implicit Invocation", *IEEE TENCON'94*, Singapore, 22-26 Aug., 1994, IEEE Press, pp. 43-49.
8. K. J. Goldman, B. Swaminathan, T. P. McCartney, M. D. Anderson and R. Sethuraman, „The Programmer's Playground: I/O Abstractions for User-Configurable Distributed Applications", *IEEE Transactions on Software Engineering* **21** (9), 1995, pp. 735-746.
9. J. Kramer, J. Magee and A. Finkelstein, „A Constructive Approach to the Design of Distributed Systems", *Tenth International Conference on Distributed Computing Systems (ICDCS'90)*, Paris, France, 26 May - 1 June, 1990, IEEE Press, pp. 580-587.
10. D. C. Luckham, „Specification and Analysis of System Architecture Using Rapide", *IEEE Transactions on Software Engineering* **21** (4), 1995, pp. 336-355.
11. Manifold home page, URL: http://www.cwi.nl/~farhad/Manifold.html.
12. G. A. Papadopoulos and F. Arbab, „Coordination of Systems With Real-Time Properties in Manifold", *Twentieth Annual International Computer Software and Applications Conference (COMPSAC'96)*, Seoul, Korea, 19-23 Aug., 1996, IEEE Press, pp. 50-55.
13. G. A. Papadopoulos and F. Arbab, „Control-Based Coordination of Human and Other Activities in Cooperative Information Systems", *Coordination'97*, 1-3 Sept., 1997, Berlin, Germany, LNCS 1282, Springer Verlag, pp. 422-425.
14. G. A. Papadopoulos and F. Arbab, „Coordination of Distributed Activities in the IWIM Model", *International Journal of High Speed Computing*, World Scientific, 1997, Vol. 9 (2), pp. 127-160.
15. G. A. Papadopoulos and F. Arbab, „Coordination Models and Languages", *Advances in Computers*, Marvin V. Zelkowitz (ed.), Academic Press, Vol. 46, August, 1998, 329-400.
16. G. A. Papadopoulos, „Distributed and Parallel Systems Engineering in Manifold", *Parallel Computing*, Elsevier Science, special issue on Coordination, 1998, Vol. 24 (7), pp. 1107-1135.
17. J. M. Purtilo, „The POLYLITH Software Bus", *ACM Transactions on Programming Languages and Systems* **16** (1), 1994, pp. 151-174.
18. M. Shaw, R. DeLine, D. V. Klein, T. L. Ross, D. M. Young and G. Zelesnik, „Abstractions for Software Architecture and Tools to Support Them", *IEEE Transactions on Software Engineering* **21** (4), 1995, pp. 314-335.
19. I. Sommerville and G. Dean, „PCL: A Language for Modelling Evolving System Architectures", *Software Engineering Journal*, IEE, March 1996, pp. 111-121.

Developing a Distributed Scalable Enterprise JavaBean Server

Yike Guo and Patrick Wendel

Department of Computing, Imperial College
180 Queen's Gate, London SW7 2BZ, UK
{y.guo,p.wendel}@doc.ic.ac.uk

Abstract. We present here approaches of a distributed scalable Enterprise JavaBean (EJB) server. The first approach is based on the idea of the resource broker. By this model, the system is composed of one or several entry points called EJB "Front" servers, a resource broker and a set of participating servers called EJB "Back" servers. The resource broker gives the system its dynamic load balancing capacity and its dynamic scalability by implementing a notification protocol with the participants and providing information to the entry point servers. An experimental version of the server has been developed and we give some performance results achieved on it. Another approach is based on using a Jini/JavaSpace as a coordination system. We present the design and the implementation of this system as well as the issues of developing this Jini-based approach towards the distributed EJB server. We finally propose a new approach based on full Jini functionality.

1 Introduction

Component servers are now commonly used in multi-tier applications. The need for these servers to become more scalable, reliable and reach better performance is growing. One possible way to improve scalability and reliability consists in using a distributed architecture for the component server. Several component servers can be gathered into one distributed and scalable component server. In that way, we can increase the fault-tolerance of the system. Concerning performance, we can implement a load balancing functionality on top of this distributed server assuming components can be instantiated or found on several nodes of the system. We present possible architectures for this distributed component server, based on the Enterprise JavaBeans component model.

M. Bubak et al. (Eds.): HPCN 2000, LNCS 1823, pp. 207-216, 2000.
© Springer-Verlag Berlin Heidelberg 2000

2 Background

As we chose Sun's Enterprise JavaBeans for our distributed component model, we explain now briefly how it works as well as some basic theoretical definitions of load balancing. Finally, we describe Sun's Jini architecture for distributed computing and spontaneous networking.

2.1 Enterprise JavaBeans

Enterprise JavaBeans (EJB) technology defines a model for the development and deployment of reusable Java server components [1]. *Components* are pre-developed pieces of application code that can be assembled into working application systems. Java technology currently has a component model called JavaBeans, which supports reusable development components. JavaBeans is just a visual component model. Components are linked by events. For instance, if a component raises an event then it launches execution of one method of another component. These links can be designed graphically with application builders. The EJB architecture logically extends the JavaBeans component model to support *server components*.

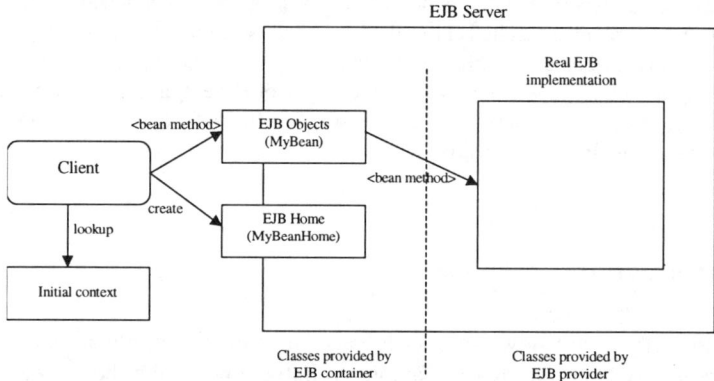

Fig. 1. Simplified diagram of the Enterprise JavaBeans Architecture. The bean implementation is never accessed directly but via the container classes.

Server components are application components that run in an application server. EJB technology, a robust Java technology environment, can support the rigorous demands of large-scale, distributed, mission-critical application systems. It supports application development based on a multi-tier, distributed object architecture in which most of an application's logic is moved from the client to the server. The application logic is partitioned into one or more business objects that are deployed in an application server.

This model also defines a set of clearly separated roles that applies for the server management, so that an organisation using it can plan and attribute a specific task for each specialist involved. These roles are: *Application Assemble, EJB component builder, Deployer, EJB container provider, EJB server provider,*

The EJB specification also defines an implementation architecture and a set of interfaces that an EJB server must implement (Figure 1).

2.2 Load Balancing

Definition

Load balancing is the result of the following observation: *If you analyse the workload of a set of computers, you often reach the situation where some computers are idle and some others are overloaded.* The idea of load balancing is to execute tasks that belong to over-loaded computers on less loaded ones. Theoretically speaking, there are two kinds of load balancing, *static* and *dynamic*. Static load balancing is defining how we will distribute tasks in order to solve a given problem. This is based on a priori knowledge of the problem and the resources available. Dynamic load balancing is the task of balancing the load whenever the system needs it by finding available resources at run time.

In a distributed object computing environment, we want to create a 'cluster' or a set of computers that will manage load balancing, in order to improve global responsiveness with respect to object activation. If one resource is busy then the new tasks it is asked to execute will be transferred to another resource for execution. This mechanism has to remain transparent from the user.

Resource Policy Issue

A number of protocols have been developed for dynamic load balancing which can be used for this implementation. A typical model is to define a central server, acting as a *resource broker*, that will keep information about the computing entities and define some balancing criteria and an evaluation function over these criteria in order to determine which computer seems to be the most suitable to carry out the client's request. In this protocol, the best computer(s) can be determined and the requests are distributed over the cluster of computers as fairly as possible. This model works well if one assumes that each request takes on average the same time to be processed. We can also wait for a participant to tell the central server if it is able or not to process a request. In that case the central server when asked for a computer will wait for a notification from a participant telling it that it can now process the request. This is more a job scheduling vision of the balancing. When applying such model, a simple scheduling policy may not be directly applicable to component scheduling. For example, in the context of EJB scheduling, each independent EJB server is multithreaded. Thus, our participant server would accept a request as long as it has remaining threads. Hence a participant can have 10 client requests to process at the same time while another one will not have any. So a simply resource policy will result in an unbalanced system due to the servers multithreading.

2.3 Jini – JavaSpaces

Jini is a Java architecture defining a small, simple set of conventions to allow services and clients to interact to form a flexible and dynamically changing network computing environment. Jini enhanced Java distributed object computing model (Remote Method Invocation, RMI) with two key components: *'Discovery/Join Protocol'* and *'Lookup Service'*.

- *Discovery and Join Protocol* is activated when a new service (or device) is to join a Jini network computing system (Jini Federation) . A D*iscovery packet* is sent to the whole net, asking for any lookup server which acts as a place where services advertise themselves and clients go to find a service. This packet is received and processed by the *Lookup* mechanism. Upon receipt of such a request packet, a lookup service will exchange proxies to register the service and advertise it to the whole federation.
- *Lookup server* stores information about all services available. For instance, when a printer joins the network then it registers with the *Lookup* that keeps the printer driver or interface. If a client wants to use this printer, it will download information (proxy) from the *Lookup*. Hence drivers do not need to be loaded in advance on the client.
- Jini provides other functionality dedicated to distributed programming. These functions are *Leasing,* which is developed specially for introducing time to resource allocation and reclamation, *Distributed events* and *Distributed transactions*.

JavaSpaces is another service provided on top of all the Jini services. A JavaSpace can realise a simple Jini system with a simple pattern-based information retrieval mechanism [2]. That is, a space can be viewed as an associate memory where information about services can be written into the space and retrieved (taken) based on its contents.

3 Design

We present the resource broker-based implementation in detail as well as the current JavaSpace implementation. We also discuss the need for a full Jini-based implementation and how it could be achieved.

3.1 Resource Broker Approach

Architecture

Our architecture is based on the resource broker approach [3]: A distributed EJB server consists of a set of ordinary EJB servers (the *participants*), a *resource broker* that gathers information about every EJB server and one or several EJB *front server(s)*, which are the entry points to the whole system. In order to set up this "front" server, we define a new server that manages connection with the client and, security issues like the basic server. Fundamentally, this server will have the same signature as the EJB server but it will behave differently and will not manage any beans. This server also contains implementations of home interfaces of all the available type of beans in the cluster of EJB servers. Hence, a client will connect to this server, asking for a particular home interface. Then, it can invoke the *create* method of the home interface. The implementation of home interface will ask the *resource broker* for the best computer that can perform a bean creation based on a scheduling policy. Then it will invoke directly the bean creation method on the best computer and return a remote reference on the new bean to the client.

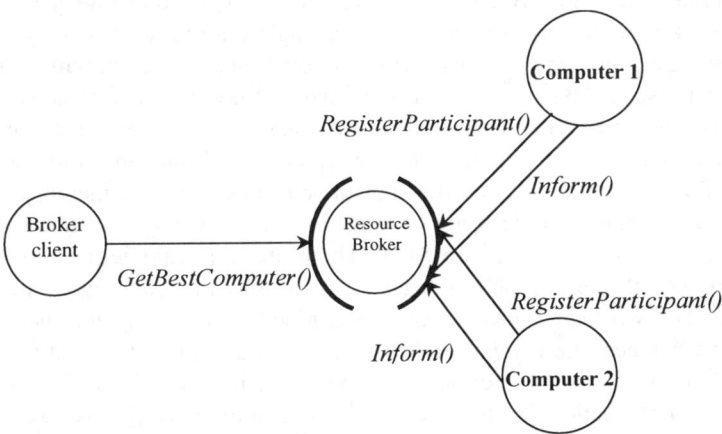

Fig. 2. Resource Broker architecture and integration with the system.

The Figure 3 represents bean creation, but it does not explain how 'entity beans' retrieval will be managed. The *resource broker* will perform this task. We will just invoke a method to find a particular bean type with a particular primary key and the *resource broker* will find it either by looking into its list of active beans or by asking the best computer to activate it.

Load indices

First we have to choose which indices we will use to decide which computer is the least loaded one or the best one for a particular task.

The Java environment provides some indices: *total amount of memory, free memory*.

The EJB server implementation also defines and provides some useful indices: *The number of active beans in a container, the number of threads currently used by bean method invocations, the maximum number of threads for bean methods invocations*.

No extra effort is required to obtain these indices, as they are already available. But one needs more information related to computing resources such as *CPU utilisation, idle time, swap space available, network bandwidth, throughput required, security constraints, priority requirement*.

Scheduling Policy

The *resource broker* will be responsible for finding the best computer. Therefore it must define a function over the load indices described, and find the computer that minimises this function. First we can use a single criteria function only depending on any index. But it is quite obvious that a multiple criteria function will form a policy to give better results. This is because a single index is not sufficient to determine whether a computer is the best one from a list of available machines. For instance, a computer may have plenty of free memory but many clients at the same time, and so it will answer slowly even if it is considered as the best one using the memory criteria. Conversely, a computer that does not have enough free memory will not answer quickly even though the other indices are good. Therefore, we could define a multiple criteria function over all the available indices. Also we may like some indices to be preponderant and only if these indices cannot determine the best computer, then we use other ones. For instance, the number of free threads is quite important and a computer that has no free threads should not be evaluated as the best one. So we will be using several chained evaluation functions to form a scheduling policy. Also we will take into account the security constraints and user's priority requirement et.al.

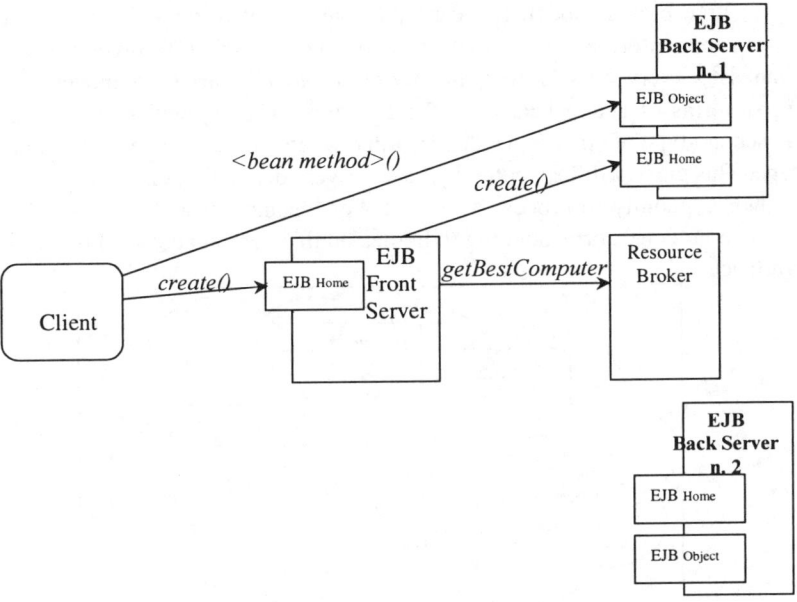

Fig. 3. Distributed EJB server architecture based on Resource Broker.

3.2 Jini/JavaSpace Approach

We now use the Jini/JavaSpace service to build the coordination system of the distributed server. The new architecture is depicted in the diagram of Fig. 4.

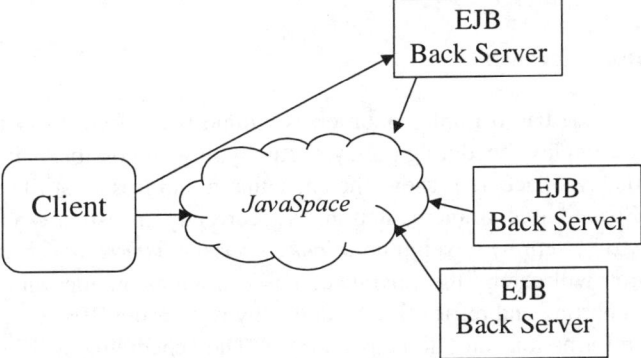

Fig. 4. Architecture using the Jini/JavaSpace service

The main design issue is therefore to define how we will use the JavaSpace and which kind of entries we are going to read and write on it. Here again different policies and protocols can be used. One approach is to choose the protocol we described for the resource broker model. It means that, instead of using a special *broker*, the JavaSpace keeps information about all the active beans. This information is kept up to date thanks to every back server that *writes* or *takes* entries whenever their state changes in terms of active beans. So the first type of entry will represent an active bean. A second type of entry is needed in order to retrieve a bean home for a particular bean type. This entry will be written by back servers depending on homes they contain and on their capability to process a request. For instance, if a Back Server is full of requests then no entry corresponding to homes on this computer should be available on the JavaSpace.

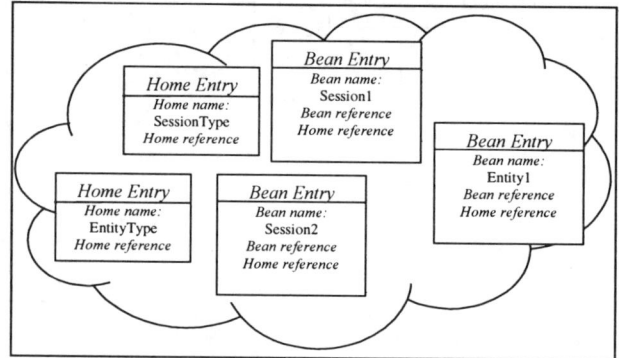

Fig. 5. Inside the JavaSpace

The client, instead of connecting to a Front Server, will connect to the JavaSpace and retrieve the home reference it is looking for. Here is a main difference with the broker approach. It means that the computer is chosen when the home is retrieved and not when a method of this home is invoked. Then the client uses this home interface as usual to create/find/remove a bean.

3.3 Full Jini Approach

The Jini/JavaSpace based implementation is simple but suffers from the problem of incorporating complex scheduling policy. Actually, it is impossible using the matching mechanism of JavaSpace, to retrieve the entry that minimises an attribute. Thus, based on the JavaSpace-based implementation, we currently investigate a full Jini-based distributed EJB system by developing a *lookup service with resource awareness*. This implementation will extend the JavaSpace based implementation with more resource information reporting and retrieval mechanism by registering EJB server together with its resource information into lookup services. The scheduling will be done when a component is required via a lookup request . Complex scheduling functions can be integrated with the lookup service.

4 Results

We based our work on an EJB Server already implemented [4] and extended it to a resource-broker based distributed EJB Server. Some simple performance tests have been performed on this implementation.

The first test measures the overhead due to the new EJB Server architecture. It compares times for creating a bean and calling an empty method on the original EJB Server and on the distributed EJB Server. As the participating servers remain pure EJB Servers as well, we also compare the invocation time on a participant directly, without using a Front Server and the resource broker.

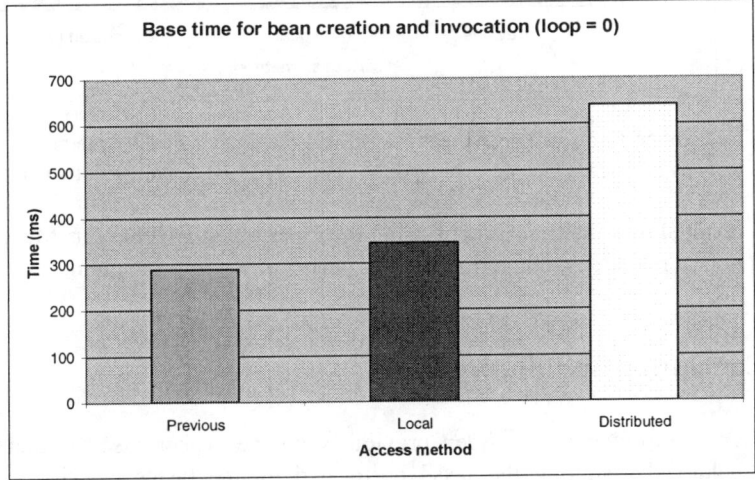

Fig. 6. Bean creation and invocation of an empty method. Times on the base EJB Server *(Previous)*, when calling a participating server directly *(Local)* and using the distributed EJB Server *(Distributed)*. The *loop* parameter is an arbitrary complexity argument of the bean method called by the client.

As expected, we can see the overhead due to additional communications between the components of the system. But the advantage of the distributed version comes up when the client performs multiple bean creations and invocations at the same time (the client simulates a multi-client environment).

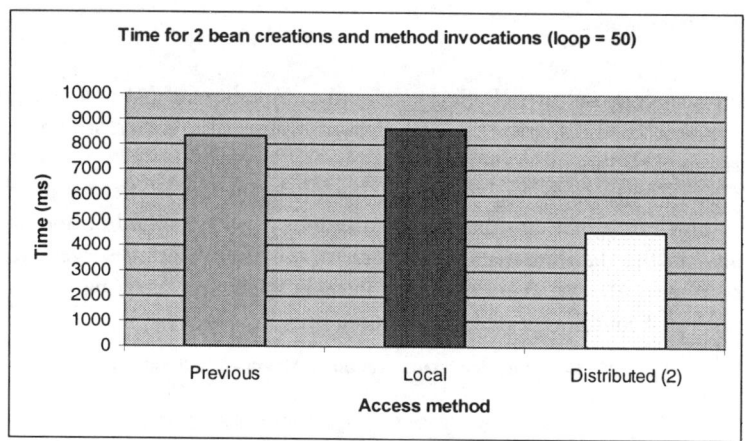

Fig. 7. The client performs two bean creations at the same time and calls a method on each bean. *Distributed(2)* means that we deployed the distributed EJB Server with two participants.

This result shows the best case of using the Distributed EJB Server. Most probably we will not, in real cases, reach this level of performance improvement.

5 Conclusion and Future Works

Performance improvement, as seen previously on the broker-based implementation, can be achieved assuming the tasks performed on the beans are time-consuming enough. Above a certain complexity threshold, the distributed architecture increases a lot the server responsiveness. On top of performance, the distributed server is also more reliable. If a participant crashes at some point, the whole system keeps working and beans can still be created and activated on the remaining participants. Conversely, we can add a new participant while the system is running.

Hence we have proven the potentiality of the distributed server. We also proved the feasibility of using the Jini architecture via the JavaSpace, to replace our broker. We now have to implement the Jini-based version and take full advantage of this environment to enhance both scalability and performance of the server.

References

1. Matena, V., Hapner, M.: Enterprise JavaBeans, ver. 1.0, Sun Microsystems (1998)
2. JavaSpaces Specification, Sun Microsystems (1998)
3. Wendel, P.: Load Balancing Mechanism for an Enterprise JavaBeans Server (1999)
4. Tsirikos, D.: Design and Implementation of an Enterprise JavaBeans Server (1998)

CFMS – A Collaborative File Management System on WWW

Ruey-Kai Sheu[1], Ming-Chun Cheng, Yue-Shan Chang[1], Shyan-Ming Yuan,
Jensen Tsai, Yao-Jin Hung, and Ming-Chih Lai

Department of Computer and Information Science, National Chiao-Tung University,
Hsin-Chu, Taiwan, R.O.C.
{rksheu,native,ysc,smyuan}@cis.nctu.edu.tw
and
NOVAS Software Inc. Hsin-Chu, Taiwan, R.O.C.
{jensen,yjh,Kelvin}@springsoft.com.tw

Abstract. The increasing complexity and geographical separation of design data, tools and teams have demanded for a collaborative and distributed management environment. In this paper we present a practical system, the CFMS, which is designed to manage collaborative files on WWW. Wherever remote developers are, they can navigate the evolutions and relationships between target files, check in and check out them conveniently on web browsers using CFMS. The capabilities of CFMS include the two-level navigation mechanism for version selection, the relationship model that conceptually manages the relationships between physical files as well as the prefix-based naming scheme that can uniquely identify requested objects.

1 Introduction

The increasing complexity and geographical separation of design data, tools and teams have demanded for a collaborative and distributed management environment That is, companies need good management tools to leverage the skills of manipulating the most talented and experienced resources, wherever in the company or the somewhere they are located.

Major challenges for collaborative file management include version control, configuration management, concurrency access control, environment heterogeneity, and the dispersion of developers [1]. Take the Electronic Design Automation (EDA) paradigm for example, design teams realize that the number of tools needed to implement complex systems is ever increasing, and such tools are generally provided by many different suppliers [2]. At the same time, developers at different sites would parallelize the developments with many variants of each component, which might contain and share many files and many versions of each file. When controlling versions, develop-

---

[1] They are also lectures of the Department of EE of Ming Hsing Institute of Technology.

ers have to deal with versions and configurations that are organized by files and directories. This is inconvenient and error-prone, since there is a gap between dealing with source codes and managing configurations. It is necessary for engineers to be capable of identifying the items which they are developing or maintaining.

In this paper we concentrate on issues of developing a web-based and visualized collaborative file management system. The web-based design has advantages of eliminating the need to install and administrate client softwares, resolving the cross-platform problems, providing the accessibility for distributed developers, and having a uniform and compatible interface for different users. Not only does the proposed CFMS have all the advantages of web-based environments, but also captures and manages change requests among team members transparently. Through the proposed two-level navigation version selection mechanism, CFMS users can define and choose the configuration of the developed component dynamically and visually. Without directly operating the bothersome chaos of physical files, users only need to manipulate the conceptual relationships between requested files on web.

The rest of the paper is organized as follows. In the next section, we discuss the related works about collaboration management tools. In section 3, CFMS system models are detailed. Section 4 briefly introduces the CFMS system architecture. Section 5 shows the implementation of CFMS. Finally, the conclusion is given in section 6.

2 Related Work

There are several collaboration management tools in many paradigms. To illustrate, RCS [3], SCCS [4], DSEE [5], and ClearCase [6] concern the activities for cooperative system development. RCS and SCCS are the most widely known version management system. These systems are built on the notion of *vault* [7] from where users must extract all of the sources, even you do not plan to change them. Besides, parallel development will result in even more copies. These copies are not under the control of configuration systems and may lead to more problems because they do not support concurrency control facilities. There are several later tools, such as CVS [8], built on top of RCS. These tools improve the management of private work areas, but do not really solve the fundamental problem inherent in vaults.

DSEE and ClearCase are tools of the second-generation configuration management environment, which advance upon earlier tools for defining and building configurations. They solve the problems inherent in vaults through the use of a *virtual file system*. Intercepting native operating system I/O calls, they hook into the regular file system at user-defined points by user-specified rules. Intercepting system calls will tight the private work areas with the central data center together and introduce problems encountered in traditional systems. Additionally, the rule-based configuration mechanism is too complex to be used for web users.

To solve problems of old systems we design the CFMS as a moderate three-tier architecture. A middle-tier is introduced to decouple the front-end users and the back-end file system, and simplify the system complexity. In CFMS, a conceptual relation model is proposed. Instead of the text-based or rule-based methodologies used in traditional tools, front-end users could traverse the relationship graph to define or navigate configurations The web-based CFMS will promote the collaboration management paradigm into the distributed heterogeneous environment.

3 System Model

Three essential system models which compose of the three-tier model, the relationship model and the prefix-based naming scheme for collaborative file management are clearly identified in this section. These three models are orthogonal and are integrated to form the building blocks of the CFMS.

3.1 Three-Tier Model

In comparison with the vault-based and virtual file design of traditional collaboration management tools, the three-tier model looses the tight-couple relation between the client side representations of requested objects and the back-end file systems. In the front-end, users of different sites could define configurations for their private workspaces and use SCCS-like check in/out operations to access the requested files. In the middle tier, the relationships between all the managed files are constructed and presented by a bi-direction graph. Front-end users can surf back-end files by traveling the relationship graph. The third tier is the vast file system, which consists of many types of files generated by different tools. Each physical file could be uniquely identified and conceptually versioned in the second tier. That is, the second tier hides the complexity of directly manipulating the physical files and provides higher flexibility and extensibility for CFMS. Figure 3.1.1 shows the three-tier architecture of the proposed web-based collaborative file management system. Front-end browsers show the user-selected configuration graphs, which might share nodes in the central relationship graph. The middle tier hides the complexity to manage the relations for the third-tier flat files.

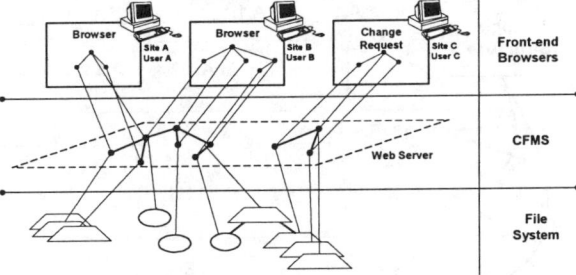

Fig. 3.1.1. Three-tier architecture of CFMS

3.2 Relationship Model

CFMS uses logic structures to manage files, which are organized into a three-level hierarchical architecture and they are projects, libraries and files from the highest level down to the lowest one. A project is the achievement of a specific objective. It involves several libraries as well as stand-alone files. A library is a frequently used group of files. The relationship model identifies the correlations between managed logic structures and classifies them into six categories: version, build, configuration, equivalence, hyper-link and sibling. The main contribution of this paper is to identify and clarify these concepts into orthogonal, rather than sequential relationships.

- **Version:** version relationships are composed of branches and derived items, and can be represented by two-dimensional graphs. In Figure 3.2-1, the generic version graph is described. Object A.1 is the root of the version history for the logic structure A. Object A.1.2 is a new version of A.1 after the creation time of A.1.1. The relationship between A.1 and A.1.2 is the branch relation. Other edges in the version history are derived relations. The logic structure A could be a file, a library or a project.

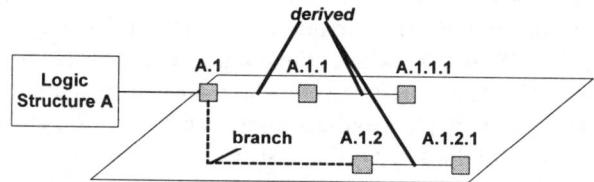

Fig. 3.2.1. The generic version graph of a logic structure

- **Build:** the build relation stands for the aggregation of a specific logic structure. Figure 3.2.2 shows the version graph as well as the build relation of the logic structure A. The logic structure A is composed of two logic structures, X and Y. Both X and Y could be any logic structures that are lower or equal to the level of A in the hierarchy of logic structures.

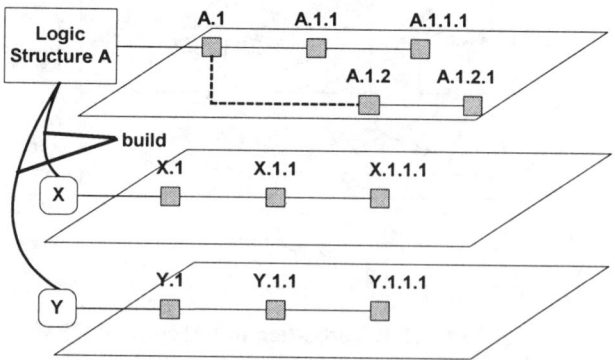

Fig. 3.2.2. The build relation of the logic structure A

- **Equivalence:** equivalence relationships are different representations of a logic structure in different development stages or platforms. For example, a C++ source program could be compiled as a PC-based or workstation-based object files. These files are of the same meaning logically, but of different representations in the real world. In Figure 3.2.3, we assume that X.1.Y.1 is created from X.1 and @.Y.1 is created from X.1.1.1 by tools which take X-versioned objects as inputs and generate outputs objects of the same type as Y. Here, edge (X.1, X.1.Y.1) and edge (X.1.1.1, @.Y.1) are two equivalence relations. Where the '@' notation is used to represent the prefix discussed in next section.
- **Sibling:** siblings are relationships between parallel-developed versions of a logic structure. In other words, siblings are parallel version history paths of a logic structure. Object Y.1, X.1.Y.1 and @.Y.1 are siblings of Y in figure 3.2.3. When someone creates an new item of Y type from X.1.1.1, @.Y.1 (@=X.1.1.1) is the most adequate identity of version name than others. Neither Y.1.1.1 nor X.1.Y.1 is the parent of @.Y.1 for the derived relation. If we connect the edge (X.1.Y.1, @.Y.1) or (Y.1.1.1, @.Y.1) with relations rather than sibling, users will be confused with them.

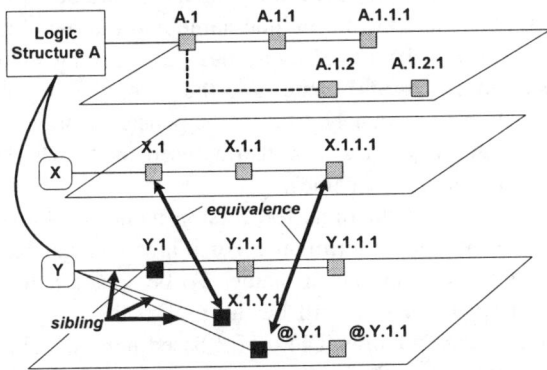

Fig. 3.2.3. The equivalence and sibling relation of the logic structure A

- **Configuration:** configuration relationships consist of snapshots of higher-level logic structures. Figure 3.2.4 illustrates the snapshots of the logic structure A. Object A.1 consists of X.1 and Y.1. Object A.1.1 is composed of X.1.1 and Y.1. Object A.1.2.1 involves X.1.1.1 and X.1.Y.1.
- **Hyper-link:** the hyper-link relation tights the manipulated files as well as the documentations that describe the related information of them. In figure 3.2.4 we shows the examples of hyper-link relations that record the hyper links to web pages describing related information, including why they are improved, when the object is updated, who did the changes, where did the author locate, how it is operated, etc.

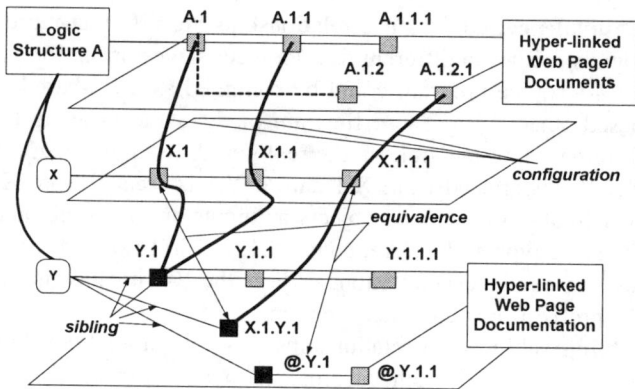

Fig. 3.2.4. The configuration and hyper-link relation of the logic structure A

3.3 Prefix-Based Naming Scheme

Engineers must be able to identify the items which they are developing or maintaining. In particular, each item must have a unique name and the name must be meaningful. In CFMS, the proposed prefix-based name has the meaning of the evolution of the item, the logic structure type and name, and the corresponding development stages. The prefix-based name is primarily used to help user to traverse and navigate the multi-dimensional relation graph. It can also be used for text-based query for version selection based on logic structure names.

The followings are the productions for version name resolution, where <> stands for non-terminals and others are terminals. Version names can be abstracted as a prefix concatenating with the relationship name. To be specific, the version name of a new item for the derived relation will be the prefix, the original old version name, concatenating with a derived name. The prefix-based naming scheme gives each relationship a unique name, respectively. Because the prefix-based naming scheme gives a unique identity for each orthogonal relation, the version name for each object will guarantee uniqueness even versioned objects have the same prefix.

<version name>	→ <prefix>.<relation>
<relation>	→ <equivalence> | <derived> | <branch>
<branch>	→ .(number of branches + 1)
<derived>	→ .1 | <derived>
<equivalence>	→ <prefix>.<base>.(number of siblings + 1)
<prefix>	→ <base>.<derived> | <base>.<branch> | <version name>
<base>	→ <logic name> | <logic name>.1
<logic name>	→ P_<name> | L_<name> | F_<name> | F_<name>_<stage>
<stage>	→ (umber of development steps + number of platforms)
<name>	→ string of character set excluding the dot

4 System Architecture

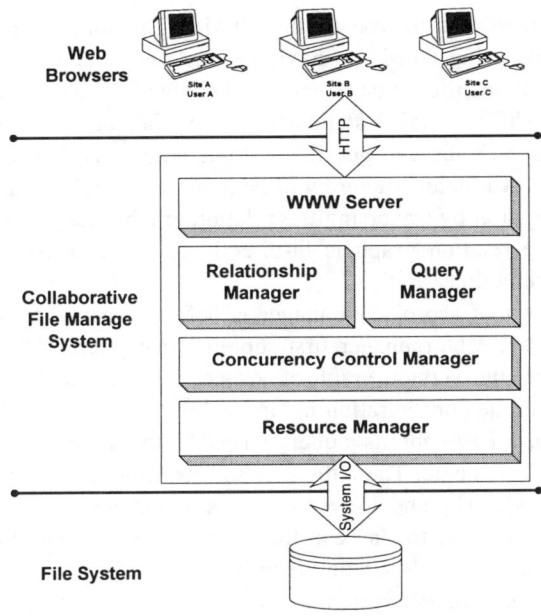

Fig. 4.1. CFMS System Architecture

The CFMS is a three-tier architecture. The front-end user interfaces are WWW browsers. Through HTTP, users download the Java Applet into the client sites. The Java Applet is responsible for drawing the partial relationship graph based on configuration for users. The reason why Java Applet is used here is that current SGML-based standard markup languages are not suitable to draw the complex multi-dimensional relationship graph. In the middle tier, the CFMS is composed of relationship manager, query manager, concurrency control manager and the resource manager. The relationship manager administrates the relations between collaborative files. The query manager provides facilities for text-based query. Users can submit search terms to select the logic structures which satisfy the search criteria. Concurrency control manager is responsible for the management of parallel developments on target files. CFMS supplies the SCCS-like check in/out operations to access files. It allows multiple read requests for a specific file at a given time. Only the user who gets the write lock can update that file. Resource manager controls all the physical files through the physical file I/O system calls.

5 System Implementation

We have implemented a prototype of the CFMS. The following figures show the front-end GUI while using CFMS to navigate or define the configuration for a collaborative purpose. In the prototype, a two-level navigation mechanism is designed for version selection. The first level stands for the initialization for version selection. Users can input a search term to get all the logic structures that contain the term in their version names. All logic structure will be returned by default. The second level is the visualized navigation by traversing the relation graph returned from the first-level search. Because the relation graph is bi-directional, users can traverse the version graph in any valid direction.

Figure 5.1 shows that a project manager is defining the configuration of version 1.1.1.1.2 for Project A. The manager first submit a query term "A" to search Project A. Then, he traverse the version graph of Project A. The selected items for version 1.1.1.1.2 are listed in the configuration list in the left side. Figure 5.2 is an example to check out files. A user first submits a query term "A" to search any logic structure that contains "A" in its logic name. Then, the user selects Project A from the list box in the corner of the right side. The check-out list shows items which he wants to check out. Figure 5.3 describes the case to check in files. While users get into the check in menu, all the checked-out items are shown on the browser by default. Users can choose items to check into the server-side file system.

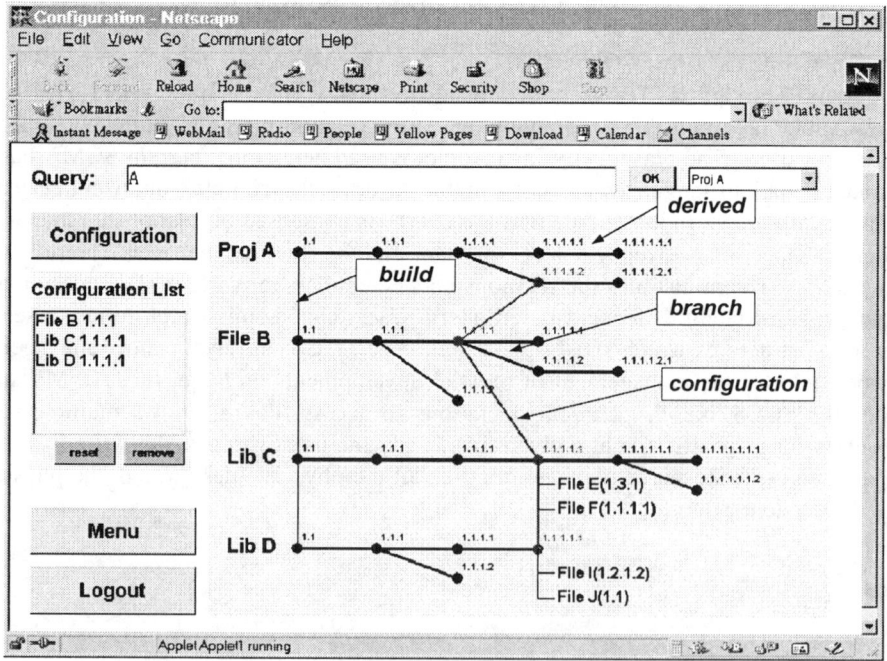

Fig. 5.1. CFMS clients traverse the relationship graph for configuration

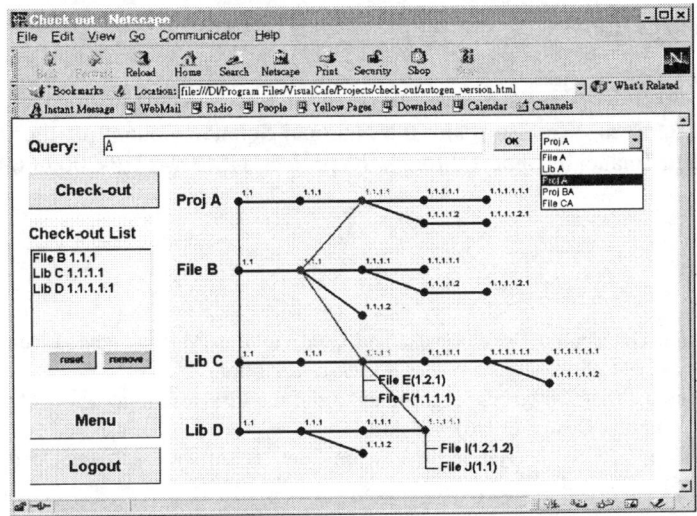

Fig. 5.2. CFMS clients traverse the relationship graph to check out files.

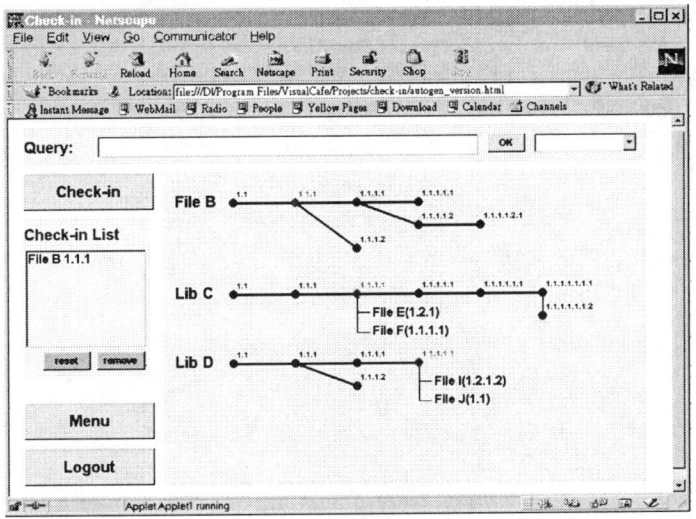

Fig. 5.3. CFMS clients traverse the relationship graph to check in files.

6 Conclusion

In this paper, the CFMS, a web-based collaborative file management, is proposed and implemented to demonstrate its feasibility. The relationship model clarifies the relations between collaborative logic structures, which represent physical files. Integrating with the prefix-based naming scheme, the two-level version navigation provides a visualized and dynamic mechanism for version selection on WWW.

Acknowledgement

We are grateful for the many excellent comments and suggestions made by the anonymous referees. We also thank the National Science Council of the Republic of China for their financial support through the project, NSC99-2213-2009-069.

References

1. Ulf Asklund: Distributed Development and Configuration Management. Licentiate Thesis, ISSN 1404-1219 (1999) http://www.cs.lth.se/~ulf/lic.html
2. L. Benini et al.: Distributed EDA tool integration: the PPP paradigm. Proc. of International Conf. on Computer Design (1996) 448-453
3. Walter F. Tichy.: RCS – A System for Version Control. Software – Practice and Experience, 15, 7, (1985) 637 – 654
4. Rochkind, M. J.: The Source Code Control System. IEEE Trans. on Software Engineering, V SE-1, N 12. (1975)
5. David. Whitgift.: Methods and Tools for Software Configuration Management. Wiley series in software engineering practice. (1991) 89-108
6. ClearCase Concepts. Atria Software Inc., Natick, Mass. (1993)
7. Walter F. T.: Configuration Management. John Wiley & Sons Ltd. (1994)
8. CVS - Concurrent Versions Systems (2000) http://www.gnu.org/software/cvs/cvs.html

Adding Flexibility in a Cooperative Workflow Execution Engine

Daniela Grigori, Hala Skaf-Molli, and François Charoy

LORIA, INRIA Lorraine, Campus Scientifique
BP 239, 54506 Vandoeuvre les Nancy, France
{dgrigori,skaf,charoy}@loria.fr

Abstract. This paper describes an approach to support cooperation in a workflow system. It is based on the combination of a cooperative transaction protocol (COO) and a traditional workflow model. This combination allows activities to exchange data during their execution. It also allows some activities to start in advance regarding the predefined control flow (anticipate) and to exchange results with their preceding activities. All these communications of draft results are done under the control of the extended cooperation protocol and under the responsibility of the users. This allows for more flexibility regarding the actual execution of the process model while keeping some control on the way the exchanges are done.

1 Introduction

Workflow management systems are gaining a wide acceptance in the service industry due to their ability to model and control business processes. The rapid evolution of information technologies and the generalisation of networking allow to consider new perspectives for different kinds of reorganisation and work organisation. However, as it has been pointed out in many papers[1], the rigidity of current workflow systems remains a problem in a context where people and organisations have to evolve constantly in a moving environment. Moreover, especially in the service industry, where activities may be creative and where people like to keep some initiative on the way they work, introducing rigid process control is considered today as counterproductive. The risk is that the user of workflow systems spend more time figuring out how to by-pass rather than to follow blindly the "yellow brick road".

To ease the acceptance of workflow management systems in organisations and to ease process adaptation to a changing environment, several paths are followed today, led by different visions of what should be workflow management, what should be the role of users in the process and how it can be executed, controlled and changed. We can consider three main points of view that should help to get a better acceptance of workflow : considering the process as a resource for action, adding flexibility to the system and evolving the process model.

Considering the process as a resource for action basically means that the process is a guide for users upon which they can build their own plan. It's not a definitive constraint that has to be enforced. Thus, the users have all the initiative to execute their activities. They are not constrained by the predefined order of activities but can be considered as being inspired by it and encouraged to do it.

The second approach uses the process as a constraint for the flow of work, but it is admitted that it may change during its life time. The process can be dynamically adapted during its execution.

These two approaches consider flexibility at the level of the process execution itself. In one case, the model is a guide to reach a goal, in the another case, the model is a path to reach the goal that may change during its course.

The third way consist in evolving the process model itself to allow for more flexible execution. In this case, flexibility has to be modelled and is anticipated during the process modelling step. This is one of the branches that are followed by the COO project [2] and by other similar work [3].

In this paper, we will consider a fourth way to provide flexibility to process that won't be based on the way the process model is used or instantiated, neither on the way it can be evolved or modelled. We propose to add some flexibility in the workflow management system execution. The model is defined as usual with a control and a data flow specification. The execution engine is based on a cooperative transaction protocol that we have developed[4]. This protocol provides an optimistic approach about concurrency control between activities. It also introduces the possibility for running activities to provide several successive states for the same objects[5]. This allows users to exchange data and start activities execution with some anticipation. This approach has the advantage to be compatible with other proposals: it does not change the way the model is interpreted, but only the way it is executed. We will see that the degree of freedom that is added to the execution engine provides users with some initiatives on the way they conduct their work, while retaining a large part of the control that is required, even in a very rigid (not evolving) process model. Our goal is to reconcile the need of freedom required by users during the execution of a process and the need of control of project managers that are accountable for the correct execution of the process.

2 Cooperative Process

A lot of work has been done to provide definition of cooperative process[2, 6]. The set of processes that we will consider in our approach that can be called cooperative is characterized mainly by their properties. Some of them are related to the process definition and some of them are related to their cooperative nature.

Processes are executed by a group of users; this is common for cooperative processes. Moreover these users are aware of participating in these processes. The processes have a goal, known from their participants. All (or at least most) of the participants acting in the process are aware of this common goal, in the achievement of which they will find benefits. They have reasons to cooperate, communicate and exchange information spontaneously.

In this paper, we limit ourselves to cases where the products can be stored on a computer and can be described in term of states so that they can be managed and controlled by the system. Typical cases of cooperative processes that are considered here are co-editing, co-engineering or web site development. Moreover, our approach will apply best to iterative processes were feedback between users is very important during all the stages of the process. These are the kind of processes where the activities are not easy to execute. Their result may change depending on the production of other activities, implying a lot of feedback and communication. This is not compatible with the rigid approach of common production process execution as shown in the following example.

This example is a simple process: write a paper for a conference. This can be presented by the following workflow:

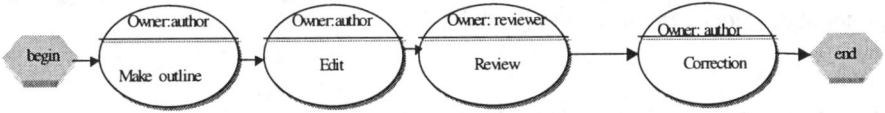

Fig. 1. Simple Edition workflow

This workflow is described by using the "begin and end" tasks dependencies[7]. It can be interpreted as follows: the Editing task starts when the task that makes the outline of the paper is finished. Then a Review task takes place followed by a correction one.

Imagine now that three researchers A, B and C from different universities want to follow the above workflow to write a paper. To do that, A and B will write the paper and C will review it. A, B and C decide to use a common repository to share their results.

The co-author A makes the outline of the paper. At the end of this task, the paper Editing tasks take place. A, B writes respectively *partA*, partB then they send their documents to C who can start the review task. If C is not satisfied of the work done by any of the co-authors, he adds comments. The correction task can then be started. A, B modifies respectively *partA, partB*. This gives the following workflow:

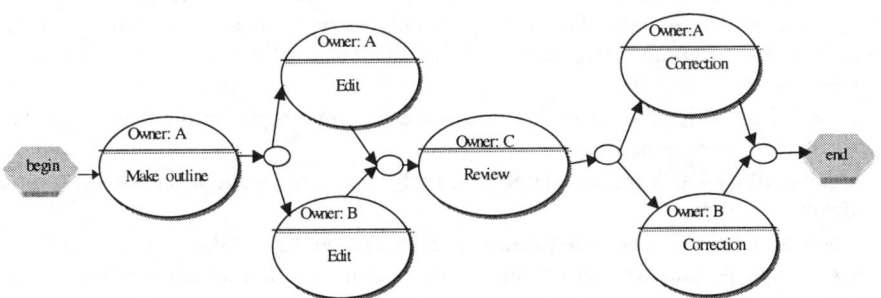

Fig. 2. Workflow execution

In fact, this scenario is a little theoretical. In reality, co-authors and people involved in creative and cooperative work in general, will not use this rigid working style[8, 9]. They adopt usually a more flexible approach. For instance, A and B will exchange drafts in order to synchronize themselves on the content and on the writing style. They will also provide early draft to C to get some feedback from him and to take his comment into account during the first phase of the work.

This kind of execution is not supported with a classical workflow management system. Following the predefined process, users would have been obliged to wait before being provided with feedback on their work. They would probably have used other communication channels to exchange information, out of the control of the system, leading to inconsistencies between the view of the process execution as controlled by the system and the actual events occurring for this process. Our proposal described in the following section allows users to exchange data during the execution of the process and to control how it is done, in order to be sure that no inconsistencies will occur because of them. We will rely on a cooperative transaction protocol developed in our earlier work [4, 5].

3 Introducing Flexibility in Workflow Execution

As we explained in the introduction, flexibility can be obtained by relaxing some of the constraints in the data flow and in the execution dependencies that exists during production workflow execution without changing the actual process. Thus we will rely on a traditional process model, describing the flow of control and the flow of data but we will provide users with some freedom on the way they can exchange data and on the way they can execute activities, based on the original specification. This will be made possible by the Coo cooperative transaction protocol [4].

3.1 The COO Approach for Cooperation

The COO protocol has been developed to allow exchange of intermediate results during cooperative transaction execution. This relaxes the isolation property of traditional transaction protocol which is considered as fundamental for long duration transactions. Of course, relaxing isolation induces the risk of inconsistencies due to dirty read or lost update. The Coo protocol provides mean to avoid this risk : it obliges users to compensate dirty read before terminating their task. The protocol works roughly as follow:
- A set of activities are executing and share a common public database. Data in this database are versioned.
- Each activity has a private database in which it can checkout data from the public database (read).
- Each activity can checkin (write) data in the public database at any time of its execution. A data written during its execution is called an *intermediate result* (a draft for instance). A data written at the end of its execution (commit time) is called a *final result*.
- The public database always contains the final result of an object and sometimes an intermediate result of the same object which is called the *current value*.

- If an activity reads an intermediate result during its execution, it must read the corresponding final result before to commit. We consider that reading the final result allows the user to compensate the dirty read done before.
- If two (or more activities) are dependent on mutual intermediate results, they are grouped and must synchronize in order to terminate simultaneously. Users have to negotiate in order to agree on the final value of both objects at the same time. A grouped transaction can be considered as one transaction in which two (or more) users interact.

It must be recalled that publishing an intermediate result will always be under the responsibility of the writer and that reading an intermediate result is also always under the responsibility of the reader. Thus, dependency and grouping can only occur if both party agree on it, and thus are knowledgeable of the risk.

3.2 Breaking Isolation between Activities

Encapsulating workflow activities in cooperative transactions allows to break the isolation between them. During their execution, activities may communicate intermediate results through the shared database (the process data container).

The COO protocol ensures the correctness of data exchange. It may impose additional constraints on activities termination order for simultaneous executing activities. For example, activities *A: Edit* and *B: Edit*, in the previous example may exchange intermediate results during their execution. They will be forced to finish simultaneously. During the terminate phase they must read each other's final results and agree on them. This kind of grouping of activities is forced by the fact that both activities are mutually dependant and could not terminate otherwise. *A* and *B* will be forced to cooperate in order to terminate in a way that was not specified in the process definition but that has been decided by users. From the inside, both activities actually execute as one cooperative activity. From the outside, the process is still going on as usual.

3.3 Allowing Anticipation of Results

Traditional Workflow management systems impose a end-start dependency between activities. This means that an activity can be started only when the preceding ones are finished. In a design workflow this may not be acceptable. Most of the time, activities may overlap and start with intermediate products. Early feedback may even be provided between two successive activities.

The flexibility we introduce in the control flow is the possibility to anticipate, i.e. to allow an activity to start its execution earlier. Even though the conditions for its activation may not be completely satisfied, the responsible of the activity may start its execution. In this case, its private database contains intermediate values of its input objets (if some of activities producing these objects have not been finished yet) or even no value at all for some of them. The intermediate results of these activities allow other subsequent activities to start anticipating and preceding activities to obtain comments on their work.

In our previous example, the Review activity can anticipate, reading an intermediate value of *partA*; it publishes some comments; thus allowing the actor in charge of activity *A:Edit* to obtain them even before he finishes his work. He may then take them into account to avoid to many comments in the correction activity.

When all activation conditions are met, an anticipating activity enters the normal executing state. All final values of the input parameters are copied into the activity's private database. These data were produced by the preceding activities in the control flow, as specified in the data flow. Having already been started, this activity will be able to finish its execution earlier. It needs only to take into account the latest modifications of its input parameters. The choice to anticipate is left to the activity responsible. It is not modeled by the process designer.

At the same time, by starting an activity in anticipating mode, the actor assumes the risk that his work may be useless. We refer to the case when an anticipating activity is in a path that is actually not taken in the given instance of the process execution (dead path).

Activities in anticipating state may even publish intermediate results in the public database. However, intermediate results of anticipating activities must not impose additional constraints for activities preceding them in the control flow. In our cooperative workflow example, Edit activity precedes Review activity in the control flow; Review may read an intermediate value of *partA* and Edit may read an intermediate value of comments. According to the COO protocol, in the absence of control flow definition, they would form a transaction group and would have to finish simultaneously. But, as the control flow defines a preceding dependency order between the two activities, we consider that the reading of the intermediate results by the Edit activity doesn't make it dependent on the Review activity. We relax this requirement of the cooperative protocol, by considering that the control flow has a higher priority over concurrency control. Since Edit precedes Review, the latter activity will make necessary adjustments and Edit activity must assume its responsibility of reading an intermediate result of anticipating activity.

Activities that do not have a predecessor-successor relationship with the anticipating activity can not read its intermediate results. This restriction avoids to make an activity dependent on an anticipating activity which is not connected to the former by a path in the control flow. Since no order is defined between them, we can neither require the first activity to wait the end of an anticipating activity, nor suppose that the logic of the control flow may compensate for this reading. This also avoids grouping between anticipating activities which may contradict the order defined in the control flow.

Intermediate results of activities placed in a conditional branch cannot be read by activities preceding the conditional node. Since an anticipating activity placed in a conditional branch for which the conditional expression was not yet evaluated, may not be executed, we prevent its influence to preceding activities. Suppose for instance that an activity in a conditional branch anticipates and publish results to preceding activities. If at the end, the activity cannot be actually executed because of its activation condition, it cannot compensate for the results that may have been taken into account by terminated preceding activities.

These constraints assure that the exchange of intermediate results does not contradict the pre-defined order of activities; it may only impose supplementary dynamic order relations.

The possibility to anticipate also requires the modification of the execution model in the execution engine. A new activity state is added: the anticipating state.

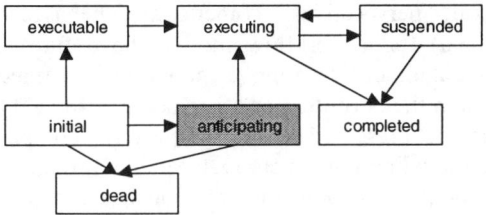

Fig. 3. State transition diagram for activities

When a process is instantiated all activities are in the initial state. They are executable if their activation conditions are satisfied. There is no condition for an activity to be in the anticipating state. The anticipating state is the same as the executing state except that the activation condition are not satisfied. When in anticipating state, an activity goes automatically in executing state as soon as its activation conditions are satisfied (not going through the executable state). If the activity appears to be in a dead path, its state becomes dead.

The execution model allowing anticipation ensures, at the same time, that the result of the process corresponds to its definition. Even though activities may be started earlier than scheduled, the order of activity termination is always respected and for an external observer, the process seems evolve as planned. It must also be recalled that flexibility is never imposed to users but decided by them. They are never obliged to produce draft result, to read others draft results or to anticipate. They are allowed to do it as they need it.

4 Related Work

Workflow flexibility is a subject of great interest in the academic research on workflows[1]. An overview and a taxonomy for this problem area are provided in [10, 11]. There are three main approaches: allowing for dynamic modification of running processes, supporting a higher degree of freedom in execution by using less prescriptive models and considering the process as a resource for action. Our proposition is to keep workflow model simple and to allow more flexibility in its execution. This proposition has the advantage of being compatible with other approaches.

Research projects like ADEPTflex[12], Chautauqua[13], WASA[14] and WIDE[15] provide explicit primitives to dynamically change running workflow instances. These primitives allow to add/delete tasks and to change control and data flow within a running workflow instance; constraints are imposed on the modifications in order to guarantee the syntactic correctness of the resulting process instance. ADEPTflex allows to work on tasks even when the conditions for their execution are not yet completely satisfied. The difference with respect to our solution is that it implies the restructuring of the process graph. Even though structural integrity is guaranteed in these approaches, *ad hoc* modifications represent serious interventions in the control flow of the workflow system.

In the second approach, less restrictive models are proposed. In Mobile[16], the authors define several perspectives (functional, behavioral, informational, organizational, operational) for a workflow model, the definitions of perspectives being independent of one another. Descriptive modeling is defined as the possibility to omit irrelevant aspects in the definition of a perspective. In [17] [2] other examples of descriptive modeling are presented as techniques for compact modeling. The authors propose simple modeling constructs that better represents real work patterns to be used, instead of a composition of elementary constructs.

Considering the process model as a resource for action implies to offer to users a set of services as a support for their awareness, helping them to treat the breakouts. In [18] the authors propose a formal model which allows to compute automatically the exceptional paths and change the process model consistently to a target behavior.

In [19], the author argues that a flexible workflow model and system may permit using the process as a resource for action or as a coordination tool, depending on the application requirements. A three-dimensional domain space is defined for a workflow model. The three axes are the amount of detail of the procedure description, the conformance required by the organization and the degree of operational abstraction. In order to address different points in this domain, the workflow model is enhanced with goal activities and regions. A goal node represents a part of the procedure with an unstructured work specification; its description contains goals, intent or guidelines. Each node in the model belongs to a region whose type is defined by a point in the domain space. Modalities of evolving the enactment system to support goal nodes are presented: including a virtual environment and providing contextual assistance.

The exchange of intermediate results between activities is supported in some workflow models[2, 20, 21]. In contrast with our approach, this exchange is included in the model, though anticipated during process modeling step.

In [20], an event-based communication mechanism allows the exchange of information between activities or sub-processes. Thus, activities in a sub-process that depend on the results of activities in another sub-process, can be started as soon as appropriate messages are received, and do not have to wait until the whole sub-process is finished. While, here, the events are used to start an activity, we allow the exchange of information all the time during activities execution.

[21] proposes a behavior definition for a task as the basis for modeling less-restrictive workflows as well as supporting dynamic workflow changes. New control flow dependencies can be specified; in particular the simultaneous dependency allows controlling data exchange between simultaneous active tasks, assuring that the dependent task does not terminate before the preceding task.

5 Conclusion

In this paper, we described how the combination of a cooperative transaction protocol and a classical workflow management system can be used to provide a more flexible system. This allows users to keep some initiative on how they work and how they exchange data during their work. The cooperative transaction protocol allows data exchanges between executing activities under the control of both the producer and the reader. The protocol ensures that these exchanges are done safely. We also rely on this approach to allow users to anticipate in their activities. It is well known that a cooperative process cannot be executed as a flowing sequence of activities, but contains many unpredictable exchanges that cannot be modeled in advance. The approach we propose allows these kinds of interactions. The users are controlled by the process model but can take some freedom on the way they exchange data during their execution and on the way activities are started. This freedom is constrained by the fact that an anticipating activity is not an actually executing activity. But we feel that this approach provides a good compromise between the rigidity of a process model and the flexibility needed by a team cooperating on a project. Moreover, the process model itself remains simple to understand and can still be controlled.

Of course this does not remove the need for a flexible model able to evolve during its life time. More specifically, we think that a cooperative process cannot be modeled entirely at the beginning of its execution. This will be the next step of this work to adapt the cooperative transaction protocol to a flexible model able to evolve and to be defined incrementally as it is the case for most cooperative processes.

References

[1] M. Klein, C. Dellarocas, and A. e. Bernstein, "Towards Adaptive Workflow Systems," presented at Workshop within the Conference on Computer Supported Cooperative Work, Seattle, 1998.
[2] C. Godart, O. Perrin, and H. Skaf, "coo: a Workflow Operator to Improve Cooperation Modeling in Virtual Processes.," presented at *9th Int. Workshop on research Issues on Data Engineering Information technology for Virtual Entreprises (RIDEVE'99)*, 1999.
[3] D. Georgakopoulos, "Collaboration Process Management for Advanced Applications," presented at International Process Technology Workshop, 1999.
[4] G. Canals, C. Godart, F. Charoy, P. Molli, and H. Skaf, "COO Approach to Support Cooperation in Software Developments," *IEE Proceedings Software Engineering*, vol. 145, pp. 79-84, 1998.
[5] H. Skaf, F. Charoy, and C. Godart, "Maintaining Shared Workspaces Consistency during Software Development," *Software Engineering and Knowledge Engineering*, vol. 9, 1999.
[6] G. De Michelis, "Computer Support for Cooperative Work: Computers between Users and Social Complexity," COMIC-MILAN-1-1, 1995.
[7] W. M. Coalition, "The Workflow Reference Model," WFMC-TC-1003 Version 1.1, 1995.
[8] E. Beck and V. Bellotti, "Informed Opportunism as strategy," presented at 3rd European Conference on Computer Supported Cooperative Work, 1993.
[9] P. Dourish, "Using Metalevel Techniques in a Flexible Toolkit for CSCW Applications," *ACM Transactions on Computer Human Interaction*, vol. 5, pp. 109-155, 1998.
[10] F. Leymann and D. Roller, *Production Workflow*: Prentice Hall, 1999.

[11] P. Heinl, S. Horn, S. Jablonski, J. Neeb, K. Stein, and M. Teschke, "A Comprehensive Approach to Flexibility in Workflow Management Systems," presented at *Joint Conference on Work Activities Coordination and Collaboration (WACC'99)*, San Francisco, 1999.

[12] Y. Han, A. Sheth, and C. Bussler, "A taxonomy of Adaptive Workflow Management," presented at *Towards Adaptive Workflow Systems, CSCW'98 Workshop*, Seattle, USA, 1998.

[13] M. Reichert and P. Dadam, "ADEPTflex - Supporting dynamic Changes of Workflows Without Losing Control.," *Journal of Intelligent Information Systems*, vol. 10, 1998.

[14] C. Ellis and C. Maltzahn, "Chautaqua Workflow System," presented at 30th Hawaii Int Conf. On System Sciences, Information System Track,, 1997.

[15] M. Weske, "Flexible Modeling and Execution of Workflow Activities," presented at *31st Hawaii International Conference on System Sciences*, Software Technology Track (Vol VII), 1996.

[16] F. Casati, S. Ceri, B. Pernici, and G. Pozzi, "Workflow Evolution," presented at *15th Int. Conf. On Conceptual Modeling (ER'96)*, 1996.

[17] S. Jablonski, "Mobile: A Modular Workflow Model and Architecture," presented at 4th international Working Conference on Dynamic Modeling and Information Systems, Noordwijkerhout, NL, 1994.

[18] S. Jablonski and C. Bussler, *Workflow management - Modeling Concepts, Architecture and implementation*: International Thomson Computer Press, 1996.

[19] G. J. Nutt, "The Evolution Toward Flexible Workflow Systems," presented at *Distributed Systems Engineerin*, 1996.

[20] C. Hagen and G. Alonso, "Beyond the Black Box: Event-based Inter-Process Communication in Process Support Systems.," presented at *9th International Conference on Distributed Computing Systems (ICDCS 99)*, Austin, Texas, USA, 1999.

[21] G. Joeris, "Defining Flexible Workflow Execution Behaviors," presented at Enterprise-wide and Cross-enterprise Workflow Management - Concepts, Systems, Applications', GI Workshop Proceedings - Informatik'99, Ulmer Informatik Berichte Nr. 99-07, University of Ulm, 1999., 1999.

A Web-Based Distributed Programming Environment*

Kiyoko F. Aoki and D. T. Lee

Institute of Information Science, Academia Sinica
Nankang, Taipei, Taiwan
{kiyoko,dtlee}@iis.sinica.edu.tw
http://iis.sinica.edu.tw/

Abstract. A Java-based system called the *GeoJAVA System* was introduced in [1]. This system allows a user to remotely compile his/her own C/C++ programs and execute them for visualization among a group of remote users. *DISPE*, which stands for DIStributed Programming Environment, expands on the *GeoJAVA System* by allowing the resulting executables to be run on systems other than the host on which they were compiled, thus making the system more versatile. *DISPE* uses Common Object Request Broker (CORBA) services to enable executables compiled on this system to invoke methods in libraries on remote sites in an architecturally heterogeneous environment. Not only does this allow users to compile and execute their programs remotely, but the maintenance and duplication of libraries is lowered since agents are used to search for symbols in libraries located remotely and to compile them with the user's source code. As long as there is an Internet connection between the hosts on which these libraries reside, the agents can search and compile with these libraries.

1 Introduction

The *GeoJAVA System* introduced in [1] is a system developed for researchers of computational geometry to enable them to develop geometric algorithms without the hassle of the administrative aspects of programming, such as downloading, setting up and maintaining libraries and compiling programs. Briefly, this system consists of web-based interfaces where users upload their C/C++ programs that visualize geometric algorithms to the web server, compile them remotely, and execute their program, thus broadcasting the results to remote users via the provided visualization tool. The programs only need to provide minimal code for the actual visualization, and the management of remote users is provided by the system.

DISPE expands on this system to provide a more generalized programming environment that any researcher can use. Whereas the *GeoJAVA System* required

---

* This work supported in part by the National Science Foundation under the Grant CCR-9731638, and by the National Science Council under the Grant NSC-89-2213-E-001-012.

that the user upload their files to a remote host and compile with the libraries there, *DISPE* allows the user to take advantage of mobile agents to compile the code with distributed remote libraries. So these agents can be dispatched from a local computer to compile the local source code with remote libraries. Work related to *DISPE* will be introduced next, followed by a brief description of the system's design. The conclusion and future work are given in the last section.

2 Related Work

DISPE is a system that encompasses several areas. It is an extension of the *GeoJAVA System*, which incorporates a visualization tool with a collaboratory, allowing remote users to interact and solve geometric problems. It is also a compilation system for distributed libraries, different from existing compilers for high performance computing or parallel computers. It takes advantage of mobile agents to search for libraries during compilation, and it uses CORBA for dynamic, distributed execution. In this section, we discuss some work related to the main components of *DISPE*, namely agents and CORBA. To our knowledge, no known system provides all of the functionality that *DISPE* does.

2.1 Agents

The terminology of agents should be discerned between agents of artificial intelligence (AI), which are more like robots, and agents as described later in this article. Projects whose foci are more in the former area include work done by the Software Agents group at MIT Media Lab[22] and Softbots at the University of Washington[2].

Quite a few agent products are being developed, as is evident on the Agent Society home page[11]. Many of these are agent systems that provide a framework for agent projects. A few major ones are listed below.

Aglets [3] are the agents provided by IBM's Aglets Software Development Kit (ASDK), which is an environment for programming mobile Internet agents in Java. Aglets are Java objects that can move from one host on the Internet to another. When an aglet moves, it takes along its program code as well as its data. *DISPE* uses Aglets in its implementation.

Voyager [24] is a 100% Java agent-enhanced Object Request Broker (ORB) that combines mobile autonomous agents and remote method invocation with complete CORBA support and comes complete with distributed services such as directory, persistence, and publish-subscribe.

MOA [5] was designed to support migration, communication and control of agents. It was implemented on top of the Java Virtual Machine and is compliant with the Java Beans component model, which provides for additional configurability and customization of agent systems and agent applications, as well as interoperability which allows cooperation with other agent systems.

2.2 CORBA

The Object Management Group (OMG) developed the CORBA standard in response to the need for interoperability among the rapidly proliferating number of hardware and software products available. CORBA allows applications to communicate with one another no matter where they are located or who has designed them.

Since the inception of CORBA, a number of different vendors have implemented their own versions of CORBA, each a little different from the other, especially where the CORBA specification was not very detailed. In implementing *DISPE*, requirements in deciding upon a CORBA implementation included adherence to version 2.2 of the CORBA specification, ease of use of the API, and availability. Based on these requirements, TAO and MICO seemed the most applicable.

TAO[23], is a real-time ORB end system designed to meet end-to-end application quality of service (QoS) requirements by vertically integrating CORBA middleware with operating system I/O subsystems, communication protocols, and network interfaces. **MICO** [20] has a clean API that supports the CORBA 2.2 specification, and it is freely available.

There are several other popular vendors who provide CORBA implementations. A full list of CORBA implementations can be found at [21]. There are several projects underway using CORBA, as can be seen in [14]. The ones that seem to be the most closely related to *DISPE* are the DOMIS Project at MITRE[15] and GOODE at the University of Lille[16], the latter of which has developed CorbaWeb[13] and CorbaScript[12].

3 Design and Implementation

The *DISPE* system is composed of the original *GeoJAVA System* plus the Compilation Agents and the *CORBAizer*. The *CORBAizer* is an application that can be downloaded and used by any user, and the Compilation Agents consist of the Java agents which reside in Agent Contexts on each remote host. These Agent Contexts provide a layer of security between the agents and the host so that (1) agents do not gain unlimited access to the host's resources, and (2) the host cannot directly manipulate the agents and its data. Figure 1 illustrates the general architecture for the system. The work involved in implementing these components of *DISPE* is given next, followed by a brief example explaining the general flow of the compilation process using agents.

3.1 The Agents

The Compilation Agents are Java objects that use native compilers. The main Compiler agent communicates with several types of agents in order to obtain information and data from remote sites and to complete the compilation process. These other agents are the Include agent, the Client Agent and the Library Proxy

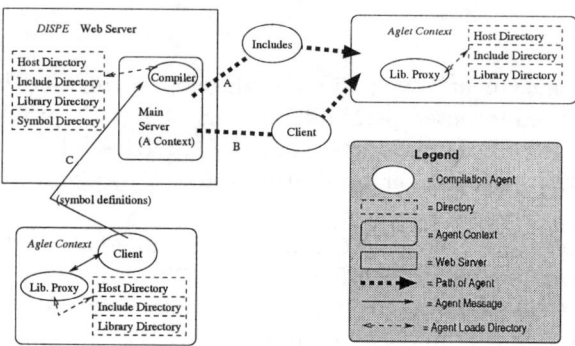

Fig. 1. Agents Architecture

Agent. While the main Compiler agent is stationary and remains at the original site at which the compilation begins, the Include Agent searches for missing include header files, the Client agent travels and searches for missing symbols during the linking phase, and the Library Proxy Agent serves as a stationary agent at each server host that responds to Include Agent requests for header file locations and to Client agent requests for library locations. Library Proxy Agents are created and disposed of by the agents who need them.

There is also a Vulture object for the Compiler agent which performs some checks for fault tolerance. For example, if an agent somehow "dies" unexpectedly, the Compiler agent would not know of it, so the Vulture object is used as a separate thread which occasionally sends a "ping" message to the dispatched agents, and if it does not receive a reply within a certain time frame, it sends a replacement agent to the same site, or, if the site has gone down, to any existing backup site.

In order to search for hosts, include header files, libraries and symbols, the Compiler agent maintains a hash table of this information which is read from disk upon startup and written to disk before being disposed of. This hash table serves as the Directory. The advantage of this approach is that the web server's agent host system keeps a central database file of this information in a simple hash table, and no new protocols or additional management software need to be introduced into the system. Java proved to be very convenient in this respect because of its serialization capabilities. So when the Compiler agent is first started, it reads four hash tables representing the Host, Include, Library and Symbol Directories containing host, header file, library, and symbol information, respectively, collected from previous compilations.

3.2 Performing the Compilation

The compilation process performed by the agents are discussed next in a step-by-step manner.

Invoking the Native Compiler. The first step is to invoke the compiler. The original gcc compiler goes through preprocessing, compilation, assembly, and linking[9], each invoked by separate executables. The linking phase was modified for the agents to use. In a normal compilation, the linker is automatically called by the compiler with various flags and options, so in order to keep consistent with the flags used during a "normal" compilation, the flags that needed to be passed to the linker were determined by running a sample compilation with the verbose (-v) flag. The same flags were then used when the agent called the linker.

The linker's main() method was modified into a method that was compiled into a shared library, which the Compiler agent loads when it is instantiated. To invoke the compiler, then, the Compiler agent first makes a system call to invoke the C++ compiler with the -c flag, indicating that it should compile up to the linking stage, but not link. Then the linker is called via the modified method in the shared library.

Finding Header Files. Before reaching the linking stage, the compiler should first handle missing header files. A parser is used to parse the messages displayed during this earlier compilation phase for any messages indicating that an include file could not be found.

For missing include header files, the Compiler agent will consult its own copy of the Directory to see if any of the include files have been found before. If so, an Include Agent is dispatched to the host where the include files are located, taking along a list of the missing filenames. If these include files are not located in the Directory, the Include Agent is sent to the Main Server, which is a central host where libraries and other host information is stored. The Include Agent will then begin travelling and searching for the include file names that it has been given.

When an Include Agent arrives at a host, it searches for a Library Proxy Agent, and if it cannot find one, it creates it. The Library Proxy Agent reads the Directory of include files located at the host and sends the Include Agent a formatted list of include files. When the Library Proxy Agent first starts up, it also informs the parent of the Include Agent, whose proxy it receives from the creating agent, of the list of hosts that it "knows" about from its own Directory. The Compiler agent will then add any new hosts from this list. In this way, new host information can be dispersed in a "natural" manner by the agents themselves.

Any include files found are sent back to the Compiler agent and incorporated into the compilation. This agent keeps track of the "repeat count" which indicates how long it should wait before it should determine that the compilation cannot continue.

Linking. Once this compilation phase completes successfully, the Compilation agent will begin the linking phase by making a call to the modified gcc linker. The linker will first attempt to link the source code with whatever libraries are available on the local host.

The linker uses a hash table to store all of the symbols in a compilation, and within the hash table is a linked list of undefined symbols. During the compilation, this linked list is used to merge in data from libraries to "pull in" symbol definitions into the compilation. The agents take advantage of this linked list for the compilation.

To illustrate a compilation procedure using Figure 1, first, the Compiler agent is instantiated upon a call for a compilation. Given the source code information, the Compiler agent dispatches an Include agent when it detects any missing header files in the compilation stage (before the linking stage). So the Include agent (A) is sent to a host (based on information in the Directory on the DISPE server), where it finds the Lib Proxy agent. The Lib Proxy agent tells the Include agent which header files are located there based on the Directory information at the host. If there is a match for the file that the Include agent is searching for, the Include agent sends the header file back to the Compiler agent. If there are missing header files remaining, the Include agent will dispatch itself to the next host. The result of the compilation determines whether or not to continue on to the linking stage. The linking stage will begin with a link in order to determine which symbols are missing. If there are symbols missing, then a Client agent (B) is dispatched to the hosts where the header files were found. These Client agents will communicate with the Lib Proxy agents in a manner similar to that of the Include agent. However, once a symbol is found in a remote library, instead of sending each symbol back to the Compiler agent, the Client agent will begin its own "mini-compilation" in order to draw in as many needed symbols as possible. This process will result in an archive of object files containing the needed symbols for the compilation, which is sent back (C) to the Compiler agent. When the necessary symbols are found, the Compiler agent re-compiles the source code with the archive(s) found, and if more missing symbols remain, a message is sent to the Client agent(s) to find the new missing symbols (thus incrementing the "repeat count"). The Client agents are told to dispose of themselves once the compilation completes successfully (i.e., no missing symbols remain), or the "repeat limit" is reached (which can be set according to the user's wishes).

3.3 The GUI

A web interface used as a GUI to the Compiler agent is described next. We use a daemon which creates a socket and listens for messages to create Compiler agents. An applet connects to the socket and simply displays any messages that it receives while listening on the socket. The daemon takes the applet's socket ID and passes it to the newly created Compiler aglet. Thus, the Compiler aglet can send messages to display on the applet to indicate the status of the compilation.

Figure 2 is an instance of this web interface. Note, however, that a user can also use these agents from their local machine as long as they have downloaded the necessary components.

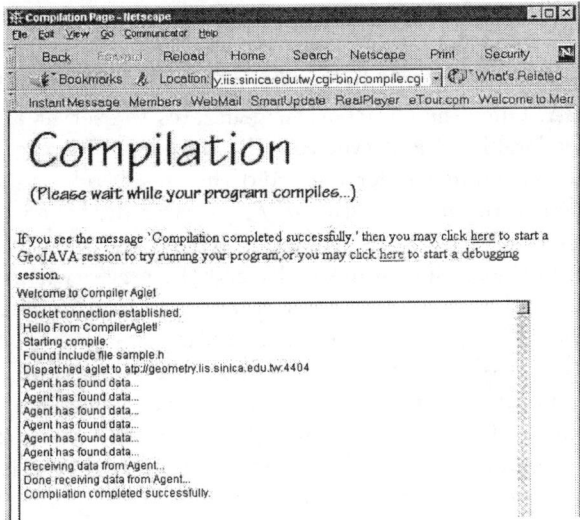

Fig. 2. Agent Compiler GUI

3.4 Registration Tool

The Registration Tool is a Java application that the user runs when he/she wishes to make a library available to the agents. Once a library is registered into the system via this Registration Tool, agents can find the new library when they arrive at the host through the Context providing the registered information. In the case that a new host is being created, the host can be added to the system during a compilation, where, in the beginning, the Compiler Agent will ask if any new hosts are to be added, and the user can input the information at that time.

3.5 The CORBAizer

Although the compilation agents can compile with static libraries, compiling with shared, or dynamic link, libraries is a different story. Because shared libraries are only useful (at the time of this writing) in a system where they are loadable into the address space of the running executable [8], they cannot be used in a distributed system. In addition, we still have architecture issues where libraries on different architectures cannot be compiled together, even if they have symbol definitions to offer. Fortunately, these issues dissolve with the *CORBAizer*.

The *CORBAizer* is based on the Exerciser Generator introduced in the RunClass tool[10]. This tool dynamically instantiates objects through the user's commands via the GUI and allows the methods within these objects to be examined. In order for a library to be inspected, the header files of the library have to be parsed and converted into a format that the system's engine can understand. The CORBAizer is based on this parser that is used in the RunClass tool.

The *CORBAizer* parses the header files of a library and generates two sets of C++ code: server code and client code. These source code files can be compiled separately, even on separate architectures, and used to communicate with each other via CORBA. Once the CORBAizer generates the server and client source files from the header files, the server code is compiled with the original library corresponding to the input header files, and the client code is used to generate a new library that is registered with *DISPE*, as was described previously. The server code can either be added to the CORBA Implementation Repository, which is a daemon that will automatically call the server when a request for a method in its library is called, or the server can be run manually to wait for requests from clients. Then, once the client library is added to the system, any program using the classes and methods defined in the original, possibly shared, library will be compiled with the client library just generated, and when the program is executed, the user's program will use CORBA to make a connection with the server program which executes the appropriate methods remotely. The user need not be concerned with any details of the CORBA implementation since it is all handled by the client library.

The basic idea behind the *CORBAizer* is that from a library's header files, a "skeleton" for the library's contents can be retrieved, which is used to generate the server and client code. The client is a copy of the skeleton with the implementations of the objects replaced by CORBA calls for searching for and connecting with the server and invoking the "real" methods. The server code takes advantage of CORBA's *tie* feature [6], which allows legacy classes to be wrapped by a CORBA class that takes the legacy class as the object of a template. The legacy class is then called when the CORBA class in the server receives a request for a method invocation.

4 Conclusion and Future Work

DISPE has great potential in paving the way for a new style of programming. Much of the latest and most solid technology in computer programming has been incorporated into the system, such as Java and CORBA. This should prove to be a plus for *DISPE* as it has been built on technology with a solid foundation.

The use of Java-based agents is indeed an innovative concept, and one that can grow and be useful for years to come. There are no signs of the C/C++ language weakening in the future, despite the growing use of Java. Thus, the integration of two of the most popular languages today in a distributed system should be very beneficial for C/C++ programmers. Remote compilation, whether using the "Traditional" or "Agent Compilation" frees users from worrying about the setup of libraries and include files and their directories, allowing them more time to focus on programming. The remote execution is all location transparent, where groups of users at remote sites can easily demonstrate geometric algorithms to one another, and the CORBAizer is especially useful in allowing remote libraries to be accessed during a program's execution. The CORBAizer should be very practical for users of existing legacy libraries written in C/C++. It is not easy to manually develop CORBA libraries from scratch, let alone to port a legacy library to CORBA. Thus, the CORBAizer will definitely be useful for all programmers in any field where there is a need to integrate legacy systems with CORBA. Finally, remote debugging capabilities, a necessity for most programmers, complete the *DISPE* package. No known system provides such a complete programming framework.

In determining how we were to implement the search for hosts and libraries ("searching" for symbols is performed once an agent reaches a remote host and access a library), at first, the lightweight directory access protocol (LDAP)[19, 18] seemed very attractive. The attractiveness was compounded by the fact that Sun Microsystems also announced the availability of the Java Naming and Directory Interface (JNDI)[17] which supports LDAP. Therefore, as more hosts and libraries are added to the system, the use of LDAP may become useful in implementing the Directories used in the system. This would address versioning and consistency issues related to the header files and libraries introduced into the system.

As in any distributed system, the issue of security needs to be addressed, especially in the compilation and execution of programs that use foreign libraries. Both the code being compiled and the libraries registered in the system need to be checked for any potentially dangerous code such as system calls. In approaching this issue, agents may use a little more intelligence in compiling the user's code, perhaps by detecting any potentially dangerous methods and checking with the user if the detected code should be compiled into the program. A similar procedure may be followed during the registration and/or CORBAization of libraries.

The CORBAizer that has been implemented generates server and client code that is compatible with MICO's CORBA implementation. So future work can be put into generating code for other CORBA implementations as well.

The Registration Tool currently reads three sets of files to determine the libraries and include files available to the compilation agents. However, in the case when large lets of libraries and include files are residing on a system, it may be more efficient to use a database. Therefore, the Registration Tool may make use of a database in the future.

References

1. K.F. Aoki, D. T. Lee, *Towards Web-Based Computing*, accepted to Int'l Journal of Computational Geometry and Applications, Special Edition, 1999.
2. O. Etzioni, D. Weld, *A Softbot-Based Interface to the Internet*, Communications of the ACM, July, 1994.
3. D.B. Lange, M. Oshima, *Programming and Deploying Java Mobile Agents with Aglets*, Addison Wesley Longman, Inc., 1998.
4. D.T. Lee, C.F. Shen, S.M. Sheu, "GeoSheet: A Distributed Visualization Tool for Geometric Algorithms", *Int'l J. Computational Geometry & Applications*, **8,2**, April 1998, pp. 119-155.

5. D. Milojicic, W. LaForge, D. Chauhan, *Mobile Objects and Agents (MOA), Design Implementation and Lessons Learned*, The 4th USENIX Conference on Object-Oriented Technologies and Systems (COOTS), Santa Fe, New Mexico, April, 1998.
6. Object Management Group. *The Common Object Request Broker: Architecture and Specification*, Revision 2.2, OMG Technical Document 98-07-01.
7. Object Management Group. 1998. *Mobile Agent System Interoperability Facility - OMG Revision Task Force Document.* Framingham, MA: Object Management Group.
ftp://ftp.omg.org/pub/docs/orbos/98-03-09.pdf.
8. *Solaris Linker and Libraries Guide*, Sun Microsystems, Inc., 1997. http://docs.sun.com:80/ab2/coll.45.4/LLM/Ab2TocView?
9. R.M. Stallman, *Using and Porting GNU CC, Version 2.8.1.* Free Software Foundation, 59 Temple Place - Suite 330, Boston, MA, 02111-1307, 1998.
10. T.R. Chuang, Y.S. Kuo, C.M. Wang, *Non-Intrusive Object Introspection in C++ – Architecture and Application*, Proceedings of the 20th International Conference on Software Engineering, pp. 312-321, Kyoto, Japan, April 1998.
11. The Agent Society. http://www.agent.org/.
12. CorbaScript. http://corbaweb.lifl.fr/CorbaScript/index.html.
13. CorbaWeb. http://corbaweb.lifl.fr/index.html.
14. Cetus Links - CORBA.
http://www.cetus-links.org/oo_corba.html#oo_corba_projects.
15. DOMIS. http://www.mitre.org/research/domis/index.html.
16. GOODE. http://corbaweb.lifl.fr/GOODE/index.html.
17. Java Naming and Directory Interface (JNDI).
http://java.sun.com/products/jndi/.
18. An LDAP Roadmap & FAQ.
http://www.kingsmountain.com/ldapRoadmap.shtml.
19. Lightweight Directory Access Protocol (LDAP) FAQ.
http://www.critical-angle.com/ldapworld/ldapfaq.html.
20. MICO. http://www.mico.org/.
21. Cetus Links - CORBA - ORBs.
http://www.cetus-links.org/oo_object_request_brokers.html.
22. Software Agents Group. http://agents.www.media.mit.edu/groups/agents/.
23. TAO. http://www.cs.wustl.edu/~schmidt/TAO.html.
24. Voyager. http://www.objectspace.com/voyager.

Track III

Computer Science Track

Performance Analysis of Parallel N-Body Codes

P. Spinnato, G.D. van Albada, and P.M.A. Sloot

Faculty of Science
Section Computational Science
Universiteit van Amsterdam
Kruislaan 403, 1098 SJ Amsterdam, The Netherlands
{piero,dick,sloot}@science.uva.nl

Abstract. N-body codes are routinely exploited for simulation studies of physical systems, e.g. in the fields of Computational Astrophysics and Molecular Dynamics. Typically, they require only a moderate amount of run-time memory, but are very demanding in computational power. A detailed analysis of an N-body code performance, in terms of the relative weight of each task of the code, and how such weight is influenced by software or hardware optimisations, is essential in improving such codes. The approach of developing a dedicated device, GRAPE [9], able to provide a very high performance for the computation of the most expensive computational task of this code, has resulted in a dramatic performance leap. We explore on the performance of different versions of parallel N-body codes, where both software and hardware improvements are introduced. The use of GRAPE as a 'force computation accelerator' in a parallel computer architecture, can be seen as an example of Hybrid Architecture, where a number of Special Purpose Device boards help a general purpose (multi)computer to reach a very high performance.

1 Introduction

N-body codes are a widely used tool for the simulation of dynamics of astrophysical systems, such as globular clusters, and galactic clusters [10]. The core of an N-body code is the computation of the (gravitational) interactions between all pairs of particles which compose the system. Many algorithms have been developed to compute (approximate) gravity interactions between a given particle i and the rest of the system [2-4]. Our research is concerned with the simplest and most rigorous method [2], which computes the exact value of the gravity force that every other particle exerts on i. Unlike the well-known hierarchical methods [3,4], this method retains full accuracy, but it implies a computational load which grows as N^2, being N the total number of particles. Consequently the computational cost becomes excessive even with a few thousands of particles, making parallelisation attractive. Recently, parallelisation of N-body codes has become an important research issue [13-15].

The huge computational requirements of N-body codes make the design and implementation of special hardware worthwhile. The goal of our research is the study of an emergent evolution in this field: Hybrid Computer Architectures.

An hybrid architecture is a parallel general purpose computer, connected to a number of Special Purpose Devices (SPDs), which accelerate a given class of computations. An instantiation of this model is presented in [11]. In the light of this, we have evaluated the performance of such a system: two GRAPE boards attached to our local cluster of a distributed multiprocessor system [1]. The GRAPE SPD [9], is specialised in the computation of the inverse square law, governing both gravitational and electrostatic interactions:

$$\mathbf{F}_i = G \frac{m_j m_i}{|\mathbf{r}_j - \mathbf{r}_i|^3}(\mathbf{r}_j - \mathbf{r}_i) \qquad (1)$$

(where m_i and m_j are star masses in the gravity force case, and charge values, in the Coulomb force case). The performance of a single GRAPE board can reach 30 GigaFlop/s. Gravothermal Oscillations of Globular Clusters cores, and other remarkable results obtained by using GRAPE, are reported in [9]. Though some fundamental differences, like electrostatic shielding, exist, this similarity in the force expression allows us in principle to use GRAPE for both classes of problems.

Our research aims at understanding how such architectures interact with a given application. For this purpose, we have used NBODY1 [2] as a reference code. It is a widely used code in the field of Computational Astrophysics. It is rather simple, but includes all the relevant functionalities of a generic N-body code. By using NBODY1, we can determine the scaling properties of various parallel versions of the code, with and without use of GRAPE boards. The data obtained are used for the realisation of a Performance Simulation model that will be used to study a more general class of hybrid architectures and their interaction with various types of N-body codes.

2 Architecture Description

The GRAPE-4 SPD is an extremely powerful tool for the computation of interactions which are a function of r^{-2}. Given a force law like (1), the main function of a GRAPE board is to output the force that a given set of particles, the j-particles, exerts on the so called i-particles. This is done in a fully hardwired way, by means of an array of pipelines (up to 96 per board). Each pipeline performs, at each clock-cycle, the computation of the interaction between a pair of particles.

A GRAPE-4 system consisting of 36 boards was the first computer to reach the TeraFlop/s peak-speed [9]. GRAPE-4 is suitable for systems of up to 10^4–10^5 particles, when running an N-body code whose computational complexity scales as N^2 ([1]). More sophisticated algorithms exist, which reduce the computing cost to $\mathcal{O}(N \cdot \log N)$, at the price of a decreased accuracy, and an increased code complexity [3,15]. The latter codes change the work distribution between

---

[1] Besides the $\mathcal{O}(N^2)$ complexity due to force computation, another term due to temporal integration has to be accounted for. See discussion in next section.

the GRAPE and the host, since many more computations not related to mere particle-particle force interactions must be done by the host. This can make the host become the system's bottleneck, and makes interesting a study of architectures where the host is a high performance parallel machine.

We connected two GRAPE boards to two nodes of our local DAS cluster. The DAS is a wide-area computer resource. It consists of four clusters placed in various locations across the Netherlands (one cluster is in Delft, one in Leiden, and two in Amsterdam). The entire system includes 200 computing nodes. A 6 Mbit/s ATM line connects remote clusters. The main technical characteristics of our DAS-GRAPE architecture are summarised in the table below:

local network	host	GRAPE	channel
Myrinet	PPro 200 MHz	2 boards 30 GFlop/s peak	PCI9080
150 MB/s peak-perf.	64 MB RAM	62 and 94 pipes per board	33 MHz clock
40 μs latency	2.5 GB disk	on-chip memory for 20.000 j-particles	133 MB/s

3 Code Description

We chose NBODY1 as the application code for our performance analysis work because it is a rather simple code, but includes all the main tasks which GRAPE has been designed to service. This allows us to evaluate the performance of our system. A number of modifications have been made on the code, in order to parallelise it, and to let it make full use of GRAPE's functionalities. An overview on the code is given in what follows. We made use of MPI to parallelise it.

3.1 The Basic: Individual Time-Step

The original version of NBODY1 uses individual time-steps. Each particle is assigned a different time at which force will be computed. The time-step value Δt depends on the particle's dynamics [2]. Smaller Δt values are assigned to particles having faster dynamics (i.e. those particles which have large values in the higher order time derivatives of their acceleration). At each iteration, the code selects that particle having the smallest $t + \Delta t$ value, and integrates only the orbit of that particle. This reduces the computational complexity, with respect to a code where a unique global time step is used. The individual time step approach reduces the temporal complexity to $\mathcal{O}(N^{1/3})$, whereas the global time step approach is $\mathcal{O}(N^{2/3})$ [7] ([2]).

An effect of individual times is that, for each particle, values stored in memory refer to a different moment in time, i.e. the moment of its last orbit integration. This means that an extrapolation of the other particles' positions to time t_i is needed, before force on i is computed.

---
[2] These figures for the temporal complexity are valid for a uniformly distributed configuration. More realistic distributions show a more complicated dependence on N, although quantitatively only slightly different.

Parallelisation. Since contributions to the gravity force on a given particle i are computed from all the other particles using eq. (1), regardless of their distances from i, an uniform distribution of particles to each processing element (PE) suffices to assure load balancing. The force computation is done by broadcasting the coordinates of the currently selected particle i. Then each PE computes the partial component to the force on i, by accumulating contributions from its own particles. Finally such components are sent back to the PE which hosts i, where the force resultant is computed, the particle's orbit is integrated, and the new values are stored.

To identify the particle i on which force will be computed, a global reduction operation is done, in order to find which particle has the least $t_i + \Delta t_i$ value, and which PE owns it. This information is broadcasted to all PEs, since they must know the extrapolation time, and the i-particle owner.

3.2 Toward a GRAPE Code: Block Time-Step

Since its introduction, NBODY1 has evolved to newer versions, which include several refinements and improvements (*cf.* [13]). In the version of NBODY1 used in our study we implemented the so called *hierarchical block time step* scheme [8]. In this case, after computing the new Δt_i, the value actually assigned is the value of the largest power of 2 smaller than Δt_i. This allows more than one particle to have the same Δt, which makes it possible to have many i-particles per time step, instead of only one. Using this approach, force contributions on a (large) number of i-particles can be computed in parallel using the same extrapolated positions for the force-exerting particles, hereafter called j-particles. Moreover, when a GRAPE device is available, it is possible to make full use of the multiple pipelines provided by such hardware, since each pipeline can compute the force on a different particle concurrently.

Parallelisation. Having many i-particles, instead of only one, makes attractive to use a somewhat different parallel code structure. If the i-particles reside on different processors, distributing the particles as in the individual time-step case could cause too convoluted communication patterns, with consequential increase of code complexity. Therefore, we chose to let every PE have a local copy of all particle data. The force computation is done in parallel by making each PE compute force contributions only from its own set of j-particles, assigned to it during initialisation. A global reduction operation adds up partial forces, and distributes the result to all PEs. Then each PE integrates the orbits of all i-particles, and stores results in its own memory. For what concerns the search for i-particles, each PE searches among only its j-particles, to determine a set of i-particles candidates. Then a global reduction operation is performed on the union of such sets, in order to determine the real i-particles, i.e. those having the smallest time. The resulting set is scattered to all PEs for the force computation. Since every PE owns a local copy of all particle data, only a set of labels identifying the i-particles is scattered, reducing the communication time.

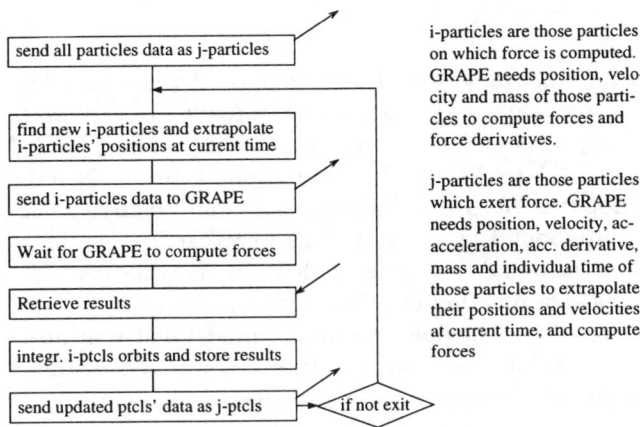

Fig. 1. Basic sketch of NBODY1 tasks. Diagonal arrows symbolise communication with GRAPE.

3.3 The GRAPE Code

The API for the GRAPE hardware consists of a number of function calls, the most relevant for performance analysis being those which involve communications of particles data to and from the GRAPE. Such communication operations are: sending j-particle data to GRAPE, sending i-particle data to GRAPE, receiving results from GRAPE. A sketch of the program flow for an N-body code which uses GRAPE is given in fig. 1.

Parallelisation. The presence of the GRAPE boards introduces a certain degree of complexity in view of code parallelisation. The GRAPE-hosts obviously play a special role within the PEs set. This asymmetry somehow breaks the SPMD paradigm which parallel MPI programs are expected to comply with. Besides the asymmetry in the code structure, also the data distribution among PEs is no more symmetric. The force computation by exploiting GRAPE boards is done, similarly to the non-GRAPE code, by assigning an equal number of j-particles to each GRAPE, which will compute the partial force on the i-particle set, exerted by its own j-particles. After that, a global sum on the partial results, done by the parallel host machine will finally give the total force. The GRAPE does not automatically update the j-particles' values, when they change according to the system evolution. The GRAPE-host must take care of this task. Each GRAPE-host holds an 'image' of the j-particles set of the GRAPE board linked to it, in order to keep track of such update. Since all force computations and j-particles positions extrapolations are done on the GRAPE, the only relevant work to do in parallel by the PEs set, is the search for i-particles candidates, which is accomplished exactly as in the code described in the previous subsection.

4 Results

Measurements for the evaluation of performance of the codes described in the previous section were carried out. They were intended to explore the scalability of parallel N-body codes. Sample runs were made scaling both N, and PEs; the former from 1024 to 16384, the latter from 1 to 24. NBODY1 does not need a large amount of run-time memory, just about 200 bytes per particle, but is heavily compute-bound [5]. Our timings were carried out in order to show the relative computational relevance of the various code tasks, and how such relevance changes as a function of N and PEs.

Our runs were started having a Plummer model distribution as initial condition (density of particles decreasing outward as a power of the distance from the cluster centre). The gravity force is modified by introducing a *softening parameter*, which is a constant term, having the dimension of a length, which is inserted in the denominator in eq. (1). It reduces the strength of the force in case of close encounters and thus prevents the formation of tightly-bound binaries. In this way very short time-steps and correspondingly long integration times are avoided. The use of a softening parameter is common practice in N-body codes. In our runs, this parameter was set equal to 0.004. As a reference, the mean inter-particle distance in the central core of the cluster, when $N = 16384$, is approximately equal to 0.037.

4.1 Individual Time-Step Code

The essential tasks of this version of the code (hereafter called IND) are basically the same as in the code-flow depicted in figure 1. That case refers to the code which makes use of GRAPE; in the present case no communications with the GRAPE device are done.

As described in the previous section, the parallel version of this code implements communications in the task regarding the i-particle search, and when i-particle's position is broadcast, and partial forces are gathered by the PE that owns the i-particle. Figure 2 shows the timings and the related performance of the parallel version of the IND code. Performance is defined as:

$$P_n = \frac{t_1}{n \cdot t_n}$$

where n is the number of PEs used, and t_n the execution time when using n PEs. Timings refer to 1000 iterations of the code. Their dependence is linear with respect to N, since the number of operations to compute the force on a given particle scales linearly with N, and in each run the same number of force computations is performed, i.e. 1000, independently of the total number of particles. An interesting super-linear speedup is visible in fig. 2b, arguably due to an optimised cache utilisation. This figure also clearly shows how this code suffers of a communication overhead when the computational work-load is light, i.e. for low values of the N/PEs ratio, but performs quite satisfactorily when this ratio is high, thanks to the compute-intense characteristics of the N-body code, and the high performance communication network of our architecture.

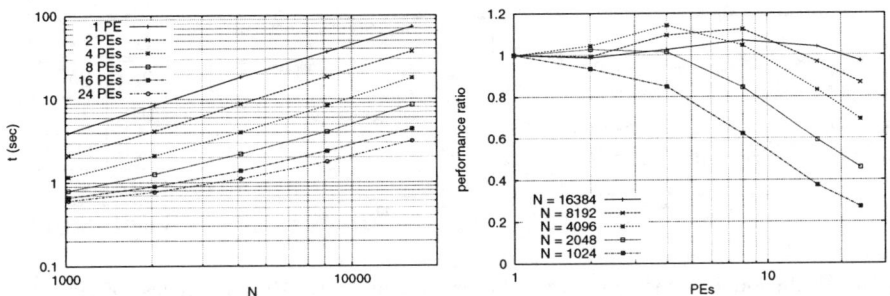

Fig. 2. *a*: Global timings for the parallel individual time-step code, running for 1000 iterations. *b*: Performance of the code.

4.2 Block Time-Step Code

The basic tasks of this version of the code (BLOCK hereafter) are the same as those described for the IND code. The only difference is that now the number of i-particles per iteration can be greater than 1. As stated, this optimises the force computation procedure, also in view of the use of GRAPE, but, on the other hand, increases the communication traffic, since information about many more particles must be exchanged each time step.

The effect of this is clearly shown in the figures presented here. Fig. 3 shows total timings and performance of this code; in this case the execution time grows as a function of N^2 because the number of i-particles, i.e. the number of force computations, grows approximately linearly with N. Since the computational cost for the force on each particle also grows linearly with N, the resulting total cost is $\mathcal{O}(N^2)$. A table listing the mean number of force computations per iteration, to show such linear scaling with N, is given in fig. 4c. Fig. 3b shows

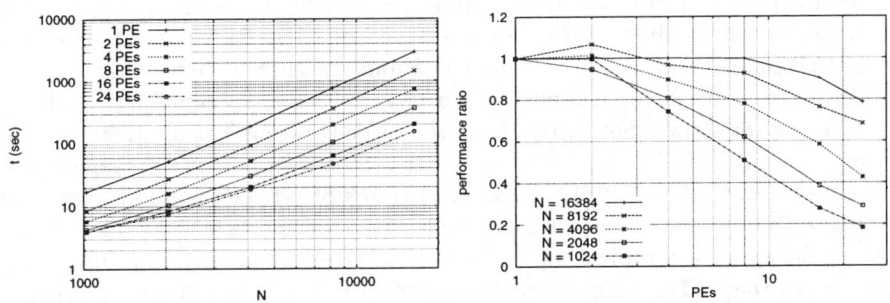

Fig. 3. *a*: Global timings for the parallel block time-step code. Here force on many particles is computed at each time-step *b*: Performance of the code.

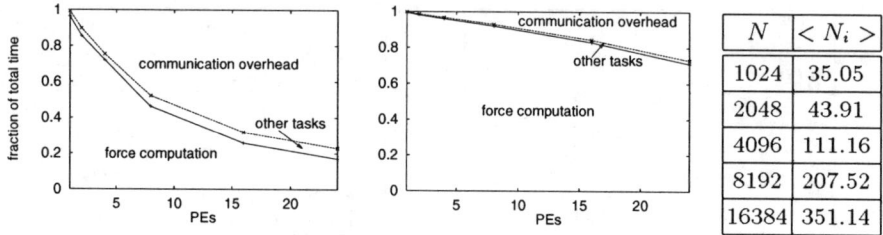

Fig. 4. Evolution of execution time shares. *a*: runs with 1024 particles; *b*: runs with 16384 particles; *c*: Mean number of *i*-particles (i.e. of force computations) per iteration in the runs of the BLOCK code.

how the performance gain of this code is less spectacular than the gain of the IND code, since communication overhead plays a larger role in the total execution time. This large overhead can be seen in figures 4*a,b*, which also show how the execution time shares evolve as a function of PEs number. These figures show that for the BLOCK code, almost all the computational part of the execution time is spent in the force computation task; the *j*-particles extrapolation, that takes roughly 25 ~ 30% of the total time in the IND code (data not shown), here is reduced to a fraction of one percent.

4.3 GRAPE Code

The code version which makes use of GRAPE boards will be called GRP hereafter. A code-flow of the serial version of GRP is sketched in fig. 1. The communication overhead of the parallel version now includes also network communications. The parallel code runs have been done by using only the DAS nodes connected to the GRAPE boards at our disposal, thus the maximum number of PEs in this case is 2.

It is clear from fig. 5 that the parallel performance is very poor. The large communication overhead, which dominates the GRP code as can be seen in fig. 6, can explain this. Here, GRAPE0 refers to the GRAPE with 62 pipelines, and GRAPE1 to the GRAPE with 94 pipelines. Figure 5 shows that runs on GRAPE1 are a bit faster, thanks to the larger number of pipelines available. The other figure shows that, apart from the large communication overhead, the time share spent in GRAPE computations (i.e. force computations) is quite low, resulting in a low efficiency of this code, in terms of GRAPE exploitation. One reason for that is of course the very high speed of the GRAPE. This device is by far faster in accomplishing its task than its host and the communication link between them. The figures clearly show that for our hardware configuration the capabilities of the GRAPE will only be fully utilised for problems of over 40000 particles (for single GRAPEs) and approximately double than that for the parallel system. This number is, however, limited by the on-board memory for *j*-particles of GRAPE.

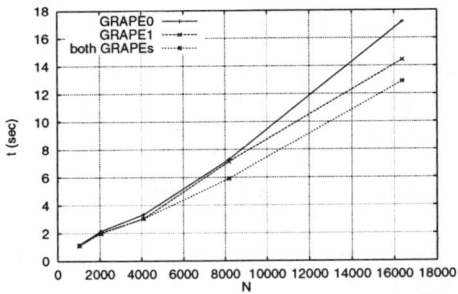

Fig. 5. Execution time for the GRP code.

Measurements [12] show that most of the time spent in communication is due to software overhead in copy operations and format conversions; analogous measurements [6], performed on a faster host, showed a higher communication speed, linearly dependent on the host processor clock speed. Nevertheless, even though GRAPE boards are not exploited optimally, the execution times for the GRP code are by far shorter than those for the BLOCK code. The heaviest run on 2 GRAPEs is about one order of magnitude faster than the analogous run of the BLOCK code on 24 PEs. A global comparison of the throughput of all codes studied in this work is given in the next subsection.

4.4 Codes Comparison

In order to evaluate the relative performance of the three versions of the N-body code studied in this work, a series of runs has been made, where both a 8192 particles system, and a 32768 particles system were simulated for 7200

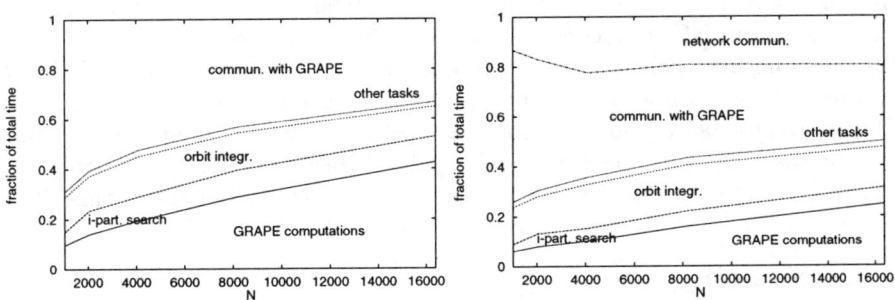

Fig. 6. Evolution of execution time shares. a: runs on GRAPE0; b: runs on both GRAPEs. Data for GRAPE1 (not shown) are qualitatively very similar to GRAPE0 data.

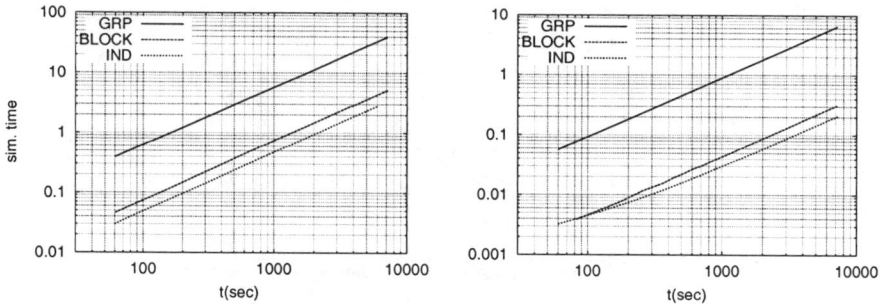

Fig. 7. Performance comparison for the three versions of the N-body code. a: runs with 8192 particles; b: runs with 32768 particles.

seconds. This will illustrate our expected better scaling of the GRP code, with respect to an increasing computational load. Initial conditions and values of numerical parameters were identical to the ones previously specified. The fastest hardware configuration was used in each case, i.e. 24 PEs for the IND and BLOCK code runs, and 2 PEs (and 2 GRAPEs) for the GRP run. Figs. 7a,b show the evolution of the simulated time, as a function of the execution time. In such a way, the performance of each code is made clear in terms of how long one should wait before a simulation reaches a certain simulated time. The figures show that the GRP code outperforms the other two codes by a factor 8, when the computational load is lighter, and by a factor 20, with a heavier computational load. In both cases the BLOCK code is 1.5 times faster than the IND code, thanks to the optimisation of the j-particles extrapolation step. Fig. 7b shows an initial overlapping of these two codes performance curves, due to a start-up phase, which is not visible in fig. 7a, because at the first timing event (after 60 s) this system is already stabilised.

These figures clearly show the large performance gain obtained with GRAPE. Using only two PEs, an order of magnitude better performance was attained compared to the BLOCK code on 24 PEs. Due to the reduction in the time needed for the force calculation, the communication overhead for the GRP code accounts for approximately 50% of the total execution time (*cf.* fig. 6). Hence an even larger relative gain may be expected for larger problems, as the relative weight of the communication overhead will become less. The difference in performance between the two cases shown in fig. 7 clearly illustrates this effect.

5 Discussion

The main conclusions from our work are, apart from the very good parallel performance of the BLOCK and especially the IND code, that the GRP code shows a dramatic performance gain, even at a low efficiency in terms of GRAPE boards exploitation. Such low efficiency is mainly due to a very high communication overhead, even for the largest problem studied. This overhead can be strongly reduced with the use of a faster host, and by the development of an interface requiring fewer format conversions. The GRAPE hosts in the system that we studied have a 200 MHz clock speed. Nowadays standard clock speeds are 3 to 4 times faster; the use of a state-of-the-art processor would reduce the host and communication times significantly. An extremely powerful machine as GRAPE, in any case, can be efficiently exploited only when the problem size remarkably increases, hence attaining the highest SPD utilisation.

The measurements described in this paper have been used to validate and calibrate a performance simulation model for N-body codes on hybrid computers. The model will be used to study the effects of various software and hardware approaches to the N-body problem.

Acknowledgements. The work related to the parallelisation of the IND code was done at the Edinburgh Parallel Computing Centre (EPCC) under the TRACS programme, supported by the EU. EPCC and TRACS are warmly acknowledged for their support. Sverre Aarseth and Douglas Heggie are also acknowledged for having made available to us the original serial N-body codes. Discussions with Douglas Heggie were unvaluable for the work related to the parallelisation of the IND code.

References

1. http://www.cs.vu.nl/das/
2. Aarseth, S.J.: Direct Methods for N-body Simulations. In Brackhill, J.U., & Cohen, B.I. (eds.): Multiple Time Scales. Academic Press (1985)
3. Barnes, J. & Hut., P.: A Hierarchical $\mathcal{O}(N \cdot \log N)$ Force-Calculation Algorithm. Nature **324** (1986) 446
4. Cheng, H., Greengard, L., & Rokhlin, V.: A Fast Adaptive Multipole Algorithm in Three Dimensions. J. Comp. Phys. **155** (1999) 468
5. Hut, P.: The Role of Binaries in the Dynamical Evolution of Globular Clusters. In Milone, E.F., & Mermilliod, J.-C. (eds.): proc. of Int. Symp. on the Origins, Evolution, and Destinies of Binary Stars in Clusters. ASP Conf. Series **90** (1996) 391 (*also available at:* astro-ph/9602158)
6. Kawai, A., *et alii*: The PCI Interface for GRAPE Systems: PCI-HIB. Publ. of Astron. Soc. of Japan **49** (1997) 607
7. Makino, J., & Hut., P.: Performance Analysis of Direct N-body Calculations. Astrophys. J. Suppl. **68** (1988) 833
8. Makino, J.: A Modified Aarseth Code for GRAPE and Vector Processors. Publ. of Astron. Soc. of Japan **43** (1991) 859
9. Makino, J., & Taiji, M.: Scientific Simulations with Special-Purpose Computers. Wiley (1998)
10. Meylan, G. & Heggie, D. C.: Internal Dynamics of Globular Clusters. Astron. and Astrophys. Rev. **8** (1997) 1
11. Palazzari, P., *et alii*: Heterogeneity as Key Feature of High Performance Computing: the PQE1 Prototype. To appear in Proc. of Heterogeneous Computing Workshop 2000, Cancun, Mexico (May 2000). IEEE Computer Society Press (2000)

12. Spinnato, P., van Albada, G.D., and Sloot, P.M.A.: Performance Measurements of Parallel Hybrid Architectures. Technical Report, *in preparation*
13. Spurzem, R.: Direct N-body Simulations, Journal of Computational and Applied Mathematics. **109** (1999) 407
14. Sweatman, W. L.: The Development of a Parallel N-body Code for the Edinburgh Concurrent Supercomputer. J. Comp. Phys. **111** (1994) 110
15. Warren, M.S., & Salmon, J.K.: A Portable Parallel Particle Program. Comp. Phys. Comm. **87** (1995) 266

Interoperability Support in Distributed On-Line Monitoring Systems[*]

Jörg Trinitis[1], Vaidy Sunderam[2], Thomas Ludwig[1], and Roland Wismüller[1]

[1] Technische Universität München (TUM)
Lehrstuhl für Rechnertechnik und Rechnerorganisation (LRR-TUM)
Technische Universität München, D-80290 München
{trinitis,ludwig,wismuell}@in.tum.de
[2] Department of Mathematics and Computer Science
Emory University, Atlanta, GA, USA
vss@mathcs.emory.edu

Abstract. Sophisticated on-line tools play an important role in the software life-cycle, by decreasing software development and maintenance effort without sacrificing software quality. Using multiple tools simultaneously would be very beneficial; however, with most contemporary tools, this is impossible since they are often based on incompatible methods of data acquisition and control. This is due largely to their relative independence, and could be overcome by an appropriately designed common on-line monitoring system. We consider three possible platforms that might be potentially capable of addressing this issue, and discuss the relative merits and demerits of each.

1 Introduction

During software development and maintenance, a substantial amount of time and effort is spent on testing, debugging, and optimizing the code. This is especially true for parallel and distributed software [11].

Performing these tasks efficiently requires elaborate tool support. Among the tools used, *on-line* tools provide the most benefits. On-line tools—in contrast to off-line tools—are applied to an application *while it is executing*, enabling the tools to adjust to the execution. In addition, such tools are capable not only of observing target application behavior, but also of manipulating the running application and guiding its execution trajectory. Numerous different classes of on-line tools exist. Among the most common are debuggers, performance analyzers, visualizers, load balancers, and steering tools.

However, there is a fundamental problem with on-line tools. On-line tools require a module termed an *on-line monitoring system* (or *monitor*) that handles low level accesses to the target system consisting of the hardware, operating system, programming environment, libraries, and finally application. These on-line

---

[*] This work is partly funded by National Science Foundation grants ASC-9527186 and CCR-9523544, and *Deutsche Forschungsgemeinschaft*, Special Research Grant SFB 342, Subproject A1.

monitors have to support *manipulation* of the target system as well as observation. This requirement plus the demand to minimize the probe effect by keeping unwanted intrusion minimal makes on-line monitors difficult and expensive to implement. In addition, they lack portability. To reduce costs and size, on-line tools usually include only a minimal on-line monitor specifically adapted to their needs. Finally, because on-line monitors access exclusive hardware and operating system interfaces and often perform modifications (so called instrumentation) to the target system, monitors from different tools are *incompatible*.

Being able to use on-line tools concurrently and utilize the synergy of cooperation would be desirable. E. g. a user might want to use a visualizer to analyze the behavior of a distributed application at a high level, and at the same time use a debugger to analyze the application's inner workings. Similarly she or he might want to use a performance analyzer and/or a steering tool while the application is executing in a production environment where a load balancing tool is active. Another example is the desire to concurrently use a debugger and a checkpoint/restart tool in order to be able to comfortable save and reset the application during debugging. The ability to concurrently use different, specialized tools concurrently would also enable tool builders to create tool environments by simply combining tool components. However, when a user tries to use two or more on-line tools concurrently, portions of the different monitors conflict, causing the tools, the application, or both to exhibit erroneous behavior, or frequently, crash. Often, it is not even possible to concurrently start the tools due to incompatibilities caused by *structural conflicts* or conflicts on exclusive interfaces among their monitors.

At the user level, this is counter-intuitive and violates the principle of orthogonality. Aspects of the implementation (e. g. modifications performed to the target system) of certain tools disrupt others and cause the tools to fail. It would be desirable to be able to use tools concurrently. This is what we will term on-line *tool interoperability*: tools that can be concurrently applied to the same application and offer the possibility to cooperate. Tool interoperability affects two levels: the monitoring level (observation and manipulation of the shared target system) and the user level (consistency and cooperation at the user interface level). The focus of this paper is on the more complex monitoring level.

2 Interoperability

The term interoperability is used in many circumstances, most of which don't relate to our area of focus. Several standards exist that address interoperability in the wider sense, but are not usable for on-line monitoring. *CORBA* (Common Object Request Broker Architecture) [10] is a middleware architecture addressing interoperability among distributed object oriented software components. *ToolTalk* [5] is a mechanism that enables different applications to communicate and thereby interoperate. However, CORBA and ToolTalk only address application level aspects and are not sufficient for the purpose of monitoring.[1] *PCTE* (Portable Common Tool Environment) [3] is a framework to support integrated software engineering tool environments. It defines how different tools may share common objects during development. Its support to access *running* applications is neglectable and it is not usable for interoperable on-line monitoring. Therefore, while technologies aimed at interoperability exist, they address scenarios different from the monitoring situation we are addressing.

To enable interoperable on-tools, *accesses* to *shared objects* (hardware, operating system, libraries, application processes, ...) among the tools have to be *coordinated*. Coordinating accesses from different independent monitoring systems is hardly possible without support from the operating system. Therefore, the only practicable way to support coordinated accesses at the user level is to base all tools on a *common monitoring system*. Thereby, structural conflicts and possible conflicts on exclusive interfaces can also be solved through careful implementation, enabling tools to concurrently *co-exist*.

Once this is achieved, *logical conflicts* may still be encountered when one tool modifies an object that is assumed to remain invariant by some other tool (e.g. when a visualizer shows a process executing on node A while a load balancer has migrated this process to node B). This leads to a *consistency problem* concerning explicit manipulations of shared objects.

On the other hand, *transparency problems* can occur, meaning that implementation dependent side effects from one tool are observed by another tool, although they should be hidden, e.g. when a tool reads a trap instruction instead of a normal program instruction, because another tool has set a breakpoint at that particular code address.

Without further measures, tools cannot *consistently co-exist* [7]: in the presence of manipulations from other tools, due to logical conflicts, they cannot preserve a consistent view of the target system. To resolve logical conflicts, the monitor as well as the tools must be enhanced. The monitor must provide a mechanism to notify the tools about other tools' accesses. The tools must process these notifications and update any internally held or assumed information about the application to regain consistency.

Three recent on-line monitoring systems are candidates that might be used as platforms to implement consistently co-existent and interoperable on-line tools: DAMS/PDBG, DPCL, and OMIS/OCM. We will examine the systems in this order. Their capabilities to solve structural and logical conflicts will serve as a guideline to determine the level of interoperability support. Abilities to overcome consistency and transparency problems will be examined.

3 DAMS/PDBG

DAMS is a distributed monitoring system infrastructure developed at the Universidade Nova de Lisboa, Lisbon, Portugal [2, 1]. DAMS is designed as a collec-

---

[1] CORBA and ToolTalk, however, could be used by interoperable tools to interact at the user level, after the problems at the monitoring level have been solved.

tion of distributed daemons communicating via standard protocols. It includes a clearly defined interface of how this infrastructure can be extended. The basic structure is depicted in figure 1.

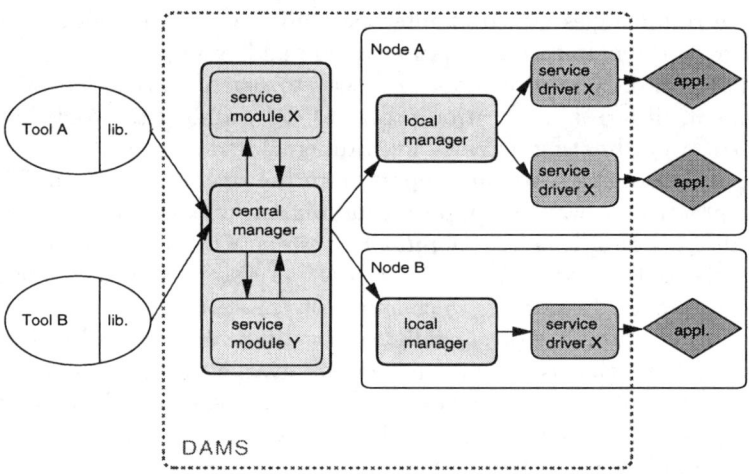

Fig. 1. Basic DAMS architecture

The standard daemons in DAMS are the *central manager* and one *local manager* per node in the distributed system. The daemons and all communication between daemons, tools, and service extensions is handled by DAMS. As such, DAMS does not provide any built-in service functionalities. To support specific types of tools, the basic infrastructure has to be extended. Functionality can be added by providing additional *services*. Services consist of a *service module* that is linked to the central manager, and a *service driver*, that is a stand-alone process connected to the local manager. Each service driver controls exactly one application process on its node.

Tools that interact with DAMS are linked with a special library. This library converts function calls to messages which are then passed to the central manager. The central manager identifies the service module that is responsible for handling the request contained in the message and forwards the necessary information. The corresponding service module inside the central manager is responsible for handling all distributed semantics. It takes appropriate actions and—via the central manager—sends messages to all local managers concerned. The local managers again forward the requests to the corresponding service drivers. Service drivers are responsible for implementing the local semantics. There is one service driver per application process, thus a service driver does not have to handle issues concerning parallelism or distribution. All replies follow the same stages in opposite order.

The first service extension implemented for the DAMS system was a distributed debugger, DDBG, later replaced by PDBG [6]. To implement the service extension, a debugging service module was implemented and brought into the service manager. In addition, a debugger service driver was written that uses gdb [14] to control the actual application process. Thus, per application process, there are two additional processes running: a debugger driver and gdb.

Since the DAMS infrastructure is very flexible and easy to extend, different classes of tools can be built on top of it, e. g. debuggers or performance analyzers. It is also possible for several tools to concurrently connect to the same application. However, DAMS does not provide any means to coordinate actions between different service extensions, thus tools built using different service extensions will not interoperate.

Nevertheless it is possible to build interoperable tools with DAMS. If the tools share the same service extension, this extension can be designed and implemented in such a way that tool interoperability becomes possible. Structural problems can be solved, enabling the tools to coexist. This has already been done in the development of the debugging service, PDBG. The PDBG extension offers services to control processes and to inspect and modify state, memory and registers. This is sufficient for a variety of tools as has already been demonstrated by basing a debugger, a test tool, and a graphical design environment on it. Besides compiler generated debugging information, no additional instrumentation like modified libraries is needed. This eliminates structural conflicts that could occur in libraries. Because only the PDBG driver accesses a process, no conflicts on potentially exclusive interfaces (like ptrace) occur.

To help resolve logical conflicts, PDBG allows tools to observe and get notified about other tools' actions on shared processes. Thus it becomes possible to implement tools that may co-exist without conflicts. DAMS supports to solve structural conflicts and logical conflicts as far as consistency problems are concerned. However, some problems remain:

- The debugging service in its current form can not support a broad spectrum of tools. E. g. its run-time overhead is far too big for the purpose of performance analysis. Extending the service to support a broader class of tools would require a complete rewrite of the service daemon since basing it on gdb is no longer feasible.
- Trying to solve transparency problems would also require a rewrite of the service.

Because transparency problems are not addressed, concurrently co-existing tools can observe implementation dependent side effects that should rather be hidden. Nevertheless, DAMS/PDBG is an excellent example of a contemporary tool infrastructure that supports interoperability among tools. Future developments will enhance its capabilities and provide a promising perspective for the future.

4 DPCL

DPCL (Dynamic Probe Class Library) [13, 12] is a project from IBM Corporation, partly in cooperation with the University of Wisconsin, Madison, and the University of Maryland, College Park. It defines a C++ class library that offers dynamic instrumentation of distributed applications.

The reference implementation of DPCL is built on the dyninst-API [4], the same API that was derived from *Paradyn* [9]. In order to support distributed applications, DPCL adds several daemon processes that can be distributed in a network. Clients then connect to these daemons. Similar to dyninst, a tool can dynamically insert pieces (called *probe expressions*) and even load modules of code into the application processes. The architecture of the DPCL implementation is illustrated in figure 2.

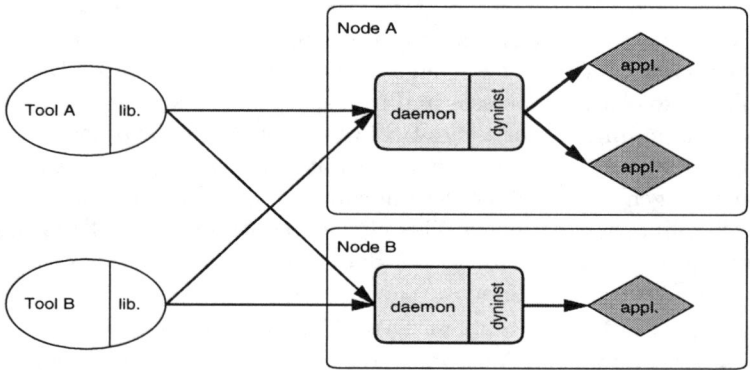

Fig. 2. DPCL architecture

On each node of the monitored system, there is a *daemon* to which tools can connect. The daemon manages all application processes on its node. A single tool can connect to multiple daemons and through them, to nodes. Also, multiple tools can connect to the same daemon and thus to the same application, concurrently. As in DAMS, communication and distribution are hidden by a library linked to the tools.

According to [13], DPCL is "designed to: ... increase interoperability among tools". Indeed, because all tools access the application through DPCL, only the shared DPCL daemon directly accesses an application process. Owing to this architecture and because the dyninst-API handles conflicts on process objects, structural conflicts as well as conflicts on implicitly shared exclusive objects like like the ptrace- or /proc-interface are solved by DPCL. Thus, tools may co-exist within this framework.

However, DPCL does not provide any means for detection and resolution of logical conflicts. Thus, consistency and transparency problems can not be solved, making it impossible to implement consistently coexistent tools based on DPCL alone.

This means that at the current time, there is no support to build consistently co-existing or cooperating tools on top of DPCL. DPCL primarily aims at improving the portability of tools. Interoperability in the sense of this survey is a possible future extension.

5 OMIS—On-Line Monitoring Interface Specification

OMIS (On-Line Monitoring Interface Specification) started out in 1995 as a joint project between LRR-TUM, Technische Universtität München, Munich, Germany, and Emory University, Atlanta, GA, USA. The goal of this project was to create a standard for runtime tool development for parallel and distributed systems [8]. The specification defines an interface of a programmable monitoring system that is powerful and general enough to support a wide variety of tools. To support this versatility, and in order to be able to support newer tools and different programming environments, OMIS is designed to be extensible.

OMIS alone is only a specification. The implementation, OCM (OMIS compliant monitor), was also started at LRR-TUM. It targets PVM environments and is termed OCM/PVM. The implementation is not yet complete; the current version serves as a platform for two tools, also developed at LRR-TUM, a debugger (DETOP), and a visualizer (VISTOP). An initial version of a checkpointing tool (CoCheck) is also running. The basic architecture of OCM/PVM is depicted in figure 3.

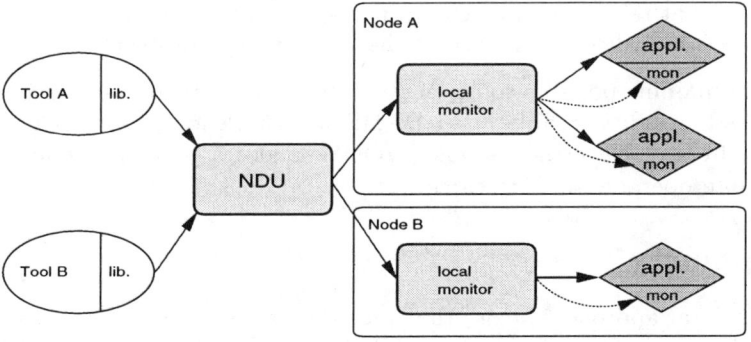

Fig. 3. Basic OCM/PVM architecture

Just like DAMS and DPCL, OCM/PVM is implemented as a distributed system of communicating processes. Tools access the monitor through a library that hides communication details. Requests issued by the tool are first sent to a daemon termed the *NDU* (Node Distribution Unit). Here, requests are parsed, global semantics are handled and the requests are broken up into sub-requests that can be handled on a single node. The NDU inserts and handles all necessary synchronization actions. Then, the sub-requests are sent to the *local monitors*, one of which runs on every node in the monitored system. The local monitor controls all application processes on its node and is responsible for handling local requests. It does not have to handle issues relating to distribution. Replies relating to tool requests are sent from the local monitors to the NDU, where they are analyzed and assembled before finally arriving at the tool.

Since performance analysis tasks have to be handled extremely efficiently, some actions are executed in the context of the observed application processes. OCM does not utilize dynamic instrumentation, therefore, parts of the monitor have to be linked to the application code before runtime. In addition, libraries are instrumented with monitoring components which could potentially result in structural conflicts.

However, OCM/PVM is designed and implemented in such a way that these and other structural conflicts, and conflicts on exclusive objects are resolved. All tools that share an object use the same controlling monitor process that coordinates accesses. Therefore, coexisting tools are possible.

Since OMIS is intended for research in interoperable tools, it goes one step further and also addresses logical conflicts. It defines events that signal accesses, which may be used to partly overcome consistency problems. However, there are also some shortcomings.

- The implementation is incomplete and some services required for consistency are missing from the current implementation.
- The defined events relate to actions performed by the tools. Because the monitor is extensible and new actions can be defined, a tool can not observe events that relate to an extension it does not know about.
- There is no defined way to solve the transparency problem.

The remaining possible consistency problems have their cause in OMIS' and OCM's extensibility. Similarly to DAMS, transparency problems are not yet being addressed. Nevertheless, OMIS/OCM is also a promising foundation for further interoperable tools research.

6 Summary

Today, several approaches exist that provide starting points for research in interoperable on-line tools. To support interoperable tools at the monitoring level, structural as well as logical conflicts have to be solved. All three systems examined in this paper can solve structural conflicts. Two of the systems also address logical conflicts and thus allow to overcome the consistency problem. However, support for the transparency problem is still unsatisfactory.

Recent developments try to address the transparency problem. While it is questionable whether all problems relating to tool interoperability can be solved satisfactorily, we expect that truly interoperable tools that further facilitate software development and maintenance will soon become available.

References

1. J. C. Cunha and V. Duarte. Monitoring PVM programs using the DAMS approach. *Lecture Notes in Computer Science*, 1497, 1998.
2. J. C. Cunha, J. Lourenco, J. Vieira, B. Moscao, and D. Pereira. A Framework to Support Parallel and Distributed Debugging. In P. Sloot, M. Bubak, and B. Hertzberger, editors, *Proceedings of High-Performance Computing and Networking, HPCN'98*, volume 1401 of *Lecture Notes in Computer Science*, pages 708–717, Amsterdam, The Netherlands, Apr. 1998.
3. European Computer Manufacturers Association (ECMA). Portable common tool environment: Abstract specification (standard ECMA-149), June 1993.
4. J. K. Hollingsworth and B. Buck. *DynInstAPI Programmer's Guide Release 1.1*. Computer Science Dept., Univ. of Maryland, College Park, MD20742, May 1998.
5. A. M. Julienne and B. Holtz. *ToolTalk and open protocols, inter-application communication*. Englewood Cliffs, NJ, 1994.
6. J. Loureno and J. C. Cunha. The PDBG Process-level Debugger for Parallel and Distributed Programs, Aug. 1998. Poster at the SPDT'98.
7. T. Ludwig, R. Wismüller, and A. Bode. Interoperable Tools based on OMIS (Abstract). In *Proc. 2nd SIGMETRICS Symposium on Parallel and Distributed Tools SPDT'98*, page 155, Welches, OR, USA, Aug. 1998. ACM Press.
8. T. Ludwig, R. Wismüller, V. Sunderam, and A. Bode. *OMIS — On-line Monitoring Interface Specification (Version 2.0)*, volume 9 of *LRR-TUM Research Report Series*. Shaker Verlag, Aachen, Germany, 1997. ISBN 3-8265-3035-7.
9. B. P. Miller, J. M. Cargille, R. B. Irvin, K. Kunchithap, M. D. Callaghan, J. K. Hollingsworth, K. L. Karavanic, and T. Newhall. The Paradyn parallel performance measurement tools. *IEEE Computer*, 11(28), Nov. 1995.
10. OMG (Object Management Group). The Common Object Request Broker: Architecture and Specification — Revision 2.2. Technical report, February 1998.
11. C. M. Pancake and C. Cook. What users need in parallel tool support: Survey results and analysis. In IEEE, editor, *Proceedings of the Scalable High-Performance Computing Conference, May 23–25, 1994, Knoxville, Tennessee*, pages 40–47, 1109 Spring Street, Suite 300, Silver Spring, MD 20910, USA, 1994. IEEE Computer Society Press.
12. D. M. Pase. An API for Run-Time Instrumentation of Single- and Multi-Process Applications: Class Reference Manual. Draft Document, Version 0.2, May 1998. IBM Corporation, RS/6000 Development, 522 South Road, MS P-963, Puoghkeepsie, New York 12601, ftp://grilled.cs.wisc.edu/DPCL/ref.bk.ps, also available from: http://www.ptools.org/projects/dpcl/.
13. D. M. Pase. Dynamic Probe Class Library (DPCL): Tutorial and Reference Guide. Draft Document, Version 0.1, July 1998. IBM Corporation, RS/6000 Development, 522 South Road, MS P-963, Puoghkeepsie, New York 12601, ftp://grilled.cs.wisc.edu/DPCL/pubs.bk.ps, also available from: http://www.ptools.org/projects/dpcl/.
14. R. M. Stallman and R. H. Pesch. *Using GDB, A Guide to the GNU Source Level Debugger, GDB Version 4.0*. Free Software Foundation, Cygnus Support, Cambridge, Massachusetts, July 1991.

Using the SMiLE Monitoring Infrastructure to Detect and Lower the Inefficiency of Parallel Applications

Jie Tao*, Wolfgang Karl, and Martin Schulz

LRR-TUM, Institut für Informatik
Technische Universität München, 80290 München, Germany
{tao,karlw,schulzm}@in.tum.de

Abstract. High computational demands are one of the main reasons for the use of parallel architectures like clusters of PCs. Many parallel programs, however, suffer from severe inefficiencies when executed on such a loosely coupled architecture for a variety of reasons. One of the most important is the frequent access to remote memories. In this article, we present a hybrid event-driven monitoring system which uses a hardware monitor to observe all of the underlying transactions on the network and to deliver information about the run-time behavior of parallel programs to tools for performance analysis and debugging. This monitoring system is targeted towards cluster architectures with NUMA characteristics.

1 Introduction

Parallel processing is a key technology both for commercial and scientific applications. Besides traditional multiprocessor and multicomputer systems, clusters of PCs are gaining more and more acceptance. Due to the synchronization and communication between processes as well as unbalanced allocation of processors, however, the parallel performance is often not as excellent as expected. This situation is more serious in fine-grained parallel systems due to the management and organization overhead. In order to obtain high system performance, it is therefore necessary to develop tools which will improve the locality of memory references and balance the load distribution among processors. Growing complexity of hardware, however, makes performance evaluation and characterization more difficult. This is especially true for PC-Cluster with DSM character as in those systems any communication is handled implicitly without the ability for a direct observation. The challenge is to design a powerful performance monitoring framework, which is able to deliver detailed information about the run-time communication behavior and to help the run-time system make correct decisions concerning the partitioning and redistribution of data and threads. Such a monitoring infrastructure has been developed for the SMiLE (Shared Memory in a LAN-like Environment) project at LRR-TUM.

The SMiLE project [2, 9] investigates in high performance cluster computing. As the limited communication performance over standard LANs restricts performance of parallel applications, the high-speed low-latency Scalable Coherent Interface(SCI) [1,

---

* Jie Tao is a staff member of Jilin University of China and is currently pursuing her Ph.D at Technische Universität München of Germany

5] is taken as interconnection technology. With hardware Distributed Shared Memory (DSM) and high performance communication characteristics SCI-based PC clusters are very suitable for High Performance Computing (HPC). Otherwise, fast data access and communication among cooperating threads within applications requires detailed information about the behavior of the system. Monitoring the dynamic behavior of a compute cluster with hardware-supported DSM like the SMiLE PC cluster, however, is very challenging as communication might occur implicitly on any read or write. This fact implies that monitoring must be very fine-grained, making it almost impossible to avoid significant probe overhead with software instrumentation. A hardware monitor is needed. The SMiLE hardware monitor provides the programmer or the system software with detailed information about any memory transactions on the network and the occurrence of user-defined events. With this information, the user or the system software is able to determine why a program runs slowly and how data and threads should be partitioned or redistributed. The hardware monitor, however, is not at its final stage. Performance tools must be developed which will analyze the observed behavior and manipulate the applications in order to achieve a more efficient execution. Based on a tight cooperation between monitor and tools, memory locality will be improved and load balancing mechanisms will be implemented. This article presents the design of the overall system.

The remainder of this paper is organized as follows. In Section 2 the SMiLE hardware monitor will be introduced, followed by an experimental evaluation using a few typical DSM applications in Section 3. The monitor infrastructure is then described in Section 4. Finally, we conclude in Section 5 with a short description of related work and a brief summary.

2 The SMiLE SCI Hardware Monitor

The SMiLE cluster consists of a number of PCs connected via SCI to a parallel system with NUMA characteristics (Non-Uniform Memory Access). Each node is equipped with a PCI-SCI adapter which can be attached by a hardware monitor. The node architecture of the SMiLE PC cluster is shown in Fig. 1.

The PCI-SCI adapter, described in detail in [2], serves as an interface between the PC's I/O bus and the SCI interconnection network. Like any other available PCI-SCI adapter the SMiLE PCI-SCI adapter consists of three logical parts: a PCI unit which interfaces to the PCI local bus, a Dual-Ported RAM (DPR) in which the incoming and outgoing packets are buffered, and an SCI unit which interfaces to the SCI network and performs the SCI protocol processing for packets. The PCI unit is connected to the DPR via the DPR bus and the SCI unit via the B-Link. Here, the B-Link, a 64-bit-wide synchronous bus, serves as the carrier of all incoming and outgoing packets to and from the SCI interface. Control information is passed between the PCI unit and the SCI via a handshake bus.

The SMiLE hardware monitor [6, 8] as part of the event-driven hybrid monitoring approach is attached to the PCI-SCI adapter as an additional PCI card. As shown in Fig.1, it consists of three modules: B-Link interface, counter module, and PCI interface. The B-Link, a central point on which all remote memory accesses can be monitored, is designed to acquire information from packets through the B-Link. The information

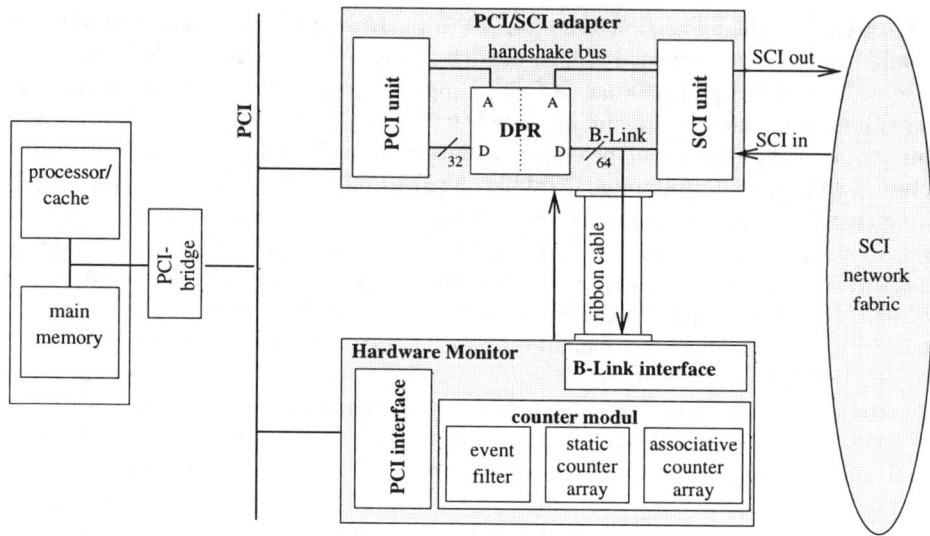

Fig. 1. The SMiLE SCI node architecture: a PCI-SCI adapter and the hardware monitor card installed in a PC

includes: transaction command (read, write, lock, unlock), source and destination IDs, memory address (page number and offset), and other characters (incoming, outgoing, response, and request).

As part of the hardware monitor, the *counter module* is responsible for the recording of the monitoring results. Its three components – an event filter, a static counter array, and an associative counter array – allow the programmer to utilize it in two working modes for performance analysis: the static and the dynamic mode. The static mode allows users to explicitly program the hardware for event triggering and action processing on definable SCI regions. This can be used to monitor given data structures or parts of arrays and will likely be applied in combination with language specific profiling extensions for detailed studies. Here, the event filter is used to allow the programmer to define events and the static counter array is used to record the frequency of events.

The *event filter* comprises a *page table* and an *event station*. The descriptors of all pages to be observed by the hardware monitor in a node are stored in the *page table*. The accurate events, which specify packets including data from certain data structures or arrays, are described in the *event station*. The *event station*'s *page frame* field points to a page descriptor in the *page table* while the *bottom* and the *top* address fields specify the range within the indexed page. The *transaction type* describes the relevant operation to that page. The *event selector* is used to decide when to begin and to end the monitoring of an event. The *counter* is used to count the event. When the information about a packet arrives from the B-Link interface, first the page number and destination ID will be compared with the page number and node number in the *page table*. With the index of the *page table* the corresponding entries of the *event station* are checked. If the event matches one of the count events defined by the programmer (and the event has been previously enabled), the corresponding counter is incremented. The programmer can also specify events in their programs to enable or disable the counting of an event defined in the *event station*. In case a counter overflow, its value will be spilled to a user-defined ring-buffer, which is also used by the dynamic mode.

The *dynamic* working mode is designed to deliver detailed information to tools for performance evaluation. In this mode, all packets passing the B-Link will be monitored. In order to be able to record all remote memory accesses to and from a node with only limited hardware resources, the monitor exploits the spatial and temporal locality of data accesses in a similar way as cache memories do. The hardware monitor contains a content-addressable counter array managing a small working set of the most recently referenced memory regions. If a memory reference matches a tag in the counter array, the associated counter is incremented. If no reference is found, a new counter-tag pair is allocated and initialized to 1. If no more space is available within the counter array, first counters for neighboring address areas are merged or a counter-tag pair is flushed to a larger ring buffer in main memory. This buffer is repeatedly emptied by the system software in a cyclic fashion. In the case of a ring buffer overflow, a signal is sent to the software process urging for the retrieval of the ring buffer's data.

With typical PC I/O bandwidths of around 80 MB/s (PCI's bandwidth of 133 MBytes per second can only be reached with infinite bursts), monitoring represents a more 0.8% of the system's bus load and shouldn't influence program execution too much.

3 First Results from the Memory Behavior Tests Using the SMiLE Monitor

We have noticed that some parallel applications don't run efficiently upon our SMiLE cluster due to too many remote memory accesses [14]. In order to understand the memory behavior of parallel programs in detail we have tested a few applications from the SPLASH-2 benchmarks suite [16]. As the physical implementation of the hardware monitor is currently under development, a simulator of the hardware monitor and its appropriate driver have been implemented. The simulator is based on the multiprocessor memory system simulator Limes [10], which allows to produce a trace file during the simulation of the parallel execution of an application. The trace file contains all packets which are transfered in the case that the application runs natively on an SCI-based cluster. The monitor simulator reads the trace file and treats every packet in the same way as a hardware implementation would. The monitoring results are collected by the driver and offered to the user. Fig. 2 and Fig. 3 provide a few examples of such statistics of the memory behavior of the FFT, LU, RADIX and OCEAN programs of the SPLASH-2 suite [16].

The results indicate that each page has a different access behavior. Some pages are accessed much more frequently by local processor while some ones are accessed more often by remote ones. There are also pages which are accessed evenly by all of the processors. The tests also reveal that for each application the remote memory accesses take a great proportion of all memory transactions. For FFT, e.g., among the total 93372

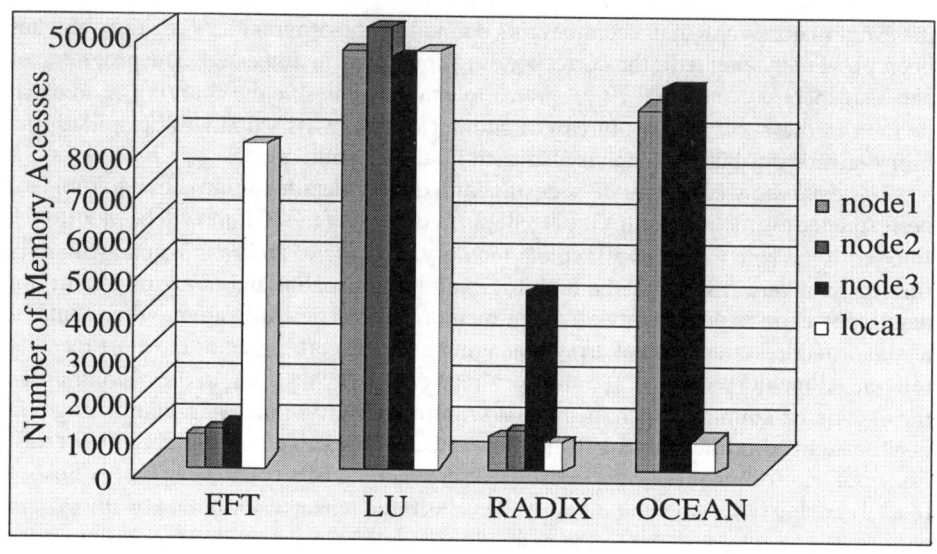

Fig. 2. The number of memory accesses of a single page on node 0 related to different applications

memory accesses to shared data by all of the four nodes the remote ones are 48139 times, i.e. 51.6% of the total accesses. For the LU program, the ratio of remote accesses to total ones is even 79.1% (4415847 remote, 5582318 total).

4 The Software Infrastructure of the Monitoring System

The results shown above dramatically demonstrate the necessity of memory locality optimization. In addition to a tool for this task, other tools like a dynamic load balancer are necessary to reach a close to optimal parallel execution. The locality optimizer needs information about memory behavior from the hardware monitor and the load balancer

source \ page	0	1	2	3	4	5	6	7	8	9	10	11
1	13890	0	0	36676	17011	58154	7489	71522	1924	81469	0	442
2	14945	0	0	47974	19473	69568	9902	91510	2343	102254	0	13
3	16269	0	0	32931	63486	17209	77154	6896	87099	511	89273	9
local	71900	0	0	34832	51884	20496	65568	10256	74556	4112	79192	16

Fig. 3. The number of memory accesses of all different pages of LU on node 0

needs the CPU states of the processors from the operating system, etc. These tools work sometimes simultaneously and the communication among them is also at times necessary. Therefore, an interface must exist to connect the tools together and the tools with the low-level environment together. OMIS / OCM is just such an interface allowing the safe interoperability between arbitrary OMIS compliant tools.

The OMIS(On-line Monitoring Interface Specification) [4] project at LRR-TUM provides the basis to integrate individual tools and the low layers of the computers into a single environment. OCM [15], an OMIS Compliant Monitoring system is a standardized interface which allows different research groups to develop tools that can be used concurrently with the same program. The OMIS/OCM system has two interfaces: one for interaction with the different tools, the other for interaction with the program and all system underlying layers which keep the program running. The interaction between tools and OMIS/OCM system is handled via asynchronous procedure calls. The tool invokes a service request and either waits for results coming back from the OMIS/OCM system or specifies a call-back to be invoked when results are available. OMIS/OCM provides means to extend these interfaces.

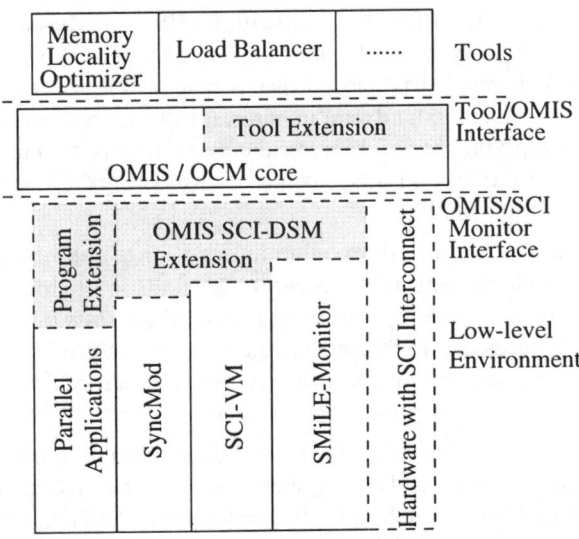

Fig. 4. The Software Infrastructure of the Monitoring System

With OMIS / OCM we have designed a whole monitoring infrastructure which is shown in Fig.4. This infrastructure consists of three main components: Tools, OMIS / OCM, and a Low-level Environment. As described above, the tools need information about the low-level environment such as node status and memory access behavior etc., which will be offered by the SMiLE-HW monitor, the program code, libraries, and the operating system. These are summarized as the Low-level Environment.

The information delivered from the SMiLE hardware monitor is purely based on physical addresses and does not suffice for the evaluation of applications. In order to make the information readable from the application level the translation of physical addresses seen by the monitor into virtual addresses visible from the application level must be realized. The translation of virtual addresses into physical address, on the other hand, also has to be supported as any event definition for the hardware monitor requires physical addresses. This translation is controlled by the DSM layer, in this case the SCI Virtual Memory(SCI-VM) [13]. This layer extends the virtual memory management of the operating system to a cross-node memory resource control and is responsible for setting up the correct virtual address mappings to both local and remote memory. It therefore is able to directly provide the required information to the monitor system and also the information which aid in the performance evaluation of applications. While the hardware monitor lacks the ability of monitoring synchronization primitives, which generally have a critical performance impact on shared memory applications, the Sync-Mod module delivers various statistical information. This ranges from simple lock and unlock counters to more sophisticated information like mean and peak lock times and peak barrier wait times. This information will allow the easier detection of bottlenecks in applications.

To connect the Low-level Environment with OMIS / OCM system so that the observed information about system behavior can be delivered to various envisioned tools an *OMIS SCI-DSM Extension* is needed that is responsible for the realization of all the service requests related to hardware monitor, for the address translation in both directions, and for supplying the information about synchronization primitives. Another module, the *Program Extension* is responsible for abstracting the virtual addresses from the symbol-table of the compiler.

Although the user can utilize the monitor information to optimize execution of their applications it is also important and necessary to develop tools to improve the efficiency of parallel programs transparently. From the results of the tests described in section 3 we notice that it is necessary to redistribute the data among processors in order to decrease the effect of remote memory accesses. A special tool for this purpose, a Memory Locality Optimizer, is currently under development.

The Memory Locality Optimizer focuses on an optimization that can significantly reduce remote data accesses in DSM system. The performance of a parallel system with DSM character depends on the efficient exploitation of memory locality. Although remote memory accesses via hardware-supported DSM deliver high communication performance compared to traditional systems, they are still an order of magnitude more expensive than local ones. Therefore, powerful tools have to be introduced to exploit data locality. The Locality Optimizer consists of a decision-making subsystem and a data- and thread-migration subsystem. The first subsystem analyzes all memory transactions in order to recognize which part of the global memory is accessed frequently by which thread. Based on this analysis and also based on information available from the compiler, the user, the SCI-VM, and the synchronization system, it decides which page or thread should be moved. The migration subsystem will then automatically and transparently migrate data and threads at run-time. In this case the communication among parallel threads will be minimized.

In the following a few strategies that aid in partitioning and redistribution of data and threads are introduced:

1. Two or more logically unrelated data objects that are used by different application components (threads) should not be placed on the same page.
2. A page that is accessed frequently from a single remote thread but rarely from others should be migrated to node executing this thread. However, when local threads also access this page frequently then the page should not be migrated and it will be taken into consideration whether the remote thread should be moved to the node on which the page is located.
3. A page that is accessed often by more than one thread can be: a) distributed among threads with that each threads has only a part of this page in its local memory; or b) placed only in one memory; or c) replicated onto all nodes. In the late case additional consistency enforcing mechanisms have to be introduced.
4. A thread which access only different pages of a memory on a processor frequently should be migrated onto this processor.

Each node of the cluster executes a Memory Locality Optimizer which is in charge of counting the accessed frequency of every active local page. For the LU program in the SPLASH-2 suite, for example, if page 9 in node 0 is moved to node 2 instead staying on node 0, the remote memory accesses of node 2 to node 0 would be lowered for 35.5% while the remote accesses of node 0 would be increased only for 0.1%.

Good locality, however, is not the sole property needed to achieve optimal performance, load balancing is of equal importance. Another tool, the Load Balancer is therefore also being developed at LRR-TUM for the SMiLE project. This tool is used to increase the utilization of system resources and improve the performance by balancing the load. The main goal is to split the work evenly among all processors. It consists of two parts: a static and a dynamic load distribution mechanism. The static mechanism is responsible for the initial assignment of load, i.e., finding the most appropriate processor for a being created parallel thread according both to the varying system capabilities and the application varying needs, while the dynamic one deals with the load redistribution after startup, i.e., transferring load from an overloaded host to an idle or an underloaded one. Static load distribution strategies are only effective when applied to problems that can be partitioned into tasks with uniform computation and communication requirements. There exist, however, a large number of problems with non-uniform and unpredictable computation and communication requirements. Therefore, it is essential that a DSM system is adaptive to dynamic changes in the environment and is able to respond to varying needs of an application by adopting dynamic load distribution mechanisms. Dynamic load distribution mainly means thread migration, but this work must be done with the consideration of data locality as there is a potential conflict between data locality and load balancing. The threads that are chosen to be migrated by the load balancing policy must be selected with care, in order to minimize the resulting remote memory accesses. The destination of the migration must be also considered carefully and the choice of the most lightly loaded processor is not sufficient as improper placement of threads might degrade performance due to the high communication overhead.

In order to connect tools by OMIS / OCM, the *Tool Extension* module must be defined adding a description of the new service requests of the tools into OMIS / OCM system.

5 Related Work and Conclusion

Over the last years, monitoring support has become increasingly available on research as well as on commercial machines. The SHRIMP hardware performance monitor has been developed to measure the running behavior of the SHRIMP PC cluster, which is connected with special network adapters capable of performing remote updates [12]. In contrast to our hybrid hardware monitor, it has a DRAM in the monitor card to record the complete access histograms and run-time measurements of running applications. Similar to the SMiLE approach, these measurement can be used to modify mutable parts of the system hardware and to help with software development and application tuning. Trinity College Dublin has designed a trace instrument for SCI-based systems [11] that allows full traces of interconnect traffic. This instrument provides hardware designers and software developers with a tool that allows a deeper understanding of the temporal behavior of their hardware and software. In contrast to the SMiLE monitoring system, however, this trace environment can not be used for on-line tools due to the large amount of trace data.

Tool environments for DSM-oriented systems, however, are less wide-spread [3, 7]. On-line tools based on hardware monitors with the purpose to improve system performance have, to our knowledge, not yet been presented. We believe that performance data provided by a hardware monitor should not only be utilized by programmer or off-line tools to the analyzation of system performance, but also by on-line tools to enable the run-time manipulation of parallel executions. Data migration based on observed memory transactions, for example, can improve the performance of an application dramatically. The combination of the hardware monitor and on-line tools therefore offer a great potential for High Performance Computing.

References

1. IEEE Standard for the Scalable Coherent Interface(SCI). IEEE Std 1596-1992,1993, IEEE 345 East 47th Street, New York, NY 10017-2394, USA.
2. G. Acher, W. Karl, and M. Leberecht. PCI-SCI protocol translations: Applying microprogramming concepts to FPGAs. In *Proceedings of the 8th International Workshop, FPL'98*, volume 1482 of Lecture Notes in Computer Science, pages 238–247, Tallinn, Estonia, September 1998. Springer Verlag, Heidelberg.
3. D. Badouel, T. Priol, and L. Renambot. A performance tuning tool for DSM-based parallel computers. In *Proceedings of Europar'96-Parallel Processing*, volume 1123 of Lecture Notes in Computer Science, pages 98–105, Lyon, France, August 1996. Springer Verlag.
4. M. Bubak, W. Funika, R. Gembarowski, and R. Wismüller. OMIS-compliant monitoring system for MPI applications. In *Proc. 3rd International Conference on Parallel Processing and Applied Mathematics - PPAM'99*, pages 378–386, Kazimierz Dolny, Poland, September 1999.

5. Hermann Hellwagner and Alexander Reinefeld, editors. *SCI: Scalable Coherent Interface: Architecture and Software for High-Performance Computer Clusters*, volume 1734 of Lecture Notes in Computer Science. Springer-Verlag, 1999.
6. R. Hockauf, W. Karl, M. Leberecht, M. Oberhuber, and M. Wagner. Exploiting spatial and temporal locality of accesses: A new hardware-based monitoring approach for DSM systems. In *Proceedings of Euro-Par'98 Parallel Processing / 4th International Euro-Par Conference Southampton*, volume 1470 of Lecture Notes in Computer Science, pages 206–215, UK, September 1998. Springer Verlag.
7. Ayal Itzkovitz, Assaf Schuster, and Lea Shalev. Thread migration and its applications in distributed shared memory systems. *The Journal of Systems and Software*, 42(1):71–87, July 1998.
8. Wolfgang Karl, Markus Leberecht, and Martin Schulz. Optimizing data locality for SCI-based PC-clusters with the SMiLE monitoring approach. In *Proceedings of International Conference on Parallel Architectures and Compilation Techniques(PACT '99)*, pages 169–176, Newport Beach, CA, October 1999. IEEE Computer Society.
9. Wolfgang Karl, Markus Leberecht, and Martin Schulz. Supporting shared memory and message passing on clusters of PCs with a SMiLE. In *Proceedings of the third International Workshop, CANPC'99*, volume 1602 of Lecture Notes in Computer Science, Orlando, Florida, USA(together with HPCA-5), January 1999. Springer Verlag, Heidelberg.
10. D. Magdic. Limes: An execution-driven multiprocessor simulation tool for the i486+–based PCs. School of Electrical Engineering, Department of Computer Engineering, University of Belgrade, POB 816 11000 Belgrade, Serbia, Yugoslavia, 1997.
11. Michael Manzke and Brian Coghlan. Non-intrusive deep tracing of SCI interconnect traffic. In *Conference Proceedings of SCI Europe'99*, pages 53–58, Toulouse, France, September 1999.
12. M. Martonosi, D. W. Clark, and M. Mesarina. The SHRIMP performance monitor: Design and applications. In *Proc. SIGMETRICS Symposium on Parallel and Distributed Tools*, pages 61–69, Philadelphia, May 1996.
13. M. Schulz. SCI-VM: A flexible base for transparent shared memory programming models on clusters of PCs. In *Proceedings of HIPS'99*, volume 1586 of Lecture Notes in Computer Science, pages 19–33, Berlin, April 1999. Springer Verlag.
14. Martin Schulz and Hermann Hellwagner. Global virtual memory based on SCI-DSM. In *Proceedings of SCI-Europe '98*, pages 59–67, Bordeaux, France, September 1998. Cheshire Henbury.
15. R. Wismüller. Interoperability support in the distributed monitoring system OCM. In *Proc. 3rd International Conference on Parallel Processing and Applied Mathematics - PPAM'99*, pages 77–91, Kazimierz Dolny, Poland, September 1999.
16. Steven Cameron Woo, Moriyoshi Ohara, Evan Torrie, Jaswinder Pal Singh, and Anoop Gupta. The SPLASH-2 programs: Characterization and methodological considerations. In *Proceedings of the 22nd Annual International Symposium on Computer Architecture*, pages 24–36, June 1995.

Run-Time Optimization Using Dynamic Performance Prediction

A.M. Alkindi[1], D.J. Kerbyson[1], E. Papaefstathiou[2], and G.R. Nudd[1]

[1]High Performance Systems Laboratory, Department of Computer Science
University of Warwick, UK
{alkindi,djke,grn}@dcs.warwick.ac.uk
[2]Microsoft Research, Cambridge, UK
efp@microsoft.com

Abstract. With the rapid expansion in the use of distributed systems the need for optimisation and the steering of application execution has become more important. The unquestionable aim to overcome bottle-neck problems, allocation, and performance degradation due to shared CPU time has prompted many investigations into the best way in which the performance of an application can be enhanced. In this work, we demonstrate the impact of using a Performance Prediction Toolset, PACE, which can be used in Dynamic (*On-The-Fly*) decision making for optimising application execution. An example application, the FFTW (The Fastest Fourier Transform in the West), is used to illustrate the approach which itself is a novel method that optimises the execution of an FFT. It is shown that performance prediction can provide the same quality of information as a measurement process for application optimisation but in a fraction of the time and thus improving the overall application performance.

Keywords: Performance Optimisation, Dynamic Performance Prediction, Performance Modeling, Application Steering, FFTW.

1 Introduction

Advances in technology, increasing user interaction with complicated systems, and the existence of powerful communication networks, have all made it easier to create high performance solutions. However, the ease in which these solutions can be formulated is highly dependent upon the complexity of the available resources, and the nature of the applications involved. A significant amount of work is being undertaken to ease this process – much of which is based on the use of performance information to guide the application execution on to the target systems. The promise of Information Power GRIDS [1] relies on the availability of systems and the utilisation of performance information to guide application execution.

Several performance tool-sets have been proposed and implemented dealing mainly with performance instrumentation and measurement. Tools such as Falcon [2], Paradyn [3], and Pablo [4] are being used to assist in the performance tuning and identification of bottlenecks in applications. These tool-sets typically include various features to allow application instrumentation at various levels, data collection, and interrogation through visualisation modules. Most of the performance monitoring tools are used for post-mortem analysis – i.e. to investigate the performance after the event has occurred. This may be in the analysis of the performance data after the end of application execution, or dynamically as the application is being executed. Other performance tools are being used to identify bottlenecks that may be inherent in the application design on particular systems. For example TASS is concerned with the relationship between parallel algorithms and architectures [5].

There has been very little work on dynamic optimisation using performance prediction while an application is being executed. Modelling approaches have the potential to provide the same performance information but without measurement on the systems. Accuracy of the performance information is important, and should be qualitatively the same as the measurements.

Several performance prediction tools are being developed including PAMELA [6,7] which enables a performance model to be constructed at compile time and combined with applications. This allows access to performance information whilst the program is executing.

In this work, we demonstrate the impact of using Dynamic (*On-The-Fly*) performance prediction to guide the execution of sequential applications. An example high performance application, the FFTW (Fastest Fourier Transform in the West) from MIT [8,9], is used to illustrate the possible use of this approach.

The toolset being developed at Warwick, PACE (Performance Analysis and Characterisation Environment) [10,11] encompasses the performance aspects of application software, its resource use and mapping, and the performance characteristics of hardware systems. It enables performance models to be constructed using an underlying performance specification language, CHIP^{3}S (Characterisation Instrumentation for Performance Predication of Parallel Systems) [11], and to be evaluated dynamically requiring only a few seconds of CPU time. The model can be compiled into a self-contained binary which can be executed and linked with an application.

In this paper we show how a PACE performance prediction model can be used to dynamically determine the execution behavior of an application. The FFTW is used to demonstrate this capability but the approach is general and could be used in systems which are dynamic in nature such as the Information Power GRIDs. The overall performance of the FFTW can be improved due to a reduction in time required to determine the best FFT calculation method for a given target processor. The outcome of this work is to provide a performance modelling approach which can be used for dynamic decision making, and also provides improvements to the FFTW performance.

The use of dynamic performance prediction is introduced in Section 2. An overview of the PACE toolset is given in Section 3. In Section 4, the FFTW package is described and the use of PACE models are illustrated in Section 5. Results comparing the default behavior of the FFTW and the FFTW with PACE performance models are given in Section 6. These show that dynamic performance prediction can be used to accurately undertake the decision making processes to optimise the FFTW execution.

2 Application Steering Using Performance Prediction

There are a number of decisions that a user has to make in order to execute an application. These are traditionally specified through the use of two types of application parameters, broadly classed into two categories:

Problem parameters - those that specify the format of the data to be processed e.g. data size, and the type of results required such as accuracy in a numerical calculation, the number of iterations required in an iterative solution etc.

System parameters – those that specify how a target system will be utilised, e.g. in specifying the number of processors to be used in a high performance system, and possibly also to specify the platform to be utilised when several are available.

The problem parameters need to be specified by the user (always) and depend on the calculation required. However, the system parameters are used to determine the mapping of the application onto the available system, and are normally used to reduce execution time. By coupling a performance model into the application, predictions can be automatically made to determine these system parameters. These *on-the-fly* decisions can be made with negligible overhead in comparison to the total application execution time if the performance model is rapid in its evaluation.

In addition, there are many applications in which numerical routines are used which may have many methods available for solution. Consider the situation depicted in Fig. 1a where an application has several methods of solution and two target systems are available. In this scenario, the user traditionally specifies the problem parameters, decides on the actual code to be used, and the target system along with relevant system parameters.

The use of a performance model in this example can be used to determine which code should be executed, on which target system, along with the relevant system parameters, based on achieving the minimum execution time. Thus, the user need only specify the problem parameters, with the performance model directing the application execution. Fig. 1b illustrates this situation, and shows the chosen code and system (solid arrows), and other available choices (dotted arrows).

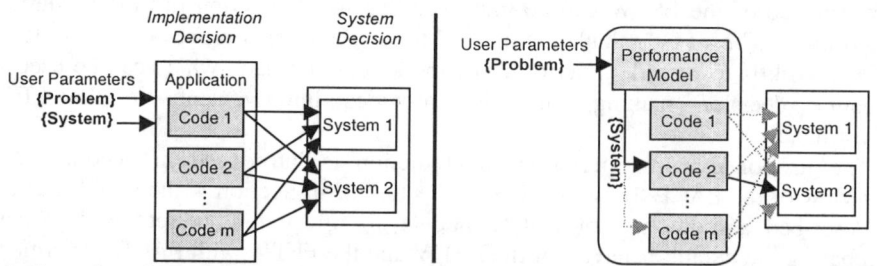

Fig. 1. Execution of an application having multiple code implementations and available systems. (a) User directed approach, (b) Performance directed approach.

The decisions made by the performance model require a number of separate scenarios to be evaluated. Each scenario is a function of the mapping, the implementation, and the available system(s). The best scenario is taken to be that which results in a minimum predicted execution time over the total number of scenarios evaluated. Thus, rapid performance model evaluation time is required.

The FFTW is an example code having multiple implementations which minimise the execution of a Fast Fourier Transform using an extensive measurement procedure [8,9]. It is shown in Section 4 how this measurement procedure can be replaced using performance prediction.

In addition, individual performance models can be used collectively within a task scheduling environment in which a scheduler is responsible for maintaining useful activity on system resources. The scheduling process is enhanced by use of predicted execution times, and mapping information from each task, e.g. [12].

3 The Performance Analysis and Characterisation Toolset (PACE)

PACE is a modelling toolset for high performance and distributed applications. It includes tools for model definition, model creation, evaluation, and performance analysis. It uses associative objects organised in a layered framework as a basis for representing each of a system's components. An overview of the model organisation and creation is presented in the following sections.

3.1 Model Components

Many existing techniques, particularly for the analysis of serial machines, use Software Performance Engineering (SPE) methodologies [13], to provide a representation of the whole system in terms of two modular components, namely a software execution model and a system model. However, for high performance computing systems, the organisation of models must be extended to take into account concurrency. The layered framework is an extension of SPE for the characterisation of parallel and distributed systems. It supports the development of several types of models: software, parallelisation (mapping), and system (hardware). The functions of the layers are:

Application Layer – describes an application in terms of a sequence of subtasks. It acts as the entry point to the performance study, and includes an interface that can be used to modify parameters of a performance scenario.

Application Subtask Layer – describes the sequential part of every subtask within an application that can be executed in parallel.

Parallel Template Layer – describes the parallel characteristics of subtasks in terms of expected computation-communication interactions between processors.

Hardware Layer – collects system specification parameters, micro-benchmark results, statistical models, analytical models, and heuristics that characterise the communication and computation abilities of a particular system.

In the layered framework, a performance model is built up from a number of separate objects. Each object is of one of the following types: application, subtask, parallel template, and hardware. A key feature of the object organization is the independent representation of computation, parallelisation, and hardware.

Each software object (application, subtask, or parallel template) is comprised of an internal structure, options, and an interface that can be used by other objects to modify its behaviour. The main aim of these objects is to describe the system resources required by the application which are modelled in the hardware object.

Each hardware object is subdivided into many smaller component hardware models, each describing the behaviour of individual parts of the hardware system. For example, the memory, the CPU, and the communication system are considered in separate component models e.g. [14].

3.2 Model Creation

PACE users can employ a workload definition language (CHIP³S) to describe the characteristic of the application. CHIP³S is an application modelling language that supports multiple levels of workload abstractions [11]. When application source code is available the Application Characterization Tool (ACT) can semi-automatically create CHIP³S workload descriptions. ACT performs a static analysis of the code to produce the control flow of the application, operation counts in terms of SUIF language operations, and the communication structure. This process is illustrated in Fig. 2. SUIF (Stanford Intermediate Format) [15] is an intermediate presentation of the compiled code that combines the advantages of both high level and assembly language. ACT cannot determine dynamic performance related aspects of the application such as data dependent parameters. These parameters can be obtained either by profiling or with user support.

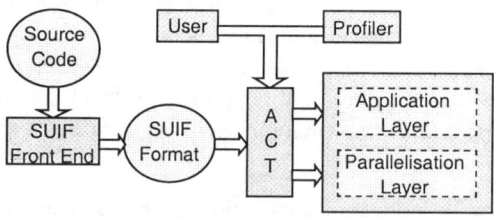

Fig. 2. Model Creation Process with ACT.

The CHIP³S objects adhere to the layered framework. A compiler translates CHIP³S scripts to C code which are linked with an evaluation engine and the hardware models. The final output is a binary file which can be executed rapidly. The user determines the system/application configuration and the type of output that is required as command line arguments. The model binary performs all the necessary model evaluations and produces the requested results. PACE includes an option to generate predicted traces (PICL, SDDF) that can then further analyzed by visualization tools (e.g. PABLO) [16].

4 The Fastest Fourier Transform in the West (FFTW)

FFTW is a Portable C package that computes one or more dimensional complex discrete Fourier Transform (DFT) [8,9]. It is claimed that the FFTW can compute the transform in a faster time than other available software due to its self-configuring nature. It requires a sequence of measurements to be made for it's software components (codelets) on the target system, and then uses a configuration of these to determine how best to compute the transform of any given size. There are three essential components that make up the FFTW, shown in Fig. 3.

Codelets: These are a collection of highly optimized blocks of C routines, which are used to compute the transform. The codelets are generated automatically. There are two types of Codelets:

Non-Twiddle: codelets that are used to solve small sized transforms (N <= 64), and

Twiddle: codelets that solve larger transforms by the combination of smaller ones.

Planner: The planner determines the most efficient way in which the codelets can be combined together to calculate the required transform. Each individual combination of codelets is called a plan. Typically 10's of plans are available for a specific FFT size.

Executor: The execution of each plan is performed and measured by the Executor. This results in the run-time cost of each plan on a specific system. From this, the best plan can be determined – i.e. the plan with the minimum execution time.

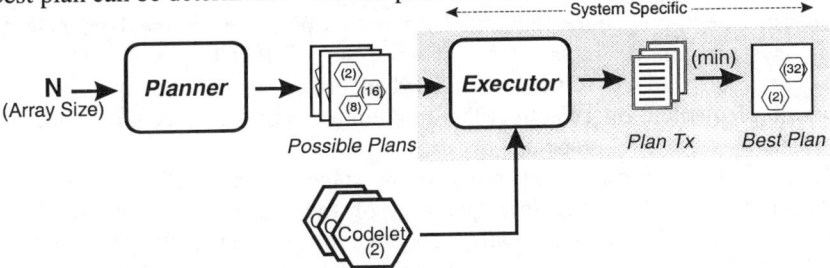

Fig. 3. Main components of the FFTW

For example, for the FFTW to compute the transform of a complex array with a size **N**, the *Executor* factors it first into $N = N_1 N_2$. Then, N_2 transforms of N_1, and N_1 transforms of N_2 are recursively computed hence calculating the exact transform of the complex array recursively.

Assuming that a complex array of size **N=64** is to be transformed, the number of possible FFTW plans is 20 as shown in Table 1. The plans are forwarded one by one to the Executor for measurement on the target system. The resulting execution times from each plan are included in Table 1 for a SUN Ultra 1. In this plans 1 to 6 contain only a twiddle codelet, and the remaining plans contain both a Twiddle and Non-Twiddle codelets. Plans that compute the full transform are numbers 1, and 16-20 (the remaining plans represent transforms of smaller sizes and may be a part of other plans). In this case, the best (fastest) plan to solve the N=64 transform is Plan 16, which is computed by the Non-Twiddle Codelet 2 and Twiddle Codelet 32.

Table 1. Example FFTW plans for N=64 (Timings for SUN Ultra 1)

Plan #	1	2	3	4	5	6	7	8	9	10	11	12	13	14	15	16	17	18	19	20
Codelet Size 1		X Non-Twiddle Codelets																		
2					X	X O	X	O	X		X	X		X			X			O
3		O Twiddle Codelets																		
4				X			O	X		X O			X	O		X		O		
5																				
6																				
7																				
8				X					O		O		O	X				X O		
9																				
10																				
16			X										O			O			X	
32		X																		X
64	X																			
Time μS	65	27	11	5	2	1	7	11	14	20	24	29	41	45	51	63	88	92	96	112

The performance of FFTW relies heavily on the successful selection of the best plan. The procedure for selecting the optimum configuration comes with a high initial cost that involves the measurement of every plan, a process that needs to be repeated several times in order to remove measurement noise induced by system load. The initial tuning stage cost is exponentially increases when the size of the problem grows due to the resulting larger number of possible transformations.

5 Dynamically Predicting the FFTW Performance

The measurement procedure required by the FFTW to select the best plan for a specific system can be replaced by using a set of PACE performance models. Separate models can be formed of individual codelets, and the operation of the planner implemented by selecting the minimum predicted execution time of an appropriate set of codelet models.

The model that represents the initial tuning stage of the FFTW includes a single application task that mimics the operation of the Planner. The models for the individual codelets are automatically generated by analyzing the existing codelet codes by ACT. A subtask represents an individual codelet called by the planner to produce a possible solution.

The PACE performance model generates and evaluates all possible plan models, instead of actually executing the codelets on the target platform. The planner uses the individual codelet model predictions to combine them to overall plan performance predictions. The planner keeps a record of all plan predicted times in order to identify the fastest combination. The predicted optimum plan is then used by the FFTW executor to perform the transformation. The evaluation of each plan can be performed for a range of systems and problems sizes. The evaluation time for this process is orders of magnitude smaller than the corresponding measurements. An overview of the prediction process is shown in Fig. 4.

The validity of the PACE optimization depends on the selection of the optimum plan. To achieve this the performance models require accurate prediction of each FFTW plan's execution time. The advantage of using PACE models is to minimize the initial measurement time that FFTW requires for each data size and each new target system. Another advantage of using PACE to predict the best plan is reduce the influence of background loading which may effect the measurement procedure.

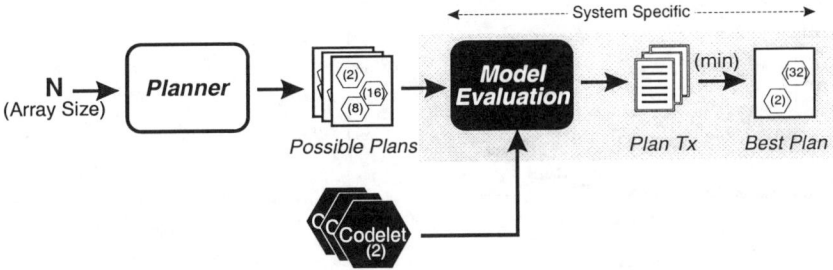

Fig. 4. Components of the FFTW using PACE performance prediction.

6 Results

In this section we present a comparison of plan execution times (measured from FFTW) versus plan predicted times evaluated from PACE models for a range of array data sizes and workstations. In all cases PACE and FFTW best plans are calculated to be the same.

Four sets of comparisons between measurements and predictions for the execution of the FFTW for different configurations are shown in Figs. 5a to 5d. In all cases, the X-axis includes all the plans generated by the FFTW planner. Note that the plans generated include combinations of codelets that might not successfully compute the entire FFT. The FFTW planner generates redundant plans assuming that they might be needed by future twiddle codelets.

Table 2. Comparison of plan execution time (s) versus model evaluation times for a range of array sizes running on SUN Ultra 10 and SUN Ultra 1 workstations.

Size	SUN Ultra 1		SUN Ultra 10	
	FFTW (s)	PACE (s)	FFTW (s)	PACE (s)
256	0.006	0.007	0.003	0.005
2 K	0.059	0.010	0.032	0.006
128 K	7.069	0.130	3.842	0.009
196 K	16.537	0.028	8.761	0.018

Table 2 shows the time required for the FFTW executor to measure all plan scenarios vs. the time required to evaluate the plan models for several problem and system configurations. The array size varies from small arrays (256 elements) to more realistic cases with 196K elements. Timings are provided for the Sun Ultra 10 and Sun Ultra 1 workstations. With the exception of the 256 elements case the PACE model evaluation times are orders of magnitude faster than the execution times. Thus using PACE models results in a performance improvement to the FFTW.

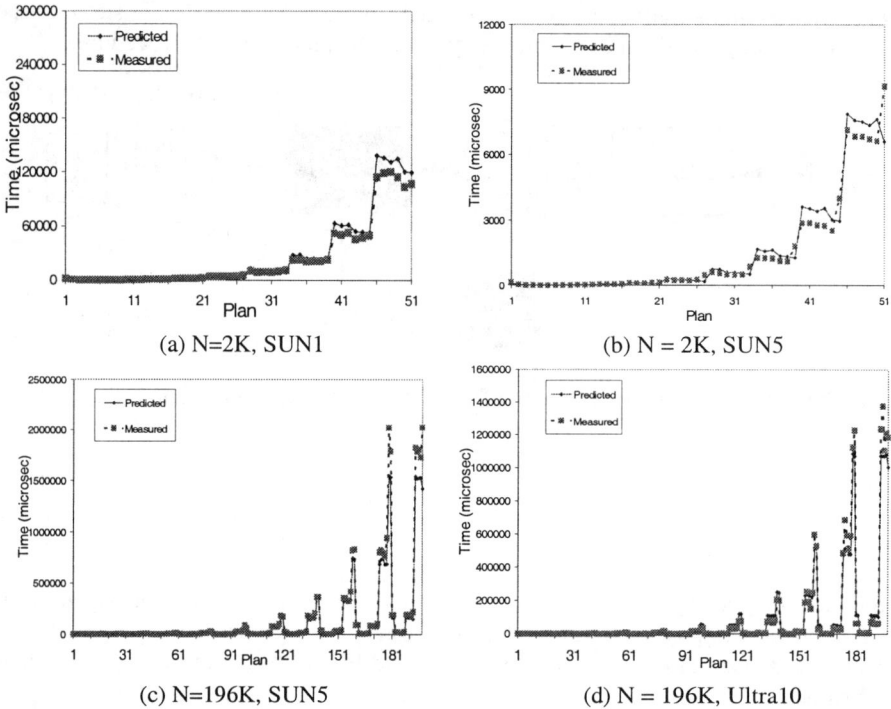

Fig. 5. *Comparison of FFTW execution times and PACE Performance Predictions.*

7 Conclusion

In this work we have shown how a novel performance prediction system may be applied for on-the-fly performance prediction to steer the execution of applications. The PACE system enables a performance model to be incorporated into an application executable and can be used in a decision making procedure to determine how the application can be executed on the target system. Accurate predictions can be produced using this toolset, and rapid evaluation of the performance models enables on-the-fly use.

The FFTW high performance application was used to illustrate the approach. The existing application contained a costly measurement process to determine how best to calculate the transform on the target system. By incorporating a PACE performance model, it was shown that the same process could be undertaken using performance prediction in a fraction of the time, but resulting in the same transform calculation.

The PACE system is currently being extended to provide support for performance prediction in computational environments which may be dynamically changing, and to aid the scheduling of multiple applications on available resources. This corresponds in part to the challenges currently posed by the development of Computational GRIDs

Acknowledgement

This work is funded in part by DARPA contract N66001-97-C-8530, awarded under the Performance Technology Initiative administered by NOSC. Mr. Alkindi is supported on a scholarship by the Sultan Qaboos University, Oman.

References

1. Foster, I., Kesselman, C.: The Grid, Morgan Kaufmann, (1998).
2. Gu, W., Eisenhauer, G., Schwan, K.: On-line Monitoring and Steering of Parallel Programs, Concurrency: Practice and Experience 10 9 (1998) 699-736.
3. Miller, B.P., Callaghan, M.D., Cargille, J.M., Hollingsworth, J.K., Irvin, R.B., Karavanic, K.L., Kunchithapadam, K., Newhall, T.: The Paradyn Parallel Performance Measurement Tools, IEEE Computer 28 11 (1995) 37-46.
4. DeRose, L., Zhang, Y., Reed, D.A.: SvPablo: A Multi-Language Performance Analysis System, in Proc. 10th Int. Conf. on Computer Performance, Spain (1998) 352-355.
5. Ramachandran, U., Venkateswaran, H., Sivasubramaniam, A., Singla, A.: Issues in Understanding the Scalability of Parallel Systems, in Proc. of the 1st Int. Workshop on Parallel Processing, Bangalore, India (December, 1994) 399-404.
6. Van Gemund, A.J.C., Reijns, G.L.: Predicting Parallel System Performance with Pamela, in Proc. 1st Annual Conf. of the Advanced School for Computing and Imaging,, Heijen, The Netherlands (1995) 422-431.
7. Balakrishnan, S., Nandy, S.K., van Gemund, A.J.C.: Modeling Multi-threaded Architectures in PAMELA for Real-time High-Performance Applications, in Proc. 4th Int. Conf. on High-Performance Computing, Los Alamitos, California, IEEE Computer Society 407-414 (December, 1997).
8. Frigo, M., Johnson. S.G.: FFTW: An adaptive software architecture for FFT, In Proc. of the IEEE Int. Conf. on Acoustics Speech, and Signal Processing, 3, Seattle (1998) 1381-1384.
9. Frigo. M.: A fast Fourier Transform Compiler. in Proc.. of the ACM SIGPLAN Conf. on Programming Language Design and Implementation (PLDI'99), Atlanta (1999).
10. Nudd, G.R., Papaefstathiou, E., et.al., A layered Approach to the Characterization of Parallel Systems for Performance Prediction, in Proc. of Performance Evaluation of Parallel Systems, Warwick (1993) 26-34.
11. Papaefstathiou, E., Kerbyson, D.J., Nudd, G.R., Atherton, T.J.: An overview of the CHIP3S performance prediction toolset for parallel systems, in: 8th ISCA Int. Conf. on Parallel and Distributed Computing Systems, Florida (1995) 527-533.
12. Perry, S.C., Kerbyson, D.J., Papaefstathiou, E., Grimwood, R., Nudd, G.R.: Performance Optimisation of Financial Option Calculations, To Appear in Parallel Computing, Elsvier North Holland (2000).
13. Smith, C.U.: Performance Engineering of Software Systems, Addison Wesley (1990).
14. Harper, J.S., Kerbyson, D.J., Nudd, G.R.: Analytical Modeling of Set-Associative Cache Behavior, IEEE Transactions on Computers 48 10 (1999) 1009-1024.
15. Wilson, R., French, R., Wilson, C., et.al.: An Overview of the SUIF Compiler System, Technical Report, Computer Systems Lab Stanford University (1993).
16. Kerbyson, D.J., Papaefstathiou, E., Harper, J.S., Perry, S.C., Nudd, G.R.: Is Predictive Tracing Too Late for HPC Users?, in High Performance Computing, Kluwer Academic (1999) 57-67.

Skel-BSP: Performance Portability for Skeletal Programming

Andrea Zavanella

Dipartimento di Informatica
Università di Pisa Italy
zavanell@di.unipi.it

Abstract. The Skel-BSP methodology provides an adaptive support for skeleton programs aiming to achieve performance portability for parallel programming. The adaptivity is obtained by choosing templates and implementation parameters according to the target machine characteristics. Each choice is made using an optimization theorem demonstrated using EdD-BSP cost predictions. The work presents several strategies to optimize the pure data parallel subset of Skel-BSP. Data parallel programs can be written in Skel-BSP combining `map, reduce` and `scan` skeletons. Analogous results have been already derived for the stream parallel subset of the language (`pipe, farm`).

1 Introduction

A renewed diffusion of parallel platforms, from symmetric multiprocessors to PCs clusters, encourages the efforts aiming to provide parallel software with traditional sequential software properties as modularity, code reusability and portability across platforms. The paper concentrates on a further aspect, specific to the parallel programming context: the so called performance portability. Parallel computers performance can not be modeled using a single parameter, as in the sequential case, and the design of parallel application should be somehow specialized to fit the characteristics of the target architecture. This means that many applications must be heavily restructured, when ported on different architectures, to run efficiently. Several abstract models have been proposed: BSP, LogP [2, 8] but they have rarely been used in connection with some high level, structured programming system. The work presents the Skel-BSP methodology as a solution providing a skeleton language, which is a subset of P3L [3], with performance portability. The methodology uses an extension of the BSP model (the EdD-BSP) developed to this purpose. The paper presents the main results on how to optimize the data parallel subset of a skeletal language made by the three skeletons: `map, reduce` and `scan`.

2 The Skel-BSP Approach

Skel-BSP forces the programmer to concentrate on exposing a "parallelization strategy" more than a parallel algorithm. The programmer knowledge and expertize is exploited in the direction of writing a strategy as the composition of already defined parallel patterns (skeletons). The optimal implementation is then devised by the Skel-BSP compiler which relies on three additional components:

- a set of implementation templates,
- a set of performance equations,
- the EdD-BSP parameters of the target architecture.

The general scheme of the Skel-BSP compiler is shown in Fig. 1. The "local"

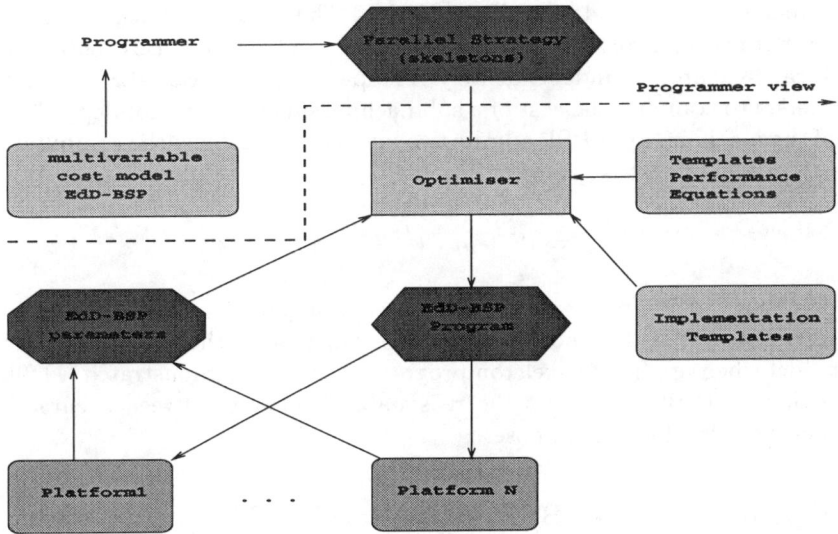

Fig. 1. The scheme for performance portability in parallel programming

optimizations are stored in a set of "already optimized" and reusable components (templates). The optimizer is charged of adapting the program structure to the target architecture using the EdD-BSP parameters and the performance equations derived for the implementation templates.

3 The Edinburgh-Decomposable-BSP

Skel-BSP makes use of the EdD-BSP model, a BSP [8] variant, as a general framework in which programs performance is predicted. A BSP computation is organized as a sequence of supersteps. Each superstep has a local computation phase and a global communication phase concluded with a barrier synchronization. The cost of each superstep is given by Eq. 1:

$$T_{sstep} = W + hg + L \tag{1}$$

Where W is the maximum amount of work performed in the local computation phase, and h is the maximum number of messages sent or received during the communication phase. The parameters g and L are the classic BSP parameters to measure the cost to send a single message and to perform a barrier synchronization. The EdD-BSP variant introduces two extensions of the BSP model:

1. the Hookney technique introduced in BSP by Miller in [7];
2. the decomposability proposed by Kruskal et al. in [4].

The first extension introduces two parameters: g_∞ and $N_{1/2}$ in place of the g parameter of the standard BSP definition. The second extension introduces the `partition` and `join` operations to create and destroy BSP submachines which can synchronize independentely. The peculiarity of EdD-BSP is that the parameters to compute the cost of a submachine superstep do not depend on the size of the machine. EdD-BSP admits two kinds of supersteps: the computational supersteps whose cost is given by Eq. 2

$$T = W + hg_\infty(\frac{N_{1/2}}{h} + 1) + L \qquad (2)$$

and join/partition supersteps which cost is simply L. The main limitation of the model is that it does not account for locality, nevertheless the accuracy of the model when applied to skeleton programs has been demonstrated [11,9]. We claim that EdD-BSP provides the reasonable trade-off between accuracy and simplicity needed for our purposes.

4 Optimizing Skel-BSP Data Parallel Programs

The Skel-BSP data parallel skeletons are: `map`, `reduce` and `scan`. A sequence of data parallel modules can be implemented using the `comp` skeleton. The `map` skeleton models a generic domain decomposition pattern in which a collection of input data structure (typically multidimensional arrays) are distributed among a grid of *virtual processors*, each virtual processor performs a sequential computation on its local partition then the results are collected. An input data structure can be distributed using three different strategies:

- *broadcast*: which replicates the structure on all the virtual processors;
- *scatter* which partitions the structure among the virtual processors;
- *multicast* which partitions the structure and then broadcasts each partition to a subset of virtual processors.

The general scheme of the implementation template written using the BSP-lib [6] is shown in Fig.2. The performance model of `map` assumes that the following parameters are known:

- t_f: time to compute the local computation in a virtual processor;
- t_{unpack}: time to extract the data partition for a virtual processor;
- n_w: number of processors utilized as parallel workers.

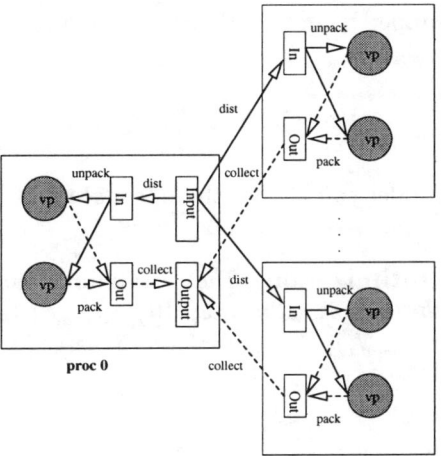

Fig. 2. The general scheme of the map template

The completion time T_{map} can be written as:

$$T_{map} = T_{dist} + T_{calc} + T_{coll} \qquad (3)$$

where:

$$T_{dist} = T_{scatter} + T_{brodcast} + T_{multi}$$

We assume that the sizes of the structures to be distributed are: d_b, d_s and d_m. The components of the distribution cost can then be written as:

$$T_{scatter}(n_w) = d_s g_\infty \left(\frac{N_{1/2}}{d_s/n_w} + 1\right) \qquad (4)$$

$$T_{broadcast}(n_w) = n_w d_b g_\infty \left(\frac{N_{1/2}}{d_b} + 1\right) \qquad (5)$$

$$T_{multi}(n_w) = \rho d_m g_\infty \left(\frac{N_{1/2}}{\rho \frac{d_m}{n_w}} + 1\right) \qquad (6)$$

Assuming that the size of the result is d_r we can write the time for collection as:

$$T_{coll} = (n_w) = d_r g_\infty \left(\frac{N_{1/2}}{d_r/n_w} + 1\right) \qquad (7)$$

and the time spent for local computation as:

$$T_{calc} = \frac{d_r}{n_w} t_p \qquad (8)$$

The parameter ρ, is the *factor of replication* associated to the particular *multicast* strategy and defined as: $\rho = \frac{d_{in} n_w}{d_m}$. Finally variable t_p in Equation 8 is the time per element to unpack and compute the results defined as $t_p = t_f + t_{unpack}$.

Since $\rho < p$ a simple upper bound for the completion time of the map module is given in Eq. 9.

$$T_{map} = g_\infty [n_w(d_m + d_b + 3N_{1/2}) + (d_s + d_r)] + \frac{d_r t_p}{n_w} + 2L \qquad (9)$$

The Skel-BSP compiler can choose the optimal number of processors using the Theorem 1:

Theorem 1 (Map Optimization) *The minimum completion time of a generic instance of the map skeleton:* map $([t_f, t_{unpack}], [\rho, d_b, d_s, d_m, d_r])$ *on a EdD-BSP computer* $(g_\infty, N_{1/2}, L, p)$ *is obtained using* n_{opt} *processors, where:* $n_{opt} = Min(p, \hat{n})$ *and:*

$$\hat{n} = \sqrt{\frac{t_p d_r}{g_\infty(3N_{1/2} + d_m + d_b)}}$$

Proof 1 *Using the approximate expression in Eq. 9, we can derive a simple expression of the derivate of the completion time and then it easy to prove that:*

$$T_{map}(\hat{n})' = 0$$

and that:

$$T''_{map}(\hat{n}) > 0$$

Finally, since $T_{map}(n_w)' > 0$ *in* $[1, \hat{n}]$, *if* $p < \hat{n}$ *the minimum of the completion time is in p and the Theorem is proved.*

The Skel-BSP **reduce** models the parallel "reduction" of an array by means of a binary associative operator **opn**. The **reduce** skeleton is implemented in Skel-BSP by choosing between two distinct templates. The compiler selects the template according to the instance of the problem and the characteristics of the target architecture. The first superstep of both the templates is a scatter distribution. Each processor receives a subset of the structure of n/p elements. The first template uses two more supersteps:

1. each processor performs the reduction of the local partition using and sends the partial results to the collector;
2. the collector computes the result reducing the p partial results.

The second template uses a number of supersteps logarithmic in the processors count. In this case the p partial results are reduced using a binary tree. The cost analysis of the **reduce** templates uses the following parameters:

- t_{op} the time to compute the binary operator;
- n number of elements of the structure to reduce;
- d size of the elements of the structure.

The cost of the distribution superstep is:

$$T_{dist} = ndg_\infty(\tfrac{N_{1/2}}{nd/p} + 1) + L \tag{10}$$

The EdD-BSP cost of the two-phase technique is given in Eq. 11:

$$T_{red1} = \tfrac{n}{p}t_{op} + dg_\infty(\tfrac{N_{1/2}}{d} + 1) + L + pt_{op} \tag{11}$$

The EdD-BSP cost of the binary-tree technique is given in Eq. 12:

$$T_{red2} = \tfrac{n}{p}t_{op} + \log p(t_{op} + dg_\infty(\tfrac{N_{1/2}}{d} + 1) + L) \tag{12}$$

The choice of the template is made exploiting the following Theorem:

Theorem 2 (Reduce Template Optimization) *A* **reduce** *($[t_{op}], [d, n]$) instance can be implemented on a EdD-BSP computer ($g_\infty, N_{1/2}, L, p$) with minimum completion time using the two-phase algorithm when:*

$$\tfrac{t_{op}}{\Pi} < dg_\infty(\tfrac{N_{1/2}}{d} + 1) + L$$

Where:

$$\Pi = \frac{\log p - 1}{p - \log p}$$

Proof 2 *Follows from Eq. 11 and 12.*

The choice of the optimal number of processor to optimize the completion time of **reduce** is given by the following Theorem:

Theorem 3 (Reduce Processors Optimization) *The minimum completion time for an instance* **reduce** *($[t_{op}], [d, n]$) on a EdD-BSP computer ($g_\infty, N_{1/2}, L, p$) is obtained using n_{opt} processors where: $n_{opt} = Min(p, \hat{n})$ where:*

$$\hat{n} = \begin{cases} \sqrt{\tfrac{nt_{op}}{g_\infty N_{1/2} + t_{op}}} & \text{if} \quad \tfrac{t_{op}}{\Pi} < dg_\infty(\tfrac{N_{1/2}}{d} + 1) + L \\ \tfrac{\sqrt{b^2 + 4ac} - b}{2a} & \text{otherwise} \end{cases}$$

and:

$$a = 2g_\infty N_{1/2}$$
$$b = t_{op} + g_\infty N_{1/2} + d_M g_\infty + L$$
$$c = nt_{opt}$$

Proof 3 *From Theorem 2 we have that when:*

$$\frac{t_{op}}{\Pi} < dg_\infty(\frac{N_{1/2}}{d}+1)+L$$

*the compiler selects the two-phase template and then cost of the computational part of **reduce** is given by Eq. 11, in this case it is easy to prove that:*

$$\begin{cases} T'_{red}(\hat{n}) = T'_{red}(\sqrt{\frac{nt_{op}}{g_\infty N_{1/2}+t_{op}}}) = 0 \\ T''_{red}(\hat{n}) = T''_{red}(\sqrt{\frac{nt_{op}}{g_\infty N_{1/2}+t_{op}}}) > 0 \end{cases}$$

Since $T'_{red}(n_w) < 0$ in $[1,\hat{n}]$ the minimum of T_{red} in $[1,p]$ is in $Min(\hat{n},p)$. In all the other cases, we have that T_{reduce} is computed using by Eq. 12 and we can prove that:

$$\begin{cases} T'_{red}(\hat{n}) = T'_{red}(\frac{-b+\sqrt{b^2+4ac}}{2a}) = 0 \\ T''_{red}(\hat{n}) = T''_{red}(\frac{-b+\sqrt{b^2+4ac}}{2a}) > 0 \end{cases}$$

Since $T'_{red}(n_w) < 0$ $[1,\hat{n}]$ the minimum of the completion time is in $Min(\hat{n},p)$ and the Theorem is demonstrated.

The Skel-BSP **scan** skeleton models the parallel prefixes (or suffixes) of an array structure using an associative operator **opn**. The **scan** skeleton is implemented by choosing between two distinct implementation templates. The compiler selects the template according to the instance of the problem and the characteristics of the target architecture. In both the templates is distributed so that each processor receives a subset of size n/p. The first template requires two more supersteps:

1. step: each processor computes the prefixes on the local portion of the structure and sends the rightmost element of the prefix array to all the processors with greater index.
2. step: processor $i > 0$ computes the ith segment of the prefix using $i + n/p$ operations. The results are gathered by processor 0.

In the second template a number of supersteps logarithmic in the number of processors is employed. During step i processor $0 \geq j < p - 2^i$ sends its rightmost value to the processor $j + 2^i$. During superstep $i+1$ processors $2^i < j < p$ computes the prefix of their array with the received value. The cost analysis of the **scan** templates uses the following problem dependent parameters:

- t_{op} the time to compute the binary operator;
- n the number of elements of the array to scan;
- d the size of the elements of the structure to scan.

The cost of the distribution and collection are:

$$T_{dist} = ndg_\infty(\frac{N_{1/2}}{\frac{nd}{p}} + 1) + L \qquad (13)$$

$$T_{coll} = ndg_\infty(\frac{N_{1/2}}{\frac{nd}{p}} + 1) + L \qquad (14)$$

The EdD-BSP cost of the two-phase computation is:

$$T_{two} = \frac{p^2 - p + 2n}{p}t_{op} + (p-1)dg_\infty(\frac{N_{1/2}}{d} + 1) + L \qquad (15)$$

The cost for the logarithmic computation is:

$$T_{log} = \log p(\frac{n}{p}t_{op} + dg_\infty(\frac{N_{1/2}}{d} + 1) + L) \qquad (16)$$

The choice of the best template is made according to the following Theorem:

Theorem 4 (Scan Template Optimization) *An instance of the Skel-BSP construct* **scan** *($[t_{op}], [d, n]$) can be implemented on top of a EdD-BSP computer ($g_\infty, N_{1/2}, L, p$) with minimum completion time using the two-phase template when $n > n_{two}$ with:*

$$n_{two} = \frac{(p-1)t_{op} + dg_\infty(\frac{N_{1/2}}{d} + 1)(p - \lambda) - \lambda L}{\lambda t_{op}}$$

and:

$$\lambda = \log p - 1$$

Proof 4 *Follows from comparing Eq. 15 and 16.*

Once the best template is selected we can optimize the number of processors using the following Theorem:

Theorem 5 (Scan Processors Optimization) *The minimum completion time for an instance* **scan** *($[t_{op}], [d, n]$) on a EdD-BSP computer ($g_\infty, N_{1/2}, L, p$) is obtained using n_{opt} processors where: $n_{opt} = Min(p, \hat{p})$ and*

$$\hat{p} = \begin{cases} p_{two} \text{ when } n > n_{two} \\ p_{log} \text{ when } n \leq n_{two} \end{cases}$$

Where:

$$p_{two} = \sqrt{\frac{2nt_{op}}{t_{op} + 3g_\infty N_{1/2} + d_M g_\infty}}$$

And p_{log} such that:

$$\frac{ap_{log}^2 + bp_{log}}{\log p_{log} - 1} = c$$

with:

$$a = 2g_\infty N_{1/2}$$
$$b = d_M g_\infty + g_\infty N_{1/2} + L$$
$$c = nt_{op}$$

5 Conclusions, Related, and Future Work

The Skel-BSP methodology to achieve performance portability for skeleton programs has been introduced. The extension of the BSP model (EdD-BSP) utilized to predict programs performance is presented. The paper describes the optimization techniques employed by the Skel-BSP compiler in the case of data parallel programs. The Theorems introduced provide an example of the power of using a simple abstraction as the EdD-BSP model when optimizing parallel programs. The optimization techniques can be easily embedded in the structure of a language compiler such as P3L or SKIE to provide a first layer of "local" optimizations. This layer is to be combined with a "global" optimization layer needed when the program is a skeletons composition. Other results on stream parallel optimization under EdD-BSP have already been published [9] while further results on optimizing data parallel composition, and nesting of stream and data parallel programs have been submitted for pubblication. A new interesting approach to optimize stream parallel skeleton has been proposed in [1] while a base for global optimization has been exposed in [5]. The next step of this work is to perform a set of validation experiments using our templates (written using the BSP-lib) [10] on a range of parallel platforms having different performance characteristics.

References

1. M. Aldinucci and M. Danelutto. Stream parallel skeleton optimization. In *proceedings of the 11th IASTED International Conference on Parallel and Distributed Computing and Systems*, MIT, Boston, USA, November 1999. IASTED/ACTA press.
2. D. Culler, R. Karp, D. Patterson, A. Sahay, K.E. Schauser, E. Santos, R. Subramonian, and von T. Eicken. Logp: Towards a Realistic Model of Parallel Computation. In *Proceedings of the 4th ACM SIGPLAN*, May 1993.
3. M. Danelutto, R. Di Meglio, S. Orlando, S. Pelagatti, and M. Vanneschi. A methodology for the development and the support of massively parallel programs. In D. B. Skillicorn and D. Talia, editors, *Programming Languages for Parallel Processing*. IEEE Computer Society Press, 1994.
4. P. De La Torre and C. P. Kruskal. Submachine locality in the bulk synchronous setting. *Lecture Notes in Computer Science*, 1124:352–360, 1996.
5. S. Gorlatch and S. Pelagatti. A transformational framework for skeletal programs: Overview and case study. In Jose Rohlim, editor, *Proc. of Parallel and Distributed Processing. Workshops held in Conjunction with IPPS/SPDP'99*, volume 1586 of *LNCS*, pages 123–137, Berlin, 1999. Springer.
6. J. M. D. Hill, B. McColl, D. — C. Stefanescu, M. W. Goudreau, K. Lang, S. B. Rao, T. Suel, T. Tsantilas, and R. H. Bisseling. BSPlib: The BSP programming library. *Parallel Computing*, 24(14), 1998.
7. R. Miller. *Two approaches to architecture-independent parallel computation*. PhD thesis, Oxford University Computing Laboratory, 1994.
8. L.G. Valiant. A Bridging Model for Parallel Computation. *Communications of the ACM*, 33(8):103–111, August 1990.

9. A. Zavanella. Optimising Skeletal Stream Parallelism on a BSP Computer. In P. Amestoy, P. Berger, M. Dayde, I. Duff, V. Fraysse, L. Giraud, and D. Ruiz, editors, *Proceedings of EURO-PAR'99*, number 1685 in LNCS, pages 853–857. Springer-Verlag, 1999.
10. A. Zavanella. *Skeletons and BSP: Performance Portability for Parallel Programming*. PhD thesis, Dipartimento di Informatica Univesita di Pisa, 2000.
11. A. Zavanella and S. Pelagatti. Using BSP to Optimize Data-Distribution in Skeleton Programs. In P. Sloot, M. Bubak, A. Hoekstra, and B. Hertzberger, editors, *Proceedings of HPCN99*, volume 1593 of *LNCS*, pages 613–622, 1999.

Self-Tuning Parallelism

Otilia Werner-Kytölä and Walter F. Tichy

Karlsruhe University, Department of Computer Science
Am Fasanengarten 5, 76128 Karlsruhe, Germany
tichy@ira.uka.de

Abstract. Assigning additional processors to a parallel application may slow it down or lead to poor computer utilization. This paper demonstrates that it is possible for an application to automatically choose its own, optimal degree of parallelism. The technique is based on a simple binary search procedure for finding the optimal number of processors, subject to one of the following criteria:
- maximum speed,
- maximum benefit-cost ratio, or
- maintaining an efficiency threshold

The technique has been implemented and evaluated on a Cray T3E with 512 processors using both kernels and real applications from Mathematics, Electrical Engineering, and Geophysics. In all tests, the optimal parallelism is found quickly. The technique can be used to determine the optimal degree of parallelism without manual timing runs. It thus can help shorten application runtime, reduce costs, and lead to better overall utilization of parallel computers.

1 Introduction

Users of parallel supercomputers are primarily concerned with speed, which means they opportunistically employ all available processors for their applications. While this heuristic often works, there are applications that would actually run faster with fewer processors. This situation arises whenever an application scales poorly, due to high communication overheads or insufficient problem size. To find the optimal degree of parallelism, users would have to manually schedule a sequence of timing runs varying both processor set sizes and problem sizes. We show that an application can automatically tune the size of the processor set it uses. If the application scales well, this strategy will, of course, choose the maximum number of available processors.

A related consideration is that parallel applications may be subject to the law of diminishing returns: Increasing the number of processors may improve application runtime only marginally, while efficiency drops. Hence, opportunistically choosing the maximum number of available processors may have a negative effect on the overall utilization of a parallel supercomputer. While poor utilization is not an issue for personal computers, it is a concern for to supercomputer administrators and users who have to pay for resource consumption. For example, suppose we run an application with n processors in time t achieving an efficiency of less than 50%. Suppose furthermore that running the same application with $n/2$ processors increases runtime only marginally, say by a factor of 1.3. However, with $n/2$ processors we can run two applications at the same time, or schedule two runs of the same application, on two different data sets, at once. When doing so, we need $1.3t$ time units to do the work that would otherwise take $2t$ time units. Thus, improving machine utilization can be beneficial for all users of a scarce computation resource.

To address these issues, we have developed a system called *ParGrad* that automatically tunes the degree of parallelism of an application according to the following criteria:

- maximum speed,
- maximum benefit-cost ratio, or
- maintaining an efficiency threshold.

The authors' research group in Karlsruhe pioneered the idea of dynamically tuning the degree of parallelism for shared memory machines [15, 9]. The present article extends this work to distributed memory. It solves the problem of automatic redistribution of data whenever the degree of parallelism changes. It also extends the set of optimization criteria.

ParGrad works for parallel programs written in C++ for both shared and distributed memory parallel computers. The technique could also be incorporated into Fortran or other languages. We assume that the application program has already been parallelized. The programmer merely indicates the optimization criterion, which parallel loops should be tuned, and the data structures that are touched by these parallel loops. *ParGrad* requires no compiler modifications. Instead, the self-tuning procedure automatically executes parallel loops with varying numbers of processors and measures the actual run times and efficiencies. A database saves the optimal processor assignments for each application and problem size. Later executions of an application thus start up with the optimal parallelism immediately.

2 Related Research

We do not address the problem of processor allocation on a multiprocessor, which is a well known problem with adequate solutions, e.g. [10], [13], and [5]. This research assumes that allocation services are available. For comparing research in choosing the degree of parallelism, we developed the classification scheme shown in Table 1. An important criterion is the *memory model*, which distinguishes between parallel computers with shared memory (Sha) and distributed memory (Dis). The *adaptation level* describes whether the degree adaptation is embedded in the operating system (Sys) or takes place at application level (Pro). A further question is whether the data for making decisions about the degree of parallelism is obtained dynamically during execution time (Dyn) or statically at compile time (Sta). The adaptation can be implemented through a library

(Lib), a runtime system (Run), or through program transformation (Tra). Finally, some research addresses the problem of *data redistribution* (+), but most work ignores it (−). The accepted *programming languages* are listed in the last column of the table.

Table 1. Classification of related research

Criterion / Work	Memory Model	Adaptation Level	Analysis Method	Implementation	Data Redistribution	Programming Language
This work	Dis	Pro	Dyn	Lib,Tra	+	C++
Reimer	Sha	Pro	Dyn	Lib,Tra	−	Fortran
Hall and Martonosi	Sha	Pro	Dyn	Run	−	C
Squillante et al.	Dis	Sys	Sta	Run	−	any
McCann et al.	Sha	Sys	Sta	Run	−	C++
Yue and Lilja	Sha	Sys	Sta	Tra,Run	−	Fortran
Tucker and Gupta	Sha	Sys	Dyn	Run	−	any with threads
Edjlali et al.	Dis	Sys	Dyn	Lib,Tra	+	HPF-like
Hänßgen	Sha	Pro	Dyn	Lib,Tra	−	C
Downey	Sha	Sys	Sta	Run	−	any

We build on the ideas of Reimer [15], who first demonstrated the feasibility of automatic parallelism tuning. We extend this work to distributed memory by triggering an automatic data redistribution whenever the degree of parallelism changes and adding a database to collect performance data. We also introduce the benefit-cost ratio as an optimality criterion and develop an execution-time model that can be used as a predictor. Finally, we provide an extensive experimental analysis of the technique using a set of kernels and full applications.

Hall and Martonosi [8] propose a run-time mechanism for shared memory machines that dynamically adjusts the number of processors in an application employing threads. The only criterion used is the speed-up and the technique is limited to shared memory.

Squillante et al. [18, 16] determine the parallelism degree statically based on heuristics instead of dynamically, as we do. Since the degree of parallelism does not change at runtime, redistribution of data is unnecessary.

The work of McCann et al. [13] differs from ours along the criteria memory model, adaptation level, and analysis method.

Yue and Lilja [22] base the adaptation on the system load and a statical forecast of the work in a parallel loop, while our work uses measurements at execution time.

Tucker and Gupta [17] assume that a program reaches its best performance when the number of processes is equal to the number of processors in the system. They develop a technique which allocates a similar number of processes to all runnable programs. This approach can lead to poor results for some programs.

The goal of parallelism tuning differs significantly from the goal of Edjlali et al. [4]. They describe a method through which a program executes in an environment where the number of available processors changes over time. According to their method, a program will allocate or release processors, depending only on the total number of available processors, without checking whether this is appropriate for the performance of the program or for the overall utilization of the parallel system.

Hänßgen [9] works with recursive programs on shared memory machines. With recursion, the available parallelism is unpredictably large and extremely fine-grain. Hänßgen uses measurements and simulations to reduce the amount of parallelism for optimizing execution time.

Downey [3] suggests adapting parallelism at the operating system level based on heuristics, without taking application characteristics into account.

In summary, our work deals with the full generality of distributed memory. Only one other approach, Edjlali's, addresses dynamic data redistribution. *ParGrad* collects actual runtime information and adapts parallelism at the application level.

3 The Technique

ParGrad makes the following assumptions: The program of interest has been parallelized for a computer with shared or distributed memory, using the SPMD or data parallel model. The tuning function relies on the assumption that speed and efficiency can be optimized by varying the number of processors assigned to parallel loops, whose extent is determined by the size of the data structures or the input data, i.e., the problem size. We assume a homogeneous parallel computer with identical processors, memory sizes, and communication channels.[1] As programming language we chose C++, but the technique could be implemented in most other programming languages. Compiler modifications are not required.

The steps to be carried out when using *ParGrad* are illustrated in Fig. 1. First, the user identifies the tuning goal: maximum speed, maximum benefit-cost relation, or maintaining an efficiency threshold. This choice is specified by setting the variable *STRATEGY* in the class *My_UserModule* (see Fig. 2). This class is a control class that collects all parameters for the tuning process. Second, the user identifies the parallel loops where adaptation takes place. These are typically the outer loops in the program, although inner loops can also be

---
[1] With a somewhat more complex search function, processors of different speeds could also be handled.

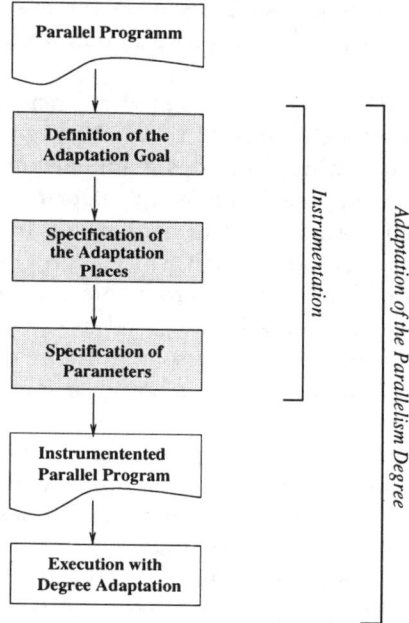

Fig. 1. Overview of the tuning steps

optimized. All that is needed for the programmer is to insert a call to *ParGrad()* at the end of each loop, or, if the program should be optimized as a whole, at the end of the entire program. Finally, a few additional parameters should be set. Chief among these is the data structures that the parallel loops work on. On a distributed memory machine, these data structures need to be redistributed whenever the number of processors changes. *ParGrad* automatically redistributes vectors and arrays. It can handle other data structures as well, but requires that the programmer code redistribution methods. The interface of these methods is specified by the object-oriented architecture of *ParGrad*. *ParGrad* then calls these methods automatically.

Figure 3 is a trivial SPMD program for adding two vectors. This program is started simultaneously on several processors. Each processor has a section of each array stored locally. The call to *ParGrad()* sets up a timer and periodically changes the number of processors until the optimum is found. Figure 2 shows that the data distribution is set to block distribution and the data structures involved are the one-dimensional arrays a, b, and c. The goal of the tuning process is to maintain an efficiency of at least 70%.

After completing this instrumentation, the parallel program is ready to compile. When executed, it will automatically find the maximum degree of parallelism in which the efficiency threshold is maintained.

```
1   My_UserModule::My_UserModule()
2   {
3     set(STRATEGY, EISS);
4     set(EFFTHRESHOLD, 0.70);
5     set(SOURCE, a, b, c);
6     set(DATA_DISTRIBUTION, BLOCK);
7     set(DATA_STRUCTURE, VECTOR);
8   }
```

Fig. 2. Instrumentation of the program above

```
1   for (i = 0 ; index < number_of_elements/_num_pes() ; i++)
2   {
3       a[i] = b[i] + c[i];
4       ParGrad();
5   }
```

Fig. 3. Simple SPMD program

3.1 Tuning Strategies

In the following, the degree tuning strategies RISS, PeSS, EISS (based on [15]), and CBISS are introduced. RISS and PeSS maximize speed, EISS maintains an efficiency threshold, and CBISS maximizes the benefit-cost ratio. All strategies start with the maximum number of available processors and successively half the size of this set until the optimum is found. The strategies start the search from the high end rather than with a single processor, because obviously this direction reduces the overall runtime. Another important consideration is that the adaption strategies consider only processor set sizes that are powers of 2 and thus may miss the actual optimum. Searching for the exact optimum is possible. However, our experience indicates that the optimum is flat, i.e., if 2^k is the processor number chosen by our algorithms, then the difference to 2^{k-1} and 2^{k+1} processors is small, and this difference is easily overwhelmed by the cost of additional data redistribution. Also, a finer optimization requires a parallel computer that allows allocation of processor set sizes other than 2^k. Finally, data access computations may become more expensive for "odd" processor numbers.

An important issue is whether the simplified binary search of halving the processor set size actually terminates at the optimal power of 2. To answer this question, we use a simple model that assumes that the overall execution time $T(p)$ is the sum of computation and communication time. Computation time is inversely proportional to the number of processors, while communication time is a polynomial of degree 2 in the number of processors:

$$T(p) = a/p + b \times p^2 + c \times p + d$$

Thus, computation time decreases and communication time increases with the number of processors. The result has the typical "bathtub" shape, where the minimum signifies maximum speed. The location of the minimum depends on the application. For applications that scale well, a minimum may never materialize for realistic processor set sizes, while for others the the minimum may be at 16 or fewer processors. In the PhD dissertation of one of the authors [20], this model is explored in depth and found to be adaptable to a wide range of applications, in particular all applications and kernels discussed in this article. When the coefficients a, b, c, d are determined, the optimal degree of parallelism can be computed perfectly. Unfortunately, this model is not practical, since the coefficients are application dependent and must be determined by extensive timing runs. However, the model justifies the use of the simplified binary search procedure, because the curve has single local and global minimum.

- **Runtime Incremental Search Strategy** (RISS): Maximizes speed by halving the number of processors as long as the speed increases while doing so. Suppose the optimum is 64 and execution starts with 512 processors. *ParGrad* then measures the runtime of a number of iterations of the parallel loop with 512, 256, 128, 64, and then 32 processors. The timing with 32 processors will be higher than with 64, so *ParGrad* will switch back to 64.
- **Prediction-Driven Search Strategy** (PeSS): Also maximizes speed. It uses the above model for prediction. It measures four points on the "bathtub" curve to estimate the coefficients for $T(p)$ and then uses it to predict the time for the next point. If the predicted runtime for this point is better, *ParGrad* switches the number of processors accordingly, measures this point, and predicts again. Otherwise, the number of processors remains unchanged. In the above example, PeSS first measures execution times for 512, 256, 128, and 64 processors and then predicts that 32 processors would increase runtime. So it would converge towards 64 processors without trying the slower 32 processor case. PeSS converges faster than RISS.
- **Efficiency-Driven Incremental Search Strategy** (EISS): Maintains an efficiency threshold. The method measures the efficiency for a given degree of parallelism and keeps halving the degree until the efficiency threshold is met. Note that this method must execute and measure the parallel loop for the single processor case, because this value is needed for computing the efficiency.
- **Benefit-Cost-Driven Incremental Search Strategy** (BCISS): Maximizes the benefit-cost ratio. The search procedure halves parallelism as long as this relation grows. Cost is computed as

$$C(p) = T(p) \times p$$

where p is the number of allocated processors and $T(p)$ the execution time with p processors. The benefit is the speed-up, i.e. $T(1)/T(p)$. The benefit-cost ratio is then

$$R_{BC} = \frac{T(1)}{T(p) \times C(p)} = \frac{T(1)}{T(p)^2 \times p}$$

BCISS searches for the maximum of R_{BC}. Note that it needs to measure $T(1)$ as well.

One of these strategies is selected by the user and is employed in the search for the optimal parallelism degree.

3.2 Granularity of Tuning

Obviously, the cost of redistribution of data structures on a distributed memory machine can be substantial. This cost must be amortized over several iterations or even over several runs of the entire application.

Intrarun Tuning: In intrarun tuning, the degree of parallelism is adjusted during a single execution of the program. This adjustment is performed separately for every parallel loop marked with a call to *ParGrad()*. Because of the cost of redistribution, the programmer can specify a period for each loop, i.e., the number of per-processor iterations of a loop before a step of the search strategy may be executed. Recall that a search step halves or doubles the number of processors and requires data redistribution.

Interrun Tuning: In this tuning mode, the degree of parallelism is adjusted between different executions of the program. This mode is necessary for programs for which the overhead of data redistribution would be prohibitive, or the entire runtime too short for tuning. Interrun tuning requires no dynamic data redistribution, because data distribution is fixed for each run.

In both cases, the runtime for an iteration of each instrumented parallel loop is stored in a database. This database is consulted every time an application is started to get the best known value so far and to continue the search from this value. Once the optimum is found, it is also stored in the database and used in subsequent runs. The runtime data is not only collected for each parallel loop, but also for different problem (data structure) sizes. Thus, this database collects a runtime profile of each application over time. The time stamp on the program is used to determine whether the program has been changed. If this time stamp is younger than the time stamp of the program's profile, then the profile is marked as invalid.

An interesting extension of this technique would be to make the choice of inter- or intrarun tuning automatic. For this to work, *ParGrad* would need a time threshold rather than an iteration count before a redistribution is allowed to take place. This threshold should be substantially larger than the redistribution time.

3.3 Data Redistribution

Every time the number of allocated processors changes, the data must be redistributed among the memory units of the distributed memory machine.

Arrays, matrices and cubes as well as sets of of objects and records can be redistributed automatically. Arrays, matrices and cubes can be redistributed block-wise or cyclic. Moreover, 2-D and 3-D arrays can be redistributed row-wise or column-wise. The redistribution functions built into *ParGrad* cover many of the standard cases and are optimized for the Cray T3E (using E-registers and the *shmem*-library). The object-oriented architecture of *ParGrad* makes it easy to add distribution functions for additional data structures. The details of this architecture are beyond the scope of this paper, but can be found in [20].

4 Experimental Results

ParGrad is written in C++ and runs on the Cray T3E/900. The following results are based on measurements carried out on a Cray T3E/900 at HLRS of the Stuttgart University, Germany. This computer consists of 512 DEC Alpha processors running at 450 MHz. We use small kernels as well as full applications as benchmarks. The kernels were chosen to test different communication patterns; the applications to show that the technique is applicable in real life situations.

4.1 Application Characteristics

Livermore Loops: three of the well known Livermore Loops [6], namely LL1, LL3, and LL13 were selected as representatives. They supply the communication patterns indexed access, reduction, and indirect indexed access, resp. These are not covered in the other programs below.

Radix Sort: a parallel sort algorithm [21], for which the key exchange between processors corresponds to an all-to-all communication pattern. The implementation of this kernel is based on the Split-C [2] benchmark.

Veltran Operator: a geophysical operator that is used for the analysis of the earth's subsurface layers. The input data is stored in a three dimensional matrix and is distributed column-wise among the processors. Each processor needs data in columns which are located in other processors. Our implementation is based on the Fortran version in [11].

Transmission Line Matrix (TLM) Method: The TLM method is an electrical engineering application used in the analysis of the electromagnetic compatibility between electronic equipment. The interprocessor communication is of neighbor-neighbor type. Our implementation is based on [7].

Partial Differential Equation (PDE) Solver: a solver from the NAS benchmark suite [1] that solves a partial differential equation using three dimensional forward and inverse FFT's. The processors communicate with each other in the transpose phase. The implementation is based on [12].

Jacobi Iteration: the classic iterative solver for linear systems. Each element of the matrix is set to the average of its four neighbors in a neighbor-neighbor interprocessor communication pattern. The implementation is based on [14].

4.2 Results

Some of the results achieved by applying intrarun as well as interrun tuning are presented in Table 2. The first column provides the name of the parallel program, while the second and third columns shows the optimization criterion and strategy used. All optimization criteria were tried on all the benchmarks. However, the table only shows the interesting cases. The optimal value (found without *ParGrad*, by trying all processor set sizes by hand) is given in the next column. The last column shows the sequence of processor set sizes that *ParGrad* tried. The last number in this sequence is the parallelism degree the automatic search settled on. It is easy to see that this is always the optimum.

Table 2. Experimental Results

Program	Goal	Strategy	Optimum	Tuning sequence
LL1	efficiency	EISS	32	256,128,64,32
LL3	benefit-cost ratio	BCISS	64	256,128,64,32,64
LL13	speed	RISS	128	128,64,128
Radix Sort	speed	RISS	16	128, 64, 32, 16, 8, 16
Radix Sort	speed	PeSS	16	128, 64, 32, 16
PDE Solver	speed	RISS	16	64, 16, 8, 16
Veltran Operator	efficiency	EISS	16	256, 128, 64, 32, 16
TLM Method	benefit-cost ratio	BCISS	32	64, 32, 16, 32
Jacobi Iteration	efficiency	EISS	16	128, 64, 32, 16

The first three rows of Table 2 show the results for LL1, LL3, and LL13. These kernels scale well for large problem sizes, so when optimized for speed, *ParGrad* always chooses the maximum number of processors. The table varies the criterion among the three loops, to see the different tuning behavior. The degree adaptation improves the efficiency as well as the benefit-cost ratio of the kernels. The kernels run quickly, so an interrun tuning was chosen.

Radix Sort. The data structure to be redistributed is a vector of 128K elements. This kernel scales poorly; a run with 128 processors takes over four times longer than a run with a single(!) processor.

Table 2 shows two cases, one with RISS the other with PeSS. Both find the optimum for maximum speed. PeSS avoids the run with 8 processors and two redistributions. This kernel runs long enough for intrarun tuning. However, timing the 256 processor case causes a high penalty. When starting closer to the optimum, the overhead of redistribution is within a factor of two of the best possible runtime with 16 processors. An interrun tuning would have been preferable.

PDE Solver. The data structures to be redistributed are four arrays which contain intermediate and final results. The RISS strategy correctly tunes the parallelism for maximum speed.

Veltran Operator. The data structures to be redistributed are two 3-dimensional matrices which contain input data and partial results. This application scales decently, so for maximum speed, 512 processors would be chosen. Given an efficiency threshold of 80%, the EISS intrarun strategy correctly finds the optimum The efficiency is actually above 80% for processor set sizes ranging from 1 to 16; EISS correctly chooses the fastest among these. For a threshold of 90%, the optimum is 2 processors.

TLM Method. The data which must be redistributed is contained in four different structures. This program also scales decently. For maximizing the benefit-cost ratio, CBISS was tried successfully. The benefit-cost relation was about 23% higher than without tuning, while the execution time was 27% longer. For cost-conscious users, this strategy delivers the best cost/performance.

Jacobi Iteration. The EISS interrun strategy correctly found the optimum when given an efficiency threshold of 70%. By going from 128 to 16 processors, the efficiency is not only above the given threshold but is also 4.5 higher than with 128 processors.

In summary, *ParGrad* finds the optimal degree of parallelism reliably. However, some judgment is required when choosing interrun vs. intrarun tuning. Finding the optimum by hand (which we had to do for the evaluation) is extremely tedious by comparison. Letting the computer do the work of determining the optimum not only saves programmer time, it is also more reliable then letting programmers guess the optimum.

5 Conclusion

This paper presented and evaluated a technique for automatically tuning the parallelism of applications on parallel computers. Three different criteria for optimization, viz. speed, efficiency, and benefit-cost were defined and implemented. The technique was designed for the difficult environment of distributed memory, but also works on shared memory computers by simply turning off the data redistribution.

The data needed for tuning is gained through automatic timing measurement during execution. No compiler support is required. The user needs to specify only a few parameters and indicate the loops to be tuned and the associated data structures. The added code for setting these parameters is minor (between 2.3 and 3.3% of the total application) and could be reduced even more by providing a simple graphical interface.

With this technique, the user can let the program itself determine the processor set sizes for maximum speed. For example, the radix sort algorithm ran up to 3.5 times faster than without tuning. Efficency and benefit-cost can also be improved, leading to better computer utilization and lower cost.

Future work should seek to fully automate this technique. First, the choice between inter- and intrarun tuning should be performed automatically. Second, the tuning code should be implanted automatically. This step would also require the automatic detection of the data structures to be redistributed. We also plan to port *ParGrad* to workstation clusters, in particular ParaStation [19].

References

1. D. Bailey, J. Barton, T. Lasinski, and H. Simon. The NAS parallel benchmarks. Technical Report 103863, NASA, July 1993.
2. David E. Culler, Andrea Dusseau, Seth C. Goldstein, Arvind Krishnamurthy, Steven Lumetta, Thorsten von Eicken, and Katherine Yelick. Parallel programming in Split-C. In *Proceedings of Supercomputing '93*, pages 262–273, Los Alamitos, CA, November 1993. IEEE Computer Society Press.
3. Allen B. Downey. Using Queue Time Predictions for Processor Allocation. In *Proceedings of the International Parallel Processing Symposium*, pages 35–57, Berlin, April 1997. Springer-Verlag.
4. Guy Edjlali, Gagan Agrawal, Alan Sussman, Jim Humphries, and Joel Saltz. Compiler and Runtime Support for Programming in Adaptive Environments. Technical Report UMIACS-TR-95-83 and CS-TR-3510, UMIACS and Department of Computer Science, University of Maryland, 1997.
5. M.S. Squillante et al. An analysis of gang scheduling for multiprogrammed parallel computing environments. In *Proceedings of the 8th ACM Symposium on Parallel Algorithms and Architectures*, pages 89–98, New York, NY, June 1996. ACM.
6. John T. Feo. An Analysis of the Computational and Parallel Complexity of the Livermore Loops. *Parallel Computing*, 7:163–185, February 1988.
7. Martin Gebhardt. Parallelisierung des 3D-TLM-Algorithmus mittels MPI. Master's thesis, Department of Electrical Engineering, Karlsruhe University, 1998.
8. Mary Hall and Margaret Martonosi. Adaptive parallelism in compiler-parallelized code. In *Proceedings of the 2nd SUIF Compiler Workshop*, Stanford University, August 1997. USC Information Sciences Institute.
9. Stefan U. Hänssgen. *Effiziente parallele Ausführung irregulärer rekursiver Programme*. PhD thesis, Department of Informatics, Karlsruhe University, 1998.
10. Hans-Ulrich Heiss. *Prozessorzuteilung in Parallelrechnern*. BI-Wissenschaftsverlag, 1994.
11. Matthias Jacob. Implementing large-scale parallel geophysical algorithms using the Java programming language - a feasibility study. Master's thesis, Department of Informatics, Karlsruhe University, 1998.
12. Honghui Lu, Sandhya Dwarkadas, Alan L. Cox, and Willy Zwaenepoel. Message passing versus distributed shared memory on network workstations. In *Proceedings of the Supercomputing'95*, pages 64–65, New York, NY, December 1995. ACM.
13. Cathy McCann, Raj Vaswani, and John Zahorjan. A dynamic processor allocation policy for multiprogrammed shared-memory multiprocessors. *ACM Transactions on Computer Systems*, 11(2):146–178, May 1993.
14. Matthias M. Müller, Thomas M. Warschko, and Walter F. Tichy. Prefetching on the Cray-T3E: A Model and its Evaluation. Technical Report 26/97, Department of Informatics, Karlsruhe University, 1997.

15. Niels Reimer, Stefan U. Hänssgen, and Walter F. Tichy. Dynamically adapting the degree of parallelism with reflexive programs. In *Proceedings of the Third International Workshop on Parallel Algorithms for Irregularly Structured Problems (IRREGULAR)*, pages 313–318, Santa Barbara, CA, USA, August 1996. Springer LNCS 1117.
16. Mark S. Squillante. On the Benefits and Limitations of Dynamic Partitioning in Parallel Computer Systems. In *Proceedings of the 8th International Parallel Processing Symposium*, pages 219–238, Berlin, April 1995. Springer-Verlag.
17. Andrew Tucker and Anoop Gupta. Process Control and Scheduling Issues for Multiprogrammed Shared-memory Multiprocessors. In *Proceedings of the 12th ACM Symposium on Operating Systems Principles*, pages 159–166, New York, NY, December 1989. ACM Press.
18. Fang Wang, Marios Papaefthymiou, and Mark S. Squillante. Performance Evaluation of Gang Scheduling for Parallel and Distributed Multiprogramming. In *Proceedings of the 8th International Parallel Processing Symposium*, pages 277–298, Berlin, April 1997. Springer-Verlag.
19. Thomas M. Warschko, Joachim M. Blum, and Walter F. Tichy. Design and evaluation of ParaStation 2. In *Proceedings of the International Workshop on Distributed High Performance Computing and Gigabit Wide Area Networks*, pages 283–296. Springer LNCS, September 1999.
20. Otilia Werner-Kytölä. *Automatische Einstellung des Parallelitätsgrades von Programmen*. PhD thesis, Department of Informatics, Karlsruhe University, 1999.
21. Steven C. Woo, Moriyoshi Ohara, Evan Torrie, Jaswinder P. Singh, and Anoop Gupta. The SPLASH-2 programs: Characterization and methodological considerations. In *Proceedings of the 22nd Annual International Symposium on Computer Architecture*, New York, NY, June 1995. IEEE Computer Society Press.
22. Kelvin K. Yue and David J. Lilja. Efficient Execution of Parallel Applications in Multiprogrammed Multiprocessor Systems. Technical Report HPPC-95-05, High-Performance Parallel Computing Research Group, Department of Electrical Engineering, Department of Computer Science, Minneapolis, Minnesota, 1995.

A Novel Distributed Algorithm for High-Throughput and Scalable Gossiping

Vincenzo De Florio, Geert Deconinck, and Rudy Lauwereins

Katholieke Universiteit Leuven
Electrical Engineering Department, ACCA Division
Kardinaal Mercierlaan 94, B-3001 Heverlee, Belgium

Abstract. A family of gossiping algorithms depending on a combinatorial parameter is introduced, formalized, and discussed. Three members are analyzed. It is shown that, depending on the pattern of the parameter, gossiping can use from $O(N^2)$ to $O(N)$ time, N being the number of communicating members. The last and best-performing algorithm, whose activity follows the execution pattern of pipelined hardware processors, is shown to exhibit high throughput and efficiency that are constant with respect to N. This translates in unlimited scalability for the corresponding gossiping service provided by this algorithm.

1 Introduction

A number of distributed applications like, e.g., distributed consensus [7], or those based on the concept of restoring organs (N-modular redundancy systems with N-replicated voters) require a base service called gossiping [4, 5]—this is the case, e.g., of the distributed voting tool described in [2].

Informally speaking, gossiping is a communication scheme such that every member of a set has to communicate a private value to all the other members. Gossiping is clearly an expensive service, as it requires a large amount of communication. Implementations of this service can have a great impact on the throughput of their client applications and perform very differently depending on the number of members in the set. This work describes a family of gossiping algorithms that depend on a combinatorial parameter. Three cases are then analyzed under the hypotheses of discrete time, of constant time for performing a `send` or `receive`, and of a crossbar communication system. It is shown[1] how, depending on the pattern of the parameter, gossiping can use from $O(N^2)$ to $O(N)$ time, N being the number of communicating members. The last and best-performing case, whose activity follows the execution pattern of pipelined hardware processors, is shown to exhibit an efficiency constant with respect to N. This translates in unlimited scalability of the corresponding gossiping service. When performing multiple consecutive gossiping sessions, the throughput of the system can reach the value of $t/2$, t being the time for sending one value from one member to another. This means that a full gossiping is completed every two basic communication steps.

The paper is structured as follows: Sect. 2 draws a formal model and introduces the family of gossiping algorithms discussed thereafter. The three case studies are reported in Sect. 3, Sect. 4, and Sect. 5. Section 6 concludes this work.

---
[1] Proofs have been omitted from the current version of this paper. They can be found in [3].

2 Formal Model and Definitions

Definition 1 (System). $N + 1$ *processors are interconnected via some communication means that allows them to communicate with each other by means of full-duplex point-to-point communication lines. Communication is synchronous and blocking. Processors are uniquely identified by integer labels in* $\{0, \ldots, N\}$; *they will be globally referred to, together with the communication means, as "the system".*

Definition 2 (Problem). *The processors own some local data they need to share (for instance, to execute a voting algorithm [2]). In order to share their local data, each processor needs to broadcast its own data to all the others via multiple sending operations, and to receive the N data items owned by its fellows. A discrete time model is assumed—events occur at discrete time steps, one event at a time per processor. This is a special class of the general family of problems of information dissemination known as* gossiping *[5]. Let us call this context as "the problem".*

It is worth mentioning that a relevant difference with respect to the basic Gossip Problem [5] is that, in the case presented in this paper, on each communication run, members are only allowed to broadcast their own *local* data, while, in general [4], they "pass on to each other as much [information] they know at the time", including data received from the other members in previous communication runs.

Definition 3 (Time Step). *Let us assume the time to send a message and that to receive a message to be constant. This amount of time be called "time step".*

Definition 4 (Actions). *On a given time step t, processor i may be:*

1. *sending a message to processor j, $j \neq i$; this is represented as relation $i\, S^t\, j$;*
2. *receiving a message from processor j, $j \neq i$; this is represented as $i\, R^t\, j$;*
3. *blocked, waiting for messages to be received from any processor; where both the identities of the involved processors and t can be omitted without ambiguity, symbol "$-$" will be used to represent this case;*
4. *blocked, waiting for a message to be sent, i.e., for another processor to enter the receiving state; under the same assumptions of case 3, symbol "$\frown$" will be used.*

The above cases are referred to as "the actions" of time step t.

Definition 5 (Slot, Used Slot, Wasted Slot). *A slot is a temporal "window" one time step long, related to a processor. On each given time step there are $N + 1$ available slots within the system. Within that time step, a processor may use that slot (if it sends or receives a message during that slot), or it may waste it (if it is in one of the remaining two cases). In other words:*
Processor i is said to use *slot t if and only if $U(t, i) = \{\exists j\ (i\, S^t\, j \vee i\, R^t\, j)\}$ is true; on the contrary, processor i is said to* waste *slot t iff $\neg U(t, i)$.*
The following notation,

$$\delta_{i,t} = \begin{cases} 1 & \text{if } U(t, i) \text{ is true,} \\ 0 & \text{otherwise,} \end{cases}$$

will be used to count used slots.

Definition 6 (States WR, WS, S, R). *Let us define four state templates for a finite state automaton to be described later on:*

WR state. A processor is in state WR_j if it is waiting for the arrival of a message from processor j. Where the subscript is not important it will be omitted. Once there, a processor stays in state WR for zero (if it can start receiving immediately) or more time steps, corresponding to the same number of actions "wait for a message to come".

S state. A processor is in state S_j when it is sending a message to processor j. Note that, by the above assumptions and definitions, this transition lasts exactly one time step. For each transition to state S there corresponds exactly one "send" action.

WS state. A processor that is about to send a message to processor j is said to be in state WS_j. In the following, the subscript will be omitted when this does not result in ambiguities. The permanence of a processor in state WS implies zero (if the processor can send immediately) or more occurrences in a row of the "wait for sending" action.

R state. A processor that is receiving a message from processor j is said to be in state R_j. By the above definitions, this state transition also lasts one time step.

For any $i \in [0, N]$, let $\mathcal{P}_1, \ldots, \mathcal{P}_N$ represent a permutation of the N integers $[0, N] - \{i\}$. Then the above state templates can be used to compose $N + 1$ linear finite state automata (FSA) with a structure like the one in Fig. 1, which shows the state diagram of the FSA to be executed by processor i. The first row represents the condition that has to be reached before processor i is allowed to begin its broadcast: a series of i couples (WR, R).

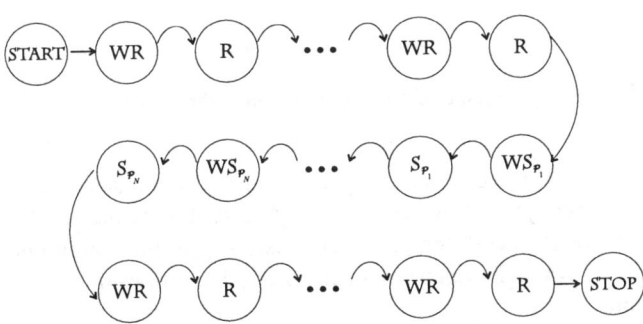

Fig. 1. The state diagram of the FSA run by processor i. The first row consists of i couples (WR, R). (For $i = 0$ this row is missing). $(\mathcal{P}_1, \ldots, \mathcal{P}_N)$ represents a permutation of the N integers $[0, N] - \{i\}$. The last row contains $N - i$ couples (WR, R).

Once processor i has successfully received i messages, it gains the right to broadcast, which it does according to the rule expressed in the second row of Fig. 1: it orderly sends its message to its fellows, the j-th message being sent to processor $\mathcal{P}_j$.

The third row of Fig. 1 represents the reception of the remaining $N - i$ messages, coded as $N - i$ couples like those in the first row.

Proposition 7. *For any permutation $\mathcal{P}$, the distributed algorithm described by the state diagram of Fig. 1 solves the problem of Def. 2 without deadlocks.*

Definition 8 (Run). *The collection of slots needed to fully execute the above algorithm on a given system, plus the value of the corresponding actions.*

Definition 9 (Average Slot Utilization). *The average number of slots used during a time step. It represents the average degree of parallelism exploited in the system. It will be indicated as μ_N, or simply as μ. It varies between 0 and $N + 1$.*

Definition 10 (Efficiency). *The percentage of used slots over the total number of slots available during a run. ε_N, or simply ε, will be used to represent efficiency.*

Definition 11 (Length). *The number of time steps in a run. It represents a measure of the time needed by the distributed algorithm to complete. λ_N, or simply λ, will be used for lengths.*

Definition 12 (Number of Slots). $\sigma(N) = (N+1)\lambda_N$ *represents the number of slots available within a run of $N + 1$ processors.*

Definition 13 (Number of Used Slots). *For each run and each time step t,*

$$\nu_t = \sum_{i=0}^{N} \delta_{i,t}$$

represents the number of slots that have been used during t.

Definition 14 (Utilization String). *The λ-tuple*

$$\nu = [\nu_1, \nu_2, \ldots, \nu_\lambda],$$

representing the number of used slots for each time step, respectively, is called utilization string.

In the next sections, three cases of $\mathcal{P}$ are introduced and discussed. In the following it is shown how varying the structure of $\mathcal{P}$ may develop very different values for μ, ε, and λ. This fact, coupled with the physical constraints of the communication line and with the number of available independent channels, determines the overall performance of this algorithm.

In the following, a fully connected, or crossbar, interconnection is assumed, allowing any processor to communicate with any other processor in one time step through the interconnection [8].

3 First Case: Identity Permutation

As a first case, let $\mathcal{P}$ be equal to the identity permutation:

$$\begin{pmatrix} 0, \ldots, i-1, i+1, \ldots, N \\ 0, \ldots, i-1, i+1, \ldots, N \end{pmatrix}, \tag{1}$$

i.e., in cycle notation [6], $(0)\ldots(i-1)(i+1)\ldots(N)$. Note that in this case only singleton cycles are present.

Using the identity permutation means that, once processor i gains the right to broadcast, it will first send its message to processor 0 (possibly having to wait for processor 0 to become available to receive that message), then it will do the same with processor 1, and so forth up to N, obviously skipping itself. This is represented in Table 1 for $N = 4$. Let us call this a run-table. Let us also call "run-table x" a run-table of a system with $N = x$.

id step →	1	2	3	4	5	6	7	8	9	10	11	12	13	14	15	16	17	18
0	S_1	S_2	S_3	S_4	R_1	–	R_2	–	–	–	R_3	–	–	–	R_4	–	–	–
1	R_0	↷	↷	↷	S_0	S_2	S_3	S_4	R_2	–	–	R_3	–	–	–	R_4	–	–
2	–	R_0	–	–	–	–	R_1	S_0	↷	S_1	S_3	S_4	–	R_3	–	–	–	R_4	–
3	–	–	R_0	–	–	–	–	R_1	–	–	R_2	S_0	S_1	S_2	S_4	–	–	–	R_4
4	–	–	–	R_0	–	–	–	–	R_1	–	–	R_2	–	–	R_3	S_0	S_1	S_2	S_3
ν →	2	2	2	2	2	2	4	2	2	2	4	2	2	2	2	2	2	2

Table 1. A run ($N = 4$), with $\mathcal{P}$ equal to the identity permutation. The step row represents time steps. Id's identify processors. ν is the utilization string (see Def. 14). In this case μ, or the average utilization is 2.22 slots out of 5, with an efficiency $\varepsilon = 44.44\%$ and a length $\lambda = 18$. Note that, if the slot is used, then entry $(i, t) = \mathcal{R}_j$ in this matrix represents relation $i\,\mathcal{R}^t\,j$.

It is possible to characterize precisely the duration of the algorithm that adopts this permutation:

Proposition 15. $\lambda_N = \frac{3}{4}N^2 + \frac{5}{4}N + \frac{1}{2}\lfloor N/2 \rfloor$.

Lemma 16. *The number of columns with 4 used slots inside, for a run with $\mathcal{P}$ equal to the identity permutation and $N + 1$ processors, is*

$$\sum_{i=0}^{N-1} \lfloor \frac{i}{2} \rfloor.$$

Figure 2 shows the typical shape of run-tables in the case of $\mathcal{P}$ being the identity permutation, also locating the 4-used slot clusters.

The following Proposition locates the asymptotic values of μ and ε:

Proposition 17. $\lim_{N \to \infty} \varepsilon_N = 0$, $\lim_{k \to \infty} \mu_k = \frac{8}{3}$.

4 Second Case: Pseudo-random Permutation

This Section covers the case such that $\mathcal{P}$ is a pseudo-random permutation of the integers $0, \ldots, i-1, i+1, \ldots, N$.

Fig. 2. A graphical representation for run-table 20 when $\mathcal{P}$ is the identity permutation. Light gray pixels represent wasted slots, gray pixels represent R actions, black slots are sending actions.

Figure 3 (left picture) shows the values of λ using the identity and pseudo-random permutations and draws the parabola which best fits to these values in the case of pseudo-random permutations. Experiments show that the choice of the identity permutation is even "worse" than choosing permutations at random. The same conclusion is suggested by Fig. 4 that compares the averages and efficiencies in the above two cases.

Table 2 shows run-table 5, and Fig. 5 shows the shape of run-table 20 in this case.

id step $\downarrow \rightarrow$	1	2	3	4	5	6	7	8	9	10	11	12	13	14	15	16	17	18	19	20	21	22	23	24
0	S_5	S_1	S_3	S_2	S_4	R_1	-	-	-	-	R_2	-	-	-	R_3	-	-	-	-	R_4	R_5	-	-	-
1	-	R_0	S_5	⌢	⌢	S_0	S_3	S_2	S_4	R_2	-	-	-	R_3	-	-	-	-	R_4	R_5	-	-	-	-
2	-	-	-	R_0	-	-	-	R_1	S_5	S_1	S_0	S_3	S_4	-	-	R_3	-	-	-	-	R_4	R_5	-	
3	-	-	R_0	-	-	-	R_1	-	-	-	-	R_2	S_5	S_1	S_0	S_2	S_4	-	-	-	R_4	R_5	-	-
4	-	-	-	-	R_0	-	-	-	-	R_1	-	-	R_2	-	-	-	R_3	S_5	S_1	S_0	S_3	S_2	-	R_5
5	R_0	-	R_1	-	-	-	-	R_2	-	-	-	R_3	-	-	-	-	R_4	⌢	S_1	S_0	S_3	S_2	S_4	
$\nu \rightarrow$	2	2	4	2	2	2	2	2	4	2	2	2	4	2	2	2	2	2	4	4	4	2	2	

Table 2. Run-table 5 when $\mathcal{P}$ is chosen pseudo-randomly. μ is 2.5 slots out of 6, which implies an efficiency of 41.67%.

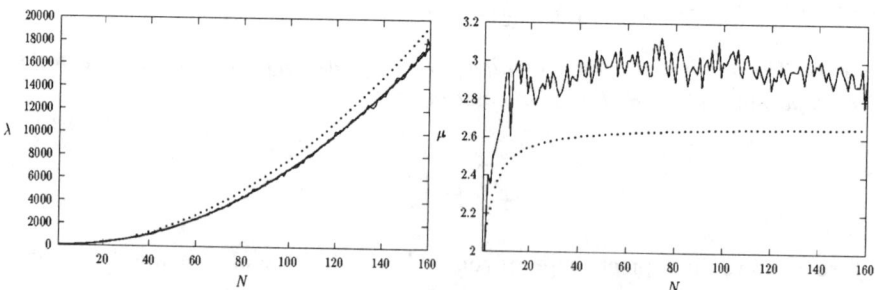

Fig. 3. Left picture: comparison between lengths when $\mathcal{P}$ is the identity permutation (dotted parabola) and when it is pseudo-random (piecewise line), $1 \leq N \leq 160$. The lowest curve ($\lambda = 0.71N^2 - 3.88N + 88.91$) is the parabola best fitting to the piecewise line—which suggests a quadratic execution time as in the case of the identity permutation. Right picture: comparison between values of μ when $\mathcal{P}$ is the identity permutation (dotted curve) and when $\mathcal{P}$ is pseudo-random (piecewise line), $1 \leq N \leq 160$. The latter is strictly over the former. Note how μ appears to tend to a value right above 2.6 for the identity permutation, as claimed in Prop. 17.

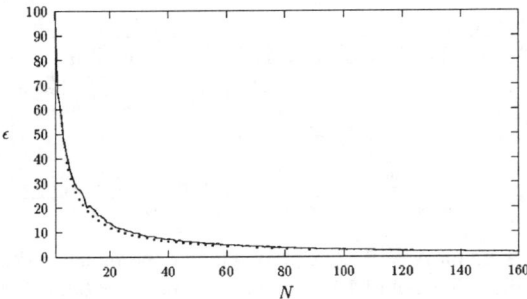

Fig. 4. Comparison between values of ε in the case of the pseudo-random permutation (piecewise line) and that of the identity permutation (dotted curve), $1 \leq N \leq 160$. Also in this graph the former is strictly over the latter, though they get closer to each other and to zero as N increases, as proven for the identity permutation in Prop. 17.

Fig. 5. A graphical representation for run-table 20 when $\mathcal{P}$ is a pseudo-random permutation.

5 Third Case: Algorithm of Pipelined Broadcast

Let $\mathcal{P}$ be the following permutation:

$$\begin{pmatrix} 0, \ldots, i-1, i+1, \ldots, N \\ i+1, \ldots, N, 0, \ldots, i-1 \end{pmatrix}. \quad (2)$$

Note how permutation (2) is equivalent to i cyclic logical left shifts of the identity permutation. Note also how, in cycle notation, (2) is represented as one cycle; for instance, $\binom{0,1,2,4,5}{4,5,0,1,2}$, viz., (2) for $N = 5$ and $i = 3$, is equivalent to cycle $(0, 4, 1, 5, 2)$.

A value of $\mathcal{P}$ equal to (2) means that, once processor i has gained the right to broadcast, it will first send its message to processor $i + 1$ (possibly having to wait for it to become available to receive that message), then it will do the same with processor $i + 2$, and so forth up to N, then wrapping around and going from processor 0 to processor $i - 1$. This is represented in Table 3 for $N = 9$.

Figures quite similar to Table 3 can be found in many classical works on pipelined microprocessors (see e.g. [8]). Indeed, a pipeline is a series of data-paths shifted in time so to overlap their execution; likewise, permutation (2) tends to overlap its broadcast sessions as much as possible. Clearly pipe stages are represented here as full processors, and the concept of machine cycle, or pipe stage time of pipelined processor, simply collapses to the concept of time step as introduced in Def. 3.

A number of further considerations brought the Authors to the name of "pipelined broadcast" for this special member of the family of algorithms described so far. They are emphasized in the following using the italics typeface.

Indeed using this permutation leads to better performance. In particular, after a start-up phase (*after filling the pipeline*), sustained performance is close to the maximum—a

Table 3. Run-table of a run for $N = 9$ using permutation of permutation (2). In this case μ, or the average utilization is 6.67 slots out of 10, with an efficiency $\varepsilon = 66.67\%$ and a length $\lambda = 27$. Note that ν is in this case a palindrome.

number of unused slots (*pipeline bubbles*) still exist, even in the sustained region, but here μ reaches value $N + 1$ half of the times (if N is odd). In the region of decay, starting from time step 19, every new time step a processor fully completes its task. Similar remarks apply to Table 4; this is the typical shape of a run-table for N even. This time the state within the sustained region is more steady, though the maximum number of used slots never reaches the number of slots in the system.

Table 4. Run-table of a run for $N = 8$ using permutation (2). μ is equal to 6 slots out of 9, with an efficiency $\varepsilon = 66.67\%$ and a length $\lambda = 24$. Note how ν is a palindrome string.

It is possible to show that the distributed algorithm described in Fig. 1, with $\mathcal{P}$ as in (2), can be computed in linear time:

Proposition 18. $\forall N : \lambda_N = 3N$.

The efficiency of the algorithm of pipelined broadcast does not depend on N:

Proposition 19. $\forall k > 0 : \varepsilon_k = 2/3, \mu_k = \frac{2}{3}(k + 1)$.

Table 5 shows how a run-table looks like when multiple broadcasting sessions take place one after the other. As a result, the central area corresponding to the best observable performance is prolonged. In such an area, ε has been experimentally found to be equal to $N/(N + 1)$ and the throughput, or the number of gossiping sessions per time step, has been found to be equal to $t/2$, t being the duration of a time step. In other words, within that area a gossiping session is fully completed every two time steps in

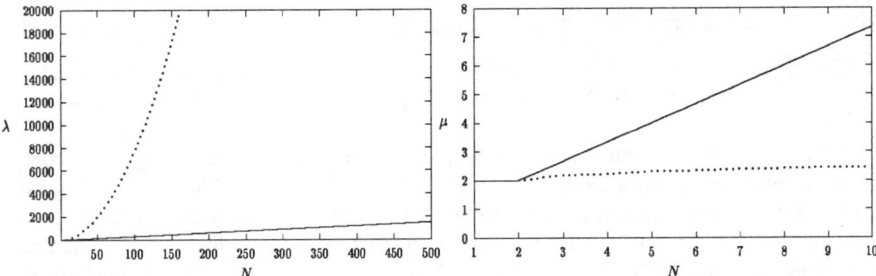

Fig. 6. Left picture: comparison between run lengths resulting from the identity permutation (dotted parabola) and those from permutation (2). The former are shown for $1 \leq N \leq 160$, the latter for $1 \leq N \leq 500$. Right picture: comparison between values of μ derived from the identity permutation (dotted parabola) and those from permutation (2) for $1 \leq N \leq 10$.

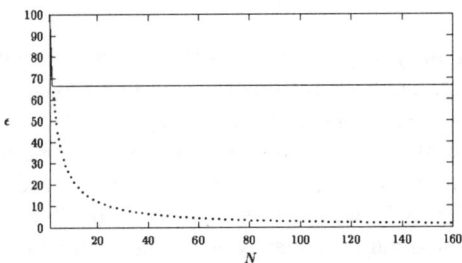

Fig. 7. Comparison of efficiencies when $\mathcal{P}$ is the identity permutation and in the case of permutation (2), for $1 \leq N \leq 160$.

the average. A number of applications that are based on multiple gossiping sessions, e.g., distributed consensus [7], may benefit from this result to reach good efficiency and scalability.

A key requirement of the model assumed herein is clearly that any communication calls for exactly one time step to take place between any two processors, as it is, e.g., in a crossbar system. This is similar to the constraint of hardware pipelined processors which call for a number of memory ports equal to n, n being the number of pipeline stages supported by that machine—in this way the system is able to overlap any two of its stages. This of course turns into requiring to have a memory system capable of delivering n times the original bandwidth of a corresponding, non-pipelined machine [8]. Note how the specularity of graphs like the one in Fig. 8 translates into a palindrome ν.

Fig. 8. A graphical representation for run-table 30 when $\mathcal{P}$ is permutation (2).

	1	2	3	4	5	6	7	8	9	10	11	12	13	14	15	16	17	18	19	...											
0	S_1	S_2	S_3	S_4	$-$	R_1	R_2	R_3	R_4	$\frown$	S_1	S_2	S_3	S_4	$-$	R_1	R_2	R_3	R_4	...	$\frown$	S_1	S_2	S_3	S_4	$-$	R_1	R_2	R_3	R_4	$-$	$-$	$-$
1	R_0	$\frown$	S_2	S_3	S_4	S_0	$-$	R_2	R_3	R_4	R_0	$\frown$	S_2	S_3	S_4	S_0	$-$	R_2	R_3	...	R_4	R_0	$\frown$	S_2	S_3	S_4	S_0	$-$	R_2	R_3	R_4	$-$	$-$
2	$-$	R_0	R_1	$\frown$	S_3	S_4	S_0	S_1	$-$	R_3	R_4	R_0	R_1	$\frown$	S_3	S_4	S_0	S_1	$-$	...	R_3	R_4	R_0	R_1	$\frown$	S_3	S_4	S_0	S_1	$-$	R_3	R_4	$-$
3	$-$	$-$	R_0	R_1	R_2	$\frown$	S_4	S_0	S_1	S_2	$-$	R_4	R_0	R_1	R_2	$\frown$	S_4	S_0	S_1	...	S_2	$-$	R_4	R_0	R_1	R_2	$\frown$	S_4	S_0	S_1	S_2	$-$	R_4
4	$-$	$-$	$-$	R_0	R_1	R_2	R_3	$\frown$	S_0	S_1	S_2	S_3	$-$	R_0	R_1	R_2	R_3	$\frown$	S_0	...	S_1	S_2	S_3	$-$	R_0	R_1	R_2	R_3	$\frown$	S_0	S_1	S_2	S_3
	2	2	4	4	4	4	4	4	4	4	4	4	4	4	4	4	4	4	4	...	4	4	4	4	4	4	4	4	4	4	4	2	2

Table 5. The algorithm is modified so that multiple gossiping sessions take place. As a consequence, the central, best performing area is prolonged. Therein ε is equal to $N/(N+1)$. Note how, within that area, there are consecutive "zones" of ten columns each, within which five gossiping sessions reach their conclusion. For instance, such a zone is the region between columns 7 and 16: therein, at entries $(4,7)$, $(0,9)$, $(1,10)$, $(2,11)$, and $(3,12)$, a processor gets the last value of a broadcast and can perform some work on a full set of values. This brings to a throughput of $t/2$, where t is the duration of a slot.

6 Conclusions

A formal model for a family of gossiping algorithms depending on a combinatorial parameter, permutation $\mathcal{P}$, has been introduced and discussed. Three case studies have been proposed, simulated, and analyzed, also categorizing in some cases their asymptotic behaviour. In one of these cases—the algorithm of pipelined broadcast—it has been shown that the efficiency of the algorithm does not depend on N.

The algorithm so far analyzed has been adopted in the "voting farm", a distributed tool for integrating N-modular redundancy and programming restoring organs [2], as a means to discipline the right of broadcast within a group of voters. An enhanced version of this tool is being designed in the framework of the TIRAN [1] project.

Acknowledgements. This project is partly sponsored by the ESPRIT-IV project 28620 TIRAN. Geert Deconinck is a Postdoctoral Fellow of the Fund for Scientific Research - Flanders (Belgium) (FWO).

References

1. O. Botti et al. TIRAN: Flexible and portable fault tolerance solutions for cost effective dependable applications. In *Proc. of Euro-Par'99, LNCS*, 1685:1166–1170, Aug.–Sep. 1999.
2. V. De Florio, G. Deconinck, and R. Lauwereins. Software tool combining fault masking with user-defined recovery strategies. *IEE Proceedings — Software*, 145(6):203–211, Dec. 1998.
3. V. De Florio and G. Deconinck. A novel distributed algorithm for high-throughput and scalable gossiping. Tech. Rep. ESAT/ACCA/2000/1, K.U.Leuven, February 2000.
4. A. Hajnal, E. C. Milner, and E. Szemeredi. A cure to the telephone disease. *Canadian Math Bulletin*, 15:447–450, 1972.
5. S. Hedetniemi, S. Hedetniemi, and A. Leistman. A survey of gossiping and broadcasting in communication networks. *Networks*, 18:419–435, 1988.
6. D. E. Knuth. *The Art of Computer Programming, Vol. 1*. Addison-Wesley, Reading MA, 1973.
7. L. Lamport, R. Shostak, and M. Pease. The Byzantine generals problem. *ACM Trans. on Programming Languages and Systems*, 4(3):384–401, July 1982.
8. D. A. Patterson and J. L. Hennessy. *Computer Architecture — A Quantitative Approach*. Morgan Kaufmann, S. Francisco, CA, 2nd edition, 1996.

Parallel Access to Persistent Multidimensional Arrays from HPF Applications Using *Panda**

Peter Brezany[1], Przemysław Czerwiński[1],
Artur Świętanowski[2], and Marianne Winslett[3]

[1] Institute for Software Science
University of Vienna, Liechtensteinstrasse 22, A-1090 Vienna, Austria
{brezany,przemek}@par.univie.ac.at
[2] Department of Statistics, Operations Research and Computer Methods
University of Vienna, Universitätsstrasse 5, A-1010 Vienna, Austria
swietanowski@bigfoot.com
[3] Database Research Laboratory, Department of Computer Science
University of Illinois, 1304 W. Springfield, Urbana, IL 61801 USA
winslett@uiuc.edu

Abstract. A critical performance issue for a number of scientific and engineering applications is the efficient transfer of data to secondary storage. Languages such as High Performance Fortran (HPF) have been introduced to allow programming distributed-memory systems at a relatively high level of abstraction. However, the present version of HPF does not provide appropriate constructs for controlling the parallel I/O capabilities of these systems. In this paper, constructs to specify parallel I/O operations on multidimensional arrays in the context of HPF are proposed. The paper also presents implementation concepts that are based on the HPF compiler developed at the University of Vienna and the parallel I/O runtime system Panda developed at the University of Illinois. Experimental performance results are discussed in the context of financial management and traffic simulation applications.

1 Introduction

The primary goal behind the development of the family of High Performance Fortran (HPF) languages was to address the problems of writing data parallel programs for architectures where the distribution of data impacts performance. This development has almost fully focused on providing support for compute intensive parts of scientific and engineering applications and only minimally addressed their input and output (I/O) aspects ([3, 4, 9]). However, no I/O language extensions proposed were included into the final HPF specification documents. Consequently, at present, HPF and the existing HPF programming environments

---

* The work described in this paper is being carried out as part of the research project "Aurora" supported by the Austrian Research Foundation, and is also supported by NASA under grant NAGW 4244 and the Department of Energy under grant B341494.

do not provide an appropriate framework for programming and implementing I/O intensive applications.

In the past years, I/O subsystems have become the focus of much research, leading to the design of parallel I/O hardware and matching system software. So far, the major software research efforts have been led along two lines, i.e., parallel file system research and parallel I/O library research. The most important results include IBM/PIOFS [1], Galley [7], Panda [8], and ViPIOS [12].

Panda (http://drl.cs.uiuc.edu/panda/) is a DBMS-style parallel I/O library providing support for most common I/O access patterns which occur in typical scientific parallel applications. From a logical point of view, Panda operates on a high performance data store called the scientific data repository [5].

The work presented in this paper describes the design and implementation of a system providing the HPF users with the opportunity to access the functionality of the Panda library at a high abstraction level. Our prototype implementation has been done in the context of an HPF compiler called the Vienna Fortran Compiler (VFC) [2] developed at the University of Vienna. Experimental performance results are discussed in the context of a real financial management application and a synthetic application derived from a real road traffic simulation application.

This paper is organized as follows. Section 2 introduces the extensions of HPF to allow the HPF programmers to access the functionality of Panda. The implementation of the VFC/Panda interface is briefly described in Section 3. Section 4 discusses the application aspects and experimental performance results. Conclusions and ideas for further research are introduced in Section 5.

2 Language Support

In this section we introduce new HPF language constructs and a machine model which their design and implementation are based on. To explicitly indicate the deviations of these language constructs from the HPF standard and to indicate the relation to the concept of the scientific data repository (SDR) applied in Panda, these new language constructs are introduced by the prefix "!SDR$".

2.1 The Parallel Machine Model

We postulate our target parallel architecture as a system $\mathcal{S} = (\ N,\ IN,\ T)$, where N is the set of nodes that perform parallel computational and parallel I/O tasks. The nodes are classified into 3 classes: (1) computation nodes - they include a processor and internal memory, (2) I/O nodes- they include a processor, internal memory, and a set of disks; the processor is responsible for controlling I/O activities only, and (3) combined nodes - they include a processor, primary memory, and a set of disks; the processor fulfills both compute and I/O tasks. IN is the interconnection network that connects the nodes. T is the set of parameters characterizing interprocessor communication and the access time to secondary storage.

2.2 Array Distribution Across I/O Nodes

Parallel I/O operations require special support at the data organization and data access level. Files are partitioned into a set of chunks (data tiles) which are distributed across the I/O nodes using a declustering algorithm (e.g., [6]) to achieve parallelism during read and write operations.

The HPF concepts related to the data distribution onto computational nodes may be extended by the mapping of data onto I/O nodes.

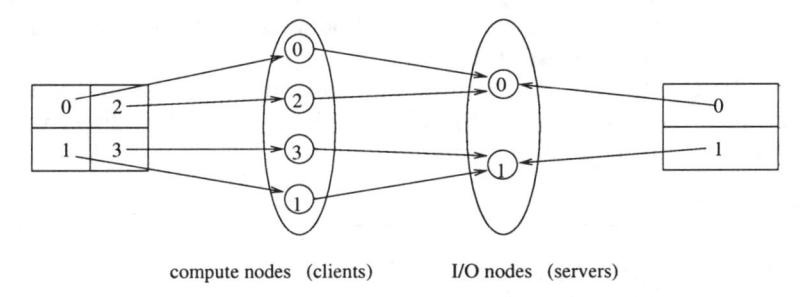

Fig. 1. Different array data distributions in memory and on disk.

Figure 1 shows a 2D array distributed (BLOCK, BLOCK) across 4 compute nodes arranged in a 2 mesh. Each piece of the distributed array is called a compute chunk, and each compute chunk resides in the memory of one compute node. The I/O nodes are also arranged in a mesh, and the data can be distributed across them using HPF-style directives.

The array distribution on disk can also be radically different from that in memory. For instance, the array in Figure 1 has a (BLOCK, *) distribution on disk. Each I/O chunk resulting from the distribution chosen for disk will be buffered and sent to (or read from) local disk at one I/O node, and that I/O node is in charge of reading, writing, gathering, and scattering the I/O chunk. For example, in Figure 1, during a write operation I/O node 0 gathers compute chunks from compute nodes 0 and 2, reorganizes them into a single I/O chunk, and writes it to disk. In parallel, I/O node 1 is gathering, reorganizing and writing its own I/O chunk. For a read operation, the reverse process is used.

In our approach, I/O nodes are explicitly introduced by the declaration of an *I/O processor array*. For example, in relation to Figure 1, the following declaration introduces a 1D I/O processor array *IOP*:

!SDR$ IOPROCESSORS :: IOP(2)

The array distribution onto the logical array of I/O processors is specified by a new directive which in relation to Figure 1 has the following form:

!SDR$ IODISTRIBUTE (BLOCK, *) ONTO IOP :: A

Mapping of logical I/O processors to physical processors and logical disks onto physical disks is implementation dependent. In the Panda architecture, the Panda data servers run on the dedicated I/O nodes, and the Panda clients run on the compute nodes. The application interfaces the clients. During each I/O request, Panda servers read or write arrays from or to disks and move data to and from clients using message passing facilities such as the MPI library.

2.3 I/O Operations

Panda's interface provides support for the following I/O operations: read and write a list of n arrays and perform a checkpoint, restart, or time step output and input operation for a list of n arrays.

In our HPF interface, the Panda operations are applied to so called *data transfer objects*. A data transfer object is specified by a block of directives of the form introduced in Figure 2.

```
!SDR$ IODEF      transfer_type :: transfer_object_name
!SDR$     INVOLVED :: array_list
!SDR$     IODISTRIBUTE ( dist_spec₁ ) ONTO IO_proc_array :: array₁
              ...
!SDR$     IODISTRIBUTE ( dist_specₙ ) ONTO IO_proc_array :: arrayₙ
!SDR$ END IODEF
```

Fig. 2. Specification of a Data Transfer Object.

In this specification, the directive INVOLVED introduces a list of arrays which will be written or read to or from the repository and the directives IODISTRIBUTE specify I/O distributions of these arrays. The way the arrays are to be transferred is determined by a keyword which stands in place of *transfer_type*. The following transfer types can be specified in this context:

- **CHECKPOINT** - For long-running production runs, it is desirable to save the state of certain arrays periodically (checkpoint) in order to resume (restart) from a previous state in case of a system failure. Because checkpoint operations are performed frequently, a good I/O data distribution is required to minimize the I/O time.
- **TIMESTEP** - For time-dependent applications, snapshots of certain arrays are output at selected intervals over time. Output data will then be read and analyzed by visualization tools.
- **TRADITIONAL** - The operations read and write are applied to a list of arrays in the traditional way.

When the HPF application does not specify any I/O data distribution, Panda automatically selects a distribution for the arrays that will be written.

The example in Figure 3 shows the specification of the operations checkpoint, restart, timestep, and timestep-read.

```
PARAMETER :: M = ..., N = ...
REAL, DIMENSION (N, N) :: A, B, C, D
!SDR$   IOPROCESSORS  :: IOP(M, M)

        ! declare transfer object of checkpoint type
!SDR$   IODEF   CHECKPOINT  :: Simulation01
!SDR$         INVOLVED :: A, B
!SDR$         IODISTRIBUTE (BLOCK, BLOCK ) ONTO IOP :: B
!SDR$   END IODEF
        ! write checkpoint
!SDR$   WRITE (Simulation01)
        ...
        ! read checkpoint (restart)
!SDR$   READ (Simulation01)

        ! declare transfer object of timestep type
!SDR$   IODEF   TIMESTEP  :: TemperatureAndPressure
!SDR$         INVOLVED :: T, P
!SDR$         IODISTRIBUTE (CYCLIC, BLOCK) ONTO IOP :: T
!SDR$         IODISTRIBUTE (BLOCK, BLOCK) ONTO IOP :: P
!SDR$   END IODEF
        ! write timestep
!SDR$   WRITE (TemperatureAndPressure)
        ...
        ! read timestep
!SDR$   READ (TemperatureAndPressure)
```

Fig. 3. Specification of the operations checkpoint, restart, timestep, and timestep-read.

The directives

   !SDR$ OPEN()   and   !SDR$ CLOSE()

specify opening and closing a repository, respectively. The paths to the files storing the repository data are not declared in the application programs; they are set within the Panda environment during the Panda installation.

3 Implementation Notes

The implementation of the interface between HPF and the Panda library consists of two main parts: extensions to the HPF compiler VFC and a wrapper library built on top of Panda.

VFC translates an HPF code into the Fortran 90 code containing calls to the runtime libraries. The !SDR$ directives are analyzed and translated into Fortran 90 data structures and sequences of procedure calls to the wrapper library.

The wrapper library provides a set of Fortran 90 subroutines that enable to call Panda operations from a Fortran 90 program. The wrapper library performs all the transformations and conversions that are necessary for coupling the C++ world of Panda to the VFC Fortran 90 environment.

The current implementation provides support both for arrays of simple and derived types. The system can accept wide range of derived types, namely, those which consist of components having INTEGER, REAL and DOUBLE PRECISION types, both scalars and arrays, and of derived types that match recursively this criterion.

4 Application Case Studies

In this section we discuss how we have applied our parallel I/O support to two model applications. The first thereof is an application module belonging to the Aurora Financial Management System. In this case study, we focus our attention on non-conventional use of the checkpoint/restart feature. The second application is a kernel of a road traffic simulator. In this case, both checkpoint/restart and time-step abilities of Panda have been employed in the traditional way.

4.1 The Aurora Financial Management System

Financial management under uncertainty is a general term for activities such as asset allocation (general portfolio management) and asset-liability management (as in insurance, pension fund management, bank management). They all involve substantial uncertainty resulting from an unpredictable nature of events influencing the value of investment over time.

The *Aurora Financial Management System* supports a decision maker in finding portfolio allocations, that will ensure meeting future obligations (liabilities), safety of the investment and a reasonable profit (the interested reader may consult [10, 11] for details).

The system models all relevant risks (both on the asset side and the liability side). The ALM (Asset Liability Management) model is developed within a framework of stochastic dynamic optimization methods and models.

The Simulation Module. Among other important data, the ALM model for a pension fund needs a credible method of proposing scenarios of future pension payments and contributions. In the optimization model [11], a random variable representing net external cash outflows relative to fund's capital is added to the model. Its value is a result of many random events taking place in the population of pension fund members (PFMs) over time, such as joining or leaving the fund, retirement, death with or without leaving dependents, etc. There is no analytical

```
      ! initialization routines, including reading input data

      !SDR$ IODEF CHECKPOINT :: POPULATION
      !SDR$       INVOLVED :: p%people
      !SDR$       IODISTRIBUTE (BLOCK) ONTO iop:: p%people
      !SDR$ END IODEF

      !SDR$ WRITE(POPULATION)

      DO s = 1, max_sym
         IF (s > 1) THEN
            !SDR$ READ(POPULATION)
            ! ... initialization of simple data structures ...
         END IF
         ! call to the main simulation routine
      END DO
      ! clean-up
```

Fig. 4. The outermost simulation loop

model for this variable's probability distribution, thus simulation is the only practical means of estimating the distribution.

A simulation model has been built that mimics the individual PFM's accounting activities taking place in the pension fund with the progress of time. Given a planning horizon, a single simulation consists of finding individual contributions and pension payments for each of the PFMs, as well as the values of all underlying random variables (e.g., number of pensioners who died in a particular month).

With the size of single risk groups in the order of one thousand to ten thousand people, a single simulation of the whole risk group takes at least a few seconds of CPU time. To obtain statistically meaningful results, thousands of simulations are needed to build just one model.

At each step of the simulation, the summary data (the random variables we want to observe, for inclusion in the optimization problem later) has to be stored for further processing by other components of the model building software.

The initial state of the population is read from an external database at the beginning of the simulation run. This operation, together with other initialization routines, usually takes a significant part of the time of the whole simulation run. Often this time is in the order of magnitude of the time used for the actual simulation. One can increase efficiency by performing initialization routines only once, then storing the initial state of simulation internal variables and restoring them at the beginning of each simulation run. We used the checkpoint/restart facility of Panda for this purpose.

After the initialization phase of the simulation that includes reading input data from an external database the checkpoint operation is executed. Before

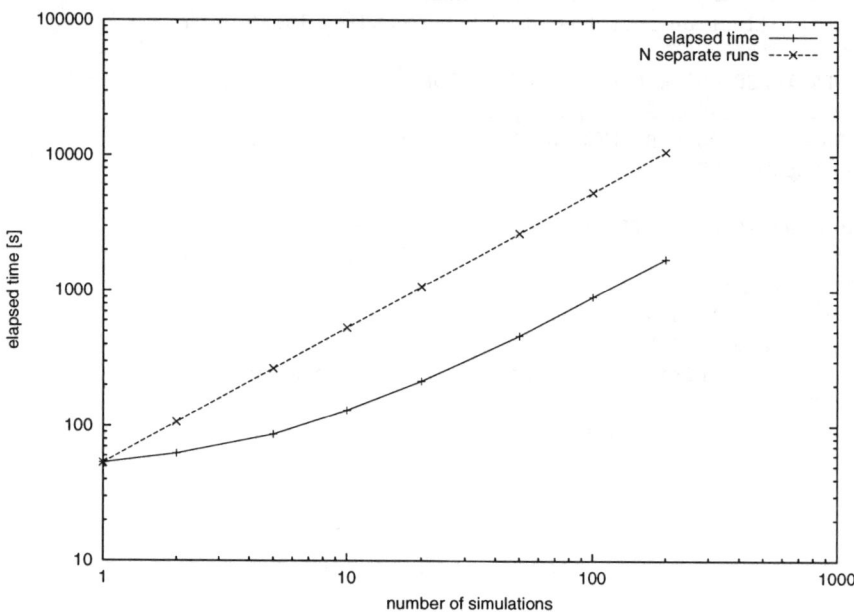

Fig. 5. Execution time of application compared to the time of a single simulation run multiplied N times (N - number of simulation runs).

each iteration of the simulation (except the first one), the restart operation is performed and the re-initialization of trivial data structures takes place. The overall structure of the outermost simulation loop is shown at the figure 4. The variable p is an object of **Population** type that entirely describes the simulation state.

Figure 5 depicts the results of the performance test in regard to increasing number of simulation runs (N). The elapsed time is compared to the time of executing a single simulation run multiplied N times. It is shown that the great efficiency improvement has been achieved.

The performance tests were carried out on the workstation cluster comprising 4 Ultra SPARC machines.

4.2 Traffic Simulation Kernel

The previous application case study has shown the expected achievement in efficiency, however, this result was quite predictable. On the other hand, the question of the amount of overhead resulting from Panda I/O operation remains unanswered. The characteristic of the application used in the previous study and the way of using our solution in it makes it difficult to perform extensive tests of overhead.

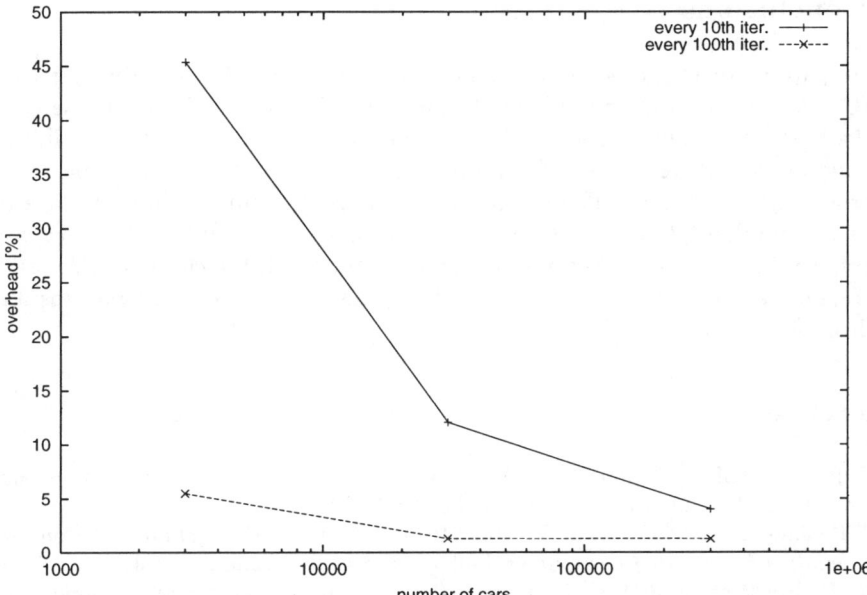

Fig. 6. Overhead of I/O operation in the synthetic test (overhead here is calculated as percentage of increase of execution time). The result are shown for two different frequencies of I/O operations. .

To evade these obstacles we used a synthetic application, that has much simpler structure and allows for controlling the balance between amounts of computation and I/O operations.

The synthetic application performs actually a simple road traffic simulation. Traffic simulators belong to the wide class of so-called "long-running applications", which can be very I/O intensive. Two main I/O patterns occur: periodical storing output of simulation after the specified number of simulation steps for further analysis and visualization and writing checkpoints.

We carried out performance tests for different data sizes (represented here by the number of cars) and different I/O operation frequencies: none, every 10th and every 100th iteration. Figure 6 displays the amount of overhead calculated as percentage of increase of the execution time.

Although for small problem sizes the overhead is substantial, for large problems (which are really interesting) it becomes little important (despite it grows in absolute values). For the test application the overhead drops to approximately one percent of the execution time. We consider this result very satisfying, especially taking into account that so far our implementation does not utilize all capabilities of Panda, especially overlapping I/O operation and computation. The utilization of this feature should lead to even less substantial overhead.

5 Conclusions

In this paper, we proposed a set of language constructs for specifying parallel I/O operations on distributed multidimensional arrays in HPF. These constructs, based on the new concept of *data transfer object*, offer great extensibility allowing, in the future, to specify complex I/O access patterns, hints and data pre-processing operations. The extensions to the HPF compiler and interface to the Panda parallel I/O library have been implemented. On two application case studies, we showed that by amending an application with the checkpoint/restart feature, one can significantly improve its functionality without risking a substantial drop of performance.

References

1. F. Bassow. IBM AIX Parallel I/O File System: Installation, Administration, and Use, IBM, May 1995. Document Number SH34-6065-00.
2. S. Benkner et al. "VFC - The Vienna HPF+ Compiler", *Proc. of the Int. Conf. on Compilers for Parallel Computers*, Linkoping, Sweden, June 29 - July 1, 1998.
3. R. R. Bordawekar and A. N. Choudhary. Language and compiler support for parallel I/O. In *IFIP Working Conference on Programming Environments for Massively Parallel Distributed Systems*. Swiss, April 1994.
4. P. Brezany, M. Gerndt, P. Mehrotra, and H. Zima. Concurrent file operations in a High Performance FORTRAN. In *Proceedings of Supercomputing '92*, pages 230–237, 1992.
5. P. Brezany and M. Winslett. Advanced Data Repository Support for Java Scientific Programming. HPCN Europe 1999, Springer-Verlag, LNCS 1593, pp. 1127–1136.
6. P. Ciaccia and A. Veronesi. Dynamic Declustering Methods for Parallel Grid Files. Proc. of the Conference "Parallel Computation", Klagenfurt, September 1996, Springer-Verlag, LNCS 1127, pp. 110–123.
7. Nils Nieuwejaar and David Kotz. The Galley parallel file system. In *Proceedings of the 10th ACM International Conference on Supercomputing*, May 1996.
8. K. E. Seamons and M. Winslett. Multidimensional Array I/O in Panda 1.0. *Journal of Supercomputing*, Vol. 10, No. 2, pages 191–211.
9. M. Snir. Proposal for I/O. *Posted to HPFF I/O Forum*, July 1992.
10. E. Dockner, H. Moritsch, G.Ch. Pflug, and A. Świętanowski. The AURORA financial management system. Techn. rep. AURORA TR1998-08, Vienna University, 1998.
11. G.Ch. Pflug and A. Świętanowski. Dynamic asset allocation under uncertainty for pension fund management. To appear in *Control and Cybernetics*.
12. E. Schikuta, T. Fuerle and Helmut Wanek. ViPIOS: The Vienna Parallel Input/Output System. In Proc. Euro-Par'98, Southampton, England, Springer-Verlag, LNCS.

High Level Software Synthesis of Affine Iterative Algorithms onto Parallel Architectures

Alessandro Marongiu[1], Paolo Palazzari[2], Luigi Cinque[3], Ferdinando Mastronardo[3]

[1]Electronic Engineering Department, University "La Sapienza" Rome
marongiu@tce.ing.uniroma1.it
[2]ENEA HPCN project – C.R.Casaccia –Rome
palazzari@casaccia.enea.it
[3]Dipartimento di Scienze dell'Informazione, Universita' "La Sapienza", Rome

Abstract. In this work a High Level Software Synthesis (HLSS) methodology is presented. HLSS allows the automatic generation of a parallel program starting from a sequential C program. HLSS deals with a significant class of iterative algorithms, the one expressible through nested loops with affine dependencies, and integrates several techniques to achieve the final parallel program. The computational model of the System of Affine Recurrence Equations (SARE) is used. As first step in HLSS, the iterative C program is converted into SARE form; parallelism is extracted from the SARE through allocation and scheduling functions which are represented as unimodular matrices and are determined by means of an optimization process. A clustering phase is applied to fit the parallel program onto a parallel machine with a fixed amount of resources (number of processors, main memory, communication channels). Finally, the parallel program to be executed on the target parallel system is generated.

1 Introduction

Automatic extraction of parallelism from algorithms and automatic parallel code generation are a key issue for wide diffusion of parallel systems. An hybrid approach to such a task is the Fortran HPF language [27],[28] which contains extensions to the Fortran 90 language to manage data parallel constructs, avoiding to the programmer the management of data communication, synchronization and allocation. Another interesting approach is constituted by the SUIF parallelizing compiler [22],[23]. In this work we present the results concerning a parallelizing environment developed for the SW implementation of iterative algorithms with affine dependencies, a significant class of algorithms in fields such signal processing or linear algebra. This High Level Software Synthesis (HLSS) environment, starting from a high level description of the affine iterative algorithm, automatically generates an optimized parallel code for a target distributed memory machine. The results will concern the Quadrics/APE100 massively parallel system [15]. We do not give a detailed description of all the theories on which the HLSS is based, but we stress the behavior of the environment describing the programming languages used to specify the program to parallelize.

The theoretical background of the HLSS relies on the System of Affine Recurrence Equation (SARE) computational model ([10], [12], [20]). A lot of studies have been devoted to the HLSS of affine iterative algorithms ([1]-[4],[6]-[12],[17]-[19], [20]). The HLSS is based on the following design flow (fig. 1):
- the affine iterative algorithm is specified through LIGHTC language (a subset of the ANSI C);
- the LIGHTC program is compiled through the LIGHTC compiler which extracts the equivalent SARE formulation;
- the SARE is allocated and scheduled on a set of Virtual Processors;
- the set of virtual processors is clustered;
- and finally the target code is generated and compiled for the target architecture;

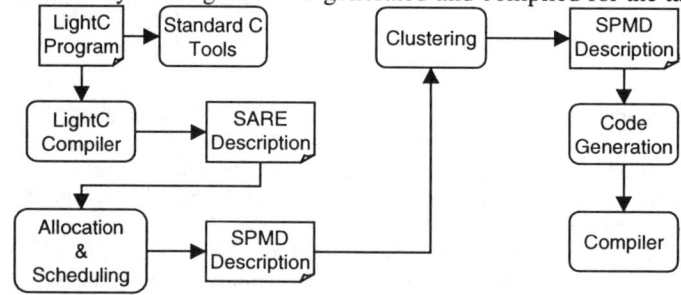

Fig. 1. Design flow in the HLSS parallelizing environment.

2 SARE Computational Model: A Review

This section briefly presents SARE computational model. Details can be found in [10],[12],[14],[20]. A SARE is described by a certain number of equations E:

$$X(z)=f(...,Y[\rho(z)],...) \text{ with } z \in I_E \tag{1}$$

where:
- X and Y are multi-dimensional array variables. Variables appearing only on the right side of equation E are input variables.
- $z \equiv (i\ j\ ...\ k)^T$ is the $N \times 1$ iteration vector. $X(z)$ is $X[i,j,...k]$. $S_X(z)$ is the statement which computes $X(z)$;
- I_E is the iteration space related to (1), i.e. it is the set of indices z where equation (1) is defined. It is described as a parameterized polytope: $I_E = \{z \in Z^N | A_E z + B_E s \geq b_E\}$, being s a parameter vector. Being SARE a single assignment model, different equations assigning $X(z)$ must be defined on disjointed iteration spaces I_E. The union of all the iteration spaces is the global iteration space polytope I.
- $\rho(z)$ is the index mapping function. It defines a flow dependence between the points $\rho(z)$ and z: execution of statement $S_X(z)$ requires $S_Y[\rho(z)]$ to be already executed. $\rho(z)$ is affine, i.e. $\rho(z) \equiv Rz+r$.
- f is a general function used to compute $X(z)$.

The main advantage of SARE model is that the analysis of algorithm dependencies, along with extraction of parallelism and mapping optimization, can be done at

compile time. Such an analysis is performed through dependence vectors.
Definition 1 *(Dependence Vector)* : Given a SARE equation (1), the dependence vector associated with a (final/intermediate) result variable Y is:

$$d_{Y,\rho(z)} = z - \rho(z); \qquad (2)$$

$d_{Y,\rho(z)}$ is the distance vector between points z and $\rho(z)$ and represents the dependence between statements $S_X(z)$ and $S_Y[\rho(z)]$. Generally $d_{Y,\rho(z)}$ depends on the iteration vector z. $d_{Y,\rho(z)}$ is uniform if it does not depend on z. Dependence vector is uniform if $R \equiv Id$ (Id is the identity matrix). In fact from (2) we have $d_{Y,\rho(z)} = z - \rho(z) = z - Rz - r = (Id-R)z - r$. Previous expression is independent from z if $R = Id$. In this case we have $d_{Y,\rho(z)} = -r$.

As example, we use the SARE expression of matrix-matrix multiplication algorithm. Given two $q \times q$ matrices A and B, it is defined by the following SARE:
- $C(z_1, z_2, z_3) = A(z_1, z_3)B(z_3, z_2)$ for $0 \le z_1 \le q-1$, $1 \le z_2 \le q$, $z_3 = 0$;
- $C(z_1, z_2, z_3) = C(z_1, z_2, z_3-1) + A(z_1, z_3)B(z_3, z_2)$ for $0 \le z_1 \le q-1$, $0 \le z_2 \le q-1$, $0 \le z_3 \le q-1$;
- Output on $C(z_1, z_2, z_3)$ for $0 \le z_1 \le q-1$, $0 \le z_2 \le q-1$, $z_3 = q-1$.

3 The *SIMPLE* Language

A first way to specify a SARE is the using of the SIMPLE (SARE IMPLEmentation) Language (SL). SL is an *equational* and a *single assignment language* used to represent SARE algorithms, similarly to *ALPHA language* ([5], [13]) and to the language used in the *tool OPERA* ([8],[9]). The following code represents previous matrix-matrix example:

```
SIMPLE PROGRAM MAT_MUL
Ind[i,j,k]
Par[q] {q-4>=0}
Input A[2] {i>=0,j>=0,-i+q-1>=0,-j+q-1>=0};
Input B[2] {i>=0,j>=0,-i+q-1>=0,-j+q-1>=0};
Result C;
C[]=f0(A[i,k],B[k,j]);
{k=0,i>=0,-i+q-1>=0,j>=0,-j+q-1>=0};
C[]=f1(C[i,j,k-1],A[i,k],B[k,j]);
{k-1>=0,i>=0,-i+q-1>=0,j>=0,-j+q-1>=0,-k+q-1>=0};
Write (C[]);
{k-q+1=0,j>=0,i>=0,-j+q-1>=0,-i+q-1>=0};
```

A typical *SIMPLE* program consists of two principal parts:
- declarations
- SARE description

Declaration part introduces all the information that specify the SARE context, i.e.
- the iteration vector declaration;
- the parameter vector declaration along with its validity domain;
- input variable declaration and their definition domain;
- result variable declaration.

SARE description part introduces the SARE equation describing the algorithm. Iteration vector declaration is composed by a sequence of N different identifiers; each identifier corresponds to a component of the iteration vector. The number N of declared indices is the SARE dimension. Iteration vector declaration is delimited by the keyword `Ind` and `.`.

Parameter declaration is optional (it is used only for parameterized SARE). It is composed by a sequence of identifiers; each identifier corresponds to a component of the parameter vector. The number of declared parameter identifiers is the dimension of the parameter vector. The parameter validity domain follows the parameter declaration. Parameters declaration, along with its definition domain, is delimited by the keyword `Par` and.

In SIMPLE language an input variable is considered to be read-only, i.e. it cannot be rewritten by an assignment equation. Input variable declaration is delimited by the keyword `Input` and. Each input variable declaration is composed by the input variable identifier followed by the input variable dimension and finally by the input variable definition domain.

In SIMPLE language a result variable is a read-write variable, i.e. it can be written and/or read by an assignment equation. Result variable declaration is delimited by the keyword `Result` and `;`. Each result variable declaration is composed by the result variable identifier. No dimension or definition domain is needed for result variables because they are obtained through SARE analysis. In fact the dimension of a result variable is always N, while its definition domain is the union of all iteration spaces corresponding to the equations which assign such a result variable.

The SARE description part, which comes after the declaration part, is composed by a sequence of SARE equations; each equation is followed by the corresponding iteration space.

Finally the SARE output is defined: for each final result variable, the set of indices, indicating where the corresponding output is stored, is given.

4 LIGHTC

One of the main advantages of the SARE model is the possibility to describe a SARE algorithm by means of a serial program constituted by a sequence of not perfectly nested affine loops. In [26] it is described a formal technique to convert a program, written in an imperative language, such as Fortran or C, into a set of affine recurrence equations. On the basis of such a technique, we have developed a tool that is able to convert an affine iterative program written in LIGHTC, into a SIMPLE program representing the equivalent SARE formulation of the original program.

Among the several ways to specify a parallel application, the possibility to use the standard C language is surely the best one. In fact
- the C language is widely diffused both in academic and industrial context;
- the specifics given through an high level language are non-ambiguous because the semantics of the language is univocally defined; moreover due to the serial nature of the C language, specifications can be given in a very straightforward way;
- it is possible to compile and execute the program describing the application with standard programming tools in order to verify program syntactic and semantic correctness at a very high abstraction level and, so, in a very fast way.

5 LIGHTC Program and Its Translation into a *SIMPLE* Program

LIGHTC language syntax is the same of the ANSI C language.
The set of LIGHTC iterative programs, which can be converted into a SARE formulation, are the ones with static control and with affine loop bounds and indices. A LIGHTC program must respect the following constraints:
- it is composed by a sequence of non perfectly nested loops;
- loop bounds (lower l and upper u) must be affine expressions of algorithm parameters and outer loop indices;
- the increment of each loop index must be 1;
- data structures are multi-dimensional arrays. Scalar values are allowed as they can be considered 0-dimensional arrays;
- the only admissible instructions are assignment to a multi-dimensional variable;
- no conditional statements, goto and while instructions are allowed.

As an example of LIGHTC language, in the following a fragment of the LIGHTC program representing the matrix-matrix multiplication algorithm is given:

```
LIGHTC PROGRAM MAT_MUL
void main(){
   /*variable declaration and allocation*/
   ...
   /*reading input matrices*/
   for (i=0;i<=q-1;i++)
      for (j=0;j<=q-1;j++) {
         a[i][j]=get_input();
         b[i][j]=get_input();
      }
   /*begin matrix product*/
   for (i=0;i<=q-1;i++)
      for (j=0;j<=q-1;j++) {
         c[i][j]=f0(a[i][k],b[k][j]);
         for (k=1;k<=q-1;k++)
            c[i][j]=f1(c[i][j],a[i][k],b[k][j]);
      }
   /* output c */
   for (i=0;i<=q-1;i++)
      for (j=0;j<=q-1;j++)
         output(c[i][j]);
}
```

LIGHTC to SIMPLE transformation is performed through the technique presented in [26] and outlined here:
- LIGHTC original program is transformed into a single assignment version by a) renaming left hand variables of the assignment statements b) expanding the indices of renamed variables according to the number of surrounding loops;
- once the program has been transformed into single assignment form, an analytical dependence analysis is performed in order to write the recurrence equation.

In the following it is shown the SIMPLE code representing the equivalent SARE expression of the LIGHTC program **LIGHTC PROGRAM MAT_MUL**:

```
SIMPLE PROGRAM MAT_MUL2
Ind[x1,x2,x3];
Par[p1] {p1-4>=0};
Input s0[2]{x1>=0,x2>=0,-x1+p1-1>=0,-x2+p1-1>=0,1>=0};
```

```
Input s1[2]{x1>=0,x2>=0,-x1+p1-1>=0,-x2+p1-1>=0,1>=0};
Result s2,s3;
s2[]=f0(s0[x1,x3],s1[x3,x2]);
{x3=0,x1>=0,-x1+p1-1>=0,x2>=0,-x2+p1-1>=0};
s3[]=f1(s2[x1,x2,0],s0[x1,x3],s1[x3,x2]);
{x3-1=0,x1>=0,-x1+p1-1>=0,x2>=0,-x2+p1-1>=0};
s3[]=f1(s3[x1,x2,x3-1],s0[x1,x3],s1[x3,x2]);
{x1>=0,-x1+p1-1>=0,x2>=0,-x2+p1-1>=0,2<=x3<=p1-1};
Write( s3[]);
{x3-p1+1=0,x2>=0,x1>=0,-x2+p1-1>=0,-x1+p1-1>=0 };
```
It was generated by the LIGHTC compiler. It is worthwhile to note that this SIMPLE program differs from the previous one but it is semantically equivalent.

6   Allocation and Scheduling of SARE

Allocation and Scheduling Module (ASM) accepts the algorithm expressed in SIMPLE form and gives, as output, the parallelized SARE algorithm (to be clustered). ASM is based on a space-time transformation which, for each point $z \in I \subset Z^N$, gives the time instant and the processor where the corresponding statement will be executed. In the following we report a brief review of the theory underlying the parallelization tool; for details, along with theorems enunciation and proofs, see [14] and [24].

The space-time transformation is composed by:
- the allocation function p(z) which returns the processor where $S_X(z)$ will be executed; p(z) is an *m*-dimensional function because we are dealing with a *m*-dimensional parallel machine.
- the timing function t(z), which returns when each statement $S_X(z)$ will be executed; t(z) is *n*=(*N*-*m*)-dimensional function.

These two functions must
- be admissible because they must preserve the semantics of the algorithm, i.e. dependence relations have to be maintained;
- be compatible, i.e. no more than one statement can be executed on a processor at the same time.

We choose t(z) as an integer and affine function

$$T(z) \equiv \Lambda z + \underline{\alpha} \qquad (3)$$

being Λ an integer *n*×*N* matrix and $\underline{\alpha}$ an integer *n*×1 vector. Most of the recent works ([1],[2],[7]-[9],[11],[12],[16],[19]) related to algorithm parallelization use one-dimensional affine timing functions (*n*=1), i.e. t(z)=λz+α, where λ is 1×*N* *timing vector*, and α is a scalar: consequently the target machine must be (*N*-1)-dimensional. Several extensions have been proposed in the literature. In the case of one-dimensional timing function the concepts of affine by statement ([17],[18]) and affine by variable ([20]) are introduced. Other authors adopt *n*-dimensional timing functions (*n*>1); for example in ([3], [10]) a general timing function is introduced without a clear investigation on the theoretical concepts involved. In [6] the special case of the projection of *N* nested loops onto a 1-dimensional linear array through two timing hyperplanes is presented. A general approach to multidimensional projection is also given in [4]. In [14] a new class of *n*-dimensional timing function was introduced: it is an extension of Lamport functions (which can be obtained as special case for *n*=1).

The condition for the admissibility of the timing function is given in the *n-dimensional Timing Function Admissibility Theorem* [14]: a *n*-dimensional t(z) is admissible iff, for every dependence vector d, $\Lambda d \gg 0$, where $\gg$ is the lexicographical ordering symbol. In [14] it was also demonstrated that, given a *n*-dimensional timing function, the corresponding processor space is *m*-dimensional. So we choose a *m*-dimensional grid as parallel architecture topology. Each processor is uniquely identified by a set of coordinates $(\pi_1 \pi_2 \ldots \pi_m)^T$. Given a point $z \in I$, p(z) returns the coordinates of the processor executing $S_X(z)$. p(z) is found by projecting the *N*-dimensional iteration space *I* along *n* directions in order to collapse *I* onto the *m*-dimensional processor space. Every projecting direction is characterized by a $N \times 1$ *projection vector* ξ_i. Projection vectors must be linearly independent. Projection directions must be compatible with t(z), i.e. statements assigned to the same processor have to be scheduled at different times. Compatibility is assured by the *Compatibility Theorem between Projection Vectors and Timing Vectors* [24]: projection vectors ξ_i and t(z) are compatible iff the set of *N* vectors $(\xi_1 \xi_2 \ldots \xi_n k_1 k_2 \ldots k_m)$ are a basis for Q^N, being vectors k_i a basis of $Ker(\Lambda)$.

p(z) is chosen in the set of affine functions and can be compactly written as

$$p(z) = \Sigma z + \underline{\beta} \quad (4)$$

where Σ is the $m \times N$ allocation matrix whose rows are σ_i and $\underline{\beta}$ is a $N \times 1$ vector whose components are β_i.

By composing timing (3) and allocation (4) functions, the space-time transformation function T(z) is obtained:

$$T(z) = \binom{t}{p} = \binom{\Lambda}{\Sigma} z + \binom{\underline{\alpha}}{\underline{\beta}} = Tz + \underline{\gamma} \text{ with } T = \binom{\Lambda}{\Sigma} \text{ and } \underline{\gamma} = \binom{\underline{\alpha}}{\underline{\beta}} \quad (5)$$

Applying T(z) to *I*, the transformed iteration space I_T is obtained. Matrix *T* must be unimodular with integer elements [19], [24]; unimodularity ensures that *I* and I_T have the same number of points with integer coordinates. Each point of I_T is individuated through a double set of coordinates: the first *n* coordinates, i.e. *geometrical time* t vector, and the second *m* coordinates, i.e. the p vector.

The new set of coordinates introduced by T(z) gives when and where a statement $S_X(z)$ is executed. Because $T(z): z \rightarrow (t,p)$, we have $S_X(z) \rightarrow S_X(t,p)$, i.e. the statement $S_X(z)$ is executed on processor with coordinates p and scheduled at time t. The space-time coordinate system also distributes variables among processors. Because $T(z): z \rightarrow (t,p)$, we have $X(z) \rightarrow X(t,p)$, i.e. variable $X(z)$ is stored in the local memory of processor p and addressed through vector t: $X(t)$.

The (near) optimal mapping function T(z) is achieved through a heuristic optimization process guided by a cost function which takes into account both processor load balancing and memory optimization [25]; as heuristic the Simulated Annealing algorithm [21] was used.

7 Clustering and Code Generation.

The number of processors, individuated through the space-time transformation, is usually too large to be implement on the target parallel machine which, generally, has few computing elements. The clustering technique is introduced in order to collapse the individuated set of processors (virtual processors) on the processors available on the target machine (physical processors). Because physical resources are less than the requirements, they will be shared in time. The clustering approach used in the HLSS environment is based on the one presented in [29]. Clustering collapses several virtual processors into one physical processor whose resources are time-shared among the virtual processors mapped onto it. The mapping function between the virtual processors and the physical processors can be done using the two classical approach:
- LSGP (Locally Serial Globally Parallel) approach. The virtual processor array is divided into subsets, each one is mapped onto a single physical processor. Each physical processor serially emulates the virtual processors mapped onto it.
- LPGS (Locally Parallel Globally Serial) approach. The virtual processor array is divided into subsets. Each virtual processor of the same subset is mapped onto a different physical processor. Each physical processor serially emulates virtual processors belonging to different subset.

The serial emulation of the virtual processors onto the physical processors is done through a static scheduling, avoiding the run-time overhead introduced by calls to the operating system scheduler. Once applied the space-time transformation and the clustering, the final parallel code is generated according to the target architecture. Details on the code generation can be found in [24].

8 Experimental Results

As example, we use the matrix-matrix multiplication LIGHTC algorithm. In the following pictures we report the actual 'wall-clock' times obtained when executing the parallel code, generated by *HLSS* methodology, on the APE100/Quadrics machine (configuration with 128 processors). Fig.2.a shows the execution times and memory requirements obtained in correspondence of different mappings produced during the space-time optimization process. In this case it is used a cost function tuned to find a space-time transformation leading to fast algorithms. The solid line represents the true execution time on the parallel machine versus the cost function, while the dotted line represents the memory requirements per processor versus the cost function.

Fig.2.b shows the execution times and memory allocation obtained with a cost function tuned to find a trade-off between execution time and memory allocation. Because of the less emphasis on time, we expect to have a bigger execution time but with less memory requirements. In fact in this case we achieve an execution time of 257 ms with a memory allocation of 3536 word/processor; this mapping allows the multiplication of matrices up to 650x650.

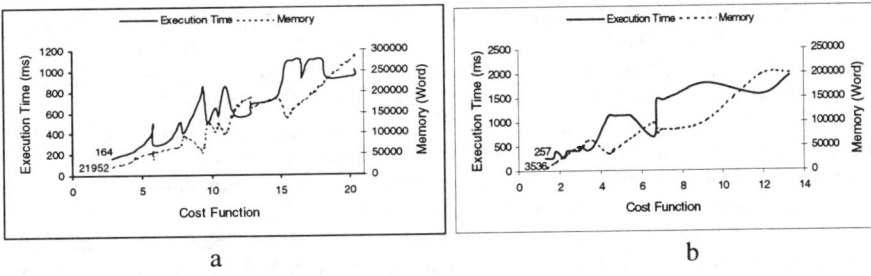

Fig. 2. Results of scheduling and allocation optimization process

9 Conclusions

In this work a High Level Software Synthesis methodology (HLSS) has been presented. This methodology deals with iterative algorithms with affine dependencies, a significant class of algorithms in the field of signal processing and numerical computations, and is based on the integration of several techniques which
- transform a C program into a System of Affine Recurrence Equation (SARE) written into the SARE IMPLEmentation language (SIMPLE);
- parallelism is extracted from the SIMPLE program through a space-time mapping function, derived through an optimization process;
- a clustering step is applied in order to implement the obtained parallel program onto a machine with a finite number of resources (memory, processors) and, finally the actual parallel code is generated.

The work is illustrated with a simple example showing the results achievable when previous steps are applied to the generation of a parallel program onto an actual parallel machine. The tests presented are executed on a SIMD massively parallel system of the family APE100/Quadrics.

References

1. Clauss P., "An Efficient Allocation Strategy for Mapping Affine Recurrences into Space and Time Optima Regular Processor Arrays", *Parcella*, Sep. 1994.
2. Clauss P., Perrin G.R., "Optimal Mapping of Systolic Algorithms by Regular Instruction Shifts", *IEEE Intern. Conf. Application-Specific Array Processors,* pp. 224-235, Aug. 1994.
3. Feautrier P., "Automatic Parallelization in the Polytope Model", *Les Menuires*, Vol. LNCS 1132 pp. 79-100, 1996.
4. Feautrier P., "Some Efficient Solution to the Affine Scheduling Problem, II Multi-Dimensional Time", *Intern. J. Parallel Programming*, Vol. 21, pp. 389-420, Dec. 1992.
5. H.Le Verge, Ch.Mauras, P.Quinton - "The alpha language and its use for the design of systolic arrays" - *Journal of VLSI and signal processing*, 3 :173-182, 1991.
6. Lee P., Kedem Z., "Synthesizing Linear Array Algorithms from Nested For Loop Algorithms", *IEEE Trans. on Computers*, Vol. C-37, No. 12, pp. 1578-1598, Dec. 1988.
7. Lengauer C., "Loop Parallelization in the Polytope Model", *CONCUR*, Vol. LNCS 715, pp. 398-416, 1993.

8. Loechner V., Mongenet C., "OPERA: A Toolbox for Loop Parallelization", International Workshop on Software Engineering for Parallel and Distribuited Systems, PDSE, 1996.
9. Loechner V., Mongenet C., "A Toolbox for Affine Recurrence Equations Parallelization",*HPCN '95*, Milan, Vol. LNCS 919. pp. 263-268, May 1995.
10. Loechner V., Mongenet C., "Solutions to the Communication Minimization Problem for Affine Recurrence Equations", *EUROPAR '97*, Vol. LNCS 1300,pp.328-337 1997.
11. Mongenet C., "Data Compiling for System of Affine Recurrence Equations", *IEEE Intern. Conf. on Application - Specific Array Processors*, ASAP, pp. 212-223, Aug. 1994.
12. Mongenet C., Clauss P., Perrin G.R., "Geometrical Tools to Map System of Affine Recurrence Equations on Regular Arrays", *Acta Informatica*, Vol. 31, No. 2, pp. 137-160, 1994.
13. D. K. Wilde - "The ALPHA language" - Technical Report Internal Publication N.827, IRISA, May 1994
14. Marongiu A., Palazzari P., " A New Memory-Saving Technique to Map System of Affine Recurrence Equations (SARE) onto Distributed Memory Systems". *Proc. Intern. Parallel Processing Symposium IPPS 99* – Puerto Rico – April, 12th,16th 1999.
15. Bartoloni et al, "A Hardware implementation of the APE100 architecture", *Int. Journal of Modern Physics*, C4, 1993.
16. Lamport L., "The parallel execution of DO loops", *Comm. Of the ACM*, 17 (2):83-93, Feb. 1974.
17. Darte A., Robert Y., "Affine-by-Statement scheduling of uniform and affine loop nests over parametric domains",*Journ. of Parallel and Distributed Computing,* Vol. 29, pp 43-59, 1995
18. Darte A., Robert Y., "Mapping uniform loop nests onto distributed memory architectures", *Research Report n° 93-03 LIP -ENS Lyon*, Jan. 1993.
19. Dowling M.L., "Optimal code parallelization using unimodular transformations", *Parallel Computing*, 16, pp. 157-171. 1990.
20. Mongenet C., "Affine dependence classification for communications minimization", *ICPS Research Report No. 96-07*, downloadable from icps.u-strasbg.fr/pub-96
21. Kirkpatrick S., Gelatt C.D., Vecchi M.P., "Optimization by simulated annealing", Science, Vol. 220, N. 4589, 13 May 1983
22. M. W. Hall, J. M. Anderson, S. P. Amarasinghe, B. R. Murphy, S.-W. Liao, E. Bugnion and M. S. Lam, " Maximizing Multiprocessor Performance with the SUIF Compiler", *IEEE Computer*, December 1996.
23. A. W. Lim and M. S. Lam, "Maximizing Parallelism and Minimizing Synchronization with Affine Transforms", Conference Proceedings of the 24th Annual ACM SIGPLAN-SIGACT Symposium on Principles of Programming Languages, January, 1997.
24. Marongiu A., Palazzari P., "Automatic Mapping of System of N-dimensional Affine Recurrence Equations (SARE) onto Distributed Memory Parallel Systems" Accepted to appear in the special issue of IEEE Trans. Software Engineering on Architecture-Independent Languages and Software Tools for Parallel Processing
25. A. Marongiu, P. Palazzari: "Optimization of Automatically Generated Parallel Programs" The 3rd IMACS International Multiconference on Circuits, Systems, Communications and Computers (CSCC'99) - July 4-8, 1999, Athens (Greece).
26. P. Feautrier, "Data Flow Analysis of Scalar end Array References", Intern. Journal of Parallel Programming, vol. 20, n. 1, 1991
27. High Performance Fortran Forum, " High Performance Fortran language Specification", Sci. Prog., vol. 2, 1993.
28. J. Merlin, A. Hey, "An Introduction to High Performance Fortran", Sci. Prog. vol. 4, 1995.
29. Zimmermann K.H., "A Unifying Lattice-Based Approach for the Partitioning of Systolic Arrays via LPGS and LSGP", Journal of VLSI Signal Processing, 17, pp 21-41, 1997.

Run-Time Support to Register Allocation for Loop Parallelization of Image Processing Programs

N. Zingirian and M. Maresca

Dipartimento di Elettronica ed Informatica
University of Padua, Italy
{panico mm}@dei.unipd.it

Abstract. When Image Processing Programs (IPP) are targeted to Instruction Level Parallel architectures that perform dynamic instruction scheduling, register allocation is the key action to expose the high parallelism degree typically present in the loops of such programs.
This paper presents two main contributions to the register allocation for IPP loop parallelization: *i)* a framework to identify the inefficiencies of the two basic approaches to register allocation – the first based on compiling techniques and the second based on hardware mechanisms for register renaming; *ii)* a novel technique that eliminates the inefficiencies of both approaches. Some experimental results show the effectiveness of this technique.
Keywords: Dynamic Register Renaming, Image Processing, Instruction Level Parallelism, Loop Parallelization, Register Allocation

1 Introduction

Most Image Processing programs consist of computing intensive loops that are typically highly parallelizable[1]. Instruction Level Parallel Processors equipped with dynamic scheduling and branch prediction mechanisms[2] take advantage of this property by scheduling together instructions taken from different iterations to overlap the execution of such iterations [3][4].

The point is that if instruction scheduling takes place after register allocation, as in dynamic scheduling approach, the parallelism degree is acceptable only if the iterations overlapped are free from false data dependencies. In fact, while true dependencies (Read-after-Write) cannot be removed because they derive from the algorithm, false dependencies (Write-after-Write and Write-after-Read) can be removed by allocating additional registers.

The state of the art provides two basic approaches to register allocation for loop parallelization. The first approach consists of unrolling loops [5][3] to enlarge their bodies, allocating registers at compile time (see, e.g. [6]) to remove false dependencies and, in most advanced schemes, combining software pipelining techniques[7] with loop unrolling[8] and "ad hoc" register allocation strategies [9][10].

The second approach, widely adopted in the latest processor generation, consists of equipping processors with special hardware for register renaming[11][12].

The attractive point of the first approach is that the register allocation algorithms executed at compile-time can be more effective than the ones implemented via hardware and executed at run-time. The attractive point of the second approach derives from the fact that physical register identifiers are not statically specified in the code. This fact is relevant when the program flow is not known at compile time, as in case of conditional statements inside loop bodies that frequently occur in image processing programs.

Unfortunately, it is worth noticing that if the two approaches are applied together, the efficiency of the register allocation done at compile time is lost when the hardware renames the registers.

In this paper we *i)* present a framework to evaluate the inherent inefficiencies of the two approaches that includes even the most advanced optimizing techniques for loops with conditional statements (e.g. [13] [8] [14]), *ii)* propose a novel solution that eliminates the inefficiency of both approaches, in change of an execution overhead, and *iii)* experimentally show that, in our solution, the benefits due to the efficient register allocation exceeds the overhead penalties for a number typical Image Processing program loops.

The paper is organized as follows. Section 2 introduces the terminology and the basic concepts about the register allocation strategies and the loop structures considered in the rest of the paper. Section 3 compares the approaches to register allocation. Section 4 proposes our novel approach. Section 5 presents a number of experiments that show the effectiveness of our solution. Section 6 concludes the paper.

2 Background

In this section we introduce the two basic existing approaches to register allocation, and outline the loop model considered in the rest of the paper.

2.1 Register Allocation Approaches

In the first approach the compiler analyzes basic blocks, or even larger code segments such as functions, and maps the program variables (virtual registers of intermediate code) contained in these blocks of code directly to the processor physical registers, solving the graph coloring problem, or some variants of it[6], through appropriate sub-optimal algorithms at compile time. At run time, the processor uses the register identifiers specified in the program to address the physical registers. Henceforth, this approach will be referred to as *static register allocation*.

In the second approach it is the processor that allocates the physical registers at run time. The compiler generates instructions that address register names (*logical registers*) that do not refer to the physical registers directly. The logical registers are mapped to the physical registers at run time by the processor, in such a way that the same logical register, during program execution, can correspond to different physical registers. At the decode stage, the processor[1] changes the logical register names specified in the instruction word into the corresponding physical register names (this change is called *register renaming*), according to the current state of a table, called the *register alias table*, that maps each logical register to a physical register. In addition, whenever the instruction currently decoded writes its results to a destination register, say r, the processor updates the register alias table entry corresponding to r, mapping r to a *free* physical register before renaming r. Instruction fetch is blocked when no *free* physical registers are available. The status of physical registers (i.e. either *free* or *in use*) changes as follows. A physical registers p is marked *in use* when a logical register r is mapped to it, and is marked back *free* when the following conditions hold: r is mapped to a new physical register p', an instruction I has written the result on p' and all the instructions before I have been executed (i.e., I is *graduated*). Henceforth, this approach will be referred to as *dynamic register allocation*.

2.2 Loop Model

A loop is a construct that controls the execution of a block of instructions, called the *body*, several times. Each execution of the body is called *iteration*. The properties of the body and of the set of iterations characterize the loop.

Loop Body. To introduce the analysis of the following section we need to model the loop body in terms of the following three parameters:

R: the number of registers initialized before the loop and accessed by loop body instructions only for reading.
W: the number of registers written by loop body instructions at each iteration.
W^*: the number of registers necessary to contain the maximum number of values that are simultaneously alive for all the admissible instruction schedules.

While R and W can be directly counted looking at the code instructions, to extract W^* it is necessary to consider the graph of true data dependencies (data dependency graph, DDG) that depends only on the algorithm structure and neither on register allocation nor on the instruction schedule. To determine W^* it is helpful to consider the following proposition.

Proposition 1 *If all the instructions of a life range, say L, precede (or follow) all the instructions of another life range, say L', in the DDG, then there exists no admissible schedule for which L and L' can be simultaneously alive.*

The proof of this proposition is given by the fact that any schedule that makes L and L' overlap would violate the DDG. All the pairs (L, L') that do not satisfy Proposition 1 can be simultaneously alive for some instruction schedule. W^* is

---
[1] The dynamic register allocation considered in this paper is based on the register renaming model of MIPS R10000 and Alpha 21264 processors

```
for (i=0; i<N; i++)
   for (j=0; j<N; j++)
      for ( k=0; k<N; k++)
         a[i][j] + = b[i][k] * c[k][j]
```

```
   r0 <- N
   r1 <- &a
   r2 <- &b
   r3 <- &c
   r4 <- 0        #inizialize i
   r5 <- 0        #inizialize j
   r6 <- 0        #inizialize k
   <... external loops..>
   r7 <- offset [r4][r5]
   r8 <- ld (r1+r7)
LOOP
A  r9 <- offset[r4][r6]
B  r10<- ld (r2+r9)
C  r11<- offset[r5][r6]
D  r12<- ld (r3+r11)
E  r13<- r10 * r12
F  r8 <- r8 + r13
G  r6 <- r6 + 1
   branch LOOP if (r4< r0)
```

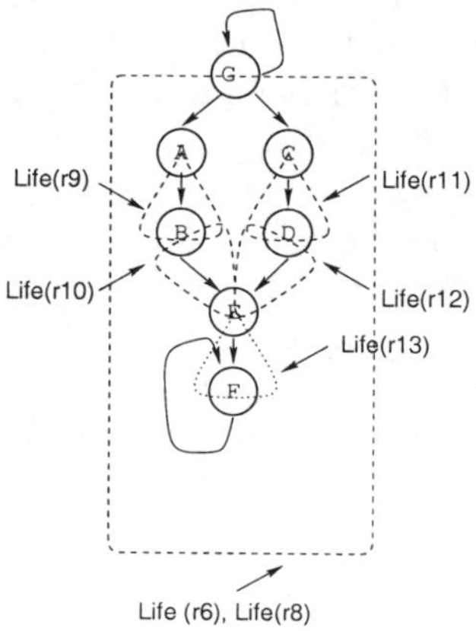

R = 3 (values in r0, r2, r3)

W= 7 (values in r9,r10,r11,r12,r13,r8,r6)

W* = 4

Fig. 1. Parameters in matrix multiply loop body

the cardinality of the the largest subset of life ranges that does not include any pair (L, L') that satisfies Proposition 1. It is immediate to show that $W^* \leq W$.

Figure 1 shows a typical loop body (matrix by matrix product) and reports the values of R,W and W^*. The DDG shows that the life range of the value in r9 cannot be overlapped to the life range of the value in r10 in any admissible schedule. Similarly r11 cannot be overlapped to r12 and r12 or r10 cannot be overlapped to r10. On the contrary values in r6 and r8 potentially overlap to all other values. As a consequence, $W^* = 4$, because no schedule can overlap the life range of more than four values (e.g. registers r9, r11, r6, r8 are sufficient to eliminate false dependencies for any schedule).

```
for each iteration I do begin
   if ( p(I) ) then begin
      body (I)
   end if
end for
```

Fig. 2. Structure of dynamic loops

Loop Iteration Set. Each iteration of the loop is identified by its own *initial status* which includes the values of the loop indices and/or depends on the values of loop indices.

In addition we say that the loop flow (or, more briefly, the loop) is

- *static* if the set of iterations to be executed is determined at compile time, while it is
- *dynamic* if the set of iterations to be executed is determined at run-time through conditional statements contained in the loop body (see Fig. 2). As in case of most image processing programs, we suppose that predicate $p(I)$ of conditional statement does not depend on data written in previous iterations.

3 Analysis of Register Allocation Efficiency

In this section we consider the register allocation efficiency of loop parallelization in the four cases obtained by applying the static and dynamic allocation approaches to static and dynamic loops, defined in the previous section. The efficiency metric ℓ adopted to evaluate each case is the maximum number of loop iterations that are made free from false data dependencies by the corresponding register allocation approach given a limited number, say $Regs$, of registers available. Table 1 reports the expressions of ℓ in the four cases of Static/Dynamic register allocations applied to Static/Dynamic Loops, in terms of R, W, W^* (see Sect. 2.2) and $Regs$ parameters.

Case	Allocation	Loop	ℓ	Techniques
S-S	Static	Static	$\lfloor \frac{Regs-R}{W^*} \rfloor$	Unrolling
D-S	Dynamic	Static	$\lfloor \frac{Regs-R}{W} \rfloor$	Hw Renaming
S-D	Static	Dynamic	$k \times \lfloor \frac{Regs-R}{W^*} \rfloor$	Unrolling + Global sched
D-D	Dynamic	Dynamic	$\lfloor \frac{Regs-R}{W} \rfloor$	Hw Renaming

Table 1. Values of ℓ in the S-S, D-S, D-S, D-D cases

In the S-S case false data dependencies are eliminated by unrolling the code and using different registers for each unrolled iteration[2]. The value of ℓ derives from the fact that each unrolled iteration does not need more than W^* registers, and from the fact that static register allocation uses only necessary registers.

In the D-S case the processor, assuming perfect branch prediction, collects the instructions belonging to several iterations in the Instruction Queue (IQ), with no need of loop unrolling. The dynamic register renaming described in Sect. 2.1 guarantees that the instructions present in the IQ are free from false dependencies. As a consequence ℓ corresponds to the maximum number of iterations whose instructions are simultaneously present in the IQ. Assuming an unbounded IQ, the lack of *free* physical registers is the only event that limits the

---
[2] We do not consider Software Pipelining because we assume that the target architecture performs branch prediction and dynamic scheduling.

number of instructions present in the IQ. The value of ℓ derives from the fact that each iteration needs W registers, because the hardware register renaming techniques allocate a new register at each write.

In the S-D Case, after unrolling the loop as in the S-S case, a *global scheduling* technique is required. The effect of *global scheduling* in Dynamic Loops is that, in the most complete case, the instructions are moved and duplicated in such a way that the structure of the unrolled program in Figure 3(a) is transformed into the structure represented by the flow graph of Figure 3(b). This structure can be obtained for instance by adopting the *if conversion* and if no predicated execution is supported also the *reverse if-conversion* (see [14]). In the case in

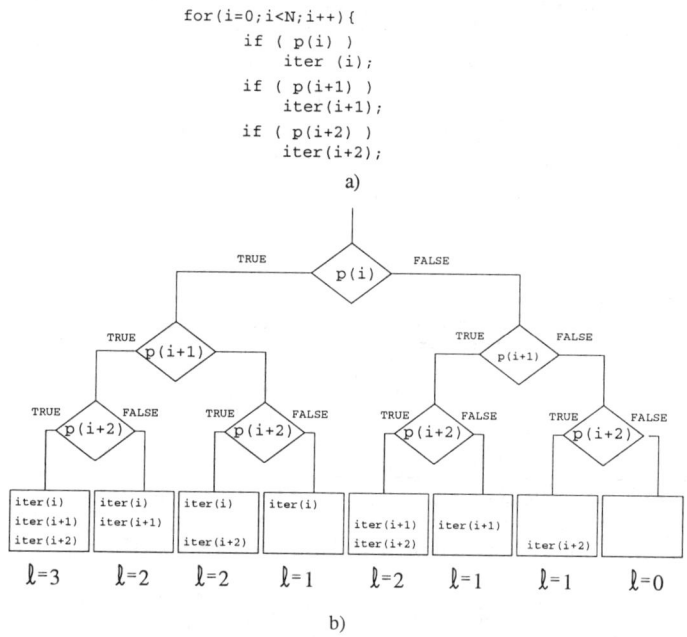

Fig. 3. Values of ℓ in case of Global Scheduling

which all the predicates of the flow graph are true, ℓ is equal to the one of the S-S case, in the case in which all these predicates are false, ℓ is null. According to the pattern of true and false predicates a factor, say k (for $0 < k < 1$), modulates the value of ℓ.

The D-D case, assuming perfect branch prediction, the value of ℓ is equal to the one of the D-S case, because the processor collects only the instructions that belong to the iterations that will be executed. As a consequence, all the registers will be allocated to the iterations actually executed, unlike the S-D case.

The above analysis reveals that in case of dynamic loops, the dynamic and static approaches to register allocations are characterized by two different types of inefficiencies if compared to the S-S case. The first inefficiency is that dynamic register allocation allocates W registers instead of only W^* registers per iteration, since it allocates a new register for each write. The second inefficiency is that static register allocation, in case of dynamic loops, introduces the inefficiency factor k. The elimination of both these inefficiencies from register allocation in dynamic loops is the objective of the solution presented in the rest of the paper.

4 Run-Time Support to Register Allocation

The solution proposed in this paper can be thought of as a generalization of loop unrolling to be applied to programs targeted to architectures that do not rename registers (e.g., the incoming IA-64 architecture).

The key of this generalization is to extend the concept of unrolled body to the concept of *cluster*. While the unrolled body is a code segment that executes a number, say U, of *consecutive* iterations of the original loop, on the contrary the *cluster* is a segment of code that executes U iterations regardless whether they are consecutive or not. This distinction is crucial because it is the constraint of the consecutive iterations that causes the inefficiency in the S-D Case. In fact, whenever the conditional statement skips an iteration of the unrolled body, the registers allocated for that iterations cannot be used by another iteration, because it is non-consecutive, and thus these registers are left unused.

On the contrary the *cluster* is free from this type of inefficiency because it contains the code of U non-consecutive iterations. The cluster is built in such a way that the k-th iteration of the cluster allocates a set of registers, say $\underline{R}^{(k)}$, to store its initial values. The attractive point is that the values in $\underline{R}^{(k)}$ and $\underline{R}^{(k+1)}$ respectively initialize iterations k and $k+1$ of the cluster even though these iterations do not correspond necessarily to consecutive iterations of the original loop. The iterations of a *cluster* are initialized by an appropriate mechanism which assigns the initial states of the next U iterations to be executed, to register vectors $\underline{R}^{(1)}, \ldots, \underline{R}^{(U)}$. This operation is a dynamic register assignment executed by additional code included in the program. This code can be regarded as a run-time support to register allocation. Figure 4(a) shows a generic dynamic loop provided with this additional run-time support (represented within frames). Whenever the conditional statement verifies that iteration I must be executed the run-time support dynamically allocates the appropriate register vector $\underline{R}^{(k)}$ to the initial state of I. Only after all the U iterations of the cluster have been initialized, the run-time support activates the execution of a cluster. This allocation can be regarded as dynamic because it cannot be determined at compile time and prevents from allocating cluster registers to iteration that will be never executed. It is immediate to verify that this approach delivers $\ell = \left\lfloor \frac{Regs - R}{W^*} \right\rfloor$ but introduces an execution overhead due to the presence of the additional run-time support code, as Fig. 4(b) shows. The experiments in the following section allow evaluating the benefits of register allocation versus the costs of the overhead.

```
                a)                    |              b)
                                      |   k = 0;
  ┌─────────────────────────────┐     |   for each iteration I do begin
  │ k = 0;                      │     |     if ( p(I) ) then begin
  └─────────────────────────────┘     |       k=k+1;
    for each iteration I do           |       case k begin
      if ( I is to be executed ) then |         1: (R_0,...,R_{c-1}) ← init st. of I
        ┌──────────────────────────┐  |         2: (R_c,...,R_{2c-1}) ← init st. of I
        │ R^(k+1) ← initial state of I │  |         ...
        │ k = k + 1 mod U          │  |         U: (R_{(U-1)×c},...
        │ if ( k == 0 ) then       │  |                 ...,R_{U×c-1}) ← init st. of I;
        └──────────────────────────┘  |       k=0;
        execute cluster (R^(1),...R^(U)) |    body(R_0,...,R_{c-1});
        ┌──────────────────────────┐  |    body(R_c,...,R_{2c-1});
        │ endif                    │  |    ...
        └──────────────────────────┘  |    body( R_{(U-1)×c}...R_{U×c-1});
     endif                            |    end case
    endfor                            |    end if
                                      |   end for
```

Fig. 4. Run-time support to register allocation: a) structure, b) implementation

5 Experiments

In this section we present some experimental results that show the effectiveness of the run-time support to register allocation described in the previous section. First we built a significant image processing benchmark suite by extracting a number of computing intensive dynamic loops from some typical image processing tasks (in this paper we present the results related to the tasks in Table 2).

Task	Algorithm

1. Affine Transform	foreach output pixel coordinates (x,y)
	  (X,Y) = inverse transform of x,y
	  if (X,Y is in input image area)
	    Out(x,y)= weighted average of pixels around (X,Y)

2. Convolution on Selected Areas	foreach pixel coordinates (x,y)
	  foreach mask coordinates (i,j)
	    if pixels (x,y) and (x+i,y+j) belong to filter area
	      Out(x,y) = Out(x,y) + In(x+i,y+j) × Mask(i,j)

3. HSV-to-RGB color conversion	foreach pixel coordinate (x,y)
	  if(saturation of In(x,y) is null)
	    Out(x,y) = gray with luminance of In(x,y)
	  else
	    Out(x,y) = combination of HSV coordinates.

4. Hough Transform	foreach pixel coordinates (x,y) and each angle k
	  if (In(x,y) is edge)
	    update Hough's Matrix according to x,y,k.

Table 2. Structure of the algorithms selected for the benchmark

Secondly, we ran the benchmark suite on a processor equipped with dynamic register renaming. Then we provided the programs with the run time support to register allocation and ran them on a second processor equipped with the same resources (see Tab. 3) as the ones of the first processor, except for hardware

Resource	Description
Integer/Logical	4 Units, 1 Cycles Latency
Floating Point	4 Units, 3 Cycles Latency
Issue/Graduation Width	6 Instructions per cycle
Instruction Queue Size	Unbounded
Registers	32 General Purpose + 32 Floating Point
Other	Out-of-order Exec., Perfect Branch Predict., No Cache Miss

Table 3. CPU Resources

register renaming. Finally we compared the results in terms of the number of instructions executed per cycle in the two experiments.

Task	W	W^*	Hw Register Renaming	Run-Time Support	Gain Factor
1	65	50	1.4 Ins/Cycl	3.9 Ins/Cycl	2.78
2	6	4	2.6 Ins/Cycl	2.9 Ins/Cycl	1.11
3	51	42	2.0 Ins/Cycl	2.4 Ins/Cycl	1.20
4	13	7	1.6 Ins/Cycl	1.8 Ins/Cycl	0.88

Table 4. Experimental Results

To perform these experiments we developed a superscalar processor simulator that allows switching on and off dynamic register renaming.

Table 4 reports the results obtained. The results show that the run time support delivers better performance than dynamic register renaming in some cases and introduces some performance penalties in other cases. The difference of performance is due to the fact that if each iterations consists of a large number of operations, then the run-time support overhead is negligible, otherwise it penalizes the overall execution.

6 Concluding Remarks and Future Work

This paper has presented a novel approach to register allocation that can be regarded as a dynamic approach because it allows assigning the registers to loop iterations at run-time. The distinctive point is that, while the dynamic register allocation usually needs appropriate hardware mechanisms, in this case it is a software run-time support included in the loop, that dynamically allocates registers. This software run-time support allows allocating registers dynamically without being constrained by the conservative assumptions typical of hardware register renaming. As a result the performance is improved to the extent that the latencies added by the run-time support turn out to be acceptable in many cases, as shown by the experiments.

Our future work will concentrate on the study of "ad hoc" hardware mechanisms aimed at removing the run-time support overheads.

References

1. C. C. Weems, E. Riseman, A. Hanson, and A. Rosenfeld. "The DARPA Image Understanding Benchmark for Parallel Computers ". *"J. of Parallel and Distributed Computing"*, 11:1–24, 1992.
2. J.L Hennessy and D.A. Patterson. *Computer Architecture: A Quantitative Approach*. Morgan Kaufmann, San Mateo, CA, second edition, 1996.
3. P. Baglietto, M. Maresca, M. Migliardi, and N. Zingirian. Image Processing on High-Performance RISC Systems. *Proceedings of the IEEE*, 84(7):917–930, July 1996.
4. N Zingirian and M. Maresca. Scheduling image processing program activities on instruction level parallel RISC through program transformations. *Lecture Notes in Computer Science*, 1225:674–686, 1997.
5. J.K. Davidson and S. Jinturkar. Aggressive loop unrolling in a retargetable optimizing compiler. In Tibor Gyimothy, editor, *Compiler Construction, 6th International Conference*, volume 1060 of *Lecture Notes in Computer Science*, pages 59–73, Linköping, Sweden, 24–26 April 1996. Springer.
6. F.C. Chow and J.L. Hennessy. The priority-based coloring approach to register allocation. *ACM Transactions on Programming Languages and Systems*, 12(4):501–536, October 1990.
7. V.H. Allan, R.B. Jones, R.M. Lee, and S.J Allan. Software pipelining. *ACM Computing Surveys*, 27(3):367–432, September 1995.
8. N.J. Warter, G.E. Haab, K. Subramanian, and J.W Backhaus. Enhanced modulo scheduling for loops with conditional branches. In *Micro*, 1992.
9. A. Aiken, A. Nicolau, and S. Novack. "Resource-Constrained Software Pipelining". *"IEEE Transaction on Parallel and Distribuited Systems "*, 6(12):1248–1271, 1995.
10. S. Lelait, G. R. Gao, and C. Eisenbeis. A new fast algorithm for optimal register allocation in modulo scheduled loops. *Lecture Notes in Computer Science*, 1383:204–213, 1998.
11. Keith I. Farkas, Norman P. Jouppi, and Paul Chow. Register file design considerations in dynamically scheduled processors. In *Proceedings of the Second International Symposium on High Performance Computer Architecture*. IEEE, January 1996.
12. Kenneth C. Yeager. The MIPS R10000 superscalar microprocessor — emphasizing concurrency and atency-hiding techniques to efficiently run large, real-world applications. *IEEE Micro*, 16(2):28–40, April 1996.
13. M. Lam. Software pipelining: An effective scheduling technique for VLIW machines. *SIGPLAN Notices*, 23(7):318–328, July 1988. *Proceedings of the ACM SIGPLAN '88 Conference on Programming Language Design and Implementation*.
14. N.J. Warter, S.A. Mahlke, W.W. Hwu, and B.R. Rau. Reverse if-conversion. In Robert Cartwright, editor, *Proceedings of the Conference on Programming Language Design and Implementation*, pages 290–299, New York, NY, USA, June 1993. ACM Press.

A Hardware Scheme for Data Prefetching

Sathiamoorthy Manoharan and See-Mu Kim

Department of Computer Science, University of Auckland, New Zealand.

Abstract. Prefetching brings data into the cache before the processor expects it, thereby eliminating potential cache misses. There are two major prefetching schemes. In a software scheme, the compiler predicts memory access patterns and places prefetch instructions in the code. In a hardware scheme, hardware predicts memory access patterns at runtime and brings data into the cache before the processor requires it.
This paper proposes a hardware scheme for prefetching, where a second processor is used for prefetching data for the primary processor. The scheme does not *predict* memory access patterns, but rather uses the second processor to run ahead of the primary processor so as to *detect* future memory accesses and prefetch these references.

1 Introduction

Prefetching brings data into the cache before it is expected by the processor, thereby eliminating potential misses. Prefetching is most effective when the total computation time in an application is about the same as the total memory access time. In this case, prefetching can successfully overlap memory accesses with computation. The maximum speed-up prefetching can achieve is 2 (or 100%).

There are two major prefetching schemes. In a hardware scheme, the hardware predicts memory access patterns and brings data into the cache before required by the processor [1, 4]. In a software scheme, the compiler predicts memory access patterns and places prefetch instructions in the code [7, 6]. Software prefetching, however, requires some hardware support, such as the provision and implementation of a prefetch instruction, and its effectiveness is mainly determined by compiler algorithms used for prefetching.

Hardware prefetching schemes make use of the regularity of data accesses. Simplest of these schemes are based on prefetching one or more cache lines adjacent to the current cache line (ie., the cache line currently in use, or currently being fetched) [9, 3, 2]. In some of these schemes, the number of cache lines prefetched can be varied dynamically [2]. Better hardware schemes detect the memory access patterns dynamically and predict on the fly what and when to prefetch. When a strided memory access is detected, these schemes prefetch the next required cache line using the stride [4, 1].

Chen proposed one of the most sophisticated hardware schemes [1]. This scheme predicts future data accesses by keeping track of past access patterns in a Reference Prediction Table. Chen proposes three strategies of increasing complexity. In the simplest strategy, during iteration i of a loop, data required for iteration $i+1$ are prefetched. For small loops, this strategy cannot hide the memory latency; and for large loops, it can prefetch more than what is required, and potentially pollute the cache. The second strategy fixes the first problem by using a lookahead program counter. This lookahead PC runs ahead of the normal PC by about the memory latency, and is responsible for issuing prefetches. For small loops, this will enable prefetches to be issued multiple iterations in advance. The third strategy extends the second one by detecting access patterns across loop levels. A major drawback of Chen's scheme is that it does not improve the performance of applications with indirect memory accesses. Besides, it requires modifications to the processor architecture, which is usually expensive.

This paper proposes an alternative hardware scheme for prefetching, where a second processor is used solely for the purpose of prefetching data for the primary processor. The scheme does not *predict* memory access patterns, but rather uses the second processor to run ahead of the primary processor so as to *detect* future memory accesses and prefetch these references. It can handle indirect memory references unlike Chen's scheme. Provided the processor can support prefetch instructions, our scheme requires no modifications to the processor architecture.

We call the primary processor the *execution engine* and the second processor the *prefetch engine*. The two engines execute images of the same binary executable. We generate the two images from a given binary executable using a binary code splitter.

At first glance, the use of a second processor solely for the purpose of prefetching may look like an overkill. With the current hardware prices, one should note that a dual processor commodity computer is only 20–25% more expensive than an otherwise comparable single processor computer. To justify the use of a second processor for prefetching, it is desirable to achieve about 20–25% speedup through prefetching. As our experimental results will show, the speedups achieved on benchmarks are up to 96% and average 45%.

It may be possible to achieve similar speedups by simply parallelizing an application. However, in this case the application developer is left with the task of parallelizing. Besides, not all applications lend themselves to parallelization. The goal of this paper is to identify a form of parallelization which separates out some of the memory fetches and issues them using the second processor as a separate thread of execution. As with classical parallelization, there are applications which do not lend themselves to prefetching: an application in which the difference between the total computation time and total memory access time is large, will not be able to benefit much through prefetching.

The rest of this paper is organized as follows. Section 2 describes the architectural details of our prefetching scheme. Section 2.1 outlines the binary code splitter that generates the program images for the two processors. Section 3 presents some experimental results comparing the performance of some benchmark applications with and without prefetching. The final section concludes with a summary.

2 The Architecture for Prefetching

Our architecture for prefetching consists of two identical processors. One processor, the *execution engine*, executes the program as normal. The other processor, the *prefetch engine*, executes a variant of the program where the memory access instructions are replaced by prefetch instructions and all other instructions, except address computation instructions and control-flow instructions, are deleted. The two engines start from the same entry point of the program, but the prefetch engine will eventually run ahead of the execution engine. This is because it does not execute any data-oriented instructions.

The architecture somewhat resembles a decoupled-architecture which distributes address generation and execution to two separate processors [10]. There is a major difference, however. The address generation engine is an integral part of a decoupled-architecture, and cannot be dispensed of. The prefetch engine in our architecture, however, is only there to boost the performance of the execution engine, so it can be safely turned off, without stalling the execution engine.

Given that the prefetch engine issues prefetch instructions, we need to use microprocessors that support a prefetch instruction. Most of the modern processors, such as the Pentium, MIPS R10K, and the Alpha, fall under this category. We choose to use the Digital Alpha microprocessor [8] since we had a cycle-level simulator available for the Alpha.

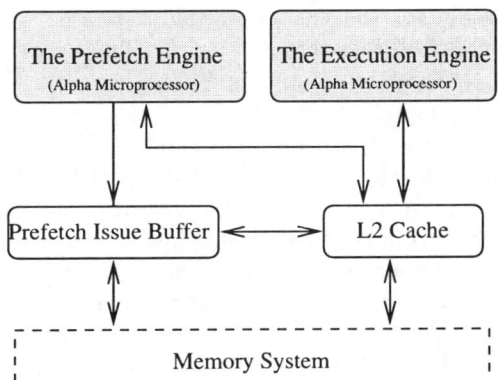

Fig. 1. Prefetching with dual processors

The organization of the execution and prefetch engines with respect to the prefetch issue buffer [5], L2 cache, and the memory system is shown in Figure 1.

The prefetch issue buffer keeps a number of prefetches and the information associated with them until the data corresponding to the prefetches is obtained from the memory and placed into the L2 cache. A prefetch is placed in the prefetch issue buffer only if the line specified by the prefetch address is not in the L2 cache; otherwise the prefetch request is simply dropped.

Note again that the execution engine executes the binary executable as normal; and that the prefetch engine executes a variant of this executable in which memory access instructions are replaced with prefetch instructions and all other instructions, except address computation and control flow instructions, are deleted. There are, however, two major issues.

1. The prefetch engine does not do any data-oriented computations. However, results of these computations are required for the control flow at the end of each basic block.
2. The prefetch engine can get far ahead of the execution engine, and thus can prefetch into L2 cache data that will not be used by the execution engine in the near future. This potentially may pollute the cache: the data the execution engine has not yet used may be replaced.

The solution is to inject synchronization instructions into the executables of both execution and prefetch engines. At a synchronization point, the contents of the registers of the execution engine are copied into the registers of the prefetch engine. This makes the results of data-oriented computations available to the prefetch engine, thus ensuring correct control flow.

By suitably placing synchronization points, we also restrict the prefetch engine from getting too far ahead of the execution engine. There is also another mechanism that limits the distance between the prefetch and execution engines: the prefetch engine can only issue prefetches when there is room in the prefetch issue buffer; when there is no room, the prefetch engine will have to wait.

The synchronization for the execution engine is a set of store instructions that copy the contents of all registers into a pre-defined area of memory, and the synchronization for the prefetch engine is a complementary set of load instructions that load into the registers contents from this pre-defined area of memory. The pre-defined memory area will be cached in the L2 cache, so that the best-case synchronization cost is proportional to the L2 cache latency rather the full memory latency. The execution and prefetch engines have a producer-consumer relationship and the ordering is guaranteed by using a boolean flag. This boolean flag is not part of the memory space, and is entirely controlled by the L2 cache controller. The flag is set when the last location of the pre-defined area is written, and cleared when this location is read. An attempt by the prefetch engine to read from the area while the flag is clear, will stall the prefetch engine until the flag is set. Similarly, an attempt by the execution engine to write to this memory area while the flag is already set, will stall the execution engine until the flag is cleared. Usually it is the prefetch engine which gets to the synchronization point first, and therefore waits for the flag to get set. However, in certain cases, it is possible that the execution engine can get to the synchronization point first. In such a case, the execution engine will wait until the flag is clear (ie., until the previous synchronization data is consumed by the prefetch engine).

A similar synchronization flag is present in the primary cache to guarantee that any item stored in the pre-defined memory area can be read only once by the processor. This essentially emulates non-cacheability for the pre-defined memory area (in the context of the primary cache).

Earlier in the section, we mentioned that the prefetch engine can be safely turned off, without stalling the execution engine. In this case, the execution engine must execute the original binary executable rather than the one that has synchronization points.

2.1 The Code Splitter

The code splitter takes as its input a compiled binary executable, and splits it into two binary images executable by the prefetch and execution engines. The binary image for the execution engine is identical to that of the original binary executable except that some synchronization instructions have been inserted. The binary image for the prefetch engine will have corresponding synchronization instructions, and will have the memory access instructions replaced by prefetch instructions, and all other instructions except address computation and control flow instructions deleted.

The main advantage of using the code splitter is that we can use existing compilers, and run existing binaries without having to re-compile. New compilers, however, can directly create separate binary images for the execution and prefetch engines.

The code splitter considers for prefetching only array memory accesses, including indirect memory addresses. Scalar memory accesses are not prefetched. In addition, the code splitter places synchronization instructions at the appropriate places in the program. Loops and conditional branches are the candidates for synchronization.

If a loop contains array memory accesses, a synchronization point is placed just before the start of the loop. Loops that contain no array memory access are simply deleted in the prefetch image.

If the execution of a loop iteration depends on the array memory accesses during the iteration, then each iteration will need to have a synchronization point. This will adversely affect the performance of the execution engine. The code splitter, therefore, leaves the execution of such loops solely at the hands of the execution engine, and deletes the corresponding loops in the prefetch image.

3 Experimental Results and Evaluation

To test our architecture and the prefetching mechanism, we use a cycle-level simulator for the Alpha architecture. The simulator simulates the EV4 architecture with the exception that it does not simulate out of order execution and super-scalar instruction issue. Neither does it simulate the instruction cache. It nevertheless simulates the primary and secondary data caches, the write buffer, a prefetch issue buffer, and a memory subsystem. System calls are also simulated, so that any statically-linked binary executable can run on the simulator.

We report results from running seven benchmark applications: MXM, CHOLSKY, MXT, VARIANCE, INRPROD, VPENTA, and BTRIX.

MXM multiples two matrices of sizes 128KB and 32KB, storing the result in a matrix of size 64KB. It uses a four-way unrolled, outer product matrix multiply algorithm. CHOLSKY performs a Cholesky decomposition on a three-dimensional array of size 160KB. MXT transposes a two-dimensional matrix of size 256KB. VARIANCE computes the variance of all elements in a one-dimensional array of size 2MB. INRPROD computes the inner product of two vectors of size 250KB. VPENTA is from the NAS kernels, and inverts three pentadiagonals simultaneously. It uses seven 2D arrays, each of size 40KB, and two 3D arrays, each of size 120KB; one of the 3D arrays is the output. BTRIX is also from the NAS kernels, and performs a block tridiagonal matrix solution along one dimension of a 4D array. It uses four 4D arrays, one of which is 540KB in size, and the rest three are 90KB.

Unless stated otherwise, the experiments use the following configuration. The size of the L2 cache is 256KB, and the memory latency is fixed at 50 cycles. The prefetch issue buffer has 8 entries.

Table 1. Speed-up resulting from prefetching

Application	MXM	CHOLSKY	MXT	VARIANCE	INPROD	VPENTA	BTRIX
Speed-up	6%	11%	28%	53%	21%	34%	19%

Table 1 compares the execution of the seven benchmarks with and without prefetching. It reports the speed-up gained by each of the benchmarks using prefetching.

3.1 Dropping Prefetches vs. Stalling the Prefetch Engine

When the prefetch issue buffer is full and there is a new prefetch to issue, there are two different strategies the prefetch engine can employ: 1. it can stall until there is room in the prefetch issue buffer, or, 2. it can drop the prefetch and continue. Table 2 compares these two strategies.

The memory locations for which the prefetches are dropped will not be prefetched. Moreover, locations that are only required for a later stage of the program may be prefetched early because of the fact that some earlier prefetches are dropped. This may result in cache pollution and can therefore lead to a less efficient program execution. A better strategy than dropping prefetches is to let the prefetch engine stall until there is room in the prefetch issue buffer. In this case, all the prefetches will be issued. Since the execution of prefetch engine is independent of the execution engine, stalling the prefetch engine does not hinder the execution of the execution engine.

The results of Table 2 show that stalling the prefetch engine is better than dropping prefetches. It shows the speed-ups when prefetches are dropped, and the speed-ups when the prefetch engine is stalled. For some applications, such as MXM and CHOLSKY, the advantage of stalling the prefetch engine is quite small, but for other applications there is a considerable gain in stalling the prefetch engine. Consider INRPROD, for example. It shows just a 0.4% speed-up when prefetches are dropped, but this goes up to 21% when the prefetch engine is stalled. For VARIANCE, there is 5% speed-up when prefetches are dropped, but the speed-up increases to 53% when the prefetch engine is stalled.

Table 2. Speed-up comparison: dropping prefetches vs. stalling the prefetch engine

Application	Prefetch strategy	
	Dropping prefetches	Stalling prefetch engine
MXM	5.8%	5.9%
CHOLSKY	10%	11%
MXT	2%	27%
VARIANCE	5%	53%
INRPROD	0.4%	21%
VPENTA	10%	34%
BTRIX	13%	19%

Note that, when the same processor is responsible for prefetching as well as execution (as is the case in software prefetching), stalling this processor generally results in slower program execution; in this case, to drop the prefetch and continue executing will be a better strategy. When there is a dedicated processor for prefetching, however, it is beneficial that the prefetch engine stalls when the prefetch issue buffer is full.

3.2 Effect of Memory Latency

The speed-up one can get through prefetching depends on the relative contributions of computation time and memory access time to the total execution time. If the memory access time is relatively small, then there is nothing to gain from prefetching. Similarly, if the relative memory access time is large, then there is not enough computation to overlap with the memory accesses, and thus prefetching cannot be effective. Prefetching is most effective when both computation and memory access equally contribute to the total execution time. In this case, a speed-up close to 100% is achievable.

This experiment shows the result of varying the memory latency. The experiment uses the application VARIANCE which has a total of 6,818,300 computation cycles when prefetching is enabled, and 6,817,396 computation cycles when prefetching is disabled. Table 3 shows the variation of total execution time with memory latency when prefetching is enabled and disabled. Note that the best speed-up occurs around a latency of 25 cycles. At this point the contribution of computation and memory access to the total execution time are roughly equal (computation time is 6,818,300 cycles, whereas the memory access time is 13,936,430 - 6,818,300 = 7,118,130 cycles). This shows that prefetching is most

Table 3. Effect of memory latency on prefetching for VARIANCE

Memory latency	Total execution time without prefetching	Total execution time with prefetching	Speedup
10	9,985,563	7,097,134	40%
20	12,619,470	7,112,596	77%
25	13,936,430	7,120,326	96%
30	15,253,390	8,178,416	87%
40	17,887,310	10,812,976	65%
50	20,521,230	13,447,536	53%
60	23,155,150	16,082,095	44%
70	25,789,070	18,716,656	38%
80	28,422,990	21,351,216	33%

effective when the total computation time is about the same as the total memory access time.

Figure 2 shows how speed-up varies with memory latency for some of the applications. A speed-up close to 100% is achievable when both computation and memory access equally contribute to the total execution time. The maximum speed-up is often less, when the applications are either computation-intensive, or memory-intensive.

3.3 Indirect Memory Access

The previous experiments considered only direct memory accesses. The prefetch engine, in these cases, did not require any memory access. However, when there are indirect memory accesses in the application, the prefetch engine will need to be able to load from the memory. For instance, consider the following code fragment.

```
for ( int i = 0; i < MAX; ++i )
    sum += a[b[i]];
```

In order to prefetch $a[b[i]]$, the prefetch engine will need to load $b[i]$. Inserting a synchronization point in the loop instead of loading $b[i]$ will be a poor choice, for it will slow down the loop rather than speeding it up.

Table 4 reports results from five applications that use indirect memory accesses. CFF2D is from the NAS kernels, and performs a complex radix fast Fourier transform on a 2D input array. VARIANCE/LL and INRPROD/LL are linked-list versions of the VARIANCE and INRPROD benchmarks. PIC-1D is one of the Livermore Loop kernels. IS is from the NAS parallel benchmark suite, and performs a bucket sort on 200,000 integers.

All five applications show reasonable speed-ups resulting from prefetching. The linked-list versions of VARIANCE and INRPROD do not gain as much speed-up as their non-linked-list versions (see Tables 1 and 4). This is because

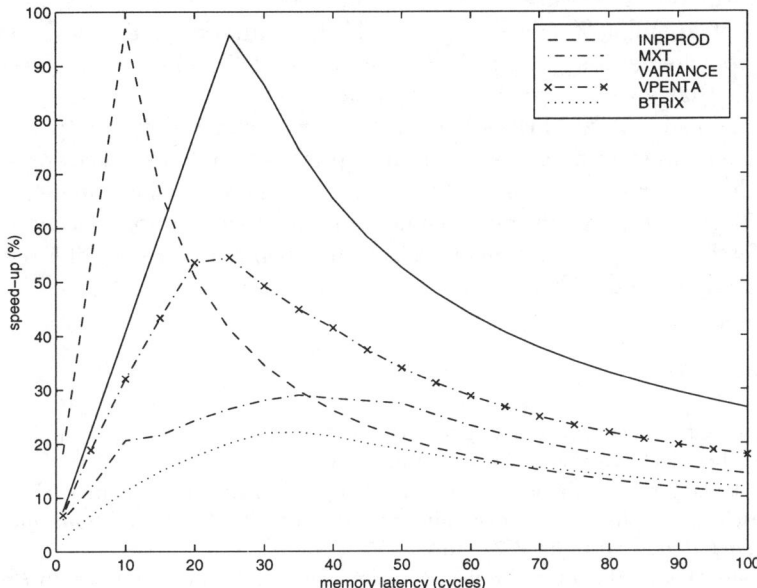

Fig. 2. Speed-up vs. memory latency

Table 4. Prefetching effectiveness and indirect memory accesses

Application	Total execution time without prefetching	Total execution time with prefetching	Speed-up
CFF2D	3,324,305	2,925,093	14%
VARIANCE/LL	35,733,775	31,543,052	13%
INRPROD/LL	6,587,186	7,467,407	13%
PIC-1D	5,109,242	4,348,065	18%
IS	28,753,396	21,623,544	33%

there is more stress on the memory system due to additional (indirect) memory accesses.

4 Summary and Conclusions

This paper proposed a hardware scheme for data prefetching, where a second processor, called a prefetch engine, is used solely for the purpose of prefetching data for the primary processor, called an execution engine. The scheme does not *predict* memory access patterns, but rather uses the prefetch engine to run ahead of the execution engine and *detect* future memory accesses and prefetch these references.

We evaluated and validated the scheme using some benchmark applications. The results indicate that hardware prefetching is a promising technique that offers worthwhile speed-ups. With a memory latency of 30 cycles, an average speed-up of about 45% can be seen in the benchmarks of Figure 2. The cost increase to acheive this speedup is only about 20–25%. This cost increase is due to the fact that a second processor is employed.

The main advantage of the scheme is that prefetching is effected without having to change an existing processor design (provided that the processor supports a prefetch instruction). Unlike other hardware schemes which *predict* memory access patterns to prefetch, our scheme *detects* future memory access locations and prefetches them. Therefore chances of not being able to use the prefetched data is smaller in our scheme.

References

1. Tien-Fu Chen. Data prefetching for high-performance processors. Technical Report PhD Thesis, University of Washington, July 1993.
2. Fredrik Dahlgren, Michel Dubois, and Per Stenstrom. Sequential hardware prefetching in shared-memory multiprocessors. *IEEE Transactions on Parallel and Distributed Systems*, 6(7), July 1995.
3. Nathalie Drach. Hardware implementation issues of data prefetching. In *Proceeding of the International Conference on Supercomputing*, pages 245–254. 1995.
4. John W C Fu and Janak H Patel. Stride directed prefetching in scalar processors. In *Proceedings of the 25th International Symposium on Microarchitecture*, pages 102–110. 1992.
5. S Manoharan and S R Bathula. Hardware support for software prefetching. In *Proceedings of the 4th Australasian Computer Architecture Conference*, pages 97–108. 1999.
6. Todd Mowry. Tolerating latency through software-controlled data prefetching. Technical Report PhD Thesis, Stanford University, March 1994.
7. Allan Porterfield. Software methods for improvement of cache performance on supercomputer applications. Technical Report PhD Thesis, Rice University, May 1989.
8. Richard Sites, editor. *Alpha Architecture Reference Manual*. Digital Press, 1992.
9. A J Smith. Cache memories. *ACM Computing Surveys*, pages 473–530, 1982.
10. J E Smith. Decoupled access/execute computer architectures. In *Proceedings of the 9th International Symposium on Computer Architecture*, pages 112–119. 1982.

A Java-Based Parallel Programming Support Environment

K.A.Hawick and H.A.James

Distributed & High Performance Computing Group
Department of Computer Science, University of Adelaide
SA 5005, Australia
Tel +61 8 8303 4519, Fax: +61 8 8303 4366
khawick@cs.adelaide.edu.au

Abstract. We have prototyped a multi-paradigm parallel programming toolkit in Java, specifically targeting an integrated approach on cluster computers. Our JUMP system builds on ideas from the message-passing community as well as from distributed systems technologies. The ever-improving Java development environment allows us access to a number of techniques that were not available using the message-passing systems of the past. In addition to the usual object-oriented programming benefits, these include: language reflection; a rich variety of remote and networking techniques; dynamic class-loading; and code portability. We are using our JUMP model framework to research some of the long sought after parallel programming goals of support for parallel I/O; irregular and dynamic domain decomposition and in particular irregular mesh support. Our system supports the usual messaging primitives although in a more natural style for a modern object oriented program.
Keywords: Message passing; multi-paradigm support; metacomputing; programming environments; Java.

1 Introduction

The Java programming language and the various environments that support it have found many uses in modern computing. Not least of these is the area of parallel and high performance programming. The JavaGrande Forum [14] has been a strong voice in the investigation of and call for improvements in the performance of Java Virtual machines. Parallel computing is an area that Java promises to aid considerably with its integrated threads package and various language features such as reflection; dynamic loading and its networks and remote methods packages. We have investigated Java as a vehicle for distributed computing systems infrastructure in some depth [12] and are now considering its capabilities for more tightly coupled parallel systems. In this paper we describe our Java Ubiquitous Message Passing (JUMP) software system for message passing.

We have taken a fresh look at many of the difficulties and requirements of message passing programming and how Java programmers would use these differently to C or Fortran programmers who have been the mainstay users driving message passing systems. This is a necessary step to achieve our main goal, which is to consider how higher level layers of libraries can be used to ease the burden of parallel programming using various decomposition paradigms. Parallel programming has come a long way in some respects since the early 1980s. It is widely accepted and used for many performance intensive applications. New generations of programmers are accepting parallel programming methods as mainstream and not some obscure part of computer science as was once the case. However, parallel computing at the message passing level remains cumbersome and error prone. We do not claim to have found a revolutionary new approach and means to solve this problem. We do however believe there is merit in considering higher level library packages that embody past experience.

The general approach we are taking with our JUMP system is to support parallel decomposition strategies in a way that does not require a lot of message passing code to be custom written for a specific application. Some of the classic decomposition strategies for parallel applications include: geometric or regular domain decomposition in an appropriate number of dimensions; scattered spatial or semi-random decomposition; unordered task decomposition. It is our thesis that many applications do not simply require one of these strategies for an optimal parallelisation, but rather many realistic codes require several. Our JUMP system is therefore designed to allow several parallel decomposition harnesses to co exist in the same running program, reusing communications infrastructure, but controlled by a different thread of execution. Multi threaded programs may therefore use different decompositions for different data structures across the same set of participating processors.

An interesting example application is a weather simulation on the globe. The atmospheric fluid flow is often modeled as a finite difference system of equations on a regular mesh or sometimes on a partially regular mesh, whereas measured data is assimilated onto this mesh from arbitrarily scattered points [10]. Time stepping the fluid mesh and assimilating data may be computationally similar in terms of performance requirement or number of raw calculations to be performed therefore no single decomposition is ideal. The same processors need to be used for both parts of the calculation but different decompositions are optimal for load balancing the two parts of the computation. Our JUMP architecture is designed to support this idea. In this paper we describe our JUMP architecture and discuss some performance insights gained. We also consider the impact of emerging technologies like Jini and JavaSpaces services as a vehicle for implementing JUMP.

2 JUMP Design and Architecture

Despite the previous work done to provide MPI support for Java [1–4, 7, 15] we believe that there is still work to be done in creating a pure Java open source MPI system. Our system is designed and implemented as a simple pure Java "MPI-like" system. Not all MPI bindings are sensible in an object-oriented programming environment such as Java. We propose a rethink of the message-passing interface specification in the light of object-oriented technology becoming more popular. As well as supporting the arguments normally used in MPI, the **send()** and **recv()** methods have been overloaded to accept fewer arguments, since providing information about data types and array sizes is available in Java through the use of reflection. This is an example of where the syntax of MPI is "unnatural" in Java. The Java language seems an obvious choice for the preferred implementation language because of both its platform-independent bytecode representation and the platform's sheer ubiquity. The Java language provides a number of other benefits too: single inheritance of classes ensures programming simplicity for new users by not requiring them to locate and compile their programs with arcane-named libraries; Java is strongly-typed throwing an exception when a received object is typecast to an invalid type, unlike C's default behaviour; Java's security managers are able to specify the permissible actions on resources, and the security managers are enforced.

JUMP builds on our past experiences in building multi-threaded daemons [12] in Java for software management. We also utilise our experience in classloading in constructing code server databases of dynamically invocable Java bytes code from databases [18]. JUMP programs can be run in a similar fashion to conventional parallel programs written in MPI or PVM. We are still experimenting with the configuration of such programs and are trying to determine where configuration information should best be located. At present we utilise a flat hostfile of participating node names in a fashion similar to PVM. Similar to both the MPI and PVM environments, the JUMP architecture is composed of three levels: user code; a user-level API; and underlying message-passing libraries (or in this case, base class).

We have measured various aspects of the behaviour of the Java 2 platform on different platforms. Object creation for simple objects takes on the order of 150 clock cycles on a modern machine, whereas the thread and socket creation overheads are on the order of millions of clock cycles [16]. This suggests that using object-oriented technologies is entirely feasible for constructing message-passing software systems. However, the high cost of thread and socket creation has driven our design objective to reuse these wherever possible [11].

The JUMP architecture is divided into two parts: the JUMP base class and the JUMP daemon. An **abstract** JUMP base class contains all the "standard" methods that the user can use for communications, initialisation and finalisation of the computation. The JUMP daemon (JUMPd) runs on each machine in the host pool, waiting to receive requests to start slave instances of user JUMP code. The daemon only exists as a bootstrap mechanism to initialize slave instances of user code. A JUMP user class extends the JUMP base class, which initiates communications through its local JUMP daemon. A network of socket connections are established between participating daemons on an as-needed basis. These can be reused to minimise socket instantiation overheads. Once a socket is established for that program between two nodes, no other will be created in our present implementation. Extending the base class allows the user's code to be used by the JUMP environment. The user is only required to supply two methods in order to extend the base class: a `public void jumpRun(String []args)` method and `public static void main(String []args)`. The former method is the main body of the user's code, run after initialisation, while the latter is used to initialise the computation. The `main` method usually consists of a call to create a new instance of the user class and a call to `jumpRun`. The user chooses where in their `main` method they will call `jumpRun` so the master instance is able to load any data it may require to distribute to the slaves. The current version of the JUMP environment communicates via TCP/IP and Internet-domain sockets.

The steps involved in the initialization of the system are shown in figure 1. Figure 1 i) shows the state of the system when the user executes the `main()` method of their program. An initialization routine in the JUMP base class contacts the local JUMP daemon (figure 1 ii) which runs in its own JVM and therefore may not share the same classpath as the user program. As shown in figure 1 iii), the local JUMP daemon sends initialization requests to an appropriate number of remote machines (the addresses of which are selected either by the user or the daemon). When each remote JUMP daemon has successfully instantiated a copy of the user program, the daemons send details of the port on which the program is listening to the master program, which distributes a complete host table (figure 1 iv). Java encapsulation and overloading allows us to provide default communicator tags to partition message space appropriately to avoid library layers colliding with each others messages. We are still considering the best syntactic form for the user to provide these explicitly when needed.

In order for JUMP to function on a truly federated cluster of compute nodes, we cannot assume the presence of a shared file system. For this reason, we use a custom classloader [18] to transfer the necessary code to the remote machine. The mechanism for remote classloading is shown in figure 2. When the user program is instantiated (through the `main()` method) a copy of the program's byte code is serialized (figure 2 i). The serialized code is sent with the run-time arguments in an initialisation request to the local JUMP daemon (figure 2 ii). The message is then distributed to each of the remote JUMP daemons that have been chosen to participate in the computation (figure 2 iii). If the class representing the user's program references any other (non-core-Java classes) not in the remote daemon's `classpath`, then when they try to instantiate a new instance of the user program, a `ClassNotFoundException` will be raised. Our classloader traps this exception (figure 2 iv) and sends a request for the class bytecode to the master user program (figure 2 v). The master user program replies to the request with a serialized copy of any requested class (figure 2 vi). Steps v) and vi) are repeated as many times as necessary to resolve all object on which the user program depends. Once all objects are resolved the slave copy of

A Java-Based Parallel Programming Support Environment

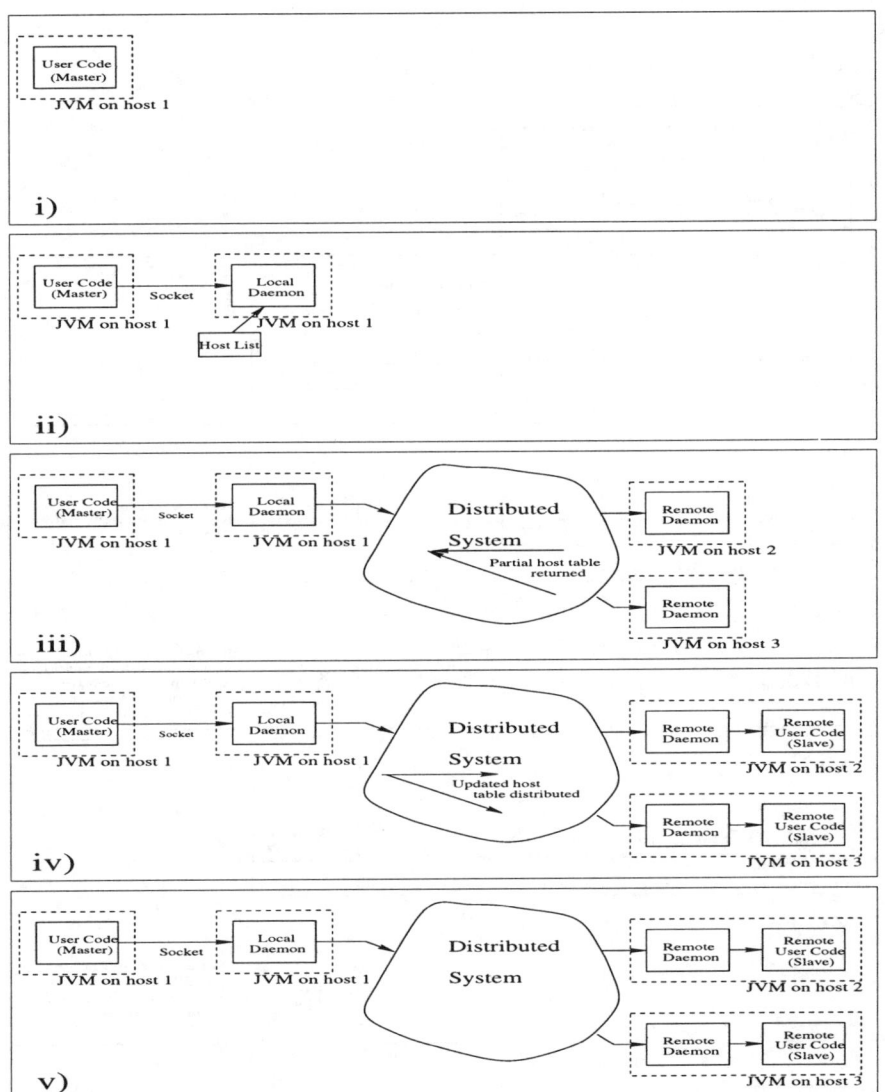

Fig. 1. Initialization of the JUMP environment. i) The user executes their program. ii) A socket is created that contacts the local daemon. iii) The local JUMP daemon selects the appropriate number of remote hosts and their JUMP daemons an initialization request. They respond with the host IP/port that the slave program listens to. iv) When all the remote daemons have responded favourably to the initialization request, the complete host table is distributed from the master JUMP daemon.

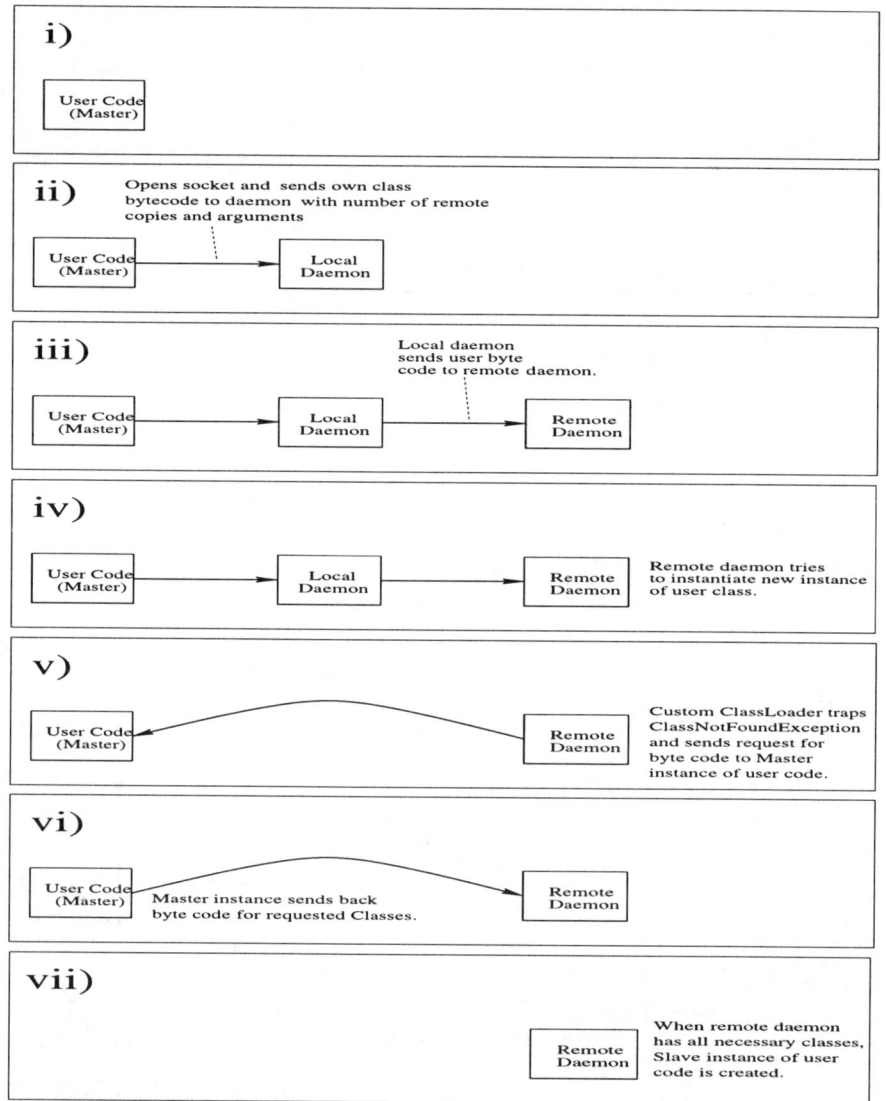

Fig. 2. JUMP distributed classloading mechanism. i) The user program loads a serialized copy of its class byte code in preparation for distribution. ii) The serialized bytecode is sent to the local JUMP daemon. iii) The bytecode is sent with an initialisation request and run-time parameters to the remote JUMP daemons. iv) The remote JUMP daemon tries to instantiate an instance of the user program on the remote machine. v) If the user program uses other non-Java core classes, a request is sent to the master instance of the user program. vi) The requested classes are sent to the remote daemon, where they are cached. Steps v) and vi) are repeated as required. vi) The slave instance is instantiated and the runJump() method is prepared with the run-time arguments.

the user program is instantiated and the port to which the slave is listening is returned through the local JUMP daemon to the master program.

To allow for the possibility of multiple, independent user programs running on the same physical machine, messages between user classes and the daemons, and between daemons are tagged with the source and destination host IP and the port on which the program is listening. With the exception of the JVM crashing or the machine running out of memory, user programs are protected from interference of other programs running on the same machine.

3 Performance Analysis

This section presents comparison of performance measurements of the current JUMP prototype against a Java version of MPI, PJMPI [17]. All tests were performed on a cluster of Digital Alphastation 2/255s with 300MHz CPUs and 128Mb of memory. Each workstation communicated via 10baseT Ethernet and ran Java version 1.2.2 with native threading. The user's jump program measured the initialization time and then the time to send a single Byte object between each of the slaves in a *ping-pong* fashion.

We measured the time, in milliseconds, to initialise the prototype JUMP system with a variable number of slave nodes on the local machine and also with the salves on remote machines. The measurements include the time taken by the Java runtime system to: search for the local daemon; serialise the user class' bytecode; send an execution request to the local daemon, where the bytecode will be uploaded if necessary; distribute an updated host table to all machines participating in the computation; and instantiate an instance of the bytecode. In the prototype, the first slave instance is always created on the same machine as the master instance. This data shows a linear progression in number of slave hosts with a least square linear fit analysis giving a gradient of approximately 507ms per extra slave and y-intercept overhead of 514 ms. The gradient, therefore represents the extra overhead that each slave adds to the instantiation overhead. A similar analysis for remote machines gives a gradient of 575 ms per slave instance and y-intercept of 498ms. This is only slightly poorer than for the local machine and we believe indicates we are obtaining good network efficiency.

A comparison is made of the time taken to send and receive a single byte between the two systems. We find that our PJMPI system takes 4.7 ms; JUMP on a local host takes 408ms; and JUMP on a remote host takes 619ms. The time taken for a single byte to be transferred in the JUMP system is many times larger than that for the PJMPI system. This can be attributed to a number of reasons: Java objects are being transferred, not primitives, and these must be serialised; a Java object is used both for the sent data and a communicator object with which a match must be made to receive the object; the data is encapsulated within a Java wrapper for identification purposes. While the comparison results are not good when data objects are small, we feel that the JUMP overheads will be amortized when large, complicated objects are to be transferred between processes. In current systems such as PJMPI the user must write each data element into a byte buffer before transfer; JUMP users may simply send a complete object to a remote process. Finally, the user of a system such as PJMPI must ensure that the code that represents their program is available on all machines in a cluster. The JUMP system alleviates this necessity by allowing Java bytecode to be serialised and downloaded by daemons on demand; the bytecode is cached so that multiple instances of the same bytecode do not incur additional download overheads.

4 Future Directions for JUMP

JUMP uses a similar approach to message communicator tags and grouping mechanisms that MPI specifies and which was we believe pioneered in the Edinburgh CHIMP system [6]. This is important to allow library layers of more sophisticated communications to operate with internal messaging that does not impact on the explicit messages sent by user program code.

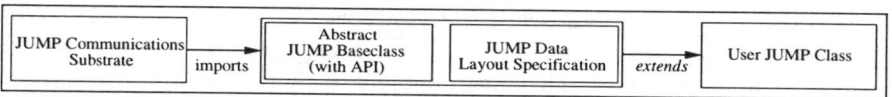

Fig. 3. Future versions of JUMP will include an extra layer of functionality to allow different data layout specifications and used by users.

We anticipate the JUMP platform will provide a platform with which we can investigate the challenging area of data layout specification. For example HPF [13] provides the ability to distribute data in block and cyclic patterns, depending on the way the data is to be used. We plan to introduce an extra layer between the user's code and the JUMP base class which will allow communications to be automatically routed to the correct user jump instance in the current data layout. For example, consider a weather simulation that combines the previous predictions with observed data (from ships, aircraft and weather balloons). The Earth is likely to be modeled as a regular grid; therefore the code that operates on the previous predictions for the surface of the Earth for which there is no observed data is likely to be a regular grid decomposition. However observational data is not spread uniformly across the surface of the Earth. It is tightly clustered around certain points (most corresponding with shipping lanes and flight paths); this observational data may be best processed using a task farming model with the ability to update the cells of the Earth grid that are affected. When this extra layer of functionality is added to the JUMP environment we expect the architecture to look like that shown in figure 3. This figure shows the proposed three-level architecture incorporating support for data layout specification.

While the current version of JUMP communicates via sockets, it is not reliant on any particular communications technology. The current version is written using TCP/IP and Internet-domain sockets; the architecture does not preclude alternate versions that use, for example, RMI and JavaSpaces [8]. We are experimenting with the tuple space approach offered by JavaSpaces. The tuple space model where messages are tagged with complex predicate objects is an attractive mechanism for building higher level systems. However, a limitation of that system is lack of support for "spaces of spaces". A single node would have to host the entire communications space, which we believe is not suitable for a high performance communications system. We are investigating ways to enable scalable tuple spaces. We have compared this with the performance that would arise if all communications in our system were brokered through a single host. This will typically present too much message congestion for any practical applications. We are waiting for the JavaSpaces technology to mature further before we use it to implement JUMP.

5 Summary and Conclusions

The Java programming language and environment is stimulating new research activities in many areas of computing, not the least of which is parallel computing. Parallel techniques are themselves finding new uses in cluster computing systems. Although there are excellent software tools for scheduling, monitoring and message-based programming on parallel clusters, these systems are not yet well integrated and do not provide very high-level parallel programming paradigm support.

We have described the design and architecture of our JUMP Java-base message passing system. We have discussed the performance and object oriented design goals that have led us in this direction. Although we envisage a need for an MPI compliant Java messaging system, we have been driven by research interests in high level layers of parallelism to develop JUMP. We have found that the performance restrictions on advanced Java technologies such as Jini and JavaSpaces and also Java RMI are too limiting for the applications we hope to support. This tradeoff may change as JVM's improve in performance, however at present we believe there is interesting work to be done in developing parallel support for various data decomposition strategies. Higher level object manipulation in a parallel Java system comes at some cost, but we believe this is worth investigating in the expectation of faster JVM's.

References

1. Mark Baker and Bryan Carpenter. Thoughts on the structure of an MPJ reference implementation NPAC at Syracuse University, Technical Note, October 1999.
2. Mark Baker, Bryan Carpenter, Geoffrey Fox, Sunh Hoon Ko, and Xinying Li. mpiJava: A Java MPI Interface. http://www.npac.syr.edu/projects/pcrc/papers/mpiJava/
3. Bryan Carpenter, Vladimir Getov, Glenn Judd, Tony Skjellum, and Geoffrey Fox. MPI for Java: Position document and draft API specification. Java Grande Forum Technical Report JGF-TR-03, November 1998.
4. Kivanc Dincer. Ubiquitous Message Passing Interface implementation in Java: JMPI. Proc. 13th Int. Parallel Processing Symp. and 10th Symp. on Parallel and Distributed Processing. IEEE, 1998.
5. Dynamic Object Group. DOGMA Home Page. http://ccc.cs.byu.edu/DOGMA.
6. Edinburgh Parallel Computing Centre. CHIMP concepts. June 1991.
7. Adam J. Ferrari. JPVM: Network Parallel Computing in Java, 1998.
8. Eric Freeman, Susanne Hupfer and Ken Arnold. JavaSpaces Principles, Patterns, and Practice, Addison Wesley Longman, June 1999. ISBN 0-201-30955-6.
9. W. Gropp, E. Lusk, N. Doss and A. Sjkellum. A High-Performance, Portable Implementation of MPI Message Passing Interface Standard, Argonne National Lab, 1996.
10. K. A. Hawick, R. S. Bell, A. Dickinson, P. D. Surry and B. J. N. Wylie. Parallelisation of the Unified Model Data Assimilation Scheme. Proc. Workshop of Fifth ECMWF Workshop on Use of Parallel Processors in Meteorology, Reading, November 1992. European Centre for Medium Range Weather Forecasting.
11. K. A. Hawick, H. A. James, J. A. Mathew and P. D. Coddington. Java Technologies for Cluster Computing Technical Report DHPC-077, Department of Computer Science, The University of Adelaide, November 1999.
12. K. A. Hawick, H. A. James, A. J. Silis, D. A. Grove, K. E. Kerry, J. A. Mathew, P. D. Coddington, C. J. Patten, J. F. Hercus and F. A. Vaughan. DISCWorld: An Environment for Service-Based Metacomputing. Future Generation Computing Systems (FGCS), (15)623–635, 1999.
13. High Performance Fortran Forum. High Performance Fortran Language Specification version 2.0. High Performance Fortran Forum, January 1997.
14. Java Grande Forum. Java Grande Forum Home Page.
15. P. Martins, L. M. Silva, and J. Silva. A Java Interface for WMPI. LNCS, 1497:p121–, 1998.
16. J. A. Mathew, P. D. Coddington and K. A. Hawick. Analysis and Development of Java Grande Benchmarks. Proc. ACM 1999 Java Grande Conference, April 1999.
17. J. A. Mathew, H. A. James and K. A. Hawick. Development Routes for Message Passing Parallelism in Java Technical Report DHPC-082, Department of Computer Science, The University of Adelaide, January 2000.
18. J. A. Mathew, A. J. Silis, and K. A. Hawick. Inter Server Transport of Java Byte Code in a Metacomputing Environment. Proc. TOOLS Pacific (Tools 28), 1998.
19. Message Passing Interface Forum. MPI: A Message Passing Interface Standard.

A Versatile Support for Binding Native Code to Java

Marian Bubak[1,2], Dawid Kurzyniec[1], and Piotr Łuszczek[3]

[1] Institute of Computer Science, AGH, al. Mickiewicza 30
30-059 Kraków, Poland
bubak@uci.agh.edu.pl
[2] Academic Computer Centre – CYFRONET, Nawojki 11
30-950 Kraków, Poland
dawidk@icsr.agh.edu.pl
[3] Dept. of Computer Science, University of Tennessee
Knoxville, TN 37996-1301, USA
luszczek@cs.utk.edu
Phone: (+48 12) 617 39 64, Fax: (+48 12) 633 80 54

Abstract. As Java is being considered an appropriate environment for high performance computing, we describe advanced techniques of highly expressive and efficient combining of Java with existing code in other languages. We present their example use in developing of Java wrappers for the lip – a runtime support library which enables easy and portable parallelization of irregular and out-of-core problems. Sample performance comparison results of out-of-core computations in Java and C are also presented.

1 Introduction

Recently, there is a growing interest in application of the Java language [1] in the high performance computing (see e.g., [2–5]). As the performance of Java is continuously increasing [3, 4] the main issue is the lack of scientific libraries designed for use in Java. At the same time, there is a good deal of precisely tuned and thoroughly tested scientific libraries written in C, Fortran and other languages. The most appropriate way to make them available in Java is to provide pertinent interfaces for them via the Java Native Interface (JNI) API [6].

Existing tools for JNI wrapper generation do not allow for human influence on the native-to-Java transition process. This results in the interfaces which do not take any advantage of Java's specific features and introduce extraordinary syntactic overhead, which could otherwise be avoided. As a solution we provide an implementation of techniques and propose guidelines which on one hand automate large amount of the aforementioned transition process while on the other hand give wide range of possibilities to hide underlying native language environment and deliver to the user the existing functionality as if it were designed and implemented in Java.

2 Main Issues

Due to the differences between Java and other languages, problems arise just in the course of interface generation. Taking the C language as representative one, the most important issues comprise

- primitive data types in C have variable sizes which may differ from standardized Java data type sizes across different platforms,
- there is no direct counterpart to C pointers in Java [8], which makes it difficult to manage array subsections,
- multidimensional arrays in C have contiguous data layout, while in Java they do not,
- C structures can be emulated by Java objects, but the layout of the fields of objects may differ across JVMs, what prohibits direct access to them.

However, Java has a lot to offer: it is a safe, object oriented and easy to use programming language, with automatic memory management and exception handling. It is reasonable to develop Java-to-native interface in such a way that these features are retained, so that the users could use an underlying native code just as they would use the Java code.

Another important problem is linking native codes to Java. In a given language, any programming library can be used by an application in a quite straightforward manner by employing either static or dynamic linking mechanism (see Fig. 1.)

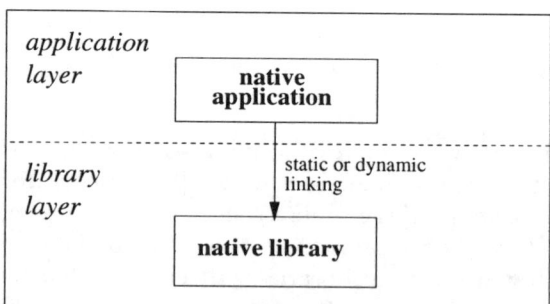

Fig. 1. Using native library in an application.

The use of the same library in Java is shown in Fig. 2. This time, it is much more complicated due to the fact that the class files generated by the Java compiler have system-independent format so they cannot be directly linked with any native programming library. The linking process must take place during the program execution and is performed by the JVM. The library ought to contain symbols that the JVM looks for. These symbols represent methods of Java classes

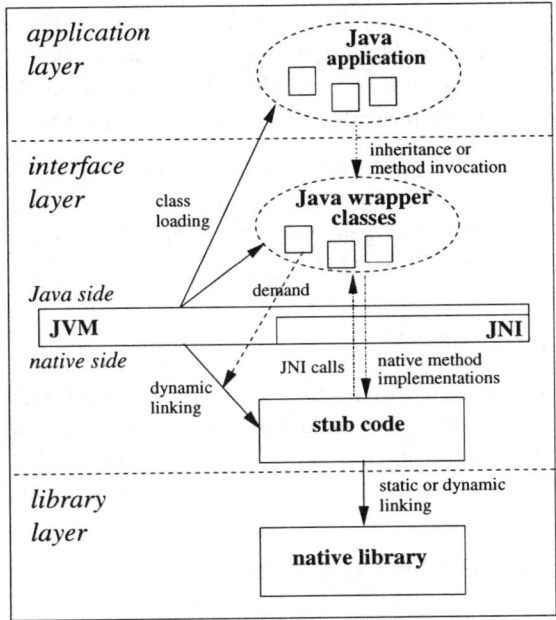

Fig. 2. Using native library in a Java application.

specified as `native` and their implementation is simply a wrapping code which calls functions of the native library.

The dynamic linkage of the wrapping code is performed by JVM with a call to the `System.loadLibrary()` method. The communication between Java and the native code is performed with functions specified by the JNI. Thus, multiple native libraries may be linked in this way at the same time, and it is also possible for them to interoperate between each other as they would in the traditional single-language application. The most notable feature of JNI is that it enables the wrapping code to have the same functionality as pure Java, i.e., it can create objects, call methods, throw exceptions, and have access to Java variables. The problem, however, may arise when performing floating-point calculations [4, 5].

3 Support for Interface Creation

The JCI tool [7, 9] is an automatic Java-to-C interface generator which enables creation of such an interface with a less effort. However, the JCI does not make any use of Java's specific features. As a result, the process of writing Java programs, which use the generated interface, is rather cumbersome. Moreover, the resulting programs are hard to understand and unsafe as they are capable of crashing the JVM upon invocation of a method with an incorrect argument.

Many of these issues can be fixed by manual refinement of the interface, but the changes must be done in a machine-generated code, which is rather inconvenient.

When a need arises for an interface which better suits the Java programming style, we provide set of sophisticated tools and techniques which

- simplify manual creation of Java-to-native interface,
- make access to the JNI features much more safe and easy,
- make frequent programming mistakes to occur as compilation errors rather than hard-to-track runtime errors.

3.1 Invoking Java from C

On the native language side, Java objects can be manipulated through the JNI calls. We provide a set of macros and functions that make the JNI calls much more convenient. For instance, they are responsible for checking whether a given JNI function call has caused an exception and acts appropriately.

3.2 Java-Style Native Programming

One set of macros, namely `JTRY`, `JCATCH`, `JFINALLY`, `JEND_TRY`, `JTHROW` and `JTHROW_NEW`, manages, on the native side, the JVM exceptions which can be thrown as a result of JNI calls. It is done simply, safely and effectively in the Java-like fashion. Another set of macros, `JSYNCHRONIZED` and `JEND_SYNCHRONIZED`, makes locking and unlocking Java monitors much safer, as it guarantees that the monitor locked in `JSYNCHRONIZED` will be unlocked in `JEND_SYNCHRONIZED` even if an exception is thrown in the code surrounded by these macros. Moreover, the lack of the `JEND_SYNCHRONIZED` macro call after `JSYNCHRONIZED` will cause the compiler syntax error rather than runtime synchronization problem.

These features help avoid some serious programming errors which are hard to track, thereby they greatly reduce the time needed for the wrapping code development. They also make the resulting interface much easier to read and understand.

3.3 Verification of Native Methods Arguments

Several macros were provided to simplify checking the correctness of arguments passed to the native methods. Examples are `CHKARG` and `CHKARG_NOT_NULL` macros, which are responsible for testing the specific conditions and causing the `IllegalArgumentException` to be thrown with the appropriate message if the conditions are not satisfied.

3.4 Arrays

Since in Java there are no pointers, there is also no way to refer to the memory area in the middle of an array using just a single variable of a primitive type. Therefore, it is difficult to wrap such C functions that take a pointer into an array as an argument. The solution we have developed are *virtual arrays*. A virtual array may represent a Java array, a portion of Java array, or even a portion of another virtual array. It is possible to use virtual arrays whenever in C the pointers into arrays are used. Virtual arrays can grow the underlying real arrays as it becomes necessary to store a larger amount of data. The growth can be enforced by native functions which are intended to write into the virtual array.

A problem arises, however, with conversion between Java and native arrays because the primitive data types in Java may not match those in C. Therefore, in some cases, automatic conversion of the whole array will result in unavoidable performance penalty.

3.5 Passing Arguments by Reference

In Java, like in C, all parameters are passed by value. In C, however, one may use a pointer to simulate the pass-by-reference semantics. Since in Java there are no pointers, special purpose classes are provided to encapsulate pointers to primitive types. Such classes were already introduced in [9].

3.6 Encapsulation of a Native Data Structures

The `DataEncapsulator` class is provided to enable type-preserving encapsulation of any C pointer. The classes derived from `DataEncapsulator` may provide additional methods that operate on the encapsulated pointer. In this way it is possible to treat underlying native data structure as an object. The `DataEncapsulator` class is also responsible for managing memory space occupied by the underlying data structure whenever necessary, e.g. when the encapsulating object is freed by garbage collector.

3.7 Multidimensional Arrays

Encapsulator classes for two- and three-dimensional native arrays with contiguous data layout were developed. Different methods of those classes enable copying portions of the encapsulated arrays into and from the Java arrays. They also allow for accessing single array elements at the specified coordinates, but this requires native method invocations each time what introduces an unavoidable overhead.

These classes are useful in common case when transformations of large arrays are performed mainly at the native side. They help to minimize the amount of the memory used and the time needed for copying of data between Java and native code.

3.8 Callback Functions

Many native libraries allow some degree of customization of their behavior with the mechanism of the *callback functions*. They are defined by the user and passed to the library as arguments. It would be desirable for Java programmers to use Java methods as callback functions.

We provide a solution of this problem in the form of an extensible set of Java interfaces and two additional C macros, namely: `DECLARE_WRAPPER_FUN` and `USE_WRAPPER_FUN`. The former is used to declare the native function responsible for wrapping the Java method, while the latter is used when the actual function pointer must be passed to the native function. Once the mechanism is used, the users may develop their own *function classes*, written entirely in Java, and pass objects of these classes to the wrapped native library functions thereby customizing their behavior, like with native callback functions.

The important issue related to this mechanism is that it requires a proper synchronization, since it involves the use of static variables and potentially may be thread-unsafe.

3.9 Java Signatures

When using JNI to access methods and fields of Java objects and classes, JVM *signatures* must be employed. Signatures are strings used internally by the JVM to choose an appropriate reference to a method or field whenever required. Since the use of signatures is error prone, we provide a set of macros to make it more convenient. For example, the signature of a Java method

```
long f(int n, String s, int[] arr);
```
which is
```
(ILjava/lang/String;[I)J
```
can be obtained by the macro call
```
JMS(long, MARGS3(int, JCLS(java/lang/String), JARR(int)))
```

Most of improper uses of these macros result in compile-time errors, whereas use of wrong signature would lead to a runtime error.

4 The lip Library Example

The lip [11] is a portable runtime support library for both in- and out-of-core irregular problems. Irregular problems are parallelized in the similar way as in the CHAOS library [12] whereas the out-of-core functions are based on the idea of *in-core section* [13]. The lip is built on top of the Message Passing Interface (MPI). The use of the MPI [14] as a communication layer makes the lip portable. At the same time, there is no support for solving irregular and out-of-core (OOC) problems in Java, whereas such a support is needed [2]. The latter, together with the portability of the lip, were the most compelling reasons to choose this library as an example usage of Java-to-native interface created using the techniques described above. In addition, there is often a need to call MPI routines directly in programs which use the lip. Thus, rudimentary Java-MPI wrappers [10] were developed conforming to MPI-2 C++ bindings specification whenever possible.

```
JNIEXPORT void JNICALL
Java_pl_edu_agh_icsr_lion_lip_Lip_localize(JNI_PARAMS,
           jobject jmaptable, jobject jg_indcs, jobject jl_indcs,
           jint jldata, jobject jgs_area, jobject jnewsched)
{
   /* LIP library data objects */
   LIP_Maptable maptable;
   LIP_Schedule newsched;

   VARRINFO v[2];          /* virtual arrays maintenance data */

   int gs_area;

   /* Verification and conversion of method arguments. */
   GET_ENCAPS_PTR(maptable, jmaptable);
   CHKARG(IS_VALID_IVARR(jg_indcs), "g_indcs");
   CHKARG(IS_VALID_IVARR(jl_indcs), "l_indcs");
   CHKARG_NOT_NULL(jgs_area, "gs_area");
   CHKARG_NOT_NULL(jnewsched, "newsched");

   JTRY {
      /* Preparing virtual arrays maintenance data. */
      VARR_PREPARE(v[0], jg_indcs);
      VARR_PREPARE_ENSURE_SIZE(v[1], jl_indcs, v[0].size);
      VARR_LOCK(v, 2);

      /* Call to LIP library. */
      LIP_Localize(maptable, v[0].data_ptr, v[1].data_ptr,
                   v[0].size, (int)jldata, &gs_area, &newsched);

      if (newsched == LIP_SCHEDULE_NULL) {
         JTHROW_NEW(LIP_EXCEPTION, "LIP_Localize() failed");
      }
      ASSIGN_ENCAPS_PTR(jnewsched, newsched);
      SET_OBJOFINT_VAL(jgs_area, gs_area);
   } JFINALLY {
      VARR_UNLOCK(v, 2);
   } JEND_TRY;
}
```

Fig. 3. Wrapper of the lip function LIP_Localize.

Fig. 3 shows one of the most sophisticated members of the Java-to-lip interface, which is the implementation of native Java method wrapping the lip function LIP_Localize. This sample demonstrates how some of the techniques aforementioned can be used. The jmaptable and jnewsched arguments are instances of classes inherited from DataEncapsulator (3.6), whereas g_indcs and l_indcs are virtual arrays (3.4). The jgs_area argument is int encapsulator (3.5).

5 Experimental Results

```
class Test extends LipNode {
  ...
  protected int startComputing() throws LipLibraryException {
    File cf = new File();                   // create lip file object
    double[] x, y;
    int[] indices, l_indices; // indirection arrays: global and local
    ...

    // Maptable describes data distribution
    mtab = new Maptable(Lip.MAP_BLOCK, ...);

    // open file with index array
    cf.open( "indices", MPI.MODE_RDONLY ...);

    for (/* all i-sections */) {
      cf.read(...);                  // read i-section data into local array
      // index translation and communication schedule generation
      schedule = Lip.localize(mtab, indices, l_indices, ...);
      Lip.gather(..., schedule); // gather data from other nodes

      for (j=0; j<is_size; j++) {  // perform compution
        y[l_indices[j]] += f( x[l_indices[j]], n );
      }
      Lip.scatter(..., schedule, MPI.SUM); // scatter computed data back
    }
    cf.close();
    ...
  }
}
```

Fig. 4. Sample Java code for OOC computing with the lip.

Fig. 4 presents a sample of Java version of an OOC test program. The code demonstrates a generic irregular OOC problem. There are two *data arrays* – x and y, together with *indirection arrays*: indices and its local version l_indices. The indirection arrays may be so large that they cannot fit into main memory and, therefore, must be stored on disk. The data arrays are distributed among the computing nodes. During the computation phase, transformations of data arrays are performed. Since the data arrays are indexed through indirection arrays which are altered during the program execution, the data access pattern is not known until runtime.

The full test programs in C and Java from Fig. 4 were run on a cluster of 10 Linux PCs each with 333MHz i686 Celeron processor and 32 MB physical memory. The machines were connected with Ethernet and each PC was equipped with 2 GB local hard disk, while approximately 600 MB of it being available for the program use. The LAM6.3b1 [15] was used as an MPI implementation together with the JVM from Java Development Kit ver. 1.1.7 for Linux. The workload incurred by the call to f() (see Fig. 4) function was controlled with parameter n. Each of the tests was run twice in C and in Java for two different values of n. The data arrays were of type double[] whereas index arrays were of type int[].

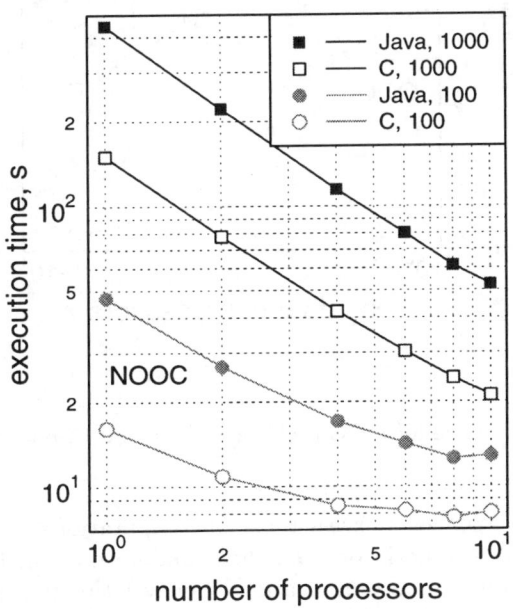

Fig. 5. Execution time for NOOC problem in Java and C, $n = 100$, $n = 1000$.

Fig. 5 shows a small test case where both data and index arrays are small enough to fit into the main memory (there are 3 arrays of 360360 values each). Fig. 6 shows an OOC case where the index array is large enough (3603600 int values) so that it does not fit into the main memory and therefore must be stored on disk. Fig. 7 presents timings for a *constant-time problem* where both data and index array sizes were proportional to the total number of the computing nodes N (the array sizes were $180180 \times N$ values).

The execution times of the Java program were about 2.5 - 3 times larger than those for C, nevertheless, we hope that the use of Just-in-time (JIT) compilers would eliminate that difference. It is worth noting that the scalability of the problem in C and Java is very similar, and it depends on the *computation-to-communication* ratio. For $n = 100$, there is a noticeable saddle where the

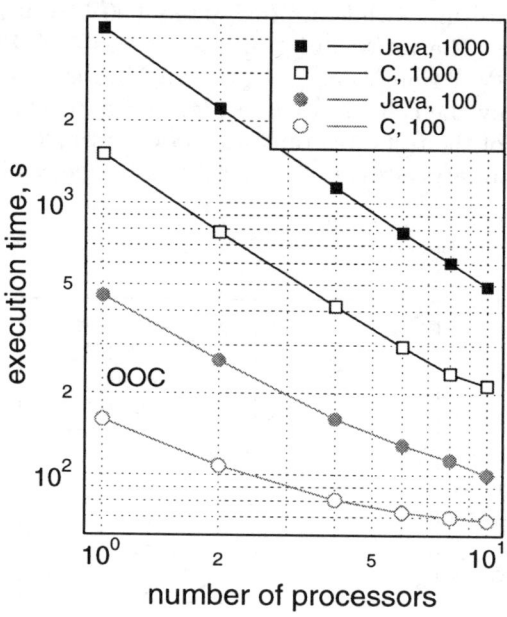

Fig. 6. Execution time for OOC problem in Java and C, $n = 100$, $n = 1000$.

communication and I/O costs excess those of computations. It is also important that our problem is a generic one and has random data distribution, thus it requires a large amount of communication whenever the data is assigned to the computing nodes.

6 Conclusions and Future Work

This paper describes some general issues which arise in the course of development of a Java-to-native interface which is used to wrap a native programming library code for its use in Java. The new techniques of mapping Java features to existing native code were proposed. It was also demonstrated with an example of the lip library that Java may be used to solve irregular and out-of-core problems while obtaining performance levels comparable to those of C/C++.

At present, we are working on an extension of the Java language that would allow to embed a native language code directly into Java programs, in a syntactic manner similar to that provided by some C/C++ compilers which allow to include assembly language in the C/C++ code. This embedded native code would be able to refer to Java variables, objects and methods directly rather than through JNI calls. Such an approach will enable the user to combine Java's rapid software development with the performance of the existing native code

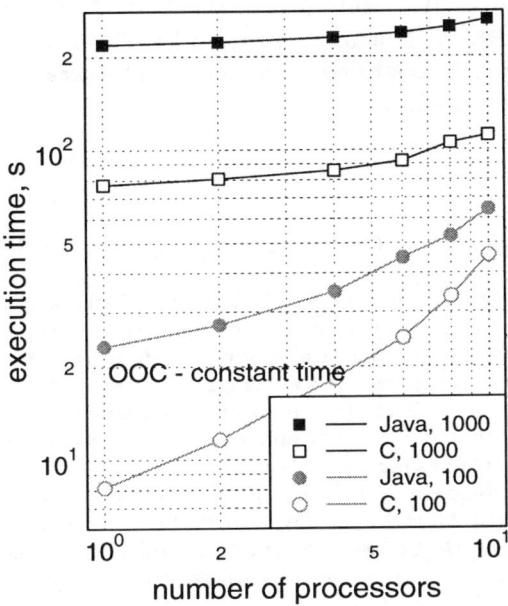

Fig. 7. Execution time for OOC problem in Java and C, constant time problem, $n = 100$, $n = 1000$.

libraries. The effort required for the interface creation will be significantly reduced since the difficult aspects of using JNI will be completely hidden from the users; moreover, they can work on a single source file per Java class rather than multiple ones. As a result, they can focus on what should be done rather than how to do it. The dedicated translator will transform a program written in such an extended Java into the pure Java code and accompanying native JNI-based source files. Such an extension could also be considered for native languages as it allows to embed the Java code into them. Translation process would transform such a mixed code into the pure native code containing proper JNI and Java Invocation API [6] calls to attach the JVM and use Java classes.

At the next stage we intend to provide a Graphical User Interface (GUI) tool, which would make the development of wrappers to native programming libraries even easier. As an output, such a tool would generate the extended Java code with an embedded native language code. This extended Java code can be further refined and processed as described above. The fully automatic wrapper generator is also under consideration, with its output being subject to potential refinement by means of the GUI tool.

Acknowledgements. We would like to thank Prof. P. Brezany for valuable discussions and to Dr. W. Funika for his remarks. This research was done in the framework of the Polish-Austrian collaboration and it was supported in part by the KBN grant 8 T11C 006 15.

References

1. Gosling, J., Joy, B., Steele, G.: The Java Language Specification. *Addison-Wesley*, 1996.
2. Zhang, G., Carpenter, B., Fox, G., Li, X., Wen, Y.: The HPspmd Model and its Java Binding. Chapter 14 in: Buyya, R. (Ed.): High Performance Cluster Computing. vol. 2: Programming and Applications, *Prentice Hall, Inc.*, 1999.
3. Java Grande Forum at: http://www.javagrande.org/
4. Philippsen, M.: Is Java Ready for Computational Science? In: Proc. 2nd European Parallel and Distributed Systems Conference, July 1998, Vienna http://math.nist.gov/javanumerics/
5. Boisvert, R.F., Dongarra, J.J., Pozo, R., Remington, K.A., Stewart, G.W.: Developing Numerical Libraries in Java. *ACM-1998 Workshop on Java for High-Performance Network Computing*, Stanford University, Palo Alto, California. http://www.cs.ucsb.edu/conferences/java98/papers/jnt.ps
6. JavaSoft. Java Native Interface. http://java.sun.com/products/jdk/1.2/docs/guide/jni/
7. Mintchev, S., Getov, V.: Towards Portable Message Passing in Java: Binding MPI. In: Bubak, M., Dongarra, J., Waśniewski, J., (Eds.), Recent Advances in Parallel Virtual Machine and Message Passing Interface, *Proceedings of 4th European PVM/MPI Users' Group Meeting, Cracow, Poland, November 1997, Lecture Notes in Computer Science* **1332**, Springer-Verlag, Berlin-Heidelberg, 1997, pp. 135-142.
8. Demaine, E.D.: Converting C Pointers to Java References. *ACM-1998 Workshop on Java for High-Performance Network Computing*, Stanford University, Palo Alto, California. http://www.cs.ucsb.edu/conferences/java98/papers/pointers.ps
9. Getov, V., Flynn-Hummel, S., and Mintchev, S.: High-Performance Parallel Programming in Java: Exploiting Native Libraries. *ACM-1998 Workshop on Java for High-Performance Network Computing*, Stanford University, Palo Alto, California, 1998; http://www.cs.ucsb.edu/conferences/java98/papers/hpjavampi.ps
10. Getov, V., Gray, P., Sunderam, V.: MPI and Java-MPI: Contrasts and Comparisons of Low-Level Communication Performance. *SuperComputing 99*, Portland, USA, November 13-19, 1999.
11. Bubak, M. and Łuszczek, P.: Towards Portable Runtime Support for Irregular and Out-of-Core Computations. In: Dongarra, J., Luque, E., Margalef, T., (eds): Recent Advances in Parallel Virtual Machine and Message Passing Interface, Proc. 6th European PVM/MPI Users' Group Meeting, Barcelona, Spain, September 26-29, 1999, Lecture Notes in Computer Science, Springer, 59-66.
12. Saltz, J., et al.: A Manual for the CHAOS Runtime Library, UMIACS Technical Reports CS-TR-3437 and UMIACS-TR-95-34, University of Maryland; March 1995 ftp://ftp.cs.umd.edu/pub/hpsl/chaos_distribution/
13. Brezany, P.: Input/Output Intensively Parallel Computing, LNCS **1220**, Springer, Berlin Heildelberg New York (1997).
14. Message Passing Interface Forum: MPI-2: Extensions to the Message-Passing Interface, July 18, 1997; http://www.mpi-forum.org/docs/mpi-20.ps
15. LAM/MPI Parallel Computing; http://www.mpi.nd.edu/lam/

Task Farm Computations in Java*

M. Danelutto

Department of Computer Science
University of Pisa, Italy
http://www.di.unipi.it/~marcod

Abstract. We describe an experiment in the development of an efficient Java support for task farm computations. The support allows Java programmers to rapidly develop parallel task farm applications starting from the plain sequential code. The target architecture we considered during the development of the support is a cluster of Unix workstations. We show experimental results that demonstrate the feasibility of the approach and we discuss the performance of this Java task farm support used on a typical workstation cluster. The task farm support discussed here is the first step towards the implementation of a full skeleton based parallel programming environment in Java.
Keywords: Task farm, cluster computing, skeletons, load balancing.

1 Introduction

Recent works on parallel programming environments demonstrated that efficient parallel applications can be developed following a "structured parallel programming model" based on the skeleton concept [19, 22, 7]. A skeleton is nothing but a known, efficient and parametric parallelism exploitation pattern that can be instantiated to derive different parallel applications [17]. Among the skeletons usually considered in these frameworks we found task farm and pipeline skeletons [4, 3, 9]. These parallelism exploitation patterns model parallel computations that employ a set of processes to compute independent tasks appearing onto some kind of input data stream (this is the task farm skeleton) as well as computations that compute a result by using a set of parallel processes each computing a "stage" (or phase) of the overall computation leading to the final result (this is the pipeline skeleton).

Despite the amount of positive features of the Java language (code portability, native concurrent and network programming features, object oriented and reflection features, etc.) there is no universal consense on the fact that Java can be used to program efficient parallel computations. This is mainly due to a couple of factors: the (relatively) poor performance achieved by Java byte-code and the minor amount of optimizations exploited in the execution of Java programs due to the machine independent features of the Java Virtual Machine. The recent availability of "Just In Time" compilers for Java [18] superseded some of

---

* This work has been partially funded by the Italian projects MOSAICO and PQE2000

these problems. A lot of work is being performed on "parallel Java", however. A whole set of different projects using or enhancing Java features to implement different parallel programming models is discussed in [13]. Some of the projects aim at embedding existing parallel processing environments such as MPI [16] or PVM [21] in the Java language. Other projects provide classes or language extensions that make popular parallel programming models such as HPF [11,6] available to the Java community.

To our knowledge, however, there are no projects trying to provide Java programmers with some skeleton parallel programming features. Therefore we started thinking at a full embedding of a skeleton programming model in Java. We wanted to exploit all the concurrent (parallel) and network features of Java in order to implement a support that allowed the Java programmer to develop parallel applications using a skeleton approach. In particular, we wanted to exploit positive network features of Java to implement an efficient support for cluster of workstations (COW) parallel machines [23,5]. These kind of parallel architectures are very popular at the moment, as they promise to deliver performances comparable to the ones delivered by Massively Parallel Architectures (MPP) at a fraction of the cost.

In this work, we discuss the design and the implementation of a Java runtime support enabling programmers to rapidly and efficiently develop parallel task farm applications in Java. The support is based on skeletons, in that the programmer is allowed to structure the parallel aspects of his application by using a proper skeleton choice. The design of our task farm support is actually the first step towards the design of a complete skeleton based parallel programming environment in Java (actually, our support already allows both task farms and pipelines to be used). Our work has many contact points with other parallelism related activity on Java. In particular, [14] discusses an approach which is similar to ours in the implementation techniques used, but aims at parallelizing shared memory, multi-threaded Java programs, rather that programs containing explicit parallel constructs (skeletons). [8] also presents a framework for parallel programming in Java which is close to ours. However, the framework is intended to provide support to higher level parallel programming models and, furthermore, because it considers plain Internet as the parallel target architecture, it deals with detail, such as security, that we do not take into account in our approach. Finally, [12] shows how complex parallel applications can be developed on COW machines, taking advantage of implicit RMI mechanism similar to the ones we used in our task farm support. Overall, the key point that distinguishes our work from these others is the usage of the skeleton abstraction to model parallelism in Java.

In the rest of the paper, we briefly discuss task farm computations (Sec. 2), then we describe an implementation of a support for task farm computations in Java (Sec. 3) and eventually we discuss the results achieved with the prototype implementation of our skeleton support on a small COW we have recently installed at our Department (Sec. 4).

2 Task Farm Computations

A task farm processes a stream of input data items, the *tasks*, in order to produce a stream of output data items, the *results*. Each result is computed out of a task in a completely independent way: neither interaction with the processes computing the other results is required, nor any knowledge of the other tasks is required to compute the result out of the task data. In a task farm computation, parallelism is exploited by using a set of homogeneous parallel processes (the *workers*) such that each worker process can compute a result out of a task. The cardinality of the worker process set is the parallelism degree of the task farm. An *emitter* process will be used to schedule tasks to the workers, by using a scheduling policy that guarantees the load balancing, and a *collector* process will be used to gather results from the workers. Provided that the emitter or collector processes do not become bottlenecks, this process structure efficiently implements a task farm. The task farm is able to process a new task every T_w/p, in theory, where T_w is the time spent in the computation of a single task and p is the parallelism degree. Actually, because of the times spent in transmitting tasks and results between emitter, worker and collector processes, this task farm implementation will deliver a slightly worst performance. Task farm computations have been often labelled as "embarrassingly parallel" [23]. Each task can be computed independently of the other ones and therefore clever implementations can be devised that achieve very high efficiency without requiring consistent programming efforts. Despite the embarrassingly parallel nature of task farm computations, many real problems can be modeled after this parallelism exploitation pattern. All the applications processing an high number of input data items (e.g. medical image post-processing applications) fit the task farm pattern, as well as simple applications partitioning each input data set into independent sub-tasks that produce results that must be recombined in order to get the final result. Therefore, the task farm parallelism exploitation pattern looks like to be a valid candidate to demonstrate the efficiency of some parallelism exploitation mechanisms: on the one side it is simple and clear to understand, on the other side it is a useful parallelism exploitation pattern.

3 Java Task Farm Implementation

We implemented task parallel computations in Java adopting a "macro dataflow" run time support, such as the one we designed for skeletons in [10]. According to this schema, a centralized pool of fireable macro dataflow instructions (tasks to be computed) is maintained, which is accessed by a set of independent processes that hold all of the code necessary to compute a single task. These processes access the centralized task pool to get a task, compute the task and deliver the result to the task pool, in a loop. Tasks are inserted in the task pool as soon as they are available on the input data stream and results are delivered to the output data stream as soon as they are deposited in the task pool. Despite the centralization of the task pool, these implementation schema for structured parallel computations has already been demonstrated effective [10].

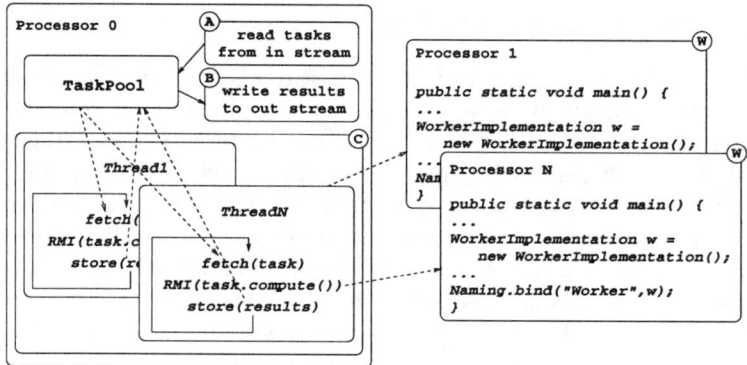

Fig. 1. Prototype task farm implementation schema

We implemented a `TaskPool` object that can hold and deliver, upon request, `Task` objects. A `Task` object is nothing but an object with a particular `void compute(void)` method which is to be provided by the user. A `Task` object can be created by providing the task input data to the `Task` constructor. The `compute` method can be called to compute the result out of the task data.

Using these objects, we set up a prototype task farm support as follows (see Figure 1). On the `root` machine (say processor 0 in the COW) three distinct processes are run. A dedicated process (marked as A in the Figure) gets input data sets from the input stream and inserts `Task` objects in the `TaskPool`; another dedicated process (B) gets computed `Tasks` (results) from the `TaskPool` and delivers them onto the output stream. [1] Finally, a task farm `main` process (C) sets up a number of threads, one for each one of the processors in the COW. Each thread, in a loop, fetches a `Task` to be computed from the `TaskPool` (the access to the `TaskPool` object is `synchronized`), computes the `Task` by invoking the `Task compute` method using RMI on one of the processors in the COW, and eventually stores the results in the `TaskPool`. As there is exactly one thread "bounded" to each one of the processors in the COW, this mechanism guarantees load balancing. The processors in the COW, in fact, behave as "task servers" and the threads deliver tasks to be computed only when these servers are ready to compute the task. On the other machines in the COW, a process (W) is run implementing the worker process in such a way that this process can accept `compute` method invocation on a task via RMI.

With such kind of support, tasks are scheduled "on demand" to the processing elements in the COW. Each PE receives a task to be computed as soon it turns out to be idle. Each PE receives the first task from the associate thread in the `main` program as soon it is included in the pool of available PEs or as soon as it has delivered the result of a previously assigned task.

---

[1] In our prototype implementation of the task farm, data streams are implemented by disk files.

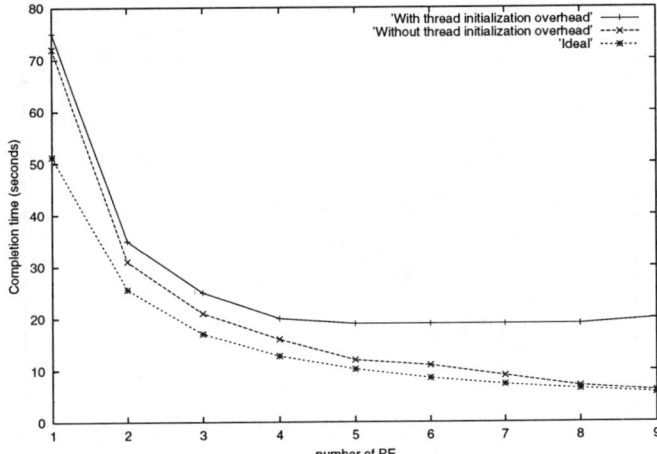

Fig. 2. Overhead in thread initialization

With this kind of support, in order to run a task farm, the user must simply supply the proper Task class and run the task farm program onto the COW.

4 Experimental Results

We run a set of experiments in order to check the feasibility of our approach. The target architecture we used is a plain workstation cluster (Beowulf class [1]). In particular, we used Backus, a dedicated Linux PC cluster with 10 233Mhz Pentium II nodes interconnected by means of a switched Fast Ethernet network, which has been installed at our Department in the framework of the Italian MURST project MOSAICO. The cluster is "dedicated" in that no other computations but our parallel ones where running on the PCs during the experiments.

In order to perform our experiments we set up a set of synthetic task farm applications. The worker code of such applications was made up of simple loops just consuming a fixed amount of processor time. Such applications allowed to achieve precise measures without incurring in the necessity of developing different code to test different aspects of the support.

First of all, we estimated the overhead involved in thread setup within our code. Figure 2 shows the typical completion times measured on our COW. The times are related to the computation of 1K task, each taking 50 msecs to compute and requiring/delivering less than 1.5 Kbyte of data (i.e. a quantity of data, including serialization data needed by RMI, less than the local network MTU). The upper curve refers to raw completion times, including the thread initialization code, while the middle curve refers to the completion times purged of the thread initialization overhead. It is clear that in this case the time spent in setting up the threads used to implement the worker self-scheduling is high.

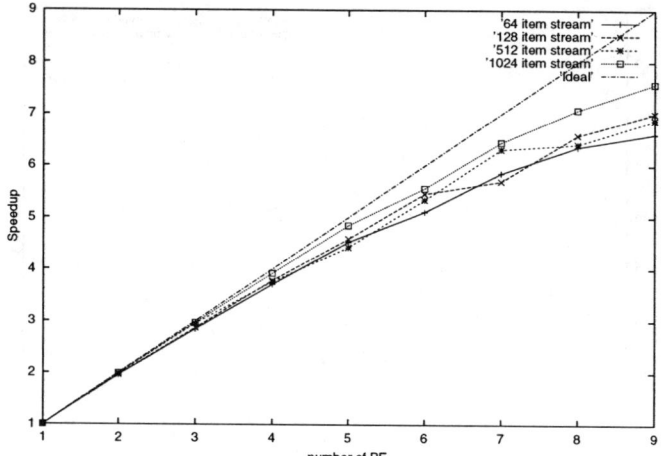

Fig. 3. Impact of stream length

However, if we do not take into account thread initialization overhead, the farm scales well with times that are close to the ideal ones.[2] As the initialization overhead is to be payed just once and for all when the task farm main process is run, it is correct to measure the times without taking into account the initialization time. In the rest of the paper, all the times displayed in the pictures do not include such initial overhead.

In task farm computations one can expect that the more tasks have to be computed the better speedup is achieved. We measured the completion times relative to the computation of task farms with a different number of input tasks (Figure 3) and we observed that the farm behaves as expected.

We also took into account the impact of the computation to communication ratio with respect to scalability. Due to the communications needed to propagate tasks to be computed to the worker processors, we expected that efficiency is higher in case the computation to communication ratio is higher. Figure 4 shows that the prototype code achieves good efficiency (90% or more) when computation to communication ratio is beyond 300, i.e. when the time spent in communicating the task input data and the task results is 300 times lower than the time spent in the task computation. Therefore we can argue that our task farm support is suitable for medium to coarse grain computations only. This result is not surprising, as other studies related to the use of Java for parallel WEB computing already demonstrated that network latencies make fine grain parallel computations unfeasible [20].

The results we achieved with the prototype implementation demonstrate to be quite good. However, we also tried to measure the effectiveness of our task

---

[2] The ideal completion time is given by the product of the time spent in the execution of a single task by the number of tasks divided by the number of worker processes.

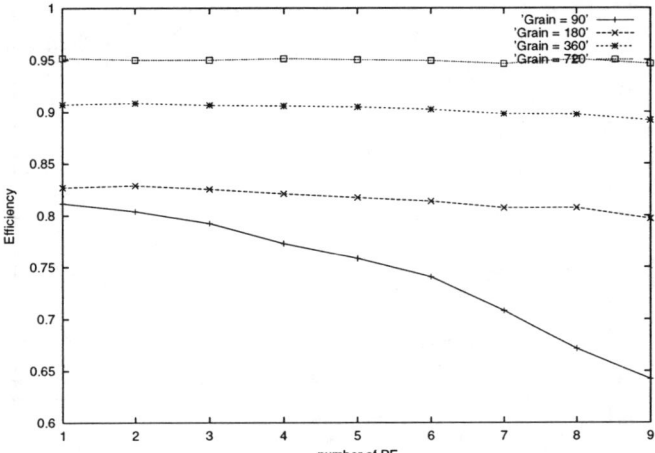

Fig. 4. Computational grain effect

farm support, with respect to both the implementation language chosen and the mechanisms used. Therefore we wrote two different versions of the support: a version entirely implemented in C and using the standard Unix TCP Sockets, and a version written again in Java but using plain TCP sockets to propagate tasks and results to a from the worker processors. Then, we measured the completion times achieved when computing a stream of tasks with the three implementations of the task farm support (see Figure 5): C task farm support is the most efficient, and the measured completion times are definitely very close to the ideal ones. TCP Socket implementation in Java is the worst one. The completion times measured in this case almost double the ideal completion times. Actually, the experiments have been performed with jdk 1.1 because our COW software configuration does not allow to run jdk 1.2 at the moment. Moving to jdk 1.2 could lower the gap between the RMI and the TCP socket implementation of the support due to the better implementation of socket mechanisms in jdk 1.2. As the Java run time support actually uses sockets to implement RMI, we argue that our Java socket implementation of the support performs poorly as we did not spent any time in trying to exploit all the options that can be used to make TCP/IP socket communication efficient. As an example, we did not use BufferedStreams over the socket streams. Furthermore, we expect that Java RMI support makes an efficient usage of the UDP/IP socket layer, which is implicitly faster that the TCP/IP layer we used in the implementation of our Java socket task farm support. Finally, the Java version of the support using RMI demonstrates to be a 15% to 20% away from the C version in terms of performance. This may be due to the overheads proper of RMI and object serialization mechanisms [15]. However, as the main goal of our work was to demonstrate the feasibility of the approach while preserving all the positive

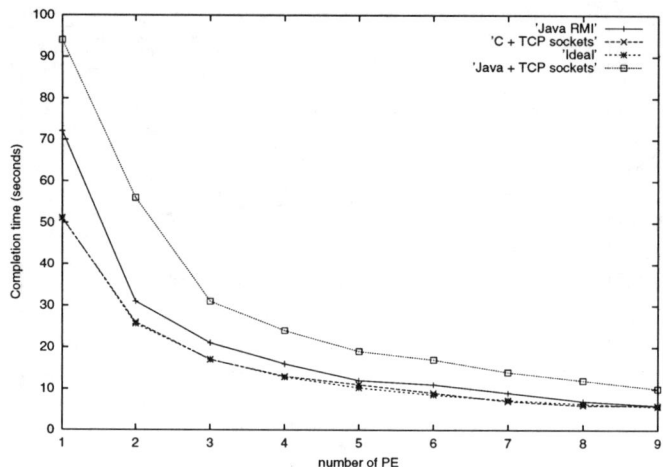

Fig. 5. Comparison between different implementations of the task farm support

features of the Java programming environment, this has to be taken as a quite good result.

All the results discussed until now have been obtained by running our Java task farm support on a dedicated COW. However, we also run the same experiments on a network of workstations (NOW) currently used by people working at our department. The processing elements of the NOW range from very simple Pentium to more sophisticated Dual Pentium III processing elements, all running the Linux operating system. On the NOW, we obtained results in terms of both scalability and performance that are comparable with the ones achieved on the COW. This demonstrates that the load balancing algorithm implemented in the task farm support is effective [3] (Figure 6 plots the tasks computed on each processing element when running a task farm computing 1K task on 7, 6 and 5 workstations of the NOW. Workstation **PE4** is the busiest one and actually participates in the computation computing the smaller number of tasks).

Despite the fact all the experiments have been performed on Pentium based COWs/NOWs, the support is perfectly suited to run task farm computations on clusters/networks hosting different processing elements (e.g. SPARCS or Alpha based machines, possibly with different operating systems than Linux). Our task form support exploits basic Java mechanisms in order to schedule computations over the COW PEs, fully relying on Java code portability across different architectures.

---

[3] the processing elements of the NOW were used, during the experiments, by people accessing Internet with WEB browsers, sending and receiving email, compiling and running other programs, editing and compiling LaTeX files, etc.

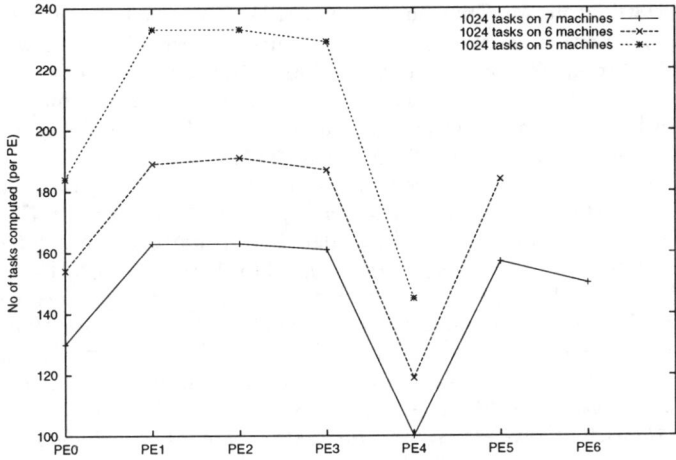

Fig. 6. Load balancing on the network of (busy) workstations

5 Conclusions and Future Work

We developed a task farm support in Java that can be used to program parallel applications exploiting this particular kind of parallelism exploitation pattern. The support has been tested on a cluster of Linux PCs interconnected by means of a switched, Fast Ethernet network. The experimental results demonstrated that the approach is feasible in that both scalability and efficiency has been achieved when running medium to coarse grain parallel computations. At the moment we are working to include in the support also the support for pipeline computations. Actually, we have already slightly modified the user interface of the support, in such a way the user can ask the parallel execution of skeleton program using arbitrary nestings of pipeline and farm skeletons. [4] The experiments performed demonstrated that scalability and performance results are achieved also in this case. We are also working in order to make full usage of the dynamic class loading features of Java in the implementation of the "worker servers" that at the moment have to be started by using a small shell script on the COW processing elements.

References

1. The Beowulf Project . http://www.beowulf.org.
2. M. Aldinucci and M. Danelutto. Stream parallel skeleton optimisations, *Proc. of the Eleventh IASTED International Conference Parallel and Distributed Computing and Systems*, pages 955–962, November 1999. MIT Boston, USA.

---

[4] This "extended" support has the same structure described in Sec. 3, but we use some known stream parallel skeleton optimisations [2] to minimize the communication overheads.

3. P. Au, J. Darlington, M. Ghanem, Y. Guo, H.W. To, and J. Yang. Co-ordinating heterogeneous parallel computation. In L. Bouge, P. Fraigniaud, A. Mignotte, and Y. Robert, editors, *Europar '96*, pages 601–614. Springer-Verlag, 1996.
4. B. Bacci, M. Danelutto, S. Orlando, S. Pelagatti, and M. Vanneschi. P^{3}L: A Structured High level programming language and its structured support. *Concurrency Practice and Experience*, 7(3):225–255, May 1995.
5. M. Baker and R. Buyya. Cluster Computing at a Glance. In Rajkumar Buyya, editor, *High Performance Cluster Computing*, pages 3–47. Prentice Hall, 1999.
6. S. Benkner and H. Zima. Compiling High Performance Fortran for distributed memory architectures. *Parallel Computing*, 25(13–14):1785–1825, December 1999.
7. George Horatiu Botorog and Herbert Kuchen. Efficient high-level parallel programming. *Theoretical Computer Science*, 196(1–2):71–107, April 1998.
8. B. O. Christiansen, P. Cappello, M. F. Ionescu, M. O. Neary, K. E. Schauser, and D. Wu. Javelin: Internet-based parallel computing using Java. *Concurrency: Practice and Experience*, 9(11):1139–1160, November 1997.
9. M. Cole. *Algorithmic Skeletons: Structured Management of Parallel Computations*. Research Monographs in Parallel and Distributed Computing. Pitman, 1989.
10. M. Danelutto. Dynamic Run Time Support for Skeletons. In *Proceedings of the PARCO'99 Conference*, August 1999. Delft, The Netherlands.
11. High Performance FORTRAN Forum. High Performance FORTRAN Language Specification. Technical Report Rice University, Houston, TX, January 1997.
12. M. Gimbel, M. Philippsen, B. Haumacher, P. C. Lockemann, and W. F. Tichy. Java ad a Basis for Parallel Data Mining in Worlstation Clusters. In P. Sloot, M. Bubak, A. Hoekstra, and B. Hertzberger, editors, *Proceedings of HPCN'99 Conference*, pages 884–894. Springer Verlag, 1999. LNCS No. 1593.
13. D. C. Hyde. Java and Different Flavors of Parallel Programming Models. In Rajkumar Buyya, editor, *High Performance Cluster Computing*, pages 274–290. Prentice Hall, 1999.
14. P. Launay and J. L. Pazat. A Framework for Parallel Programming in Java. In P. Sloot, M. Bubak, and B. Hertzberger, editors, *Proceedings of the HPCN'98 Conference*, pages 628–637. Springer Verlag, LNCS, 1998.
15. M. Migliardi and V. Sunderam. Networking Performance for Metacomputing in Java. In *Proc. of the IASTED International Conference on Parallel and Distributed Computing and Systems*, pages 220–225, November 1999. MIT Boston, U.S.A.
16. P. S. Pacheco. *Parallel Programming with MPI*. Morgan Kaufmann, 1997.
17. S. Pelagatti. *Structured Development of Parallel Programs*. Taylor & Francis, 1998.
18. M. P. Pletzbert and R. K. Cytron. Does "Just in Time" = "Better Late than Never". In *Proceedings of the POPL'97: The 24th ACM SIGPLAN-SIGACT Symp. on Principles of Programming Languages*, pages 120–131, 1997.
19. J. Serot, D. Ginhac, and J.P. Derutin. SKiPPER: A Skeleton-Based Parallel Programming Environment for Real-Time Image Processing Applications. In *Proceedings of the 5th International Parallel Computing Technologies Conference (PaCT-99)*, September 1999.
20. L. M. Silva. Web-Based Parallel Computing with Java. In Rajkumar Buyya, editor, *High Performance Cluster Computing*, pages 310–326. Prentice Hall, 1999.
21. V. S. Sunderam and G. A. Geist. Heterogeneous parallel and distributed computing. *Parallel Computing*, 25(13–14):1699–1721, December 1999.
22. M. Vanneschi. PQE2000: HPC tools for industrial applications. *IEEE Concurrency*, 6(4):68–73, 1998.
23. B. Wilkinson and M. Allen. *Parallel Programming. Techniques and Applications Using Networked Workstations and Parallel Computers*. Prentice Hall, 1999.

Simulating Job Scheduling for Clusters of Workstations

J. Santoso[1,2], G.D. van Albada[1], B.A.A. Nazief[3], and P.M.A. Sloot[1]

[1] Department of Computer Science, Universiteit van Amsterdam
Kruislaan 403, 1098 SJ Amsterdam, The Netherlands
{judhi,dick,sloot}@wins.uva.nl
[2] Department of Informatics, Bandung Institute of Technology
Jl. Ganesha 10, Bandung 40132, Indonesia
[3] Department of Computer Science, University of Indonesia
Jakarta, Indonesia
nazief@cs.ui.ac.id

Abstract. In this paper we study hierarchical job scheduling strategies for clusters of workstations. Our approach uses two-level scheduling: global scheduling and local scheduling. The local scheduler refines the scheduling decisions made by the global scheduler, taking into account the most recent information. In this paper, we explore the First Come First Served (FCFS), the Shortest Job First (SJF), and the First Fit (FF) policies at the global level and the local level. In addition, we use separate queues at the global level for arriving jobs, where the jobs with the same number of tasks are placed in one queue. At both levels, the schedulers strive to maintain a good load balance. The unit of load balancing at the global level is the job consisting of one or more parallel tasks; at the local level it is the task.

1 Introduction

With the advent of large computing power in workstations and high speed networks, the high-performance computing community is moving from the use of massively parallel processors (MPPs) to cost effective clusters of workstations. Furthermore, integrating a set of clusters of workstations into one large computing environment can improve the availability of computing power, e.g Globus (Czajkowski et al. [3]), andthe Polder Metacomputing Initiative [9]. However, to efficiently exploit the available resources, an appropriate resource management or scheduling system is needed. Scheduling systems for large environments can be divided into two levels: scheduling across clusters (*wide area scheduling*) and scheduling within a cluster. The wide area scheduler queries the local schedulers to obtain the candidate resources to allocate the submitted jobs [11].

Scheduling systems within a single cluster of workstations have been described in the literature [1], [5], [7], [10], [12]. Various strategies are used to achieve the scheduling objectives. For example two-level time-sharing scheduling for parallel and sequential jobs has been introduced by Zhou et. al [12] to attain a good performance for parallel jobs and a short turn-around time for sequential jobs. At the upper level, time slots are used. Each time slot is divided into 2 slots (one for sequential jobs and a second for tasks of parallel jobs). However at the local level a task of a parallel job may share its time slot with sequential jobs. K.Y Wang et al. [10] use hierarchical decision scheduling, implemented using global and local schedulers. The global scheduler is responsible for long term allocation of system resources, and the local scheduler is responsible for short term decisions concerning processor allocation.

In this paper we use two level scheduling in a different way: global scheduling across clusters and local scheduling within a cluster. The global scheduler has a number of functions. One of these is the matching of the resources requested by a job to those available in the participating clusters. Another is to obtain the best utilisation of the available clusters. The local scheduler is responsible for scheduling jobs to a specific resource. The global scheduler and local schedulers have various scheduling policies that can be used independently. To observe our scheduling strategies' performance, we have developed a simulation environment. By using this environment, the influence of scheduling strategies on the overall performance will be shown.

In this paper, we will not be concerned so much with the architecture and implementation of the scheduling system, but rather with the scheduling strategies that can be used to attain the best possible performance in our environment. We will specifically address the prioritisation of parallel jobs at the global scheduling level. To support this, we use multiple queues at the global level.

This paper is organised as follows: In section 2 we discuss the the relevant aspects of our scheduling model as well as several scheduling policies that we will use. In section 3 we explain our simulation environment, including job and resource descriptions. The simulation results are presented in section 4, after which we present our conclusions.

2 The Scheduling Model

The scheduling model is based on a hierarchical approach, Figure 1. We distinguish two levels of scheduling: scheduling at the top level (global scheduling) and scheduling at the local level. The global and local scheduler frequently interact with each other to make an optimal schedule. The local scheduler will refine the scheduling decisions made by global scheduler to adapt to changes in resource availability.

2.1 The Global Scheduler

The global scheduler schedules the arriving jobs to the available clusters. Once a job has been scheduled by the global scheduler, it is passed on to the local cluster scheduler. For low loads, the global scheduler's task is to find the best offer, e.g. in terms of performance or cost. Jobs are handled on an individual

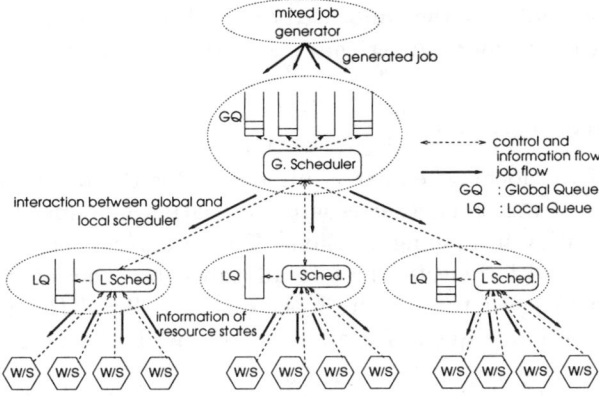

Fig. 1. Global and Local Scheduler interaction

basis, and no global queue needs to be maintained. For high system loads (typical for high-throughput computing), the scheduler should also try to maximise the system utilisation and reduce the over-all turn-around time. To this end, the scheduler should know the available capacity in each participating cluster. As, due to locally submitted work, this capacity may vary in an unpredictable manner, it may be attractive to delay scheduling until the needed capacity actually becomes available. We assume that jobs delayed in this way are maintained in a global scheduling queue. We have studied two distinct queuing strategies at the global level: a single global queue and separate global queues. In the single queue model, one of three scheduling strategies is used to assign jobs to clusters: First Come First Served (FCFS), Shortest Job First (SJF) and First Fit (FF) policies. FF policy maintains a queue in FCFS order, but searches that queue for the first job that can be scheduled on the available resources. SJF and FCFS wait until the first job in the queue can be scheduled.

When using separate queues, each type of job (distinguished by the number of tasks), is placed in a separate queue. The number of queues is equal to the number of job types. Dispatching a job from the queue uses the SJF policy and starts either from the queue of jobs with the smallest number of tasks or, alternatively, from the queue with the largest number of tasks. We call the first strategy the S-SJF policy and the second strategy, L-SJF.

The global scheduler interacts with the local scheduler to make the best schedule. The interaction between the local and global scheduler is different when a local queue is used. When no local queue is used, the global scheduler will match the job's resource request to the currently available local resources. When a local queue is used, the maximum capacity of the cluster as well as the current and maximum queue length are considered.

In all cases the scheduling is done by sorting the clusters in order of decreasing capacity (free slots on nodes when no local queue is used, free slots in the local queue otherwise). Jobs are then allocated to the cluster with the largest capacity, until it can accept no more. After that the next cluster is filled.

2.2 The Local Scheduler

The role of the local scheduler is important for the overall system performance. It tries to maximise the resource utilisation by using the most recent resource information in making decisions. The resource (workstation) with the largest capacity will be selected first to allocate jobs.. One of three scheduling policies is used to select the jobs from the queue: FCFS, SJF or FF. The local scheduler uses only one queue for all types of job. The maximum local queue size is limited.

The local scheduler also performs task scheduling for parallel jobs. It will schedule only a limited number of tasks on a specific node. Tasks are assigned to available nodes in a Round-Robin fashion starting from the most lightly loaded node.

2.3 Task Dispatching

The runnable tasks on each node are dispatched in a Round-Robin fashion. As the tasks in a parallel program are assumed to synchronise after every time step. Tasks block at each synchronisation point and become runnable again only when all other tasks in the job have advanced to the same synchronisation point. T he communication itself is assumed to be instantaneous.

3 The Simulation Environment

In this section, we describe the components of our simulation model and its implementation aspects, as well as statistical characteristics of some simulation variables. Our simulation model uses a discrete-event approach.

3.1 Components of the Model

We define six basic components to build our simulation environment. Each component has various attributes to identify its state and several functions / methods to modify that state as well as to interact with the other components. Every component has an object association with one or more object instances. For example the generator component and the global scheduler have only one object instance. The components of the simulator are listed below.

3.2 Implementation

Our simulation environment was developed using the simulation package C++SIM [2]. We use two additional random number generators to enhance the quality of randomness for some independent variables. The first generator *job*   : The job object contains information about the attributes that describe the characteristics of a job.

resource : The resource object contains information about machine attributes and several methods that represent the machine capabilities.

queuing : The queuing object is responsible for managing the jobs in a queue. Each queuing system is represented by a single instance.

generator: The generator object generates jobs of various types. There is only one generator object instance in the system.

g-sched : The global scheduler object is responsible for scheduling the newly arriving jobs. There is one g-sched object instance in the system.

l-sched : The local scheduler object schedules jobs received from the global scheduler. There is one l-sched object instance in each cluster.

is the *drand48()* generator (standard C library), and the second one is *mtGen* generator from Matsumoto [6]. *Drand48()* is used to generate the random variates from two-stage hyper-exponential distribution used for the execution time of the job. *MtGen* is used to generate the random variates from the exponential distribution used for the inter-arrival time of jobs. To generate the job types, we use the internal generator of C++SIM.

3.3 Configuration

To use the simulation environment, we need to set the system configuration. For this purpose we have to model the resource, job structure, and the arrival processes. Furthermore, a set of performance criteria is needed. As the total number of parameters in the system already is quite large (global scheduling model, length of local queue, arrival rate, and local scheduling policy), we decided to keep the remainder of the model as simple as possible.

The resources consist of four identical clusters, and each cluster has four identical nodes (workstations). Every cluster is running a local scheduler and each node belonging to a cluster has a CPU scheduler. The system model is shown in Figure 1.

We distinguish two basic types of jobs, sequential jobs and parallel jobs. Sequential jobs consist of a single task. Parallel jobs can have between two and four identical tasks. For parallel jobs, we use a synchronised model where each task is divided into sections taking one unit of CPU time. After each section, all tasks in the job synchronise.

The arrival process generates jobs of each type (sequential or two, three or four parallel tasks) together with their attributes (execution time and number of tasks) in a certain proportion. For each type, a separate hyper-exponential distribution is used for the execution time. As experience shows that massively parallel jobs tend to run longer [4], we have chosen the distributions accordingly.

Table 1 shows the parameters of the job generator. The amount of work in a job is the product of the number of tasks and the execution time. As the

Table 1. Distribution of generated jobs

Number of tasks	Job fraction	Mean exec.	Std.Dev	Avg. Work
1	0.7	4	5	2.8
2	0.1	4	5	0.8
3	0.1	8	10	2.4
4	0.1	16	18	6.4

simulated system contains 16 nodes, 16 units of work can be processed per unit time. This corresponds to a minimum average job inter-arrival time of 0.775 time units. An important simplifying assumption is that the actual CPU time needed is available to the scheduler.

4 Simulation Results

In this section, we show our simulation results of the selected measurements with a brief discussion. A more comprehensive discussion of all the results is given in the next section.

4.1 Measurement Strategy

We have used the following parameters as the variables in our simulations :

- *Global queue model* : separate queue or single global queue.
- *Global scheduling policy*: FCFS, SJF, or FF (when a single queue is used; SJF otherwise.)
- *Local queue size* : (zero, four, and eight)
- *Local scheduling policy* : FCFS, SJF, or FF (when a local queue is used)
- *Job inter-arrival time* : from 0.8 to 1.5 time unit

We do not use all parameter combinations. By re-seeding the random generators, we made 16 independent runs of 4000 jobs each, with a start-up period of 200 jobs for each parameter combination. Except for their spacing in time, the same set of 16 times 4200 jobs was used for each parameter setting. System utilisation and run time were measured from the time the 201st job was submitted to the moment that the 4200nd job was finished. Performance statistics for each job were obtained at completion for the 201st job up to the 4200nd job. Job generation continued until the end of the simulation. As the scheduling system influences the order in which jobs are executed, there are differences in the population sampled.

For each run we have measured the average turn-around time for all types of (completed) jobs together and for each type of job separately, and the standard deviation of the turn-around times. We also measured the overall maximum turn-around time, the system utilisation, the number of jobs remaining in the global queue at the end, and the average of the slowdown factor (defined as the ratio between the response time of a job run on the shared resources and that on the dedicated resource).

We present the measured values for the performance estimators for runs with mean jobs inter-arrival times of 0.8, 0.9, 1.1, 1.3 and 1.5 time units, corresponding to loads of about 97 %, 86 %, 71 %, 60 %, and 52 %, respectively.

4.2 Turn-Around Times

The average turn-around time is a good indicator for the level of service provided by a system. We have measured the average turn-around times for each type of job (sequential, and two to four parallel tasks) and the average over all types.

Tables 2 through 4 show the results for the average turn-around time for all jobs, that for sequential jobs and that for four task jobs. The tables show that, for the highest arrival rate (inter-arrival time 0.8 time units), the smallest queue first policy (S-SJF) with no local queue gives the shortest turn-around times. When a local queue is used, the SJF policy has the shortest turn-around time, and FF appears to be the next best, but its bias against large parallel jobs results in longer turn-around times for those jobs.

Table 2. Reference model turn-around times T as a function of inter-arrival time. The errors given are the r.m.s. errors is the averages as derived from the spread in the computed averages for each of the 16 independent runs. They have not been computed from the internal values in each run, as those values may be highly correlated.

Global policy	Local policy	Local queue	0.8 T	error	0.9 T	error	1.1 T	error	1.3 T	error	1.5 T	error
S-SJF		0	24.3	0.5	18.2	0.6	11.6	0.2	9.1	0.2	7.9	0.1
S-SJF	FCFS	4	36.9	1.5	22	1	12.0	0.3	9.1	0.2	7.9	0.1
S-SJF	FCFS	8	46.7	2.3	25.0	1.7	12.1	0.3	9.1	0.2	7.9	0.1
S-SJF	SJF	4	26.5	0.8	18.8	0.7	11.7	0.2	9.1	0.2	7.9	0.1
S-SJF	SJF	8	26.2	0.7	18.7	0.7	11.6	0.2	9.1	0.2	7.9	0.1
S-SJF	FF	4	34.3	1.5	20	1	11.7	0.2	9.1	0.2	7.9	0.1
S-SJF	FF	8	41.2	2.5	22	1	11.8	0.2	9.1	0.2	7.9	0.1
L-SJF		0	53.7	5.6	22.2	1.6	11.6	0.2	9.1	0.2	7.9	0.1
L-SJF	FCFS	4	89	10	30	3	12.0	0.3	9.1	0.2	7.9	0.1
L-SJF	FCFS	8	90	10	29	3	12.1	0.3	9.1	0.2	7.9	0.1
L-SJF	SJF	4	31	2	18.9	0.7	11.7	0.2	9.1	0.2	7.9	0.1
L-SJF	SJF	8	26.2	0.7	18.7	0.7	11.6	0.2	9.1	0.2	7.9	0.1
L-SJF	FF	4	71	8.9	24	2	11.8	0.2	9.1	0.2	7.9	0.1
L-SJF	FF	8	60	6.6	22.7	1.8	11.8	0.2	9.1	0.2	7.9	0.1
G-FCFS		0	100	12	28.8	2.8	11.9	0.3	9.1	0.2	7.9	0.1
G-SJF		0	25.0	0.5	18.7	0.7	11.7	0.2	9.1	0.2	7.9	0.1
G-FF		0	87	11	25.1	2.3	11.8	0.3	9.1	0.2	7.9	0.1

4.3 Standard Deviation and Slowdown

The average turn-around times are not a good indicator for the ways the scheduling delays are distributed over the jobs. The standard deviation of the

Table 3. Reference model turn-around times T_s for sequential jobs as a function of inter-arrival time. Errors as in Table 2.

Global policy	Local policy	Local queue	0.8 T_s	error	0.9 T_s	error	1.1 T_s	error	1.3 T_s	error	1.5 T_s	error
S-SJF		0	13.3	0.2	10.5	0.3	6.8	0.13	5.42	0.07	4.86	0.05
S-SJF	FCFS	4	23.4	0.8	14.5	0.8	7.1	0.19	5.46	0.08	4.85	0.05
S-SJF	FCFS	8	32	1	17	1	7.2	0.22	5.45	0.08	4.86	0.05
S-SJF	SJF	4	14.9	0.3	11.0	0.3	6.8	0.12	5.42	0.07	4.85	0.04
S-SJF	SJF	8	14.4	0.3	10.8	0.3	6.7	0.13	5.42	0.07	4.85	0.04
S-SJF	FF	4	20.0	0.6	12.7	0.6	6.9	0.15	5.43	0.07	4.85	0.04
S-SJF	FF	8	26	1	14.1	0.9	6.9	0.16	5.42	0.07	4.85	0.04
L-SJF		0	57	7	16.5	1.9	6.8	0.15	5.44	0.07	4.86	0.05
L-SJF	FCFS	4	105	14	26.2	3.8	7.2	0.21	5.46	0.08	4.85	0.05
L-SJF	FCFS	8	102	14	23.8	3.3	7.2	0.23	5.45	0.08	4.86	0.05
L-SJF	SJF	4	23.0	2.8	11.1	0.4	6.8	0.12	5.42	0.07	4.85	0.04
L-SJF	SJF	8	14.4	0.3	10.8	0.3	6.7	0.13	5.42	0.07	4.85	0.04
L-SJF	FF	4	78	2	18.2	2.5	6.9	0.16	5.43	0.07	4.85	0.04
L-SJF	FF	8	58	8	15.3	1.7	6.9	0.16	5.42	0.07	4.85	0.04
G-FCFS		0	93	12	22.4	2.7	7.1	0.21	5.45	0.07	4.85	0.05
G-SJF		0	14.4	0.2	11.0	0.3	6.8	0.14	5.42	0.07	4.85	0.05
G-FF		0	79	11	17.8	2.1	6.9	0.17	5.42	0.07	4.85	0.05

Table 4. Reference model turn-around times T_4 for four task jobs as a function of inter-arrival time. Errors as in Table 2.

Global policy	Local policy	Local queue	0.8 T_4	error	0.9 T_4	error	1.1 T_4	error	1.3 T_4	error	1.5 T_4	error
S-SJF		0	100	3.7	66.7	3.1	41.8	0.9	32.1	0.7	27.1	0.5
S-SJF	FCFS	4	120	6.9	68.8	3.9	42.0	0.9	32.0	0.7	27.1	0.5
S-SJF	FCFS	8	131	9.1	69.2	3.9	42.1	1.0	32.1	0.7	27.1	0.5
S-SJF	SJF	4	100	4.4	67.4	3.1	42.0	1.0	31.9	0.7	27.0	0.5
S-SJF	SJF	8	104	4.2	68.5	3.2	41.7	0.9	32.0	0.7	27.0	0.5
S-SJF	FF	4	120	7.5	68.2	3.4	41.6	0.9	32.1	0.7	27.1	0.5
S-SJF	FF	8	130	9.8	68.7	3.3	41.9	0.9	32.1	0.7	27.1	0.5
L-SJF		0	74	1.5	60.9	2.0	41.4	0.9	32.1	0.7	27.1	0.5
L-SJF	FCFS	4	80	1.3	63.1	2.0	42.0	0.9	32.0	0.7	27.1	0.5
L-SJF	FCFS	8	89	2.0	65.8	2.5	42.2	1.0	32.1	0.7	27.1	0.5
L-SJF	SJF	4	96	3.6	66.8	3.2	42.0	1.0	31.9	0.7	27.0	0.5
L-SJF	SJF	8	105	4.5	68.5	3.2	41.7	0.9	32.0	0.7	27.0	0.5
L-SJF	FF	4	89	1.9	65.8	2.4	41.6	0.9	32.1	0.7	27.1	0.5
L-SJF	FF	8	102	3.8	68.0	2.8	41.9	0.9	32.1	0.7	27.1	0.5
G-FCFS		0	143	12	68.4	3.6	41.6	0.9	32.0	0.7	27.1	0.5
G-SJF		0	97	3	66.8	3.2	41.8	1.0	32.1	0.7	27.1	0.5
G-FF		0	137	12	68.3	3.3	41.4	0.9	32.0	0.7	27.1	0.5

turn-around time is a measure that (table 5) increases when jobs experience large differences in delays. This may indicate that the behaviour of the system is unstable, but it can also be due to a strong preference for one type of job above another.

Slowdown (table 6) is a measure for the delay experienced by a job, as compared to its minimal execution time. It tests the common expectation that small jobs should only experience little delay.

Table 6 for the slowdown indicates that for the highest arrival rates, short jobs receive a much better service from the smallest job queue first with SJF and FF. Table 5 shows that, the spread in turn-around times also is somewhat smaller for these policies. For slightly lower arrival rates, these differences readily disappear.

Table 5. Standard deviation σ of the distribution of the turn-around times. The σ values are based on the internal variation in each simulation run, combined with the variation between the runs. The errors again are the estimated r.m.s. errors in the result.

Global policy	Local policy	Local queue	0.8 σ	0.8 error	0.9 σ	0.9 error	1.1 σ	1.1 error	1.3 σ	1.3 error	1.5 σ	1.5 error
S-SJF		0	54	11	35	7	22.7	1.9	17.6	1.68	15	1
S-SJF	FCFS	4	56	13	35	6	22.8	1.9	17.5	1.64	15	1
S-SJF	FCFS	8	60	14	36	5	23.2	2.1	17.6	1.65	15	1
S-SJF	SJF	4	54	12	35	6	22.8	2.0	17.5	1.61	15	1
S-SJF	SJF	8	58	12	36	7	22.6	1.9	17.6	1.64	15	1
S-SJF	FF	4	54	12	34	5	22.7	1.9	17.6	1.70	15	1
S-SJF	FF	8	57	14	35	4	22.8	2.0	17.6	1.69	15	1
L-SJF		0	153	72	49	20	22.6	1.9	17.6	1.61	15	1
L-SJF	FCFS	4	197	89	61	27	23.0	2.0	17.5	1.64	15	1
L-SJF	FCFS	8	183	86	51	20	23.2	2.2	17.6	1.65	15	1
L-SJF	SJF	4	63	19	35	7	22.8	2.0	17.5	1.61	15	1
L-SJF	SJF	8	57	11	36	7	22.6	1.9	17.6	1.64	15	1
L-SJF	FF	4	166	80	46	17	22.7	1.9	17.6	1.70	15	1
L-SJF	FF	8	131	67	38	10	22.8	2.0	17.6	1.69	15	1
G-FCFS		0	85	18	38	6	22.8	1.8	17.5	1.63	15	1
G-SJF		0	56	10	36	8	22.8	2.0	17.5	1.59	15	1
G-FF		0	78	16	36	5	22.6	2.0	17.5	1.61	15	1

Table 6. Reference model slowdown for all jobs as a function of inter-arrival time. Errors as in Table 2.

Global policy	Local policy	Local queue	0.8 s.d.	0.8 error	0.9 s.d.	0.9 error	1.1 s.d.	1.1 error	1.3 s.d.	1.3 error	1.5 s.d.	1.5 error
S-SJF		0	8.4	0.4	5.5	0.3	2.6	0.1	1.8	0.08	1.5	0.06
S-SJF	FCFS	4	42	5	15.0	1.8	3.3	0.3	1.8	0.07	1.5	0.06
S-SJF	FCFS	8	61	5	22	3	3.5	0.4	1.8	0.06	1.5	0.06
S-SJF	SJF	4	13.3	0.9	6.8	0.6	2.7	0.1	1.8	0.09	1.5	0.06
S-SJF	SJF	8	11.3	0.8	6.5	0.5	2.6	0.1	1.7	0.06	1.5	0.06
S-SJF	FF	4	30	2	12.1	1.6	2.9	0.2	1.8	0.09	1.5	0.06
S-SJF	FF	8	44.6	3.9	15.6	2.5	3.0	0.2	1.7	0.06	1.5	0.06
L-SJF		0	16.9	1.3	7.4	0.5	2.5	0.1	1.8	0.08	1.5	0.06
L-SJF	FCFS	4	80	9	22	4	3.4	0.2	1.8	0.07	1.5	0.06
L-SJF	FCFS	8	93	11	28	5	3.7	0.5	1.8	0.06	1.5	0.06
L-SJF	SJF	4	18	2	6.5	0.5	2.7	0.1	1.8	0.09	1.5	0.06
L-SJF	SJF	8	11.7	0.9	6.5	0.5	2.6	0.1	1.7	0.06	1.5	0.06
L-SJF	FF	4	74	22	14	2	2.8	0.2	1.8	0.09	1.5	0.06
L-SJF	FF	8	54.5	4.7	16	3	3.0	0.2	1.7	0.06	1.5	0.06
G-FCFS		0	200	31	33	7	3.4	0.3	1.8	0.06	1.5	0.06
G-SJF		0	9.7	0.6	6.2	0.4	2.7	0.1	1.8	0.07	1.5	0.06
G-FF		0	165	28	23	5	3.0	0.2	1.8	0.09	1.5	0.06

4.4 Throughput

Our measurements of the system utilisation and jobs remaining in the queue (data not shown) indicate that for the highest arrival rate, the processing lags behind the arrival of work. This can be indicative of saturation. We have, therefore, performed additional experiments over 40 000 jobs for all models. For an inter-arrival time of 0.8 time units, the performance measures deteriorate in all cases, indicating that after 4000 jobs, the system has not yet reached its equilibrium state. Yet, average turn-around times, jobs remaining in queue and other indicators for saturation do not appear to grow out of bounds for any of the scheduling policies. At this arrival rate, the FF policy leads to the best system utilisation both at the global and the local levels, closely followed by FCFS. SJF continues to show short average turn around times and small slowdown values.

5 Discussion

In this paper, we have reported on simulation experiments on the effectiveness of various strategies for the global scheduler in our system using single and separate queues. The purpose of splitting the global queue is to observe more closely the effects of the scheduling policies on the performance of a certain type of job.

For mixed jobs table 2 the results indicate that turn-around time for the S-SJF policy (no local queue) is slightly shorter than that obtained for a single global queue with the same policy, G-SJF. For four-task jobs, the L-SJF policy gives a better service than does the S-SJF policy, which on the other hand, gives the best performance for sequential jobs.

In general, the performance of S-SJF is better than L-SJF for high arrival rates. The SJF and FF policies show better performances than FCFS policy almost for all jobs. If a good global policy (S-SJF) is used, the configuration without local queue has a shorter turn-around than that using a local queue. From the experiments, we conclude that the best strategy strongly depends on the desired performance characteristics.

In our paper, we have made a variety of simplifying assumptions that may influence the results.

- We have assumed the job execution time (CPU time) is known by the scheduler when the jobs are submitted to the system. This assumption is also used to calculate the slowdown for each job.
- We have assumed no (variable) background loads for the clusters, and also that the expected work-loads for a job, as used by the scheduler, is a perfect predictor of the actual work-load. Dropping these assumptions should make delayed global scheduling (with a minimal local queue) even more profitable.
- We have assumed a homogeneous system. Dropping this assumption leads to a far more complex scheduling problem, particularly if jobs with inhomogeneous resource requirements are allowed.

6 Conclusions

From our simulation results by using a separate global queue and more scheduling policies at the local level, we conclude that :

- For global queue ordering, the shortest job queue first strategy (S-SJF) gives a better performance than that of the largest job queue first strategy (L-SJF).
- For a good global policy (S-SJF), the performance with local queue is worse than that of without local queue. For a bad global policy (L-SJF), a local queue is advantageous.
- For a single global queue strategy, the SJF policy has the shortest turn-around time, and FF policy gives the highest utilisation.

6.1 Future Research

The simulation environment and the results of these simulation experiments will be used in our research on run-time support systems, including our work on "Dynamite" [8], which supports dynamic load balancing for PVM and MPI jobs in clusters through the migration of tasks.

Acknowledgments

We gratefully acknowledge support for this work by the Dutch Royal Academy of Science (KNAW) under grant 95-BTN-15.
We wish to thank T. Basaruddin of the University of Indonesia and Benno Overeinder of the University of Amsterdam for fruitful discussions.

References

1. D. L. Clark, J. Casas, W. Otto, M Prouty, and J. Walpole. Scheduling of Parallel Jobs on Dynamic, Heterogeneous Networks.
   http://www.cse.ogi.edu/DISC/projects/cpe, 1995.
2. C++SIM Simulation Package, Univ. of Newcastle Upon Tyne,
   http://cxxsim.ncl.ac.uk/, 1997.
3. K. Czajkowski, I. Foster, N. Karonis, C. Kesselman. A Resource Management Architecture for Metacomputing Systems. In *Proc. Job scheduling strategies for Parallel Processing*, vol. 1459 of *LNCS*, pages 62-82, Springer-Verlag, 1998.
4. D.G. Feitelson, and L. Rudolph. Metrics and benchmarking for parallel job scheduling. In *Proc. Job Scheduling Strategies for Parallel Processing*, vol. 1459 of *LNCS*, pages 1-24, Springer-Verlag, 1998.
5. A. Hori, H. Tezuka, Y. Ishikawa, N. Soda, H. Konaka, and M. Maeda. Implementation of gang scheduling on workstation cluster. In *Proc. Job Scheduling Strategies for Parallel Processing*, vol. 1162 of *LNCS*, Springer-Verlag, 1996.
6. M. Matsumoto and T. Nishimura. "Mersenne twister: A 623-dimensionally equidistributed uniform pseudo-random number generator". In *ACM Transactions on Modeling and Computer Simulation*, vol. 8 of *ACM*, January, 1998.

7. P. G. Sobalvarro, S. Pakin, W.E. Weihl, and A.A. Chien. Dynamic Coscheduling on Workstation Cluster. In *Proc. Job Scheduling Strategies for Parallel Processing*, vol. 1459 of *LNCS*, Springer-Verlag, 1998.
8. G.D. van Albada, J. Clinckemaillie, A.H.L. Emmen, J. Gehring, O. Heinz, F. van der Linden, B.J. Overeinder, A. Reinefeld, and P.M.A. Sloot. Dynamite - blasting obstacles to parallel cluster computing. In *High-Performance Computing and Networking (HPCN Europe '99)*, vol. 1593, Springer-Verlag, April 1999.
9. A.W. van Halderen, B.J. Overeinder, P.M.A. Sloot, R. van Dantzig, D.H.J. Epema, and M. Livny. Hierarchical Resource Management in the Polder Metacomputing Initiative. *Parallel Computing*, pages 1807-1825, November 1998.
10. K. Y. Wang, D. C. Marinescu, and O. F. Carbunar. Dynamic Scheduling of Process Groups. *Concurrency: Practice and Experience*, pages 265-283, April 1998.
11. J. B. Weissman, and A. S. Grimshaw. A Federated Model for Scheduling in Wide-Area Systems. In *Proceedings of the Fifth IEEE International Symposium on High Performance Distributed Computing*, pages 542-550, Syracuse, NY, Aug. 1996.
12. B.B Zhou, R.P. Brent, D. Walsh, and K. Suzaki. Job Scheduling Strategies for Networks of Workstations. In *Proc. Job Scheduling Strategies for Parallel Processing*, vol. 1459 of *LNCS*, pages 143-157, Springer-Verlag, 1998.

A Compact, Thread-Safe Communication Library for Efficient Cluster Computing

M. Danelutto and C. Pucci

Dept. Computer Science, University of Pisa, Italy
{marcod,puccic}@di.unipi.it

Abstract. We describe a compact, thread-safe communication library for cluster computing. The library provides the most used communication functions, such as point-to-point send/receives and broadcast, scatter and gather collective operations. In addition, our library allows processes running on different processing elements of a cluster to share portions of memory. We show experimental results that demonstrate the library outperforms classical implementations of MPI (mpich) in collective operations and achieves comparable performance in point-to-point communications on Beowulf class workstation clusters.

1 Introduction

Cluster of workstation (COW) architectures are becoming more and more interesting as they promise to deliver supercomputer performance at a fraction of the supercomputer cost [1]. Commonly available tools for parallel computing on COWs, such as PVM and MPI [2,3], provide the user with tens to hundreds of different functions that can be used to implement communications. In order to implement interprocess communication the user may choose among a variety of combinations of the functions provided.

Common implementations of the MPI and PVM standards do not exploit architectural features provided by a PC cluster. They use a COW environment like an unoptimized network, using TCP – and relying on the TCP reliability – to implement communications. As a consequence, collective operations such as a broadcast, that could be performed exploiting UDP/IP primitive broadcast and/or multicast features, are performed by using collections of point-to-point TCP/IP operations. Furthermore, due to the fact that common implementations of MPI and PVM are supposed to be used on heterogeneous environments, data exchange formats are used in communications. This is not necessary, in general, when using COWs, as the processors of a COW are usually homogeneous and they use the same data formats; therefore the usage of external data formats introduce unnecessary overheads.

Nowadays MPI and PVM implementations are not thread-safe [4,5]: they do not use threads in their implementation and threads cannot be used in user code. Every process that is generated (explicitly by the user or implicitly by the tools) is a heavyweight Unix process.

In this work we describe a new communication library we designed at our Department, aimed at solving some of the problems stated above. The library implements the most common communication functions plus a particular "flavor" of shared memory system. In our work, "shared" memory has to be intended as "remotely accessible" memory. You can use this memory (for reading and writing data) as fast as possible from the same machine that physically holds the memory or from a remote machine, but you must explicitly issue specific library calls both to read and to write shared memory portions (i.e. the access is not transparent to the final user).

Our library has been developed in C under the Linux OS, using all positive features of this operating system: sockets, UDP/IP transport protocol, fast interprocess communication when allowed (System V Shared Memory, Memory Queues) [6], POSIX threads [7], etc. In this paper, we will first discuss the library design issues (Sec. 2), then we will discuss some of the library relevant implementation details (Sec. 3) and, eventually, we will discuss some performance results that demonstrate the effectiveness of our approach (Sec. 4).

2 Library Design Choices

The structure of our communication library is basically similar to the structure of other communication library packages. We provide the user a linkable library, that gives execution-time support (`libste.a`), a server (`Demone`), that is used to maintain the computation status (this is a plain Unix daemon), and a support command, to initiate and terminate parallel computations using the library. Therefore, in order to run parallel applications using our communication library, the user code of the different processes participating to the computation (and issuing calls to our communication library) must be linked with the library, first. Then, a `Demone` process must be started on each one of the processing elements involved, and the `Par` command may be used to start the parallel computation. Point-to-point computations simply involve the user process code and the associated library code, in the general case. Collective operations and shared memory accesses involve also the `Demone` processes. User code interacts with `Demone` processes by using plain UNIX (System V) message queues.

In the design of our communication library, we decided to use UDP/IP – at transport level of the OSI model – to implement interprocess communications because it implements faster communications than TCP/IP. Furthermore the communication model of UDP/IP is connectionless, while the one of TCP/IP is connection-oriented. This is another point that can be exploited to achieve a better performance in communications (processes do not need to connect before communicating). The UDP protocol is not reliable. However, when COW architectures are taken into account, due to the dedicated (Ethernet) network used and to the small distance between the nodes of the network, UDP data are usually lost only when a receive socket buffer happens to be full. Reliability is shifted at the application (library) level in our software and it is guaranteed by an appropriate high level protocol (see Sec. 3).

Support functions	
`int InitAll(int *argc, char **argv);`	`char *SteStrError(int Error);`
Process Naming	
`int DeclareProcessId(int Pid);`	`int MyProcessId(void);`
`int NumProcessId(void);`	`int RemProcessId(int Pid);`
Groups	
`int DeclareGroupId(int Gid, int Kind);`	`int JoinGroup(int Gid);`
`int AddToGroup(int Gid, int Elem);`	`int LeaveGroup(int Gid);`
`int RemFromGroup(int Gid, int Elem);`	`int NumProcessInGroup(int Gid);`
Point-to-point communications	
`int Send(int Dest, void *Buf, int Len);`	
`int Receive(int Sender, void *Buf, int *Len);`	
`int ReceiveAny(int *Sender, void *Buf, int *Len);`	
Collective communications	
`int Barrier(int Gid);`	
`int Broadcast(int Gid, int Root, void *Buf, int *Len);`	
`int Scatter(int Gid, int Root, void *Buf, int *Len, int elem_size, int Mode);`	
`int Gather(int Gid, int Root, void *Buf, int *Len, int elem_size, int Mode);`	
Shared Memory	
`int DeclareShRAM(int Shid, int Len);`	`int RemShRAM(int Shid);`
`int LockShRAM(int Shid);`	`int UnlockShRAM(int Shid);`
`int WriteToShRAM(int Shid, void *Buff, int Len, int offset);`	
`int ReadFromShRAM(int Shid, void *Buff, int Len, int offset);`	
`int BroadcastShRAM(int Gid, int Shid, int Len, int offset);`	

Table 1. Function's prototypes

In order to achieve good efficiency, we decided to implement collective communications and synchronizations using the low-level, primitive broadcast operation that comes with UDP. Therefore our library implements collective communications by single UDP broadcast communications plus some (un-avoidable) memory copies. We implemented daemon processes on each one of the COW processing elements to support the implementation of collective communications as the UDP protocol does not allow different processes to manage the same port number [6]. The daemon processes (one per processing element in the COW) manage all the broadcasts performed in the COW by the processes of the same run or of different, simultaneous runs.

We decided to implement a minimal number of primitives within our library. Each primitive is associated with a well-known semantics. We preferred to implement a small set of simple and intuitive primitives rather than a wide set of infrequently used functions. The set of primitives we implemented in our library covers 99% of the function calls issued to MPI library in production software in a classical parallel computer center [8]. As far as point-to-point communications are concerned, our library just provides asynchronous `Send` and synchronous `Receive` (both symmetric and asymmetric). Concerning collective communications, instead, the library provides synchronous `Barrier`, `Broadcast`, `Scatter` and `Gather` functions only.

We make a deep use of threads in our library and in the **Demone** process code. The library and the server code use threads to achieve efficiency (due to easy sharing of data and low cost context switch typical of threads) by parallelizing computations and communications (i.e. we adopted typical *excess parallelism* programming techniques in the implementation of both library and server code). Furthermore, threads are spawned every time a new process is added to the computation in such a way that a single thread in the library is responsible of the management of a single user process directly involved in the parallel computation (i.e. of a process that sends or receives data to (from) other processes). Finally, in the server, threads are extensively used to handle different requests coming from client (user) processes concurrently.

Threads can also be used in user computations, i.e. in the code of user processes calling our library. Different threads can call all the library functions concurrently as a single process or as different entities. A thread can obtain a name (i.e. an identifier valid within the current parallel computation) and may receive and send data to other processes (identifiers), or a process can obtain a name and allow its threads to call library functions concurrently. This is possible as every function implemented in our library is definitely thread-safe and can be called concurrently from threads preserving its semantics.

Blocks of shared, remotely accessible memory[1] can be declared and used with our library. Such blocks are implemented as System V Shared Memory [6]. Every time a shared memory block is declared in the user code, a thread is created within our library and associated to the block. This thread logically belongs to the local daemon process (the **Demone**) run by our library on each machine and is in charge of managing all the read and write activities from the remote processes that refer to that particular block. A mutual exclusion mechanism is implemented within the library, to allow consistency in concurrent accesses to the "shared" memory blocks, in such a way that accesses to the blocks are guaranteed to be atomic[2]. Reading and writing to the shared blocks are performed as standard memory read and writes in case the block is local (the process issuing the read/write call runs on the same machine where the shared memory block has been allocated) as the blocks are implemented using Unix System V shared memory facilities. The accesses are implemented by means of remote communications in case the process issuing the calls is running on a machine different from the one physically holding the shared block.

Our library supports read and write operations involving shared memory blocks (`ReadFromShRAM`, `WriteToShRAM`), as well as lock and unlock operations (`LockShRAM`, `UnlockShRAM`). In addition to these operations, a function is provided (`BroadcastShRAM`) implementing an efficient copy operation that distributes the contents of a shared memory block to a set of different shared memory blocks. It's worth pointing out that the shared memory primitives included in the library just allow processes to explicitly read or write memory

---

[1] That will be named *ShRAM = Shared Remotely Accessible Memory*

[2] Both operations involving the whole block and operations involving just a portion of the block are guaranteed to be atomic.

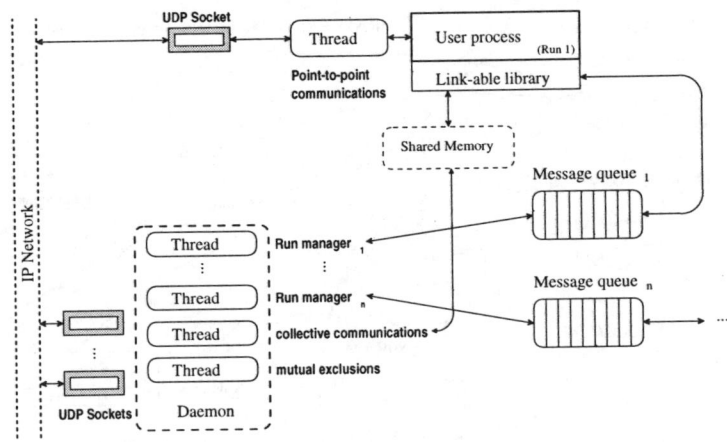

Fig. 1. Thread subdivision and other resources usage.

blocks allocated on different processing elements. Neither caching is performed, nor copies of the remote memory blocks are maintained by the support processes on the target architecture processing elements.

3 Library Implementation

The server process, which is run by our library on each node participating to the parallel computation, maintains the computation status, actively performs collective communications, and activates threads associated to ShRAM blocks; servers and clients (user) processes communicate by using UNIX System V message queues. A message queue is created for every Run[3] started, in such a way that the library code linked to the user processes involved in that particular parallel computation could use the queue to communicate with the server process allocated on the node. Obviously the linkable part of support hides these details to the user.

The whole system is organized in threads and threads are used to allow high concurrency between the different activities performed by the library and server code. A thread is associated to every process (or thread) that obtains a name (and therefore becomes eligible to participate in a communication and/or in an access to a shared memory block). Each Demone process is also divided in threads: a thread manages mutual exclusion over the network, another thread manages the collective routines (sending and receiving data), another one manages general initialization and starting of other tasks, and the last one is associated with every Run that will be started. This thread will wait over the message queue associated with the Run (see Figure 1).

---

[3] A Run is a parallel computation, exactly like a $task$ in a PVM parallel computation.

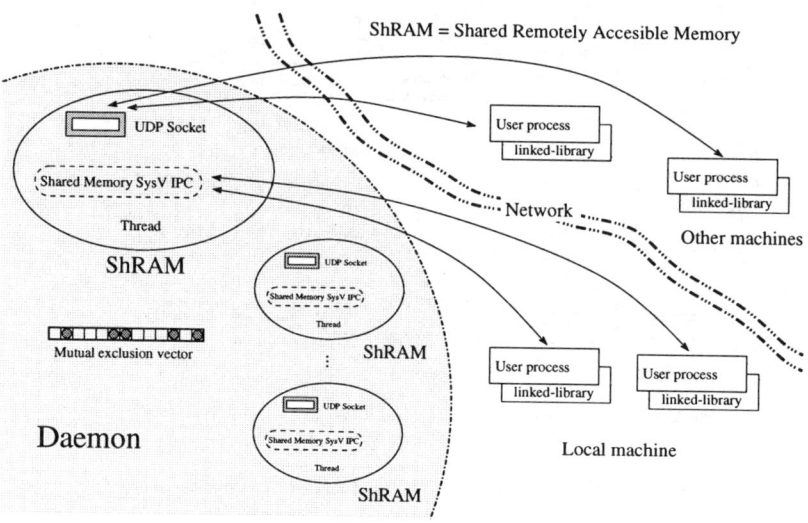

Fig. 2. Organization of the ShRAM system.

The library uses exactly one **Demone** process for each processing node. This process is only used for collective and shared memory operations, as these operations require a centralization point to properly handle the ports used in the UDP collective operations. It is not a bottleneck for point-to-point communications because it is not involved in these primitives at all.

Every object (process, process group and shared memory segment) participating in a computation is identified by an integer. The process-names (Pid) are used to select processes in point-to-point communications, the names of process groups (Gid) are used to identify the processes involved in collective communications. Finally, the ShRAM object names (Shid) are used to perform operations on shared memory blocks. Groups can be set up with Pid or with Shid (but not both, the Kind parameter is used to identify groups of processes versus groups of ShRAM). Process groups are used to perform collective communication operations, while shared memory block groups are used to perform BroadcastShRAM operations (see below). A process (or a thread) can join a parallel computation (formerly a *Run*) by calling an initialization routine and then by getting a name. All the implemented functions are shown in Table 1 along with their C prototypes.

When a process obtains a name the support spawns a thread that manages the sending part of any point-to-point communication. When the user process calls the Send function the support just waits while the associated thread copies data into the local send buffers. After the last copy – data is split into packets – the Send library call issued by the user process actually returns. The thread sends data to the destination process using an UDP socket. It completely manages the high level protocol needed to implement reliable communications using unreliable

UDP sockets. The `Receive` and `ReceiveAny` functions are implemented directly by the library code linked to the calling process and are always blocking. The big problem here is how to offer reliability over an unreliable protocol, without degrading communication performance. We used a variant of a standard protocol (i.e. the *Sliding Windows* protocol [9]). This variant was used because data loss (and consequentially performance degrading) mainly happens in a COW when the socket receive buffer becomes full. Due to the blocking semantic of the `Receive` and `ReceiveAny` functions, continuous data sending[4] is not a great solution (this is the policy currently implemented in the TCP protocol [6]). In our library data is initially sent just a few times (the number of times can be configured during library compilation) and then if the destination process does not send acknowledgements the sender will wait for an extended rendez-vous before actually start re-sending data continuously. This approach works better with algorithms using synchronous communication modes (best communication performance). When algorithms use asynchronous communication modes, data rate can be halved.

When a `Broadcast` is performed, the `Demone` processes are actively involved because the UDP low-level broadcast needs a centralization point. This is because the UDP/IP low-level broadcast distribute data at all receipts waiting on a single UDP port [6]. In turn, this implies that a single process on every machine can wait and receive broadcasted data. A process just passes the request to the local `Demone` and waits for completion; data is exchanged between processes and daemons (the centralization points) by System V shared memory blocks (request and answers are exchanged usually with System V message queues). Therefore the cost of a broadcast operation is just the cost of a single network communication plus the cost of some memory copies (one for each process, copies on different machines are obviously done in parallel). The cost of a broadcast operation is independent of the number of the machines involved and is proportional to the number of processes that are involved in the broadcast on a single machine.

Even with our kind of "shared" memory we implemented a small but complete set of primitives [10]. Using our library the programmer is allowed to create and destroy a ShRAM block, acquire and release a mutual exclusion on a block, and read and write data from/to the block. It is also possible to refer a group of ShRAM blocks and distribute data – in a single collective operation – to all the blocks involved (`BroadcastShRAM`).

Writing data in a remote shared memory block implies a remote communication via UDP sockets. The communication target, in this case (i.e. the process that actually owns this socket) is the thread associated with the ShRAM block. Any mutual exclusion possibly needed is managed by servers (one lock for every ShRAM block, managed by the daemon that holds the block). Figure 2 outlines the structure of ShRAM system.

Finally this library is easily configurable. A single source file is provided with the definitions of all the parameters that could be configured by the final li-

---

[4] i.e. sending data again and again until an ack is eventually received

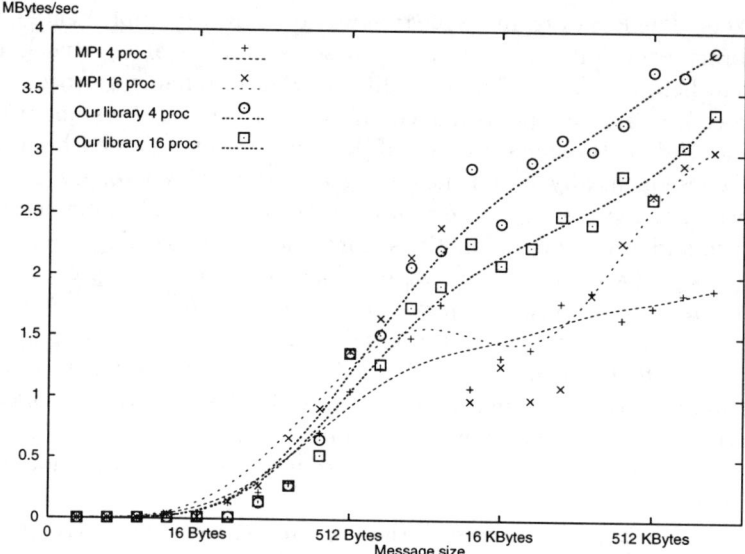

Fig. 3. Point-to-point performance with our library and MPI.

brary user (`Limits.h`). By changing the default parameter values in that file and with a simple re-compilation of the library code, users can adapt the library to the architecture at hand. The configuration parameters that can be usefully changed basically manage the communication protocols and memory consumption (e.g. they concern buffer length and number, number of sends performed by the transmitting protocol before actually starting the extended rendez-vous phase, maximum number of machines involved in the parallel computations, and so on).

4 Performance Results

Our communication library has been developed using a workstation cluster installed at the Department of Computer Science of the University of Pisa. The name of the COW is *Backus* and it is currently equipped with 11 standard PCs, with Intel Pentium II processors running at 266MHz, 128MBytes of RAM and an IDE disk on every machine. Those computers are interconnected by a fast Ethernet network by means of a 3Com SuperStackII switch (crossbar). The switch guarantees that communications are performed without interferences between different machines. Only one machine is connected to the external network, other machines use a private, class B, IP subnet. On this system both our library and the `mpich` implementation of MPI [11] are available.

We performed a complete set of experiments to validate the library design choices and the effectiveness of the approach. Most of the functional tests where

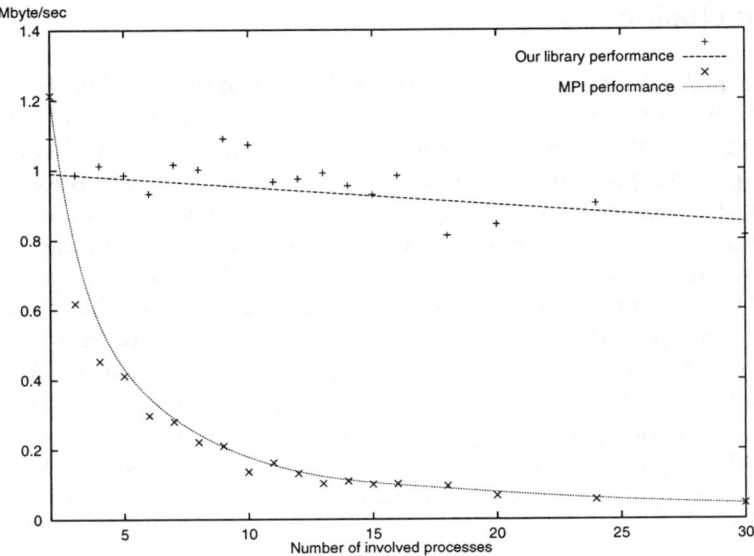

Fig. 4. Broadcast performance comparison between MPI and our library.

performed with simple, ad hoc code. The performance tests, instead, has been performed adapting some of the test source code provided along with the mpich MPI source code. In particular, point-to-point communication performance was evaluated with the systest program, that comes with the mpich distribution. This program has been rewritten in such a way that our library functions where used instead of the standard MPI ones to test the communication performance of our library. Figure 3 summarizes the results achieved. Our library is slightly faster than mpich in point-to-point communications. To be honest, we must say that in those cases where really asynchronous point-to-point communications are performed by the processes involved in the parallel computation, the times we achieved by using our library are slightly worst than those achieved using mpich. The performance of collective operations (broadcast), instead, was compared using simple programs that broadcast some data to a group of processes. In this case, results achieved are definitely better than those of mpich (see Figure 4). This is an expected result, as we use plain UDP broadcast to implement library broadcasts, whereas mpich emulates the broadcast by using a set of TCP point-to-point communications.

5 Conclusions

We developed a communication library for cluster computing which is very compact (the library just provides the commonly used communication primitives of MPI), thread-safe, and that provides the programmer of parallel applications with the possibility to efficiently access remote node memory. The library demonstrates to be as efficient as existing, state-of-the-art communication libraries in point-to-point communications and achieves much better performances in collective communications. This results are due to specific design choices we made. Most of the communication library algorithms and code have been specifically optimized to exploit the architectural features of cluster of workstations interconnected by means of switched Fast Ethernet networks. Thread safeness is a particular advantage with respect to other similar tools: the programmer may use simpler communication primitives in conjunction with threads to achieve efficiency in parallel computations that, without threads, could be (possibly) achieved by using more complex communication primitives and/or program structures.

References

1. M. Baker and R. Buyya. Cluster Computing at a Glance. In Rajkumar Buyya, editor, *High Performance Cluster Computing*, pages 3–47. Prentice Hall, 1999.
2. W. Gropp and E. Lusk. The MPI communication library: its design and a portable implementation. In *Proceedings of the Scalable Parallel Libraries Conference*, pages 160–165. IEEE Computer Society Press, October 1993. Mississippi.
3. Al Geist, Adam Beguelin, Jack Dongarra, Weicheng Jiang, Robert Manchek, and Vaidy Sunderam. PVM 3 Users Guide and Reference Manual. Technical report, Oak Ridge National Laboratory, September 1994.
4. William D. Gropp and Ewing Lusk. User's Guide mpich, a Portable Implementation of MPI. Technical Report ANL-96/6, Mathematics and Computer Science Division, Argonne National Laboratory, 1996.
5. A. Beguelin, J. Dongarra, A. Geist, R. Manchek, and V. Sunderam. A User's Guide to PVM (Parallel Virtual Machine). Technical Report TM-11826, Oak Ridge National Laboratory, June 1991.
6. W. Richard Stevens. *UNIX Network Programming*. Prentice Hall, 1999. Second Edition, Vol. 1 and 2.
7. S. J. Norton and M. D. DiPasquale. *THREADTIME The multithreaded programming guide*. Prentice Hall, 1997.
8. R. Rabenseifner, P. Gottshling, W. E. Nagel, and S. Seidl. Effective performance problem detection of MPI programs on MPP systems: From the global view to the detail. In *Proceedings of the ParCo'99 Conference – Parallel Computing*, 1999.
9. Andrews S. Tanenbaum. *Computer Networks*. Prentice Hall, 1997.
10. Cray Research Inc. Cray C/C++ Reference Manual. Technical Report SR-2196, Cray Research Inc., 1998.
11. W.D. Gropp, E. Lusk, N. Doss, and A. Skjellum. An high-performance, portable implementation of the MPI message passing interface standard. *Parallel Computing*, September 1996.

EPOS and Myrinet: Effective Communication Support for Parallel Applications Running on Clusters of Commodity Workstations*

Antônio Augusto Fröhlich[1], Gilles Pokam Tientcheu[1], and Wolfgang Schröder-Preikschat[2]

[1] GMD FIRST, Kekulésraße 7
D-12489 Berlin, Germany
{guto|pokam}@first.gmd.de
http://www.first.gmd.de/~guto
[2] University of Magdeburg, Universitätsplatz 2
D-39106 Magdeburg, Germany
wosch@cs.uni-magdeburg.de
http://ivs.cs.uni-magdeburg.de/~wosch

Abstract This paper presents the EPOS approach to deliver parallel applications a high performance communication system. EPOS is not an operating system, but a collection of components that can be arranged together to yield a variety of run-time systems, including complete operating systems. This paper focuses on the communication subsystem of EPOS, which is comprised by the *network adapter* and *communicator* scenario-independent system abstractions. Like other EPOS abstractions, they are adapted to specific execution scenarios by means of scenario adapters and are exported to application programmers via inflated interfaces. The paper also covers the implementation of the *network adapter* system abstraction for the Myrinet high-speed network. This implementation is based on a carefully designed communication pipeline and achieved unprecedented performance.

1 Introduction

Undoubtedly, one of the most critical points to support parallel applications in distributed memory systems is communication. The challenge of enhancing communication performance, especially in cluster of commodity workstations, has motivated numerous well succeeded initiatives. At the hardware side, high-speed network architectures and fast buses counts for low-latency, high-bandwidth inter-node communication. At the software side, perhaps the most significant advance has been to move the operating system out of the communication pathway. In this context, several forms of active messages, asynchronous remote

---

* This research has been partially supported by the Federal University of Santa Catarina, by CAPES Foundation grant no. BEX 1083/96-1 and by Deutsche Forschungsgemeinschaft grant no. SCHR 603/1-1.

copy, distributed shared memory, and even optimized versions of the traditional send/receive paradigm, have been proposed. Combined, all these initiatives left the giga-bit-per-second, application-to-application bandwidth barrier behind [7].

Nevertheless, good communication performance is hard to obtain when dealing with anything but the test applications supplied by the communication package developers. Real applications, not seldom, present disappointing performance. We believe many performance losses to have a common root: the attempt to deliver a generic, all-purpose solution. Most research projects on high performance communication are looking for "the best" solution for a given architecture. However, a definitive best solution, independently of how fine-tuned to the underlying architecture it is, does not exist, whereas parallel applications simply communicate in quite different ways. Aware of this, many communication packages claim to be "minimal basis", upon which application-oriented abstractions can (have to) be implemented. One more time, there cannot be a best minimal base for all possible communication strategies. This contradiction between generic and optimal is presented in details in [8], and serves as motivation for the project EPOS.

In EPOS [5], we intend to give each application its own run-time support system, specifically constructed to satisfy its requirements (and nothing but its requirements). EPOS is not an operating system, but a collection of components that can be arranged together in a framework to yield a variety of run-time systems, including complete operating systems. Besides application-orientation, the project aims on high performance and scalability to support parallel computing on clusters of commodity workstations. The following sections describe EPOS communication system design, its implementation, and its performance.

2 Communication System Design

EPOS has been conceived following the guidelines of traditional object-oriented design. However, scalability and performance constrains impelled us to define some EPOS specific design elements. These design elements will be described next in the realm of the communication system.

2.1 Scenario-Independent System Abstractions

Granularity plays a decisive role in any component-based system, since the decision about how fine or coarse components should be have serious implications. A system made up of a large amount of fine components will certainly achieve better performance than one made up of a couple of coarse components, since less unneeded functionality incurs less overhead. Nevertheless, a large set of fine components is more complex to configure and maintain.

In EPOS, visible components have their granularity defined by the smallest-yet-application-ready rule. That is, each component made available to application programmers implements an abstract data type that is plausible in the application's run-time system domain. Each of these visible components, called *system abstractions*, may in turn be implemented by simpler, non application-ready components.

In any run-time system, there are several aspects that are orthogonal to abstractions. For instance, a set of abstractions made SMP safe will very likely show a common pattern of synchronization primitives. In this way, we propose EPOS system abstractions to be implemented as independent from execution scenario aspects as possible. These adaptable, scenario-independent system abstractions can then be put together with the aid of a *scenario adapter*.

Communication is handled in EPOS by two sets of system abstractions: *network adapters* and *communicators*. The first set regards the abstraction of the physical network as a logical device able to handle one of the following strategies: datagram, stream, active message, or asynchronous remote copy. The second set of system abstractions deals with communication end-points, such as links, ports, mailboxes, distributed shared memory segments and remote object invocations. Since system abstractions are to be independent from execution scenarios, aspects such as reliability, sharing, and access control do not take part in their realizations; they are "decorations" that can be added by scenario adapters.

For most of EPOS system abstractions, architectural aspects are also seen as part of the execution scenario, however, network architectures vary drastically, and implementing unique portable abstractions would compromise performance. As an example, consider the architectural differences between Myrinet and SCI: a portable active message abstraction would waste Myrinet resources, while a portable asynchronous remote copy would waste SCI resources. Therefore, realizations for the *network adapter* system abstraction shall exist for several network architectures. Some abstractions that are not directly supported by the network will be emulated, because we believe that, if the application really needs (or wants) them, it is better to emulate them close to the hardware.

2.2 Scenario Adapters

EPOS system abstractions are adapted to specific execution scenarios by means of *scenario adapters*. Currently, EPOS scenario adapters are classes that wrap system abstractions, so that invocations of their methods are enclosed by the **enter** and **leave** pair of scenario primitives. These primitives are usually inlined, so that nested calls are not generated. Besides enforcing scenario specific semantics, scenario adapters can also be used to "decorate" system abstractions, i.e., to extend their state and behavior. For instance, all abstractions in a scenario may be tagged with a capability to accomplish access control.

In general, aspects such as application/operating system boundary crossing, synchronization, remote object invocation, debugging and profiling can easily be modeled with the aid of scenario adapters, thus making system abstractions, even if not completely, independent from execution scenarios.

The approach of writing pieces of software that are independent from certain aspects and later adapting them to a given scenario is usually referred to as *Aspect-Oriented Programming* [2]. We refrain from using this expression, however, because much of AOP regards the development of languages to describe aspects and tools to automatically adapt components (*weavers*). If ever used in EPOS, AOP will give means but not goals.

2.3 Inflated Interfaces

Another important decision in a component-based system is how to export the component repository to application programmers. Every system with a reasonable number of components is challenged to answer this question. Visual and feature-based selection tools are helpless if the number of components exceeds a certain limit —depending on the user expertise about the system, in our case the parallel application programmer expertise on operating systems. Tools can make the selection process user-friendlier, but certainly do not solve the user doubt about which selections to make. Moreover, users can usually point out what they want, but not how it should be implemented. That is, it is perhaps straightforward for a programmer to choose a mailbox as a communication end-point of a datagram oriented network, but perhaps not to decide whether features like multiplexing and dynamic buffer management should be added to the system.

The approach of EPOS to export the component (system abstraction) repository is to present the user a restricted set of components. The adoption of scenario adapters already hides many components, since instead of a set of scenario specific realizations of an abstraction, only one abstraction and one scenario adapter are exported. Nevertheless, EPOS goes further on hiding components during the system configuration process. Instead of exporting individual interfaces for each flavor of an abstraction, EPOS exports all of its flavors with a single *inflated interface*. For example, the datagram, stream, active message, and asynchronous remote copy *network adapters* are exported by a single `Network_Adapter` inflated interface as depicted in figure 1.

An inflated interface is associated to the classes that realize it through the *selective, partial realize* relationship. This relationship is partial because only part of the inflated interface is realized, and it is selective because only one of the realizations can be bound to the inflated interface at a time. Each selective realize relationship is tagged with a key, so that defining a value for this key selects a realization for the corresponding interface. The way this relationship is implemented enables EPOS to be configured by editing a single key table, and makes conditional compilations and "makefile" customizations unnecessary.

The process of binding an inflated interface to one of its realizations can be automated if we are able to clearly distinguish one realization from another. In EPOS, we identify abstraction realizations by the signatures of their methods. In this way, an automatic tool can collect signatures from the application and select adequate realizations for the corresponding inflated interfaces. Nevertheless, if two realizations have the same set of signatures, they must be exported by different interfaces.

The combination of *system abstractions, scenario adapters* and *inflated interfaces*, effectively reduces the number of decisions the user has to take, since the visual selection tool will present a very restricted number of components, of which most have been preconfigured by the automatic binding tool. Besides,

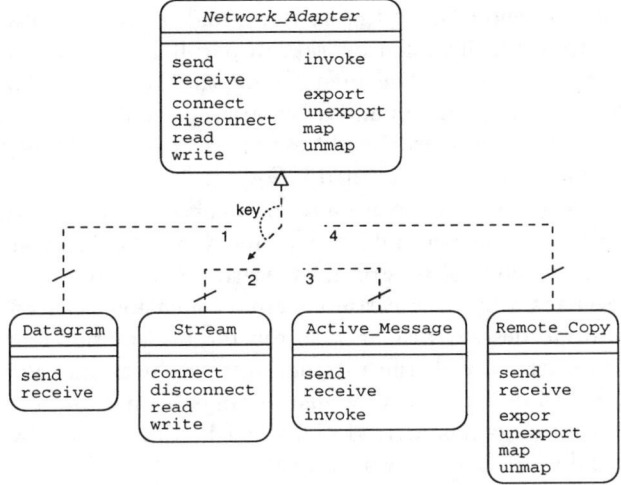

Figure1. The Network_Adapter inflated interface and its partial realizations.

they enable application programmers to express their expectations concerning the run-time system simply by writing down well-known system object invocations.

3 Communication System Implementation for Myrinet

EPOS is coded in C++ and is currently being developed to run either as a native ix86 system, or at guest-level on Linux. The ix86-native system can be configured either to be embedded in the application, or as μ-kernel. The Linux-guest system is implemented by a library and a kernel loadable module. Both versions support the Myrinet high-speed network. EPOS communication system implementation is detailed next.

3.1 Platform Overview

EPOS is currently being implemented for a PC cluster available at GMD-FIRST. This cluster consists nowadays of a server and 16 work nodes, each with an AMD Athlon processor running at 550 MHz, 128 MB of memory on a 100 MHz bus, and a 32 bits/33 MHz PCI bus in which a Fast-Ethernet and a Myrinet network adapter are plugged. This platform and some performance figures have been introduced in [4], however, it is important to recall some of its characteristics in order to justify implementation decisions.

The Myrinet network adapter present in each node of our cluster has a processor, namely a LANai 4.1, 1 MB of memory and three DMA engines, respectively for transferring data between main memory and the memory on the network adapter, to send data to the network, and to receive data from the network. These DMA controllers can operate in parallel and perform two memory accesses per processor cycle. The memory on the Myrinet adapter is used to store the LANai control program and as communication buffer as well; it is also mapped into the main processor's address space, thus enabling data transfers without DMA assistance (programmed I/O).

A simple message exchange can be accomplished by using programmed I/O or DMA to write the message into the memory on the Myrinet adapter, and then signaling to the control program, by writing a shared flag, that a message of a given size is available in a certain memory location. The control program can then generate a message header with routing information and configure the send DMA controller to push the message into the network. The receiver side can be accomplished in a similar way, just adding a signal to the main processor to notify that a message has arrived. This can be done either by a shared flag polled by the main processor or via interruptions.

If the memory management scheme adopted on the node uses logical address spaces that are not contiguously mapped into memory, additional steps have to be included in order to support DMA. EPOS can be configured to support either a single task (the typical case for MPI applications running on single processor nodes) or several tasks per node. The ix86-native, single-task version does not need any additional step, since logical and physical address spaces do match. The multi-tasking and Linux-guest version, however, allocate a contiguous buffer, of which the physical address is known, and give programmers two alternatives: write messages directly into the allocated buffer; or have messages copied into it.

Figure 2 depicts a message exchange between two applications (including the additional copies). The data transfer rate for each stage has been obtained and is approximately the following: 140 MB/s for the copy stages 1 and 5; 130 MB/s for the host/Myrinet DMA stages 2 and 4; and 160 MB/s for the send and receive DMA stages 3.1 and 3.2. Therefore, the total data transfer rate is limited to 130 MB/s by the host/Myrinet DMA stages.

3.2 Communication Pipeline

In order to deliver applications a communication bandwidth close to the 130 MB/s hardware limit the software overhead must be reduced to an insignificant level. Fortunately, a careful implementation and several optimization can help to get close to this limit. To begin with, the DMA controllers in the Myrinet adapter are able to operate in parallel, so that stages 2 and 3.1 of figure 2, as well as stages 4 and 3.2, can be overlapped. However, these stages are not intrinsically synchronous, i.e., there is no guarantee that starting stage 3.1 just after starting stage 2 will preserve message integrity. Therefore, overlapping will only be possible for different messages or, what is more interesting, different pieces of a message. We took advantage of this architectural feature to implement a communication pipeline for EPOS.

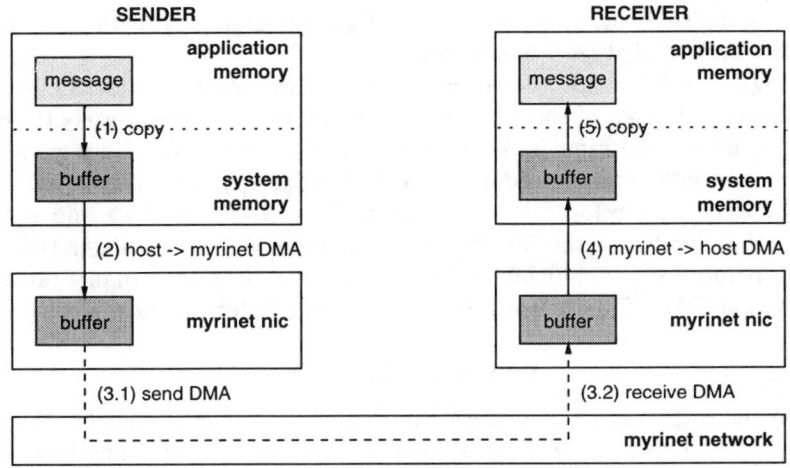

Figure2. Steps involved in a message exchange.

EPOS communication pipeline for Myrinet has been designed considering the time messages of different sizes spend at each stage of figure 2. This time includes the overhead for the stage (per-packet cost) and its effective data transfer rate (per-byte cost). It is important to notice that the overhead includes synchronization operations and the waiting time for the next stage to become available. According to Myrinet documentation, the delay between stages 3.1 and 3.2 is of 0.5 µs per switch hop. As this latency is much smaller then any other in the pipeline, we will consider stages 3.1 and 3.2 to completely overlap each other, thus yielding a single pipeline stage 3.

A message sent through the network is now split in small packets that move through the stages of the pipeline. In order to sustain a transfer rate close to the maximum, at least two requirements must be fulfilled: first, the number of packets must be at least equal to the depth of the pipeline (five in our case), and second, the packet length must be such as to minimize the total message transmission time. To determine the optimal packet length for the pipeline, we modeled it according to the following rules:

- As soon as a packet n leaves a pipeline stage i, it enters the next stage $i+1$;
- A packet n enters stage i of the pipeline as soon as packet $n-1$ leaves it.

In this way, we can compute the optimal packet length $P_{optimal}$ by looking for a maximum of the function $T(L, P)$, which represents the time a message of size L, split in packets of length P, takes to move through the pipeline [3]. This function can be expressed as the sum of the time a packet spends on each of the pipeline stages. Using a linear cost model [1], the time spent by a packet n in a stage i can be expressed as:

$$T_i^{(n)} = \beta_i + P\tau_i \qquad (1)$$

where β_i is the constant per-packet cost assigned to stage i, P is the packet size, and τ_i the inverse of stage i bandwidth.

Taking in consideration the characteristics of the platform described earlier, we can assume the cost of the copy stages 1 and 5 to be approximately the same, since they involve the same hardware resources, i.e., processor and memory. For the same reason we can consider the cost of stage 2 to be approximately the same of stage 4. Therefore, $\beta_1 \cong \beta_5$ and $\tau_1 \cong \tau_5$, while $\beta_2 \cong \beta_4$ and $\tau_2 \cong \tau_4$. Since the bandwidth of the Myrinet network (160 MB/s) is greater than that of main memory copy (140 MB/s), which in turn is greater than that of the host/Myrinet DMA (130 MB/s), we can write the following inequality:

$$\tau_2 \cong \tau_4 > \tau_1 \cong \tau_5 > \tau_3 \qquad (2)$$

From the inequality 2 above, we can deduce an upper bound for the delay experienced by a packet traveling through the stage 3 of the pipeline (inequalities 3 and 4); and also the delay experienced by the last packet of the message at stage 4 (inequality 5)[1].

$$\beta_1 + \beta_2 + \beta_3 + P\tau_2 > \beta_3 + P\tau_3 \qquad (3)$$

$$\beta_1 + \beta_2 + \beta_3 + P\tau_2 > \beta_1 + P\tau_1 \qquad (4)$$

$$\beta_1 + \beta_2 + P\tau_2 > \beta_1 + P\tau_1 \qquad (5)$$

Assuming that the message size L is an integer multiple of P, the function $T(L, P)$ could be expressed as:

$$T(L, P) = \sum_{k=1}^{2} T_k^{(1)} + \sum_{i=1}^{\lceil \frac{L}{P} \rceil} T_3^{(i)} + \sum_{j=4}^{5} T_j^{(\lceil \frac{L}{P} \rceil)} \qquad (6)$$

From the inequalities above, and from the approximations suggested earlier, we can express $T(L, P)$ as:

$$T(L, P) \cong 2(2\beta_1 + \beta_2) + 2(\tau_1 + \tau_2)P + \lceil \frac{L}{P} \rceil (\beta_1 + \beta_2 + \beta_3 + P\tau_2) \qquad (7)$$

and the optimal packet size can be computed by solving the following equation:

$$\frac{dT(L, P)}{dP} = 0 \Rightarrow P_{optimal} = \sqrt{\frac{L(\beta_1 + \beta_2 + \beta_3)}{2(\tau_1 + \tau2)}} \qquad (8)$$

By applying a linear regression approximation [6] to each stage of the pipeline on the set of predictor variables, which has been obtained from the non-pipelined

---

[1] The overhead of upstream pipeline stages has to be considered because there is no analytical way to determine a relation among them. Even measurements are not possible, since execution time is not predictable in our CPUs.

transfer of messages of various sizes, we computed the values for β and τ to be approximately the ones given in table 1.

Table1. Computed values for the per-packet cost β and inverse bandwidth τ of each pipeline stage.

stage	β_i(in μs)	τ_i(in $\frac{\mu s}{KB}$)
1, 5	3.0	5.9
2, 4	0.6	7.6
3	1.3	7.5

As an example, the calculated optimal packet length for a 64 KB message is 3490 bytes. In practice, 4 KB packets are to be used. We solved equation 8 for several message lengths in order to determine the optimal packet length for ranges of message lengths. With this information in hand, we implemented an adaptive pipeline that automatically selects the appropriate packet length according to the message length, thus minimizing the message transfer latency.

3.3 Short Messages

Although the pipeline described above has a very low intrinsic overhead, programming DMA controllers and synchronizing pipeline stages may demand more time than it is necessary to send a short message via programmed I/O. In order to optimize the transfer of short messages using programmed I/O, which usually has a mediocre performance on PCs, we instructed our processors to collect individual write transactions that would traverse the PCI bridge to form 32 bytes chunks. Each chunk is then transferred in a burst transaction. This feature is enabled by selecting a "combine" cache policy for the pages that map the memory on the Myrinet adapter into the address space of the process. For the current implementation, messages shorter than 256 bytes are transferred in this way.

3.4 Performance Evaluation

We evaluate the performance of EPOS *network adapter* system abstraction, by measuring the latency and the bandwidth experienced at the application level. A single-task, ix86-native and a Linux-guest version have been considered. The one-way tests have been executed in the platform previously described and consist of one node sending messages of different sizes to an adjacent node, i.e., one connected to the same Myrinet switch. The results are presented in figure 3, and, when compared to the 130 MB/s limit, give an efficiency rate of 85% for 64 KB messages in the Linux-guest version and 92% in the native version.

4 Conclusion

In this paper we presented the EPOS approach to deliver a high performance communication system to parallel applications running on clusters of commodity

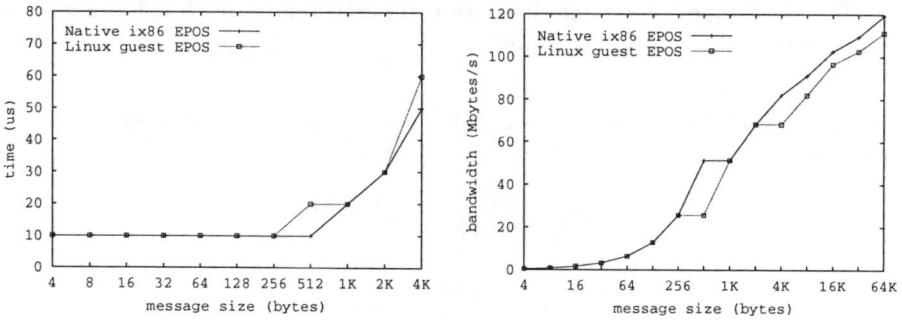

Figure3. EPOS *network adapter* one-way latency (left) and bandwidth (right).

workstations. We demonstrated how *system abstractions, scenario adapters* and *inflated interfaces* can simplify the process of run-time system configuration, mainly by reducing the number of decisions the user has to take. We also describe the *network adapter* system abstraction implementation for the Myrinet high-speed network that interconnects the nodes in our PC cluster.

The results obtained so far are highly positive and help to corroborate EPOS design decisions. The evaluation of EPOS *network adapter* abstraction revealed performance figures that, as far as we are concerned, have no precedents in the Myrinet interconnected PC cluster history. However, EPOS is a long term, open project that aims to deliver application-oriented run-time systems. Many system abstractions, scenario adapters, and tools are still to be implemented in order to support a considerable set of applications.

References

[1] J. de Rumeur. *Communications dans les Réseaux de Processeurs*. Masson, 1994.
[2] G. Kiczales et al. Aspect-Oriented Programming. In *Proc. ECOOP'97*, Springer-Verlag, 1997.
[3] R. Wang et al. Modeling Communication Pipeline Latency. In *Proc. SIGMETRICS'98*, June 1998.
[4] A. Fröhlich and Schröder-Preikschat. SMP PCs: A Case Study on Cluster Computing. In *Proc. First Euromicro Workshop on Network Computing*, August 1998.
[5] A. Fröhlich and Schröder-Preikschat. High Performance Application-Oriented Operating Systems – the EPOS Aproach. In *Proc. 11th SBAC-PAD*, September 1999.
[6] R. Jain. *The Art of Computer Performance Analysis*. John Willey & Sons, 1991.
[7] L. Prylli and B. Tourancheau. BIP: a New Protocol Designed for High Performance Networking on Myrinet. In *Proc. PC-NOW'98*, April 1998.
[8] W. Schröder-Preikschat. *The Logical Design of Parallel Operating Systems*. Prentice-Hall, 1994.

Distributed Parallel Query Processing on Networks of Workstations

Cyrus Soleimany[1] and Sivarama P. Dandamudi[2]

[1] IBM Canada, Toronto, Canada
cyruss@ca.ibm.com
[2] Centre for Parallel and Distributed Computing, School of Computer Science
Carleton University, Ottawa, Canada
sivarama@scs.carleton.ca

Abstract. Network of workstations (NOW) is an attractive alternative to parallel database systems. Here we present a distributed architecture for parallel query processing on networks of workstations. We describe a comprehensive design that synthesizes knowledge from parallel and distributed systems area on scheduling, load sharing and loop scheduling. To study the performance of this architecture, we have implemented the distributed architecture using PVM on a Pentium-based NOW system. Our results show that the distributed architecture achieves good speedups and scaleups.

1 Introduction

A typical database system is based on the familiar client-server model. Client nodes submit queries/transactions and receive responses from the database server. Thus, in this model, all the work is done by the server while the client node acts as an interface to the user. With even recent PC-based client nodes providing traditional workstation-class performance, performance improvements can be obtained by offloading some of the processing typically done on the server to these powerful client nodes. Parallel query processing takes advantage of the idle cycles on the client nodes to process a query. The use of networks of workstations has been proposed for parallel query processing [1].

The implications of this proposal are two fold. One, we can use the existing client workstations to speedup query processing. The other implication is that we can build parallel database systems based on a LAN-connected network of workstations (NOW).

The NOW-based architecture presented here is a shared-disk architecture. This is in contrast to the popular shared-nothing architecture used for parallel database systems. Even though the shared-nothing architecture was preferred a few years ago [4], recent technological advances (increased communication bandwidth, low latencies, large memories and faster processors) have shown that shared disk is probably preferred [6]. Our proposed architecture falls into this category. The goal of our project is to explore various architectures that are suitable for parallel query processing on NOW-based systems [2, 12, 13]. In this paper, we present a distributed architecture for parallel query processing on NOWs.

2 Distributed Architecture

This section describes the proposed distributed query processing policies for query scheduling and load sharing. These policies have been adapted from the parallel and distributed systems areas [3, 10, 8].

Due to dynamic query and non-query (background local jobs) arrivals, query complexity, and heterogeneity of the system, some nodes might be overloaded while others are under-loaded [10]. In the distributed architecture, we consider the weighted combination of the non-query and query loads as an indicator of the system load. Then, a simple threshold can be used to determine the state of a node. Threshold values classify the state of a node as under-loaded (i.e., receiver state), overloaded (i.e., sender state), or OK (i.e., normal state) [10]. Several techniques and policies have been proposed in the literature to study load sharing in distributed systems [10]. These policies (sender/receiver-initiated policy, adaptive policy) have been presented to provide load sharing, but not in the context of database query processing.

In our architecture, each node maintains three job queues: a query queue for local and remote queries, a non-query queue for background loads, and a pending queue for partial results of queries or join operations. As in the previous load sharing policies, each node dynamically operates in one of the three states. The state of a node is determined by comparing with two predefined threshold parameters (lower bound and upper bound threshold) [11]. The current load of a node is defined as:

$$l = a.f_1(B) + \sum_{i=1}^{|Q|} g(q_i) \qquad (1)$$

where l = load of the node (expressed in units of job)
$f_1(B)$ = number of jobs in the non-query queue
a = a constant that represents weight of the non-query job queue
$|Q|$ = number of queries in local-query queue
$g(q_i)$ = cost of the ith query in the local-query queue

Another parameter, which increases communication overhead, is the maximum number of probes. P_l represents the maximum probes a workstation uses to locate an under-loaded workstation for sharing the load. Each workstation probes up to P_l nodes to find a workstation to share the load. If such a workstation is found, work is transferred to that node for remote execution [10]. However, granularity of work is related to the current load of the sender workstation. Sending an entire query puts less overhead on the sender workstation than sending an operation or a partial operation. Thus, the granularity of work transferred is as follows:

- An entire query along with the relations information, if the sender workstation is over-loaded, or
- An operation of a query or sub-query along with the relations information, if the sender workstation is in the OK status, or

- A partial operation along with fragments of relations, if the sender workstation is under-loaded [1].

Scheduler, one of the components of the system, runs on each node and has three sub-components: *Load Information Gathering, Transfer Policy,* and *Location Policy*. The main responsibilities of the *Scheduler* are to establish connections with other Schedulers, to communicate with local processes, to respond to all inquiries by external and internal processes, to allocate idle nodes, to reply to the load request of other nodes, and to schedule jobs with appropriate operation levels. Also, in order to maintain consistency among workstations, the following rules have to be taken into account by all workstations:

1. If an under-loaded node receives a request from another node for sharing jobs and then becomes overloaded, it will accept the request and will not transfer the remote job to any other nodes[2].
2. If an under-loaded node receives a remote partial operation, it will not transfer the remote job to any node under any circumstances, due to the cost of transferring.
3. If an overloaded node could not find any under-loaded node to share the job, it would start processing the job until under-loaded nodes become available. While it is processing the job, it could send a probe message to locate an under-loaded node.

The *Transfer Policy* component uses the current load of the system and determines whether to transfer the query or to process it locally. If the load is below the lower bound threshold, the node is in the receiver state (under-loaded state). Thus, the *Scheduler* processes the query in intra-operation (partial-operation) level by splitting a query into several operations and scheduling each operation on as many nodes as possible. This level of operation requires more management and supervision since, for each join operation, the Location Policy component has to be called. The number of nodes involved in each partial-operation level is determined as:

$$e_s = \frac{e_1}{M_w} + (M_w - 1)[C + C^R + m] \qquad (2)$$

where

e_s = estimate of the total execution time with load sharing
$\frac{e_1}{M_w}$ = estimate of the execution time on M_w nodes
C = estimate of the communication time for sending fragments of relations to another node
C^R = estimate of the communication time for receiving the result

---

[1] When the sender workstation is under-loaded, it first sends one partial operation to itself and while it is working on the join operation, it sends a probe message to some other nodes to share the load in order to reduce the response time of a query. Note that the probing is done in parallel with other activities.
[2] A node can send an accepted remote job to another node, if the *number of remote limit* associated with the job is less than the prespecified *transfer limit*.

from remote node
m = estimate of the merging time of the results
M_w = maximum number of nodes that can be employed for load sharing
e_1 = execution time of the operation without load sharing

Assume that a query Q_1 that requires four relations R_1, R_2, R_3, and R_4 to process is submitted to workstation W_1. If W_1 is overloaded then it would transfer the entire query Q_1 to another workstation (assume that, after probing, one idle workstation is located and assigned to Q_1). But if the load of W_1 is OK it tries to send one join operation to an under-loaded one. If W_1 is under-loaded, the *Scheduler* of W_1 calculates e_1 (estimate execution time of a query). If we assume W_1 can employ one workstation to share the load, $M_w = 2$, the last equation gives an estimate of the execution time for two workstations. If $e_1 > e_s$ then W_1 is able to send the partial operation to another workstation. By incrementing M_w and applying the same equation, we can find the maximum number of required nodes for the join operation. After determining M_w, workstation W_1 sends the partial operations to each of the available workstations. All these cases assume that after probing, W_1 is able to find an under-loaded workstation to share the load. Otherwise, it has to process the query locally. Note that in the last equation the communication time, speed of workstations, and the granularity of the jobs are assumed to be the same for all workstations.

Probing for the target workstation stops as soon as the maximum number of under-loaded workstations needed is satisfied or the number of probes reached P_l. The maximum number of workstations, M_w, is the estimated number of workstations on which the *Scheduler* can schedule the current job. However, if the scheduler could not get M_w under-loaded workstations, it will schedule the job on a smaller number of workstations.

The *Location Policy* used here is the sender-initiated policy. Also, for better performance, it first contacts the *Information Gathering* process, which maintains the recent load information on the whole system, to get information on nodes that are likely to be in the receiver state [11].

Our implementation uses the nested-loop join, which is the simplest join technique. The simplicity of this algorithm has made it a popular choice for hardware implementation in database machines. Also, it has been found that this method can be parallelized with great advantage. The parallel version of this algorithm is more efficient than the most other methods [7, 9].

3 Experimental Environment and Implementation

We have conducted our experiments on a Pentium-based network of workstations. The current configuration consists of four types of nodes: Pentium 133 with slow cache, Pentium 133 with fast cache, Pentium 166, and Pentium Pro 200. All nodes have 32 MB of memory. We have used Parallel Virtual Machine (PVM) for message passing [5].

For the purpose of our simulation, a set of tables is created. These tables are stored as regular files under the Linux file system, and all nodes have access to these tables through the DBMS node (one dedicated workstation to database activities). The size of the tables varies between 5000 to 9000 tuples, each tuple consisting of two attributes[3]. There are a total of 16 tables; two of these tables have unique key numbers, which are sequentially generated. Thus, there is no duplicate keys in these two tables. The rest of tables have some common key values, which are generated by the normal distribution with the same mean value and different variance. We also maintain the *selectivity factor* among tables for the purpose of the join operation, which represents the fraction of tuples that qualify compared to the number of tuples in the Cartesian product of the two tables (see [11] for detail).

4 Performance Evaluation

In this section we study the performance of the distributed architecture for parallel query processing. For all the experiments, a batch strategy has been used to compute the confidence intervals (at least 30 batches were used). This strategy produced 95% confidence intervals that are less than 1% of the mean response times when the inter-arrival rate is low to moderate (i.e., lightly loaded system) and less than 5% for moderate to high inter-arrival rates (i.e., heavily loaded system).

4.1 Selection of Parameters

The performance of our system relies on several parameters that have to be predefined in order to run the simulation. After conducting several experiments, we observed that a probe limit P_l of 3 provided the best performance. Another parameter T_s refers to the reinitiation period [3]. When probing fails to identify an underloaded node, the originator of the probes waits T_s seconds before initiating another probe. It can be observed that decreasing the value for T_s results in increasing the overhead of *Transfer Policy* since it reinitiates probing shortly after each probe failure. In the experiments, the value for T_s is set to 2 seconds.

A second preliminary experiment was conducted to study the performance sensitivity of the system to probe limits. These results suggest that the best performance is achieved when the probe limit is 3. When the probe limit increases beyond 3, performance of the system degrades rapidly due to an increase in communication and waiting times.

For the purpose of finding the upper and lower threshold values, we have conducted several experiments. From these experiments, we have observed that a lower threshold limit of 1 and an upper threshold limit of 4 provided the best performance. More details on these experiments are available in [11].

4.2 Performance as a Function of Inter-Arrival Time

Figure 1 shows the performance sensitivity to query load on an 8-node system. The fragmentation size is set to 200,000 bytes, which means no partial operation

---

[3] For simplicity of our experiments, we consider only *key* and *value* as two attributes for each tuple. However, the model does not impose any restrictions on the size of tables, tuples, and attributes.

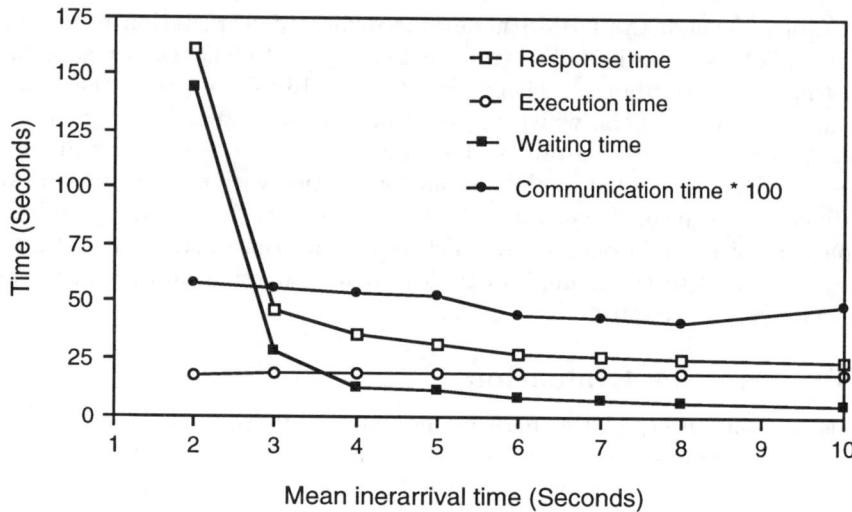

Fig. 1. Performance sensitivity to query load (Note that the communication time shown is 100 times the actual value)

is permitted at least at the first level of join operations. As we shall see in Section 4.4, communication time is minimized at this fragmentation size. It can be observed that when the system is lightly loaded the average response time is 24.31, which is 6.62 times less than 161.04 at heavily loaded status. It is clear that waiting time forms a significant part of the total response time, whereas the execution time, as expected, is almost constant in all the cases.

We also note that the communication time is relatively small compared to the response time (or even execution time). For example, communication time is about 2% of the response time at light query loads (when the inter-arrival times are in the range between 6 and 10 seconds). It becomes even less significant, at moderate to high query loads. However, we will see in Section 4.4 that communication time can be a significant portion of the response time for smaller fragmentation sizes.

Communication time varies from a high of 0.58 second when the inter-arrival time is 2 seconds to 0.48 second when the inter-arrival time is 10 seconds. Communication time is minimum (0.41 second) at inter-arrival time of 8 seconds. Communication time increases with query load (low values of inter-arrival times) because the system would have to do more probing to find nodes in the receiver state. On the other hand, the reason for the increase in the communication time at low query loads is that we send partial operations, which induces more communication overhead.

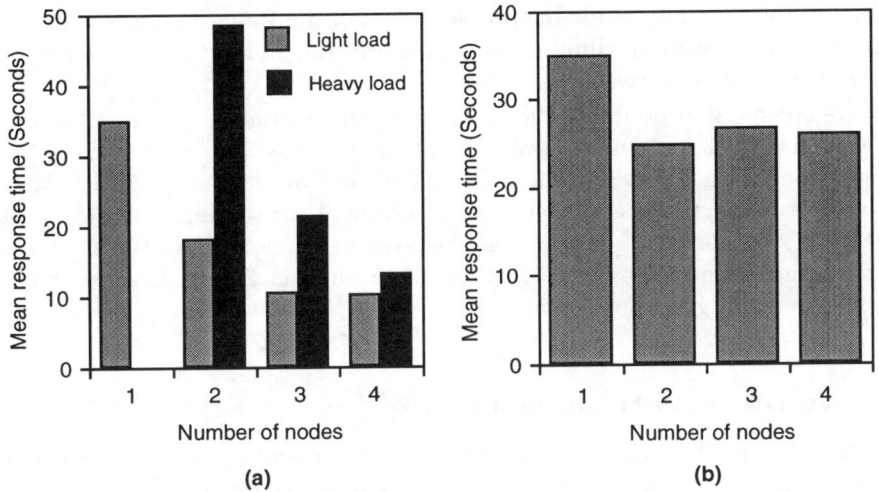

Fig. 2. System speedup and scaleup results (a) Speedup at light and heavy loads (b) Scaleup results

4.3 System Speedup and Scaleup

Ideal parallel systems demonstrate two key properties, namely *linear speedup* and *linear scaleup*. This section presents the speedup and scaleup results for the distributed architecture.

Speedup refers to the characteristic of a parallel system to reduce the execution time as we increase the number of processors. If increasing the number of processors results in a proportional improvement in performance (e.g., response time), the parallel system is said to be successful in achieving linear speedup.

Two experiments are performed to evaluate the speedup of the system under light and heavy loads. The number of processors is varied while the fragmentation size is fixed at 200,000 bytes. Figure 2a shows the performance of a lightly loaded system (the shaded bars). It can be observed that the speedup of a 2-node system compared to a 1-node system is 1.92, and the speedups of 4-node and 8-node systems are 3.37 and 3.44 respectively. Note that there is no performance difference between 8-node and 4-node systems since the system is lightly loaded.

Figure 2a also shows the performance of a heavily loaded system (the black bars). The job inter-arrival time is fixed at 10 seconds. The number of processors is varied from 2, 4, and 8 nodes (no response time is reported for the 1-node system due to saturation). It can be seen that the 4-node and 8-node system can service the same amount of load 2.27 and 3.7 times faster than the 2-node system. From these two experiments, we conclude that the distributed architecture achieves good speedups under both light and heavy workloads.

The third experiment deals with the system scaleup. Linear scaleup refers to the characteristic that with increasing system size (e.g., number of nodes), the system can perform proportionally larger amount of work during the same time period. While speedup holds the problem size constant and increases the system size to reduce execution time, scaleup measures the ability to increase both the system and problem sizes.

We assume that doubling the mean inter-arrival time (e.g., 6 to 12 seconds) results in half the amount of work to be done. Thus, for the 8-node, 4-node, 2-node, and 1-node systems, the inter-arrival time of job is set to 6, 12, 24, and 48 seconds respectively. Another way of looking at the scaleup is to increase the database size. However, due to practical system-imposed limitations, we could not conduct such scaleup experiments. The results in Figure 2b show that the system provides good linear scaleups.

4.4 Sensitivity to Fragmentation Size

In the previous sections, we assumed no partial operation, at least at the first level of a join operation. However, the load sharing aspect of the system can be exploited in intra-operation parallelism. Generally, a larger fragmentation size results in fewer partial operations. There is a trade-off between the fragmentation size and the response time. In this section we study the impact of the fragmentation size on the performance of the system. Increasing the number of partial operations increases the communication overhead, which becomes a significant part of the total response time as shown in Figure 3. It can be observed that as the size of the fragment decreases the response time also decreases until the fragment size reaches 20,000 bytes. After this point the response time increases. The reason for this behavior is that the fine-grain tasks introduces more overhead and increases communication, which directly affects the response time.

The second observation is that as the fragmentation size decreases, the communication cost increases irrespective of the system workload. This is due to the fact that without partial operation the system transfers the entire relations to other nodes, which is the only cost. However, by allowing partial operations, the system has to calculate the size of each partial operation, send a probe message, wait for the response, and transfer the fragment. These overheads in addition to the communication cost are responsible for increasing the response time. Thus, the smaller the fragment size the higher the communication cost.

Note that the response time of the heavily loaded system is higher than the lightly loaded system, whereas the communication cost of the heavily loaded system is lower than the lightly loaded system. The reason is that the system under light workloads has a higher possibility to share the load with some other nodes, while under heavy workloads, the system goes above the lower threshold limit and tries to balance the load by sending the join operation (sends only the execution plan of the join operation, not the relations) to another under-loaded node rather than sharing the load by sending partial operation (sends join operation along with fragments of relations) to some other under-loaded nodes.

A further observation from these results is that the best fragmentation size is 20,000 bytes whether the query load is low or high.

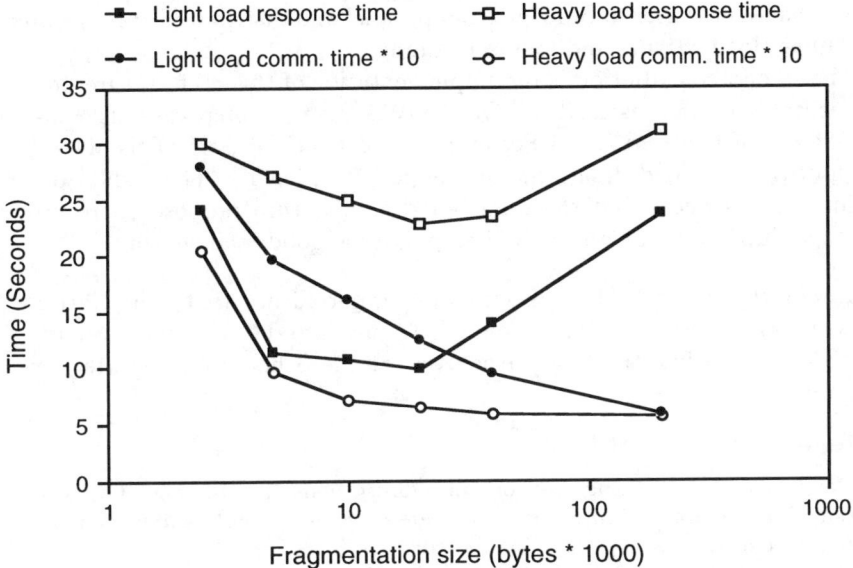

Fig. 3. Sensitivity to fragmentation size (Note that the communication times shown are 10 times the actual value)

4.5 Impact of Background Load

Results in the last three sections are obtained under the assumption that there is no background load (i.e., non-query local load) in the system. In a NOW environment, evaluating performance under background load is important due to the dynamic nature of the background load. We have conducted several experiments to study the impact of the background non-query load on the performance of the distributed architecture. These results show that the distributed architecture is able to provide performance improvement even in the presence of the background load (these results are not presented due to space limitation). Details of these experiments are available in [11].

5 Summary

Network of workstations present a great opportunity to speedup query execution. Query processing can benefit from a NOW-based system because it not only provides additional processor cycles but also adds substantial main memory, which reduces the number of disk access. The improved performance is often obtained without any additional cost by using the existing client workstations.

We make two main contributions to the area of parallel query processing. First, we have presented the design of a distributed architecture for parallel query processing on NOW-based systems. We have described a comprehensive design that synthesizes knowledge from parallel and distributed systems area on scheduling, load sharing and loop scheduling.

The second contribution is the implementation of this architecture on an experimental Pentium-based NOW. Our NOW system is heterogeneous, consisting four classes of Pentium (from Pentium 133 to Pentium 200). This allows us to test how the proposed design handles node heterogeneity. The results reported in this paper suggest that the distributed architecture proposed here provides good speedups and scaleups as well as providing good load sharing.

Acknowledgements. This research was supported in part by the Natural Sciences and Engineering Research Council of Canada and Carleton University. This work was done while the first author was at the School of Computer Science, Carleton University.

References

1. Dandamudi S.P., "Using Network of Workstations for Database Query Operations," *Proceedings of International Conference on Computers and Their Applications*, Tempe, Arizona, 1997, pp. 100-105.
2. Dandamudi S.P. and Jain G., "Architecture for Parallel Query Processing on Networks of Workstations," *Proceedings of International Conference on Parallel and Distributed Computing Systems*, New Orleans, 1997, pp. 444-451.
3. Dandamudi S.P., "Sensitivity Evaluation of Dynamic Load Sharing in Distributed System," *IEEE Concurrency*, July/September 1998, pp. 62-72.
4. Dewitt D. and Gary J., "Parallel Database System: The Future of High Performance Database Systems," *Communications of The ACM*, 1992, Vol. 35, No. 6, pp. 85-98.
5. A. Giest, A. Beguelin, J. Dongarra, W. Jiang, R. Mancheck, and V. Sunderam, "PVM: Parallel Virtual Machine — A Users' Guide and Tutorial for Networked Parallel Computing", *The MIT Press*, 1994.
6. Invited Lecture, Workshop on High Performance Computing on Very Large Datasets, 7th Int. Conf. High Performance Computing and Networking, Amsterdam, April 1999.
7. Harris E.P. and Ramamohanarao K., "Join Algorithm Costs Revisited," *VLDB Journal*, Vol. 5, No. 1, 1996, pp. 64-84.
8. Lilja D.J., "Exploiting the Parallelism Available in Loops," *IEEE Computer*, Vol. 27, No. 2, 1994, pp. 13-26.
9. Mishra P. and Eich M., "Join Processing in Relational Databases," *ACM Computing Surveys*, Vol. 24, No. 1, 1992, pp. 63-113.
10. Shivaratri N.G., Krueger P., and Singhal M., "Load Distributing for Locally Distributed Systems," *IEEE Computer*, Vol. 25, No. 12, 1992, pp. 34-44.
11. Soleimany C., *Distributed Parallel Query Processing on Network of Workstations*, M.C.S. Thesis, School of Computer Science, Carleton University, Ottawa, 1998.
12. B. Xie and S. Dandamudi "Hierarchical Architecture for Parallel Query Processing on Networks of Workstations," *Proceedings of the Int. Conf. High Performance Computing*, Chennai, India, 1998, pp. 351-358.
13. S. Zeng and S. Dandamudi "Centralized Architecture for Parallel Query Processing on Networks of Workstations," *High Performance Computing and Networking*, 1999, Amsterdam, pp. 683-692.

Track IV

Industrial and End-User Applications Track

High Scalability of Parallel PAM-CRASH with a New Contact Search Algorithm

Jan Clinckemaillie[1], Hans-Georg Galbas[2], Otto Kolp[2], Clemens August Thole[2], and Stefanos Vlachoutsis[1]

[1]ESI SA, Rue Saarinen 20, Silic 270, F-94578 Rungis Cedex, France
[2]GMD-Institute for Algorithms and Scientific Computing (SCAI)
Schloss Birlinghoven, D-53757 St. Augustin, Germany

Abstract. The numerical simulation of crashworthiness plays an important role in the vehicle design-cycle. PAM-CRASH, from ESI, has been successfully exploited for industrial simulation on massively parallel systems. The most time critical part of the parallel simulation is the contact handling. ESI and GMD have developed a new contact search algorithm (CSA) [1]. The implementation of this algorithm into PAM-CRASH leads to a better scalability. Results of measurements on a 128-processor SGI machine demonstrate the good performance of PAM-CRASH with CSA.

1 Introduction

PAM-CRASH is an explicit time-marching, finite-element program used for the numerical simulation of the highly non-linear, dynamic phenomena arising in short-duration contact-impact problems. Its major domain of application is in the crashworthiness assessment of passenger cars [2]. During the last years a message-passing parallel PAM-CRASH program has been constructed which includes all features necessary for a full, industrial front or off-set car-crash simulation. The effectiveness of the code migration has been demonstrated on a range of parallel platforms, with benchmarking being performed by two industrial end-users (Audi & BMW) using industrially relevant benchmark models. Preliminary results [3] highlight the crucial role that the parallel contact-impact algorithms play in determining the overall performance of the parallel crashworthiness simulation. In the project KALCRASH, funded by the German Ministry for Research and Education (BMBF), a new contact search algorithm (CSA) has been developed for handling contact-impact algorithms on distributed-memory machines. First simulations of CSA with input data of PAM-CRASH showed encouraging results [1].

The integration of CSA into parallel PAM-CRASH has taken place during the project SIM-VR [4]. PAM-CRASH with CSA (PCSA), is an experimental version of PAM-CRASH restricted to contact type 6. In the moment it can handle one slideline, i.e. a subset of the structure within which contact may occur. But the approach is more general such that multiple slidelines can be integrated effectively. Also other contact types such as contact type 36 are possible.

The input data set was the BMW-CRASH50, a 60 000 element model of BMW. The BMW-CRASH50 was considered for one slideline concerning contact. This slideline of about 21 000 contact segments contains the part of the structure most affected in a 40 % off-set crash.

The integration task has been performed in several steps:
- Definition of an interface between PC and CSA.
- Implementation of the interface and testing on an IBM-SP2 of GMD.
- Improvement of the interface and implementation on different parallel systems.

Furthermore the integration of a static load balancing strategy was performed. To solve the problem of load balancing an over-partitioning concept has been developed [1]. The main idea here is to combine parts of the structure with high contact intensity (in the front of the car) with parts of low contact intensity (in the rear) to achieve a good balance for the contact part of the problem while keeping the remaining part of the problem balanced. To integrate this strategy into PAM-CRASH, partition files were constructed for the BMW-CRASH50 model.

After these steps performance measurements were carried out on a 34-processor IBM-SP2 of GMD and on a 128-processor SGI-Origin 2000 of SGI. The results generally show that a good scalability is obtained for PCSA.

2 CSA a Fast Contact Search Algorithm

The parallel, multi-level strategy for the proximity search addresses two of the major causes of overhead involved with the inclusion of contact-impact algorithms: by minimising the amount of non-local data provided for the proximity search, overheads arising from both communication and repetition of computation related to the search (for example, sorting) are minimised.

Within the macroscopic part of the strategy, a hierarchy of levels is constructed for each process which partitions all elements, local and non-local, according to their distance from the set of locally held nodes. Only those elements in the first level could possibly be involved in contact with the locally held nodes. Following the macroscopic phase, the microscopic part of the strategy determines, for each local node, the nearest elements within the first level in case contact is possible.

Each level-step is followed by a fast step, which is repeated n times. During such a fast-step the co-ordinates of nodes which are candidates for contact pairs and which are determined during the previous level-step are exchanged between processes if necessary.

In each time step the pairs of nodes and segments which may be involved in contact will be given to the part of PAM-CRASH which handles the impact calculations.

How often a fast-step is repeated depends on different parameters as relative speed of the nodes or distance of the levels. But considering the BMW-CRASH50 application the number of repetitions of the fast-loop is between 100 and 200, i.e. the number of time consuming level-steps is between 0.5 – 1 % of the total number of time steps.

CSA involves a complex communication structure with a very dynamic behaviour. On each level a process can communicate with a different number of processes. This number is growing with higher levels. The number of communications may be different for the different processes. If a process comes to the same level again, then the number of communications and the partners may have changed.

3 Integration of CSA into PAM-CRASH

The goal of porting CSA into PAM-CRASH was to substitute the proximity searching contact algorithm of PAM-CRASH by the new algorithm CSA [1].

The contact computing includes two important operations; the first is the proximity search that results in a contact candidates list of nodes or segments and the second is the impact computing which operates a final geometrical contact detection and the application of contact forces. The first phase generally needs a high computing time. Using CSA for this phase results in considerable reduction of computing time.

The parallel platforms were the MPI (Message Passing Interface) platform, the IBM SP2 machine of GMD with 34 processors and the SGI-Origin 2000 with 128 processors. On the SP2 machine and on the Origin the native form of MPI on each machine was used.

In the contact computing part the first contact phase, i.e. the proximity search, was substituted by the new algorithm. New routines were created in order to adapt the CSA algorithm results into the PAM-CRASH arrays. The impact computing itself remains unchanged. All this work was made for the contact type 6 of PAM-CRASH for which CSA was originally developed.

After this implementation phase, PCSA was tested on the SP2. The first performance results did not fulfil the expectations, which are obtained by earlier simple simulations [1]. However it followed an improvement phase. The main result of this phase was to avoid doubled computing and doubled communication, for example all communication necessary for contact search was included in CSA and could be avoided in the PAM-CRASH part. After this phase and encouraging results on SP2, it was decided to perform test runs for larger systems on the 128- processor SGI-Origin 2000 of SGI.

4 Load Balancing

To get a good performance from a parallel system a good distribution of the work to the different processors is of great importance. In [1] a static load balancing strategy (over-partitioning) was developed which is based on domain decomposition tools. With the help of a domain decomposition tool the car will be divided into parts of equal size. Such a tool makes a well balanced partition for the non contact part but in general not for the contact part.

The over-partitioning concept works as follows: Instead of using the partition of the domain partition tool directly a refined partition is considered. A refined partition is a partition containing a multiple number of blocks than parallel processors; for example, if 8 processors are taken a partition of 32 blocks will be considered. Then each processor is assigned the same number of blocks, (i. e. 4). This has the effect that the non contact part remains well balanced, on the other hand the distribution of blocks to the processors are done in such a way that the contact part of the car is well distributed among the processors. This results in a new partitioning file which is given to the system in place of the file developed by the domain decomposition tool. The combination of the blocks of the finer partition (32) to form blocks of the coarser partition (8) has to be done very carefully, since the number of neighbours of a processor increases. The resulting increase in communication has to be kept as low as possible. Work to develop such partitions automatically for a given application model is done in the ongoing Autobench project funded by the BMBF.

Starting from the over-partitioning concept the development of a dynamical load balancing approach has been considered in [5].

5 Performance of PCSA and Load Balancing

The BMW-CRASH50 model consists of about 60 000 elements and 80 000 time steps are needed for a simulation. The contact regions are defined by slidelines. In this investigation one slideline with about 21 000 segments is used for contact.

We made measurements on two systems on the 34-processor IBM-SP2 RS/6000 of GMD and on the 128-processor SGI-Origin 2000 with 195 MHz.

On IBM-SP2 we considered systems with 1, 8, 16 and 32 processors. On the Origin we made measurements for 1, 32, 48, 63, 96 and 126 processors.

For the same number of processors several partitions of elements are considered. The o-partitions based on the over-partitioning concept as described above generate a better performance up to 48 processors. But they became less efficient for larger systems. The advantage of the o-partitions is the better distribution of the contact segments to the processors which on the other hand involves more communication between the processors. In the studied test case there are only 20 700 possible contact segments. From this fact it is natural that a further 'better distribution' of the contact nodes to the processors does not lead to a better performance if a certain number of processors is exceeded. Another disadvantage of the o-partitions is that the different blocks belonging to one processor create a higher communication in the finite element phase of the PAM-CRASH program. If more processors are used the finite element work per processor is reduced but communication for this work is growing relatively.

The best result with o-partitions is obtained with 32 processors. The gain for such a partition was about 15% compared to the best PAM-CRASH partition.

In Figure 1 we have shown the speedup of PAM-CRASH with CSA and without CSA on SP2 and on the Origin of SGI. The speedup on SP2 is given for 8, 16 and 32 processors whereas the speedup on the Origin is given for 32, 48, 63, 96 and 126 processors. The best speedup for PCSA is 20 for 32 processors on SP2 and the best speedup for PCSA on the Origin is 50 with 126 processors which results in a run time of about 27 minutes.

Considering efficiency (Figure 2) the 'break-even point' of 0.5 is reached with 96 processors for PCSA on the Origin. From this a good scalability of PAM-CRASH can be derived. Besides we have to consider that the BMW-CRASH50 model with about 60 000 elements and one slideline of about 21 000 elements for contact is rather small for such a large number of processors.

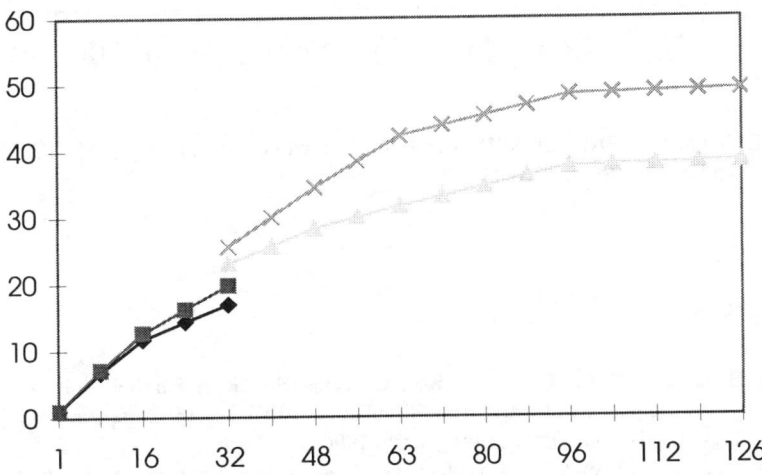

Fig. 1. Speedup of PAM-CRASH with CSA and without CSA on IBM-SP2 for 8 – 32 processors and on the Origin of SGI for 32 – 126 processors.

6 Summary

The scalability of parallel PAM-CRASH could be improved by integrating a new contact search part. Moreover it was shown that new partitions constructed with the concept of over-partitioning for the distribution of workload reduce the problem of imbalance. In the whole a speedup of 50 for a relative small application as the BMW-CRASH50 model shows the possibility of PAM-CRASH to obtain high scalability for larger applications on large parallel systems.

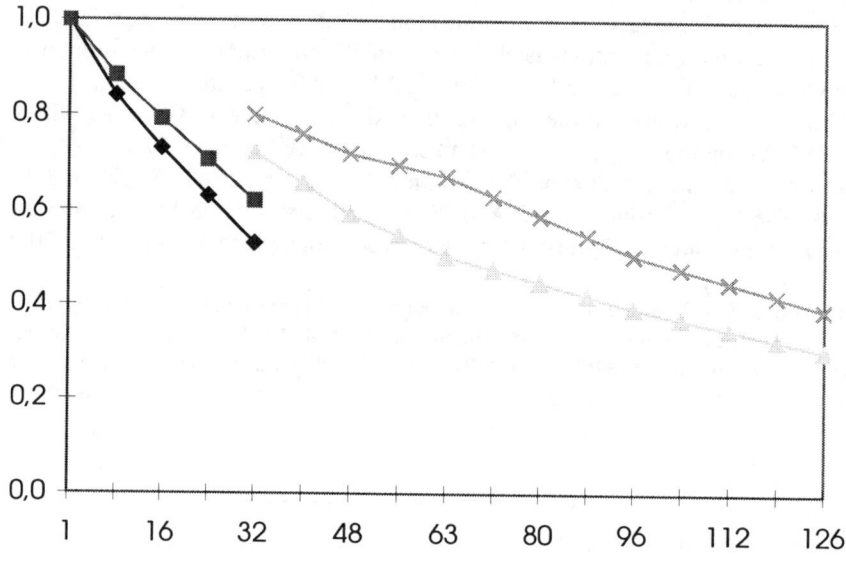

Fig. 2. Efficiency of PAM-CRASH with CSA and without CSA on IBM-SP2 for 8 – 32 processors and on the Origin of SGI for 32 – 126 processors

References

1. Elsner, B., Galbas, H. G., Goerg, B., Kolp,O., Lonsdale, G.: A Parallel Multi-level Contact Search Algorithm in Crashworthiness Simulation. Advances in Computational Structures Technology, Civil- Comp Press, Edingburgh, 1996.
2. Clinckemaillie, J., Elsner, B., Lonsdale, G., Meliciani, S., Vlachoutsis,S., de Bruyne, F., Holzner, M.: Performance Issues of the Parallel PAM-Crash Code. The International Journal of Comp. Applications and High Performance Comp., Vol. 11, No. 1, 1997.
3. Lonsdale, G., Elsner, B., Clinckemaillie, J., de Bruyne, F., Holzner, M.: Experiences with Industrial Crashworthiness Simulation Using the Portable, Messing-Passing PAM-CRASH Code, Lecture Notes in Computer Science 919, Springer-Verlag, 1995.
4. Thole, C.-A., Kolp, O., Galbas, H.G., Vlachoutsis,S., Werner,H., Wind, J., Clinckemaillie, J., Goebel, M.: SIM-VR Interactive Crash Simulation, Lecture Notes in Comp. Science and Engineering, Springer-Verlag, to appear.
5. Galbas, H.G., Kolp, O.: Dynamic Load Balancing in Crashworthiness Simulation. Lecture Notes in Computer Science, 1573, Vector and Parallel Processing-VECPAR'98, Springer-Verlag, Berlin, 1999.

Large-Scale Parallel Wave Propagation Analysis by GeoFEM

Kazuteru Garatani[1], Hisashi Nakamura[1],
Hiroshi Okuda[2], and Genki Yagawa[3]

[1] Research Organization for Information Science & Technology (RIST)
1-18-16 Hamamatsucho Minato-ku Tokyo, Japan
k-garatani@tokyo.rist.or.jp, nakamura@tokyo.rist.or.jp,
[2] Yokohama National University, 79-5 Tokiwadai Hodogaya-ku Yokohama, Japan
okuda@typhoon.cm.me.ynu.ac.jp
[3] University of Tokyo, 7-3-1 Hongou Bunkyo-ku, Japan
yagawa@garlic.q.t.u-tokyo.ac.jp

Abstract. GeoFEM is solid earth simulator software, which is under development to be used by STA's super parallel computer, "Earth Simulator (GS40)". GeoFEM has already been used to analyze large-scale static linear problems up to 100M (100,000,000) degrees of freedom (DOF) and demonstrated high performance computing ability by parallel processing. Therefore, in the present study, we apply the same data structure and a similar parallelization method to wave propagation problems. We describe the formulation, parallelization procedure and benchmark results of the wave propagation analysis function of GeoFEM. According to the simple benchmark problem, GeoFEM demonstrated more than 98% of the rate of CPU usage performing parallel wave propagation analysis of 100M DOF problem using 1,000 PEs and also shows almost linear scalability on the Hitachi SR2201.

1 Introduction

GeoFEM[1],[2],[3] is parallel FEM code developed to simulate solid earth phenomena. The system is aimed to be used by the super parallel computer, "Earth Simulator" which is planned to be completed in the 2001 fiscal year. GeoFEM has already been applied to analyze large-scale static linear problems up to 100M (100,000,000) degrees of freedom (DOF) and demonstrated high performance computing ability by parallel processing[4],[5]. The system is composed of several sub-systems which depend on each analysis target. Structure sub-system covers large-scale and complex solid earth phenomena by means of constructing a structural problem. The function of wave propagation analysis which is one of the structure sub-system function is to analyze the dynamic response of solidus behavior using an explicit method. This analysis function is responsible for the smallest spatial scale and the shortest time scale of the GeoFEM analyses[2],[6], and is targeted at the seismic wave propagation problem.

In the present study, first we describe the formulation of the wave propagation analysis function of the GeoFEM. The method used in this analysis, is a discrete-space system by the finite element method (FEM), and a discrete-time system by the central differential method. Next, the wave propagation analysis function, implemented by explicit time integration is used to examine a dynamic response problem. In order to investigate the effect of parallel processing, we adopt an efficient data structure, which is also used in static analysis[7],[8] for the wave propagation analysis function. This data structure and partitioning idea is one of the best suited for parallel FEM. Finally, the results of the benchmark analysis are discussed.

2 Formulation

The present study, the GeoFEM wave propagation feature is analyzed by evaluating the behavior at time $t+\Delta t$ of the equation of motion at current time t using the central differential method. This procedure is called an "explicit method" because the method does not require simultaneous equations[9].

2.1 Equation of Motion

Applying the FEM to a discrete-space system, the equation of motion is expressed by mass matrix $[M]$, dumping matrix $[C]$, internal force vector $\{p\}^{(n)}$, external force vector $\{f\}^{(n)}$, displacement vector $\{u\}^{(n)}$, and its first and second order time derivation, velocity and acceleration vector, $\{\dot{u}\}^{(n)}, \{\ddot{u}\}^{(n)}$ respectively (1). This discretization does not depend on element shape or on a continuous model. Here $^{(n)}$ means at time t.

$$[M]\{\ddot{u}\}^{(n)} + [C]\{\dot{u}\}^{(n)} + \{p\}^{(n)} = \{f\}^{(n)} \tag{1}$$

2.2 Explicit Time Integration

The next central differential equations concerned with the 1st and 2nd derivations of the displacement vector $\{u\}^{(n)}$ are given by the Taylor expansion of displacement at time; t up to the Δt^2 term at time; $t + \Delta t$ and $t - \Delta t$, denoted by $\{u\}^{(n+1)}, \{u\}^{(n-1)}$.

$$\{\dot{u}\}^{(n)} = \frac{1}{2\Delta t}(\{u\}^{(n+1)} - \{u\}^{(n-1)}) \tag{2}$$

$$\{\ddot{u}\}^{(n)} = \frac{1}{\Delta t^2}(\{u\}^{(n+1)} - 2\{u\}^{(n)} + \{u\}^{(n-1)}) \tag{3}$$

Substituting equations (2) and (3) and rearranging for $\{u\}^{(n+1)}$ equation (1) becomes,

$$\{u\}^{(n+1)} = \left[[M] + \frac{\Delta t}{2}[C]\right]^{-1}$$

$$\left\{\Delta t^2\left[\{f\}^{(n)}-\{p\}^{(n)}\right]+2[M]\{u\}^{(n)}-\left[[M]-\frac{\Delta t}{2}[C]\right]\{u\}^{(n-1)}\right\} \tag{4}$$

In order to solve equation (4) without using simultaneous equations, the mass matrix $[M]$ and dumping matrix $[C]$ must be diagonal matrices. GeoFEM adopts a lumped mass matrix concentrated at a nodal point and also adopts a proportional dumping, which is assuming the dumping matrix is proportional to the mass matrix. Under these assumptions, equation (4) can be rewritten as a a scalar equation, composed of i-th vector components denoted by $_i$ and diagonal matrix components denoted by $_{ii}$.

$$u_i^{(n+1)}=\frac{\Delta t^2(f_i^{(n)}-p_i^{(n)})+2M_{ii}u_i^{(n)}-(M_{ii}-\frac{\Delta t}{2}C_{ii})u_i^{(n-1)}}{M_{ii}+\frac{\Delta t}{2}C_{ii}} \tag{5}$$

Equation (5) is conditionally stable, since it is limited by the time increment Δt. This restriction is known as the "Courant condition" which is expressed by the minimum nodal distance of an element l_e and elastic wave speed c as follows.

$$\Delta t<\frac{l_e}{c} \tag{6}$$

2.3 Internal Force

First, in order to solve equation (4), the internal force vector $\{p\}^{(n)}$ must be evaluated. The internal force vector is given by the integral of the transpose matrix of the displacement-strain relationship $[B]^T$ and stress $\{\sigma\}$ as follows.

$$\{p\}^{(n)}=\int [B]^T\{\sigma\}^{(n)}dv \tag{7}$$

Usually, the internal force is computed using "element by element" approach, that is by evaluating equation (7) for each element. However for a linear problem, equation (7) can be rewriten as follows.

$$\{p\}^{(n)}=[K]\{u\}^{(n)} \tag{8}$$

Here, $[K]$ is the stiffness matrix, same as a linear static problem, and is composed of the superposition of the elemental stiffness matrix $[K^e]$ as expressed below.

$$[K^e]=\int_{elm}[B]^T[D][B]dv \tag{9}$$

In GeoFEM, the internal force is evaluated using equation (8) assuming a linear problem. Equation (8) implies that after the stiffness matrix is composed, only multiplication is required to obtain the internal force. This method has both merit and demerit. The merit is a reduction in the computation process, because only one stiffness matrix need to be composed compared with the element by element approach requires evaluating equation (7) for each step. However, this method requires a large amount of memory in order to store the stiffness matrix $[K]$. The amount of memory required is similar to that of static analysis, and can be estimated using equation (10).

$$(\text{Memory requirement})[\text{Byte}] = 1,000 \times (\text{Degrees of freedom})[\text{DOF}] \quad (10)$$

Equation (10) means approximately 120 variables (double precision real type variable; 8 bytes/variable) are required per one DOF. However in comparison, if the internal force is evaluated using equation (7) and the element by element approach, then only approximately 10 variables are required per DOF. This requires approximately 1/10 of the amount of memory required by current method. As a result, GeoFEM method can be used to analyze linear wave propagation problems with high speed computation.

3 Parallel Processing

Under the assumptions described in the previous section, equation (4) becomes a scalar equation, equation (5) composed of independent displacement terms. However, matrix multiplication is required in order to compute the internal force term as given by equation (8). This means, to parallelize these processes, only the computation process of equation (8) requires communication. The GeoFEM system adopts node based partitioning as shown in figure 1 in accordance with network theory. As can be seen from figure 1, overlapped elements exist at each domain boundary. For this type of partition, every internal node must belong to one subdomain that external node is defined as a node that is connected to an internal node on a subdomain, but that is not an internal node. So, internal nodes in a subdomain has all of the connected node information as for internal or external nodes. In the case of matrix multiplication of equation (8), for a matrix as shown in figure 2 composed of internal and external nodes, the value of the internal nodes can be calculated correctly. After the time step computation of equation (5), the value of the external nodes by communication from the value of the internal nodes which belong to the other subdomain. This alternate nodal value transfer realizes more effective communication than the element based partitioning system. The time progress procedure and communication necessary to allow parallel processing is as follows.

1. Prepare partial matrix $[\bar{K}]$ and vector $\{\bar{u}\}^{(n)}$ composed of internal and external nodes in each subdomain as shown in figure 2.
2. Compute matrix multiplication concerning the internal nodes in each subdomain.
3. After evaluation of equation (5), communicate the value of external nodes $\{\bar{u}\}^{(n+1)}$ corresponding to internal nodes in the other subdomain.
4. Repeat steps 2 and 3 above.

This type of partitioning, communication and data structure is defined as the GeoFEM solver subsystem[7],[8] which allows efficient FEM parallel processing.

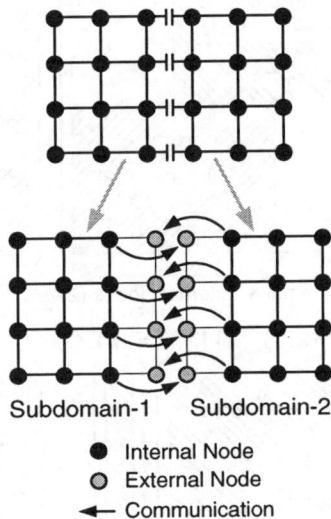

Fig. 1. Node Based Partition

4 Benchmark Analysis

4.1 Problems

In this section, various elapsed times are measured to evaluate the performance of the wave propagation analysis for a large-scale parallel problem. The benchmark problem is a unit cube model as shown in figure 3. The model allows the scale of the problem and subdomain or PE count to be changed easily and is suitable for a parameter study to evaluate parallel performance. This type of benchmark problem is also used for static linear analysis. The analysis condition is to apply an impacted load on the $X = 1$ plane at time $t = 0$, under appropriate material constants. The analysis is executed for 100 time integration steps and the elapsed time is measured. The parallel computation was executed using in Hitachi SR2201 of the University of Tokyo. The SR2201 has 1,024 PEs and 256MB memory in each PE.

4.2 Measurement of Elapsed Time

Figure 4 shows the elapsed time measured for the following two parameters, the number of nodal points per line of subdomain, n_{sn}; and the subdomain division

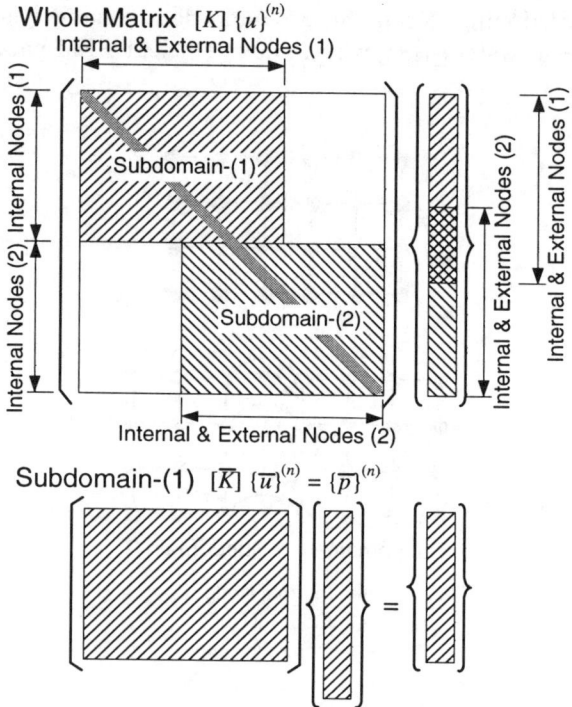

Fig. 2. Parallel Matrix Operation

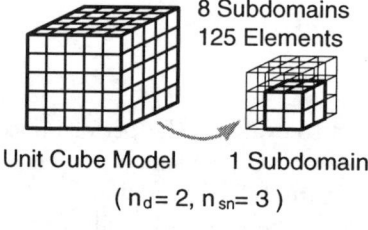

Fig. 3. Cubic Mesh for Benchmark Analysis. This cube has a unit length in the X, Y and Z directions. The model can be described by the following two parameters: the number of subdomains on each side of the model (n_d) and the number of internal nodes on each side of the subdomain (n_{sn}).

count per line of model, n_d. In this benchmark, n_{sn} varies from 2 to 33 or 39 due to the memory capacity limitation of one PE, and n_d changes from 1 to 10. The maximum DOF case is $n_{sn} = 33$ and $n_d = 10$ (1,000PE). This gives a total node count of $330 \times 330 \times 330 = 35,937,000$ nodes or $107,811,000$ DOF, that is, approximately 100 million DOF. The solid line in figure 4 represents a prediction of the elapsed time in one PE. In this type of analysis, the total floating point operation count is proportional to DOF, and so it follows that elapsed time is also proportional to DOF. The elapsed time can be decomposed into two parts; mass or stiffness matrix composition process and the time integration process. Therefore, a predicting equation for measured elapsed time, expressed by equations

(11) and (12) can be obtained applying a least minimum square approximation. The sum of the two equations gives total elapsed time with 100-time integration as given by equation (13). As apparent from equations (11) and (12), the elapsed

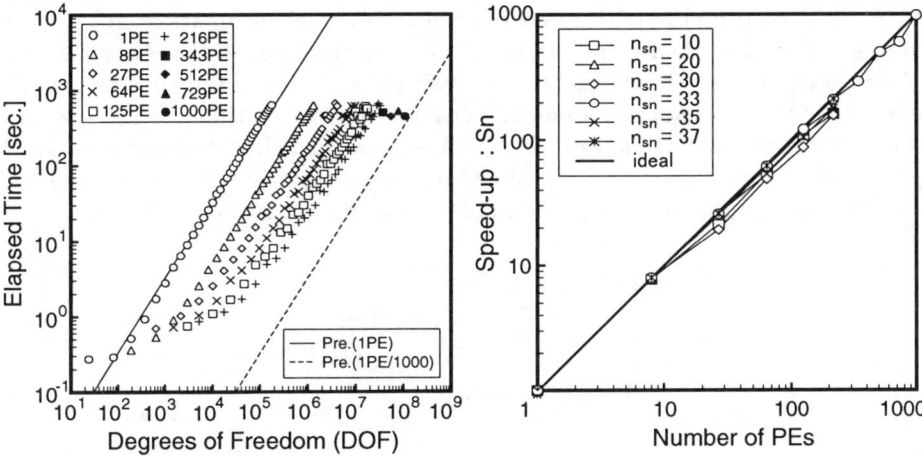

Fig. 4. Elapsed Time by Cubic Model

Fig. 5. Scalability

time to compose the mass or stiffness matrix is approximately 100 times larger than the time required to execute one integration step. In the former, the time to compose the stiffness matrix is dominant. Hence, if an element by element approach is applied without composing stiffness matrix, it takes approximately 100 times longer to compute this type of analysis, because the stiffness matrix re-evaluates each time step. The current GeoFEM is approximately 100 times faster than the method traditionally used in this field[9], but uses ten times as much memory and limiting linear analysis.

$$\text{(Time to Compose Mass and Stiff. Mat.)} = 1.51 \times 10^{-3} \times \text{DOF} \quad [\text{sec.}] \quad (11)$$
$$\text{(Time of one Integration Step)} = 1.61 \times 10^{-5} \times \text{DOF} \quad [\text{sec.}] \quad (12)$$
$$\text{(Total Time (100 Integration Steps))} = 3.12 \times 10^{-3} \times \text{DOF} \quad [\text{sec.}] \quad (13)$$

The dotted line in figure 4 shows 1/1000 of equation (13). This means if the parallel efficiency is 100 % using 1000 PEs, the computation will be finished in this elapsed time. Actually the elapsed time using 1000 PEs is very near to the dotted line, which shows that high parallel efficiency is achieved by this computation.

4.3 Parallel Performance

In general, in order to obtain the parallel efficiency P_n or speed-up S_n, not only the elapsed time by parallel processing T_n is required, but also the time taken by one PE T_1, as in equation (14). Usually, it is very difficult to compute a large-scale parallel problem using only one PE, because of the memory limitations. At the case of static analysis, the elapsed time prediction equation is used instead of T_1. However in the present study, this method is not suitable due to the less of accuracy in the prediction as can be seen in figure 4. Therefore, the speed-up S_n is obtained from the CPU usage of equation (15), as in equation (16). Here n means the number of subdomains or PEs. In equation (15), $\{(Working\ Time) + (Idle\ Time)\}$ means the elapsed time in each PE and $(Idle\ Time)$ can be directly evaluated by the sum of communication time and awaiting time. The reason why this type of approximation is possible, is due to the fact that in this explicit wave propagation analysis, operation count differs only slightly between parallel and serial computation, and so equation (16) holds approximately.

$$S_n = T_1/T_n\ ,\quad P_n = S_n/n \tag{14}$$

$$W_n = \frac{\sum(Working\ Time)}{\sum\{(Working\ Time) + (Idle\ Time)\}} \tag{15}$$

$$S_n \simeq W_n \times n\ ,\quad (W_n \simeq P_n) \tag{16}$$

Figure 5 shows the scalability of S_n by evaluating W_n. Table 1 contains the calculated values for ($n_{sn} = 33$). We have to point out on these figure and table, keeping the node count of one PE in constant, so it becomes total scale of problem larger as PE count increasing. This means keeping n_{sn} constant, and varying only n_d, what is called a scaled speed–up under constant granularity. For the largest scale problem (107,811,000 DOF,1000PE) the rate of CPU usage W_n=98.23% and speed-up S_n=982. A reduction in efficiency can be observed in several PE counts in table 1. The reason for this phenomenon was investigated and an abnormally long message time in some cases was found. However, the occurrence of this delay is not regular, and therefore, it may be the effect of an other job or waiting for an input or output. In this performance evaluation, a constant time increment was assumed for all cases, even though for actual analysis, when the space discretization decreases due to large-scale analysis, the time increment must be reduced to maintain the numerical stability.

5 Conclusion

The present study, we focused on the large-scale wave propagation analysis feature of the GeoFEM system. The formulation, method of parallel processing and results of performance analysis using a benchmark problem have been described. In conclusion, the present study has shown the scalable feature for this type of analysis and high performance in parallel processing. In a future study, the authers intend to apply this GeoFEM method to an actual problem to simulate solid earth phenomena.

Table 1. Rate of CPU Usage ($n_{sn} = 33$)

n_d	PEs	Nodes	DOF	W_n	S_n
1	1	35,937	107,811	100.00%	1.00
2	8	287,496	862,488	99.54%	7.96
3	27	970,299	2,910,897	95.32%	25.74
4	64	2,299,968	6,899,904	96.30%	61.63
5	125	4,492,125	13,476,375	97.08%	121.35
6	216	7,762,392	23,287,176	97.36%	210.29
7	343	12,326,391	36,979,173	85.32%	292.63
8	512	18,399,744	55,199,232	97.88%	501.15
9	729	26,198,073	78,594,219	82.80%	603.59
10	1,000	35,937,000	107,811,000	98.23%	982.31

Acknowledgement

Authors would like to thank Professor G.Yagawa, of the University of Tokyo, the supervisor of the "GeoFEM" project and all the members of the GeoFEM team for their discussion and advice.

References

1. http://geofem.tokyo.rist.or.jp
2. G.Yagawa, H.Okuda, H.Nakamura GeoFEM: Multi-purpose parallel FEM system for solid earth, WCCM IV 1998;Vol.II:1048-1058
3. M.Iizuka, H.Nakamura, K.Garatani, K.Nakajima, H.Okuda, G.Yagawa GeoFEM: High-performance parallel FEM for geophysical applications, ISHPC II 1999; LNCS-1615: 292-303
4. K.Garatani Feasibility study of GeoFEM for solving 100 million DOF problems, 1st ACES Workshop Proceedings; 331-336 1999
5. K.Garatani, K.Nakajima, H.Okuda, G.Yagawa Three dimensional elasto-static analysis of 100 million degrees of freedom, *contributed to Computers and structures* 1999
6. K.Garatani, H.Nakamura, H.Okuda, G.Yagawa GeoFEM: High performance parallel FEM for solid Earth, HPCN Europe 1999;LNCS 1593:132-140
7. R.S.Tuminaro, J.N.Shadid, S.A.Hutchinson Parallel sparse matrix vector multiply software for matrices with data locality, SAND 95-1540J, 1995
8. K.Nakajima and H.Okuda, Parallel Iterative Solvers with Localized ILU Preconditioning for Unstructured Grids on Workstation Cluster, 4-th Japan-US Symposium on FEM in Large-Scale Computational Fluid Dynamics Proceedings, 25-30, 1998
9. D.R.J. Owen, E.Hinton , *Finite Elements in Plasticity*, Pineridge Press, 1980.

Explicit Schemes Applied to Aeroacoustic Simulations The RADIOSS-CFD System

Dimitri Nicolopoulos[1] and Alain Dominguez[2]

[1] M^3, 58 Avenue des Caniers,13400 Aubagne, France
dimitri@mcube.fr
[2] SGI, 21 rue Albert Calmette, 78350 Jouy en Josas, France
aldo@sgi.com

Abstract: Explicit schemes applied to the resolution of compressible Navier Stokes equations and structure conservation laws resolution can be used efficiently to derive acoustic behavior of complex industrial systems. High CPU cost leads to a careful parallelization of the code and yields good results on NUMA computer systems.

1   RADIOSS-CFD System Presentation

1.1   Background

Since 15 years, the R&D team of the RADIOSS group has developed a wide expertise in the Explicit Finite Element Analysis field. The first solver was dedicated to highly non-linear structural phenomena simulation. It is now a worldwide leader for crashworthiness analysis. In 1992, it was decided to develop a Navier Stokes compressible solver [1], fully coupled with the existing structural software in order to simulate complex fluid structure interaction problems involving unsteady compressible flows. Since 1996, a strong emphasis has been put on aero-acoustics. A strong collaboration with industrial clients worldwide allowed the development of an original methodology in order to solve with only one code what used to be split between at last two to three different packages. The RADIOSS-CFD system involves two main solver modules, M-Explicit and M-Implicit.

1.2   Methodology Description

In order to be able to compute in a single simulation the aero-acoustic behavior of complex industrial devices, the following methodology has been developed [2]:
   a-   Define a FE model using unstructured mesh,
   b-   Run a first simulation in order to reach a converged unsteady flow,
   c-   Run a second simulation (continuation of run b) to record the time domain data:

-In the computational domain if the virtual microphone is located within the domain boundaries.
-On the domain boundaries with enough points to resolve the highest frequency to be resolved if the noise in the far field is sought.
d- Post process the results to get the classical frequency domain results:
-Global Sound Pressure Level at specific locations,
-Distribution along the different frequencies (frequency resolution is classically limited for the Fourier Transform to the sampling rate divided by the number of recorded samples).

Remarks about the Methodology:
The definition of the acoustic characteristics for a given problem will constrain some of the basic characteristics of the numerical model:
-Mesh size:
A careful research work has been conducted to determine the minimal number of elements per wavelength in order to accurately capture the acoustic noise generation [1]. A rectangular domain meshed with an unstructured grid has been loaded with a white noise propagating horizontally. The transfer function of the numerical algorithm and of the discretization has been estimated by studying acoustic wave propagation over 200 elements. As a result, we have a good level of accuracy (less than 2dB attenuation and 6% dispersion after 200 elements travel) as soon as:

$$\text{Element size} < \lambda_{min}/6$$

Where, if we assume c is the sound speed in the considered medium, u the flow speed and f the frequency.

$$\lambda = c/f \text{ in the far field, } \lambda = u/f \text{ close to the sources}$$

-Simulation duration / sampling
The simulation total duration for the recording phase should be $1/F_{min}$ in order to be able to treat properly the lowest considered frequency. In order to get a fine enough resolution when switching to the frequency domain, sampling should be at least $4*F_{max}$.

In air, where a [200Hz, 3000Hz] domain is often considered and velocities of 20 to 50m/s are often seen, Typical elements sizes range from less than a 1mm close to noise sources to about 10 mm further apart and the frequency resolution after the FFT of 6 Hz.

1.3 Key Features of M-Explicit

In order to be able to efficiently apply this methodology, the solver must have the following key features:

Arbitrary Lagrange Euler (ALE) Formulation.

The structures are modeled using a Lagrangian representation, where the grid and the material are tight together. However, the traditional Eulerian representation for fluids will not be suited to represent coupled fluid structure phenomena where the structural parts are vibrating/moving, leading to a time dependent shape of the fluid domain. Therefore, ALE techniques are used, where the fluid mesh is attached to the vibrating/moving structures while the material can still flow through the element boundaries. The Navier Stokes equations are then rewritten taking into account a new velocity term, w that is the mesh velocity:

$$\rho\left(\frac{\partial u}{\partial t} + [(u-w)\cdot \nabla]u\right) - \nabla \sigma = 0$$

$$\frac{\partial \rho}{\partial t} + [(u-w)\cdot \nabla]\rho + \rho \nabla u = 0$$

$$\frac{\partial \rho e}{\partial t} + [(u-w)\cdot \nabla]\rho\ e + (\rho\ e + p)\nabla u = 0$$

Large Eddy Simulation (LES) Turbulence Model.

The modelization of the turbulent behavior of the flow that occurs in the class of problems involving complex geometry like rotating fans is of primary importance. After a first implementation of the classic k-ε turbulence model, it appeared that the need for a more refined one was needed. RANS methods solve average component variables, by modeling the turbulent fluctuations. Unlike such turbulent models, that by nature act as zero frequency cut-off devices, LES considers the mesh as a primary spatial filter whose cut-off wave length is function of its element size.

In LES, the grid is able to resolve the larger eddies, whereas smaller grains of turbulence must be modeled (Sub Grid Scale model). The very simple Smagorinsky sub-grid scale model has been implemented in M-Explicit. And lately refined to handle better the damping of the acoustic waves within turbulent boundary layers [3].

$$Vsgs = (C_S \Delta)^2 \sqrt{e_{ij}.e_{ij}}$$

Non-conforming Fluid-Fluid Interface.

In order to be able to perform simulations of flow around rotating mechanical part, embedded in a non-symmetric environment (e.g. centrifugal fans), it is critical to have a non-conforming fluid-fluid interface. This option allows a conservative momentum transfer between he rotating fluid elements attached to the rotor and the non-rotating fluid elements themselves connected to structural parts.

Structure Modelization.

Structures can be modeled as rigid bodies or as deformable materials. Several structural material laws are available ranging from simple isotropic elastic material to complex multi-layered orthotropic composites. Large deformations of the structural parts can be handled as well as structure/structure contacts.

2 Parallel Implementation

2.1 Implementation

Users have the choice between a shared memory (SMP) and a distributed memory (SPMD) version [4]. Both versions share the same source code. A considerable amount of work has been conducted in the RADIOSS group since 1991 on parallel and massively parallel systems. It appeared that it is possible to reach a fairly high level of performance with an SMP implementation provided a dynamic load balancing technique is carefully implemented. . Today, most of M-Explicit installed versions are SMP. The M-Explicit SMP architecture key features are the following:

Granularity. In order to reach a satisfactory parallel performance, one has to design the code parallelization at the highest possible level. M-Explicit SMP threads are launched from the main program, leading to one M-Explicit executable operating on each processor with different data and synchronizing only a couple of times per time step.

Dynamic Load Balancing. It is extremely difficult to reach a correct scalability if it is not used. The goal is to reach a parallelism degree higher than 99% in order to scale well on 32 to 128 processor systems. The cost of various items being non-constant during the simulation, the best way to minimize load-unbalancing problems is to dynamically feed the processors with the next available task. Tasks are computation blocks which size is fitted to processor cache (or vector pipe length). For instance, groups of similar elements, or a rigid body are tasks. The potential downside of dynamic load balancing is of course the remote memory access latencies since most of the modern systems are NUMA designs. Chapter 2.3 (performance) shows that this potential threat is not practically damaging.

Single Source Code. This applies for mono and parallel versions, SMP and SPMD. It is critical, both for clients and developers to have this uniqueness of source code. It avoids the plague of having one version (the parallel one) being more efficient but trailing the standard one for new options. We learned this lesson the hard way in 1993 when we developed our first SIMD version for MasPar and Thinking Machine CM-5 systems [4].

Results Uniqueness: Running explicit codes in parallel can lead to some drift due to round off errors as soon as the number of cycles becomes important. Those round off errors come from the non-deterministic way of performing the assembling of the accelerations at each cycle. On a 64 bits executable, statistically, one can observe that after about 3,000 cycles for 1,000,000 element models, solutions start to drift from each other if several runs of the very same input deck are compared. Although no run is theoretically better, this phenomenon makes it difficult to understand design modifications, version to version comparisons and even debugging. Performing a fixed order assembling solves this problem. Besides providing an independent solution from the number of processors or users on the system, it scales much better. The cost being a higher memory need.

M-Explicit SMP Algorithm:

```
While T<Tend
    Do in parallel, dynamic scheduling
    Compute elements and nodal accelerations
    Parallel assembling of accelerations
    Synchronize on accelerations
    Integrate v = ∫ ∂v/∂t dt
    Compute kinematic constraints
    Synchronize on velocities
    Compute new grid position w
    Integrate X = ∫ wdt
    Compute time step for next iteration
    T = T + dt
End While
```

2.3 Performances

Case 1: Airflow around a car body 2.5 Million elements

Computer: SGI Origin 2000@195MHz 16 processors

# Processors	Speedup	Elapsed/elt./cycle
1	1	78.6 E-06 s
8	6.5	12.1 E-06 s
16	9.4	8.40 E-06 s

Case 2: Airflow around a car model, 2 Million elements

# Processors	Speedup	Parallelism	Elapsed/elt./cycle
1	1	NA	77.3 E-06 s
4	3.8	98%	20.2 E-06 s
8	8.1	100%	9.52 E-06 s
16	14.3	99.2%	5.41 E-06 s
32	26.5	99.3%	2.92 E-06 s

3 Applications

To date, M-Explicit has been used successfully by industrial clients worldwide to simulate aeroacoustic problems. Here is a list of the most common applications today:

Problem	Location	Validation Status
Axial Fan noise analysis	Europe, USA	Validation in Progress
Centrifugal Fan noise analysis	Europe, USA	Validated
Exhaust noise analysis	USA, Japan	Validated
Air Intake noise analysis	USA	Validated
Full car aeroacoustic	Europe, USA	Validation in progress

4 Future Developments

We have seen that M-Explicit can successfully simulate complex fluid structure interactions and thanks to compressibility and its explicit integration scheme derive noise analysis of the problems. However, some developments are still needed in order to reach a higher level of performance and accuracy:
-Acceleration of the convergence toward the unsteady flow regime. Explicit is a handicap for reaching such a converged regime since it has to fulfill the CFL time step condition. Therefore, a new module, M-Implicit has been added to the RADIOSS-CFD system to quickly initialize velocity fields [5].
-The next major version of the code will feature SMP & SPMD in a transparent way for the user. A new data structure will be defined to fit better the need of modern NUMA systems and reach even higher parallelism levels.

References

[1] M. Hsi, F. Périé, *"Computational Aeroacoustics for Prediction of Acoustic Scattering"* Second Computational Aeroacoustics Benchmark Problems Workshop Tallahasee Florida , November 4-5, 1996.
[2] Subrata Roy & al. "Designing Axial Flow Fan for Flow and Noise" SAE 99F-34, 1999
[3] Eric Lenormand "Contribution à la simulation des grandes échelles d'écoulements turbulents compressibles en milieux confinés" PhD dissertation, University of Paris VI
[4] Dimitri Nicolopoulos "A comparison of Parallel Paradigms Applied to an Industiral Code: RADIOSS" HPCN 1996 Proceedings, Springer
[5] Farzin Shakib "AcuSolve, A powerful Solver Technology" ACUSIM Software Inc. internal document

Creating DEMO Presentations on the Base of Visualization Model

Elena V. Zudilova and Denis P. Shamonin

Visualization Systems Lab
Institute for High Performance Computing and Data Bases
Of. 55, Fontanka 118, St. Petersburg 198005, RUSSIA
Tel.: +7 (812) 2519092, Fax: +7 (812) 2518314
http://www.csa.ru/visual
E-mail: zudilova@fn.csa.ru; sham@fn.csa.ru

Abstract. The paper is devoted to the complex process that begins by obtaining numerical data while solving a scientific problem and completes by the creation of the demo for illustrating the final solution in the graphical form. The represented visualized results were obtained after the simulation of the particle movement in the system of three bodies on the LaGrange surface and the simulation of mature convective clouds development. All calculations were carried out using supercomputers of CONVEX C-series, PARSYTEC systems and Cray J-90. Then the numerical data was converted using Geomview - special viewing program oriented for 3D visualization under IRIX 6.2 (SGI Octane workstation). It became possible thanks to the additional module written in C/C++ and aimed to converting data of special formats into the series of graphical images. As a result, 3D animation films were elaborated.

1 Introduction

Supercomputers of different architectures are becoming more and more popular while solving different mathematical and physical problems demanding large data volumes. The choice of this modern and powerful technique is caused by the possibility to decrease the solution period for complicated tasks from several months up to several weeks or even hours. The data obtained during such numerical experiments is usually too large, so sometimes the only way to analyze it is to work with the graphical interpretation of results.

The paper is devoted to the process of visualization based on converting the huge numerical data volumes into 3D animation films. 7 stages of the following process are distinguished and form the Visualization Model.

The results presented in the paper are based on more than four years experience in the sphere of numerical simulation and visualization of different physical problems. The visualization of two different physical processes - the particle movement in the system of three bodies on the LaGrange surface and convective cloud development in extreme conditions - were selected for the illustration material of the paper. All the results were obtained using the modern computational and visualization resources of the St. Petersburg Institute for High Performance Computing and Data Bases. Institute is one of the most powerful supercomputer centers in Russia and equipped by supercomputers of HP Convex series, Sun Enterprise cluster, Parsytec massive-parallel computers, special SGI graphic stations Octane, etc. [1].

2   Main Stages of Creating Demo Presentation

The methods of visualization are very important for both analyzing numerical simulation results and popularization of supercomputer methods. The dynamics of different physical and chemical processes are better observed in graphical form. Moreover, there are many special information systems based on numerical simulation methods, such as: weather broadcasting, monitoring, etc. where only the visualized data can be used as the output result.

As a whole the complex process of creating a DEMO presentation for the physical process can be introduced as the Visualization Model on the fig.1. The Visualization Model includes three main phases: numerical simulation, data conversion and graphical interpretation.

Numerical Simulation: 3D visualization means simulating research conditions in three coordinates (x,y,z) through establishing choreography or movement for the 3D-objects step by step, setting up lights and applying textures [5]. The following approach needs additional computer memory. To obtain the data necessary for 3D visualization by means of common computer resources (simple PC) sometimes is not a simple task as this process may cover the period of several months, especially if the investigated phenomenon is too complicated. So the best way to obtain data for 3D visualization is to organize the numerical simulation on the base of supercomputer resources, of course, if these resources are available.

The visualization of the particle movement in the system of three bodies on the LaGrange surface and convective cloud development in extreme conditions demands huge data arrays for creating 3D animation, so the numerical simulation was organized on HP Convex and Parsytec supercomputers, later on Cray J-90. As for the software, two special program complexes QUANTUM and CONVEC were developed using programming languages C/C++ and Fortran.

Data Conversion: Data volumes obtained during the numerical simulation are usually of the special formats and the majority of visualization tools does not understand them. So the next step is to convert data to the appropriate format(s). In accordance with software selected for the graphical interpretation a special converter should be designed and implemented. QUANTUM provides two files of data - one for the surface zones and the second defining the particle movement trajectory. The output of CONVEC contains special file of strings, divided on several parts equal to the number of zones, where each line contains three coordinates and the volume of the investigated function - distributions of different cloud parameters for different time moments (velocity, cloud droplets content and rain drops, dry and wet aerosol particles concentrations, etc.).

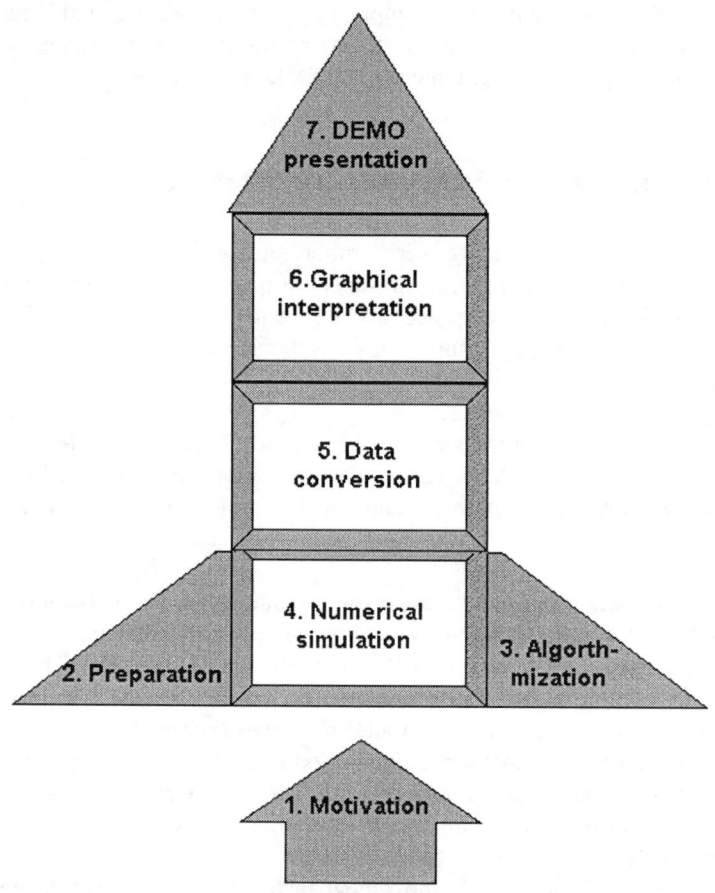

Fig. 1. Main stages of creating DEMO presentation - Visualization Model

Graphical Interpretation: Special visualization tools for different operating systems make it possible to get graphical images from the numerical data of the corresponding format. Some tools are aimed to the creation of static images, others provide animation films from static images or even from the numerical data. Different colour palettes, textures, lighting effects, etc. are the additional tools to make the DEMO presentation more attractive for the audience. For the graphical interpretation of the numerical simulation results the SGI Octane workstations were selected as hardware and Geomview program package as software. The description of how it was made is contained in the third section of the paper.

Preparation and Algorithmization stages are necessary for the organization of the first main phase of the Visualization Model - Numerical Simulation.

The *Preparation* stage is very complicated as it covers the elaboration of mathematical models, development of numerical schemes for effective simulation, definition of the investigated zones and fields, selection of scales and criterions.

Algorithmization is necessary for the organization of the computing process. It includes selection of a computer (PC or supercomputer if available) and programming language, creation of the algorithm adapted to the architecture of a computer (if it is a supercomputer of vector-parallel or massive-parallel architecture, then parallel algorithm is to be developed), program implementation, testing and validation.

Motivation includes factors defining the importance and necessity to visualize the scientific tasks or phenomena.

The next section of the paper contains the description of creating DEMO presentations of two different physical processes for SGI Octane workstations after the results of numerical simulation were obtained for each phenomenon.

3 Visualization of Numerical Simulation Results

The visualization package Geomview 1.6.1 (Geometry Center) was selected for the graphical interpretation of the data obtained during the numerical simulation.

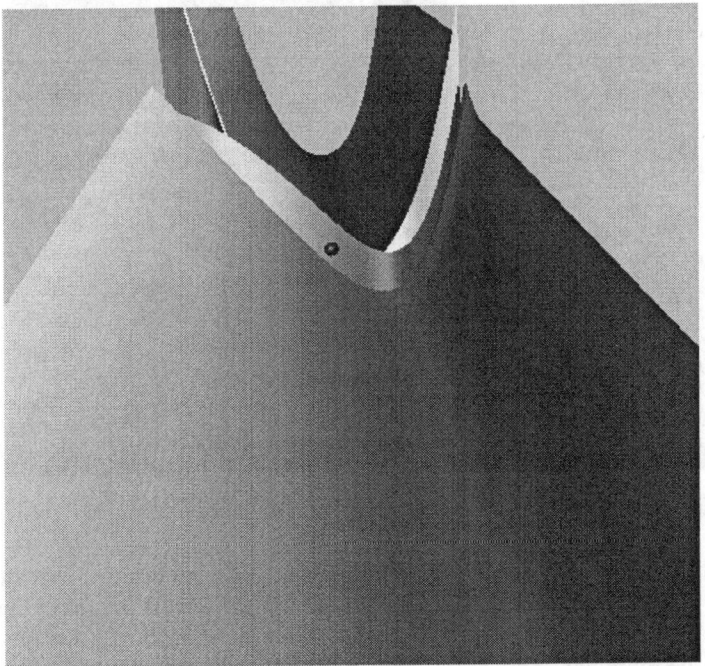

Fig. 2. Particle movement in the system of three bodies on the LaGrange surface

Geomview is an interactive 3D viewer that allows to demonstrate and to manipulate geometric objects. The package was installed on the SGI Octane workstation, operating system IRIX 6.5. The real power and flexibility of Geomview comes from its ability to work in conjunction with external software. It is possible to connect Geomview to another program that runs concurrently and generates data for Geomview to display. As the source program changes the data, the Geomview image reflects the change instantly.

There were two main difficulties of using Geomview for the visualization of results obtained during the numerical simulation. First, that the following package is orientated to work with the static geometry objects only. And the second, that like the majority of existing visualization tools, it works only with OOGL (Object Oriented Graphics Library) File Formats such as QUAD, MESH, OFF, etc.

Fig. 3. Convective cloud development in process: buoyant thermal is forming

The additional external module DYNVIS (Dynamic Visualizator) written in C/C++ for Geomview package makes it possible to obtain the 3D animation movies from a set of several files of COFF format. Each file describes one zone of the process at the concrete time moment. Those files are processed by gcl (Geomview Command Language) commands defined separately in a file used for indicating the display engine components - windows sizes and locations, cameras positioning, characteristics of light sources, etc. This program module interacts with the

Geomview through gcl commands. As a result it is possible to obtain the animation of the same process in several projections and with different visualization effects [6].

Fig. 2 and 3 contain the screen-shorts of 3D animation films produced for particle movement in the system of three bodies on the LaGrange surface and the process of convective cloud forming in extreme conditions correspondingly. The data for the visualization of the first process was obtained through the numerical experiments carried out on the massive-parallel computer Parsytec CCe. As for the second process, the numerical simulation was carried out on the supercomputer HP SPP-1600/8 of the vector parallel architecture and later on Cray J-90 [2, 4]. Special parallel algorithms permitting to solve the same task on several processing nodes at the same time were elaborated and implemented.

The organization of each 3D movie is different. As for the film that screen short is presented on the fig. 2, the surface is static and can be moved manually and the particle is moving chaotic in accordance with defined trajectory.

The screen short of the movie on the fig. 3 represents three projections of the process development at the current time moment. Each episode illustrates the process at the current time moment in one of three different projections. For building the graphical images the data of all selected zones at certain time moments was invoked. Then the isosurfaces of the selected layers for each zone were built. And at last, these 3D surfaces were overlapped in one image. Each layer is of the different colour and can be seen through the other layers, as they all are transparent.

4 Conclusion

Using the modern computational and visualization hardware and software gives the ability to create the attractive 3D movies for the very complicated mathematical and physical processes, even with the elements of Virtual Reality - next step in 3D visualization when the stereo virtual objects can be interacted during the presentation process.

The introduced in the following paper approach of creating DEMO presentations based on the Visualization Model is aimed to make the methods of 3D visualization more popular in different research spheres and, especially in the scientific researches where the numerical simulation is the only way of obtaining information.

For example, catastrophic natural and anthropogenic phenomena, such as volcanic eruptions, large forest fires, burning of oil and gas wells, large industrial explosions accompanied by the formation of mature convective clouds may considerably change atmospheric conditions in the region where the phenomenon takes place and also can exerts essential negative influence both upon the global climate and upon the ecological situation in the nearest regions. So the prediction of the following catastrophic events and their possible consequences is becoming more and more important and actual. Based on the results of numerical simulation and then interpreted in the graphical form those predictions can be used as a powerful tool for solving many ecological problems.

References

1. Bogdanov A.V., Gevorkyan A.S., Gorbachev Yu.E., Zudilova E.V., Mareev V.V., Stankova E.N.: Supercomputer Cluster of the Institute for High Performance Computing and Data Bases, Proceedings of CSIT Conference, Yerevan, Armenia, pp. 353-354 (1999).
2. Bogdanov A.V., Gevorkyan A.S.: Reactive scattering in the three-body system as imagining point quantum dynamics on 2-D manyfolds, Proceedings of the International Workshop on Quantum Systems, Minsk, Belarus (1996).
3. Gorbachev Y., Zudilova E: 3D-Visualization for Presenting Results of Numerical Simulation. Proceedings of HPCN Europe - 99, pp.1250-1253 (1999).
4. Stankova E. N., Zatevakhin M. A.: Investigation of aerosol-droplet interaction in the mature convective clouds using the two-dimensional model. 14th International Conference on Nucleation and Atmospheric Aerosols. Helsinki, Finland, 26 - 30 August 1996, pp. 901-904 (1996).
5. Stankova E., Zudilova E. Numerical Simulation by Means of Supercomputers. Proceedings of HPCN Europe - 98, pp. 901-903 (1998).
6. Zudilova E.: Using Animation for Presenting the Results of Numerical Simulation. Proceedings of International Conference of Information Systems Analysis and Synthesis) / Conference of the Systemics, Cybernetics and Informatics, Orlando, Florida, USA, pp. 643-647 (1999)

Very Large Scale Vehicle Routing with Time Windows and Stochastic Demand Using Genetic Algorithms with Parallel Fitness Evaluation

Matthew Protonotarios, George Mourkousis, Ioannis Vyridis, and Theodora Varvarigou

National Technical University of Athens, Department of Electrical and Computer Engineering, Division of Computer Science
9 Iroon Polytechniou, 157 73 Athens, Greece
{matprot,georgemr,jvs,dora}@telecom.ntua.gr

Abstract. This paper deals with a real-life vehicle routing problem concerning the distribution of products to customers. A non-homogenous fleet of trucks with limited capacity and allowed travel time is available to satisfy the stochastic multiple product demand of a set of different types of customers with earliest and latest time for servicing. The objective is to minimize distribution costs while maximizing customer satisfaction and respecting the constraints concerning the vehicle capacity, the time windows for customer service and the driver working hours per day. A model describing all these requirements has been developed as well as a genetic algorithm to solve the problem. High Performance Computing has been used to allow the pursuit for a near-optimal solution in a sensible amount of time, as the parallel chromosome fitness evaluation counterbalances the increased size and complexity of the problem.

1 Introduction

This work has been developed in the framework of the ESPRIT project DYNALOG. The DYNALOG project develops a dynamic, integrated system to meet the current needs in the field of logistics. As transportation costs represent a large percentage of the total costs in distribution companies the amount of time needed for high quality routing decisions is required to be as short as possible, in order to provide superior reactivity and cost-effectiveness. The use of HPCN enables the use of advanced software systems to fulfil these requirements within sensible amount of time.

The end-users in the project are dairy companies that have complementary problems, which reflects the diversity of needs across Europe. MEVGAL (Greece) produces and delivers ultra-fresh dairy products while demand is not always known at departure time. FIEGE (Germany) offers large volume logistic services throughout Europe and they are of the largest logistics companies in Europe. In their case the demand is known just a few hours before route scheduling.

The Vehicle Routing Problem with Time Windows (VRPTW), besides retail distribution as in our case, applies in a variety of situations like mail delivery, newspapers delivery, waste collection and consequently has taken up the time of many researchers. However, there is still a lot of space for further research due to the high complexity of the problem, which allows only search for better and computationally faster approximations of the optimal solution. It is worth mentioning that even finding a feasible solution for a given number of vehicles is NP-hard (see Savelsberg [2]). The use of Genetic Algorithm techniques to encounter the problem seems to be a very promising. Thangiah [4] has followed this approach with GIDEON to face VRPTW with deterministic customer demands.

We used a genetic algorithm in order to form an assign and route method to heuristically solve the VRPTW with stochastic demands and an implementation of the Message Passing Interface standard for the parallel chromosome fitness evaluation. The MPI standard is available through several sets of libraries and was used due to its stability, simplicity and interoperability. MPI instructions are easily available to applications and most implementations make possible the interaction between processes in a cluster of mixed Win32 and UNIX workstations over a TCP/IP network. Thus, most of the commonly available hardware resources can be used for cost effective parallel computing under the MPI standard.

2 Problem Description

A dairy company is engaging a non-homogenous fleet of trucks (some owned by the company and some are franchised trucks) in order to distribute a variety of dairy products to a large number of customers situated in a dense geographic region. The trucks, initially located at a given depot(s), have to serve (some of) the customers within specified periods of time. Each customer's demand is stochastic, respecting historical data and current trend as well as external factors that influence the demand. The target is to create a number of baseline routes on a weekly basis for the vehicles, within a user-defined period that respect certain additional constraints like driver working hours per day, franchised drivers' revenue, predetermined or non-compatible pairings amongst customers and vehicles. Hence, our objective is to consider all expected fluctuations in demand across the predefined period, while keeping customer satisfaction high and distribution costs to a minimum.

3 Mathematical Formulation

Here, we present a mathematical formulation of the problem on which our approach was based. This model is an initial simplified mathematical description as a mixed integer problem. However, it might prove useful for a better understanding of the problem and indicate of the difficulties that would arise if analytical methods were to be used.

$$\min \underbrace{\sum_{i,j,k} x_{ijk} \cdot c_{ij}}_{\text{Travelling Cost}} + \underbrace{\sum_{j,d,k} [y_{jkd} - \tilde{v}_{jd}]^+ \cdot c_j^1}_{\text{Returned Products Cost}} + \underbrace{\sum_{j,l,k} [\tilde{v}_{jd} - y_{jkd}]^+ \cdot c_j^2}_{\text{Unsatisfied Demand Cost}}$$

subject to:

- A vehicle arrives to a customer j at most once:

$$\sum_i x_{ijk} \leq 1, \quad \forall j \in N, k \in K \quad (1)$$

- At most one vehicle k arrives to a customer j only once:

$$\sum_{i,k} x_{ijk} \leq 1, \quad \forall j \in N \quad (2)$$

- A vehicle arrives to customer j only if departs customer j and vice versa:

$$\sum_i x_{ijk} = \sum_i x_{jik}, \quad \forall j \in N, k \in K \quad (3)$$

- Extra constraint in order to avoid sub-tours:

$$u_i - u_j + n \cdot \sum_k x_{ijk} \leq n-1, \quad \forall i,j \in N = \{1,2,\ldots,n\} \quad (4)$$

- Sum of product quantities loaded to any truck k any weekday l, planned for delivery to each customer j, must not exceed truck capacity:

$$\sum_k y_{jkd} \leq V_k, \quad \forall k \in K, l \in L \quad (5)$$

- Total time spent travelling and servicing customers for any truck k must be less or equal to total permitted working time reduced by the total waiting time:

$$\sum_{i,j} x_{ijk} \cdot (t_{ij} + s_j) \leq \left[T - \sum_j w_{jk}\right]^+, \quad \forall i,j \in N, k \in K \quad (6)$$

- Waiting time of truck k at customer j cannot be greater than maximum permitted waiting time and this holds only if truck k arrives to customer j:

$$w_{ik} \leq \sum_j x_{jik} \cdot W, \quad \forall i \in N, k \in K \quad (7)$$

- The time of arrival of any truck to customer j must greater or equal to the time of arrival to customer i plus the waiting time plus the service time plus the travelling time from customer i to customer j, given that at least one truck is travelling from i to j (note that according to 2 at most one truck can travel from i to j):

$$t_j \geq t_i + w_{ik} + s_i + t_{ij} - (1 - x_{ijk}) \cdot T, \quad \forall i,j \in N, k \in K \quad (8)$$

- The time of arrival of any truck to customer i plus service time of customer i must be less or equal to latest service time for customer i:

$$t_i + s_i \leq L_i, \quad \forall i \in N \quad (9)$$

- The time of arrival of any truck to customer i plus service time plus waiting time at customer i must be greater or equal to earliest service time for customer I (note that the waiting times at customer i for all trucks but one are equal to zero according to 7):

$$t_i + s_i + \sum_k w_{ik} \geq E_i, \quad \forall i \in N \quad (10)$$

- Non-negativity constraints for all variables. Finally note that (1),(2) and (3) are slightly modified for node 0 (the depot) in order for the trucks to be able to return in the end of the day.

where:
- $i, j \in N = \{1,2,...,n\}$ set of customers
- $k \in K = \{1,2,...,m\}$ set of trucks
- $d \in D = \{1,2,...,7\}$ set of weekdays
- V_k is truck's k volume capacity
- W is maximum permitted waiting time at any customer
- T is total permitted working time for any truck any day
- $TW_j = [E_j, L_j]$ is the time window for customer j
- c_j^1 is the penalty for each unit of volume of returned products for customer j
- c_j^2 is the penalty for each unit of volume of unsatisfied demand for customer j
- c_{ij} is the travelling cost from customer i to customer j
- t_i is the arrival time at customer i (continuous)
- t_{ij} is the travelling time from customer i to customer j
- s_i is the service time at customer i
- w_{ik} is the waiting time of truck k at customer i (continuous)
- u_i is an auxiliary continuous variable
- Decision variables:
  $x_{ijk} = 1$, if vehicle k goes from customer i to customer j
  $= 0$, otherwise,
  and,
  y_{jkd} = Product volume loaded in truck k at day d for customer j (continuous)
- $\tilde{v}_{jd}$ is the expected demand for customer j on day d (note that it is a stochastic variable that must be discretized in order to reduce the model to a mixed integer programming model)

- $[f]^+$ denotes the positive part of f, that is $[f]^+ = \begin{cases} f & ,if\ f \geq 0 \\ 0 & ,otherwise \end{cases}$

4 The Use of a Genetic Algorithm Model

In contrast to classical search, with GAs the possibilities are not explored in an exhaustive way in some order, but a current set of candidate solutions is improved by stochastically selecting one or more parents and using them to generate one or more children by applying a genetic operator. The advantage of a GA is that of genetic diversity: each step in the search process considers a set of candidates, instead of a single one, selected on the basis of their quality, for the creation of new candidates. Hence, a GA explores and exploits the relationship between the inner structure and the quality of candidate solutions automatically. The so-called *building block hypothesis* states: a GA detects simple patterns within candidates, the building blocks, that contribute to the quality, and composes candidate solutions consisting of high quality building blocks.

A traditional genetic algorithm can be generally described in the next few steps:
1. Initialization of the population. Initialization requires the creation of sufficient individuals to full the population. The term individual is used to group together the two forms of a single solution: the chromosome, which is the raw 'genetic' information (genotype) that the GA deals with, and the phenotype, which is the expression of the chromosome in the terms of the model. During the initialization of the population the fitness of the initial individuals is evaluated. The fitness of an individual in a GA is the value of the objective function for its phenotype.
2. If search termination criteria are met then bring search to a halt else proceed. The termination techniques can be as simple as a maximum number of generations or more complicated, customary devised for the needs of each specific problem.
3. Select two or more chromosomes for crossing. Selection allows (some of) the building blocks of chromosomes that satisfy certain criteria (usually high fit values) to survive into the next generation. Note that the method for determining which individuals should be allowed to interchange their genetic information has great bearing on the rate and quality of the algorithm convergence.
4. Crossover and Mutation. Crossover splices chromosomes at random points and exchanges the splices among the spliced chromosomes. Mutation is the direct change of a gene's value (usually to a bit's complementary value).
5. Offspring's Fitness Evaluation.
6. Replacement. This is the last stage of the breeding cycle were it is determined which individuals stay in the population and which are replaced. The technique of choice is on a par with the selection in influencing convergence.
7. Go to step 2.

Traditional GAs are primarily applicable for problems where there is no interaction (epistasis) between the variables of a candidate solution. However, as it is also obvious from the mathematical model presented earlier, this is not the case with

DYNALOG, which can be characterized as a Constraint Optimization Problem (COP). Hence, the original COP problem was transformed into its GA-suited counterpart by relaxing the constraints that could not be embedded into the operators (capacity, route time, time windows constraints), penalizing them and finally including them into the objective function. In this case an infeasible solution might become feasible but it gets penalized in the objective so that it becomes a "bad" feasible solution. Finally, higher weights were assigned to the coefficients related to tardiness and truck capacity in order to first obtain a feasible solution and then minimize over the rest of the factors of the GA-suited counterpart of the objective function.

The DYNALOG system uses an assign and route method to solve the problem. That is, given a set of customers and central depot(s), the system clusters the customers assigning one truck for each cluster's service.

Chromosomes represent valid solutions and consist of n genes. Gene i in the chromosome represents the i_{th} customer in which is assigned a vehicle from $h_1, h_2, ..., h_m$. Then the vehicle is routed within each sector by solving the corresponding TSP problem using the cheapest insertion method [6].

Each genome has three primary operators: initialization, mutation, and crossover. With these operators you can bias an initial population, define a mutation or crossover specific to your problem's representation, or evolve parts of the genetic algorithm as your population evolves. The operation of these operators and therefore the algorithm's evolution depends on the random number generator applied as we use random numbers to perform action with a predefined probability. It is worth mentioning that for every seed the random number generator produces a unique sequence of random numbers. Every time that a specific seed is used the evolution of the algorithm (when using the same parameters) is identical, having the potential to derive experimental results.

The type of the genetic algorithm used is the Steady State GA (SSGA) [7],[8]. The SSGA follows the steps of a traditional GA as previously described and it is similar to Simple GAs (SGA) in the way the genetic operators are applied but it differs in the replacement step. More specifically, the SGA starts by randomly generating a population of N individuals. Then selects the individuals for breeding using fitness proportionate reproduction. That is, individuals with higher fitness scores are selected, with replacement, to create a mating pool of size N. The genetic operators of crossover and mutation are applied in this stage in a probabilistic manner which results in some individuals from the mating pool to reproduce. The selection procedure of the SSGA is identical to SGA. However, in SSGA, the offspring are inserted into the population, thus replacing two individuals, soon after they are generated (successive replacement strategy) whereas the SGA generates N offspring prior to replacing the entire population (simultaneous replacement strategy). Finally, as its name implies during the evolution the size of the population remains constant.

The solution algorithm can be summarized in the next few steps:
1. Initialization of the population (including original chromosome fitness evaluation).
2. If search termination criteria are met then bring search to a halt else proceed. In this a case a maximum number of generations was used.
3. Select the individuals for breeding using fitness proportionate reproduction.

4. Mutate and Combine (via crossover) the selected chromosomes in a probabilistic manner.
5. Express the chromosome in the terms of the model (phenotype) that is Truck to Customer Assignments.
6. For each phenotype solve the corresponding TSP Problems.
7. Find optimal truckloads for each one of the truck routes.
8. Compute the Objective Function of each phenotype and use its value as the chromosome's fitness value.
9. Replace Chromosomes in population.
10. Go to Step 2.

5 Parallel Chromosome Fitness Evaluation

In order to reduce the data exchange between the various processes, the genetic part of the algorithm remains the same at each process, while the evaluation process that is the most computationally expensive is spliced.

During the evolution, all processes have exactly the same main population at the same time. This is obtained by initializing all processes with the same seed. The identical seed forces all processes to create the same initial population. Then, we perform in parallel the evaluation of the chromosomes in every generation. Due to the fact that for a specific population all processes acquire the same chromosome objective scores for every individual in the population and perform exactly the same replacement method, the next generation will be the same in all processes. Generations after the initial one are created by the application of selection and chromosome operators to the population. The initial identical seed used by all processes, guarantees that selections and mating operators among processes are performed in exactly the same way.

Throughout the evolution all processes have approximately the same utilization and execution time but by applying the chromosome fitness evaluation in parallel we obtain significantly lower execution times than the serial version.

In the serial algorithm the chromosomes' fitness evaluation is performed sequentially. During the evaluation of each chromosome, we solve the corresponding TSP Problems and find optimal truckloads for each one of the truck routes computing the Objective Function of each phenotype and use its value as the chromosome's fitness value.

Taking into account the sequential nature of the evolution, there are two points where the computation can be sliced. We can either share in the available processors the evaluation of the entire chromosomes of fragment of the population or share in the available processors the evaluation of fragment (as far as it concerns truck routes) of every chromosome in the population.

When parallelizing the population evaluation we divide the chromosomes of the population in groups. The number of groups is equal to the number of processes involved in the computation and each process is responsible for evaluating the chromosomes in the group it owns. After the computation is completed the chromosomes' scores are distributed among process so that all of them have the scores for the entire population.

It is obvious that if the number of processes does not divide the size of the population, some processes will evaluate more chromosomes than the others will.

The second method is derived by parallelizing the chromosomes' fitness evaluation. In this case chromosomes in the population are evaluated in serial order. Each chromosome evaluation procedure is spliced in the processes participating in the parallel computation. More precisely, what is shared among processes are the chromosome's phenotype corresponding TSP problems and optimal truckloads solution for each one of the truck routes described in it. After each process has finished its partial computation the results are distributed among them and summed up to give the chromosome's fitness value.

Similarly to the previous case, if the number of processes performing the parallel computation does not divide the number of trucks described in the chromosome's phenotype, some processes will solve more TSP problem optimal truckloads solutions than the others will. Moreover in this particular case of parallelization it must always be guaranteed that the number of processes is smaller than the number of trucks described in the phenotype of any chromosome of the population, or else we can't divide trucks to processes.

6 Computational Results

The methodology described in the previous sections was tested both on 1-CPU on-board machines connected through a TCP/IP LAN network and 2-CPU on-board machines. We measured the computation time of the serial and parallel versions of the algorithm. The tests were performed for 180, 500, 1000 and 2000 customers on 1,2 and 3 processors. The computation time in each case is the amount of time (clock time) in seconds that are required by the GA to evolve for 100 generations (see table 1).

The parallel version performs better than the serial in most cases (see figure 2). The exceptions are observed when many processes try to perform the computation through a LAN. Examining the speedup figures (figure 2) it's noticeable an opposite performance pattern for the two parallelization methods. For small customer sets, better results are obtained by the parallel population evaluation. Increasing the number of processes realizing the computation the method's performance decreases proportionately to the number of processes. In the contrary, when dealing with large customer sets, the best results are obtained by the parallel chromosomes' fitness evaluation. In the parallel population evaluation the speedup figures decrease as the number of customers increase. In the other hand, parallel chromosomes' fitness evaluation show a clear increase in the speedups obtained as the number of customers increase. Shared memory implementations outperform those based on distributed memory. Moreover among shared memory implementations the parallel chromosome's fitness evaluation performs better than the parallel population evaluation as the number of customers increases.

Table 1. Computation times / Speedup figures (in bold)

		Number of customers /trucks			
	Processes	180/4	500/12	1000/22	2000/44
Serial execution	1	824.094	2151.59	4980.95	10121.723
Parallel population evaluation	2 (shared memory)	532.318 **(1.548)**	1401.251 **(1.535)**	3420.943 **(1.456)**	7023.283 **(1.441)**
	2 (LAN)	559.534 **(1.473)**	1455.283 **(1.478)**	3547.398 **(1.404)**	7119.314 **(1.422)**
	3 (LAN)	323.225 **(2.550)**	1098.172 **(1.959)**	2651.624 **(1.878)**	5363.592 **(1.887)**
Parallel chromosome fitness evaluation	2 (shared memory)	565.469 **(1.457)**	1284.534 **(1.675)**	2884.84 **(1.727)**	5663.967 **(1.787)**
	2 (LAN)	1819.12 **(0.453)**	2614.56 **(0.823)**	4459.391 **(1.117)**	5781.51 **(1.751)**
	3 (LAN)	2587,220 **(0,319)**	2867,926 **(0,750)**	3318.587 **(1.501)**	4639.872 **(2.181)**

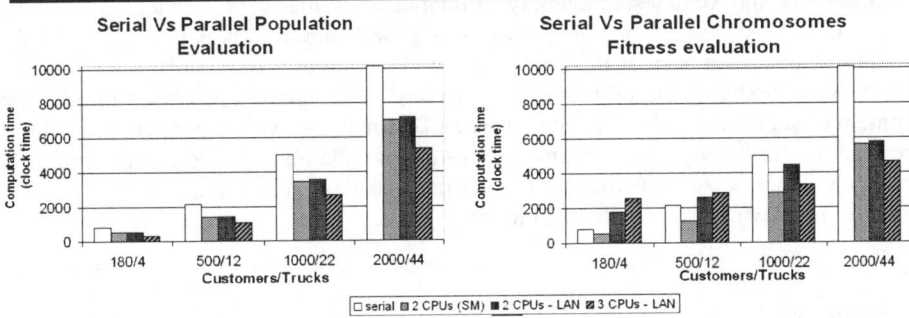

Fig. 1. Comparison of execution times between the serial algorithm and the parallel versions.

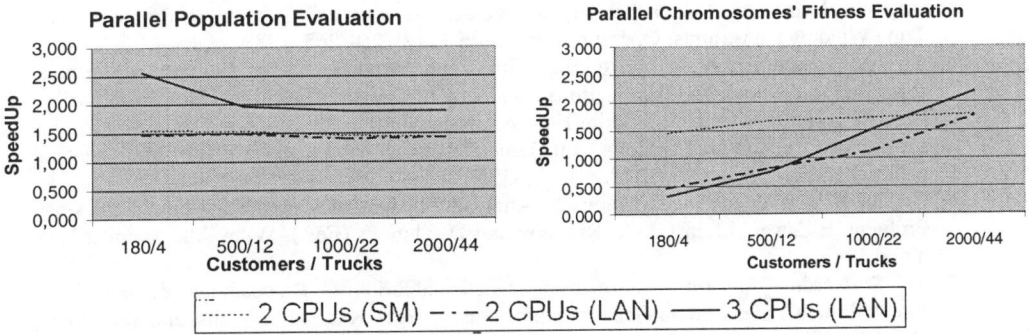

Fig. 2. Diagrams of speedup figures obtained for various

7 Conclusion

In this paper we have presented a model for solving a real-life vehicle routing problem which meets the requirements of a dairy company. However, the same model applies to many other situations where similar needs and restrictions occur. The latter include: limited vehicle capacity, time windows of serving customers, restricted duration of each route, combination of customers in the same or in different routes, allocation of customers to specific vehicles/drivers, exclusion of customers from specific vehicles/drivers. A significant feature of the problem is the fact that the demand of each customer is not known in advance. It may vary according to expected factors (e.g. seasonality) and unexpected ones (e.g. strikes) and so it is given probabilistically. The possible fluctuations in the demand should be taken into account as they affect the return cost for the remaining products as well as the satisfaction of the customers, and subsequently, in connection with the capacity constraint for the vehicles, they may dramatically influence the scheduling of the routes. In the last section, a strategy that uses genetic algorithms is presented. As the problem depends on a great deal of factors and has quite a few constraints, the size and the complexity of the problem are quite large. The method proposed implies the segmentation of the problem in smaller parts. Distinct, almost independent procedures are required for each part so that parallelization can play a key-role in our approach. Problems of this size and this form constitute an ideal area for the exploitation of capabilities provided by HPCN techniques

References

1. Michalewicz, Zbigniew, Genetic Algorithms + Data Structures = Evolution Programs (3rd edition) Springer-Verlag, 1996
2. Savelsbergh M.W.P., Local Search for Routing Problems with Time Windows. Annals of Operations Research 4, 285-305, 1985.
3. Solomon, Marius M., Algorithms for the Vehicle Routing and Scheduling Problems with Time Window Constraints. Operations Research 35 (2), 254-265, 1987.
4. Thangiah, Sam R., Vehicle Routing with Time Windows using Genetic Algorithms, submitted to the book on "Applications Handbook of Genetic Algorithms: New Frontiers".
5. Thangiah, Sam R. – Osman, Ibrahim H. – Vinayagamoorthy, Rajini – Sun, Tong, Algorithms for the Vehicle Routing Problems with Time Deadlines. American Journal of Mathematical and Management Sciences, 13 (3&4), 323-355.
6. Golden B. and W. Stewart, Empirical Analysis of Heuristics in the travelling Salesman Problem, E.Lawer, J.Lenstra, A. Rinnooy and D. Shmoys (Eds.), Wiley-Interscience, New York, 1985.
7. G. Syswerda, Uniform crossover in genetic algorithms, Proceedings of the Third International Conference on Genetic Algorithms, Morgan Kaufmann Publishers, San Mateo, CA, June 1989.
8. D Whitley, GENITOR: a different genetic algorithm, Proceedings of the Rocky Mountain Conference on Artificial Intelligence, Denver Colorado, 1988.

Extracting Business Benefit from Operational Data

Terence M. Sloan[1], Paul J. Graham[1], Kira Smyllie[1], and Ashley D. Lloyd[2]

[1] EPCC, The University of Edinburgh, EH9 3JZ, UK
tms@epcc.ed.ac.uk
WWW home page: http://www.epcc.ed.ac.uk/
[2] Management School, The University of Edinburgh, EH8 9JY

Abstract. EPCC is a technology transfer centre within the University of Edinburgh in the United Kingdom. It has assisted a number of organisations to extract extra business benefit from operational data. This paper outlines some of these data intensive, knowledge discovery projects undertaken by EPCC. It highlights issues common to these projects despite their diversity in solution methods, computational requirements and application area.

1 Introduction

With the increasing integration of information technologies into business processes, many commercial and industrial organisations are now able to collect vast amounts of data relating to their internal operations and the behaviour of their markets. Many of these organisations now wish to actively exploit this information for competitive advantage. However, analysing these data volumes is a significant task that carries with it an opportunity cost for staff time. If the analysis requires expertise and perhaps even computing platforms that are not present within the organisation, then the opportunity cost of staff time spent on this task may not exceed the benefits that arise!

EPCC is a technology transfer centre within the University of Edinburgh in the United Kingdom. Its mission is to accelerate the effective exploitation of high performance computing (HPC) in industry, academia and commerce. It has assisted a number of organisations to extract extra business benefit from operational data by acting as a bridge between commercial domain expertise (e.g. The Management School) and expertise in high performance computing tools, using the former to structure the problem, and the latter to analyse it. Solutions, and real business value can then emerge if both sets of expertise collaborate in an iterative process. From national financial organisations to fishing vessel skippers it has helped address the common desire of wishing to quickly and effectively analyse operational data and act upon it. In some cases HPC-enabled solutions have been utilised, in others, workstation and PC-based solutions have been employed. Where HPC has not been applied directly, previous experiences and techniques used in HPC projects have proved very beneficial. In particular, many HPC projects involve the management and manipulation of vast amounts of data. EPCC's experience in this was therefore important.

This paper outlines some of these data intensive, knowledge discovery projects undertaken by EPCC describing the methods employed. These projects are in quite diverse application areas : resource planning in the prosecution service; detecting customer trends in retail banking; condition monitoring of connected machinery; and staff scheduling in automotive repairs. The paper highlights the issues common to these projects despite their diversity in solution methods, computational requirements and application areas. These projects are described in section 2. In section 3 those potentially time-consuming tasks involved in obtaining and formatting the operational data to be analysed are discussed. Section 4 discusses the data analysis process, and finally section 5 offers some conclusions.

2 Data Intensive Projects

EPCC has collaborated with a range of industrial and commercial organisations on data intensive projects. These projects have used a range of techniques : data mining with decision trees, statistical methods and knowledge based systems; and a range of computational platforms : shared memory enterprise servers through workstations to high end personal computers. Four of these projects are described further below. The order the projects are listed is indicative of their use of high performance computing. This first being the most HPC-intensive.

2.1 Resource Planning

In this project, the partners were trying to determine if HPC-enabled data mining can help in the allocation of resources for the prosecution of criminal cases. The project partners were the Crown Office, Scotland's prosecution service, and Quadstone, a provider of scalable high performance decision support software [1]. The project formed part of EPCC's activities under the European Union's High Performance Computing and Networking Technology Transfer Node Network.

The Crown Office retains records on all the criminal cases in Scotland. At the time of the project this totalled 2.6 GBytes of data on approximately 1 million cases. EPCC used Quadstone's Decisionhouse data mining product running on a Sun 4 processor Enterprise 3000 with 1 GByte of memory to effectively analyse a dataset containing over 40000 criminal cases (100 MBytes of data). This analysis involved data cleaning, visualisation, data extraction for operational management and data modelling to help understand the characteristics that determine the length of a criminal case. Such analysis is currently beyond the realms of a single user operating a spreadsheet on a personal computer.

This project therefore successfully demonstrated the effectiveness of HPC-enabled decision support in resource planning for the prosecution service.

2.2 Retail Banking

EPCC in collaboration with the University of Edinburgh's Management School (UEMS) and Lloyds TSB Bank Scotland plc used data mining techniques to evaluate and predict customer trends within Lloyds TSB's mortgage business. Lloyds TSB group is one of the largest financial services operations in the UK, with a consolidated revenue of £7,872m in 1999. Globalisation of the financial services market has intensified competition from traditional players and new entrants, where low-cost operators are seen to have a competitive advantage. Revenue growth is seen as critically dependent on improving customer relationship management and increased segmentation of the customer base [2]. EPCC and UEMS used various techniques to attempt to answer business questions posed by Lloyds TSB. This principally involved using decision trees and statistics to identify customer characteristics. Rulequest Research's C5.0 product [3] was used for the decision tree analysis while SAS [4] was used for the statistical analysis. The project involved three distinct stages:

1. Database Evaluation - The TSB mortgage database was evaluated to determine the nature of the information contained within it. This involved direct cooperation with TSB staff to enable understanding of the business processes used in the mortgage division and the requirements of a data mining project. Through this the appropriate and relevant data for analysis could be identified within the mortgage database.
2. Data Preparation - Having examined the TSB mortgage database in its "raw" form, the data was manipulated so data mining techniques could be applied. This involved several different steps. Firstly, the fields in the database identified by TSB as areas of interest needed to be extracted. Many of these fields then had to be adjusted before meaningful analysis could take place, for example the conversion of calendar dates to lengths of time. Then the data had to be formatted for use with the data mining software.
3. Data Mining - This stage involved the application of data mining techniques to analyse and determine associations within the mortgage data. There are many tools available ranging from those that include artificial intelligence techniques to more traditional database query and report methods. In this project, the decision tree approach was used. This involved searching the data for patterns or regularities and using these results to infer new information about the data.

These three stages are typical in any data mining project. It is interesting to note that stages 1 and 2 are generally the most time consuming and this project was no exception to that rule. This was mainly due to the size and complexity of databases which have been maintained over several years.

A 4 processor Sun Enterprise 3000 with 1 GByte of memory was used in the manipulation and analysis of the data.

2.3 Condition Monitoring

Integriti Solutions Ltd produce plant health assessment systems which are typically used on oil platforms and power stations to monitor the condition of the machinery. These systems are able to save operators large amounts of money by detecting potential problems before they result in expensive failures. The systems collect data from sensors located on connected plant such as compressors, turbines, generators and gearboxes. These systems allow alarm levels to be set for each sensor and provides tools for analysing the data. The system therefore enables the machine operators to monitor the connected machinery checking that the components are within their operational range. Large amounts of data are collected, all of which can be useful when diagnosing the cause of alarms and for optimising processes. However, this information requires great skill and experience to analyse and due to the volumes of data this can be a time-consuming task. Integriti therefore wished to examine novel computational methods for analysing this collected data.

Initially, in a proof of concept study, EPCC used knowledge discovery methods based on decision trees (C5.0 [3]) and statistical techniques (SAS [4]) to analyse this collected operational data from a client site. This proved successful and Integriti subsequently employed some of these techniques into their systems. The study was carried out on a UNIX workstation. Following from this, EPCC and Integriti designed and implemented a research prototype knowledge based system to help detect and diagnose faults when sensors on connected machinery go into alarm [5]. This knowledge based system analyses the sensor data as it is collected, identifying potential fault causes and the machinery in distress. An operator no longer has to examine large volumes of data thus its reduces the time to detect and resolve problems.

This project was funded by Scottish Enterprise, the UK government's development body. The knowledge-based system utilised FuzzyCLIPS [6], the fuzzy logic rule-based expert system shell and runs on a UNIX workstation.

2.4 Staff Scheduling

Staff scheduling is a complex problem encountered in almost all large businesses. Available manpower and skills must be deployed as effectively as possible, while being flexible enough to cope with the unexpected. Otherwise potential business may be lost, while valuable staff time is squandered unnecessarily. EPCC and Kwik-Fit GB Ltd collaborated to develop a staff scheduling model using high-end PC technology in data mining and simulation.

Kwik-Fit have around 600 centres in the UK alone. At each centre there are between 4 and 20 fitters of varying skills levels. The seven-day-a-week operation in Kwik-Fit combined with highly variable demand, means staffing is an important and complex issue. Kwik-Fit need to ensure there are sufficient staff to meet customer demand or else business is lost. The staff rota therefore directly affects business efficiency and also staff morale since staff prefer to have consecutive days off.

The project made use of vast quantities of data keyed in daily from every Kwik-Fit centre to the main Kwik-Fit database. Data was extracted from the database and analysed to uncover the staff-related factors affecting business performance. Kwik-Fit also needed to know about the effects of the existing rota situation, to identify areas where the current approach to rostering was successful and where it was falling short. Data-mining and statistical techniques were used to assess the performance of centres [7]. In addition, key results from the analysis were used to develop the model of the Kwik-Fit operation.

HPC could be applied both in the data cleaning prior to data mining, and in the possibility of simulating centres in parallel, using up-to-date simulation parameters generated by a data mining layer.

The system will help Kwik-Fit develop staff rotas better suited to their business needs. Just as importantly it will allow them to plan for the future and make assessments based on a changing commercial environment.

3 Getting the Data

Table 1 contains a breakdown of the amount of the time each project spent in getting the data and performing the actual analysis. As highlighted by Bigus [8] when discussing data mining projects, up to 80% of project time can be spent preparing the data for analysis. The projects described in this paper were no exception to this rule and it is important to realise that this time is not all computational time.

Project	% of Project Time	
	Acquiring and preparing data	Analysis and presentation
Resource Planning	62%	38%
Retail Banking	48%	52%
Condition Monitoring	40%	60%
Staff Scheduling	56%	44%

Table 1. Percentage of total project time expended in data preparation and analysis.

In the projects described in this paper, the various tasks which comprised most of the overhead in this acquisition and preparation of data are listed below.

Establishing the Focus. In all the projects, the goals of the proposed analysis had to be established. This helped determine the data to be collected for analysis and the areas of interest to the client. This also involved agreement on the the steps to be taken to ensure that a client's data and the analysis results were not made available to potential rivals.

Locate and Obtain Relevant Data. Determining which parts of an organisation are responsible for the relevant data. It can be the case that while the project is being driven by the marketing department, the data is the responsibility of the IT department. A further level of negotiation and explanation is therefore likely in order to collect the data. In all the projects concerned, this type of scenario occurred. In the condition monitoring project, data had to acquired from an operational site. In the remaining projects, typically higher management or marketing were driving the projects and so links had to be developed by EPCC with data providing departments to ensure the appropriate data was transferred in suitable formats.

Sensitive Information. When dealing with data containing private or sensitive information eg. personal details, it was necessary for the data provider to remove such fields before supplying the data. This added a further data processing overhead before the data was supplied to EPCC to be prepared for analysis. In dealing with commercial organisations, as in the retail banking and staff scheduling projects, and with public bodies as in the resource planning projects, these factors were of prime concern.

Data Formats/Transfer Method/Legacy Systems. Organising the format and method of data transfers can be costly. When dealing with vast quantities of data, suitable tape drives have to be available for data transfer between organisations. This is potentially an area for misunderstandings. It is not unusual for this data transfer process to involve a number of iterations before the relevant data in the correct format is obtained. Other issues, for example, conversions between ASCII and EBCDIC can further disrupt the process. Moreover when dealing with legacy systems, further data conversion may be necessary, for example when dealing with packed COBOL structures. In such circumstances it is very important to have historical knowledge of the data and its process to ensure that the conversions are appropriate and correct. An important point to realise is that different organisations often use the same terms for different techniques. In such circumstances it is all too easy to make assumptions about terminology. Such assumptions can lead to costly delays to the data transfer process.

Data Conversion and Cleaning Scripts. In all the projects, such scripts had to be developed or customised. The development of such scripts can be a significant programming exercise particularly when dealing with historical data spread across a range of systems.

Data Cleaning. In addition to the actual collection of the data being a time-consuming task, the preparation of data is recognised as a computationally intensive task. Here, HPC can be used to clean and pre-process data. Such preparation typically involves the removal of bad or irrelevant information, conversions of calendar dates and field reclassifications. In one of the previously mentioned projects, a flat file conversion took 12 hours of compute time, with a further hour to extract the data to be analysed. The actual data mining analysis took minutes to perform. Before embarking on a large-scale data cleaning operation it is worthwhile establishing if at the end of this process there is likely to be sufficient quality data for the ensuing analysis. This may prevent a costly, ultimately worthless data cleaning operation being undertaken.

Business Processes. As important as the data itself are the business data processes. Without collecting the knowledge of the processes the data has undergone and how it is used, a knowledge discovery project is more likely to be unsuccessful. This is where good communications and the involvement of the data provider is critical to the project.

Regardless of the size of the project or the solutions being employed, the process of collecting and preparing the data is one fraught with potential problems and time delays. It is therefore important to plan and discuss this process fully with the clients and data providers to help minimise these issues. In particular assumptions concerning any aspect of this process should not be made. Table 2 suggests a possible division of responsibilities among the participants at this stage of a knowledge discovery project. By establishing these responsibilities at the start of the project, this may go some way to alleviating the opportunity for problems and delays. The table makes the distinction between client and data provider to account for those common circumstances where the project is say, driven by marketing, with the data being under the care of the IT department.

A final point to be aware of in this context is the iterative nature of knowledge discovery projects means that the previously mentioned 80% of project time is not spent in a single collection and preparation task at the beginning of a project. Rather the collection and preparation is ongoing throughout the length of the project and continues as more knowledge is gained and the discovery process becomes more focused.

Client	Data Provider	Data Analyst
Establish goals	Data Collection	Security
Areas of interest	Data Transfer	Script development
Business knowledge	Removal of Sensitive Information	Data conversion
Relevance of Results	Data processing knowledge	Data cleaning

Table 2. A possible division of responsibilities among participants during the acquisition and preparation of data.

4 Analysing the Data

4.1 Tools

A variety of tools and techniques were applied in the projects and some of their characteristics are discussed below.

Rule-Based. FuzzyCLIPS[6] was employed in an expert system to analyse masses of sensor data and to identify faults.

Statistical. SPSS and SAS were used for basic statistics and statistical modelling. When dealing with off-the-shelf packages, it is important to be aware of their limitations. In one of the projects the data consisted of more than 350000 records each containing more than 800 fields. Such volumes cannot always be handled by off-the-shelf packages.

Decision Trees. The Decisionhouse product from Quadstone, with its parallel processing capability and its graphical user interface was successfully employed. At the other end of the extreme, the less sophisticated and less expensive, C5.0 was also very successfully employed on projects with very large amounts of data. C5.0 has been developed by J.R.Quinlan who was responsible for the development of two other popular decision tree algorithms ID3 [10] and C4.5 [11].

4.2 The Analysis Process

According to Fayyad et al [9] a large part of the application effort in knowledge discovery projects is towards formulating the problem, that is asking the right question rather than optimising the algorithmic details of a particular data mining method. They also state that domain knowledge is important in all the steps of the knowledge discovery process and that blind application of data mining methods is a dangerous activity leading to the discovery of meaningless patterns. In all the projects described in section 2, knowledge of the application area was critical to their success. EPCC was able to provide skills in knowledge discovery and data intensive applications but this was essentially useless without the additional domain knowledge provided by the project partners. In addition to the application knowledge, the knowledge of the data history is again paramount.

This cross-fertilisation of skills and knowledge across the participants in each of the projects was achieved by regular meetings and exchange of information on the latest analysis results. In a number of the projects the initial analysis results appeared very promising but closer examination revealed that these were trivialities or process artefacts. An example of such a process artefact is that in one of the projects there appeared to be a direct correlation between the absence of a particular record type in a database and a particular event occurring. When the result was presented to all involved parties, it was revealed that this record was actually removed when the event occurred. Thus the correlation was merely a data processing artefact. In light of this extra information, the input data was prepared once again to account for this. A further example is that in the staff scheduling project, staff at the centres in addition to the higher management were often consulted to help shed light on the results. One such consultation was used to define the time to perform particular tasks at a centre. This information helped reveal a better correlation between staffing levels and sales at a centre.

These types of situation were common across all the projects throughout their lifetime and so illustrate a number of useful points.

- Data preparation and collection is ongoing throughout such projects.
- The importance of the business processes and knowledge of the data processing history.
- The necessity for ongoing dialogue between all participants during the life of the project.

The skills both EPCC and the project partners brought to the projects were critical in formulating the problem to ensure meaningful results were produced and that they were produced in a timely fashion. This analysis of the data was an iterative process. With each iteration a clearer picture was built up which helped improve the clarity of subsequent questions and analysis. A common factor to the projects was where initial data mining did not reveal any correlations with a particular class of data. This led to further discussions between EPCC and the project partners to more tightly define this data class. Subsequent analyses were able to better identify correlations across the different classes of data.

As an example, in the retail banking project, we were looking for behaviour characteristics associated with a particular type of customer. In the first instance, decision tree analysis with C5.0 [3] revealed very little. As the project progressed, the definition of the customer became more focused and so the behaviours became much more apparent and C5.0 was better able to identify this customer type. Similarly in the condition monitoring project, the factors affecting corrosion only really became apparent when using input data consisting of a full range of conditions at the operational site. It transpired that initial investigations with C5.0 were on a dataset collected during a particularly stable period with low corrosion at the operational site. When further data was collected during less stable periods of operation, then C5.0 was better able to identify the factors affecting corrosion.

4.3 Assessing Results

In all the projects criteria for assessing the results from the analysis of the data were very helpful. The following proved to be useful criteria for all parties involved when examining the latest results and in determining future directions for the analysis process.

Feasibility. Trivial correlations which bore no relation to the process under investigation have to be discounted. Occasionally what appear to be feasible results can be found but it is important to verify such results to ensure they are not merely artefacts of the business process. Section 4.2 contains a description of one such occurrence on a project.

Accuracy versus Simplicity. It is important to understand if the results are broad generalisations or if they are so accurate that they classify the data into groups which are too focused. Further, it may be possible to produce very accurate data classifications however the overhead involved in applying the results may outweigh the benefits. Moreover results may generate rules which are not easily understood by domain experts. For example, in one of the projects, decision tree analysis with C5.0 produced, firstly 26 rules which could correctly classify 77% of records in a database, and secondly, 231 rules which could correctly classify 80% of the records. The overhead of applying the extra rules has to be weighed against the extra accuracy obtained.

Applicability. Results may be feasible and not trivialities but they may not be relevant to the aim of the project. Interesting business results can be generated which can lead investigations off at a tangent to the project objectives. In the staff scheduling project, the data revealed certain staff behaviours which concerned the management. This led to further investigation by the management about these behaviours. However, this result had no relevance to the goals of the project itself.

5 Conclusions

To successfully exploit operational data in knowledge discovery project requires:

- A clear division of responsibilities across project participants to ensure that all necessary tasks are performed.
- Frequent exchange of information between all participants including the data providers. This ensures that subtleties or issues associated with any of data acquisition, preparation and analysis can be identified and resolved early.
- Information on business processes and data history must be included in the data acquisition. Without such information the relevance of results cannot be clearly established.

References

1. Quadstone: http://www.quadstone.com/
2. Ellwood, P., Group Chief Executive Lloyds TSB: *Delivering Value to Shareholders.* Lloyds TSB Analysts Presentation (1999)
3. RuleQuest Research: http://www.rulequest.com/
4. SAS Institute: http://www.sas.com/
5. Sloan, T.M., Hall, H.J., Hopcraft, R.: *Exploiting Soft Computation in SMEs.* Proceedings of the 6th European Congress on Intelligent Techniques and Soft Computing. ELITE-European Laboratory for Intelligent Techniques Engineering (1998)
6. Orchard, R.A.: *FuzzyCLIPS Version 6.04 User's Guide.* Knowledge Systems Laboratory, Institute for Information Technology, National Research Council Canada (1995)
7. Smyllie, K..: *Data Mining and Simulation Applied to a Staff Scheduling Problem.* High-Performance Computing and Networking (Eds. Sloot, Bubak, Hoekstra, Hertzberger). Lecture Notes in Computer Science Vol. 1953. Springer (1999).
8. Bigus, J.P.: Data mining with neural networks. McGraw-Hill(1996)
9. Fayyad, U., Piatesky-Shapiro, G., Smyth, P.: *The KDD Process for Extracting Useful Knowledge from Volumes of Data.* Communications of the ACM Volume 39(11) (1996)
10. Quinlan, J.R.: *Discovering rules by induction from large collections of examples.* Expert Systems in the micro-electronic age (Ed. D.Michie). Edinburgh University Press (1979).
11. Quinlan, J.R.: *Improved use of continuous attributes in c4.5.* Journal of Artificial Intelligence Vol 4 (1996).

Considerations for Scalable CAE on the SGI ccNUMA Architecture

Stan Posey, Cheng Liao, and Mark Kremenetsky

SGI, Mtn View, CA, 650.933.1689
sposey@sgi.com

Abstract. Recent breakthroughs in CAE technology have lead to a high degree of parallel scaling of CAE simulations in a distributed shared-memory environment. A variety of research and commerical CAE software demonstrate scalable levels that are linear beyond 256 processors on the ccNUMA system architecture developed by SGI. Parallel algorithms typical of contemporary CAE software offer efficient strategies that overcome bottlenecks for moderate levels of parallelism. However, to achieve efficient parallelism on 100's of processors, additional consideration must be given to topics such as performance of system software and awareness of communication architectures. This paper examines the requirements for highly-scalable CAE simulations on an assortment of industrial application examples.

1 Introduction

Traditional industries such as automotive, aerospace, and power generation are challenged with an increasing need to reduce development cycles, while satisfying global regulations on safety, environmental impact and fuel efficiency. They must also appeal to demands for high-quality, well-designed products in a competitive business environment. Continuing technology advances in computer-aided engineering (CAE) simulation provide industry with a design aid that is a relavent step towards achieving these goals.

Historically, CAE simulation provided limited value as an influence on industrial design owing to excessive modeling and solution times that could not meet conventional development schedules. During the 1980's vector architectures offered greatly improved CAE simulation turn-around, but at a very high cost. RISC computing introduced in the 1990's narrowed this gap of cost-performance, however bus-based shared-memory parallel did not scale sufficiently beyond 8 processors.

Recent advancements in parallel computing have demonstrated that vector-level performance can be easily exceeded with proper implementation of parallel CAE algorithms for distributed shared-memory systems like the SGI 2800. Perhaps even more appealing is that the increased performance is offered at a fraction of the cost. These trends have influenced recent increased investmestments by users of CAE technology throughout a range of industries.

The automotive industry in particular has made substantial investments over the past three years in scalable sytems and parallel CAE software. It is estimated that during 1999 alone, total GFlop compute capacity within the three Detroit automotive OEM's increased more than 2-fold. This is in large part due to efficient parallel implementations of commerical CAE software for applications such as crashworthiness simulation, analysis of noise vibration, and harshness (NVH), and computational fluid dynamics (CFD) simulation. During a one year period, a total of 740 processors of SGI 2800 servers including two systems with 128 processors each, were deployed within the Detroit Big 3, representing about a 375 GFlop increase to existing compute capacity.

A system architecture's ability to achieve high parallel efficiency becomes increasingly important as algorithms for CAE software applications are developed towards such capability. From a hardware and software algorithm perspective, there are roughly three types of CAE simulation "behavior" to consider: implicit and explicit finite element analysis (FEA) for structural mechanics, and CFD for fluid mechanics. Each have their inherent complexities with regards to efficient parallel scaling, depending upon the parallel scheme of choice.

Most commercial CAE software employ a distributed-memory parallel (DMP) implementation based on domain decomposition methods. This method divides the solution domain into multiple partitions of roughly equal size in terms of required computational work. Each partition is solved on an independent processor, with information transferred between partitions through explicit message passing software (usually MPI) in order to maintain the coherency of the global solution.

Other choices for efficient parallel methods are shared-memory parallel (SMP) coarse grain, and hybrid parallel schemes that combine DMP and SMP within a single simulation. Hybrid parallel schemes, which are becoming increasing popular for applications that contain a mix of Eulerian and Lagrangian mechanics such as combustion, are particularly well suited to the SGI 2800 distributed shared-memory architecture.

The distributed shared-memory SGI 2800 system is based upon a cache-coherent non-uniform memory access (ccNUMA) architecture. Memory is physically distributed but appears logically as a shared resource to the user. Motivation for ccNUMA evolved at SGI as conventional shared-bus architectures like that of the SGI CHALLENGE would exhibit high-latency bottlenecks as the number of processors increased within a single system. During this same time, non-coherent distributed-memory architectures emerged, but application development for most designs was considered too difficult for commercial success.

The SGI ccNUMA implementation distributes memory to individual processors through a non-blocking interconnect design, in order to reduce latencies that inhibit high bandwidth and scalability. At the same time, a unique directory based cache-coherence provides a memory resource that is globally addressable by the user, in order to simplify programming tasks. A single image SGI 2800 system offers up to 512 processors and can expand to 1 Tbyte of memory, which is the largest SMP system currently available in industry.

2 Industrial Application Examples

A variety of industrial CAE application examples are provided that demonstrate the requirements and recent achievements towards highly scalable CAE. All examples present models from industry and represent some of the largest currently in practice. These industry examples highlight applications from commercial CFD, research CFD, explicit FEA, and implicit FEA.

Algorithms and applications for CFD have led the advance in parallel scalability. CFD models are growing in size such that the the ratio of work to overhead is rapidly increasing. Also, fluid domains have a natural load-balance advantage over structural domains owing to uniformity of fluid properties – typically air or water, such that model entities for CFD typically offer uniform computational expense from one to the next.

Structures often contain a mix of materials and finite elements that can exhibit substantial variations in computational expense, which creates load-balance complexities. The ability to efficiently scale to a large number of processors is highly sensitive to load balance quality. The examples illustrate these differing charateristics between the two disciplines.

2.1 Commercial CFD – FLUENT

Commercial CFD software such as FLUENT from Fluent Inc. (www.fluent.com) is among the most efficient parallel software in industry. Fluent in particular has provided leadship in the recent trend of CFD simulations reaching essential higher resolutions. FLUENT is a multi-purpose, unstructured-mesh CFD solver that employs an algebraic multi-grid (AMG) scheme for solution of the 3D, time-dependent Navier-Stokes equations.

The parallel implementation of FLUENT can accomodate homogenous compute environments such as SMP systems or clusters, as well as a heterogeneous network of various types of systems. A domain decomposition method with various graph partitioning schemes is used as the parallel strategy. The parallel programming model used is DMP with MPI as the message passing software.

For classic DMP implementations, several factors can inhibit parallel scalability. CFD solver algorithms affect the frequency and amount of information that must be shared across partitions. Similarly, efficient planning of what data to share (the message content), and when to do so, is important. Quality domain decomposition schemes are critical since they affect load balancing, and determine the size of partition boundaries and consequent message passing requirements.

Efficient algorithmic and partitioning strategies are provided in FLUENT that overcome the scaling issues noted. Linear scaling to about 16 processors is routinely achieved with well-designed parallel CFD software. Beyond that level, efficient scaling can be sensitive to the performance of hardware communication architectures and system software. In particular, the choice and implementation of the system software for message passing between partitions is critical.

Parallel efficiency of the AMG solver is restricted only by hardware and software communication latency of a particular system. FLUENT supports proprietary versions of MPI, some with latency improvements of nearly 3-fold over public domain MPICH. During 1997, SGI released an MPI based on standard MPICH source that was tuned to properly account for performance issues related to the SGI ccNUMA architecture.

The ccNUMA-aware MPI provided by SGI exhibits a latency of about 13 micro-seconds on the SGI 2800 system. Further experiments conducted on the SGI 2800 with a one-sided MPI implementation based upon the MPI-2 standard, show latency in the range of 3 mirco-seconds. These results suggest that further improvements of FLUENT parallel efficiency are possible in the near future.

A recent investigation on the parallel efficiency of a large automotive case involved an aerodynamics study of a full vehicle on a stationary ground plane. The flow conditions were steady-state and isothermal, and a K-epsilon turbulence model was used to resolve shear effects. The most critical aspect of this model was the size – 29M cells, which exceeds recent automotive aerodynamics modeling by 3-fold.

The system used for this aerodynamics study was a SGI 2800 with 256 MIPS processors at 195 MHz, and 33 Gbytes of memory. A segrated solver strategy in FLUENT 5.1.1 was used for solution of the 29M cell model, which required 28 Gbytes of memory. As with performance evaluations on other large automotive examples, this case exhibited a remarkably high degree of parallel efficiency for FLUENT.

Parallel performance results from 10 to 240 processors produced a 21-fold speed-up, close to ideal which occurs at 24-fold. Solution on 10 processors required 381 seconds per iteration and was reduced to just 18 seconds for 240 processors. The speed-up from 10 to 120 processors was 13-fold, or greater than ideal at 12-fold. This "super-linear" behavior can occur when domain partitions are a size where they become local-memory resident. Local memory is 256 Mbytes and the domain contains 120 partitions of roughly equal size, each about 233 Mbytes.

The study measured performance starting with 10 processors rather than one, in order to reduce the artificial scaling beneifts observed for large remote-memory access times. With 10 processors, some of the threads are distributed to nodes with data in local memory. This reduces the overall number of remote-memory hops since 10 threads require less hops than a single thread. Using a 10 processor result as a reference time is more realistic for a scalability study of this model size.

This achievement is significant for the automotive industry since aerodynamic characteristics are important during early developlment efforts that fix the vehicle shape. CFD simulation is generally not influential at this early stage. Instead, conventional aerodynamic evaluation of vehicles rely primarily on expensive wind tunnels since an equivalent CFD simulation would require among other things, models with resolution levels near the 29M cell case.

For its part, the performance limitations of large model turn-around for vehicle aerodynamics has been addressed with this study. Additional CFD modeling features, mostly related to turbulence are required before CFD simulation can effectively replace wind tunnels. Still, CFD capability has reached an effective level of accuracy and economics today as an established experimentation alternative, or at least a compliment for a wide range applications.

In another FLUENT example involving combustion simulation for an industrial burner, the parallel benefits of hybrid SMP and DMP methods are illustrated. The model contains 155K cells, and in addition to resolution of momentum and energy distribution, simulations of this class require accurate modeling of spray and chemical combustion phenomena. Combustion for this model includes chemical reactions for 6 species, and tracking of 500 disperse phase particles.

Prior to the release of FLUENT 5.1.1, models of this type would exhibit very poor, parallel scalability since the highly parallel Eulerian continuum gas phase was coupled with the serial Lagrangian disperse phase. Implementation by FLUENT of an SMP particle decomposition with m_fork threads now permits models such as these to achieve near-ideal parallel speed-up. For the 155K cell burner example, the hybrid scheme shows more than a 7-fold speed-up on 8 processors.

2.2 Research CFD – OVERFLOW

Research CFD software is typically developed for special purpose modeling. OVERFLOW from NASA Ames, is a structured over-set grid CFD code that is applied by both research and industry for investigation of aircraft aerodynamics. It solves the 3D time-dependent Reynold's-Averaged Navier-Stokes equations. Unlike most commercial CFD software that use a DMP approach, OVERFLOW uses an SMP approach that is multi-level parallel (MLP), meaning both coarse and fine-grain SMP is employed within a single simulation.

The MLP approach uses fine-grained, compiler-generated parallelism at the loop level, and at the same time performs parallel work at a coarser level using standard Unix fork system calls. During execution, load balance adjustments are fully automatic and dynamic in time. Global data is shared among the forked processes through standard shared-memory arenas. This eliminates the need to design code for synchronization and communication as would be required for DMP, and as such greatly simplfies code development.

The shared-memory MLP approach has performance advantages over DMP since it offers coarse-level parallel without a need for message passing. Since DMP requires an additional software interface such as MPI, latencies and other overhead are imposed that affect scalability. These messaging subroutines are used to communicate boundary values and other data between partitions as the solution progresses, which is not required for MLP since data is shared globally.

Beginning in 1997, NASA Ames began deployment of several SGI 2800 systems, the largest of which is a 512 processor single system image (SSI) and currently the largest SMP system in industry. Several large aerodynamic simulations have been conducted with OVERFLOW-MLP up to the maximum 512 processors. One example is a model of a full aircraft configured for landing that contains 35M grid points distributed over 160 grid zones. At the time, this model was considered among the largest ever attempted at NASA Ames.

Performance results for OVERFLOW-MLP 1.8 on the 35M grid-point model were obtained for three NASA systems: CRAY C90/16, SGI 2800/256 (250 MHz), and SGI 2800/512 (300 MHz). The level of achieved GFlops during the simulation was measured for each system and the C90 was measured at 4.6, the 2800/256 at 20.1, and the 2800/512 at 60.0 – a performance level that required only 2.5 seconds per time step. Parallel performance for both SGI 2800 systems achieved linear scaling to their maximum number of processors.

The 35M grid-point OVERFLOW study demonstrates several significant computing advances. Most important is the practicallity for model resolutions as high as 100M grid points – a level considered neccessary for "virtual wind tunnel" simulations. Another advance is the cost-performance improvements of a a new generation of parallel architectures. NASA reports the cost of the SGI 2800/512 to be 2.6-fold less than the CRAY C90. With a performance advantage of 13-fold, this provides the 2800/512 system with a 33-fold cost-performance advantage.

2.3 Explicit FEA – PAM-CRASH

Explicit FEA software PAM-CRASH from developer Engineering Systems International (www.esi.fr), is a multi-purpose structural analysis tool for high-transient, short duration dynamics. This includes such applications as automotive crash simulation, airbag deployment, and turbine rotor burst containment among others. In addition, simulations of this classs often require predictions of surface contact and penetration.

Parallel implementations of both SMP and DMP have been developed for PAM-CRASH. The SMP version exhibits moderate parallel efficiency and can be used with SMP systems only. The DMP, based on domain decomposition with MPI for message passing, exhibits high scalability and can accomodate homogenous compute environments such as SMP systems or clusters, as well as a heterogeneous network of various types of systems.

A recent performance investigation of the DMP version, PAM-CRASH 99D was conducted with simulation of a vehicle frontal crash traveling at 35 mph. The vehicle model contained 120,000 elements with 117,500 contact segments split into 11 contact interfaces. The simulation was completed for 70 milli-seconds of real time, which required about 58,000 time steps. The system used for this study was an SGI 2800/128 (300 Mhz).

Results for this study demonstrate a sustained performance of 3.2 Gflops with 32 processors and 5.4 Gflops with 64. Parallel speed-up between 32 and 64 processors yielded 1.7-fold which is nearly ideal speed-up at 2-fold. Further parallel scalability of PAM-CRASH V99 was observed up to 96 processors with performance of 6.2 Gflops. These levels of performance are the some of the highest observed for Explicit FEA applications.

Evaluation of an vehicle's crashworthiness is currently the fastest growing application for CAE simuation in the automotive industry. Proper crash management and occupant safety is a mandate by local governments, but its also viewed as a competitive advantage by many automotive developers. Further performance improvements are forthcoming with PAM-CRASH such that parallel scalability will keep pace with the growing demand.

2.4 Implicit FEA – MSC.Nastran

Implicit FEA software MSC.Nastran from developer MSC Software[1], is a multipurpose structural analysis tool with a range of linear and nonlinear capabilities. It is most popular with industrial applications that require evaluation of the dynamic response. of structures. For example, the automotive industry uses MSC.Nastran for design applications such as vehicle weight reductions, improved durability, and reductions in noise, vibration and harshness (NVH).

Beginning with MSC.Nastran 70.7, a DMP capability was developed with MPI as the message passing software. NVH applications were the primary motivation for parallel MSC.Nastran since they require a more than 10-fold increase in compute resources over other applications. Body NVH in particular can require a day of processing on a vector system such as the Cray T90.

In general, two methods are used for frequency response evaluation in NVH applications; modal and direct. The modal method performs two steps; a Lanczos eigenvalue analysis that computes natural frequencies over an excitation range of interest, and later a frequency response on the generalized modal coordinates from the Lanczos step. The direct method is a solution of the equations of motion for each frequency in the range of interest.

The modal method is conventional practice for practically every automotive NVH application today. However, a shift is underway towards increased use of the direct method since high levels of parallel scalability are observed that can provide faster turn-around over modal. MSC.Software developed parallel capability for both modal and direct methods, but the direct has a naturally parallel algorithm that exhibits much higher scalability.

An example of the parallel scaling that is possible with the modal method is provided with an automotive body-in-white (BIW) model that contains 1.5M degrees-of-freedom (DOF). The excitation frequency range of interest is 0 to 200 Hz, whereby 1076 modes were extracted during the Lanczos step. Note also that only the Lanczos step was executed in this study. The system used for this study was an SGI 2800/64 (300 MHz).

Parallel performance results for the Lanczos step from 1 to 8 processors produced a 3.2-fold speed-up, which reduced the turn-around time from 17.1 hours to 5.5 hours. Note that scalability is model dependent, since the parallel Lanczos scheme divides the frequency range, in this case 200 Hz, equally among processors, then performs the eigenvalue search. In this example, processor 1 was assigned the segment of 0 to 25 Hz, processor 2 from 26 to 50 Hz, and so on.

---

[1] http://www.mscsoftware.com/

The parallel Lanczos scheme can experience load-balance inefficiencies since it is rare for a structural to have an equal number of natural frequencies within each uniform frequency segment. The result is that some processors perform more work than others for a given segment. The direct method, however offers good load-balance since parallelisation is implemented for the independent frequency steps which each perform the same amount of work.

A second example illustrates the potential for high parallel scalability with the direct frequency response method. Performance was investigated on the same SGI 2800/64 for a vehicle BIW model that contained 536K DOF and 96 frequency steps. Results from 1 to 32 processors produced a 23-fold speed-up, which reduced the turn-around time from 31.7 hours to only 1.4 hours. On 16 processors the speed-up was 14-fold, demonstrating very good efficiency.

NVH analysis provides essential benefits to the automotive industry towards designing vehicles for improved ride comfort and quietness – clearly an increasingly competitive advantage. An NVH capability that is highly scalable is critical to the automotive industry as NVH modeling expands to higher excitation frequencies that are necessary for modeling within a higher audible range. Further noise reductions for the occupant are possible with a better understanding of noise sources at these higher frequencies.

Higher frequencies require that NVH model parameters grow substantially larger from current practice. NVH models today for trimmed BIW are typically limited to frequencies of 250 to 300 Hz, yet many automotive companies want an increase to 600Hz. The conventional modal method requires vector systems for overnight turn-around, but even next generation vector designs will not deliver the performance required for future targets of NVH modeling. Also, accuracy concerns for the modal method at higher frequencies do not exist for the direct method.

The automotive industry "standard" for NVH has been the modal method and vector systems for several years. It will be necessary to characterize the cross-over point whereby the parallel direct method with RISC systems becomes less expensive computationally. A third example demonstrates this with a comparison of a vehicle BIW using the direct method on the same SGI 2800/64, and the conventional modal method on a single processor Cray T90. The model contains 525K DOF and extracts 2714 modes for a response analysis of 96 frequency steps.

Results demonstrate roughly equal performance between the two techniques when 4 processors are used for the SGI 2800. At this level, the Cray T90 time is 8.8 hours compared with the SGI 2800, 4 processors at 8.9 hours. The 4 processor time exhibits near-ideal speed-up at 3.8-fold over 1 processor at 33.5 hours. Results for the SGI 2800 at 8 processors further reduces the turn-around time to 4.8 hours for a 7-fold speed-up over that for 1 processor.

The model used in the comparative study exhibits a higher modal density than what is typically observed today, and is perhaps more representative of future modeling practice. Modal density will increase nonlinearly with increasing excitation frequency. These results offer an encouraging alternative to conventional NVH modeling, based upon both performance, and cost-performance. An additional benefit is the improved solution accuracy of the direct over the modal method.

3 Future Directions

In the wake of recent breakthroughs for scalable CAE simulation, research and industry will continue to increase their investments in CAE technology as a product and process design aid. The motivation is simply a matter of economic benefits and improved quality that scalable CAE brings to the development process. Efficient turn-around of CAE simulations means increased modeling resolution and more comprehensive evaluation during early development stages.

Advancements will continue well into the new decade for improved CAE scalability as emerging algorithm developments and new hardware architectures lead a path towards enhanced CAE methodologies. These enhancements will encourage an increase in CFD modeling for transient flow conditions, wide-spread implementation of probabilistic structural mechanics, and production capability for multi-discipline fluid and structure coupling, among others.

CFD simulations for transient conditions have reached their potential for industrial application. Important simulation capability for transient flows include applications of automotive powertrain in-cylinder combustion, vehicle aerodynamics with rotating wheels and moving ground plane, commercial aircraft aerodynamics for off-design (non-cruise) conditions, and a variety of turbomachinery aerothermal flows for compressor, combustor, and turbine designs.

Structural FEA simulations are currently undergoing a historic transition from determinstic to probablistic. Turn-around for a single FEA analysis is considered small such that highly parallel stochastic techniques are being applied to better manage design uncertainty of scatter observed in sources such as as material properties, test conditions, manufacturing, and assembly. The discipline of explicit FEA in particular is suited to benefit from probablistic techniques.

The high-transient nonlinear modeling of dynamic events with explicit FEA, such as automotive vehicle crash, airbag-occupant interaction and aircraft bird-strike all exhibit substantial parameter scatter. This trend towards stochastic simulation is a relavent step towards an overall trend towards single and multi-discipline optmization. Lately, monte carlo stochastic methods have been applied to automotive vehicle design for improved NVH and crashworthiness.

Computer hardware architectures will continue to improve. During 2000, SGI will introduce a new generation of ccNUMA that will provide a 2-fold increase in system bandwidth with a latency decrease of 2-fold. Other performance improvements include a processor clock speed increase by 33L2 cache. These and other features will improve both single processor turn-around and parallel efficiency over the current SGI 2800 ccNUMA system.

4 Conclusions

Effective implementation of highly parallel CAE simuations must consider a number of features such as CAE parallel algorithm design, system software performance issues, and hardware communication architectures. Several examples of commericial and research CAE simulations demonstrate the possibilities for efficient parallel scaling on the SGI ccNUMA system.

Development of increased parallel capability will continue on both software and hardware fronts to enable modeling at increasingly higher resolutions. Indeed, the demonstrated benefits of cost-effective, highly scalable CAE simulation has potential to shift modeling practices within research and industry on a global basis, and continue to further advancements in industrial design applications.

Acknowledgments

The authors wish to acknowledge certain SGI colleagues for their contributions to this paper:

- Herve Chevanne, CAE Applications Engineer, SGI France
- Jeff Konz, CAE Applications Consultant, SGI Detroit
- Takahiko Tomuro, SE Manager for HPC Applications, SGI Tokyo

References

1. Laudon, J. and Lenoski, D. The SGI Origin ccNUMA Highly Scalable Server *SGI Published White Paper*, March 1997.
2. Yeager, K. The MIPS R10000 Superscalar Microprocessor *IEEE Micro*, Vol. 16, pp. 28-40, April 1996.
3. Komzsik, L. MSC/NASTRAN Numerical Methods, Version 70 1998.

An Automated Benchmarking Toolset[1]

Michel Courson[2], Alan Mink, Guillaume Marçais[3], and Benjamin Traverse[3]

100 Bureau Drive, Stop 8951
Gaithersburg MD 20899-8951, USA
{michel.courson,amink,guillaume.marcais}@nist.gov

Abstract. The drive for performance in parallel computing and the need to evaluate platform upgrades or replacements are major reasons frequent running of benchmark codes has become commonplace for application and platform evaluation and tuning. NIST is developing a prototype for an automated benchmarking toolset to reduce the manual effort in running and analyzing the results of such benchmarks. Our toolset consists of three main modules. A Data Collection and Storage module handles the collection of performance data and implements a central repository for such data. Another module provides an integrated mechanism to analyze and visualize the data stored in the repository. An Experiment Control Module assists the user in designing and executing experiments. To reduce the development effort this toolset is built around existing tools and is designed to be easily extensible to support other tools.

Keywords. Cluster computing, data collection, database, performance measurement, performance analysis, queuing system, visualization.

1 Introduction

The drive for performance in parallel computing and the need to evaluate platform upgrades or replacements are major reasons frequent running of benchmark codes has become commonplace. By "benchmark code" we refer not only to well known codes designed to evaluate platforms, but also to any piece of code used for performance evaluation or in need of tuning.

NIST is developing a prototype of an automated benchmarking toolset to reduce the manual effort involved in running benchmarks, to provide a central repository for all the collected data, current and archival, and to provide an integrated mechanism to analyze and visualize that data. To reduce the development effort and leverage the work of other tool developers, we intend to build this benchmarking toolset around existing tools and only produce "glue" and "gap" code as necessary.

---

[1] This NIST contribution is not subject to copyright in the United States. Certain commercial items may be identified but that does not imply recommendation or endorsement by NIST, nor does it imply that those items are necessarily the best available.
[2] Visiting scientist from University of Maryland, UMIACS.
[3] Guest researcher from the Institut National des Télécommunications, France.

Such an automated benchmarking toolset will support both production environments and research environments. For a production environment such a toolset can reduce the manual effort required and speed up the overall benchmark and evaluation process. In addition, the database will provide a means to produce unplanned comparisons over a much longer epoch, limited by the age of the database. Through its database of performance information, such a toolset also provides a rich environment for standard and innovative performance analysis, as well as a means to produce unplanned comparisons.

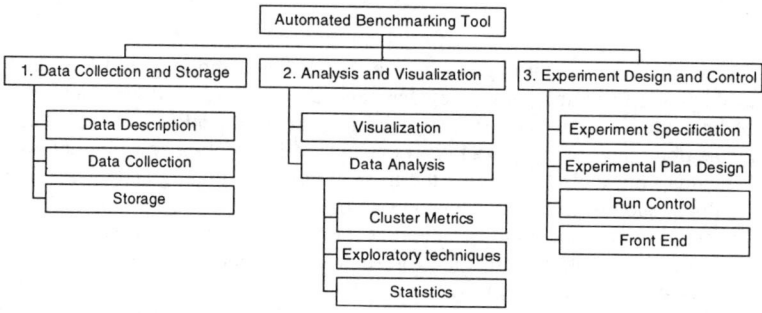

Fig. 1. Project Organization Chart

1.1 Design Overview

The proposed benchmarking toolset consists of three major modules as shown in Figure 1: (1) the Data Collection and Storage module, (2) the Analysis and Visualization module, and (3) the Experiment Design and Control module. The Data Collection and Storage module addresses the runtime collection and storage of performance data for later retrieval. The data storage organization is based on an abstract model of our computing prototype environment, a computing cluster of PCs. This abstract model is mapped on a regular SQL database. We use the PostgreSQL [POS] database management package. The collection of data is done with the SNMP protocol, using the Scotty package [TNM]. However, the tool is general enough to handle virtually any other packages implementing the SQL and SNMP protocols.

The Analysis and Visualization module provides the means to produce standard plots and charts of the raw data from the database or composite values using well known tools such as MatLab [MAT] or IDL [IDL]. This module is intended to be user extensible to allow novel and innovative techniques such as data mining to identify hidden patterns in the data.

The Experiment Design and Control module automates the definition and execution of a set of experiments. This is accomplished from a combination of information supplied by the experimenter and information extracted from the database. The culmination of this module is the execution of a set of run-time scripts; these carry out the benchmark code executions and subsequent data collection and storage of the associated performance data.

Each component is built modularly using existing open software for more flexibility and may be easily modified for experimentation with alternate features and capabilities.

1.2 Existing Work

Many tools have been developed that focus on performance measurement. Some of the most well known include: VAMPIR [VAM99], from the German company Pallas GmbH, for MPI programs, AIMS [YAN96] from NASA Ames Research Center, Paradyn [MIL95] that features performance tracing down to the statement level through dynamic instrumentation, and Pablo [AYD96] which is known for its visualization and self-defining format, among other features.

These tools focus much more extensively than our toolset on instrumenting codes to collect and compute performance measurement metrics and on locating computational bottlenecks. These and other tools also provide various visualization capabilities.

Although our toolset offers performance measurement and visualization, it does so by incorporating other existing tools, possibly some of the tools mentioned above. In addition our toolset focuses on an integrated database that spans all the "benchmark" codes over an extended time frame. This differs from current effort of the Paradyn team [MIL99], which tracks the performance of a single code with the intention of better tuning of that code. Our toolset will also provide the capability to design and manage the evolution of experiments consisting of multiple codes as well as their input and output files.

2 Performance Data Model

This section describes an abstract data model of a parallel computing environment that represents the performance data available from our cluster in a compact but accurate fashion. To implement this model, we selected a set of collection mechanisms and tools as well as a storage platform using a relational database.

2.1 Performance Data Storage Platform

We designed a formal data model to represent both abstract data structures, mechanisms and relationships within the data. This model is described in detail on our website. Although specifically designed to represent a computing cluster, this abstract model is easily changeable for other platforms. The data stored in the database is a set of "snapshots" (runs). The key feature of the model lies in the ability to create *virtual experiments:* such experiments are conducted without actually running the application, but by combining instead existing runs from previous experiments, promoting reuse of existing performance data.

Although an object-oriented database management system (OODBMS) may seem best suited for our data model, we did not find any OODBMS that meets our requirements for convenient querying and easy retrieval for the user. Such features are readily available from most database management systems supporting SQL, at the cost of a little extra administration overhead. The actual database is implemented using PostgreSQL [POS], a freely available SQL relational database.

The entity-relationship diagram (fig. 2) translates to a set of corresponding tables. Each table is the representation of an entity type; each column represents attributes, each row is one occurrence. Primary keys (unique identifiers) are used to ensure uniqueness and to allow cross-references between tables. As noted before, this makes the database administration less straightforward due to complex key management. However this complexity is only a burden for the tool administrator, not the user.

Our reference database implements most of the model and provides two other interesting features. First of all, it supports time series using a variable-granularity scheme similar to Paradyn [MIL95]; common statistics such as average are also cached in the database to facilitate simple queries. Our toolkit also provides a generic registry mechanism to store information without administrator privileges to create new tables, for example application-specific tracing information.

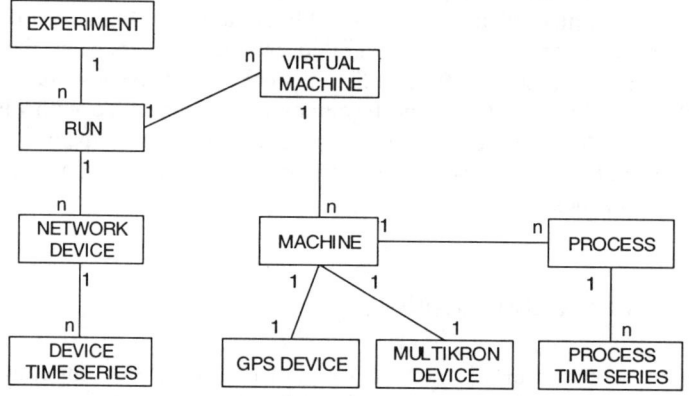

Fig. 2. Entity-Relation Diagram

2.2 Collection Tools

The automated benchmarking toolset is designed to handle data collection from most sources. In order to demonstrate the collection mechanism, our reference implementation currently supports several mechanisms at different levels in the parallel machine, as described below.

For low-level data such as memory usage, paging activity or network usage at the process level, we use a modified version of UCD-SNMP [UCD], a popular, freely available, easily extensible SNMP agent from the UC Davis. The agent already supports a very wide range of system-level data such as network activity, via standard interfaces. It was extended to give fast access to common statistics such as load average, process activity and resource usage, as well as our custom high-resolution tracing hardware, Multikron [MIN95], and GPS-synchronized timing facility. This information is collected and stored as time-series data.

Application-level data such as response time or number of processors is collected directly from the run control module (see below).

Our toolset is designed to support most third-party measurement tools that follow our data model. We have demonstrated this by collecting high-level performance data from the MPI Profiling library developed at NIST [IND99]. This library provides transparent timing and call-count information for most MPI functions (send, receive, etc.) as well as synthetic metrics including computation, communication and response time for any MPI application.

3 Experiment Design and Control

The Experiment Design and Control module provides a means to quickly and efficiently design and carry out experiments. The experiment design is shown in Figure 3. The user specifies the desired parameters that will control the execution of the application and their different values. The tool then builds an experimental plan that will analyze these parameters in an efficient way, relying on advanced statistical techniques as well as existing runs in the database to minimize the overall cost of the experiment (both in terms of time and resources). The Run Module then automatically carries out the experimental plan by executing the necessary program instantiations.

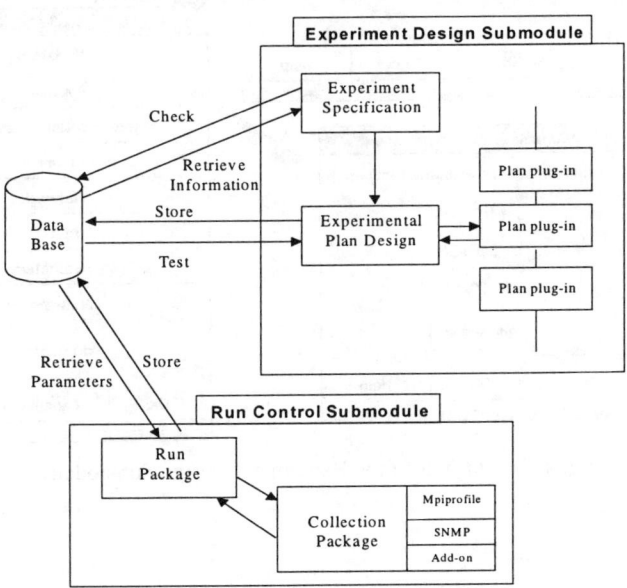

Fig. 3. Experiment Control and Design Module

3.1 Experiment Specification

We define an experiment by a set of variables (e.g. application, number of processors, etc.) and a set of values they can take. This mechanism is general enough to describe most any situation the experimenter may want to describe. Our tool provides a graphical interface that guides the experimenter through the experiment specification process. Figure 4 illustrates some of the steps followed to define the experiment. First the user selects the set of applications he wishes to study. For each application, he defines a set of variables that will control the execution of the experiment, such as the number of processors or the input file. For each variable, the user declares a set of values that these variables can take during the experiments, for example a range of 2 to 16 for a scalabilty test or different communication libraries. At each step the tool interacts with the database to propose choices from previous runs. The sample experiment shown below simultaneously tests the scalability of two applications as well as their performance using different communication libraries.

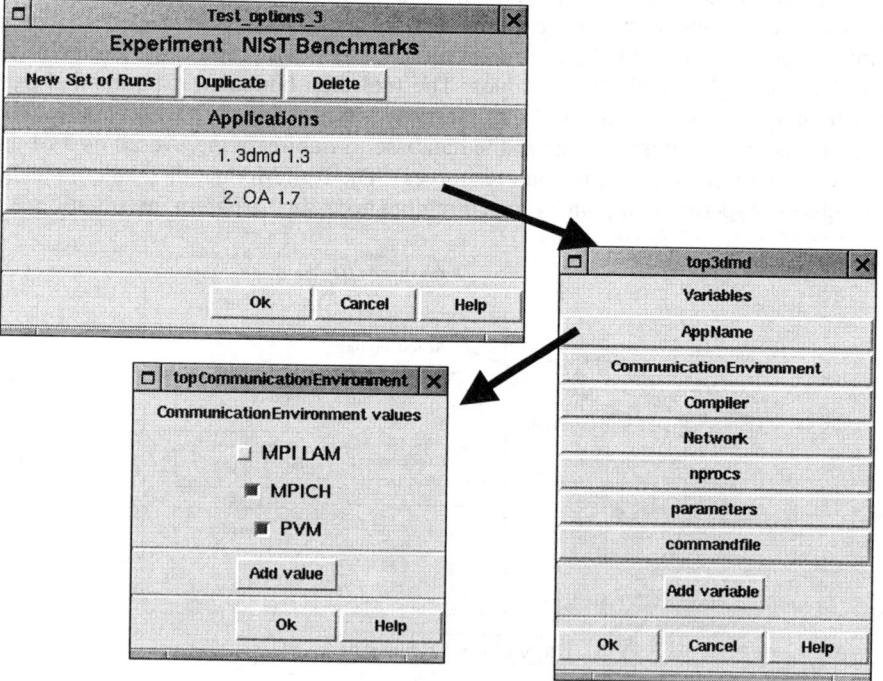

Fig. 4. Screenshots of the Experiment Design Submodule

3.2 Experimental Plan Design

An experimental plan describes the way an experiment should be carried out. The simplest case is a set of fully described runs, that is, an instance of each application run with each combination of the controlling variables' specified values declared in the experiment. Designing such an experimental plan follows a well-known process [BOX78], widely used in most scientific disciplines. Based on the requirements (e.g. standard error, interactions), such techniques define the optimal set of runs that match the requirements at the lowest cost (in terms of time and resources). A number of plan-design plug-ins can drive this package. Our initial prototype provides full-factorial plans, that select all the possible combinations of the variables ($O(e^n)$), but we expect future support of fractional-factorial plans (up to $O(n)$) with no interactions).

The next step consists of storing the fully described experiment in the database as an experiment structure pointing to a list of runs. If some of the runs needed for the experiment already exist in the database, they can be reused to save time if the experimenter desires; records for the missing instances are created in the database and flagged for execution by the Run Package. A graphical interface helps the user through the selection of existing runs, providing the choice to hand-pick them, to let the system chose one or to forcibly re-execute the application.

3.3 Run Control

The Run Package, as shown in Figure 3, establishes the link between the previously described collection tools and the database. The run package carries out the individual executions (runs) required in the experiment and manages the performance data collection process.

This component is fully automated. It relies on an existing queuing system for most of the job control. Our prototype uses DQS [DQS], a free, open-source job control system from the Florida State University. As with most queuing systems, a job is submitted in the form of a single shell script containing special control directives. For a given experiment, it accesses the database and retrieves the list of runs to be executed. One of the variables stored in the database has a special meaning: it points to a command file for the job control system, provided by the user as a template for the tool. For each run, the run package builds a new job by parsing the command file for variable substitution and instrumenting it with data collection commands. The created jobs are then submitted in a single batch to the job control system for execution. The instrumentation code within the job file is responsible for starting and stopping the collection tools as well as updating the records in the database once the job is completed.

4 Visualization and Analysis

The toolset provides an extensible visualization and analysis package. This package initially offers basic capabilities to visualize the performance data collected as well as to perform standard statistical analysis. We also devised several computed metrics adapted to get insight of performance issues specific to parallel computing architectures. As with the rest of the toolkit, it can be easily extended to support new analysis techniques, for example data mining.

4.1 Visualization Tools

The requirements for our visualization module are the capability to plot any pair of raw data items from the database (e.g. memory usage against time), the capability to display graphical tests and finally the capability to plot values from computed metrics.

The visualization tool is built on top of a third-party software package. We have chosen to use MatLab [MAT] or IDL [IDL] software, which are commercial software. They both offer good drawing capabilities in 2D and 3D. Moreover, they are able to conduct standard statistical analysis and to perform fast array/matrix computation. Both of the tools can access a database using the ODBC protocol. This feature simplifies the interaction between the visualization module and our database hosting the experiments' performance results. Finally, these two pieces of software are available for the Win95/98/NT platform, MacOS and many UNIX platforms. This allows our toolset's visualization package to run on many platforms and not only UNIX, our primary development system.

The use of free software such as Octave [OCT] rather than commercial software is also possible. It offers the same basic interface, computation and drawing capabilities as MatLab, but it has no database connectivity embedded and is only available for Win95/98/NT and UNIX.

4.2 Analysis

The analysis package helps to interpret the meaning of the performance data and get some insight into the application and/or parallel machine behavior. It uses an integrated visualization tool (see above) as a graphical output and computing platforms as needed. It will assist in performing essential statistical analysis such as standard deviation, standard error, etc. It will also provide a set of less common graphical tests to validate the results of the experiment: independence between runs, fixed location of measurements, etc.

We also developed a suite of new metrics focusing on different aspects specific to cluster computing, summarized in Table 1. These metrics rely on a simple "Communication v. Computation" model. Although this model has its limitations (it cannot describe particular cases, such as superlinear speed up), it is rich enough to help understand the internal behavior of parallel codes.

These metrics are all computed from high-level runtime computation and communication information. Each one focuses on a different aspect of the execution of a parallel job. We use the standard speedup (SU) and the efficiency (EF) as general indicators of the gain in speed. We define several overhead metrics: the global overhead (GOH) as the overhead introduced in the process of parallelizing the code compared to the sequential version and the computation overhead (POH) as the extra computation cost alone. We also define the average response time on the parallel program, which gives an indication of the load balancing, the granularity (computation-communication ratio) and the mean number of working processors. This set of metrics helps in determining what can be modified in the program or the architecture to improve performance. The analysis package will feature an automatic diagnostic of the application using such metrics to pinpoint potential improvements.

We present this set of metrics as a basis to demonstrate the functionality of the module. The user can extend the toolset with other analysis schemes and techniques to highlight particular points of parallel programs.

Definition	Description
$SU = \dfrac{R_s}{R_p}$	Speedup.
$EF = \dfrac{SU}{n}$	Efficiency.
$\overline{R_p} = \dfrac{\sum_{i=1}^{n} R_i}{n}$	Average response time of the parallel version.
$GOH = \dfrac{n\overline{R_p} - R_s}{R_s}$	Global Overhead: overhead of parallel section of the code compared to the sequential version.
$POH = \dfrac{\sum_{i=1}^{n} P_i - R_s}{R_s}$	Computation Overhead: part of the global overhead due only to computational operations.
$\eta = \dfrac{\sum_{i=1}^{n} X_i}{\sum_{i=1}^{n} P_i}$	Granularity: Communication vs. Computation ratio.
$\overline{n} = \dfrac{1}{R} \sum_{i=1}^{n} P_i$	Mean number of working processors
X_i: Communication time; P_i: Computation time;	$R_i = X_i + P_i$: response time of i-th process out of n R_s: response time of the sequential version

Table 1. Parallel Metrics

5 Summary and Conclusions

Our automated benchmarking toolset is intended to provide an extensible environment for conducting traditional computing platform evaluations as well as research-based computer/application performance analysis. The key mechanism is a central database that incorporates an abstract model of the computing platform. The database stores all available runtime performance and configuration information for each execution of each program and provides current and archival information for analysis and visualization. Various existing tools, covering such functions as data collection, analysis, visualization and runtime control, will be integrated with this database. Our goal is to integrate any performance, visualization or analysis tool into this toolset, treating these modules as interchangeable "plug-ins." To accomplish this we need a methodology to access these various possible disjoint tools. One possible approach would be to specify an access API to these tools that would activate commands or processes and exchange information.

Our initial prototype already incorporates an SQL-based database along with SNMP data collection from the various data agents on the executing platforms. We are using Tcl/Tk to handle most of the integration between the various tools and for the user interface. We are currently working on the Experiment Control Package, the access APIs and a front-end to bind the different components into a single tool.

Additional information, downloads and current status of this project is available on our website at http://www.cluster.nist.gov.

References

[AYD96] Ruth A. Aydt, "The Pablo Self-Defining Data Format",http://www.pablo.cs.uiuc.edu.
[BOX78] George E.P. Box, William G. Hunter, "Statistics for Experimenters, an introduction to design, data analysis, and modeling building", 1978.
[DQS] "DQS - Distributed Queueing System", http://www.scri.fsu.edu/~pasko/dqs.html
[IDL] "IDL - The Interactive Data Language", http://www.rsinc.com/idl/index.cfm
[IND99] "MPIProf", http://www.itl.nist.gov/div895/cmr/mpiprof
[MAT] "The MathWorks - MATLAB Introduction". http://www.mathworks.com
[MIL95] B.P. Mille et al., "The Paradyn Parallel Performance Measurement Tools", *IEEE Computer*, 28, 1995.
[MIL99] Barton P. Miller, Karen L. Karavanic. "Improving Online Performance Diagnosis by the Use of Historical Performance Data", SC'99, Portland, Oregon (USA) November 1999.
[MIN95] Alan Mink, "The Multikron Project", http://www.multikron.nist.gov
[OCT] "Octave Home Page", http://www.che.wisc.edu/octave
[POS] "PostgreSQL Home Page", http://www.pyrenet.fr/postgresql
[TNM] "Scotty - Tcl Extensions for Network Management", http://wwwhome.cs.utwente.nl/~schoenw/scotty
[UCD] University of California at Davis, "The UCD-SNMP project home page", http://www.ece.ucdavis.edu/ucd-snmp
[VAM99] Pallas Gmbh, "VAMPIR", http://www.pallas.de/pages/vampir.htm, 1999.
[YAN96] J. C. Yan and S. R. Sarukkai, "Analyzing Parallel Program Performance Using Normalized Performance Indices and Trace Transformation Techniques", *Parallel Computing* Vol. 22, No. 9, November 1996. pages 1215-1237, 1996.

Evaluation of an RCube-Based Switch Using a Real World Application

Ersin Cem Kaletas, A.W. van Halderen, Frank van der Linden,
Hamideh Afsarmanesh, and Louis O. Hertzberger

University of Amsterdam, Kruislaan 403, The Netherlands
http://www.wins.uva.nl/research/arch/

Abstract. In order to adress the problems in fulfilling the requirements of the high performance systems, the IEEE-1355 standard has been developed. A number of ESPRIT projects were initiated to support the technology within the scope of this standard. However, it is also important to promote the technology, i.e., to prove and demonstrate its applicability and efficiency. Targeting this, the ARCHES project has developed Gigabit Ethernet switch and demonstrator applications. This paper briefly addresses the efficiency issues related to the communication, and descibes the tests performed for evaluating the technology developed during the ARCHES project and the results obtained.

1 Introduction

In order to address the problem of reliable, efficient, low-latency communication within high-performance systems, the IEEE 1355 standard [3] has been developed. The scope of this standard is a range of interconnect technologies which were supported through a number of ESPRIT projects [2]. For a better deployment of the technologies developed in these projects, a better utilization through interfacing with communication standards such as PCI, SCI, ATM and Ethernet is needed.

In the ARCHES project [1], the ESPRIT developed interconnect technologies have been taken up and interfaced with SCI and Gigabit Ethernet (GE). Not only developing, optimizing, and promoting the heterogeneous interprocess communications (HIC) technology are important, but also proving the usefulness through the demonstration of applications on large scale test-beds is important to expand the deployment for the interconnect technology. The role of the University of Amsterdam (UvA) in the ARCHES project was the demonstration of efficient system software and application usage of GE technology, in order to validate the technology developed in the ARCHES project as an expert end-user of this technology.

Traditional network performance benchmarking methods are quite simple; testing only the parameters like latency, throughput, etc. without considering, for example, the effect of the applications using the network. Hence, these methods usually result in unrealistic, mostly optimistic results about the system under test. For more realistic network performance evaluation, demonstration applications should resemble the behaviour seen in the real world domain. The communication patterns of the application should be complex, irregular and should not be specifically adapted or optimized for the benchmarking. Also, standard software technologies must be used as much as possible while developing the applications. The demonstration application, at the same time, must be relevant to the interest of other potential customers to draw their attention.

The objectives of the application demonstration were to provide early user feedback, promote technology by showing actual applications using the platform, and demonstrate the benefit of the new communication technology. A demonstrator application in the field of web technology was established which uses the Gigabit Ethernet test-bed. This application considers the emerging trends and needs to distribute and enhance data management. Not only the proper validation of the new communication technology developed for the ARCHES project was considered while developing this demonstration application, but also the future of web server technology has been taken into account.

2 Application Specification

The demonstration application developed at the UvA combines web server technology with database server software in a parallel/distributed environment. As the Internet evolves and Internet applications become more diverse and complex, sophisticated data management as part of the web server system becomes more and more important. Additionally, web servers running popular web applications must deal with ever increasing numbers of client requests. This leads to an increasing demand for performance and requires a good scalability of the web server architecture, which translates to applying parallelism. The objective for deploying this server software is to demonstrate the benefit of the new communication technology as developed in the ARCHES project from an application point of view, as well as to investigate the problems and opportunities in this research area and build software based on this knowledge.

The database part of the application is based on the commercially available Matisse DBMS 4.0 for Solaris/x86. The web server part is based on the freely available Apache web server.

2.1 Overall View

The application is a distributed web server incorporating a parallel database system framework. The system runs on a cluster of high-performance homogeneous machines interconnected by a high-speed network, and can be divided into two subparts: the distributed database framework consisting of interconnected database nodes (back-end) and the distributed web server consisting of multiple HTTP daemon instances (front-end) (see Figure 1). Clients accessing the web server can connect to any of the HTTP daemons, making them able to access all the documents in the collective web server. Each HTTP daemon has the ability to connect to one or more database nodes.

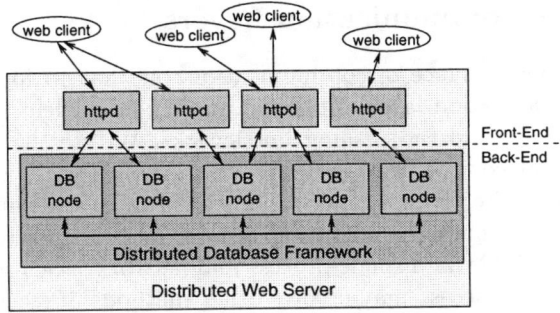

Fig. 1. Application overview.

2.2 Distributed Web Server

Several HTTP daemons run in parallel on different machines. The Apache HTTP daemon consists of several processes running in parallel. The incoming client HTTP requests are distributed over the available processes. The number of processes is dynamic, depending on the number of client requests. The dedicated plug-in, WEBDB module, connects every HTTP daemon process to a database node. The HTTP daemon process is configured in such a way that it redirects every request for static data to the WEBDB module which retrieves documents from the database replacing the normal filesystem storage.

2.3 Distributed Database Framework

Every node in the distributed database framework consists of a Matisse database server, a communication handler supporting efficient inter-node communication, and an application extension supporting access from the WEBDB module.

Each database server in the framework can perform the role of entry server for the HTTP daemons that are currently connected. The communication handler takes care of efficient communication between the nodes. It supports broadcasting requests and gathering replies over the high-speed network connecting the nodes. Therefore, the capacity of this "internal" network imposes a maximum throughput restriction. From a throughput perspective, the maximum throughput of the network to the outside world (to the HTTP daemons) and the maximum throughput of the disks in the system are important as well.

The application extension offers the functionality to let a node co-operate with other nodes in such a way that together they form a distributed database system and offers a uniform interface to the database clients (HTTP daemons), enabling HTTP daemons to connect to a single arbitrary database node to perform all its actions on the entire database.

3 Low-Level Communication Layers

To examine the performance of communication, one has to look at the structure of the communication layer implementation, the efficient mapping to the underlying hardware, and the ensuing bottlenecks. Within the field of parallel computing, different communication layers are in use. In some implementations application libraries directly access networking hardware, while in others network access is performed through the operating system. In direct hardware access, the network is only available for a single process and network security is lost. Dependent on details, direct access may involve other methods of receiving data than the interrupt-based approach, which has its impact on the possible application programming models. Using normal (TCP/IP) layers we have the traditional, full-blown network semantics which induce the necessary overhead.

Network communication therefore involves more than just the transport medium. Some of the issues that were found to play an important role in the performance and capabilities of a network communication layer will now be addressed.

Context Switching: Since the difference between special-purpose HPC systems and cluster of workstations is disappearing, programs are now running in a multi-process environment. This makes context switching overhead hard to avoid, especially since we like to overlap communication with computation.

System Call Overhead: It is normally the task of the kernel to access the actual networking hardware via a system call, shielding the hardware from the application for portability and security reasons, but also to be able to share the resource between different applications. By forfeiting the access to the network by multiple applications and allowing a user process to access low-level hardware significant speed up can be obtained. This is however not acceptable for a large number of application domains that use the network.

Interrupt/Signal Latency: Network interface cards usually signals the CPU through interrupts when it has finished sending data, or when new data was received. Handling interrupts causes overheads, which gets worse when the interrupt needs to be propagated to the user space.

Data Copying: Data to be sent out from user space is rarely stored in such a way that it can be readily handled by the networking hardware. It is almost inevitable that data needs to be copied, which has a big effect on performance.

Semantics: The semantics expected by programs or interface definitions often do not match the most optimal way of sending or receiving data. Convenience often does not match efficiency. While near optimal communication speed can be reached for simple operations, this will not be possible in practice for more sophisticated operations, like multicasting and non-blocking operations.

Reliability: Software communication layers must deal with packet loss on the hardware level. This can be implemented using sequence numbers, acknowledgments, etc., which leads to overhead.

4 Experimental Setup

The ARCHES test-bed at the UvA consists of 20 dual Pentium II-300 MHz machines running the Sun Solaris operating system each with a SCSI disk and 512 MBytes of memory. The nodes are connected to different type of networks to allow the comparison of the different technologies. Next to a common Fast Ethernet (FE) switch there is a Myrinet switched network both of which serve as a reference to the Gigabit Ethernet RCube-based [5] switch developed in ARCHES. Within the ARCHES project, only a 5 port switch was finally delivered, limiting the scalability evaluation.

The SPECweb96 benchmark [4] is used to benchmark the web-server application on the testbed system. SPECweb96 includes a client load generator, simulating the activity of a client, a data set generator to generate the data stored at the web server, and a result document generator producing output files describing the run and the results. The basic property the SPECweb96 benchmark measures is the response time of retrieving an arbitrary document from the server. Document sizes in the workload data set and their access frequencies are based on measurements on the Internet. The load generator classifies the documents according to size into groups of less than 1, 10, 100 and 1000 Kbyte. Documents in these groups are accessed with frequencies of 35%, 50%, 14% and 1%, resp.

The test environment, as depicted in Figure 2, indicates in which environment the application benchmark is executed. A configurable amount of front-end machines run the HTTP daemons, a configurable amount of back-end machines run the DB nodes.

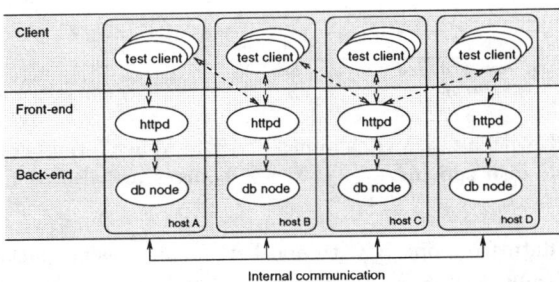

Fig. 2. Test environment

5 Experiments and Results

5.1 Low-Level Communication Layer Test Results

In order to better understand the performance of more elaborate applications we will study the results of a basic test program, a streaming point-to point application. Streaming provides a basic method for benchmarking the throughput of a system. Streaming entails the sending of data from one machine to another without further interaction between both machines on application level. The sender continuously sends packets of data of a certain size in a tight loop without waiting for acknowledgments or other events. The receiver accepts these packets of data also without any other processing. Streaming is a simple one-way communication benchmark to test the maximum throughput of a system related to the packet size used.

In our first experiment, shown in figure 3, this basic throughput is measured using a low-level driver, which is embedded into the benchmark application. The GE is directly accessed by the program without using the operating system. This is also true for the receiving application, where a polling operation checks and accepts received packets by the GE, which means that interrupts are not used. Furthermore, all packets that are being sent are already present on the network card and received packets are purged from the receive queue without transferring them to main memory.

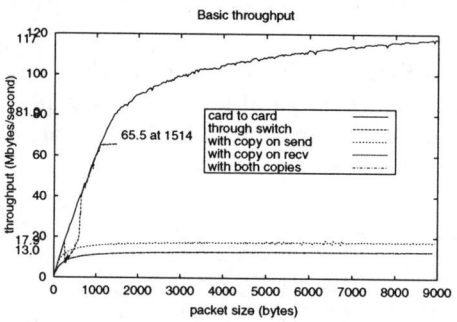
(a) With and without a switch and without switch with copying.

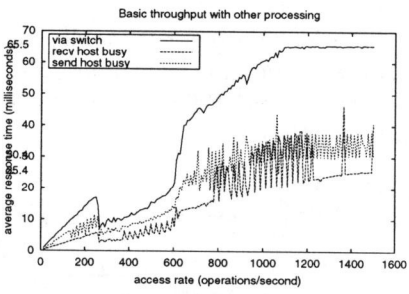
(b) When the computing resources are also used.

Fig. 3. Steaming data from one NIC to another, for increasing packet size. The total Ethernet packet frame size is set on the x-axis, and the y-axis shows the obtained throughput in Mbytes/second.

The experiment was performed with two machines connected directly to each other via a crosslink cable, and with both machines hooked up to the ARCHES RCube switch (figure 3(a)). The RCube switch does not implement extended frame sizes, which means that packets over 1514 bytes (the standard maximum Ethernet frame size) are not allowed by the switch. We see that at just above 1000 bytes per packet the throughput limit of the RCube switch has been reached at roughly 65.5 Mbytes/s. This is consistent with the maximum possible performance of the HS-Link technology, i.e. 65 Mbytes per second, upon which the switch has been built.

At the maximum standard Ethernet packet frame size, the difference between the optimum (direct coupling of two machines) amounts to 16 megabytes/s (20%). However, a near 100% performance of optimum performance is reached by some of the commercially available switches, which have been tested. The advantage of the RCube design should be its better ability to scale to a very large number of nodes.

The maximum throughput for two cards directly connected with each other climbs towards 117 Mbytes/s as the packet size increases. The current Ethernet standards define the frame size to be 1514 bytes at most. For FE this provides a proper balance between network availability, computer speed and packet overhead. On the other hand, by using GE this balance is disturbed: Only about 70% of the network performance can be achieved. The overhead for fragmenting data into packets has become too large in comparison with the time needed to send the packet. This is a known problem with the GE. Jumbo frames (Ethernet frames > 1514 bytes) were introduced, but are still not generally supported.

Figure 3(a) also shows the results when data is transferred by a memcpy operation from CPU to the network card's memory on the sender side (copy on send), with the transferral of data from the receiving NIC to the main memory of that machine (copy on recv) or with both copies. (In the figure the copy-on-recv and with-both-copies coincides). Clearly a naive copy operation will introduce a large bottleneck into the system, where the receiver side will be the most limiting factor in the performance.

Figure 3(b) shows the effect when other processes (in this case on average one) use the receiving host or the sending host to perform computation, compared with an otherwise idle system. From this we find that obtaining high throughput comes at a cost; also computing resources will be seriously affected and vice versa.

These experiments show that using a dedicated communication program will result in a speed-up close to the raw communication speed of the GE medium. However, this performance is quickly reduced when the communicating hosts try to incorporate the network into a normal computer organization setup.

There is more to communication in a computer network system than just networking hardware alone. In the final analysis, what matters is that the data must be transported inside the computer also. For example, data from the memory of one application must arrive at the memory of another application, or data from the main memory of the computer must be transferred to the network card. In our low-level benchmarks, the analysis of the data transfers inside the computer was not performed and therefore these tests say very little about the actual usage of the network.

We have seen in the measurements that as soon as this communication overhead is included, a severe drop in throughput is observed. A clear bottleneck in the internal bandwidth of the machine is present, even though this bottleneck cannot be explained from the PCI-bus, memory or CPU speed alone. The interaction between the components and the driver software needs to be seen as the cause for this bottleneck.

5.2 Application Results

The standard SpecWeb benchmark tests the performance of a web server. It does so by retrieving documents from the server and looking at the service or response time of the server in handling the request and sending all data in the document back. The clients which retrieve documents from the system coordinate their actions to put a certain load onto the system. This load is the number of documents that are being requested over a period of time. A test sample of the SpecWeb benchmark is a run over a fixed period of time. The clients divide the test period into slots according to the requested number of documents per second. In each time slot a document will be retrieved from the server, and when service time is less than the length of the time slot no other operation is started.

The system may not be able to handle the requested number of operations over the test period. The obtained access rate may therefore be less than the requested data access rate. The highest number of operations that the system can manage is the final rating according to the SpecWeb benchmark of the system. Normally, the obtained number of operations per second are used in the plottings of SpecWeb (as shown in 4(a)), but we find it more clear to study the performance using the requested number of operations in our figures, to prevent the cluttering of lines.

For our performance measurements, the response time of the system for each of the requested access rates may also be interesting. Figure 4(b) shows most results at a glance. Four different data set sizes were tested on a cluster of three nodes, interconnected through either Fast Ethernet, Myrinet or Gigabit Ethernet.

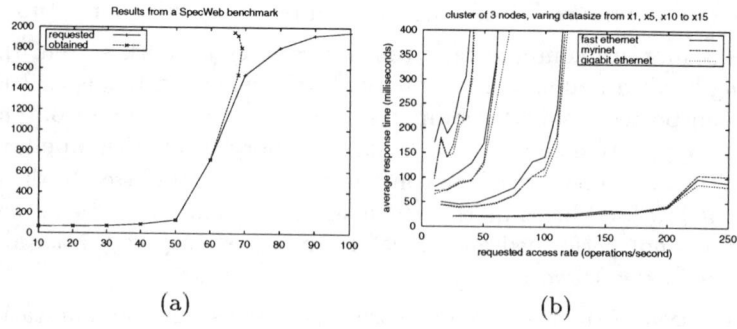

Fig. 4. Overview of application benchmark results

For the smallest data set size, where an average data object is 15 kilobytes, the system is mostly latency bound. We find that the higher latency of the Myrinet driver and the large set-up time for establishing the communication is more important than the actual speed of the network. For five times larger data objects, we do find that the performance of the system becomes dependent upon the speed of the network, which shows even more clearly in data object sizes of 10 and 15 times the default size. Here the Specweb benchmark was modified to use larger data sets. Furthermore the dataset was modified to fit always into main memory, making these results not conforming to the benchmarking rules.

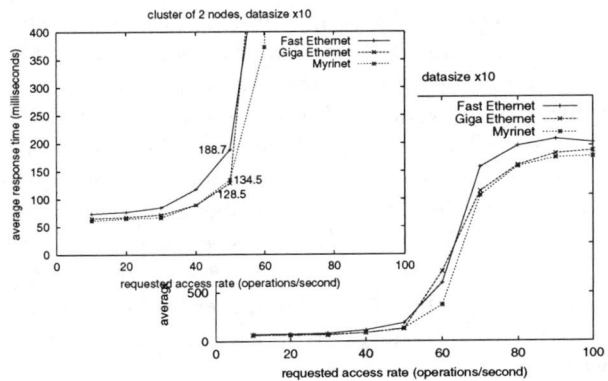

Fig. 5. Application benchmark results with varying network types as overview and enlargement of most important range as an inset.

Figure 5 shows the results of running the benchmark on all three networks using just two machines with a data size of 10 times the default. ¿From this we can see that Gigabit Ethernet is about 25% faster than Fast Ethernet. However, this does not show a significant increase in the maximum number of obtained documents per second, which is in all cases around 50.

The faster networks deliver the documents faster at this data-rate than Fast Ethernet (FE). This shows that the application software becomes the bottleneck when the system is overloaded. Even though the network is not fully exploited and the application performs no significant computation, the internal transport of data and accepting requests becomes the bottleneck.

6 Conclusions and Future Perspective

High Performance Computing (HPC) requires ever more computing power and faster networking technology to build more powerful applications. It is sometimes assumed that high-speed networks like GE can directly supply the HPC community with this increased network performance. The current networking intrastructure and communication software cannot deliver this low-latency, high bandwidth connection between individual programs on different machines in a network, as needed for parallel programs in HPC.

Next to the fast host-to-host communication, there is a growing need to provide better networking infrastructure on a secondary level. The technology by which multiple -local area- networks are connected with each other is often based on the same technology as the LANs themselves. One single or a few connections bind two networks together. This link from one network to another is shared between all the hosts in a cluster. When the interconnected networks are used as one big network, the single link connecting the networks potentially causes a bottleneck, which prevents an easy computing as within the local network.

Using a high-speed network like GE to interconnect the clusters of machines makes it possible for the combined set of machines in one cluster to exploit the high total bandwidth between the clusters. High-speed communication between individual computing nodes remains however important for HPC applications.

The network speed achieved with todays networking technology has become so high that the computing resources of a system can largely be occupied with the transfer of data. Considering the currently available networking technology and the networking software methods, we have reached the limit of how efficiently the computer as a whole can be used, because the overlapping activities (communication and computing) in the computer no longer have the desired effect. The administrative overhead of preparing communication is too large with these high network speeds.

Our results have shown that the application of GE technology is not only an issue of raw hardware performance. Instead, the net end gain for the real world usage needs to be reviewed. Fundamental better approaches in networking software need to be pursued. The performance is limited by issues that play a role on all levels of the networking infrastructure. This includes drivers, operating systems, communication libraries and programming paradigms (e.g. latency hiding techniques). All levels of communication need to be better adapted to each other to alleviate the bottlenecks that have arisen in the interaction between the layers. Such a solution should be structural and generic. It should not be targeted to suit the specific needs of a single institution or application and not just applicable to the specific network switch for which they are developed.

References

[1] Arches group. The arches project. http://www.1355.org/projects/arches.htm.
[2] IEEE. IEEE projects. http://www.1355.org/projects/projects.htm.
[3] IEEE. IEEE 1355: Standard for heterogeneous interconnect (HIC) (low-cost, low-latency scalable serial interconnect for parallel system construction). http://www.1355.org/projects/projects.htm, http://www.ieee.org/, 1995.
[4] Jason Levitt. Measuring web-server capacity. *Information week*, January 1997. http://open.spec.org/osg/web96/.
[5] Vincent REIBALDI. Rcube specification. Laboratoire MASI / Institut Blaise Pascal Universite Pierre et Marie Curie.Electronicly available at http://mpc.lip6.fr/hard.html.

MMSRS - Multimedia Storage and Retrieval System for a Distributed Medical Information System

Renata Słota[1], Harald Kosch[3], Darin Nikolow[1], Marek Pogoda[1,2], Klaus Breidler[3], and Stefan Podlipnig[3]

[1] Institute of Computer Science, AGH, Cracow, Poland {rena, darin,pogoda}@uci.agh.edu.pl
Phone: (+48 12) 6173 964, fax: (+48 12) 6338 054
[2] Academic Computer Centre CYFRONET AGH, Cracow, Poland
[3] University Klagenfurt, Institute of Information Technology
Klagenfurt, Austria
{harald,kbreidle,spodlipn}@itec.uni-klu.ac.at
Phone: (+43 463) 2700 513, fax: (+43 463) 2700 6216

Abstract. The Multimedia Storage and Retrieval System described in this paper is aimed at storing and retrieving medical multimedia data such as images and videos. There are three significant problems that automatically need to be addressed: providing enough capacity to fit Terabytes of MPEG files, efficient extraction of video fragments from MPEG files and quality of service issues. We are working on two different approaches to build MMSRS. The first one, described in this article, utilizes commercial HSM software for managing the tertiary storage hardware. The second approach that we have started to investigate is an attempt to build a specialised storage management system from scratch.

1 Introduction

The authors of this article take part in research on Distributed Medical Information System within the PARMED project [1]. We cooperate with a clinic which possesses large stuff of images and video clips (videos) illustrating typical and non-typical medical cases. In order to share this valuable material with other similar institutions for scientific and educational purposes, we provide a large virtual database [2,3] of medical video and image data, rather than video-conferencing or using medical video and image data within decision support for single patient's diagnostics [4,5]. This article presents the strategies and implementation details to store and access the medical video data.

Users of the medical multimedia database usually do not need to view the whole videos. They access more frequently fragments of them, perhaps composed in a presentation issuing from different videos. For example, a user wants to compare different methods of implantation of artificial hip joint depending on the type of the artificial limb employed. Users may need to view only the fragment showing the moment of implantation (which is only a small fragment of the whole video) from different medical operations (thus different videos).

Even if the user requires whole videos to be delivered, one can take advantage of accessing parts of it in order to make up the mind if it's worthwhile to fetch the complete one or not. This is typical for content-based retrieval applications [6] as described in this paper for the medical context. For example, a user may be interested to view a demonstration video of an "Artificial Hip Joint Replacement", then it is probable that for this very general query, a large number of videos matches the description. Transferring and viewing all of the videos is time consuming and sometimes even impossible. Therefore if the user could define a fragment of the operation essential for her/him, it would be able to transfer and view all of the fragments corresponding to this definition and then make the decision for the complete video to be fetched.

Processing requests like in the above examples is made possible by adding annotations to the videos and storing them in an object-relational meta-database [7]. The annotations are grouped in four main annotation classes: persons, events, objects and locations. Subclasses of them describe detailed annotation semantics, e.g. an operation has the different events of preparation, anaesthesia, etc. Users issue their request to the meta-database via a graphical interface which translates these requests to SQL-queries for the database (at this time we are using PostgreSQL). The database returns a set of relevant hits for which the user can either refine their demands or trigger the display of the videos.

In the current implementation [8], the user gets back links to the video fragments satisfying the query. Each link is an URL address of an MPEG file coupled with a frame (time) range.

One of the subsystems of the PARMED project is the Multimedia Storage and Retrieval System (MMSRS). The system is aimed at storing and retrieving medical multimedia data such as images and videos. There are three significant problems that automatically need to be addressed: providing enough capacity to fit Terabytes of MPEG files, efficient extraction of video fragments from MPEG files and quality of service issues.

The next section 2 presents the concept of MMSRS. Section 3 explains the design of the system. Section 4 introduces the strategies for remote access to MMSRS and handles quality of service issues. Finally, section 5 presents results, future plans and conclusions.

2 Concept of the MMSRS System

Many organisations like universities, hospitals and companies need to store large volumes of video material used for education, training, marketing or archiving (documentation) purposes. Tertiary storage [9, 10] is required because the volume of the video material cannot be stored economically on magnetic disks.

We briefly discuss some related work as follows. Brubeck et al. in [11] describe a distributed Video-On-Demand (VOD) system with videos stored permanently on tertiary storage. They concentrate on algorithms to manage the distributed cache in the video servers. Suzuki et al. in [12] study the delay issue of a hierarchical storage system in which some files are permanently stored on disk, while some others reside almost entirely in tertiary storage, with only the initial parts stored on disk. Chan et al. in [13] study the required secondary storage and bandwidth to meet a delay goal for VOD hierarchical storage system. In our research we focus on building a video storage system which will efficiently use the tertiary storage for video fragments retrieval.

There is a lot of software dealing with tertiary storage, like automated backup or archiving software and HSM (Hierarchical Storage Management) systems. The most appropriate for our purpose seems to be the HSM system [14, 15]. HSM systems utilize a combination of magnetic disk and automated media (tape or optical disk) library (AML). HSM systems move data between magnetic disk (disk cache) and AML devices depending on some migration scheme. Some HSM systems can provide access to a fragment of a file without copying the whole file into the disk cache. But when the file is an MPEG video, accessing a time or frame based fragment of the video is not possible without analysing the contents of the file.

We are working on two different approaches to build MMSRS. The first one, described in this article, utilizes commercial HSM software for managing the mass storage hardware. The second approach that we have started to investigate is an attempt to build a specialised AML management system from scratch and will be presented in the future.

When building MMSRS based on HSM, we need another piece of software, which will enhance the system functionality, so that it can store videos and access fragments of them. We call this enhancement MPEG Extension for HSM (MEH). The implementation of MEH depends on the underlying HSM functionality. In section 3 we present our implementation of MMSRS with the main focus on efficient extraction of video fragments from MPEG files.

3 MMSRS Implementation

In our implementation of MMSRS we employ an ATL 4/52 tape library managed by UniTree Central File Manager (UCFM) software. UCFM has the following features essential for the MEH implementation:

- UNIX-like filesystem (UTFS - UniTree filesystem) accessible via NFS or FTP,
- practically unlimited expandability of the filesystem's capacity,
- access to a file causes all its contents to be brought into the disk cache,
- flexible volume management mechanisms.

Files stored on UTFS generally reside on tapes. When a file is requested it is first copied to the Disk Cache (DC), where it can be read from. This operation is called *staging*. The files are removed from the DC using a LRU (Least Recently Used) algorithm. This operation is called *purging*.

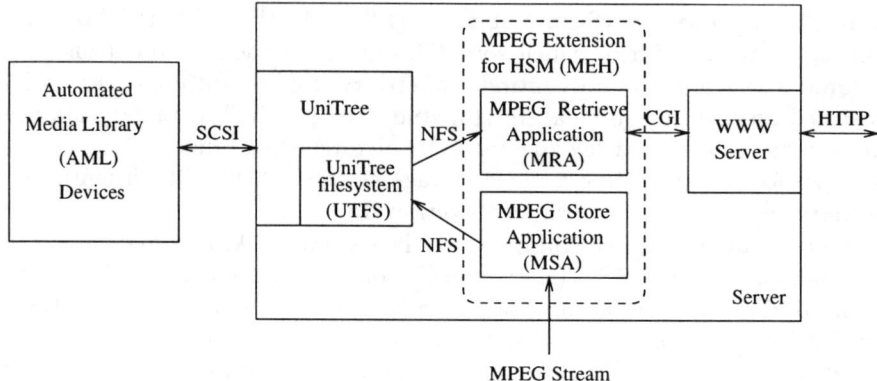

Fig. 1. Architecture of MMSRS - Multimedia Storage and Retrieval System.

Because of the features imposed by the UniTree software and our requirements it was necessary to carefully design the way in which videos are written to the UniTree filesystem. If videos were stored in whole as usual files, a user requesting a file that exists only on the tertiary storage would experience some start-up delay, because the whole file must first be staged. When the file is big the delay might be unacceptable.

The architecture of MMSRS based on HSM software is shown in Fig.1. The system consists of Automated Media Library (AML) and the following software components: AML managing system (UniTree HSM), MPEG Extension for HSM (MEH) and a public domain WWW server.

In order to workaround the difficulties mentioned above we decided to cut the videos into pieces of similar size and store them as different files. The cutting substantialy reduces the start-up delay. Each piece starts with an I-frame. The application for splitting and storing the videos into UTFS we called MPEG Store Application (MSA). Due to the chosen method of storing the videos another application for retrieving them is needed. We called it MPEG Retrieve Application (MRA). MRA is executed whenever there are client requests waiting to be served, while MSA is used by the administrator when a new video is to be stored.

MRA receives from the client the name of the video and the frame range (start frame and end frame). According to the range, it computes which video pieces (files) will be needed to produce the output MPEG stream. Then it takes the first piece, scans it to find the last I-frame whose number is less or equal to the start frame and starts transmitting the stream from there. When the piece is over the next one is taken and in this way it proceeds up to the last piece. The transmission ends when the last frame to be transmitted is out.

In order to keep the MPEG stream as smooth as possible (when the pieces reside on tapes), a simple prefetching technique is used. It is based on staging the next file to be transmitted while transferring the current one. We stage a file stored in UniTree by reading the first byte of the file.

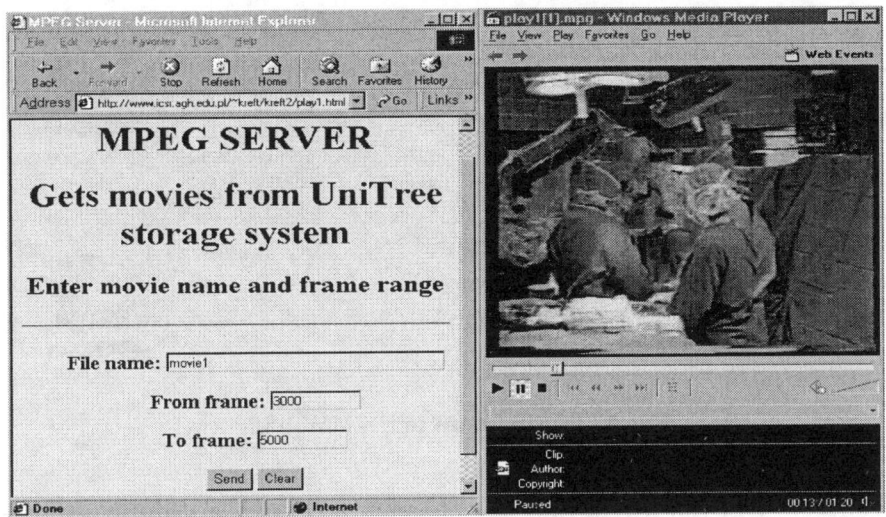

Fig. 2. WWW access to MMSRS.

MRA and MSA are written in C. MRA uses the Common Gateway Interface (CGI) to communicate with the WWW server. The video fragment is requested by the client using HTTP. In order to make the access to the MMSRS easy we provide a WWW page as shown in Fig. 2.

4 MMSRS and Quality of Service

The client of MMSRS can be a subsystem of the PARMED project or any standalone application (i.e. netscape). These direct clients do not receive Quality of Service (QoS). Anyway, depending on the current state of MMSRS (idle drives, system load, etc.) and the currently available network bandwidth they may receive contiguous video stream and view it simultaneously. Otherwise, the client should receive and save the video to the local disk and view it after that.

Storing multimedia data on tertiary storage meets some serious problems when quality of service is concerned. This is why some proxy cache guaranteeing QoS should be used.

The idea of proxy caches has been exploited for retrieving Web documents and can be successfully employed in our context for video retrieval, i.e. videos are accessed similar to text and images using the download-and-play mode. However, the requirements in space are different from that of the Web proxies, i.e. the most popular combination of size and bit-rate (320x280, 30fps, 1.6 Mbits/sec - VHS quality) takes about 12 MB/min.

We are planning to extend a previously developed traditional Video Server (VS) [16, 17], so that it behaves as a proxy cache to MMSRS. This modified VS has been called Proxy Cache VS (PCVS). Fig.3 shows typical access to the MMSRS.

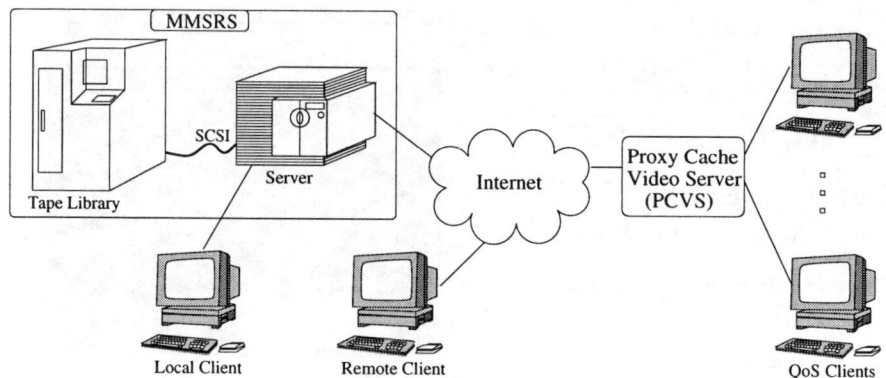

Fig. 3. Typical access to the MMSRS.

The essential point here is that we can assume that such an infrastructure is available for departments or the hospital already practising electronic exchange of multimedia data. Another advantage of using PCVS is reusability of the video for further processing or for other users who do not need to access the original server.

Let us consider an example of PCVS use in the case of a lecturer preparing video examples to support a computer based training. The lecturer first searches for the necessary video fragments (using the meta-database), then accesses them via PCVS, which causes that the fragments are cached into the PCVS, and locked until the end of the classes. Then, during the classes many students may watch different examples on their displays with corresponding QoS.

4.1 Architecture of the Proxy Cache VS

The general idea of a Proxy Cache VS (PCVS) is to support quality of service at, or near to the client. With this approach it is possible to overcome unreliable network bandwidth and give a support for interactive applications accessing repetitive times the same set of video fragments. Additionally, a proxy cache can be used to smooth the delivery of the videos.

In the sample application, we are currently using a simple video server, following some techniques employed in the Stony Brook VS [18], to store incoming video and to deliver them to the clients. We developed a mechanism to adapt the outgoing stream to the requirements of the clients, as shown in Fig.3, and to the average arrival video rate. In principle, it performs like this: after a specific amount of data is accumulated, the proxy begins to transmit the video data to the requesting client. Depending on the current network situation between the server of the corresponding video and the proxy, the proxy chooses the start-up latency for the transmission to the client. After transmission the video is cached for further requests. At the time being we are implementing this methodology into our server. The replacement policy for the cache mechanism will follow a simple LRU replacement for whole videos.

In future work, we want to combine the LRU removal policy with a strategy for quality adaptation, i.e. if a video is selected by LRU for replacement, we perform a quality adaptation of that video and not a total replacement like in typical replacement policies.

5 Conclusion and Future Works

The presented MMSRS allows to store a large volume of videos and to access them efficiently via the fragment-access feature, which decreases the latency. Additionally, the prefetching increases the quality of stream (making it smoother). The size of the pieces was adapted to achieve smooth output stream and low latency when the pieces are on tapes.

The following parameters influence the size:

- drive's transfer rate,
- head's positioning time,
- other parameters like system I/O throughput, software efficiency, disk speed etc.,
- bit-rate of the MPEG video.

To provide low latency, small size should be used, but for smoother stream the size should be bigger. On the other hand, retrieving small files on UTFS is not efficient.

We started to adopt VS techniques for accessing MMSRS with QoS. The VS acts as a proxy-cache near to the client and allows for QoS between proxy and client and the support of interactive applications as defined in the requirements of the PARMED project. The VS shall be extended with:

- client and MMSRS interfaces,
- efficient cache management (e.g. filtering instead of page replacement policies).

Some initial work has been started concerning the design of dedicated system for managing the tertiary storage. For the time being we have implemented a set of low level functions for manipulating the AML devices and the data being stored on them. When the dedicated system will be implemented it will be tested against the existing system based on UniTree.

Our goal is to design an efficient AML managing system and to explore and improve various techniques for efficient store and retrieval of multimedia data on tertiary storage.

Acknowledgements. We would like to thank L. Jarlaczyński and M. Kreft for their help with implementing the system, J.Otfinowski from Cracow Rehabilitation Centre for his help with the medical videos and J.Kitowski for valuable discussions. This research was partially supported by the KBN grant No. 8T11C00615 and by the Scientific and Technological Cooperation Joint Project between Poland and Austria: KBN-OEADD grant No. A:8N/1999.

References

1. Böszörmenyi, L., Kosch, H., Słota, R., "PARMED - Information System for Long-Distance Collaboration in Medicine", in: Conf. Proc. of Software for Communication Technologies'99 (Third International Austrian-Israeli Technison Symposium with Industrial Forum), Linz, April 26-27, 1999, Austria.
2. Zhang, A., Chang, W., Sheikholeslami, G., "NetView: Integrating Large-Scale Distributed Visual Databases", IEEE Multimedia, vol.5, No.3, pp.47-59, 1998.
3. Fernandez, I., Maojo, V., Alamo, S., Crespo, J., Sanandres, J., Martin, F., "ARMEDA: Accessing Remote Medical Databases over the World Wide Web".
4. "The Isat-Telematics project", University Lille2, France, http://www.univ-lille2.fr/isart/.
5. " The EUROMED project", Institute of Communications and Computers, National Technical University of Athens, Athens, Greece, http://euromed.iccs.ntua.gr.
6. Chang, S., Smith, J., Meng, H., Wang, H., Zhong, D., "Finding Images/Video in Large Archives", D-Lib Magazine, Feb., 1997.
7. Tusch, R., "Content-based Indexing, Exact Search and Smooth Presentation of Abstracted Video Streams", Master Thesis, University Klagenfurt, Institute of Information Technology, December 1999.
8. Kosch, H., Słota, R., Böszörményi, L., Kitowski, J., Otfinowski, J., Wójcik, P., "The PARMED Project - The first year's Experience", submitted to: The 8th International Conference on High. Performance Computing and Networking, Amsterdam, May 8 - 10, 2000, in this volume.
9. Chervenak, A.L., "Challenges for Tertiary Storage in Multimedia Servers", Parallel Computing Journal, vol.24, 1998, http://www.cc.getech.edu/fac/Ann.Chervenak/papers.html.
10. Hillyer, B.K., Silberschatz, A., "Storage Technology: Status, Issues, and Opportunities", AT&T Bell Laboratories, June 25, 1996, http://www.bell-labs.com/user/hillyer/papers/tserv.ps.
11. Brubeck, D.W., Rowe, L.A., "Hierarchical Storage Management in a Distributed Video-On-Demand System", IEEE Multimedia, Vol.3, No.3, pp.37-47, 1996.
12. Suzuki, H., Nishimura, K., "Performance analysis of a storage hierarchy for video servers", Systems and Computers in Japan, Vol.28, pp.300-308, March 1997
13. Gary Chan, S.-H., Tobagi, F., A., "Designing Hierarchical Storage Systems for Interactive On-Demand Video Services", Procedings of IEEE multimedia applications, services and technologies, Vancouver, Canada, June 1999
14. Nikolow, D., Pogoda, M., "Experience with UniTree Based Mass Storage Systems on HP Platforms in Poland", in: Bubak, M., Mościński, J., (Eds.), Proc. Int. Conf. High Performance Computing on Hewlett-Packard Systems, Cracow, Poland, November 5-8, 1997, CYFRONET KRAKOW, pp.189-19, 1997.
15. "ADIC Software", http://www.adic.com/US/English/Products/Software/.
16. Mostefaoui, A., "The Design of a High Performance Video Server for Managing TV Archives", Advances in DataBases and Information Systems (ADBIS'97), St. Petersburg, Russia, September, 1997. Springer Verlag Electronic Publications.
17. Moustefaoui A., Kosch H, Brunie L., "Semantic Based Prefetching in News-on-Demand Video Servers", accepted for publication in Multimedia Tools and Applications, Kluwer Journal.
18. Chiueh, T., Venkatramani, C., Vernick, M : "Design and Implementation of the Stony Brook Video Server", SP&E, vol. 27(2), pp.139-154 1997.

Track V

Posters

Dynamically Transcoding Data Quality for Faster Web Access

Chi-Hung Chi[1], Xiang Li, and Andrew Lim

School of Computing
National University of Singapore
Singapore 119260

Abstract. To survive in the "world-wide-wait" Internet environment, web surfers are often forced to tradeoff Web object quality for bandwidth by turning off the graphics or Java related objects through the browser configuration. Alternatively, the Web site owner might provide a lower bandwidth version of the Web pages statically. These solutions are not satisfactory because they can not make the best use of the available network bandwidth. To balance between the quality and the bandwidth, we propose a novel layered data model for Web multimedia data. Together with dynamic bandwidth negotiation and a best-fit mechanism, client can get the most appropriate version of a Web request. To make all these transparent to the clients, the layered data model and its associated best-fit mechanism will be implemented in the proxy server. This result is especially important to mobile devices for its limited bandwidth.

1. Introduction

The variance of the Internet client's bandwidth is tremendous. Different users have different network speeds to access Internet. For each user, he also faces the fluctuation of network bandwidth during different times of the day. Users' preference to different kinds of information also changes from time to time. There lies the challenge to provide appropriate quality according to network variance and different users' preference.

To balance between the quality and the bandwidth, we propose a novel layered model for Web multimedia data in this paper. The quality of a Web object presentation will be directly proportional to the number of layers it has. To make all these transparent to the clients, the layered data model and its associated best-fit mechanism will be implemented in the proxy server. One additional advantage of the proxy solution is its support of the simple, effective recovery mechanism for Web content. This result is especially important to mobile device for its limited bandwidth.

2. Related Work

There are research efforts in the area of the layered data model and the static solutions in proxy. Some researchers apply the concept of layered objects in the proxy to solve the problem. [1] suggests a framework for streaming media retransmission based on

---

[1] This work is partially supported by Advanced Research Fund RP960686 and RP974999.

layered media representations. [2] suggested the approach of SCUBA: scalable consensus-based bandwidth allocation. In this paper, the author proposes a protocol called "SCUBA" that enables media sources to intelligently account for the receiver's interest in its rate-adjustment algorithm. [3] [4] discuss how data should be delivered according to the user's requests. They propose the data distillation and refinement processes gifmunch and Transend in the proxy server [5] [6]. These works provide good foundation and valuable experience to the optimization of network bandwidth usage. However, there are still rooms for further improvement. For the layered data concept, most of the proposed models are designed for multimedia database. None of them is designed to handle Web objects in proxy. For the proxy solution, Berkeley's work can be considered as the representative one. However, it is a static solution instead of a dynamic one. Their solution always performs the transformation, no matter whether the instantaneous network bandwidth is enough. Although clients have the refined solution to get back the original quality, the solution is not transparent to the users.

3. Layered Data Model

The problem is to provide client with the most suitable presentation under different network condition and different user requirements. With the potentials of dynamic content transformation in proxy, it raises the need for a good, scalable data model for web information. We can find this trend to this kind of new data model already appeared in both protocol design and data format standard definition.

3.1. Basic Layered Data Graph

In this paper, we propose a kind of layered data model. With this model, we will map a URL into a graph. And with this graph, we can choose an appropriate presentation according to the bandwidth and users' preference. There are four types of vertices in the basic layered data graph. They are defined as follows:
- *Layer*: Under the transformation framework, the original file is coded into multiple layers. Among the layers, higher layers are the refinements to the lower ones. There will not be any transformation within one layer.
- *Presentation*: A presentation is made up of one or multiple layers.
- *Composite*: A composite is made up of one or more presentations. All the presentations under one composite are different ways of representing an object.
- *Group*: A group is made up of composites, links and even groups.

Each presentation has a **weight**. The weight is the time cost to fetch the presentation. This will be discussed in details later.

An edge in the layered data graph is a directed link that connects two vertices together. Although the starting and ending vertices can be any of the kind (at least in theory), only four types of edges are actually possible. They are *Group-Group, Group-Composite, Composite-Presentation, Presentation-Presentation.*

An edge in the layered data graph has exactly one of these three types of properties: inclusive, exclusive, and compulsory. At the presentation level, all edges have inclusive property. Each presentation is comprised of levels. The lowest level gives the basic information of the presentation; the upper levels are the refining ones. This inclusive property states that the presentation of the level *i* data of an object can only be possible if all its previous (more basic) layers are also presented to the client. A link from composite to presentation has exclusive property. When a composite has more than one child, only one will be chosen in the final request view. A link from group to group or from group to composite is compulsory. Compulsory means that edge must be included in the final request view. This certificates the integrity.

In the layered data model, we map a Web document structure of a user request to a graph, which consists of those vertices and edges mentioned in the last section. In this graph, the terminal vertices are the presentations and the non-terminal vertices are either composites or groups. Furthermore, all children of a composite are its possible presentations. For groups, their children are either composites or groups (i.e. defined recursively, as in frames). The edges in the graph describe the relationship in the final view of the Web request. With this layered data graph, information about the Web page objects and the network are stored as the attributes of the data structure for either the vertices or the edges.

3.2. Construction of the Graph

We define the layered data model in order to apply it into the practical Internet world. In other words, we need consider the HTTP protocol and find out a practical way for our model to live in such environment. Consider a client-server structure of Internet as below (in Figure 1):

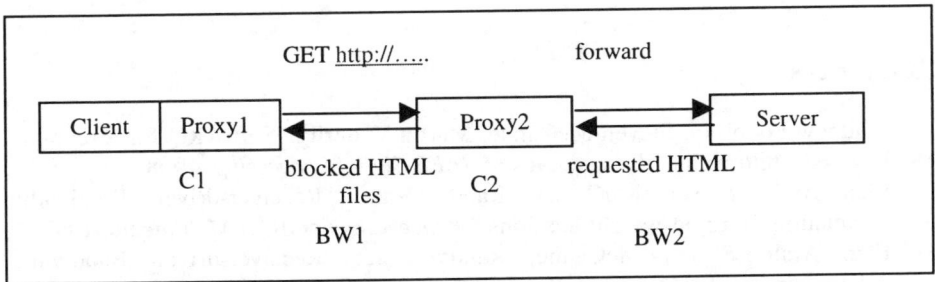

Figure 1: Typical Client-Server Structure of Internet

This table will be added into the HTML file by the server. After Proxy2 gets this table, it will mark the "Cache Where" area according to its local cache, C2. So when Proxy1 receives the first block of the HTML, it can easily construct the graph from this table. The weight area of the graph components can be calculated by using the following formulas:

- If the object is to get from server: $weight = size(\frac{1}{BW1} + \frac{1}{BW2})$
- If the object is cached in C2: $weight = \frac{size}{BW1}$
- If the object is cached in C1: $weight = 0$
- If the object is get from the server and need transformation in Proxy2:

$$weight = \frac{size0}{BW2} + \frac{size1}{BW1} + Transformation\_time$$

Where size0 is the original size and size1 is the size of the transformed objects. The Transformation_time is the transformation time for Proxy2 to do the needed transformation.
- If the object is in C2 and need transformation:

$$weight = \frac{size}{BW1} + Transformation\_time$$, where the Transformation_time is the transformation time for Proxy2 to do the needed transformations.

4. Conclusion

Bandwidth fluctuation is a major problem which most Internet users suffer from. Up to now, no good solution has been suggested. In this paper, we define a layered model for Web multimedia information presentation. The unique feature of this model is that it will deliver the best quality request view to a user, subject to the currently available network bandwidth. That is, the model is designed to work in the environment where fluctuation of network and a wide variety of user's requests are expected. Our experiment shows that the potential of this layered data model is very good and the overhead involved in the transformation and selection process is negligible.

References

1. Matthew Podolsky, Steven McCanne, Martin Vetterli , "Soft ARQ for Streaming Layered Multimedia," *Proceedings of IEEE Signal Processing*, 1998.
2. Elan Amir, Steven McCanne, Randy Katz, "Receiver-driven Bandwidth Adaptation for Light-weight Sessions," *Proceedings of ACM Multimedia*, 1997.
3. Elan Amir, Steven McCanne, Randy Katz, "Receiver-driven Bandwidth Adaptation for Light-weight Sessions," *Proceedings of the ACM Multimedia*, 1997.
4. Armando Fox, Steven D. Gribble, Yatin Chawathe, Eric A. Brewer, "Adapting to Network and Client Variation Using Infrastructural Proxies: Lessons and Perspectives," *Proceedings of ASPLOS-VII*, 1996.
5. Armando Fox, Eric A Brewer, "Reducing WWW Latency and Bandwidth Requirements by Real-Time Distillation", *Proceedings of Fifth International World Wide Web Conference*, 1996.
6. Elan Amir, Steven McCanne, Randy Katz, "An Active Service Framework and its Application to Real-time Multimedia Transcoding", *Proceedings of ACM SIGCOMM*, 1998.

Easy Teach & Learn(R): A Web-Based Adaptive Middleware for Creating Virtual Classrooms

Thomas Walter, Lukas Ruf, and Bernhard Plattner

Computer Engineering and Networks Laboratory (TIK)
Swiss Federal Institute of Technology Zürich
Gloriastrasse 35, 8092 Zürich, Switzerland
{walter,ruf,plattner}@tik.ee.ethz.ch
http://www.tik.ee.ethz.ch

Abstract. We present an architecture of a teleteaching system which supports direct interactions between all participants, i.e., lecturer and students and among students. To establish a kind of "virtual classroom" atmosphere, high-quality audio and video streams as well as teaching material are distributed between participants. For the implementation of high-quality audio and video channels, the system is based on emerging high-speed networks such as cable TV networks and any kind of digital subscriber line. A novel adaptive middleware takes care of system and network resources such that every participating user receives a teleteaching session in best quality possible. A first prototype is discussed.

1 Introduction

Asynchronous distance education refers to an environment where lecturers and students are separated in time and space. In synchronous distance education, lecturers and students are separated in space only; such an environment is also referred to as a teleteaching environment (see [1] for an overview on research in distance education).

The Easy Teach & Learn(R) project is aiming at the design and implementation of a communication and information middleware for creating virtual classrooms at the university level where students are remote from lecturers and remote from other students. The term virtual classroom emphasizes the aspect that lecturers and students are separated in space but should perceive a teleteaching session as being in a single physical classroom. In particular, everyone should see, hear and have the chance to interact with everyone else. In this project, the necessary communication technologies are developed and implemented for best possible and cost-effective transmission of multimedia streams.

2 Interactive and Synchronous Teleteaching Scenario

With Easy Teach & Learn(R), a user, either a lecturer or a student, has access to a high-performing communication infrastructures. Upon joining a virtual classroom session all applications required to receive audio and video, to view teaching aids and to communicate with the lecturer and other students are started automatically. If necessary, i.e., because a specific piece of software has not been installed on the user's PC, this software is downloaded and started without involvement of the user (see steps 1 to 5 in Figure 1). These functions are supported by a virtual classroom middleware. In addition, after joining a virtual classroom session, an adaptation process takes place that adapts network and system resources and performance so that best user perception of virtual classroom sessions is enabled. This holds true for communication between lecturer and student, student and lecturer and among students.

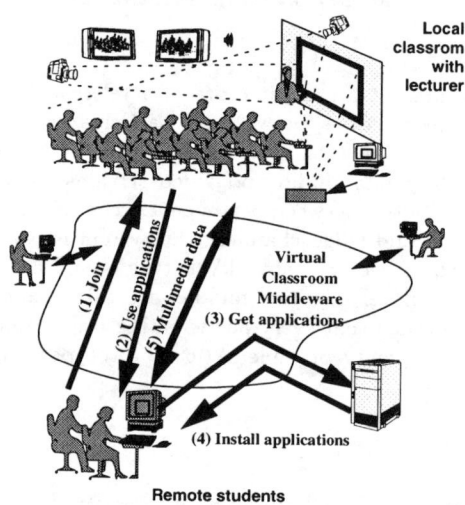

Fig. 1. Easy Teach & Learn Scenario

3   System Architecture

We present our generic virtual classroom system architecture which comprises three layers; a brief description follows. Different instances of the system architecture will be created for students, lecturers and administrators.

– Application and user interface: Within this layer the user accessible functions of the Easy Teach & Learn[(R)] system are implemented. Supported functions are: administration of different virtual classrooms, joining of announced and subscribed virtual classroom sessions via a standard Web-browser, and management of used applications for audio and video coding and decoding as well as management of teaching aids used within a virtual class. The control of interactions between lecturer and students and between students is implemented as a separate application which is supported by a specific protocol.

- Virtual Classroom Middleware: Processing video and audio streams is performed by standardized CODECs (coder/decoder devices). A system control component provides all functions to install, manage and run virtual classroom sessions. Functions to be defined and to be implemented are: access control to virtual classrooms, call control for starting and terminating virtual classroom sessions, capability exchange of systems joining a classroom, quality-of-service management, resource reservation, and interaction protocol. Packetization, framing, numbering and error detection and correction are low-level functions implemented. For some of the mentioned functions, protocols have already be defined and are in use [4], [5] and [6], which will be developed further.
- Communication: The specific functions within this layer are the adaptation of multimedia streams to the specific characteristic of the underlying communication network. Filtering of multimedia streams and adaptation to varying network properties is done transparently and cost-efficiently (with respect to resources available) within the communication layer at appropriate nodes [9] [10]. Multicast communication [3] support is an essential function of the communication system. The communication layer is part of end-systems as well as any intermediate system.

4 Prototype Implementation

In the current state of the project (Figure 2), teleteaching sessions between groups can be simulated, new system components can be tested and evaluated. We have set up a test bed with the following components: A number of PCs with MPEG [7] hardware and software CODECs. PCs are connected to different networks: Ethernet (10 and 100 MBit/s) and ATM (Asynchronous Transfer Mode) [2]. A PC equipped with a hardware MPEG CODEC is sending site. Audio and video data as well as shared teaching aids are communicated towards students sites. For the latter, the CoBrow (Collaborative Browsing) software [8] is used. A feedback channel between students and lecturer and among students is implemented by a chat.

This initial prototype will be further developed into a fully-functional Easy Teach & Learn[(R)] system with the following functional profile: joining of virtual classroom sessions, active download of software components, full interactive participation, adaptation of system resources and network capabilities, and multicast and QoS support in the underlying networks.

5 Conclusion

In this paper we have presented Easy Teach & Learn[(R)] which is designed as a system for high-quality and easy to use synchronous and interactive teleteaching. The system integrates advanced communication and information processing technologies with a novel adaptation process. The latter configures end-user systems and network for best user-perception of teleteaching sessions. The current implementation of the system has been discussed.

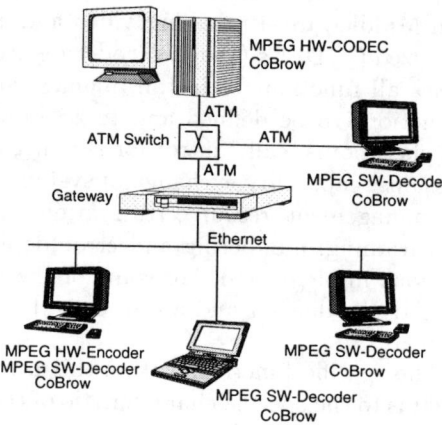

Fig. 2. Prototype Implementation

Acknowledgement. This project is funded by Stiftung Hasler-Werke and ETH Zürich.

References

1. Association for the Advancement of Computing in Education: ED-MEDIA World Conference on Educational Multimedia, Hypermedia & Telecommunications. http://www.aace.org/conf/edmedia/index.html (2000)
2. de Prycker, M.: Asynchronous Transfer Mode. 3rd Edition. Prentice Hall (1995)
3. Deering, S.: Host Extensions for IP Multicasting. Network Working Group, Request for Comments 1112 (1989)
4. Handley, M.: Session Announcement Protocol. Internet Engineering Task Force. Internet Draft draft-ietf-mmusic-sap-01.txt (1998)
5. Handley, M., Jacobson, V.: SDP: Session Description Protocol. Network Working Group, Request for Comments 2327 (1998)
6. Handley, M., Schulzrinne, H., Schooler, E., Rosenberg, J.: SIP: Session Initiation Protocol. Network Working Group, Request for Comments 2543 (1999)
7. ISO/IEC.: Information technology - Generic coding of moving pictures and associated audio. ISO/IEC Standard (1994)
8. Sidler, G., Scott, A., Wolf, H.: Collaborative Browsing in the World Wide Web. In: Proceedings of the 8th Joint European Networking Conference, Edinburgh, May 12.-15. (1997)
9. Yeadon, N., Garca, F., Hutchison, D., Shepherd, D.: Filters: QoS Support Mechanisms for Multipeer Communications. IEEE Journal on Selected Areas in Communications, Vol. 14, No. 7 (1996)
10. Zhang, L., Deering, S., Estrin, D., Shenker, S., Zappala, D.: RSVP: A new Resource ReSerVation Protocol. IEEE Network, Vol. 7, No. 5 (1993)

A Beowulf Cluster for Computational Chemistry

K.A. Hawick[1], D.A. Grove[1], P.D. Coddington[1],
H.A. James[1], and M.A. Buntine[2]

[1] Department of Computer Science, University of Adelaide, SA 5005, Australia
[2] Department of Chemistry, University of Adelaide, SA 5005, Australia

Abstract. We have constructed a Beowulf cluster of networked PCs that is dedicated to solving chemistry problems using standard software packages such as Gaussian and GAMESS. We describe the economic and performance trade-offs in the design of the cluster, and present some some selected benchmark results for a parallel version of GAMESS. We believe that the Beowulf we have constructed offers the best price/performance ratio for our chemistry applications, and that commodity clusters can now provide dedicated supercomputer performance within the budget of most university departments.

1 Introduction

In an effort to significantly increase the high-performance computing (HPC) resources available for computational chemistry at the University of Adelaide, we have designed and constructed a Beowulf cluster that is dedicated to chemistry applications [1]. We first put together a prototype system of 16 dual Pentium PCs (8 with 400 MHz Pentium II processors and 8 with 450 MHz Pentium IIIs), and we present some benchmark results obtained on this machine. We have now constructed the full Beowulf cluster, which consists of an additional 100 dual 500 MHz Pentium III PCs (i.e. a total of 232 Pentium processors), with the machines connected by a commodity Fast Ethernet network using Intel 510T switches. Our system has a peak performance of over 100 GFlops at a cost of approximately US$200,000, which is considerably less than a traditional HPC system of comparable performance from a commercial vendor. We believe our Beowulf cluster offers the best possible price/performance ratio for computational chemistry applications.

2 Computational Chemistry Requirements

Computer usage requirements will vary between different research groups and applications, however the requirements we have used to design our cluster are fairly typical for computational chemists. High-performance computers are primarily used by chemists at the University of Adelaide to determine molecular structure, using standard software packages such as Gaussian [2] and GAMESS [3,4]. At any given time, there are typically many users, each of whom may be running several different jobs. Each job typically takes several days or weeks on a single processor to produce a final result. In this situation it is more important to maximise *throughtput* [5] of the compute resource over all jobs, rather than maximising the performance, efficiency and scalability of a single job. Commodity (or Beowulf [6]) clusters provide the best price/performance for these kinds of applications requiring high throughput of many sequential jobs.

It is also useful to be able to utilise parallel processing across the nodes of the cluster. All of the major computational chemistry software packages have parallel implementations available. Parallel versions of GAMESS [3,4] are available using message passing software such as MPI. These give good speedups on a variety of distributed memory parallel computers, and should scale reasonably well on Beowulf clusters. The parallel version of Gaussian is written using a shared memory programming model. It will run in parallel on a dual processor PC, but can only be run across multiple nodes of a Beowulf cluster using Linda [7], which emulates shared memory on a distributed memory architecture. However this is only efficient on a relatively small number of processors.

3 Design Issues for the Beowulf Cluster

One of the most difficult tasks in designing and commissioning a Beowulf cluster is considering the price/performance trade-offs from the multitude of possible configuration options. There are four crucial hardware parameters to choose in the design of a Beowulf cluster: the type of processor to use in the nodes; the amount of memory installed in each node; the amount and type of disk installed in each node; and the network infrastructure that is used to connect the nodes. The best options will depend on the particular application.

Beowulf systems are typically built from commodity desktop computers, usually PCs with Intel Pentium processors or workstations with Compaq Alpha processors. Benchmarks on a variety of chemistry programs show that Alpha 21164 processors are around 1.5 times faster than Pentiums of the same clock speed, and the more recent Alpha 21264 processors are around twice as fast as the 21164 [8,9]. Dual processor PCs deliver almost twice the performance of a single processor, and currently cost less than single processor Alpha 21164 machines, while workstations based on the Alpha 21264 are currently at least twice the price of those with comparable Alpha 21164s. Hence dual processor PCs offer the best price/performance for our application requirements.

Memory is a non-trivial part of the overall node cost. We have chosen to provide 256 MBytes of memory for most of the nodes, which is adequate for virtually all of our simulations. However memory can be a limiting factor in some applications with very large molecules, so we have provided 1 GByte of memory in some of the nodes.

There are a number of technologies available for linking together a set of compute nodes in a Beowulf cluster. These include commodity networks such as conventional Ethernet running at 10 Mbit/s and Fast Ethernet running at 100 Mbit/s, as well as less common, non-commodity technologies such as Gigabit Ethernet, Myrinet, Scalable Coherent Interconnect (SCI), and implementations of the Virtual Interface Architecture (VIA) such as ServerNet and GigaNet cLAN. These non-commodity network technologies are around an order of magnitude faster (higher bandwidth and lower latency) than Fast Ethernet, but are currently around an order of magnitude more expensive, so that the cost of the network becomes comparable to the cost of the nodes.

Choosing the appropriate communications hardware for a Beowulf depends heavily on the types of problems that will be tackled. Guest [10] makes a good distinction between "grand challenge" problems and "throughput problems". While computational chemistry certainly does have grand challenge problems which require very large jobs that can only feasibly be run on large numbers of processors, there are many situations (such as ours) where job throughput is the more important criteria. In these situations it is clearly better to purchase more nodes and use cheaper commodity networking such as Fast Ethernet. However even in these cases, some support for parallel processing is desirable, although it is not necessary that the programs scale to large numbers of processors. Our benchmark results show that for many problems using parallel versions of standard chemistry packages such as GAMESS-US, reasonable scalability over multiple processors can be achieved with commodity Fast Ethernet networks.

Chemistry programs can make considerable performance improvements by storing intermediate results to disk in order to avoid re-computation. It is therefore advantageous to provide the compute nodes with a generous amount of disk space for temporary data. Cheaper EIDE hard disk drives proved adequate for the nodes, while a SCSI disk subsystem is used on the front-end machines which serve home directories and program binaries to the compute nodes.

4 Performance Benchmarks

In Figure 1 we present some speedup curves for several different calculations using GAMESS-US [3] running on our prototype Beowulf cluster. The results are for conventional Self-Consistent Field (SCF) computations for the benzene molecule using both the Restricted Hartree-Fock (RHF) and Second-Order Moller-Plesset (MP2) levels of theory. The RHF calculations do not scale particularly well, although they still give reasonable efficiencies up to around 10-15 processors. In contrast, the much more computationally intensive MP2 calculations give very good speedups. The Beowulf can therefore be efficiently used for parallel jobs on modest numbers of processors, as well as multiple sequential jobs.

Acknowledgments

This work was supported by The University of Adelaide and an Australian Research Council (ARC) Research Infrastructure Equipment and Facilities Program (RIEFP) Grant. We thank B. Hyde-Parker, J.D. Borkent, F.A. Vaughan and Andrew Naismith for technical assistance and F.A. Vaughan for discussions on system design.

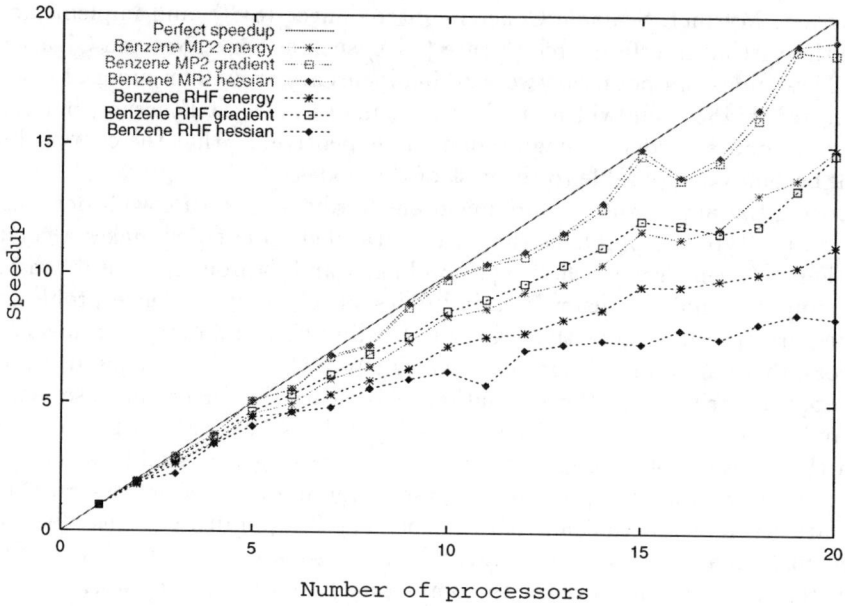

Fig. 1. Comparison of conventional speedups for benzene RHF and MP2 single-point energy, gradient and hessian calculations using GAMESS-US. The solid line represents the ideal speedup for a parallel computer.

References

1. Distributed and High-Performance Computing Group, University of Adelaide, Perseus: A Beowulf for computational chemistry,
   http://dhpc.adelaide.edu.au/projects/beowulf/perseus.html.
2. Gaussian, Inc., Gaussian, http://www.gaussian.com/.
3. Gordon Research Group, Iowa State University, GAMESS-US,
   http://www.msg.ameslab.gov/GAMESS/GAMESS.html.
4. Computing for Science Ltd, GAMESS-UK, http://www.dl.ac.uk/CFS/.
5. The Condor Project, University of Wisconsin, Condor: High Throughput Computing, http://www.cs.wisc.edu/condor/.
6. NASA CESDIS, Beowulf Project, http://www.beowulf.org/.
7. Scientific Computing Associates, Linda, http://www.sca.com/linda.html.
8. M.F. Guest, Performance of Various Computers in Computational Chemistry, Proc. of the Daresbury Machine Evaluation Workshop, Nov 1996. Updated version available at http://www.dl.ac.uk/CFS/benchmarks/compchem.html, June 1999.
9. M.C. Nicklaus *et al.*, Computational Chemistry on Commodity-Type Computers, Journal of Chemical and Information Computer Sciences, 1998, 38:5, 893-905.
10. M.F. Guest, P. Sherwood and J.A. Nichols, Massive Parallelism: The Hardware for Computational Chemistry?, http://www.dl.ac.uk/CFS/parallel/MPP/mpp.html.

The APEmille Project

Emanuele Panizzi and Giuseppe Sacco
(For the APE Collaboration)

INFN, Sezione di ROMA I - p.le A. Moro, 5 I-00185 Roma (Italy)
INFN, Sezione di PISA, v. Livornese, 1291, I-56010 S. Piero a Grado (Italy)
Università dell'Aquila, Dip. di Ing. Elettrica, Monteluco, I-67040 L'Aquila (Italy)
DESY, Platanenalle 6, D-15636 Zeuthen (Germany)
{Emanuele.Panizzi,Giuseppe.Sacco}@roma1.infn.it

Abstract. The APEmille high performance computer is a parallel SIMD machine, with local addressing, suited for massively parallel homogeneous problems, and with goal peak performance of 1 Teraflop. Several APEmille subsystems are currently running in Italy and Germany for a global amount of 240 Gflops.

1 Introduction

APEmille [1] is the third generation of a line of machines developed by INFN, dedicated to computational physics and optimized for Lattice Gauge Theories (LGT). This project, a collaboration of INFN Rome and Pisa, University of L'Aquila (Italy) and DESY Zeuthen (Germany), is meant to develop a teraflop range, parallel supercomputer oriented to floating point intensive computation and suitable for massively parallel homogeneous problems. The APEmille architecture grows from APE100 experience, and as such it has been strongly influenced by detailed features of Q.C.D. lattice problem, especially the project requirement that communication bandwidth between memory and processor should be well suited to this particular problem. APE is one of a small number of similar projects in Europe, Japan [2] and the US [3].

2 Architectural Overview

An *APEmille* machine can be viewed as a 3-D array of processing elements (PEs) with periodic boundary conditions. Each PE is directly connected to its owns 6 first neighbours through synchronous data communication channels. The grid of APEmille PEs is organized as a set of adjacent cubes of size 2 x 2 x 2 PEs.

 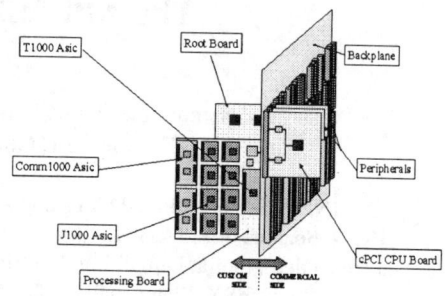

Fig. 1. The *APEmille tower* and the *crate* schematic diagram. The tower can host up to 256 PEs for 128Gflops of peak performance. The crate has 128 PEs for 64 Gflops of peak power.

In the actual implementation, the machine is based on a basic building block called Processing Board (PB). Each PB contains one control processor (called T1000), eight processing nodes (called J1000) and the support for both intra and inter board communication (a bit-sliced chip called Comm1000). In addition, each PB has its own PCI-based host-interface implemented on a FPGA. A larger APEmille machine is assembled by replicating these PBs, connected each other by communication lines and global control signals. The replication of the control processor and host interface on each board not only allows easy software partitioning of the machine into multiple SIMD partitions, but also provides a highly simplified and modular system-level hardware which improves reliability of the global system.

T1000, the control processor for the whole PB, manages the program flow (branches, loops, subroutines and function calls etc.) and broadcasts half of its very long instruction word (VLIW) stream to the J1000s. T1000 is provided with separate data (512 KB, SRAM) and program (2 Mword, SDRAM) memory.

J1000 is the arithmetic processor. It is pipelined, it has a register file of 512 word x 32 bit, and supports several data types: floating-point *normal* operations ($a \times b + c$) are implemented for 32 and 64 bit IEEE format, as well as for single precision complex and vector operands (pairs of 32 bit IEEE). Arithmetic and bit-wise operations are available for integer operands. J1000 manages its own local data memory, a SDRAM with 4M of 64 bit words (32 Mbytes).

Comm1000 handles all communication tasks needed by the machine. *Comm1000* basically interfaces with the eight nodes of the *cube*, and with 6 links connecting to nearby PB's in the *X*, *Y* and *Z* directions. The communication network automatically exploits multiple data paths for communications along more than one coordinate axes and it allows broadcast over the full machine, as well as a long lines and planes of nodes.

The performance of the different configuration of an *APEmille* system (numbers are in GigaFlops) are : Node **0.528**, Board (2*2*2) **4.2**, SubCrate (2*2*8) **16.8**, Crate (2*8*8, ~ 0.5 m^3) **66**, Tower (8*8*8, ~ 2 m^3) **264**, APEmille (32*8*8, ~ 8 m^3) **1056**. Comparing these performances, with an APE100 machine we find out that it's possible to compare an entire APE100 tower (25 GigaFlops) with an *APEmille* Sub-Crate.

A cluster of networked PCI-based PC's running Linux acts as the host for *APEmille*. Four PB's together with one host PC are plugged into the same compact-PCI bus of an APE backplane and form an *APEUnit* with 2 x 2 x 8 nodes.

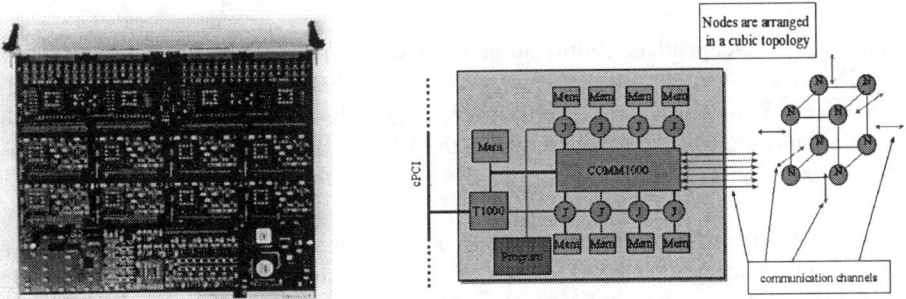

Fig. 2. The Processing Board printed circuit and its schematic diagram. In the printed circuit, from top, can be seen the two Comm1000 modules, the eight J1000 modules and the T1000 control processor module.

3 Software and Simulation Environment

The software developed for *APEmille* has three main parts: the operating system (*APEOS*), the compilation chain, and development and simulation tools.

The *APEmille* Operating System (*APEOS*) runs on the hosts. It
- provides the user interface
- downloads and uploads the APE executable programs
- drives the PB via the CPCI and accesses their memories and registers
- manages the interrupts coming from the PBs
- serves the I/O requests providing mass memory storage and file system
- allows to run programs on *APEmille* from remote network nodes.

The *APEOS* is a distributed Operating System which heavily uses Remote Object Technologies as the main networking paradigm. The user uses a front-end PC and is shielded by the underlying complexity. To allow compatibility and portability of source TAO programs the *APEOS* system services will comprise all the Ape100 system services (but one, namely the broadcast one, which is implemented in hardware). New system services are provided to manage new hardware facilities (for example high-speed networking or storage systems connected to the *APEUnit* PCI bus).

The *APEmille* processor is provided with two compilers. The first is the compiler for the *TAOmille* language, a language fully compatible with the Ape100 TAO language but with many extensions not only related to the new *APEmille* architecture. The *TAOmille* compiler is based on the *Zz* dynamic parser and has a hardware independent optimizer. The second one is a C++ compiler. The C++ language is being properly extended to support the parallel features of *APEmille*.

The hardware-dependent optimization and the micro-code scheduling is performed by FLINT, a low-level optimizer which reads assembly code and produces executable VLIW (Very Long Instruction Word) micro-code. The most important phase of this optimization is the *code shaking*, by which the VLIW patterns get re-arranged in order to fill *J1000*'s and *T1000*'s pipelines.

Two *APEmille* simulators are provided. A functional simulator, based on an "instruction set" concept, developed in C++. A gate level simulator, developed in VHDL, whose major advantage is that electrical loads, some parasitic capacitance, delays, race conditions and other hardware hazards are properly simulated. The possibility to simulate the architecture at two different level is important to produce *test vectors* and allowed a very deep test of the whole hardware and software project.

4 Status Report

As of January 30 the status of the project is summarized as follows. Several APEmille subsystems are running: 128 Gflops (tower) in APE/Rome, 64 Gflops (crate) in APE/Pisa, 32 Gflops (SubCrates) in APE/DESY, 16 Gflops (SubCrate) in Bielefeld (Germany) and the INFN is planning to build other 860 Gflops.

References

1. Tripiccione, R.: APEmille. Parallel Computing (1999) 1297-1309
2. Ukawa, A.: The CP-PACS parallel computer. Computer Physics Communications 110 (1998) 220-224
3. Mawhinney, R.D.: The 1 Teraflops QCDSP computer. Parallel Computing 25 (1999) 1281 – 1296
4. Aglietti, F. et al.: The Teraflop Supercomputer APEmille: architecture, software and project status report. Computer Physics Communications 110 (1998) 216-219

A Distributed Medical Information System for Multimedia Data – The First Year's Experience of the PARMED Project

Harald Kosch[1], Renata Słota[2], László Böszörményi[1], Jacek Kitowski[2,3], Janusz Otfinowski[4,5] and Przemysław Wójcik[2]

[1] Institute of Information Technology, University Klagenfurt, Austria
[2] Institute of Computer Science, AGH, Cracow, Poland
[3] ACC CYFRONET, Cracow, Poland
[4] Collegium Medicum Jagellonian University, Cracow, Poland
[5] Cracow Rehabilitation Centre, Thraumatology Dep., ul., Cracow, Poland

Abstract. The aim of the presented PARMED Project is to provide medical stuff with a large database of image and video data, distributed over several medical centers in Europe. This database is queried with user-friendly and intelligent interfaces. They are also responsible for processing the incoming video streams for the later use, i.e. for diagnostics or teleeducation. Other aspects of the project is to develop services for multimedia data management and archiving with quality of service requirements and the modeling of the video and image data.

1 Introduction

The principal goal of the PARMED Project [1] is to take advantage of multimedia telematics services, especially delivery of medical images and video sequences, to support long-distance collaboration and cooperation of several medical centers and medical teleeducation. The delivery of the video and image data is enabled by content-based retrieval which relies on descriptive information (also called indexes). The indexes are stored in a meta-database which is intended to be implemented in a high-performance Database Management System (DBMS). This database acts as a server for incoming video and image queries.

The fields of applications of the PARMED system are usage and exchange of scientific medical information, as well as, teleeducation rather than standard use of databases for storing and retrieving patients' information. The usefulness of exchanging scientific video and audio data on demand is not limited to the medical field. Concepts and techniques of the PARMED Project can be easily adopted in many other application fields, where collaboration is required.

This article presents the advances and the first year's experience of the PARMED project.

2 Video and Storage Server Utilities

There are high storage requirements concerning the medical data. Due to the huge amount of data it is necessary to make use of tertiary storage, like an automated tape library with suitable software management. In our implementation we employ an ATL 4/52 tape library and the UniTree Central File Manager software. The library itself is equipped with a disk cache in order to assist data streams. No QoS guarantee could be maintained with this Storage Server (SS), therefore a strategy of a Video Server (VS) and the SS cooperation is being developed [2].

The VS behaves as a proxy cache between the SS and the clients in order to provide easonable start-up times and smooth streaming of video data. Such a caching and buffering proxy is able to compensate fluctuating network bandwidth and to take advantage of interactive applications accessing repetitively the same set of video fragments.

In the sample application, we are currently using a simple video server, following the techniques employed in the Stony Brook VS [3], to store incoming video and to deliver them to the clients. We developed a mechanism to adapt the outgoing stream to the requirements of the clients and the average arrival video rate. In principle, it performs like this: after a specific amount of data is accumulated, the proxy begins to transmit the video data to the requesting client. Depending on the current network situation between the server of the corresponding video and the proxy, the proxy chooses the start-up latency for the transmission to the client. After transmission the video is cached for further requests.

3 Database Management

At present, two kinds of data are of interest – digitized Roentgen pictures and MPEG files. The content of the digital images and of the videos is modeled by an object-oriented schema. This schema contains semantically rich information about objects, events, persons, and locations, about their appearance in structural components of the video (or regions of the images), about their temporal dependencies and about their relationship to other semantic objects.

Consider, for example, the event of the Anesthesia in an operation video. This event occurs in some structural component of the video which can be either a shot or a scene: it involves surely the persons anesthetist and the patient and occurs before the Incision event.

The initial implementation of the database schema was carried out on the PostgreSQL system [4]. Due to some severe performance problems of PostgreSQL, mainly in the optimization module, we have moved to Oracle 8i. Although this step causes some additional difficulty, Oracle 8i offers higher flexibility and has the advantage of being a commonly used platform.

The architecture of the meta-database is based on three medical operations: artificial hip joint replacement, lumbar interbody fusion, screwing in of the titanium block, for which the annotation has been developed in collaboration with orthopedic surgeons.

As mentioned beforehand, four types can be indexed in the database model [4]: objects, events, persons and locations. The **object** is an *operation* or a *patient*. The patient is a passive subject of the operation. For the *patient* we need his name tag and his case history. Additional information like sex and age are also useful. The *operation* attributes are more informative: a patient name tag, the date and the method of operation, the part of the body that is operated, and some information about complications arising during and after operation. Name tags of medical staff participating in the operation are also included. The central content annotation represents the **event**. The **event** is an action which is undertaken during the operation stages (an example of operation stages concerning the artificial hip joint replacement is shown in Table 3).

Stage	Description	Stage	Description
1	Anesthesia	8.2	The femoral shaft
2	Placement of the patient	8.2.1	Preparing the femoral canal
3	Ischemia	9	Implantation
4	Preparation of operating field	9.1	Inserting the acetabular component
5	Covering of the field	9.2	Inserting the femoral stem
6	Incison	10	Attaching the femoral head
7	Joint dislocation	11	Implant repositioning
8	Preparation of bony placenta	12	Insertion the drain
8.1	The hip joint acetabulum	13	Suturing
8.1.1	Removing the femoral head		
8.1.2	Reaming the acetabulum		

For instance, ncision, is a part of each operation. In order to describe this event, the following information is required: the name tag of the physician who made the cut, the type of approach, the course of incision, the incision area, the layer, the names of the tools that have been used, and (optional) the complications. Another example can be the preparation of the bony placenta, for which similar information is required. At present, about thirty types of events are implemented. The current database model will be enhanced with new events for other types of operations.

Since the full scope of the information included into the database can be available for users with the highest privileges only, we work on methods of access privileges for different kinds of users, e.g. only restricted access to the name tags of patients.

4 Teleeducation

Although many standards concerning teleeducation and the virtual university have been developed [5] we are adopting a rather simple approach. There have been defined two functions of the teleeducation subsystem : semi-automatic preparation of the lectures and courses to support the lecturer and individual and group learning with verification of knowledge acquisition.

We have defined a lecture or a course as a set of presentation material consisting of multimedia information, logically organized by the lecturer, and presented electronically to the participants. The presentation is assembled with data already introduced into the system (and typically existing on the storage server). Besides the physical data from medical cases, some written material like slides with text and other graphical objects are also considered.

We organized the teaching material as graph in which nodes define lecture elements (atomic data) and edges – relations between them. Two versions are being developed: portable, which can be taken away to be installed on a local computer elsewhere and a stationary version making use of the VS. The graph representation is also implemented for computer based learning, which is being adopted in PARMED based on [6]. The subsystem incorporates some learning methods like preliminary learning recognition or self pace [7]. In the graph of knowledge representation the nodes determine atomic problems (knowledge elements), while edges – problem relations. The relations specify sequence of knowledge learning.

5 Future Developments

The following activites are concerned: refinement of the implementations of the meta-database and the storage server as well as development of further client's applications and conception of client's application framework.

Acknowledgments. We gratefully acknowledge the contribution of Klaus Breidler, Marian Bubak, Roman Krasowski, Jacek Mościński, Darin Nikolow, Stefan Podlipnig, Marek Pogoda and Roland Tusch. Without them, the initiation and then the advances in the PARMED project would not have been possible.

The work has been supported by Scientific and Technological Cooperation Joint Project between Poland and Austria: KBN-OEADD grant No. A:8N/1999.

References

1. Böszörmenyi, L., Kosch, H., Słota, R., "PARMED - Information System for Long-Distance Collaboration in Medicine", in: Conf. Proc. of Software for Communication Technologies'99, Hagenberg-Linz, Austria, April 26-27, 1999, Austrian Technion Society, pp. 157–164.
2. Słota, R., Kosch, H., Nikolow, D., Pogoda, M., Breidler, K., "MMSRS - Multimedia Storage and Retrieval System for a Distributed Medical Information System", appears in HPCN 2000, Amsterdam, May 8 - 10, 2000.
3. Chiueh, T., Venkatramani, C., Vernick, M : Design and Implementation of the Stony Brook Video Server. SP&E 27(2), pp. 139-154, 1997.
4. Tusch, R.: Content-based Indexing, Exact Search and Smooth Presentation of Abstracted Video Streams. Master Thesis, University Klagenfurt, Institute of Information Technology, December 1999.
5. Farance, F., Tonkel, J., "Learning Technology Systems Architecture (LTSA) Specification", http://www.edutool.com/ltsa/ltsa-400.html
6. Cetnarowicz, K., Marcjan, R., Nawarecki, E., Zygmunt, M., "Inteligent Tutorial and Diagnostic System.", Expersys'92 - Proceedings, pp. 103–108, Huston - Paris,October 1992.
7. Goldsmith, D.M., "The Impact of Computer Based Training on Technical Training in Industry", Australian Journal of Educational Technology, 1988, 4(2), pp. 103-110.

Airport Management Database in a Simulation Environment

Antonello Pasquarelli[1] and Tomáš Hrúz[2]

[1] ALENIA MARCONI Systems, via Tiburtina km 12.400, 00131 Rome – Italy
apasquarelli@lti.alenia.it
[2] Slovak Academy of Science, Mathematical Institute, Dúbravská 9, P.O. Box 56, 84000 Bratislava – Slovakia
tomas@ifi.savba.sk

Abstract. This paper presents the SEEDS simulation environment for the evaluation of distributed traffic control systems. The description starts with a general overview of the simulator, completely developed in C++ according to the CORBA standard, and then focuses on the Airport Management Database (AMDB) implemented in the prototype as an add-on module. The emphasis of the AMDB is on the wide area network operation, and this leaded to the architecture centered on the Java system. The paper shows the AMDB architecture, how it has been integrated in the core simulator and the interaction with it. The evaluation shows the ability of co-operation between Java and CORBA based C++ subsystems and the scalability of the whole system.

1 SEEDS Overview

SEEDS[1] (Simulation Environment for the Evaluation of Distributed traffic control Systems) is a distributed HPCN simulation environment composed of powerful workstations connected in a local network and it is targeted to the evaluation of Advanced Surface Movement Guidance and Control Systems (A-SMGCS). The simulation environment allows the definition and evaluation of technologies and performances needed to implement new functions and procedures of A-SMGCS, to mould new roles in the airport, to introduce new automatic tools and interfaces, to support A-SMGCS operator decisions.

The SEEDS consortium is composed of Alenia Marconi Systems (I), as coordinator, Sogitec (F), Artec (B), as industrial partners, University of Siena (I), LRR-TUM at Technische Universität München (D), Slovak Academy of Science (SK), as associated partners, and Sicta (I) as final user-partner. An European User Group, composed by the flight assistance administrations Sogel (L) and SEA (I), participated to all the phases of the project.

The software architecture of the simulator, defined using the Unified Modeling Language (UML) notation, is based on CORBA (Common Object Request Broker Architecture) as communication middleware; the DIS (Distributed Interactive Simulation) protocol has been used for the image generation and distribution. This

---

[1] SEEDS (European Project Number 22691) has been partially funded by EC DGIII in the area of HPCN. SEEDS started the first January 1997 and finished the 30th September 1999.

choice assures the scalability of the system, allowing the mapping of the different objects on heterogeneous workstations or PCs. Techniques for load balancing are used to reduce computing power, data rate and latency on the network. A distributed architecture for traffic and image generation was adopted for performance reasons. Subscription and notification mechanisms are used to synchronize processes in the distributed environment.

The simulation environment is composed of commercial off-the-shelf components and of some proprietary software modules, and it is open to be connected to other ATM (Air Traffic Management) simulators. The main software modules of the SEEDS architecture, completely developed in C++ and reported in Fig. 1, are: Scenario Generation and 2D-3D Visualization; Sensor Models; Airport Database; Surveillance, Control, Guidance and Planning Modules, Controller, Pilot and Driver models, Administration Station modules.

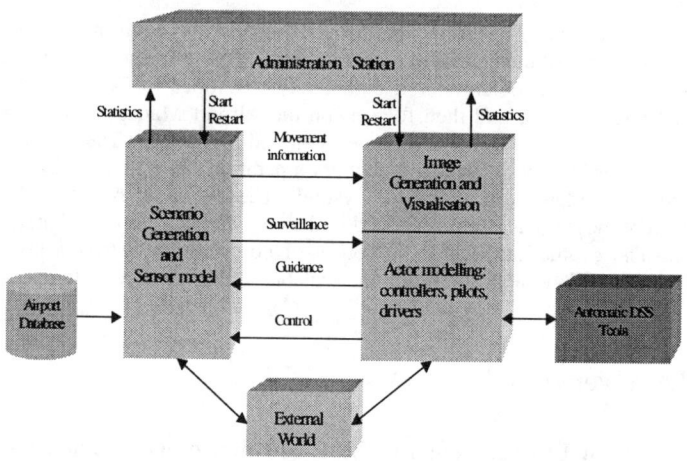

Fig. 1. SEEDS Simulator Architecture

The scenario generator is in charge to animate the scene according to the stimuli coming from the actors of the simulation and from the External World. The actors can be simulated (software processes) or real (human beings). The 3D visualization reproduces the scene as seen by the actors' eye (eye model); the 2D visualization reproduces the scene as seen by the sensors present in the airport (ASDE radar, GPS, DGPS, Magnetic/Dynamic sensor error models). The controllers have a set of decision support tools which help them to plan the aircraft surface movements. The other functions of A-SMGCS (Surveillance, Control and Guidance) are also implemented. An Administration Station is responsible to configure, start-up, stop, restart the simulation, and it collects application level and system level statistics.

2 Airport Management Database Architecture

The purpose of the Airport Management Database (AMDB) module is to add a database system to the core simulator, which implements the main aspects of a real airport database model, such as: Meteorological situation, flight data list, initial climb procedures, instrument approach charts, standard instrumental departure procedures, standard approach routes. For each procedure, the database contains textual description and visual information.

In this manner, all information centralized in the AMDB can be retrieved and updated by the airport agencies (Meteo Services, Central Flight Management Unit, etc.) and, at the same time, they are available to the SEEDS core simulator.

The main aspect of the AMDB module design is a wide area network (WAN) operation emphasis, which leaded to the architecture centered on the Java system. This solution provides the possibility of WAN access to the external aspects of the airport simulations.

In order to achieve an high level of flexibility, the AMDB architecture is designed as a three-tier, or three level, system whose main modules are the SQL database server, the application server and several clients (See Fig. 2). Considering that the AMDB is a complex system containing various hardware and software platforms, open software tools where adopted for the development of this module.

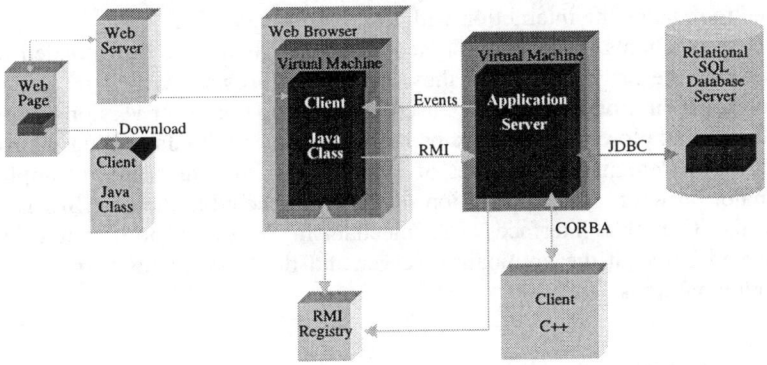

Fig. 2. Airport Management Database Architecture

The data repository has been modeled and implemented using a relational database subsystem able to handle large data sets, which occur if the airport simulator developed during the SEEDS project were used for simulation of traffic comparable to real airports. In our case the database server is the PostgreSQL system.

The AMDB system has two main kind of clients: the Java clients of the AMDB module and the C++ clients of the core simulator. The Java clients can handle the meteorological information, changing the meteo situation on the airport and storing all changes and operations to relational database for further on-line or off-line processing; they also offer support for chart maps processing and administration. The Java clients handle also flight data list for editing, processing and storing the flight information into the database. AMDB system contains also the database tables to describe the airport structures, the navigation charts, the aircraft description etc.

The application server is a Java application which manages the data contained in the database. It receives the requests coming from the set of currently running clients and it is the only agent allowed to mediate them towards the database, in order to assure an high data consistency. It exhibits also a CORBA interface in order to exchange data with the C++ modules. The event notification has been implemented in order to have a correct situation awareness in all kind of clients.

The application server manages the connections with the clients using a time-out mechanism. Considering that the AMDB module is designed with a WAN computing oriented structure, security aspects have also been taken into account.

3 Interaction with the Core SEEDS Environment

The core of the SEEDS simulator is written in C++ and uses the CORBA standard to communicate between different system modules. The AMDB module is entirely written in Java. This situation has been an excellent occasion to test the ability of co-operation between Java and CORBA based C++ subsystems. The introduction of new platform into the project has stressed the necessity to stick very closely to CORBA standards otherwise the reusability of the software modules could be decreased. In the prototype the AMDB is hosted on a Digital Alpha station. The Apache web server has been used to manage the interaction with the Java clients.

The code of clients is stored in signed archives. Clients are downloaded from the web server to the browsers where they are run on browser virtual machine. In the prototype configuration the browsers used are Netscape browser version 4.x or higher and Microsoft Explorer version 4.x or higher together with Java Plug-in from Sun. The Plug-in implements the features of Java 1.1 system, which are not implemented by the major browsers. The application server is connected to the core SEEDS system through the CORBA interface. The mechanism of event notification has been implemented between the application server and the C++ clients using the existent subscription methods.

4 Conclusions

This paper has presented the SEEDS simulation environment with particular attention to Airport Management Database, which is an add-on module of the core simulator. The chosen approach using Java to develop the module has been shown appropriate. In fact it has been a good occasion to see whether the speed of recent Java Virtual Machine (JVM) implementations are able to cope with tasks arising in such complex applications as is an airport simulator. The results of these tests have been very encouraging and they provide a good basis to extend this approach in future. Furthermore, the use of Java allowed to release any special requirement on the client machines, other than the presence of a standard web browser.

References

1. S. Bottalico. SEEDS (Simulation Environment for the Evaluation of Distributed Traffic Control Systems). A Simulation Prototype of A-SMGCS. In *Proceedings of the Internationa Symposium on Advanced Surface Movement Guidance and Control System.* Stuttgart, Germany, 21-24 June 1999.
2. Official SEEDS Web page. http://www.lti.alenia.it/EP/seeds.html

Different Strategies to Develop Distributed Objects Systems at University of La Laguna

A. Estévez[1], F.H. Priano[1], M. Pérez[2], J.A. González[3], D.G. Morales[3], and J.L Roda[3]

[1] Centro de Comunicaciones y Tecnologías de la Información (CCTI)
[2] Centro Superior de Informática
[3] Dpto. Estadística, I.O. y Computación
Universidad de La Laguna, La Laguna, Tenerife, Spain
{amesteve, fpriano, jlroda}@ull.es

Abstract. The use of *Java Remote Method Invocation* and *Java Servlets* allow the development of distributed systems. The University of La Laguna is made up of sixty departments on three separate sites. There are some 30,000 people involved in its day-to-day organisation, including students, teaching staff and service and administrative personnel. The geographical distribution of the University is a problem as far as administrative matters are concerned. Our developments in Web technology are giving users access to different services from their place of work or study. While the use of the Java version of CORBA, *Java Remote Method Invocation*, permits high interactive distributed applications; the *Java Servlets* provides an easy way to create three tier client/server applications over the Internet. In this work, we present end-user Web applications implemented using both techniques. The evaluation can be extended to any other organisation.

1 Introduction

Distributed Object Applications appear as a natural behaviour of different information systems. In distributed applications, the components that form the system are distributed: data, computation and the users. Data is distributed due to administrative and security reasons. Computation is distributed due to the use of specific computers, as parallel supercomputers, to execute particular problems. Users communicate with each other via the application. The conjunction of Internet, Web technologies and Distributed Object Applications conform a great environment for the next generation of cross-platform applications where exists highly interactive functions between servers and clients.

The University of La Laguna has decided to reorganise the services around the Web and Distributed Object Systems technologies. The University administration deals with more than 30,000 people including students, teaching staff and service and administrative personnel. Within the organisation there are different information systems. Some of the most important include the system for the Student Registration, the

Records of University Meetings, the Human Resources Administration, the Library Services and the Sports Services Administration.

The main resources of the Centre of Communications and Information Technologies (CCTI) are a powerful IBM/AS400 for administrative proposes, many IT workstations, a 4 processors computer Digital Alpha Server 4100, PCs with Linux and Apache, etc. Administrative personnel work using Windows NT Workstation with Internet browsers. Java platform will permit a fast integration and software development within all this resources.

2 Distributed Object Systems Paradigms and the Internet

Distributed Object Systems (DOS) are distributed systems in which all elements are modelled as objects. *DOS* extends an object oriented programming system by allowing objects to be distributed across a heterogeneous network. These objects may be distributed on different computers of a network, with their own address space outside an application, and they still appear as part of the application. The most popular *Distributed Object Paradigms* are the OMG's Common Object Request Broker Architecture (CORBA) [1], the Microsoft 's Distributed Component Object Model (DCOM) [2] and the JavaSoft's Java Remote Method Invocation (Java-RMI) [3].

On the other hand, all the advances in Web Technology permits the use of many different API's to develop information systems. Java provides a specific API for Web based applications named *Servlets* [4]. Java Servlets provides an easy and fast way to create 3 tier (or in general n-tier) client/server applications over the Internet. Servlets can be implemented as Distributed Object Systems.

3 Implementation of Distributed Object Systems in the Internet

In the transition to the three tiered architectures, our organisation has to evaluate the different efforts and benefits. The personnel have extensive experience with relational databases, advanced intranet language programming and techniques, value added services, AS400 and Unix, RPG, internet script languages, visual languages, HTML, protocols software, firewalls, etc., but most of the people lack knowledge about distributed applications, object technology, network protocols, programming by interfaces, Java platform. The first phase to move to a distributed architecture is the training of the computer staff. In this period, the training must live together with the day to day work.

In this way, the directive staff of the Computer Centre has planned to evaluate two different technologies in two different services. The Student Registration Service was analysed to study the capacities, efforts and benefits of the Java Servlets technology. The Teaching Plan Service was the example used to evaluate the Java RMI technology.

3.1 Java RMI Implementation of the Teaching Planning Service

An Internet browser with capabilities for Java applets and SSL security is needed in the client computer. All the input format checking is done in the applet to avoid unnecessary communications.

In this case, we have define the following three tiers. The Presentation tier composed by HTML pages with applets that check all the input data. The applet uses the Java Swing API to be user-friendly. The Business Rules involves all the computations and prepared statements necessary to carry out the request data. The applet invokes the remote method in the server object. In Java RMI, all the communications between server and client are performed using the stub/skeleton proxies. In the server, the underlying skeleton "talks" with the client stub that return the request data to the applet. Data Base tier deals with all data management.

The main points of the Java RMI evaluation are: maintenance of the new application is easy due to the new organisation layers. Database independence from new rule business. Frequently, the organisational rules, as teaching hours, student's requirements, etc., make changes to the application. With the tier division, we can modify or add new functionality. Automatic software distribution is a critical point in large organisations. The application has been divided in components that can be reused in other distributed applications. We have reused code from the servlets application.

3.2 Servlets Implementation of the Student Registration Service

The *service()* method in the server side receives a request from a client HTML form. Most of the forms performs the POST action and the *service()* method calls the *doPost()* method. The first data that is received is the username and the password. After the security process, the server responds with an HTML page with all the options a student must fill and can consult. There is mandatory data that must be filled until next page can be reached.

Within the three-tier model we define the Presentation tier composed by HTML pages with form checking using JavaScript. We try to minimise useless transactions over the network. The Business Rules tier, where the servlets Java programs define the computations to execute in the different machines. The last level is the Database Management tier.

The database management is independent of the other two tiers so if any change is done in the business rules or the user interface, this tier provides all the necessary functionality. The SQL queries formed from the client data forms are executed by the servlets over the database management.

The experience obtained can be remarked as follows. Servlets are an efficient technique to develop distributed applications under the Internet when no interactions are necessary inside an HTML page. With the three-tier model, computer staff has differentiated the database, the user interface and the business rules. New developments could be done reusing the objects of other applications. Is easier to maintain and to add new functionality. Software distribution of the application is not necessary. With the Java concept of software on demand, we take off the problem of having the same and latest versions of the application. Software distribution is done automatically as we are working in a web environment. We only need a web browser. Servlets performance is one of the primary keys. Under this model, the clients (students, administrative staff or computer staff) only need a computer with an Internet browser with SSL security capacity.

4 Conclusions

Distributed object oriented applications take advantage of the scalability and heterogeneity of the computer network associated. We have integrated different machines and operating systems under the Java Platform, the Internet and the Web. We have evaluated two different techniques to develop distributed systems: Java Servlets and Java Remote Method Invocation. Java RMI offers high interactive distributed applications and the Java Servlets provides an easy way to create efficient and fast client/server applications.

The Student Registration application was used to analyse the features, benefits and problems of Java Servlets. The Teaching Planning application was selected to study the Java RMI. The three-tier architecture used in both applications allows explicitly separating data management, logic business rules and user interface. Java's software on demand model allows the application changes to be made without physically updating each client. Client applications are distributed via the HTTP and Web browsers without the expense of installing it on each computer.

The experience obtained from the evaluation of these two techniques gives the computer directive staff the necessary feedback for the development of the rest of applications in our organisation. The applications will be created using the Internet as the communication medium and Distributed Object Systems as the platform to develop and reuse components. The results of this evaluation can be extended to any other organisation.

References

[1] CORBA: Common Object Request Broker Architecture. *www.corba.org*
[2] DCOM: Distributed Component Object Model. Microsoft Corp. *www.microsoft.com/com/tech/dcom.asp*
[3] JAVA RMI: Java Remote Method Invocation. Sun Microsystems Inc. *java.sun.com/products/jdk/rmi/index.html*
[4] JAVA SERVLETS: Java Servlets. Sun Microsystems Inc. *java.sun.com/products/servlet/index.html*
[5] PERL: *www.perl.org*
[6] CGI: *www.cgi.org*
[7] ASP: Active Server Pages. Microsoft Corp. *msdn.microsoft.com/library/tools/aspdoc/iiwawelc.htm*
[8] CCTI: Centro de Comunicaciones y Tecnologías de la Información. Universidad de La Laguna. *www.ccti.ull.es*

DESIREE: DEcision Support System for Inundation Risk Evaluation and Emergencies Management

Giovanna Adorni

E.N.E.A. H.P.C.N. Via Martiri di Monte Sole, 4, 40128 Bologna, Italy
adorni@bologna.enea.it

Abstract. The work described within this document forms an activity within the NOTSOMAD TTN. The main objective of the DESIREE project is the implementation of a Decision Support System for the Inundation Risk Evaluation and Emergencies management aimed at analysing and anticipating catastrophic flood events and at preventing and mitigating their effects on the social environmental and cultural heritage. The public administrations are highly sensitive to these aspects and are looking forward to the availability of Decision Support Systems allowing for the planning and the definition of areas at risk as well as for the forecasting of possible catastrophic events in order to define the most appropriate intervention and remedial strategies. The final result of the DESIREE project is a software package, fully integrated into a GIS and a Data Bank, running on a MIMD parallel computer specifically devoted to the simulation of flooding and related phenomena. The developed software is the result of a porting by the addition of OpenMP directives.

1 The Problem

The problem of flooding is as old as time. However, while natural flooding of large areas did not create situations more dangerous than others in the prehistoric world, the expansion of human activity and cities has made preventing damage caused by floods or harnessing over-bank flows for one's own purposes, as in ancient Egypt, a necessity that remains vital to this day.

From the end of the eighteenth century onwards, with the advent of the industrial age, there have been two courses of action: hydraulic works on the territory, such as land reclamation works, which in many cases upset a land's balance dependent on overflow, and the channelling of watercourses, especially in mountain and foothill sections, with the result that the problem of flooding is brought downstream even to areas that were originally protected.

Lastly, recent years have seen booming population and indiscriminate urbanisation create extremely dangerous situations, with floodplain areas that are inhabited, or that even house entire neighbourhoods sheltered by levees that are not particularly safe during prolonged periods of flooding, especially in the case of suspended bed rivers.

2 Flood's Effect Forecasting

The main objective of the DESIREE project is the implementation of a Decision Support System (DSS) for Inundation Risk Evaluation and Emergencies Management aimed at analysing and anticipating catastrophic flood events and at preventing and mitigating their effects on the economical, social, environmental and cultural heritage.

DESIREE responds to the need of an integrate tool for planning and management that, taking advantage of available High Performance Computer platforms, allows to locate areas at risk and to estimate expected damages; to forecast floods and inundation phenomena on the basis of real-time analysis of the present meteorological situation and of forecasts available at different time and space scales; to evaluate the effects of decisions aimed at reducing social, economical and environmental damages on the basis of planned or real-time forecasted scenarios; to allow continuous training of personnel.

3 A New Powerful Decision Support System

Following the findings of AFORISM, DESIREE will allow quantitative impact analyses on the basis of environmental, social, economical criteria by combining flood maps (generated by simulation models) with geo-referenced data (land use maps, cadaster maps, road maps, traffic information, etc.) using several tools developed in the ODESSEI project DESIREE will include at full development:

- A set of relational and mathematical models: statistical model of rainfall extremes; hydrological semi-distributed rainfall-runoff model; a one-dimensional flood routing model, a combined one-dimensional/two-dimensional flood plain inundation model; a socio-economic and environmental impact assessment model.
- A decisional support system, which manages data, flows to and from the database; organises and compares the performed scenarios; guides the user in making decisions regarding a particular issue or problem.
- A database management system, which includes data treatment procedures, information concerning watershed entities (rivers, sub-basins, etc.) and the socio-economical and environmental data.
- A geographical information system, which allows the management of geo-referenced data, with the typical recent GIS spatial analysis tools, while the data interchange between the GIS and the models, is optimised by knowledge-based system procedures.
- A user interface, which is extremely advanced and, at the same time, easy-to-use.

4 The Relevance of the High Performance Computing

DESIREE integrates different models:
- statistical models of rainfall extreme values,
- hydrological seem-distributed rainfall-runoff model (ARNO),

- one-dimensional flood routing model (PAB),
- combined one-dimensional/two dimensional flood plain inundation model (CVFE) (Fig.1).

The application of this complex system presents two kinds of problems:
- At the beginning, when the area is analysed to draw the emergency plan, many different basins must be studied; for each basin several hydrograms have to be considered; to simulate all the cases a large amount of computing time is necessary.
- In the operative emergency the Decision Support System must forecast in quasi real time the inundation events and the models have to run faster than real time.

For those reasons one of the models (CVFE), that must be run many times absorbing by itself computing hours, is modified and parallelized. The parallelization consisted on the porting of the CVFE model by the implementation of OpenMp.
CVFE is a finite element code with a main program and a tree of subroutines, a large use of DO Loops characterises the code.

By the use of tools, installed on the SGI Hw, the computing time was determined for every subroutine and detailed for each DO Loop, the heavier of them were implemented with OpenMP. Some DO Loops, that were intrinsically sequential, must be modified before the implementation.
The addition of OpenMP directives provided an interesting approach offering the chance of parallelizing the single parts of the code and obtaining the most convenient solution.

The development and tests are done using MIMD machines (SGI and IBM). The result is:
- a parallel code easier to maintain,
- a reduction of overhead and load imbalance (the CVFE code uses DO structures too heavily and in some cases depending on the time).

The parallel version of the CVFE code is validated on different Hw configurations to verify the portability and to define the scalability.

In the end we obtained a code that
- offers a good ratio between performance and portability,
- is easy to use - easy to maintain.

References

1. D. Dent, G: Mozdzynsky, D. Salmond, B. Carruthers: Implementation and Performance of OpenMP in ECMWF'sIFS Code. Fifth European SGI/Cray MPP Workshop
2. R. Ansaloni, P. Malfetti, T. Paccagnella: A weather forecast model on SGI Origin 2000. Fifth European SGI/Cray MPP Workshop
3. OpenMP Portland User Group: OpenMP Fortran Application Program Interface, Oct 1997 1.0

Database System for Large-Scale Simulations with Particle Methods

Danuta Kruk[1] and Jacek Kitowski[2,3]

[1] Institute of Physics, Jagellonian University, Reymonta 4, 30-059 Cracow, Poland
[2] Institute of Computer Science, AGH, al. Mickiewicza 30, 30-059 Cracow, Poland
[3] ACC CYFRONET, ul., Nawojki 11, 30-950 Cracow, Poland
kito@uci.agh.edu.pl

Abstract. In the paper a structure of the database system for simulation results which concern raw and multimedia data is presented.

1 Introduction

Simulation methods using particles constitute a worth field of research. Due to the progress in simulation models cross-scaling is an important element, e.g. [1]. Performing such a kind of studies one is often confronted with problems of analysis, results comparision and storing.

Two kinds of the results could be of interest: basic results (particle positions and momenta with additional quantities) and data obtained from numerical analysis of the basic results (like time snapshots, 2D/3D diagrams and movies).

From our previous experience [2] it follows that the object-oriented approach could be less efficient than a traditional relational approach. For the reported implementation Oracle8 system management has been chosen.

2 Database Description

For each individual simulation experiment we define the first order attributes (e.g. system type, calculation method) and the second order ones (i.e. interactions, initial conditions). Properties like mass, size, particle physical properties and contribution (for heterogeneous systems) are included. The next attributes are related to the problem of external forces and internal interactions in the physical systems and to initial and boundary conditions.

The kind of results defines basic results or some aggregated results obtained from the basic ones for the defined experiment. Other links give possibility for cross search of the entire database – for similar experiments and for other kinds of results obtained from the same experiment. The aggregated results represent plots of macroscopic quantities, snapshots of chosen simulation timesteps or movies presenting development of the simulation. Typical formats are being adopted, like giff, tiff, postscript and mpeg. The publication database is also included with separately defined auxiliary keywords.

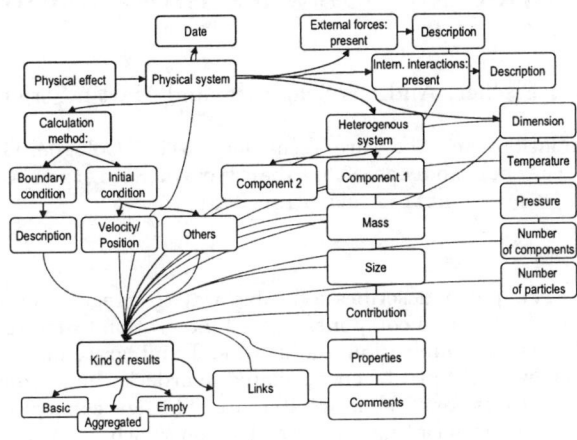

Fig. 1. Example of query flow

In Fig. 1 an example of query flow through the database is presented, with links to the **results** which can be followed at any stage of browsing, taking the rest of unspecified attributes as defaults.

3 Conclusions

Using the component programming paradigm the database system can be used as an element of the simulation environment (under development at present). It will support long-distance cooperation in the field of particle simulations. The proposed data relations offer an access to the results in aggregated forms which are easy to interpret.

The work has been sponsored by Polish Committee for Scientific Research (KBN) Grant No. 8T11C 006 15.

References

1. Dzwinel, W., Alda, W., and Yuen, D.A., Cross-Scale Numerical Simulations Using Discrete Particle Models, *Molecular Simulation*, **22** (1999) 397-418.
Dzwinel, W., Alda W., Kitowski, J., and Yuen, D.A., Using discrete particles as a natural solver in simulating multiple-scale phenomena, *Molecular Simulation*, in print.
2. Kitowski, J., Trzeciak, P., and Wajs, D., Object-oriented database system for large-scale molecular dynamics simulations, in Sloot, P., Bubak, M., Hoekstra, A., and Hertzberger, B. (eds.) Proc.Int.Conf. on High Performance Computing and Networking, April 12-14, 1999, *Lecture Notes in Comput.Sci.*, **1593** (1999) 693-701.

Script Wrapper for Software Integration Systems

Jochen Fischer, Andreas Schreiber, and Martin Strietzel

DLR, Simulation- and Software Technology, 51170 Cologne, Germany
{Jochen.Fischer,Andreas.Schreiber,Martin.Strietzel}@dlr.de
http://www.sistec.dlr.de

Abstract. This paper describes a flexible way to integrate existing applications and tools into a component-based integration framework which uses CORBA as communication middleware. To integrate an application into this framework it must be encapsulated in order to build components which conform to its component architecture. The new concept described here is based on the scripting language Python which is used to control the execution of the encapsulated application. The CORBA DSI is used to create the application dependent component interface which can be extended by the Python script.

1 Introduction

The development of frameworks, integration, or problem solving environments (PSE) is focussed in many scientific projects, such as COVISE [1] or Legion [2]. Another approach is the integration environment TENT [3].

TENT is a component-based framework for the integration of tools belonging to typical workflows in computer aided engineering (CAE) environments. The user should be enabled to configure, steer, and interactively control his personal process chain. The workflow components can run on arbitrary computing resources within the network. It is possible to combine the components to workflows in a graphical user interface. Due to the usage of CORBA, TENT supports all state-of-the-art programming languages, operating systems, and hardware architectures. A necesssary task in TENT and other integration systems is a suitable wrapping or encapsulation of the applications conforming to the component architecture of the system. Here we will describe a new concept for integrating existing applications and tools.

2 Why Script Wrappers?

Up to now all application wrappers in TENT are static, i.e. they are written in a compiler language like C++ or JAVA and are based on the generated skeletons of static IDL interfaces. The development process of such a wrapper is relatively complicated and requires a profound knowlegde of TENT and the CORBA mechanisms behind it. This applies also to the enhancement of an existing wrapper. Furthermore, it is not possible to extend a wrapper with additional functionality without a compilation step. In view of these problems an application wrapper based on a scripting language, like Python [4], has the following advantages:

- scripting languages are designed for gluing applications together [5],
- no compilation step is necessary,
- scripts can easily be modified and extended,
- the code is 3-10 times shorter compared to other high level languages,
- the developer must be experienced with the scripting language only,
- the internals of TENT and the CORBA mechanisms are completely hidden.

3   Design of a Dynamic Wrapper

A script wrapper has a layered architecture, where every layer consists of several building blocks. The architecture is shown in figure 1. The first layer represents

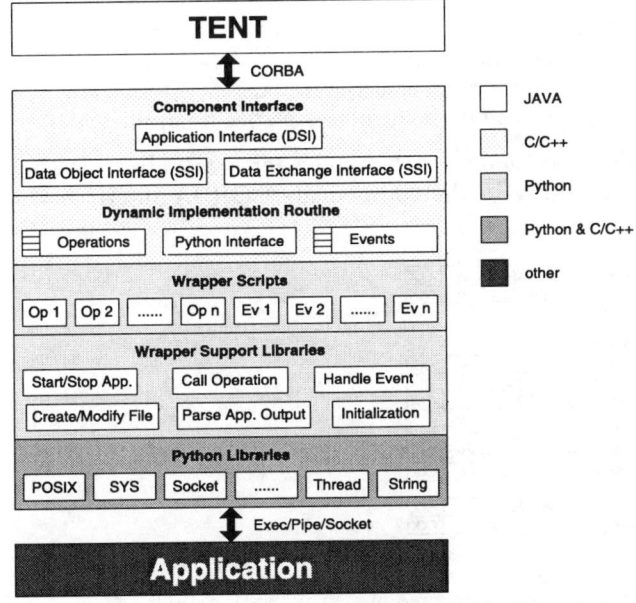

Fig. 1. Layered architecture of a script wrapper.

the component interface which consists of three parts as in every other TENT component. This layer is written in C++ because this provides greater flexibility in the choice of the convenient CORBA platform.

The difference to the static components is the application interface, which is not based on generated skeletons (SSI). In contrast to the other interfaces it is implemented using the Dynamic Skeleton Interface (DSI). This is done with a so-called Dynamic Implementation Routine (DIR). The Dynamic Implementation Routine analyses the incoming requests and delegates them to the appropriate handler routines which are registered in the DIR. There are two different types of handler routines. One for handling explicit method calls, and one for handling events (implicit method calls). The main difference between the two types is that method calls are normally executed synchronously while event handlers must return immediately and execute the triggered action asynchronously.

All available handler routines are kept in different lists in association with their signature. These lists can contain a large number of entries depending on the features of the encapsulated application. Therefore it is reasonable to structure the available set of operations and events using CORBA namespaces. It is not necessary to adopt every operation of an application to the component interface, but every component of a specific type must provide at least the core operations that are characteristic for this type. In addition to these core operations the developer can add further operations to the component interface. These operations should be grouped according to the following categories in order to facilitate the replacement of one component through another component of the same type:

- basic (for all components),
- core (for all components of the same type),
- package (additional operations for a specific type),
- special (only for the component that encapsulates a specific application).

Figure 2 shows an example of an operation hierarchy for a visualization tool. These informations must be added to the components CORBA interface and

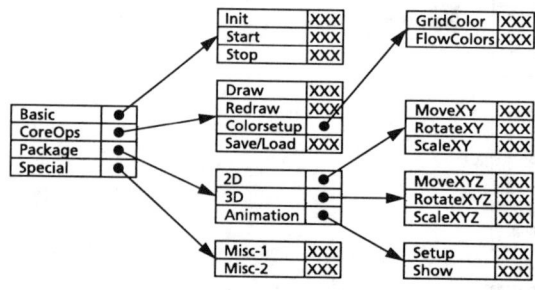

Fig. 2. Example of an operation tree of a visualization tool.

are used to determine which handler must be called to handle requests. Because the component interface is extended at runtime, the Dynamic Implementation Routine must be able to generate an appropriate IDL file or to feed the information directly into the Interface Repository. The Python interface represents the bridge between Python and C++ and provides fundamental functionalities for the interaction between C++ and Python. It consists of Python modules written in C/C++.

The handler routines written in Python are the only parts of a wrapper that must be created or modified for a new application. Hence the script is the only part in the wrapper which depends on the encapsulated application. The wrapper library provides easy to use functions for different tasks. At least the following capabilities have to be provided:

- setting up interprocess communications via socket or pipe,
- invocation of the wrapped application(s),
- generating startup and configuration files,
- context sensitive modification of created files,
- parsing the application output.

In the lowest layer of the wrapper the communication to the encapsulated application is done via the Python POSIX library which is an abstraction to the UNIX environment.

4 Conclusions

The concept presented for embedding applications has several advantages compared to the method used so far. The main advantages are due to the use of the Python scripting language. The portability of Python allows to use the wrapper scripts on every relevant platform without changes. The scripts itself can be as simple as possible because the complexity of the integration system interfaces are completely hidden from the developer who benefits from the large Python library which offers functions for almost all operation purposes. Like C/C++ Python allows the integration of every application type. Furthermore script wrappers are easier to implement and maintain. It is not restricted to applications with a programming api, in contrast to other projects with a similar approach [6].

Several applications and tools from the CFD field have already been wrapped and integrated in TENT. Some legacy and commercial CFD codes of which the source code is not available are also included. The experiences made using the described approach show that the new concept meets our requirements.

References

1. Lang, U., Rantzau, D.: *A Scalable Virtual Environment for Large Scale Scientific Data Analysis.* Future Generation Computer Systems 14 (1998) 215-222
2. Foster, I., Kesselman, C. (eds.): *The Grid: Blueprint for a New Computing Infrastructure.* Morgan Kaufmann Pub. (1999)
3. Breitfeld, T., Kolibal, S., Schreiber, A., Wagner, W.: *Java for controlling and configuring a distributed Turbine Simulation System.* Workshop Java for High Performance Network Computing, Southampton (1998).
http://www.cs.cf.ac.uk/hpjworkshop/papers/okpapers/javatent.ps
4. Python-Homepage: http://www.python.org
5. Rossum, G.: *Glue It All Together With Python.* Position Paper for OMG-DARPA-MCC Workshop on Compositional Software Architecture, Monterey (1998).
http://www.objs.com/workshops/ws9801/papers/paper070.html
6. Sistla, R., Dovi, A.R., Su, P.: *A Distributed, heterogeneous computing environment for multidisciplinary design and analysis of aerospace vehicles.* 5th National Symposium on Large Scale Analysis, Design, and Intelligent Synthesis Environments, Williamsburg, VA (1999)

Implementation of Nested Grid Scheme for Global Magnetohydrodynamic Simulations of Astrophysical Rotating Plasmas

Takuhito Kuwabara[1], Ryoji Matsumoto[2], Sigeki Miyaji[1] and Kenji Nakamura[3]

[1] Graduate School of Science and Technology, Chiba University
Inage-Ku, Chiba, 263-8522 Japan
{takuhito, miyaji}@c.chiba-u.ac.jp

[2] Faculty of Science, Chiba University, Inage-Ku, Chiba, 263-8522 Japan
matumoto@c.chiba-u.ac.jp

[3] Japan Science and Technology Corporation, kawaguchi, 332-0012 Japan
kenji@c.chiba-u.ac.jp

Abstract. We have developed an astrophysical rotating plasma simulator by which we can simulate global dynamics of magnetized rotating plasmas in 2 and 3 dimensions. One of the major applications of this simulator is a differentially rotating disk around a black hole (accretion disk) which is believed to be the central 'engine' of active galactic nuclei(AGN) such as quasars. We incorporated nested-grid scheme in order to simulate spacial scale ranging from the observed 100 pc ($\sim 10^{18}$ m) torus of AGN to the central region of the accretion disk ($\sim 10^{13}$ m). By the calculation of 3 CPU vector-parallelized code we get total CPU time of calculation about 1/2 that of non-parallelized code.

1 Introduction

Recently, many kinds of simulation schemes have been developed for various necessities of scientific field. Especially, in the field of astrophysics, wide range of simulation space needs special scheme to simulate a model from the global range to the tiny area where the catastrophic phenomena take place. The nested-grid scheme is the one to cover the wide range of evolution of the astrophysical object. Tomisaka in [1] applied this scheme to the fragmentation process of a cylindrical magnetized cloud. Norman and Bryon in [2] applied adaptive mesh refinement in computational cosmology. In this paper, we examined the calculation of 5-levels nested grid scheme and compared their efficiency. Here, we used $100^2 \times 5$ meshes. By using VX/4R at National Astronomical Observatory of Japan (NAOJ), we evaluated the performance of our codes.

2 Parallelized Nested-Grid Scheme

Fig. 1 shows a sketch of our model. Because of deep gravitational potential of the black hole, inner region should be simulated by higher resolution. The basic domain used in our simulation is a square of size L_0. We call this domain as level

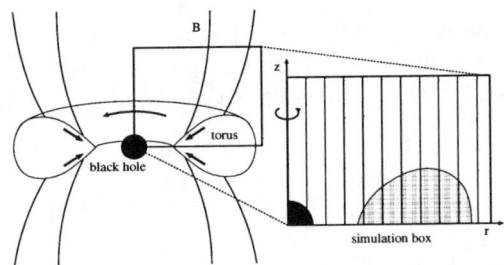

Fig. 1. A schematic picture of the simulation model and simulation box

0 domain. Nested domains reside at the inner corner of a lower level domain as shown in Fig. 2(a). Therefore the domain size of level N is $L_N(= 2^{-N}L_0)$. Domains of each level have the same grid size of (100×100). In our simulation, we want to solve the Euler equations for an inviscid flow like this,

$$\frac{\partial U}{\partial t} = \frac{\partial F(U)}{\partial x} + S , \qquad (1)$$

where F is the flux vector and S is the source term. Here, the simulation algorithm is as follows (For ease of explanation we take into account only 2 levels which is level 0 and 1).

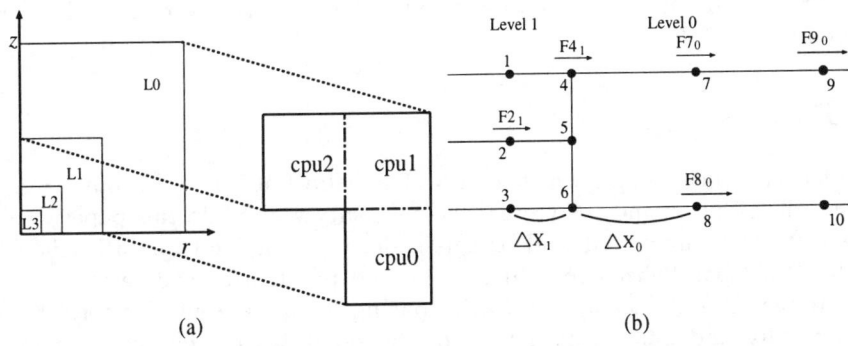

Fig. 2. (a) The nested levels and how to division of domain for parallel computing. (b) The 2D plane at boundary between level 0 and level1 and the grid structure

1)- Determine the time step Δt_0 by the CFL condition on level 0. Using this time step Δt_0 we define time step at level 1 as $\Delta t_1 = \Delta t_0/2$.

2)- Integrate level 1 with time step Δt_1 (see Fig. 4(a)).

3)- Calculate boundary values between level 0 and 1.

Fig. 2(b) shows the 2D plane at boundary between level 0 and level 1. The filled circles are grid points, and ΔX_0 and ΔX_1 are grid lengths on level 0 and level 1, respectively. The $F*_0$ and $F*_1$ are fluxes on level 0 and level 1, where the '*' reveals the number of grid point. Here, we describe one direction flow as shown in Fig. 2(b) for simplicity. For example, the integration of the quantity U at point 5 on level 1 with time step Δt_1 needs fluxes $F2_1$, $F7_0$ and $F8_0$ as follows (see Fig.2(b)),

$$U_1^1 = U_1^0 + \frac{\Delta t_1}{\Delta X_1 + \Delta X_0}\left(\frac{F7_0 + F8_0}{2} - F2_1\right) + S\Delta t_1 . \qquad (2)$$

Here, the superscript 1 shows integrated value and the superscript 0 shows the value before integrated value and S is the source term.

4)- Overwrite boundary values by those taken in procedure 3).
5)- Integrate level 1 again with the same time step Δt_1 as in procedure 2) (see Fig.4(a)).
6)- Calculate boundary values again same as procedure 3).
7)- Integrate level 0 with time step Δt_0 (see Fig. 4(a)).
8)- Overwrite boundary values by those given at procedure 6).
9)- Return to procedure 1).

For parallelization, we divide each level into three regions (see Fig. 2(a)) and compute each region parallelly. In our parallel calculation, the CPU number must be $3 \times 2 \times m$. Moreover, it is not needed so many processors for calculation on vector processor machine as on non-vector processor machine for us to get good performance (commonly, $10 \sim 15$ CPUs are used). Because of enjoying the benefit of the vector performance, we could not reduce the mesh size of each CPU less than several tens meshes. For an actual machines available, 32 CPUs would be maximum. As a test case of our scheme, here, we presented 3 CPUs case only.

3 Results

Simulation results are shown in Fig. 3. The solid lines show isocontour of the density distribution and the vectors show velocity vectors. In this paper we used 5 levels to test our nested grid scheme code. Fig. 3(a) and Fig. 3(b) show the result of non-parallelized code at $t = 2.3$ and parallelized code at $t = 3.0$. As the number of level decreases, the grid spacing becomes large. The torus rotates differentially and one rotation time of the point where the density becomes maximum inside the torus is $t \sim 6.28$.

In the panels of level 1 and level 2, it is easily to recognize the propagation of mass outflow. In other word, as shown on panels of level 1 and level 2 in Fig. 3(a), Fig. 3(b), we can successfully simulate that the information is transmitted continuously between different levels. Fig. 4(b) compares the total computational times of non-parallelized code and parallelized code. Because of the complexity in the procedure to communicate and integrate at the boundary and because of decrease in vector length (100 to 50), the parallel performance is still 66%. If we need more fine mesh in 2D, because of the increase in vector length would result better performance. For 3D code, $7 \times 2 \times m$ CPU parallelism ensure higher parallel performance.

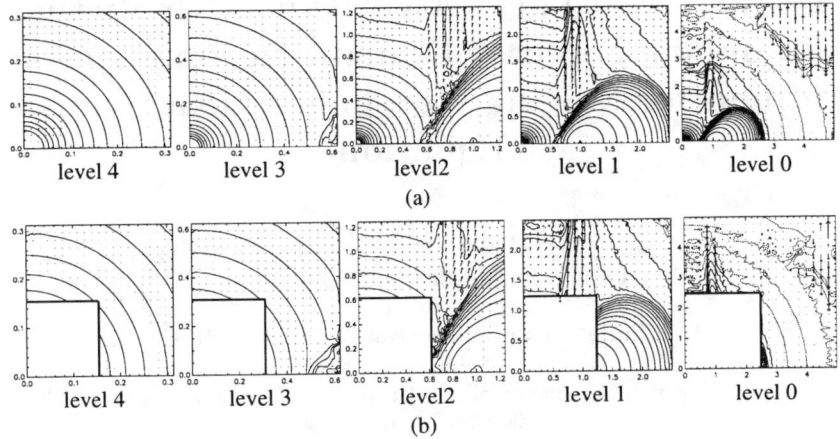

Fig. 3. Simulation result of non-parallelized code (a) and of parallelized code (b)

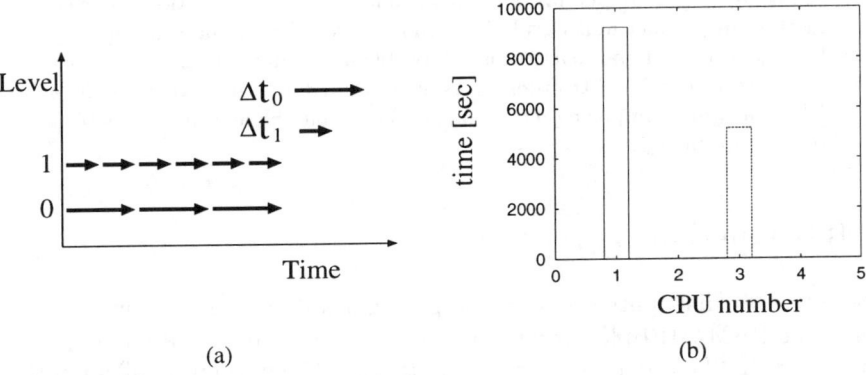

Fig. 4. (a) The integration time difference. (b) The total simulation time of non-parallelized code and parallelized code

This study is supported in part by the grant for "Astrophysical Rotating Plasma Simulater" by Japan Science and Technology Corporation.

References

1. Tomisaka, K.: Collapse and fragmentation of cylindrical magnetized clouds: simulation with nested grid scheme. Publ. Astron. Soc. Japan **48** (1996) 701–717
2. Norman, M. L., Bryan, G. L.: Cosmological adaptive mesh refinement. numa.conf (1999) 19–28

Parallel Multi-grid Algorithm with Virtual Boundary Forecast Domain Decomposition Method for Solving Non-linear Heat Transfer Equation

Guo Qingping[1], Yakup Paker[2], Zhang Shesheng[1],
Dennis Parkinson[2], and Wei Jialin[1]

[1] Transportation University, Wuhan 430063, P.R. China
qpguo@public.wh.hb.cn
[2] Queen Mary & Westfield College, University of London, E1 4NS U.K
paker@dcs.qmw.ac.uk

Abstract. This paper proposes a virtual boundary condition forecast method in parallel multi-grid algorithm design with domain decomposition for solving non-linear transient problem. Numerical results of non-linear transient heat transfer problem obtained by the algorithm in a PVM network computing environment show that the algorithm has high parallel efficiency.

1 Introduction

In recent years, distributed parallel computing based on a local network environment using PVM/MPI platform is a booming research area. But communication costs are crucial factors in network computing.[1] The normal parallel multi-grid method involves many communication costs between sub-domains that reduce algorithm efficiency and make speedup worse[2]. This paper proposes a virtual boundary forecast (VBF) method for domain decomposition in parallel multi-grid computing. Generally speaking, a parallel algorithm with the VBF method composes of three main parts: pre-calculation of initial value on coarse grid, sub-domain virtual boundary value forecast and parallel calculation of the independent sub-domain with multi-grid method without communications between sub-domains on coarse grids, smoothing on the finest grid with communications between sub-domains. This scheme substantially reduces the com munication costs of parallel MG[2]. Measured numerical results for a non-linear transient heat transfer problem shows very good performance in efficiency and speedup.

2 Key Ideas of the Scheme

Kernel ideas of the virtual boundary forecast algorithm of the parallel MG method can be outlined as follows: (1) Forecast virtual boundary values using historic values at boundary points by some methods, In this stage no communications between sub-domains; (2) Performing multi-grid calculation on each sub-domain separately, independently in parallel without any communications between sub-domains; In this stage every sub- domain solves its own independent boundary problem; (3) Smoothing results on the finest grid of whole domain in a few cycles to reduce errors derived by (2), in this stage there are communications between corresponding sub-domains, and updating the virtual boundary values; (4) Using error criterion to determine a sub-domain calculation should be terminated or not. If it can not be terminated the procedure returns to the (1) step; otherwise a termination token is sent to the neighbour computers, informing them the calculation on the sub-domain is finished, and there are no more communications between it and neighbours; meanwhile results on the sub-domain are sent to master computer for final results assembling. (5) The whole calculation is terminated when all sub-domains are finished their own computing.

In fact the virtual boundary value update happens at step (3) as well as step (1). The latter step is more important than the first one in some sense.

3 Case Study of Non-linear Transient Equation

3.1 Efficient Difference Form with Pre-processing Item

In order to explain our algorithm, we consider following non-linear heat transfer initial-boundary equation:

$$u_t = \frac{\partial}{\partial x}(k\frac{\partial u}{\partial x}) \quad 0 < x < 5; \quad t > 0; \tag{1}$$

here boundary conditions are $u(0,t) = \cos(2\pi t)$, $u(5,t) = 0$, and initial condition is $u(x,0) = 0$ $(0 < x < 5)$. Eq(1) may be rewritten as

$$u_t = ku_{xx} + k_x u_x \quad or \quad u_t = ku_{xx} + G \tag{2}$$

here $G = G(x,t) = k_x u_x$ And

$$k = (1 - f_m)k_c + f_m k_m \tag{3}$$

$$k_c = 0.1 - 0.01u + 0.001u^2, \quad k_m = 1 + 0.1u + 0.01u^2$$

$$f_m = x^2 \quad (x \leq 1) \quad f_m = 1 \quad (x > 1)$$

Where the k_c represents feature of the ceramic, and the k_m feature of the metal in ceramic/metal composite material. The eq(1) is a non-linear equation since k is a function of u. Its difference formula could be written as an efficient difference form such that

$$(2 + \rho)u_j^{(n+1)} = u_{j-1}^{(n+1)} + u_{j+1}^{(n+1)} + d_j \tag{4}$$

here

$$\rho = \frac{h^2}{k\Delta t}, and \quad d_j = \rho u_j^{(n)} + \frac{h^2}{k}G^{(n+1)} \tag{5}$$

The boundary and initial conditions are $u_0^{(n)} = \cos(2\pi t^{(n)})$ $u_M^{(n)} = 0$ $u_j^{(0)} = 0$
There $j = 0, 1, 2, \ldots, M$ represent space nodes and $n = 0, 1, 2, \ldots$ represent time steps. The h and Δt are step length of x and t respectively. We call the item d_j as a pre-processing item. If the k as well as ρ has been calculated and the d_j is pre-determined, the eq(4) then become a convergent pseudo linear iteration format. For the linear pseudo format of eq(4) the multigrid method may be used to solve it. Suppose u^* is the approaching value of u and $u = u^* + \varphi$, it is easy to show that φ satisfies following formula

$$\varphi_t = k\varphi_{xx} + r \quad here \quad r = ku_{xx}^* - u_t^* + G \tag{6}$$

Its difference formula is:

$$(2 + \rho)\varphi_j^{(n+1)} = \varphi_{j-1}^{(n+1)} + \varphi_{j+1}^{(n+1)} + \frac{h^2}{\Delta t} r_j \tag{7}$$

Therefore 2-level or multi-level MG method could be used.

3.2 VBF Algorithm Outline of Non-linear Transient Equation

There are three main nested loops in the algorithm. The outside one is time-step loop; the median one is non-linear loop; and the inner one is linear loop with the MG computing. For the non-linear and linear loop the algorithm can be described as following: (a) In each time-step on the finest grid $\Omega^{(h)}$, virtual boundary values are forecasted using last three subsequent time-step historic values at the virtual boundary; (b) using last non-linear iteration value $u_j^{(nl-1)}$ in the time-step (or initial values $u_j^{(0)}$ at the loop beginning, determined by the initial value pre-calculation stage or last time-step calculation or a mixture of them) calculate k, ρ and d_j from eq(2),(3) and (5) on each computer for each sub-domain; For values of k_x and u_x at boundary using corresponding left or right difference format calculate them to avoid some relevant communications. (c) Compute pseudo-linear equation eq(4) with V-cycle multigrid method in a few fixed times (e.g. one or two times) as a semi-independent MG on each sub-domain without communications between sub-domains in local smoothing on fine grid and all other calculations on all coarse grids. For two-level MG a procedure should be: 1).Local smoothing in a few iteration times with the forecasted virtual boundary on the fine grid (in our program the iteration times is one); 2). Calculating residues and restricting them onto the coarse grid; 3).Iterating the mapped residues on the coarse grid for a few times (e.g. 4 times); 4).Interpolating the residues of the coarse grid into fine grid and revising the corresponding unknowns; 5).Global smoothing in a few iteration times on the fine grid to reduce errors caused by the sub-domain independent MG; at this step all the sub-domains are iterated as a whole domain, and there must be communications between corresponding sub-domains. (d) Using distributed convergence control technique to determine termination of the non-linear computing [3]. If calculation for a time step is finished the results will be sent back to master computer. After the master receiving the results the slave computer then processes next time step.

3.3 Computation Speedup

Denote Nc as number of used computers and η as speedup. The measured results are shown in Table1 which clearly shows that the VBF method has very good speedup. Where 100,000 nodes are calculated. obsevered using Pentium II local

N_c	1	2	3	4	5	6	7	8	40	48
η	1	1.95	2.78	3.62	4.50	5.40	6.30	7.21	23.2	31.0

Table 1. Speedup of VBF Parallel MG Method Algorithm

network with Linux+PVM environment of QMW, London University.

4 Conclusions

Virtual boundary forecast method is an efficient way for domain decomposition-based MG parallel computing, which dramatically reduces communication overheads. Philosophy in the VBF method is simple and direct. If all virtual boundaries are known then each sub-boundary becomes an independent boundary problem, which could be calculated in parallel using any efficient sequential method; therefore the VBF method forecasts values of virtual boundaries using its historic values could speedup iteration convergence rate and reduce communication overheads between sub-domains. In fact in the non-linear transient problem this method uses a little extra calculations to reduce massive communications between processors, which are a bottleneck in MG parallel computing.

References

1. Guo Qingping et al, Network Computing Performance Eva luation in PVM Programming Environment, PVM/MPI 98, Liverpool, UK September 1998
2. W. Hackbusch, Multi-Grid Methods and Applications, Springer Series in Computational Mathematics 4, Springer Verlag, Berlin Heidelberg New York Tokyo, 1985.
3. Guo Qingping et al, Convergence Models for Parallel Multi-Grid Solution ofNon-linear Transient Equation Using Virtual Boundary Forecast Method, to be published.
* Research supported by the UK Royal Society joint project (the Royal SocietyQ724) and the Natural Science Foundation of China (NSFC Grant No.69773021)

High Performance Computing on Boundary Element Simulations

José M. Cela[1] and Andreu Julià[1]

Centro Europeo de Paralelismo de Barcelona (CEPBA)
c/ Jordi Girona 1-3, 08034 Barcelona, Spain
cela@ac.upc.es

Abstract. In this paper we describe the parallelisation of a Boundary Element Package. We exploit different levels of parallelism using threads and processes. Moreover, the threads are used to overlap communications and I/O operations with computations.

1 Introduction

The boundary element method (BEM) has several advantages over the classical finite element method in some specific simulations, like crack grow analysis. However, BEM has an important drawback, the linear systems generated by this method are dense. Then, the computational complexity grows with N^3 and the storage requirements with N^2, where N is the amount of boundary elements. Although, some tricks can be used to mitigate this problem, the complexity of the BEM limits seriously its use. Then, high performance computing is specially critical in BEM simulations.

Inside the ESPRIT project HIPSID, we have developed a new software structure of the commercial boundary element package BEASY [1]. The new software has a speed-up between 5 to 7 using just one CPU. Several techniques has been used to obtain this benefit: blocking algorithms, asynchronous I/O, thread parallelism, and message passing parallelisation.

A boundary element simulations must solve several linear systems like $Ax = b$, where A is a no symmetric full matrix, although sometimes A could be a block sparse matrix. A block LU factorisation is used as linear solver. Partial pivoting in the same column is required in some simulations. The matrix is stored in a file and organised in squared blocks, so the whole application is out of core. In several simulations the different linear systems to be solved are very similar, just some rows in the bottom of the matrix change from one system to other one. Then, it is required to reuse as much as possible a previously factorised matrix in the next factorisation. The sequential algorithm to be parallelised is the following:

```
/* if this is a partial factorisation, update the submatrix (rowStart, N) x (1, N) */
for p= 1 to rowStart-1
    for i= rowStart to N
```

```
        for j= p+1 to N
            A_ij = A_ij - A_ip A_pj

for p= rowStart to N
    if (A_pp is singular) then
        for i= p+1 to N
            if (A_ip is not singular) then
                SWAP (row i-th, row p-th) and break this search loop
    A_pp = A_pp^-1
    for j= p+1 to N       /* Update the row of the pivot */
        A_pj = A_pp A_pj
    for i= p+1 to N       /* Update the remain submatrix */
        for j= p+1 to N
            A_ij = A_ij - A_ip A_pj
```

2 Parallel Solver

We have parallelised the linear solver mixing two different techniques: message passing and threads. With the message passing technique, we obtain a parallel program which is able to run both in shared memory and distributed memory machines. With the threads, we obtain a parallel program with asynchronous I/O operations and asynchronous communications. Both parallelisation schemes, message passing and threads, can be combined in any way to adapt the program execution to the parallel computer architecture. The number of parallel processes and the number of threads are controlled by environment variables.

Moreover, we have applied blocking to the factorisation loops in order to increase data locality. The blocking factor is computed automatically depending on the available memory. The number of I/O operations is divided by two each time that the blocking factor is multiply by two.

The message passing parallelisation of the linear solver is based in a master-slave scheme. There is a master process that performs all the operations of the original version, except the factorisation and the forward and backward substitutions. Both the slaves and the master execute these algorithms in parallel. A cyclic block column distribution is used, i.e. each process stores and computes the block columns such as, the column index module the number of processes is equal to the process number. This distribution guarantees an optimal load balancing if the matrix is full dense. If the matrix is block sparse some little load unbalance will appear. Moreover, the column oriented distribution guarantees that pivot search is a local operation without any communication.

The thread parallelisation is based on a decomposition of the algorithm in basic tasks. A basic task is a matrix block operation like, $A_{pp} = A_{pp}^{-1}$ or $A_{pj} = A_{pp}^{-1} * A_{pj}$ or $A_{ij} = A_{ij} - A_{ip} * A_{pj}$. Also, the communications in the factorisation algorithm are basic tasks. A basic task requires some I/O operations and some computations. The execution of a basic task is done in a pipeline of threads.

Priority queues are used to pass a basic task from one thread to the next thread in the pipeline. Every thread in the pipeline has one input and one output queue. Each parallel process is decomposed in the following threads:

1. The **Main Thread**: This thread executes the solver algorithm. But, when a basic task must be performed, this thread just puts it in the *I/O Pending queue* or in the *Communication Pending queue*. This thread also gets tasks from the *Done queue* in order to know that the task is completed.
2. The **IO Thread**: This thread gets a task from the *I/O Pending queue*, performs the needed IO operations, and puts the task in the *Ready to Execute queue*.
3. The **Net-IO Thread**: This thread gets a task from the *Communication Pending queue*, performs the required communications, and also the required I/O operations to write(read) the received(sent) data to(from) the local storage. Finally it puts the task in the *Executed queue*.
4. The **Worker Threads**: It is possible to have one or several threads of this kind. If several ones are started the computations associated with basic tasks are done in parallel. This thread gets a task from the *Ready to Execute queue*, performs the needed computations, and puts the task in the *Executed queue*.
5. The **Eraser Thread**: This thread gets a task from the *Executed queue*, performs some operations related with the management of the memory buffer, and puts the task in the *Done queue*.

The software was developed using the object-oriented philosophy, but the source code is written partially in C and partially in FORTRAN 77. The main software objects are:

1. **Memory-Disc Object**: This object encapsulates all the details about the management of the memory buffer used to store temporally the matrix blocks. The management of the memory buffer uses the Least Recently Used (LRU) strategy to release blocks from the buffer.
2. **Thread Scheduler Object**: This object encapsulates all the details about the management of threads. Internally it uses the queue object to pass tasks between threads.
3. **Version System Object**: This object guarantees that a matrix block is in the proper state (version) before a basic operation is performed with it. In this way data dependencies between iterations of the factorisation external loop are guaranteed without barrier statements.

This software requires a message passing library and a thread library. The access to both libraries is done using an interface. In this way the code could be ported to other message passing or thread libraries with a few effort. At the present the message-passing interface is implemented using the PVM library [2], and the thread interface is implemented using both P-threads and WIN32 threads.

3 Performance Analysis

The new software has been tested for a complete set of examples, but here we present just one test case. Time measures has been obtained in a SGI O2000 using a memory buffer of just 32 Mbytes. It is a full dense matrix with 141 block rows, and each block has 120 DOFs. Then, there are 16,920 DOFs, and the matrix file size is 2.13 Gbytes. The execution time of the original solver was 11:51:52. The next table shows the present execution times for different number of worker threads and processes. The figure shows the speed-up with respect to the original solver.

Table 1. Execution times of the new solver (hh:mm:ss)

	1P	2P	3P	4P	6P	8P
1 Th	1:56:41	1:04:23	0:41:40	0:31:06	0:23:22	0:20:22
2 Th	1:04:44	0:32:51	0:24:21	0:19:18	0:17:27	0:17:26
3 Th	0:42:15	0:25:13	0:19:21	0:16:34	0:15:15	0:15:56
4 Th	0:33:54	0:19:26	0:14:59	0:13:39	0:13:37	0:13:40

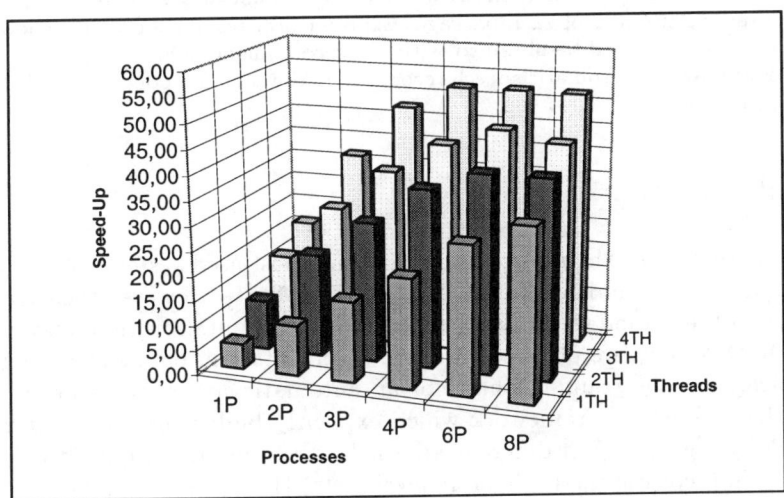

Fig. 1. Speed-up

References

1. Trevelyan, J.: Boundary Elements for Engineers. Computational Mechanics Publications (1994).
2. Al Geist et al: PVM 3 User's Guide and Reference Manual. ORNL/TM-12187, Oak Ridge National Laboratory (1994).

Study of Parallelization of the Training for Automatic Speech Recognition*

El Mostafa Daoudi, Abdelouafi Meziane, and Yahya Ould Mohamed El Hadj

Université Mohammed Ier, Faculté des Sciences
Départ. Maths — Informatique, LaRI, 60 000 Oujda, Morocco
{mdaoudi,meziane,h.yahya}@sciences.univ-oujda.ac.ma

Abstract. In this work we study the parallelization of the training phase for an automatic speech recognition system using the Hidden Markov Models. The vocabulary is uniformly distributed on processors, but the Markovian network of the treated application is duplicated on all processors. The proposed parallel algorithms are based on two strategies of communications. In the first one, called regrouping algorithm, communications are delayed until the training of all local sentences is finished. In the second one, called cutting algorithm, packages of optimal sizes is firstly determined and then asynchronous communications are performed after the training of each package. Experimental results show that good performances can be obtained with the second algorithm.
Keywords: Automatic speech recognition, Markovian modeling, parallel processing.

1 Introduction

At the present time, the most efficient and the most used systems of recognition are based on the Hidden Markov Models (HMM) [1]. However the algorithms relating to these models are very expensive in computation time and memory space. In this work, we propose parallel implementations on a distributed memory machine of the training phase for an automatic speech recognition system using the widespread framework which explicitly builds the global Markovian network by regrouping of different knowledges relating to all linguistic levels of application (acoustic level, phonetic level ...etc) [1].

To our knowledge, a few works related to the parallelization of the speech recognition are proposed in the literature [4, 2, 5].

2 Structure of the Model

A Markovian network of the our application is built in an hierarchical way [1] by calling upon linguistic knowledge structured at various levels. At the syntactic level, the sentence is dealt with like concatenation of models of words. At the lexical level, each word is described by a sequence of phonetic units and handled

---

* Supported by the European Program INCO-DC, "DAPPI" Project

like concatenation of acoustic models. At the acoustico-phonetic level, a Markovian acoustic model is associated with each phonetic unit. By compiling all these knowledges and by connecting the models of the sentences by a common input and output, we obtain the global Markovian model of this application, which will be noted by λ [1]. The common input and output are HMM representing the silence of the beginning and the end of sentences. Once the global model is built, the parameters of the underlying Markovian models are estimated from a training set composed by various pronunciations of each sentence of the vocabulary. The training is done by an iterative re-estimation procedure. It consists in determining, from an initial model λ_0, a new model λ_1 which maximizes the likelihood of the observations conjointly to the optimal path (in the probabilistic meaning). This procedure is repeated until a stop criterion is reached (convergence of the model or a maximum number of iterations). The optimal paths are determined by the Viterbi algorithm [3].

3 Parallelization

We consider a distributed memory architecture composed of p processors numbered by $(P_i)_{0 \leq i \leq p-1}$. Each processor has its own local memory and communicates with the others via an interconnected network. Since the structure of our problem does not impose any particular architecture, we will choose for the theoretical study, the ideal case where all processors are directly connected.

3.1 Parallelization Strategy

We adopt a strategy of duplication of the network. This technique consists in assigning to each processor, a copy of the global Markovian network. The set of m sentences representing the vocabulary is uniformly distributed on the processors. Then, each processor carries out the training of the local corpus composed of $\frac{m}{p}$ sentences, where each one is pronounced n times. Information resulting from the training of each sentence are not directly integrated in the network, but they are combined and stored locally. After simultaneous local training, an exchange between all processors is then carried out to re-actualize the global network. This communication, which is of *all-to-all* type, can be done either after the training of one sentence or a group of sentences. However the best means consists to find an optimal cutting which ensures overlapping of the communications and the computations. Thereafter, we give two parallel algorithms for the training according to different schemes of communications.

3.2 Regrouping Algorithm

In this algorithm the information relating to the local training in one processor are gathered and diffused in the same time to other processors. During this communication, a processor can anticipate the re-estimation of the arc probabilities of the models of its local sentences. The re-estimation of the acoustic laws associated with these models as well as the rest of the global markovian network can only be done after the termination of this communication.

Complexity: Since the global network is duplicated on all processors, the computation time, for one iteration, for the re-estimation remains identical with the sequential case. If we assume that the training cost is independent of the learned sentence, then the computation time, for one iteration, to learn the local sentences is equal to the computation time to learn sequentially all sentences divided by p. The communication time of the *all-to-all* procedure, in a totally connected architecture, is given by $\beta + L\tau$, where β is the startup time, τ is the elementary transfer time of one data and L is the message size. Since the messages are composed by information concerning the use of the transitions and the acoustic laws on the optimal paths, which is not easy to estimate, we only determine an upper bound of the communication times using the maximum number of laws and transitions on an optimal path. If a pronunciation of the sentence i is composed of T_i^j observations, then the optimal path associated with this pronunciation contains, at most, $min(T_i^j, 2N_i - 1)$ different transitions, where N_i is the number of states of the sub-network of this sentence. A transition is characterized by some parameters (its number will be noted by Cte) and by an acoustic law. The acoustic laws, that we use, are multi-gaussian of mean vectors of length μ and diagonal covariance matrices of μ sizes. It follows that the communication time performed after the local training is, at most, equal to

$$\simeq \beta + \left(\sum_{i=1}^{\frac{m}{p}} \sum_{j=1}^{n_i} \left(\left(2\mu + C^{te}\right) \times min(T_i^j, 2N_i - 1) \right) \right) \tau$$

3.3 Optimal Cutting Algorithm

In this algorithm, the corpus of the local training will be subdivided into a set of packages, each one is composed of k sentences. The training is, then, performed package by package. The obtained information by the training of each package will be diffused in an asynchronous way to other processors so that this communication is hided by the training of the next package. The main problem is to determine the optimal size of the packages which ensures a good communications and computations overlapping. Since the determination of the theoretical size of packages requires an exact evaluation of computation and communication costs, which is not easy, we have only determined the experimentally optimal size.

4 Experimentations

The evaluations are done on a vocabulary composed of 50 sentences where each one is pronounced by 6 speakers. These data are recorded at the CNET Lannion with a sampling rate of 16 kHz. We have used the pseudo-diphone as basic unit to build the global Markovian network. The parallel programs are developed under the PVM (Parallel Virtual Machine) environment on the distributed memory parallel machine TN310 composed of 32 T9000-Transputer. In table 1, we report the average execution times, in second, of one iteration of the training algorithms on different processors. It shows the improved sequential execution time. In table 2, we give the computation time, without communication, of the fastest and the slowest processors obtained by execution on 25 processors. The results show a great load unbalance between processors.

Table 1. Average execution time for one iteration of the training

p	Sequential	Regrouping	Cutting
1	117.11		
2		86.41	84.69
5		34.64	33.92
10		19.51	19.17
25		10.72	10.72

Table 2. Average computation time in second of some processors.

Processors	Computation time
P_6	4.61
P_{24}	9.71

5 Conclusion

In this work we have proposed two parallel implementations for the training. The first one gathers the communications and performs only one after the training of all local sentences. In the second one we propose a most elaborated strategy which gives a good communications and computations overlapping. Now we are working to adopt a more elaborated data distribution technique, which takes into account the training complexity of each sentence, in order to reduce the problem of the load unbalance.

References

1. E.M. Daoudi, A. Meziane, Y.O. MOhamed El Hadj, "Parallel HMM Model for Automatic Speech recognition", RR-Lari, Oujda, 1999, Submitted for publication.
2. M. Fleury, A. C. Downton, A. F. Clark, "Parallel Structure in an Integrated Speech-Recognition Network", In Proceedings of Euro-Par'99, pages 995-1004, 1999.
3. D. R. Forney, "The Viterbi Algorithm", Proc IEEE, Vol 61, $n°3$, Mai 1973.
4. H. Noda, M. N. Shirazi, "A MRF-based parallel processing algorithm for speech recognition using linear predictive HMM", ICASSP '94, pages I-597 - I-600, 1994.
5. S. Phillips, A. Rogers, "Parallel Speech Recognition", EUROSPEECH-97.

Parallelization of Image Compression on Distributed Memory Architecture

El Mostafa Daoudi*, El Miloud Jaâra, and Nait Cherif

Université Mohammed 1er, Faculté des Sciences
Département de Mathématiques et d'Informatique
60 000 Oujda, Maroc Morocco
{mdaoudi,jaara}@sciences.univ-oujda.ac.ma

Abstract. In this work we propose two parallel algorithms, for image compression, based on multilayer neural networks, by subdividing the image into blocks. The first parallel technique is based on a static distribution of blocks to processors. The advantage of this distribution is that the training phase (construction of the compressor-decompressor network) does not need any communication but its drawback is the load balancing problem. The second parallel technique improves the load balancing problem by using a dynamic distribution of blocks but it requires communication between processors. These two implementations are tested and compared on a distributed memory machine under PVM.

1 Introduction

The field of image processing has known a great interest during the last years thanks to the large number of applications such as computer vision, medicine,... etc. However, the main difficulty lies in the processing of a great number of information since the digital image is represented by a large quantity of data [1]. Therfore the cost of the transmission and the storage of images are very expensive, that's what the compression techniques are very useful. The digital compression consists of
- describing the image in its more condensed shape by reducing the size of data files which represent the image. This is done by annihilating the redundancy in the information to transmit or to store.
- allowing the reconstruction of image.

2 Image Compression Using Neural Networks

It is shown that the neural networks can be a useful tools for the image compression [1] [3] for their architecture and parallel flexibility [2] [4]. The principle of image compression using neural networks consist in realizing Compressor-Decompressor network (see Fig.1.). It imposes that the input and the output layers must be identical, while the number of cells in the hidden layer is lower than the number of cells in the input one.

---
* Supported by the European Program INCO-DC, Project "DAPPI"

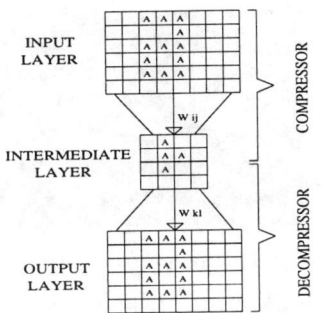
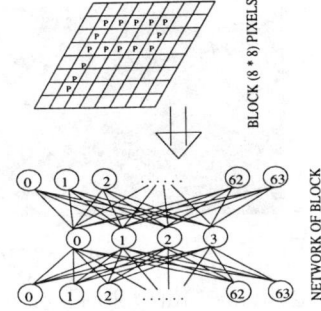

Fig.1. Representation of character a **Fig.2.** Multilayer network

In order to have a high compression ratios, it is necessary to use the same network for compressing several images. However this is limited by the training phase which becomes very expensive (several thousand of iteration) on sequential machine. Therefore, it is necessary to use a parallel computer. The training phase (compression phase) by neural technique consists in subdividing the image into blocks of equal size and then compressing each block independently by a neural network.

The compression is realized block by block. Each block is affected, independently, to one network. In our applications, images are of 304 × 304 pixels and they are subdivided into blocks where each block is of size 8 × 8 pixels [3]. The used network for compressing each block is composed of 3 layers (see Fig.2.) where:
- the input and the output layers are composed of 64 cells which represent a block of 8 × 8 pixels.
- the intermediate layer is composed of $N < 64$ cells. We notice that when the number of cells decreases, the number of iterations increases and the quality of decompressed image decreases.

3 Parallel Training

For the parallelization techniques of the compression pahse:
some blocks are assigned to each processor
processors work simultaneously on local blocks: training for each block
if necessary, exchange of data between processors
In the remaining, we present two parallel techniques based on a static and a dynamic distributions.

3.1 Parallel Static Technique for the Training Phase

Our goal in this parallel implementation is to eliminate the communication. First, image is subdivided into blocks of equal size, then blocks are uniformly distributed on processors. Each processor deals only with its local blocks. The training for local blocks is done sequentially by the same processor. So, when a processor terminates the training of its local blocks, it becomes idle even if there exist blocks in other processors which have not been yet processed.

For our experimental tests, images are of 304 × 304 pixels where each image is

Fig.3. Image before compression **Fig.4.** Image after compression

subdivided into blocks of 8×8 pixels and then we assign B bloks to each processor where B is at most $\lceil \frac{1444}{p} \rceil$ blocks and p is the number of processors. For the parallel implementations we have used the PVM environment on the distributed memory machine $TN310$ composed of 32 processors. Fig.3. and Fig.4. show the original and the obtained images after the compression and decompression. It appears that the two images are identical but in the reality these images present a little difference. We can see some spots (for example in the area [200,250]x[100,150]) which represent the blocks where the training phase has not converged: the convergence has not been obtained by the expected precision but it has been obtained by the imposed maximum number of iterations (100000 iterations).

	number of processors							
	4		8		16		32	
P	B	T	B	T	B	T	B	T
1	360	3463	180	1607	90	773	45	489
2	360	3421	180	1811	90	789	45	391
3	360	3261	180	1615	90	883	45	444
4			180	1794	90	1006	45	390
6			180	1626	90	786	45	389
7			180	1822	90	819	45	526
10					90	995	45	377
11					90	780	45	487
12					90	797	45	389
15					90	963	45	430
17							45	469
18							45	595
19							45	572
22							45	401
23							45	352
24							45	398
28							45	455
31							45	519

Table 1. Times/s of static method

	number of processors							
	4		8		16		32	
P	B	T	B	T	B	T	B	T
1	478	4286	205	1830	97	850	45	414
2	485	4289	207	1835	96	850	47	416
3	477	4285	207	1834	95	847	47	411
4			205	1829	96	847	48	417
6			205	1833	93	846	46	410
7			207	1836	94	854	47	414
10					97	850	46	414
11					97	848	46	413
12					97	851	46	414
15					95	847	46	413
17							46	409
18							46	410
19							48	417
22							44	410
23							45	415
24							46	414
28							47	414
31							45	415

Table 2. Times/s of dynamic method

P	1	2	4	8	16	32
time/s	13342	6981	3587	1824	1009	595
Efficiency	1	0.955	0.929	0.914	0.826	0.700

Table 3. Times and Efficiency of static method

P	2	4	8	16	32
time/s	12901	4289	1836	854	417
Efficiency	0.517	0.777	0.908	0.976	0.999

Table 4. Times and Efficiency of dynamic method

In table 3 we give the efficiency and the execution time obtained for 1000 iterations, for training of two images following the number p of processors. The results show that we obtain good performances but they are different to the expected ones, because, theoretically, we expected that the efficiency will be equal to one, independently of the number of processors, since all processors work independently on the same number of blocks and do not need any communication. The decrease in performances is explained by the load unbalance between processors (see table 1 where T represent the time in second of training for B bloks), even if all processors have treated the same number of blocks. For example with 32 processors, a great difference in execution time between the latest processor (processor 18: 595 seconds) and the faster one (processor 23: 352 seconds).

3.2 Parallel Dynamic Technique for the Training Phase

In order to remedy to the problem of the load unbalance between processors in the previous technique, we propose, in the following a dynamic parallel technique. It is based on the same idea as the static technique, but with a different distribution of blocks to processors. Instead of initially partitioning the bloks uniformly to processors in a static way, we adopt the following distribution:
- first, each processor deals with one block
- then, assign blocks to processors as soon as a processor becomes idle

In order to implement this technique, we assume that each processor dispose of a private memory sufficiently large in order to stock all images of the training basis. This constraint can be avoided by using an adapted environment such Athapascan. The blocks are treated on only $(p-1)$ processors (slave processors) whereas the master processor imposes, by sending a message, to each processor which becomes idle the number of the block to be treated. On the other hand each processor, as soon as it becomes idle, informs the master of its state. Table 2 shows that good improvements of the load unbalance between processors are obtained for large number of processors. It also shows that the blocks are not uniformly distributed on processors.

In table 4 we report the execution times (by 1000 iterations) obtained in seconds for the training of two images follwing the number of processors as well as the efficiencies calculated for different processors. The efficiency is calculated with regard to the number p of processors, knowing that only $(p-1)$ processors which treat the blocks whereas the master processor orders to other processors the number of each block which will be treated. To this effect, we remark that the efficiency is more better when the number of processors increases.

4 Conclusion

In this work we have proposed two parallel techniques for image compression using neural networks. The first one is based on a static distribution of blocks to processors and does not need any communication. The drawback of this distribution is the load unbalance between processors, even if all processors treat the same number of blocks, since the training time of each block is different. The second implementation, more difficult than the first one, allows to improve the load unbalance by assigning blocks to processors in a dynamic manner.

References

1. O. ABDEL-WAHHAB et M.M. FAHMY, " Image Compression using Multilayer Neural Network " *IEEE Proc - Vis. Image Signal Process, Vol. 144. N^0. 5. October 1997*
2. E.M. DAOUDI et E.M. JAARA, " Parallel Methods of Training for Multilayer Neural Network " *Euro-Par'99, Lecture Notes in Computer Science 1685, 1999*.
3. H. NAIT CHARIF, " A Fault Tolerant Learning Algorithm for Feedforward Neural Networks " *Conférence FTPD 1996, Hawaï*.
4. H. PAUGAM-MOISY, "Réseaux de Neurones Artificiels : Parallélisme, Apprentissage et Modélisation " *Habilitation à Diriger des Recherches, ENS-Lyon, 1997*.

Parallel DSMC on Shared and Hybrid Memory Multiprocessor Computers

Gregory O. Khanlarov, German A. Lukianov,
Dmitry Yu. Malashonok, and Vladimir V. Zakharov

Institute for High Performance Computing and Data Bases
P.O. Box 71, St.Petersburg 194291, Russia
{greg,luk}@fn.csa.ru, mal@csa.ru, zvv@fn.csa.ru
http://www.csa.ru

Abstract. A new algorithm of two level parallelization for direct simulation Monte Carlo is elaborated for solving unsteady problems of molecular gasdynamics on shared and hybrid memory multiprocessor computers. The first parallelization level (parallel statistically independent runs) is implemented with the aid of MPI library. For the second level (data parallelization) the standard UNIX interprocess communications are employed. Two versions of static load balancing are used for the second level. The study on speedup is carried out by solving one typical test problem of molecular gasdynamics. The computation is performed on shared memory HP/Convex SPP-1600 (8 processors) and hybrid memory PARITET (4 two processor nodes).

1 Introduction

There are quite a lot of parallel algorithms for the direct simulation Monte Carlo (DSMC) (see, for example [1, 2]). The majority of these algorithms use domain decomposition (DD) technique. The main problems of DD implementation are load imbalancing and increasing number of links between subdomains as the number of subdomains increases. The existing load balancing (LB) for DD algorithms are not efficient for solving unsteady gasdynamic problems. The present work being the continuation of [3, 4] uses different parallel algorithms designed specially for solving unsteady problems.

The direct simulation Monte Carlo [5] of unsteady flows requires several statistically independent runs to perform in order to get the necessary sample size. The work [4] describes the two level parallel (TLP) algorithm designed for solving unsteady gasdynamic problems. The first level corresponds to parallel execution of statistically independent runs. Inside each run the data parallelization (DP) is employed — this is the second level. This algorithm can be used only on shared memory computers.

At present there is a tendency towards building clusters of SMP-machines, PCs, or workstations [6]. The advantage of this approach is that a cluster can be constructed with mass produced equipment providing high cost-performance parameter.

Recently it is observed wide usage of cluster type hybrid memory systems (several shared memory multiprocessor nodes). Such computers ideally suit for implementation of TLP algorithms for which inter-node communications are rare and the parallelization inside each node is implemented with the aid of shared memory model. The present paper describes a hybrid two level parallel algorithm (HTLP) for DSMC of unsteady flows which can be implemented either on shared memory computers or on hybrid memory ones.

The study on speedup and efficiency is carried out on HP/Convex SPP-1600 and new cluster system PARITET. HP/Convex SPP-1600 is the SMP system with 8 processors sharing 2 GB of RAM. The PARITET[1] system is a new computer designed by specialists from the Institute for High Performance Computing and Data Bases, Russia. PARITET being the MPP-system allows to increase the number and power of computing nodes. The nodes of the system are connected via high-speed communication environment Myrinet. The configuration of each of 4 nodes is 2 Intel Pentium II-450 MHz processors sharing 512 Mb of RAM.

2 Parallel Algorithm

The HTLP algorithm is designed in such a way that single statistically independent runs are executed on single nodes (the first parallelization level). The communications between runs are quite rare. The MPI tool is employed for the first level implementation.

The structure of DSMC method has two consecutive stages at each time step: particle motion and collision simulation inside each computational cell. Inside these stages the data used for the simulation process is comparatively independent. To speedup the computation of a single run the DP algorithm is employed inside each run (the second level) [4]. The standard UNIX interprocess communications (IPC) technique is used for the implementation. The principal requirement of DP algorithm is the presence of shared memory processors which compute the specific run. For DP algorithm it is used two versions of static load balancing (LB).

The first version of LB equally distributes all the tasks among processors which are responsible for a single run. The task at the first stage is the simulation of particle movement, the second stage is the modeling of collision process in a cell. The distribution of tasks is carried out according to the processor's unique number, the data located in shared arrays is selected every other interval equal to the number of processors. The drawback of such LB scheme is that in some problems (for instance, expanding flows) the time required for every subsequent cell processing becomes less if the numbering of cells is performed in direction of flow expansion (the number of collisions decreases). Hence, the total computational time is maximal for the first processor and is minimal for the last one.

The second version of LB also equally maps all the tasks among processors but the distribution of tasks is carried out randomly by shuffling the data arrays.

---
[1] http://www.csa.ru/CSA/MICRO/mikro1.html

In this case the task numbers are not periodic. This version of LB makes it possible not to take into consideration the flow structure.

The presented LB schemes provide approximately equal processor load if the average time required for task processing is approximately the same.

3 Results

The unsteady monatomic gas outflow from the sonic nozzle into vacuum is chosen as a test problem. The Knudsen number defined by the gas quantities at the nozzle edge is 0.05. The problem has axial symmetry. The rectangular grid having 1816 cells is superimposed on the computational domain being 10 × 10 nozzle diameters in size. The number of particles in the computational domain varies from 0 (at the initial moment) to 100000. The number of time steps is 300. The number of statistically independent runs is 8.

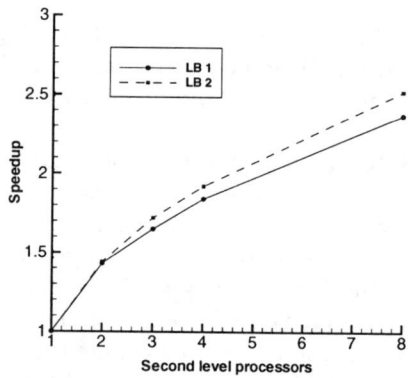

Fig. 1. Speedup on the second level for two versions of LB v.s. number of processors

The study on speedup for the second level of HTLP algorithm with two versions of load balancing is carried out on SPP-1600 computer (the number of runs is equal to unity, the number of second level processors is varied from 1 to 8). The speedup as a function of the number of processors is presented in the fig. 1. The second version of LB gives higher speedup (approximately 6% increase running on 8 processors).

The table 1 gives execution time (CPU time) of the test problem for two versions of LB schemes running on two computers. The examination of the tables clearly suggests that for all cases the execution time on the PARITET computer is 2.7–3 times less than that on SPP-1600. The PARITET computer is approximately 30 times cheaper than SPP-1600 computer. Due to this fact the cost-performance parameter of PARITET is 90–100 times lower than that of

Table 1. CPU time in seconds for two versions of LB schemes (N is the number of the first level processors, n is the number of the second level processors)

Computer	$n \backslash N$	LB 1				LB 2			
		1	2	3	4	1	2	3	4
SPP	1	1936	970	732	491.5	1940	974.5	735	493.5
PARITET		706	359	268.5	—	702	358	268.5	—
SPP	2	1352	668	499.5	336.5	1348	667	498.5	341
PARITET		452	222	166.5	—	450	220.5	165.5	—

SPP-1600 in the whole range of presented computations. The estimation of this parameter is given as the product of the computer price and CPU time. Thus, it has been shown that the new cluster-type PARITET system is more preferred than SPP-1600 for solving unsteady problems of molecular gasdynamics with the help of the HTLP algorithm.

The study was carried out in the Center of Supercomputing Applications[2] of the Institute for High Performance Computing and Data Bases under support of the Russian Foundation for Basic Research (project code 99-07-90451).

References

1. Ivanov M., Markelov G., Taylor S., Watts J. Parallel DSMC strategies for 3D computations. Proc. Parallel CFD'96. North Holland, Amsterdam, 1997, pp.485–492.
2. Wilmoth R.G. Adaptive Domain Decomposition for Monte Carlo Simulations on Parallel Processor. In. 17th Rarefied Gas Dynamics Symposium. AIAA, July 1990.
3. A.V.Bogdanov, N.Y.Bykov, G.A.Lukianov. Distributed and Parallel Direct Simulation Monte Carlo of Rarefied Gas Flows. Lecture Notes in Computer Science, Vol. 1401. Springer-Verlag, Berlin Heidelberg New York (1998)
4. A.V. Bogdanov, I.A. Grishin, Gr.O. Khanlarov, G.A. Lukianov and V.V. Zakharov. Algorithm of Two-Level Parallelization for Direct Simulation Monte Carlo of Unsteady Flows in Molecular Gasdynamics. Lecture Notes in Computer Science, Vol.1593. Springer-Verlag, Berlin Heildelberg New York (1999).
5. G.A.Bird. Molecular Gasdynamics and Direct Simulation of Gas Flows. Clarendon Press. Oxford. 1994
6. P.Kuonen, R.Gruber, A. De Vita, P.Volgers. Parallel computer architectures for commodity computing. HPCN Europe 99. Late Papers. pp. 1-10.

---

[2] http://www.csa.ru/CSA

Population Growth in the Penna Model for Migrating Population

A.Z. Maksymowicz[1], P. Gronek[1], W. Alda[2], M.S. Magdoń-Maksymowicz[3],
M. Kopeć[1], and A. Dydejczyk[1]

[1] Department of Physics and Nuclear Techniques, AGH
Mickiewicza 30, 30-059 Kraków, Poland
[2] Institute of Computer Science, AGH
Mickiewicza 30, 30-059 Kraków, Poland
[3] Department of Mathematical Statistics, Agriculture University
Mickiewicza 21, 31-120 Cracow, Poland
amax@agh.edu.pl, Phone: (048-12)617 2974

Abstract. We present computer simulation of population evolution at different locations, with migration between the sites. Calculations are based on the Penna model, with suitable modifications for the migration. We present some examples of the population growth for different scenario of the migration rules. Calculations requires about $100MB$ memory for 10^6 population which is a minimum necessary to get reliable statistics. Typical running time for 3000 iteration steps is several hours for HP S2000 machine. The problem is very suitable for parallelization with geometrical decomposition, especially for small migration limit.
Keywords: Population growth, Penna model, migration, computer simulations

1 Introduction

The standard Penna model [1,2] of the population growth may be considered as extension of the classical model of evolution presented by the logistic iterative equation

$$x(t+1) = (1+B) \cdot x(1-x), \qquad (1)$$

for normalized population $x(t) = n(t)/N$ at time $(t+1)$, where the right hand side variables are at time t. Here n is the actual population and $n < N$, the environmental capacity N. Logistic equation describes evolution rule for $t \to t+1$ iteration cycle: first eliminate fraction $x = n(t)/N$ of the population, then allow for B offsprings from each individual.

Penna model assumes that elimination of an individuals may also result from too many bad mutations already activated at given individuals age. Each individual gets an inherited *genome* which is a computer word (integer *genome*), storing information in each bit position on presence (bit '1') or lack (bit '0') of bad mutation, a disease. When getting older in one iteration step, the individual goes from age a to $a+1$ and the next bit is disclosed. In the Penna model we kill an individual with *active* mutations above a threshold T. The survival gives birth to B babies if the reproduction age R is reached. The offspring is then offered a copy of *genome* from the parent, enriched by M additional mutations randomly picked over its whole lifespan.Thus the model has 5 input parameters (N, B, R, M, T) that control population structure $n(t, a, \mu)$, the number of members at time t, of age a and number of active mutations μ. We get $n(t, a, \mu)$ from simulation. From this we may extract required information such as age structure of the population, survival rate etc. In the next section we propose model that allows for migration between locations.

2 Model

As example we consider migration between three locations labeled by i. After carrying out given number S of iterations, we stop evolution and allow for migration. The picked out items go into the buffer first, followed by further move into their new locations. This completes *one cycle* when entering new era $(t+1)$, with new $n_i(t+1)$. For simplicity, we use same set of model parameters for each site. Then we expect all $x = n/N$ same at all sites, even if $N's$ are different, for isolated locations, or for neutral migration of the random character. For non-random migration we may get new results for the equilibrium population n_i after many cycles. Preferences are defined by the probability $p(i,j)$ of a transfer from location i to j and proportional to a mobility coefficient q_i which may be different for different locations i. The model is described more in detail in our earlier papers [3, 4]. In the following section we discuss results for some scenario of such non-random migration.

3 Results and Conclusions

In calculations we used the growth rate $B = 0.5$, mutation rate $M = 1$, threshold $T = 3$ and the minimum reproduction age $R = 4$. In each case we aimed for population n of order of 10^6 or so on a 32 bit machine to gain a sensible statistical accuracy. Therefore we applied N_i of the right order. The model was tested for $q = 0$ and for random migration. Calculations were carried out for clustering and anti-clustering tendencies, this tendencies are defined by choice of probability $p(i,j) = q(i) \cdot p(j)$.

In figure 1 an example of stationary population $x = n/N$ versus mobility q is shown for three locations. The three sites have different environmental capacities N, $N_1 : N_2 : N_3 = 1 : 2 : 3$, yet we assume more intensive migration from sites of smaller population n. As result, even if we start simulation from same population density x (following same set of the evolution parameters at all sites), the tendency is that sites of smaller population would evolve to still decrease population there. This tendency is overcome if the reproduction rate is

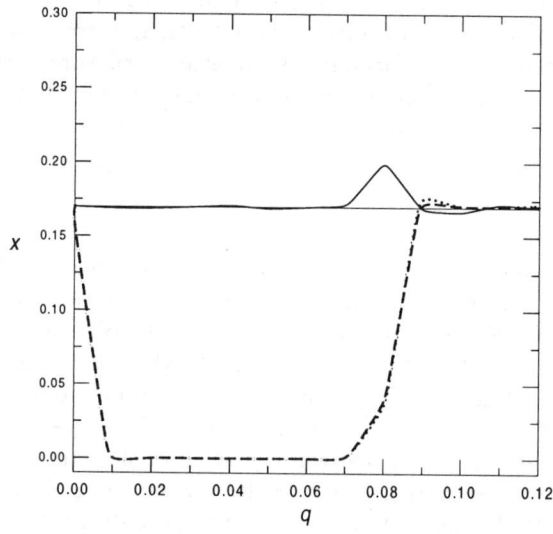

Fig. 1. Normalized population $x_i = n_i/N_i$ at 3 locations *versus* mobility q.

sufficiently high to upkeep the population, perhaps at smaller level x. If, however, larger is the mobility q forcing intensive migration to escape small sites, they may become totally deserted.

With increasing migration mobility $q_i = q$, we get 1-st order transitions of population on one of the locations and population there drops to zero. The transition from all 3 occupied sites to 2 occupied ones, 3→2, takes place at $q = 0.001$. Further increase in $q = q_i$ leads to another instability of the 2 nearly equally occupied locations. In one of them the population drops to zero as it is expected for the assumed scenario. This is the 2→1 transition at $q = 0.005$. Migration becomes so intense that small locations are completely deserted by the outgoing migrants at one iteration step, yet at the same time this very location is flooded by migrants from outside, entering presumably for short only. As result, we gain a sort of dynamical equilibrium. This is the last transition 1→3 near $q = 0.07$.

Apart from many other scenario for migration preferences and/or limitations for the health (number of mutations) or age or other factors, migration between locations may be used as a nice example to evaluate timing of such calculations in the parallel version of calculations. We did appropriate statistics for program developed within the MPI packet. All calculations were performed for typical set of model parameters as discussed above, and in the limit of small migration mobility q. We run program for different number of locations (up to four), different number of processors assigned to each location and suitably chosen to guarantee reasonable balance. We deliberately choose parameters such as population distribution among locations, number of iteration and others to extract from the CPU times important components of the timing. We were able to evaluate the total time (summed over all processors) of the population growth itself of order of 1.87 seconds per iteration cycle and independent of the number of processors. The time required to launch communications links proved to be proportional to the number of processors. Then time for passing over items between processors was again split into component responsible for a preparatory action for the transmission, independent of the number of migrants, and part directly proportional to this number. Actually, the last component was controlled by the migration mobility parameter q, and we get this component of the CPU time for migration proportional to $q^{3/2}$.

We conclude that migration between locations is important and modifies population growth. Model is much suitable for geometrical decomposition of the task for parallelization.

Acknowledgments. This work was partly supported by a grant for cooperation with University of Cologne, and we are grateful to Prof. Stauffer for his help and guidance. Main simulations were were carried out at the Academic Computer Center CYFRONET-KRAKÓW on HP Exemplar S2000.

References

1. Penna, T.J.P., A Bit-String Model for Biological Ageing, *J. Statist. Phys.* **78** (1995) 1629.
2. Penna, T.J.P. and Stauffer, D., Efficient Monte Carlo Simulations for Biological Ageing, *Int. J. Mod. Phys.* **C6** (1995) 233.
3. Magdoń, M.S. and Maksymowicz, A.Z., Penna model in migrating population - effect of environmental factor and genetics, *Physica*, **A273** (1999) 182-189.
4. Magdoń, M.S., Effect of migration on population dynamics, *Int. J. Mod. Phys.*, **10** (1999) 1163-1174.

Use of the Internet for Distributed Computing of Quantum Evolution

Alexander V. Bogdanov, Ashot S. Gevorkyan,
Armen S. Grigoryan, and Elena N.Stankova

Institute for High-Performance Computing and Data Bases,
194291, Russia, St-Petersburg, P/O Box 71
bogdanov@hm.csa.ru

Abstract. The experiment of complex physical problem solution on two remotehigh-performance systems is described. The new algorithm of the quantum evolution computation made it possible to organize computation on a cluster with three subsystems and low exchange of information between them.

1 Introduction

Up to now the calculation of the evolution of the complex quantum systems was out of the possibilities even for the most powerful supercomputers. The problems were connected both with the great number of equations and with the strong coupling of the equations resulted in large dimension of the problem, with the elements of large values far from the diagonal in the operator matrix. In the Institute of High Performance Computing and Data Bases (IHPCDB) new algorithm was proposed [1,2] which enables to solve this problem by making the evolution operator of the problem almost diagonal, that is the parallel algorithm of the problem solution was created [3]. However even the computing facilities of IHPCDB are not sufficient for the computation of all processes interesting for the problem applications.

Existing agreement between IHPCDB and GMD (German National Research Center for Information Technology, Bonn, Germany) makes it possible to joint in the regime of distributed calculations the computing facilities of those centers, comprising in one computational cluster more than 120 nodes. Practical realization of such clustering is provided by means of special tools of PVM (Parallel Virtual Machine) and MPI (Message passing interface).

M. Bubak et al. (Eds.): HPCN 2000, LNCS 1823, pp. 592-596, 2000.
© Springer-Verlag Berlin Heidelberg 2000

2   Algorithm Description

So, we propose the new approach to the solution of complex evolution equations on supercomputers and computer clusters. The main idea of the approach is to use the intrinsic properties of the pertinent physical system to make the evolution operator of the problem almost diagonal [4]. For that purpose we introduce the functional transformation of the physical variables of the system, that turns it into a set of weakly coupled equations that are afterwards solved by means of some iteration - perturbation method [5].

The proposed approach makes it possible to increase drastically the speed of computations by decreasing the exchanges between processors, to increase the dimensionality of the problem and its scalability on many processor systems. In such a way we introduce some highly scalable algorithms for several industrial and scientific applications.

Let us consider the proposed algorithm using the problem of quantum multi-channel scattering taking into account dynamic quantum chaos and stochastic processes.

Schematically the algorithm of such process computation can be presented in the following way [3]:

I - Lagrangian surface construction for the system. The curvilinear coordinate system, within which all the further calculations are performed, is derived in it;

II - Classical trajectory problem solution. At this stage the system of four ordinary non-linear differential equations of the first order is being solved numerically. The problem's parameters are collision energy E and oscillation quantum number of initial configuration n. This system is being solved by one-step method of 4th-5th order of accuracy. This method is conditionally stable (by initial deviation and right part), that's why the standard automatic step decreasing method is implied to provide its stability. It's worth mentioning that initial system degenerates in certain points. To eliminate this degeneration, the standard -procedure with differentiation parameter replacement is performed.

III -- The results of classical trajectory problem calculation are used for quantum calculations performing and complete wave function obtaining in its final state. At this stage, the numerical problem represents solution of an ordinary non-linear differential equation of the second order. Calculating this equation is a difficult task due to non-trivial behavior of differentiation parameter [6]. Differentiation algorithm consists of two stages: 1) construction of differentiation parameter values system using the results of classical problem calculation and 2) integration of initial differential equation on non-uniform system obtained by means of multi-step method. Choosing such integration step in a classical problem provides integration stability, while control is performed by means of step-by-step truncation error calculation. The obtained solution of differential equation is approximated in a final asymptote in a form of falling and reflected flat wave superposition;

IV - The results of quantum problem solution are used for obtaining the values for matrix elements of transitional probabilities of a reaction and their corresponding reaction sections. Calculation of matrix elements for initial oscillation quantum

number n and final oscillation quantum number m is performed with the use of expressions presented in [2]. Let's note that transition probability matrix obtained corresponds to one value of collision energy, stipulated at stage II;
V -velocity constant calculation. At this stage, the values for reaction
truncation matrix calculated for different collision energies are integrated
by Maxwell's distribution.

Let's remind that calculations for steps II and III are made for specific values of collision energy E and oscillation quantum number of initial state. Results of these calculations allow to obtain one line of a reaction truncation matrix, which corresponds to n. In order to obtain the entire truncation matrix, calculations at stages II and III need to be repeated as many times as dictated by the size of transitional probability matrix. As a result the entire probability matrix is obtained. The procedure described needs to be repeated for many values of collision energy in order to enable further integration and velocity constants finding.

It is clear, that most time consuming are the stages II and III and that they can be carried to large extend on independent computational systems, using one of them just to collect all the results and work out the statistics. Since from each of such computation we needed only the value of the kernel of transition functional it was possible to exchange such information over the regular Internet. All the computations of the stages IV and V were carried out on MPP system Parsytec CCe-20 of IHPCDB and the individual computations for different trajectories on MPP IBM SP-2 of GMD. We found, that MPP architecture, although old-fashioned, is very well suited for the architecture of proposed algorithm.

The parallelization was performed for the values of collision energy. Calculation of classical trajectory problem, quantum calculation and transition probability matrix calculation is performed in each of the parallel branches. Let's note that just as in the case on non-parallelized algorithm all calculations from stages II and III are performed as many times as it is dictated by the size of transition probability matrix. Due to the fact that calculation in each of the parallel branches represents a separate problem and does not interact with other branches of calculation, the effectiveness of using this parallelization algorithm vs. relatively unparallelized algorithm is nearly proportional to a number of calculation branches, i.e. to the amount of computation nodes.

As a reaction on which the algorithm was tested, a well studied bimolecular reaction Li + (FH) (LiFH)* (LiH) + H was taken. The results of testing have shown the calculation effectiveness to be nearly proportional to the number of computation nodes. We have proposed the variant of our approach for the shared memory systems. However now we have no technical possibilities to unite in large clusters systems with shared memory emulation in the regime of NUMA architecture. But this problem solution is one of the main items in the program of joint activities with GMD in the nearest years Finally we would like to stress one of the peculiarities of parallelization algorithms demonstrated - their scalability. Due to the fact that integration of transition probability matrix and rate constants calculation during stage V requires the values of matrix elements for many energy values, one can hardly find a supercomputer with an excessive number of computation nodes. As illustration we show first exact converging results of computation of reaction probability and properties of the system.

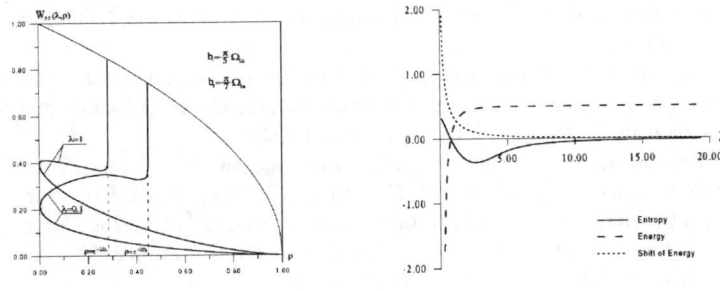

Fig. 1. The results of the first exact computation of the probability dependencies for reaction
Li + (FH) (LiFH)* (LiH) + H

3 Conclusion

Scalability of these algorithms was used for conducting distributed computing runs on the supercomputers of GMD and SARA. Internet possibilities allowed to obtain access for most difficult part of the problem – trajectories calculations - to the far more powerful computing resources that are available in IHPCDB, and so to conduct distributed computations in the different regimes including X terminal regime. In the future we are planning to provide in the similar approach the visualization of the numerical results, data preparations and preliminary tests for remote compilation on the cluster of workstations Octane and Sun Ultra, available in IHPCDB and sending them for further processing to the ONYX visualization supercomputers situated in GMD and SARA.

References

1. Bogdanov A.V. Calculation of the Quntum-Mechanical Inelastic Scattering Amplitude through Solution of the Classical Dynamical Problem. Sov.Phys.Tech.Phys. 31(7), 1986, pp. 833-834.
2. A.V. Bogdanov, A.S. Gevorkyan, G.V. Dubrovsky, Theoretical Quasi-Classical Analytical Approximation of the S-matrix for the Three-Particle Collinear Rearrangement Reaction, and Mathematical Physics, 1996, v.107, No.2, pp.609-619.
3. A.V. Bogdanov, A.S. Gevorkyan, A.G. Grigoryan and S.A. Matveev, Investigation of High-Performance Algorithms for Numerical Calculations of Evolution of Quantum Systems Based on Their Intrinsic Properties in Proccedings of 7th Int. Conference on High Performance Computing and Networking Europe (HPCN Europe '99), Amsterdam, The Netherlands, and April 12-14, 1999, pp.1286-1291.
4. Bogdanov A.V., Gevorkyan A.S.: Reactive Scattering in the Three-Body System as Imagining Point Quantum Dynamics on 2-D Manyfolds, Proceedings of the International Workshop on Quantum Systems, Minsk, Belarus, June 3-7, 1996, pp.34-40.
5. A.V. Bogdanov, A.S. Gevorkyan, A.G. Grigoryan, S.A. Matveev, Internal Time Peculiarities as a Cause of Bifurcations Arising in Classical Trajectory Problem and Quantum Chaos Creation in Three-Body System, in Proceedings of Int. Symposium "Synchronization, Pattern Formation, and Spatio-Temporal Chaos in Coupled Chaotic Oscillators", Santyago de Compostela, Spain, June 7-10, 1998.
6. A.V. Bogdanov, A.S. Gevorkyan, A.G. Grigoryan, First principle calculations of quantum chaos in framework of random quantum reactive harmonic oscillator theory, in Proceedings of 6th Int. Conference on High Performance Computing and Networking Europe (HPCN Europe '98), Amsterdam, The Netherlands, April, 1998.

Debugging MPI Programs with Array Visualization

Dieter Kranzlmüller, Rene Kobler*, Rainer Koppler, and Jens Volkert

GUP Linz, Johannes Kepler University Linz
Altenbergerstr. 69, A-4040 Linz, Austria/Europe
kranzlmueller@gup.uni-linz.ac.at
http://www.gup.uni-linz.ac.at:8001/

Abstract. Debugging message passing programs is accepted as one of the major difficulties of parallel software engineering. Besides various problems with communication and synchronization of concurrently executing processes, one of the big obstacles is the amount of data that is processed in parallel applications. Yet, inspecting these data is a basic necessity to verify the correctness of a program. Therefore the MAD environment includes an array visualization component, which displays arbitrary arrays distributed on parallel processes in MPI programs. Although the current implementation restricts arrays to HPF-like distributions, the usefulness of this first prototype already indicates how vital such a visualization feature can be for parallel program debugging.

1 Introduction

Debugging a program means to gather knowledge about its execution in order to detect and correct errors. Thus, a basic principle of debugging is to stop the program at specified breakpoints during execution and inspect the state of the processes and the values of their data [6]. While this is relatively easy for fundamental data types (e.g. char, int, float), evaluating derived types like arrays or complex data structures may impose severe problems. These problems scale with the amount of data processed by the target application, and are thus generally difficult in the parallel computing domain. Thus, there is a certain demand on debugging tools to support users in inspecting such data types.

Some solutions for this problem are provided for data-parallel programs. For example, the Connection Machine debugger *Prism* offers data visualization capabilities for programs implemented in CM Fortran [1]. Similarly, *PDT* (Parallel Debugging Tool) offers displays for distributed data values and their locations for HPF programs (High Performance Fortran) [3]. Another comparable system in this area is *DAQV*, which provides a framework for accessing data of HPF programs during runtime [4]. Finally, the Graphical Data Distribution Tool *GDDT* combines sophisticated facilities of a data distribution tool - including both regular and irregular distributions - with inspection of in-core and out-of-core data, again for applications written in HPF [2].

---
* Presenting author

So far, all these approaches have been based on data-parallel HPF-like languages. The reason for this is the way those languages treat distributed arrays in general. Firstly, the user specifies only directives on how the distribution is done during runtime. The actual distribution is then carried out by the compiler and its corresponding runtime system, respectively. Secondly, HPF data distributions are restricted to a limited number of possibilities. For instance, distributions may be partitioned in blocks or cyclic, with different variations and combinations for the dimensions of the data array.

In contrast to HPF, programs based on the MPI (Message Passing Interface) standard are much more flexible. In fact, there is no restriction of any kind on how to distribute the data onto the MPI processes of a program. Consequently, a one-fits-all solution for visualizing distributed MPI data may not be feasible. Yet, for certain applications a valid and useful solution can be provided.

In this paper we present such an approach for visualizing distributed arrays within the Monitoring And Debugging environment *MAD* [5]. The idea of this array visualization component is based on the fact, that some of HPF's data distribution schemes are equally important for the average MPI programmer. Actually, for regular problems many MPI applications implicitly adopt a HPF-like distribution, simply because it is well-suited for these particular problems. Thus, our distributed array visualization component is currently restricted to HPF-like data distributions, although there are some ideas for more convenient and flexible distribution schemes.

2 Array Visualization in MAD

The array visualization in MAD is applied with three steps, (1) instrumentation, (2) monitoring, and (3) visualization. The first step is the instrumentation phase, where the user has to decide which data to analyze and when to generate this data during the target program's execution. Upon deciding these two questions, the user places the following monitoring macro at the corresponding position in the source code of the program.

monARRAYTRACE(arrAddress,arrInfo)

During execution of the program, the monitor gathers all the needed data whenever this macro is executed. The description of the data is determined by the two parameters `arrAddress` and `arrInfo`, where the former contains the address of the array in memory, while the latter contains information about the array itself. This array information describes, how the global array is distributed onto the participating processes and thus consists of element size and type, number of dimensions, size of the global array, size of the local portion on each process, number of processes, type of distribution for each dimension (NONE, BLOCK, CYCLIC), overlapping information, and possible remainders.

After the data has been generated by the monitor, it can be visualized within the debugging environment (currently only post-mortem analysis is supported).

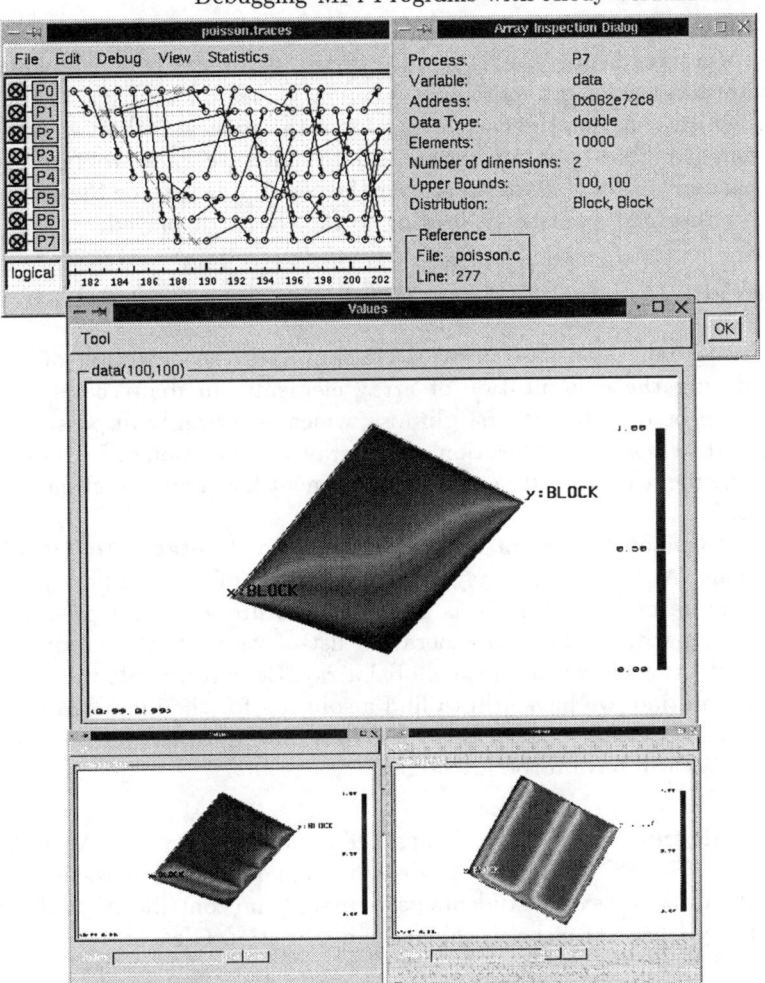

Fig. 1. Debugging session with distributed array visualization

The screenshots in Figure 1 show an example debugging session in MAD[1]. In the top-left corner is an event graph window, which serves as the main debugging interface and displays the communication pattern of the MPI processes. The highlighted symbol across all processes indicates that array data has been traced during execution. When selecting this symbol, the top-right array inspection window is opened, which displays textual information about the array contents, like data type, number of elements and dimensions, and the distribution (in this case BLOCK, BLOCK).

In addition to the textual information, the user may open array visualization windows as shown in the three screenshots in the middle and bottom of Figure 1.

---

[1] The original colors have been replaced by different shades of grey and the two lowest windows have been down-scaled by 50 percent due to printing matters.

All these windows display the data of the distributed array from a global point of view, with different colors indicating different values. In contrast to the middle window, which contains the correct results of the program, the two windows at the bottom of Figure 1 contain the same data structure for a program with erroneous behavior. In both examples, either the communication or the computation has been performed incorrectly, leading to the effects displayed.

3 Conclusion and Future Work

The array visualization system of MAD simplifies the detection of errors occurring during the computation of array elements. In many cases, erroneous behavior can be detected at first glimpse, which is certainly impossible without tool support. Thus, the integration of an array visualization system in a debugging environment represents a vital improvement for error detection in parallel programs.

The problem of the current implementation is its limitation to HPF-like data distributions. Although this may be sufficient for many parallel applications, seeking a more general solution is one of our future goals. At present, we are experimenting with an idea to generate a list of values that contains the data together with their location in the global array. However, while this is certainly a possible solution, we have still to find a solution for the increased complexity of the instrumentation task in this case. In addition, other derived data types besides arrays will have to be investigated.

Acknowledgements. This work has been sponsored by the Austrian FWF project P11157-TEC "Integrating Visualization into Parallelization Environments". In addition, several students participated and contributed to this project, most notably Christian Schaubschläger and Christian Glasner.

References

1. D. Allen, R. Bowker, K. Jourdenais, J. Simons, S. Sistare, R. Title, *The Prism Programming Environment* Proc. Supercomputer Debugging Workshop '91, Albuquerque, NM, USA, pp. 1-7 (Nov. 1991)
2. P. Brezany, P. Czerwinski, K. Sowa, R. Koppler, J. Volkert, *Advanced Visualization and Data Distribution Steering in an HPF Parallelization Environment* World Scientific, (1999).
3. Ch. Clemencon, J. Fritscher, R. Rühl, *Visualization, Execution Control and Replay of Massively Parallel Programs within Annai's Debugging Tool*, Proc. of HPCN'98, Amsterdam, Netherlands, (1998).
4. S.T. Hackstadt, A.D. Malony, *DAQV: Distributed Array Query and Visualization Framework*, Special Issue on Parallel Computing, Vol. 196(1–2), pp. 289–317 (1998).
5. D. Kranzlmüller, S. Grabner, J. Volkert, *Debugging with the MAD Environment*, Parallel Computing, Vol. 23, No. 1–2, pp. 199–217 (Apr. 1997).
6. J.B. Rosenberg, *How Debuggers Work: Algorithms, Data Structures, and Architecture* John Wiley & Sons, New York (1996).

An Analytical Model for a Class of Architectures under Master-Slave Paradigm

Yasemin Yalçınkaya and Trond Steihaug

University of Bergen
Department of Informatics
Bergen, Norway

Abstract. We build an analytical model for an application utilizing master-slave paradigm. In the model, only three architecture parameters are used: latency, bandwidth and flop rate. Instead of using the vendor supplied or experimentally determined values, these parameters are estimated using the analytical model itself. Experimental results on Cray T3E and SGI Origin 2000 indicate that this simple model can give fair predictions.

While building a performance model, it is crucial to catch the main factors of behavior of the program in question. These factors are the parallelization strategy used, the amount of communication and computation, and the architecture of the parallel computer. The software employed combined with the chosen message passing paradigm plays a significant role in the effective values of the architecture parameters. A promising approach is to build a simplified model for a "real" application on a target architecture. The purpose of this paper is to build an analytical model to predict the behavior of iterative numerical algorithms on a class of architectures using master-slave paradigm. Under the master-slave paradigm, the execution of the parallel program can be seen as a sequence of parallel and purely sequential phases. In the parallel phase the slaves compute concurrently and in the sequential phase only the master does computation. Between these phases there is communication between the master and slaves either in form of single node *broadcast* from the master, or single *sends* and *receives* between the master and any one of the slaves. We are going to analyze one of many possible parallel programs under this paradigm to show that in certain cases, it is possible to quantify the influence of the factors mentioned above on the performance, and a small number of architecture parameters can be defined to be used in an analytical model. In order to estimate the architecture parameters and their interactions we will make some assumptions.

The iterative algorithm used in the analysis is a block Jacobi algorithm for the solution of linear least squares problems outlined by Dennis and Steihaug [2]. One important aspect of this algorithm is that the amount of communication and computation changes with different values of an algorithmic parameter p [4].

For the performance analysis of the algorithm we need to estimate the run time of one loop and the total number of loops. The run time of one loop, t_{loop}, is estimated by two "components", computation and communication time.

When the number of processors at hand is smaller than the number of tasks, we assign more tasks to each slave. Assuming that when a slave finishes sending its result to the master the data is received instantaneously by the master and the slave continues with its next task until it runs out of tasks, t_{loop} is:

$$t_{loop} = \max_{\forall \text{ slaves}} \{ \sum_{i=1}^{k_{slave}} (t_{c\_slave}(i) + t_{s\_send}(i)) \} + t_{c\_master} + t_{broadcast}, \quad (1)$$

where $t_{c\_slave}(i)$ is the computation time of task i on *slave*, $t_{s\_send}(i)$ is the time to send computation results from *slave* to the master, t_{c_master} is the computation time on the master, $t_{broadcast}$ is the time used by the master to *broadcast* data to the slaves and k_{slave} is the number of tasks assigned to *slave*. The computation time is assumed to be proportional to the number of floating point operations (flops). The communication time is assumed to consist of two parts: a startup time (latency) and a part proportional to the amount of data sent. The proportion factor depends on the number of processors in use and the intercommunication topology.

To analyze the computation time, we count the number of flops, additions and multiplications, for each task. We assume that the time it takes to do a multiplication operation is equal to the time for an addition operation. The values of t_{c_master} and t_{c_slave} are estimated by multiplying the respective flop counts with the inverse of the flop rate of the architecture in use.

In the implementation, MPI is employed as the message passing library. When the slaves begin sending their results, the master is already ready to accept. We assume that the time used by the master to retrieve the arrived data from the buffer is negligible.

Let us define $t_{send} = l + \tau\beta\pi$, where τ is the message size, $\beta = 8/\text{BW}$ is the transfer time per byte, l is latency, and π is a variable depending on the number of hops between the sender and receiver. BW is the bandwidth of the parallel computer architecture and 8 is the number of bytes in one data element. For simplicity, bandwidth is taken as constant in the model. When the number of slaves is only one, we assume that the master and slave processors are neighbors, and $\pi = 1$. On SGI Origin 2000 the interprocessor communication network has hypercube topology and on Cray T3E, each PE has 6 neighbors [4]. The topological structure of the network, hence the number of network links traversed by a message, is critical. When the number of slaves is increased we do not know the topology of the partition assigned to our program but we will assume that the partition is "dense". We can approximate the average distance between any two processors by $1/2 \log_2 n_s$ on Origin 2000, and with $3/4 n_s^{1/3}$ on Cray T3E, where n_s is the number of slaves. Hence, π is $1/2 \log_2 n_s$ and $3/4 n_s^{1/3}$ [1] on Origin 2000 and Cray T3E respectively when $n_s > 1$. Single node *broadcast* operation in MPI is implemented using a binomial tree-structure approach [3]. Therefore, the *broadcast* time is proportional to $\log_2 n_s$.

We use only three parameters in the analytical model: flop rate, latency and bandwidth. However, instead of using the vendor supplied or experimentally

Table 1. Estimated values of architecture parameters

architecture	l	BW	β	α
SGI Origin 2000	22.69 μs	13.38 Mbyte	0.59 μs	7.42 ns
Cray T3E	74.16 μs	16.54 Mbyte	0.48 μs	8.67 ns

determined values, these parameters are estimated using the analytical model itself. The variable π is not considered as a parameter since it is determined by the topology of the communication network. To estimate the values of architecture parameters, we measure the execution time of an actual implementation on increasing problem sizes using one slave. Using only one slave enables us to avoid the effect of other interference from the system. Execution time on different problem sizes reflect the possible change in bandwidth. We take algorithmic parameter p to be zero. The implementation is run for a couple of times for different problem sizes, and the average of execution times for each problem is taken. We calculate the number of flops, count the number of *send*, *receive* and *broadcast* operations and estimate the latency, bandwidth and flop rate of the architectures in question as the solution of a least squares problem.

The architecture parameters estimated to be used in the model are given in Table 1. The parameter α in the table is the time to do one flop. To validate the model we look at the difference between actual execution time measurements and model predicted values. In the figures, increased test problem number means increased problem size [4]. Figures 1 and 2 depict the model validation for execution times on Cray T3E and SGI Origin 2000. We see that on Cray T3E, the predictions are quite accurate, whereas on SGI Origin 2000, when the problem size is increased we are slightly overestimating the execution time. The average deviation between the actual execution times and model predictions is 8.6% of the actual execution times on SGI Origin 2000, and 6.0% on Cray T3E.

Figure 3 displays the predicted versus actual execution time on Cray T3E using two slaves and $p = 1$. When $p = 1$ the amount of communication remains the same as when $p = 0$ but the computation on slaves is increased.

In Fig. 4, we predict the execution time of the application with a different choice of p, $p = Cs$ using 4 slaves on SGI Origin 2000. We see that for the two largest test problems the overestimation observed in the model is magnified. There are two main differences between this case and the model problem: computation on the slaves is increased along with the size of the messages sent by the slaves, and the number of the slaves is quadrupled. These size increases in the system results in increased error in the prediction.

The model can also be used in the analysis of scalability of the application [4]. The estimated parameters do not reflect the architectural characteristics only. The network load, the message passing paradigm used, the program code, the size of the problems are all factors that cause changes in the vendor supplied values for the parameters. This implies that the "estimated" latency is higher than the "machine constant latency". Similarly, the bandwidth is lower [4].

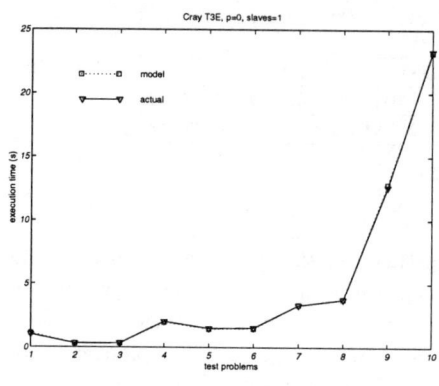

Fig. 1. On Cray T3E, using one slave, $p = 0$

Fig. 2. On SGI Origin 2000, using one slave, $p = 0$

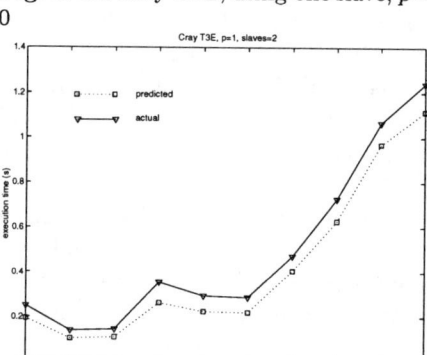
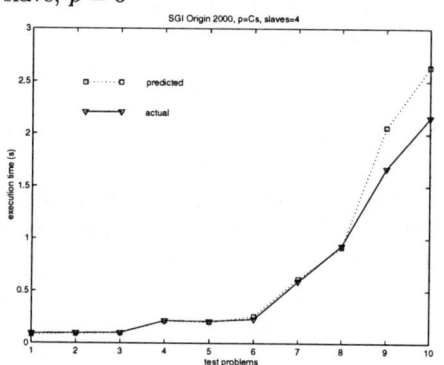

Fig. 3. On Cray T3E, using two slaves, $p = 1$

Fig. 4. On SGI Origin 2000, using 4 slaves, $p = Cs$

References

1. Culler, D. E., Karp, R., Patterson, D., Sahay, A., Schauser, K. B., Santos, B., Subramonian, R., von Eicken, T.: LogP: Towards a Realistic Model of Parallel Computation. In: Proceedings of the Fourth ACM SIGPLAN Symposium on Principles and Practice of Parallel Programming. ACM (1993) 1–12
2. Dennis, J. B. Jr., T. Steihaug, T.: A Ferris-Mangasarian Technique Applied to Linear Least Squares Problems. Tech. Rep. No. 150. Dept. of Informatics, Uni. of Bergen, Norway (1998)
3. Touriño, J., Doallo, R.: Performance Evaluation and Modeling of the Fujitsu AP3000 Message-Passing Libraries. In: Amestoy, P. et. al. (eds.): Euro-Par '99. Lecture Notes in Computer Science Vol. 1685. Springer Verlag, Berlin Heidelberg New York (1999) 183–187
4. Yalçınkaya, Y., Steihaug, T.: Are Three Parameters Enough to Represent a Parallel Computer Architecture?. In: Sixth Meeting of the Nordic Section of the Mathematical Programming Society. Proceedings. Opuscula; 49. ISSN 1400-5468, Mälardalen University, Västerås, Sweden (1999)

Dynamic Resource Discovery through MatchMaking

Omer F. Rana

Department of Computer Science
University of Wales, Cardiff, POBox 916, Cardiff CF2 3XF, UK
o.f.rana@cs.cf.ac.uk

Abstract. A decentralised resource management technique is described that makes use of 'resource capabilities' and 'task requirements' to find suitable allocations. The technique extends the 'classified advertisement' approach adopted within the CONDOR [3] system, to include resource usage history and variable resource granularity. Our approach can deal with a dynamically changing environment, which could include new tasks created at run time, or new devices added to existing clusters. A managed cluster can consist of identical or heterogeneous resources, with usage dictated by a local usage policy. The approach makes use of various algorithms that use similarities between devices to find approximate matches. Each resource administrator is required to describe the resource using a policy description scheme, which is subsequently used by a MatchMaker to find suitable allocations.

1 Introduction and Related Work

The resource management problem in the simplest form consists of, (1) selecting a set of resources on which to execute jobs generated from an application, (2) mapping tasks to computational resources, (3) feeding data to these computations, (4) ensuring that task and data dependencies between executing tasks are maintained. Resource selection or discovery generally involves identifying suitable computational engines from a pool, mostly homogeneous, based on criteria ranging from licensing constraints to processor(s) capability(ies) and background workload. In task-parallel programs, different tasks may need to be mapped to resources, whereas in the data parallel case, data decomposition becomes significant. Existing resource management systems, such as the Load Sharing Facility (LSF), for instance, involve a queuing facility to which application tasks are submitted. Such systems are primarily aimed at managing a homogeneous cluster, rather than a heterogeneous resource pool. In addition, the process of identifying a suitable queue to which tasks must be submitted is delegated to the user (either directly or via a job control language). A good overview of such cluster management systems can be found in [1]. The proposed approach uses the object-oriented description mechanism in Legion [2], but is most closely related to the 'class advertisement' mechanism in CONDOR [3].

2 Proposed System

Let $\Re$ be a set of resources, and $\Im$ be a set of tasks, where: $\Re(t) = \{R_1(t), ..., R_n(t)\}$, $\Im = \{T_1, ..., T_m\}$. Each T_i can be obtained from a task graph, or may be specified directly by the application developer. Every $R_j(t)$ specifies the state of resource 'j' at time 't'. Associated with each T_i is an execution time, such that the total execution time of an application T_{App} can be specified as: $T_{App} = \sum_{i=1}^{k} T_i$, then the objective of the resource manager is to find a mapping function Φ, such that $\Phi : T_i \mapsto_{min(T_{App})} R_j(t), \forall (i,j), t > 0, (1 < i < m), (1 < j < n)$.

We propose a de-centralised mechanism, which can cater for variable 'n' and 'm', based on the notion of a MatchMaker. Each member of the set $\Re(t)$ is wrapped as an object, with the state of the object at time 't' accessible via an interface, and methods identifying services that can be invoked on the object by a user. In this sense, the approach is similar to the technique employed in Legion. Each R_j sends an asynchronous message to a pre-defined matchmaking service 'M' (running on a given host) to indicate its availability within a cluster. Each message is tagged with the resource type: (1) computational resource 'C', (2) data storage resource 'D', (3) visualisation resource 'V', or (4) scientific instrument 'I'. The message contains no other information, and is sent to the local 'M'. On receiving the message, the local 'M' responds by sending a document specifying the required information to be completed by the resource manager at R_j. This information is encoded in an XML document, and contains specialised keywords that correspond to dynamic information that must be recorded for every device in the pool. The document also contains a time stamp indicating when it was issued, and an IP address for 'M'. The document can be automatically completed using scripts, or it can be completed manually by a systems administrator. The manager for R_j completes the document, and sends it back to 'M', maintaining a local copy. The document contains the original time stamp of 'M', and a new time stamp generated by R_j. Some parts of the document are static, while others can be dynamically updated. Once this has been achieved, the new device is now registered with the resource manager, and will continue to be a suitable candidate for task allocation until it de-registers with 'M'. If the device comes off-line or crashes, 'M' will automatically de-register it when it tries to retrieve a new copy of the document. We define each resource document as $D_\Re$. Similarly, a user wishing to execute an application issues a request document based on requirements of each task T_i within the application, and classed using the 'C','V','S','I' annotation to the matchmaking service 'M' within the local cluster. This results in a set of documents being sent to the user, for each T_i in the application. The user has complete control over the granularity of T_i, and tasks may be grouped based on known dependencies. Each document must now be completed by the user, either using pre-defined scripts or manually. The issued document contains a time stamp, and on subsequent return to 'M', contains a time stamp from the user. We define each task document as $D_\Im$. 'M' now tries to find a match between each $D_\Re$ and $D_\Im$, based on pre-defined matchmaking criteria. Each time a suitable match is achieved, 'M' sends the generator of D_{T_i} and D_{R_j} their corresponding identities. The matched participants must now activate a separate protocol to complete the allocation, and this process does not involve 'M'. The matched R_j must now de-register itself, or request a new D_{R_j}. If a local 'M' within a cluster cannot fulfill a request based on the submitted $D_\Re$, then it can forward this request to an 'M' within another cluster. The matchmaking services are therefore federated, and register with each other using a pre-defined document D_M, which identifies their IP address and start time.

2.1 System Architecture

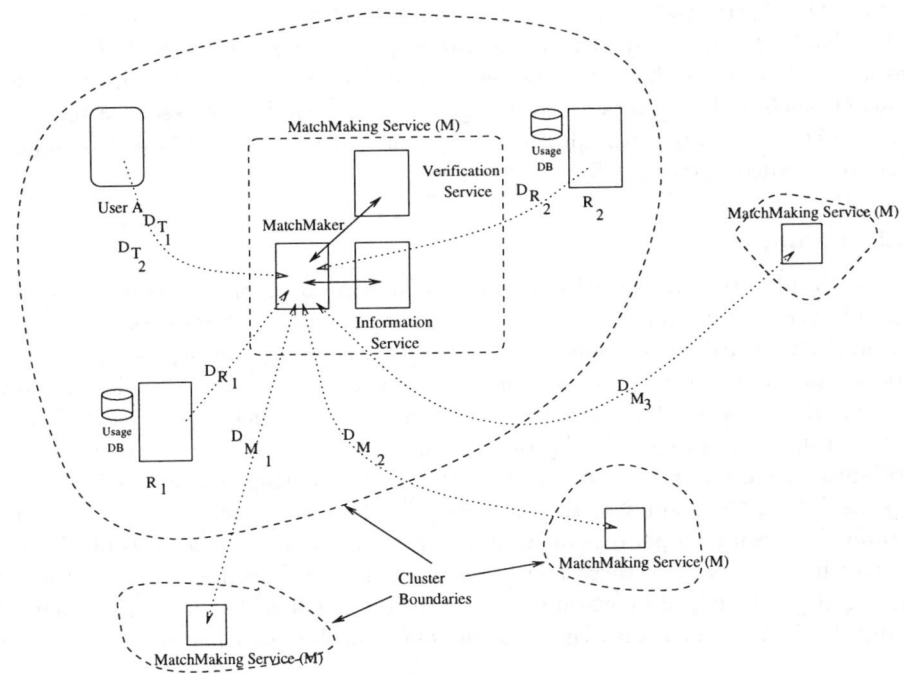

Fig. 1. *MatchMaking architecture*

Figure 1 illustrates the MatchMaking architecture, comprising of a single MatchMaker within a cluster. 'M' consists of three core components, (1) an information service, (2) a verification service, (3) the matchmaking service itself. The information service is responsible for obtaining dynamic parameter values within documents, and can interact with a resource or a user to obtain these parameters. At any given time, the final version of a document is always maintained with the MatchMaker, and the information service merely acts to facilitate the gathering process. The verification service is used to check information maintained on a given resource by invoking the information service, and is used to check submitted documents to ensure that all necessary information has been supplied. As illustrated, a user or resource can submit multiple documents, corresponding to differing granularities within the application or the resource.

The verification service contains an XML parser and a Document Type Definition (DTD) for D_R, D_T and D_M. All submitted documents can be verified prior to processing by the MatchMaker. Further, modifications may be made to any of these documents, and the next resource to register will get the newer version of the document, with a change in the document version number sent to the resource or task.

Each resource can also store a usage history in a local database, based on the $D_\Re$ schema, at intervals determined by the resource administrator. In addition, based on a local policy, an administrator may refuse to record certain parameters within $D_\Re$. MatchMakers in different domains/clusters can interact with each other, and only do so if locally generated tasks cannot be executed on local resources. The usage history database can also be used to maintain composite metrics, such as load averages over a given time period. The resource manager can either extend $D_\Re$ with additional tags, or query the MatchMaker to supply a more detailed document for completion.

2.2 Example

The example uses an execution environment composed of 4 workstations connected over an Ethernet, running a backpropagation neural network. The workstations share disk space, and have a local cache each. The neural network is divided into four tasks: (1) conditioning data prior to processing, which requires a data file to be read from a local disk and each numeric value scaled from 0.0 to 1.0, (2) running a backpropagation neural network in batch mode, (3) post-processing the results of the execution, and writing output back to disk, (4) displaying the results to the user. Hence, there are explicit scheduling constraints between the operations. We use a syntactic match to find suitable devices on which to run each of these tasks, and identify relationships between tasks to ensure that precedence constraints are satisfied. Each of these tasks is defined using the XML model, task three for instance can be described as:

```
<task>
<name short="neural3">parian.cs.cf.ac.uk/neural3</name>
<type compound="false">C</type>
<owner>scmofr</owner>
<executable value="ref://parian.cs.cf.ac.uk/neural/post.exe">
                                        post.exe</executable>
<arguments type=output value="file"
    location=ref://parian.cs.cf.ac.uk/neural/data/results.txt" />
<mmtimestamp>14.01.2000.10.15</mmtimestamp>
  <submittimestamp>15.01.2000.10.22</submittimestamp>
  <docversion>1.0</docversion>
  <constraint>
    NOT (<task name="neural2"> || <task name="neural1">)
  </constraint>
</task>
```

In this case, the `constraints` tag suggests that if documents for tasks `neural2` or `neural1` exist in the MatchMaker, then the `constraint` evaluates to False. This is therefore an alternative way to cater for data dependencies when an application has been divided into sub-tasks. A prototype using JKQML [4] has been implemented.

3 Conclusion and Future Work

An approach based on MatchMaking is described for dynamic resource discovery and management, based on the use of XML documents for describing resources and tasks. MatchMaking services can be federated, which enables the exchange of documents across cluster boundaries, where local resources are unable to meet demand. Each MatchMaker has complete control over the parameters that need to be specified for each resource, however, the MatchMaker does not participate in the subsequent scheduling and allocation process once a suitable match has been found. Each resource administrator can also provide a usage policy based on specialised tags within a document. The complexity of the MatchMaking can vary, and is based on a similar theme to that employed by web based search engines. The approach is also quite general, and can be implemented within existing meta-computing systems such as Globus.

References

1. M.A. Baker, G.C. Fox, and H.W. Yau. Review of Cluster Management Software. *NHSE Review*, 1(1), May 1996.
2. A. S. Grimshaw. Campus-Wide Computing: Early Results Using Legion at the University of Virginia. *Int. Journal of Supercomputing Applications*, 11(2), 1997.
3. Raman R, M. Livny, and M. Solomon. Matchmaking: Distributed Resource Management for High Throughput Computing. *Proceedings of the Seventh IEEE International Symposium on High Performance Distributed Computing*, july 1998.
4. IBM Research. A KQML implementation in Java, 1998. See web site at: http://www.alphaworks.ibm.com/. Last visited: February 2000.

A New Approach to the Design of High Performance Multiple Disk Subsystems: Dynamic Load Balancing Schemes

A.I. Vakali, G.I. Papadimitriou, and A.S. Pomportsis

Department of Informatics
Aristotle University of Thessaloniki, Greece
{avakali,gp,apombo}@csd.auth.gr

Abstract. The performance of storage subsystems has not followed the rapid improvements in processors technology, despite the increased capacity and density in storage medium. Here, we introduce a new model based on the idea of enhancing the I/O subsystem controller capabilities by dynamic load balancing on a storage subsystem of multiple disk drives. The request servicing is modified such that each request is directed to the most appropriate disk drive towards servicing performance improvement. The redirection is performed by a proposed algorithm which considers the disk drive queues and the disk drives "popularity". The proposed request servicing has been simulated and the load balancing approach has been shown quite effective as compared to conventional request servicing.

1 Introduction

Modern I/O subsystems are equiped with efficient policies such as scheduling, reordering of I/O requests or read-ahead. The management of a multiple disk subsystem is usually governed by a controller which is responsible for request servicing and storage system functionality. Nowdays most controllers in disk subsystems come with self-managing techniques for request servicing and are easily adapted standard systems without major software modifications [1].

We propose a load balancing approach towards reducing the difference between processor speed and disk servicing time. A similar model was proposed in [5] where the request redirection has been proven quite beneficial. Here, we formulate and extended the earlier approach by proposing an effective load balancing scheme under a specific probability updating scheme. The next section presents the multiple disks I/O storage subsystem model whereas Section 3 introduces the proposed load balancing approach Section 4 presents the simulation results and conclusions are summarised in Section 5.

2 The Storage Model

One of the most common memory models is the hierarchical memory model proposed in [6] where an abstract machine consists of a set of processors interconnected via a high-speed network and each processor access an appropriate

I/O controller. Each of these I/O controllers manages a set of disk drives [2]. The storage subsystem comprises of D individual and independent disk drives. Disk drives have associated queues that contain requests waiting to be serviced. Requests arrive to the system randomly by various independent processes. The total service time of a request in the disk mechanism is a function of the seek time(ST), the rotational latency(RL) and the transfer time(TT) whereas queue delay must be considered also for the evaluation of the overall service time [3, 4]. The most widely used formula for evaluating the expected service time involves these time metrics and it is expressed by :

$$E[ServiceTime] = E[ST] + E[RL] + E[TT] \qquad (1)$$

where $E[ST]$ refers to the expected seek time, $E[RL]$ refers to the expected rotational delay and $E[TT]$ refers to the expected transferring time. Formulae for these times have been given in [3, 4].

3 The Dynamic Load Balancing Approach

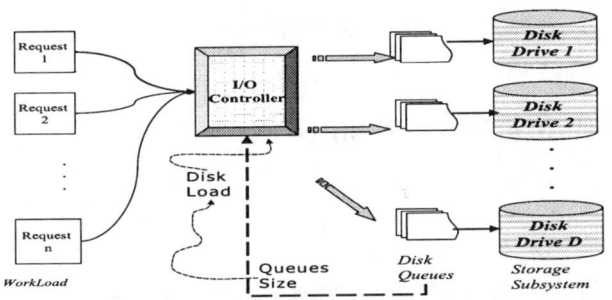

Fig. 1. The Dynamic Load Balancing Model.

In our approach the controller doesn't remain a static design where data are placed as directed by the file system. Instead the controller becomes a dynamic tool which efficiently re-directs the requests to the physical medium according to the load information. Figure 1 presents the structure of our dynamic load balancing model which alters the conventional multiple disk drives model accordingly. Writing is performed by indirecting request pattern within the controller to allow data relocation such that the service time is improved. At each time slot t, the dynamic load balancing based controller contains a probability distribution $P(t)$ over the set of disks. Thus, $P(t) = \{p_1(t), \ldots, p_D(t)\}$, where D is the number of disks. The probability distribution $P(t)$ is updated at each time slot by taking into account the estimated load of each disk drive. The load is estimated by using the disk queue length since the queue size is an indicative metric for each disk load.

Definition : The load $L_i(t)$ of disk d_i $(i = 1, \ldots, D)$ is evaluated in by :

$$L_i(t) = \max\{Q_i(t), \epsilon\} \qquad (2)$$

where $Q_i(t)$ is the number of requests waiting in the queue of disk d_i, while ϵ is a positive real number in the neighborhood of 0.
The probability $p_i(t)$ of redirecting a write request to a disk d_i is inversely proportional to the load $L_i(t)$ of this disk.
At any time slot t, for any two disks d_i and d_j, we have:

$$\frac{p_i(t)}{p_j(t)} = \frac{L_j(t)}{L_i(t)}$$

since the choice between two disks i and j for servicing a write request will be made according to their loads $L_i(t)$ and $L_j(t)$. The disk with the heavier load has smaller probability than the other disk with the lighter load. Since at any given time slot t, it holds $\sum_{i=1}^{D} p_i = 1$, we are expressing all p_is $(i = 1, 2, \ldots, D)$ in terms of one specific disk, namely the i disk drive. From the above, we derive that the choice probability of each disk d_i $(i = 1, \ldots, D)$ will be :

$$p_i(t) = \frac{\frac{1}{L_i(t)}}{\sum_{k=1}^{D} \frac{1}{L_k(t)}} \qquad (3)$$

4   Experimentation - Results

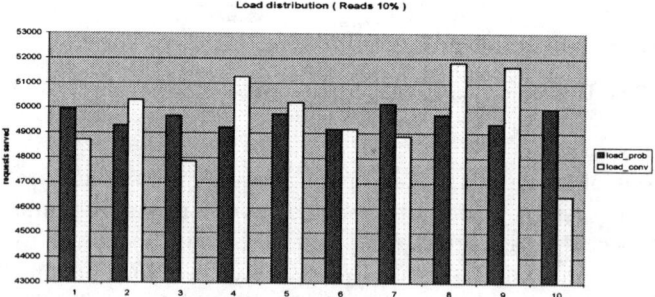

Fig. 2. Load distributions, 10% reads (Probabilistic versus conventional).

In order to study the performance of the proposed model we have simulated the conventional model and two types of models for the dynamic load balancing model, namely the deterministic and the probabilistic model. The simulation model was studied for an I/O subsystem of $2, 4, \ldots, 10$ disk drives. Each disk is configured by the characteristics proposed in [2, 3] for the *HP 97560* disk drive. The read/write ratio is a parameter varying in the range $0.1, \ldots, 0.9$. The proposed model has been showed to be beneficial in all cases when compared to the

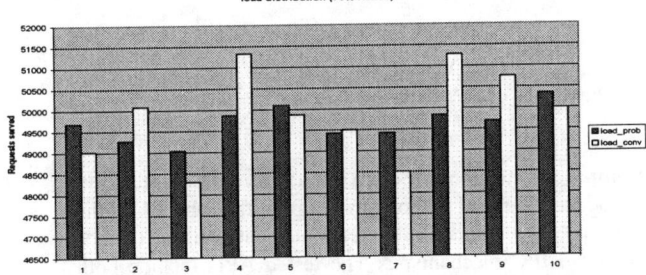

Fig. 3. Load distributions, 70% reads (Probabilistic versus conventional).

conventional I/O servicing model. Figures 2 and 3 depict the load distribution under indicative loads serviced by both the conventional and the proposed dynamic load balancing probabilistic model. Figure 2 depicts the load distribution for a parallel I/O subsystem of 10 disk drives, under both the proposed and the conventional model under 0.1 read ratio and Figure 3 presents the same bars for a 0.7 read ratio. These histograms depict the load variation for each of these models and they represent the beneficial load distribution of the proposed dynamic load balancing model.

5 Conclusions

The presented paper provided a new I/O servicing model in a parallel multiple I/O subsystem. The proposed model have introduced a request servicing redirection, based on disk queue information used as the load estimation and characterization metric. The redirection concerns write requests and the I/O controller is responsible for the new model implementation. Simulation runs for heavy disk loads have been presented and indicative results are demonstrated.

References

1. English R., Stepanov A. : Loge : A Self-Organizing Disk Controller. HP Labs, Technical Report, **HPL-91-179**, (1991).
2. Ruemmler C., Wilkes J.: 'An Introduction to Disk Drive Modeling, IEEE Computer, **27** (1994) No.3, 17-28.
3. Shriver E. : Performance modeling for realistic storage devices, Department of Computer Science, New York University, **Ph.D. Thesis** (1997).
4. Shriver E., Merchant A. and Wilkes J. : An Analytic model for disk drives with readahead caches and request reordering. ACM SIGMETRICS'98, Conference Proceedings, (1998) 182-191.
5. Vakali A.I., Papadimitriou G. I. : An Adaptive Model for Parallel I/O Processing. Proceedings of the IASTED International Conference, Parallel and Distributed Computing and Systems, **PDCS 99** (1999) 139-142.
6. Vitter J., Shriver E.: Algorithms for Parallel Memory I,II, Department of Computer Science, Brown University, Technical Report **CS-90-21** (1990).

Embarrassingly Parallel Applications on a Java Cluster

Brian Vinter
University of Southern Denmark

Abstract. This paper proposes a new benchmark suite for use with research in Java for HPC, and introduces the first set of applications that are all of the embarrassingly parallel type. The applications are implemented using threads and four kinds of networked IPC mechanisms. The tested APIs include both native Java NIPC's and interfaced to native code NIPC's. The results show that there is huge differences in the scalability of the tested API's, even for embarrassingly parallel applications.

1 Introduction

There is currently a large interest in Java for high performance computing, especially as a language base for cluster architectures. Java has a wide array of attractive features for HPC, among the more important are the facts that Java is designed for communication, has a multi-threaded model at its core, is easy to debug and is designed for code reuse.

A large number of cluster communication mechanisms have already been proposed in the spawn of a few years. There are interfaces to native communication API's; mpi-Java[1] and jPVM[2], Java implementations of NIPC's such as JPVM[3] and RMI[4] and new and interesting API's such as JavaSpaces[5] and TSpaces[6]. All of these API's are described in the literature, but are rarely compared to each other or described using metrics that allow potential users to decide which API would be the fastest for their use.

Thus a benchmark suite is proposed. The goal of this work is to provide a benchmark suite that is representable for a large variety of parallel HPC applications. The applications should be well enough documented and small enough, that researchers can easily port the suite to a new API and thus easily compare their work to that of others.

DANISH

The proposed benchmark suite is named **DANISH**, **D**anish **A**pplications for **N**etworked **I**PC and **SH**ared memory. It is important to stress that DANISH is not designed to benchmark Java Virtual Machines, nor is it designed to test the performance of physical machines. The idea behind DANISH is to provide means to easily test parallel programming API's for Java, whether they are targeted for distributed or shared memory execution. The primary measure for DANISH will be scalability rather than actual execution time or some measure of computations per timeframe. Speedup is chosen to allow for the best possible comparisons between published results, across different platforms.

2   Results

This section presents the results that were achieved from running the embarrassingly parallel section of the DANISH benchmark suite, on a CLUMP. All numbers are the mean of five executions to ensure stability, all speedups is calculated from comparing execution time to that of the sequential version.

The test machine is a cluster of 16^1 dual Pentium III, 450 MHz nodes, each with 128-MB memory. The machines are connected via Fast-Ethernet through a switch with a back-plane capacity of 2.1 Gb/sec, e.g. enough to service all NICs at the same time.

The JVM used is IBM's Java for Linux version 1.1.8 and the OS is Linux 2.1.15.

Monte Carlo PI

The Monte Carlo Pi simulator is a well-known parallel test-application. The idea is to let a square of side-length two be inscribed by a unitary circle. One then takes a large number of darts and throws them at the square, hitting random positions within the square, the fraction of the darts that hit within the circle is then one-fourth the value of Pi.

The DANISH implementation of the Monte Carlo Pi throws 10^8 simulated darts towards the square, and the paralleization of this process is done by simply splitting the darts amongst the workers and then collect the number of hits from all workers into a global sum. The Monte Carlo PI is the simplest of the applications, and with hardly any data being communicated one would expect linear speedup with this application.

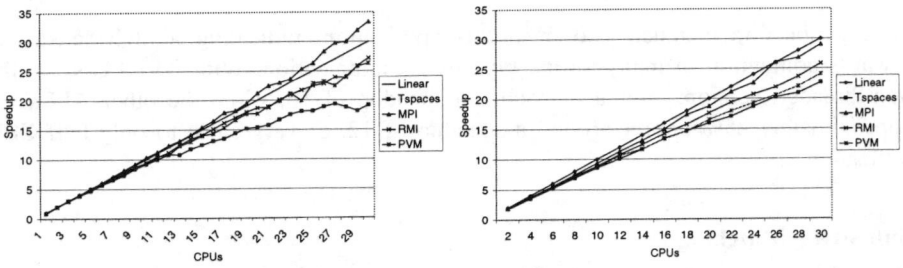

Fig. 1. Speedup of Monte Carlo PI; (a) one process per CPU, (b) one process per node

As figure 1 (a) shows, all APIs provide good speedup until approximately 10 CPUs, at this point RMI and PVM starts to loose ground while TSpaces start to perform significantly poorer than one would expect. The mpi-Java API behaves rather strangely, and shows signs of super-linear speedup from the beginning and significant super-linear speedup when using more than 16 CPU. While this behavior is hard to

---

[1] Unfortunatly one node is currently down, limiting results to 30 CPUs.

explain precisely, it is the result of a positive interference between the multi-threaded MPICH application and the multi-threaded JVM, which somehow ends up with better locality for the JVM itself within the dual CPU nodes. This can be verified by the results of running the mixed NIPC and multi-threaded versions, where only one process is running on each node, but with two worker-threads in each, one for each CPU.

Mandlebrot

The execution of the Mandlebrot benchmark turned out to introduce problems for some of the testes API's. JPVM became very unstable and TSpaces as well as RMI had problems related to memory, due to the nature of the models, in which the complete rendered picture, with size 25MB, is stored in an object handler and in the master process. This was handled by having the master process and the object server/TSpaces server run on one node and workers on other nodes. Figure 3 does not reflect this, e.g. with RMI and TSpaces there was always one more node used than reflected in the graph. However since this is due to a memory problem and is not likely to skew performance significantly to the advantage of the two API's, the CPU's on this host-node is not included in the CPU-count.

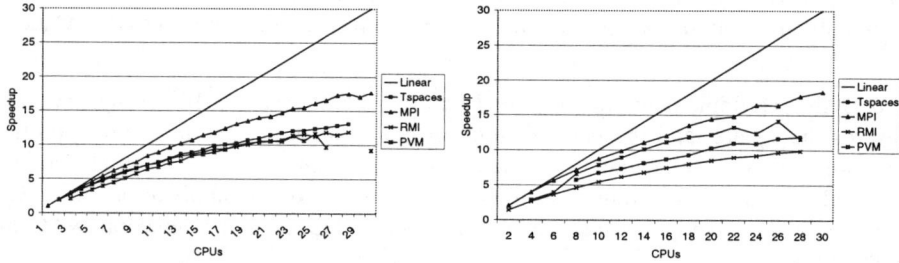

Fig. 2. Mandlebrot benchmark performance; (a) one process per CPU, (b) one process per node

Overall speedup is much poorer in this application, indicating a high sensitivity towards contention in many-to-one type of communication. With 25 CPU's, on the non-threaded version, mpi-Java reaches a speedup of 16, while the other API's are approximately 25% poorer around a speedup of 12, or very close to only half CPU utilization.

Sub-string Matching

The sub-string matching application seeks to locate the best-fit position of a given sub-string within another, much longer, string. The actual string matching is trivial, e.g. test if each character in the source string matches a position in the target string. The Sub-string matching allow us to test against contention with one-to-many type of communication. The test-data is approximately 1/3 the size of the result data from the Mandlebrot application, and are now being distributed, rather than collected. The sequential execution time is similar to the sequential versions of the previous two applications, approximately 90 seconds on our equipment.

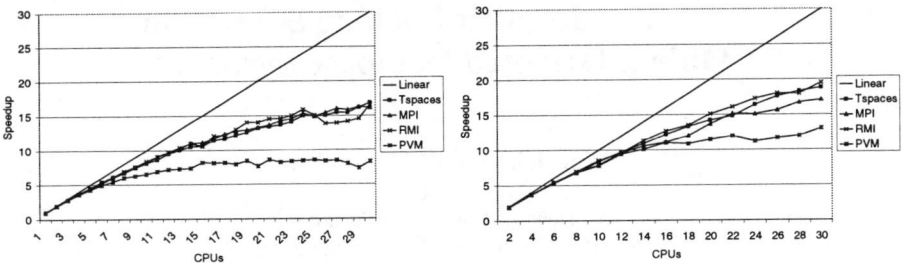

Fig. 3. Sub-string matching speedup; (a) one process per CPU, (b) one process per node

In the non-threaded version, TSpaces follows the linear speedup line more closely in this example, but hits a saturation point of speedup 15 around 20 CPU's after which point there is no significant speedup. Both MPI and RMI follow the same behavior very closely, while PVM performs very badly with this one-to-many scheme, reaching a saturation point of a speedup of 8 with 16 CPUs.

3   Conclusions

This part of DANISH found that even for absolutely embarrassingly parallel applications, with a minimum of communication, linear speedup could only be attained by one of the tested API's, mpi-Java. The remaining API's did well too though, except for TSpaces and RMI which could not break the speedup of 25 barrier.

Once real communication is introduced, the picture changes and speedup is no longer close to linear.

Only one of the three applications, the sub-string matching, shows an advantage from using multi-threaded workers, which can justify the added complexity for programming for both distributed and shared memory.

References

[1] Mark Baker, Bryan Carpenter, Geoffrey Fox, Sung Hoon Ko, and Sang Lim. mpiJava: An Object-Oriented Java interface to MPI. Proc of the Int. Workshop on Java for Parallel and Distributed Computing, IPPS/SPDP 1999.

[2] http://www.chmsr.gatech.edu/jPVM/

[3] Adam J. Ferrari, JPVM: Network Parallel Computing in Java, ACM 1998 Workshop on Java for High-Performance Network Computing .

[4] Ken Arnold, James Gosling, The Java Programming Language – second edition, Addison Wesley.

[5] Eric Freeman, Susanne Hupfer, Ken Arnold, JavaSpaces(TM) Principles, Patterns and Practice, SUN MicroSystems

[6] http://www.almaden.ibm.com/cs/TSpaces/

A Revised Implicit Locking Scheme in Object-Oriented Database Systems

Woochun Jun[*] and Kapsu Kim[**]

Dept. of Computer Education
Seoul National University of Education
Seoul, Korea
[*]wocjun@ns.seoul-e.ac.kr; [**]kskim@ns.seoul-e.ac.kr

Abstract. In this paper, we present a locking-based concurrency control scheme for object-oriented databases (OODBs). Our scheme revises the implicit locking so that incurs less locking overhead than implicit locking for any types of access to OODBs. Especially, our scheme only uses structural information of OODBs so that extra overhead to reduce locking overhead is little. In this work, we prove that our scheme performs better than the implicit locking.

1 Introduction

OODBs have been popular for many non-traditional database environments such as computer-aided design, artificial intelligence, etc. In a typical OODB, a class object consists of a group of instance objects and class definition objects. The class definition object consists of a set of attributes and methods that access attributes of an instance or a set of instances [2].

A concurrency control scheme is used to coordinate multiple accesses to the multi-user database so that it maintains the consistency of the database. A concurrency control scheme allows multi-access to a database but incurs an overhead whenever it is invoked. This overhead may affect on the performance of OODBs where many transactions are long-lived. Thus, reducing the overhead is critical to improve overall performance.

The implicit locking has been used for controlling access to OODBs ([4],[5],[6],[7]). It is based on intention locks [3]. The purpose of an intention lock on a class indicates that some lock is set on a subclass of the class. Thus, when a lock is set on a class C, it is required to set extra locking on a path from C to its root as well as on C. In implicit locking, when an operation such as class definition modification is accessed on a class, say C, locks are not required for every subclass of the class C. It is sufficient to set a lock only on the class C (in single inheritance) or locks on C and its subclasses having more than one superclass (in multiple inheritance) [4]. Thus, it can reduce locking overhead than explicit locking. But, implicit locking requires more locking overhead when a target class is near the leaf in a class hierarchy due to intention lock overhead.

For example, consider a class hierarchy in Fig 1. Assume that the class definition of the class C4 should be modified. Based on the implicit locking in Orion [4], locks are set as follows. Intention locks IWs corresponding to W (Write) locks are required for all superclasses on the path from C4 to the root C1. Also, W lock is required on the class to be accessed (called target class) C4.

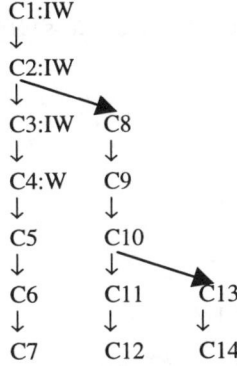

Fig. 1. Example of the implicit locking

This paper is organized as follows. In Section 2 we propose new implicit locking. In Section 3 we prove that our scheme performs better than the implicit locking. The paper concludes with future research issues in Section 4.

2 The Revised Implicit Locking

For simplicity, in this work, we only consider single inheritance. The basic idea is that some redundant locks can be reduced without affecting the correctness of the scheme. Assume that a class C is accessed so that it needs to be locked. For implicit locking, an intention lock is set on every superclass of C. On the other hand, the proposed scheme does not have to set intention locks on every superclass of C. That is, the proposed scheme sets intention locks on the root class of C and on the superclasses, which have more than one direct subclass.

Based on the idea explained above, the proposed scheme is as follows. Assume that a lock is requested on class C. Also, it is assumed that the strict two-phase locking is adopted [1].

Step 1) locking on the root class
   For the root class of C, check conflicts and set an intention lock.
Step 2) locking on superclasses
   For each superclass, which has more than one direct subclass of C, check conflicts and set an intention lock.
Step 3) Locking on the target class

Check conflicts with locks set by other transactions and set a lock on only the target class C and set an a lock on the instance to be accessed if necessary.

For example, consider the class hierarchy in Fig. 1. Also, assume the following access by transaction T_1. Fig 2.a and Fig. 2.b show locks by the implicit locking and the proposed scheme, respectively.

T_1: class definition update operation on class C14.

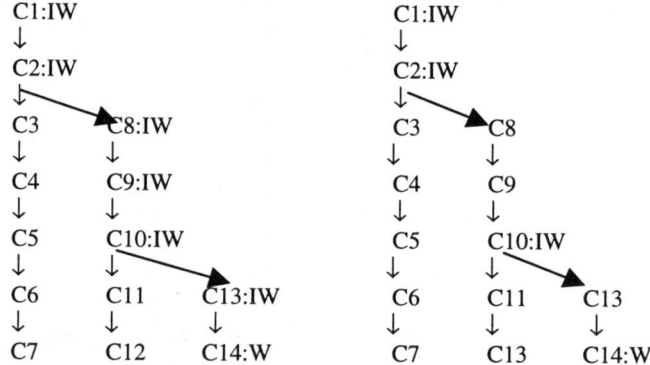

Fig. 2.a. Locks by implicit locking   Fig. 2.b. Locks by our scheme

3 The Proof of Correctness for the Proposed Scheme

In this Section, we will show that the proposed scheme performs better than implicit locking. As we discussed in Section 3.2, the proposed scheme incurs equal or fewer number of locking overhead than implicit locking for any type of accesses to OODBs. Thus, in this Section, it is sufficient to prove that the proposed scheme is correct, that is, it satisfies serializability [1]. We prove this by showing that, for any lock requester, its conflict with a lock holder (if any) is always detected. With this proof, since our scheme is based on two-phase locking, it is guaranteed that the proposed scheme satisfies serializability.

Claim) for any lock requester, its conflict with a lock requester is always detected.

We prove that there exists at least one common class in which both the lock requester (R) and the lock holder (H) set a lock.

Case I) the R and the H have the same target class.

In this case, regardless of intention locks by both requester and holder, the conflict can be detected on the target class.

Case II) Otherwise

If the R and the H have different target class, the conflict can be detected on the nearest superclass, S, from both R and H. In this case, conflict can be also detected on any superclasses of S. But, detecting conflict on those superclasses may degrade concurrency so that conflict should be detected as near as possible to both R and H. For example, if target class of R and H are C12 and C14 in Fig. 1, respectively, conflict must be detected on class C10, which is the nearest superclass of R and H.

From case I) and case ii), we can conclude that, for any lock requester, it is guaranteed that its conflict with a lock holder (if any) is always detected. Also, since our scheme is based on two-phase locking, serializability is guaranteed.

For multiple inheritance, additional locks are required. That is, locks are required for each subclass of the target class, which has more than one superclass [4]. Due to lack of space, we omit the formal proof.

4   Conclusions and Further Work

In this paper, we present a locking-based concurrency control scheme for object-oriented databases (OODBs). Our scheme revises the implicit locking so that incurs less locking overhead than implicit locking for any types of access to OODBs. Also, our scheme only uses structural information of class hierarchy in OODBs. Thus, extra overhead to reduce locking overhead is little. Finally, we prove that our scheme performs better than the implicit locking

Currently we are preparing simulation work in order to compare the performance of our scheme with implicit locking.

References

1. Bernstein, P., Hadzilacos, V. and Goodman. N., *Concurrency Control and Recovery in Database Systems,* Addison-Wesley, (1987).
2. Cart, M. and Ferrie. J., Integrating Concurrency Control into an Object-Oriented Database System, 2nd Int. Conf. on Extending Data Base Technology, Venice, Italy, Mar. (1990), 363 - 377.
3. Date, C., *An Introduction to Database Systems*, Vol. II, Addison-Wesley, (1985).
4. Garza, J. and Kim, W., Transaction Management in an Object-Oriented Database System, ACM SIGMOD Int. Conf. on Management of Data, Chicago, Illinois, Jun., (1988), 37 - 45.
5. Jun, W. and Gruenwald, L., An Effective Class Hierarchy Concurrency Control Technique in Object-Oriented Database Systems, Journal of Information And Software Technology, Vol. 40. No. 1, Apr. (1998) 45-53.
6. Lee, L. and Liou, R., A Multi-Granularity Locking Model for Concurrency Control in Object-Oriented Database Systems, IEEE Trans. on Knowledge and Data Engineering, Vol. 8, No. 1, Feb., (1996), 144 - 156.
7. Malta, C. and Martinez, J., Controlling Concurrent Accesses in an Object-Oriented Environment, 2nd Int. Symp. on Database Systems for Advanced Applications, Tokyo, Japan, Apr., (1992) 192 - 200.

Active Agents Programming in HARNESS

Mauro Migliardi and Vaidy Sunderam

Emory University, Dept. o f Math & Computer Science
Atlanta, GA, 30322, USA
om@mathcs.emory.edu

Abstract. HARNESS is an experimental Java-centric metacomputing system based on distributed object oriented technolgy. Its main features are the capability to dynamically assemble, manage and dismantle virtual machines that are reconfigurable both in terms of resources enrolled and in therms of services offered, and the capability to mix and match heterogeneous programming models on demand. In this paper we describe how the HARNESS system supports the active, mobile agents programming model.

1 Introduction

Harness [1][2] is an experimental metacomputing system based upon the principle of dynamically reconfigurable object oriented [3] networked computing frameworks. Harness supports reconfiguration not only in terms of the computers and networks that comprise the virtual machine, but also in the capabilities of the VM itself. These characteristics may be modified under user control via an object oriented "plug-in" mechanism that is the central feature of the system. At system level, the capability to reconfigure the set of services delivered by the virtual machine allows overcoming obsolescence related problems and the incorporation of new technologies. In fact, a virtual machine model intrinsically incorporating reconfiguration capabilities addresses these issues in an effective manner. At application level the reconfiguration capability of the system allows the incorporation of new capabilities into applications directly at run-time. This capability is extremely useful for long running applications (e.g. simulations) that need to evolve to adapt to new data and constraints.

The native programming model of HARNESS is distributed, Object Oriented components. However, the modular, Object Oriented nature of the HARNESS system allows on demand mix and match of components or complete applications adopting programming models that are different from its native one (e.g. PVM [4]). This process is supported by means of compatibility suites that can be plugged into the system if the need for such a programming model arise. The feasibility of this approach has been already proved in past works [5].

In this paper we describe the compatibility suite for the active, mobile agents programming model [6][7].

The paper is structured as follows: in section 2 we give an abstract overview of the

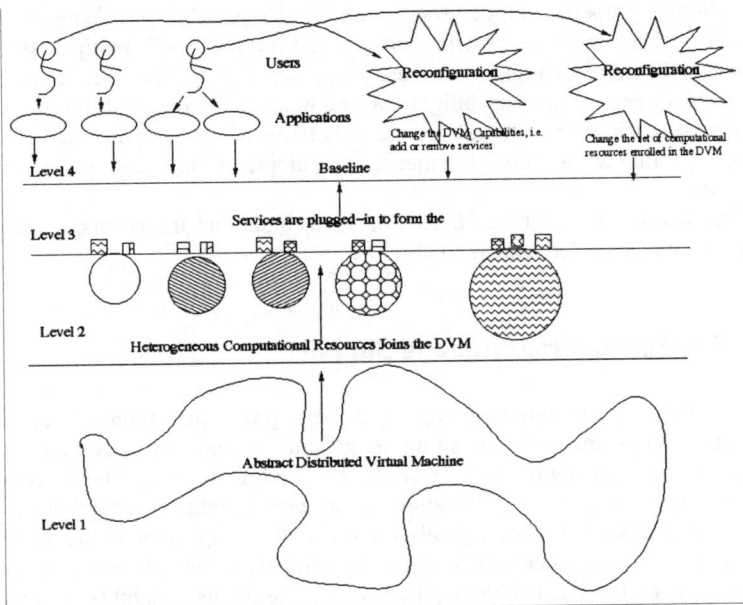

Fig. 1. Abstract model of a Harness virtual machine

system architecture; in section 3 we describe our compatibility suite for active agents programming; finally, in section 4, we provide some concluding remarks.

2 The HARNESS Metacomputing System

The fundamental abstraction in the HARNESS metacomputing framework is the **Distributed Virtual Machine** (DVM) (see figure 1, level 1). Any DVM is associated with a symbolic name that is unique in the HARNESS name space, but has no physical entities connected to it. **Heterogeneous Computational Resources** may enroll into a DVM (see figure 1, level 2) at any time, however at this level the DVM is not ready yet to accept requests from users. To get ready to interact with users and applications the heterogeneous computational resources enrolled in a DVM need to load **plug-ins** (see figure 1, level 3). A plug-in is a software component implementing a specific **service**. By loading plug-ins a DVM can build a **service baseline** (see figure 1, level 4) that is consistent with applications' requirements. Users may **reconfigure** the DVM at any time (see figure 1, level 4) both in terms of computational resources enrolled by having them **join** or **leave** the DVM and in terms of services available by **loading** and **unloading** plug-ins.

The availability of services to heterogeneous computational resources derives from three different properties of the framework: *i)* the portability of plug-ins, *ii)* the portability of the runtime system and *iii)* the presence of multiple searchable repositories. HARNESS implements these properties leveraging two different features of Java technology, i.e. the capability to layer a homogeneous architecture such as the Java Virtual Machine [8] over a large set of heterogeneous computational resources, and the capability to customize the mechanism adopted to load and link new objects and libraries.

For further details about the HARNESS metacomputing framework we direct the attention of the interested readers to [2].

3 Active Agents Programming Support

Active mobile agents programming is a very promising approach to Internet computing and distributed computing in general. Among the advantages of the mobile agents paradigm with respect to traditional distributed computing programming such as message passing we can cite its inherent adoption of Object Oriented programming and its capability to move the computation to the data instead of the traditional reverse approach. The latter property offers advantages both in the minimization of the bandwidth required as well as in its capability to survive to intermittent network partitioning by moving the computational entity at the site where the data reside. For further details about the agents paradigm we direct the intersted reader to [6] and [7].

To support the agents programming paradigm in HARNESS we have developed the HARNESS active mobile agents compatibility suite. A compatibility suite is a pluggable execution environment that can be loaded into the HARNESS system on demand to support additional programming paradigms. The Java-centric nature of HARNESS together with its capability to soft install both Java classes and native code retrieved from trusted repositories suits very well the mobile active agents programming paradigm. The HARNESS active mobile agents compatibility suite provides to users:

- the capability to load, initialize and start an agent;
- a system-wide unique naming for agents;
- the capability to perform a weak migration of agents;
- the capability to track the movements of an agent through an event system;
- the capability to obtain an RMI reference of an agent;
- a coordination system based on Sun's JavaSpaces [9].

We adopted a weak approach to the problem of agents migration in our compatibility suite because the current definition of the Java platform does not provide an architecture independent format for the execution stack of a thread. Thus our compatibility suite does not service a request for migration until all the threads of the agent have acknowledged that they are ready to migrate.

The capability to track the movements of an agent allows other agents and applications to retrieve RMI references of RMI enabled agents and interact with them in a direct, synchronous way. The coordination system based on JavaSpaces allows agents to interact asynchronously.

The compatibility suite is currently in alpha testing, however, our first tests showed that the time required to perform migration between two Sun Ultra60 workstations connected over a 100 Mbs ethernet LAN is 100 milliseconds.

4 Concluding Remarks

The modular, object oriented nature of the HARNESS system allows on demand mix and match of components or complete applications adopting programming models. This process is supported by means of compatibility suites that can be plugged into the system if the need for such a programming model arise. In this paper we have described the HARNESS active mobile agents compatibility suite. This compatibility suite allows the execution of agents based applications on a HARNESS virtual machine. The mobile agents programming paradigm is one of the most promising new approaches to distributed programming and is extremely well suited to Internet computing. Our plug-in based approach leverages the capability of the system to support weak migration of agents and multiple agents interaction mechanisms. The initial testing of our implementation shows a very low overhead for agents migration.

References

1 M. Migliardi, V. Sunderam, A. Geist and J. Dongarra, Dynamic Reconfiguaration and Virtual Machine Management in the Harness Metacomputing Framework, proc. Of ISCOPE98, Santa Fe (NM), December 8-11, 1998.
2 M. Migliardi and V. Sunderam, Heterogeneous Distributed Virtual Machines in the Harness Metacomputing Framework, Proc. of the Heterogeneous Computing Workshop, IEEE Computer Society Press, pp. 60-73, 1999.
3 G. Booch, Object Oriented Analysis and Design with Applications, Second Edition, Rational, Santa Clara, California, 1994.
4 Geist A., A. Beguelin, J. Dongarra, W. Jiang, B. Mancheck and V. Sunderam, PVM: Parallel Virtual Machine a User's Guide and Tutorial for Networked Parallel Computing, MIT Press, Cambridge, MA, 1994.
5 M. Migliardi, V. Sunderam, PVM Emulation in the Harness Metacomputing System: a Plug-in Based Approach, Proc. of 6th EuroPVM-MPI99, Lecture Notes in Computer Science, Vol. 1697, pp. 117-124, Springer Verlag, 1999
6 J. White, Mobile Agents Whitepaper, MIT Press, Menlo Park, 1996.
7 Proc. of the International Workshop on Mobile Agents, Lecture Notes in Computer Science, N. 1219, Springer Verlag, 1997.
8 T. Lindholm and F. Yellin, The Java Virtual Machine Specification, Addison Wesley, 1997.
9 Sun MicroSystems, Jini Starter Kit documentation package, available on line at http://developer.sun.com/developer/products/jini/jinidoc.html

Track VI

Workshops

LAWRA Workshop
Linear Algebra with Recursive Algorithms
http://lawra.uni-c.dk/lawra/

Fred Gustavson[1] and Jerzy Waśniewski[2]

[1] IBM T.J. Watson Research Center, P.P. Box
218, Yorktown Heights, NY 10598. USA
gustav@watson.ibm.com

[2] The Danish Computing Centre for Research and Education
(UNI•C), Technical University of Denmark
Building 304, DK-2800 Lyngby, Denmark
jerzy.wasniewski@uni-c.dk

1   Introduction to the LAWRA Project

A good quality numerical software is very important in many industrial applications. Here "quality" means:

- the practical problem is solved as fast as possible, and
- with an acceptable error (as is well-known computers make roundoff errors which can spoil the result significantly).

Therefore, some companies develop libraries of computer programs which are ready to use, and where the "best" present algorithms are implemented. One such library is LAPACK (Linear Algebra PACKage) [4] which is the basis for highly tuned libraries on different architectures, and BLAS (Basic Linear Algebra Subroutines) [5] in which some basic matrix and vector operations are implemented, and which is heavily used in LAPACK.

Modern computer architectures have a hierarchical memory (with 1 or more levels of cache memory). This allows faster algorithms but only when the algorithm is designed appropriately. There is some development in this field (for example in LAPACK and ATLAS) [4, 17] but it seems that the memory is not used to the full possible extent, and there is a reason to do further research.

In the last 3 years a new idea emerged which leads to faster algorithms on modern processors. This is the recursive formulation of all basic algorithms in numerical software packages. It turns out that recursion leads automatically to better utilization of the memory, and so faster algorithms result. Sometimes better algorithms emerge. There is a number of algorithms where recursion has been applied successfully. In one case the recursive algorithm was 10 times faster than the existing one but here the speed up was additionally due to the SMP environment. The recursive algorithms automatically benefitted from the parallelism and the traditional methods did not.

The second important issue is that recursive programs "capture" very concisely the mathematical formulation, and are easy to read and understand (but only if one knows the subject area).

New research on recursive algorithms started at the IBM Watson Research Center, USA in 1997. Then, in the beginning of 1998, two more centers joined the team working on these algorithms: UNI•C (Danish Computing Center for Research and Education), and the University of Umeå, Sweden. Also, UNI•C works in cooperation with the University of Rousse, Bulgaria on these problems. The ATLAS [17] library is extended by the recursive algorithms. At present there have been developed and tested the following recursive algorithms:

- LU decomposition for general matrices [10, 15, 2],
- Cholesky decomposition for symmetric and positive definite matrices [10, 3, 16, 2],
- QR factorization for general matrices [8],
- Several algorithms for symmetric indefinite system [2] and
- Recursive BLAS [9, 2, 13]. [1]

We have improved significantly some of the existing algorithms in LAPACK. This happens because the recursive formulation uses the cache memory more efficiently and sometimes also allows the use of larger block operations (i. e. more efficient use of BLAS).

There are many widely used algorithms in practice for which the recursive formulation has not been studied (e. g. algorithms for eigenvalues, singular values, generalized eigenvalues, generalized singular values, etc.). Based on our preliminary experience, we think that recursion will also benefit these algorithms.

Another important consequence of this research is that as computer vendors improve the hardware of parallel high performance computers, the memory hierarchies are tending to become deeper. Thus recursive algorithms, since they automatically block for them, will be able to take advantage of the new memories. In this way our research has impact on both software and hardware.

2 Outline of the Workshop

New Generalized Data Structures for Matrices Lead to a Variety of High Performance Dense Linear Algebra Algorithms

First we give a very brief overview of the Algorithms and Architecture Approach of as means to produce high performance Dense Linear Algorithms. The key idea developed is blocking for today's deep memory hierarchies.

Next we develop how recursion relates to dense linear algebra. Specifically we show it leads to automatic variable blocking which is more general than conventional fixed blocking. New concise algorithms emerge. The machine independent

---

[1] See also the LAWRA papers on http://lawra.uni-c.dk/lawra/papers/ and on http://lawra.uni-c.dk/lawra/references/

design of dense linear algebra codes is challenged. By doing so we are lead to the conclusion that new data structures are required.

Next we describe new data structures for full and packed storage of dense symmetric/triangular arrays that generalize both. Using the new data structures one is led to several new algorithms that save "half" the storage and outperform the current blocked based level 3 algorithms in LAPACK. We concentrate on the simplest forms of the new algorithms and show for Cholesky factorization they are a direct generalization of LINPACK [6]. This means that level 3 BLAS's [7] are not required to obtain level 3 performance. The replacement for Level 3 BLAS are so-called kernel routines, see [1], and on IBM platforms they are producible from simple vanilla codes by the XLF FORTRAN /citeibmlanref compiler. The results for Cholesky, on Power3 for $n \geq 200$ is over 720 MFlops and reaches 735 MFlops out of a peak of 800 MFlops. Using conventional full format LAPACK xPOTRF[2] with ESSL [11] BLAS's, one first gets 600 MFlops at $n \geq 600$ and only reaches a peak of 620 MFlops. Before leaving Cholesky, we describe a recursive formulation of Cholesky Factorization of a matrix in packed storage due to the authors and Bjarne S. Andersen [3] of UNI•C. The key to this algorithm is a novel data structure that converts standard packed format into $n-1$ "square" full matrices and n scalar diagonal elements. This formulation only use conventional xGEMM[3], a surprising result in itself.

We have also produced simple square blocked full matrix data formats where the blocks themselves are stored in column major (FORTRAN) order or row major (C) format. The simple algorithms of LU [10, 2] factorization with partial pivoting for this new data format is a direct generalization of LINPACK algorithm xGEFA[4]. Again, no conventional level 3 BLAS's are required; the replacements are again so-called kernel routines. Programming for squared blocked full matrix format is accomplished in FORTRAN [12, 14] through the use of three and four dimensional arrays. Thus, no new compiler support is necessary. As in the Cholesky case above we mention that other more complicated algorithms are possible; e.g., recursive ones. The recursive algorithms are also easily programmed via the use of tables that address where the blocks are stored in the two dimensional recursive block array.

References

1. R.C. Agawal, F.G. Gustavson, and M. Zubair. Exploiting functional parallelism on power2 to design high-performance numerical algorithms. *IBM Journal of Research and Development*, 38(5):563–576, September 1994.
2. B.S. Andersen, F. Gustavson, A. Karaivanov, J. Waśniewski, and P.Y. Yalamov. LAWRA – Linear Algebra with Recursive Algorithms. In R. Wyrzykowski, B. Mochnacki, H. Piech, and J. Szopa, editors, *Proceedings of the 3th International Conference on Parallel Processing and Applied Mathematics, PPAM'99*, pages 63–76, Kazimierz Dolny, Poland, 1999. Technical University of Częstochowa.

---

[2] LAPACK [4] subroutine name; x can be S for single precision, D for double precision, C for complex or Z for double complex.
[3] BLAS Level 3 [7] subroutine name.
[4] LINPACK [6] subroutine name.

3. B.S. Andersen, F. Gustavson, and J. Waśniewski. Formulation of the Cholesky Factorization Operating on a Matrix in Packed Storage Form. http://lawra.uni-c.dk/lawra/pspapers/bjarnepaper99.ps. The paper is sent to be published in ACM TOMS.
4. E. Anderson, Z. Bai, C. Bischof, S. Blackford, J. Demmel, J. Dongarra, J. Du Croz, A. Greenbaum, S. Hammarling, A. McKenney, and D. Sorensen. *LAPACK Users' Guide*. Society for Industrial and Applied Mathematics, Philadelphia, PA, third edition, 1999.
5. J. Dongarra et al. BLAS (Basic Linear Algebra Subprograms). http://www.netlib.org/blas/. Ongoing Projects at the Innovative Computing Laboratory, Computer Science Department, University of Tennessee at Knoxville, USA.
6. J. J. Dongarra, J. R. Bunch, C. B. Moler, and G. W. Stewart. *LINPACK Users' Guide*. Society for Industrial and Applied Mathematics, Philadelphia, PA, USA, 1979.
7. J. J. Dongarra, J. Du Croz, I. S. Duff, and S. Hammarling. A set of Level 3 Basic Linear Algebra Subprograms. *ACM Trans. Math. Soft.*, 16(1):1–28, March 1990.
8. E. Elmroth and F. Gustavson. New Serial and Parallel Recursive QR Factorization Algorithms for SMP Systems. In B. Kågström, J. Dongarra, E. Elmroth, and J. Waśniewski, editors, *Proceedings of the 4th International Workshop, Applied Parallel Computing, Large Scale Scientific and Industrial Problems, PARA'98*, number 1541 in Lecture Notes in Computer Science Number, pages 120–128, Umeå, Sweden, June 1998. Springer.
9. F. Gustavson, A. Henriksson, I. Jonsson, B. Kågström, and P. Ling. Recursive Blocked Data Formats and BLAS' for Dense Linear Algebra Algorithms. In B. Kågström, J. Dongarra, E. Elmroth, and J. Waśniewski, editors, *Proceedings of the 4th International Workshop, Applied Parallel Computing, Large Scale Scientific and Industrial Problems, PARA'98*, number 1541 in Lecture Notes in Computer Science Number, pages 195–206, Umeå, Sweden, June 1998. Springer.
10. F.G. Gustavson. Recursion leads to automatic variable blocking for dense linear-algebra algorithms. *IBM Journal of Research and Development*, 41(6), November 1997.
11. IBM. *Engineering and Scientific Subroutine Library for AIX*, Version 3, Volume 1 edition, December 1997. Pub. number SA22-7272-0.
12. IBM. *XL Fortran AIX, Language Reference*, first edition, Dec 1997. Version 5, Release 1.
13. B. Kågström, P. Ling, and C. Van Loan. GEMM-based level 3 BLAS: High-Performance Model Implementations and Performance Evaluation Benchmark. *ACM Trans. Math. Software*, 24(3):268–302, 1998.
14. M. Metcalf and J. Reid. *FORTRAN 90 Explained*. Oxford University Press, Oxford, UK, first edition, 1990.
15. S. Toledo. Locality of Reference in LU Decomposition with Partial Pivoting. *SIAM Journal of Matrix Analysis and Applications*, 18(4), 1997.
16. J. Waśniewski, B.S. Andersen, and F. Gustavson. Recursive Formulation of Cholesky Algorithm in Fortran 90. In B. Kågström, J. Dongarra, E. Elmroth, and J. Waśniewski, editors, *Proceedings of the 4th International Workshop, Applied Parallel Computing, Large Scale Scientific and Industrial Problems, PARA'98*, number 1541 in Lecture Notes in Computer Science Number, pages 574–578, Umeå, Sweden, June 1998. Springer.
17. R.C. Whaley and J. Dongarra. Automatically Tuned Linear Algebra Software (ATLAS). http://www.netlib.org/atlas/, 1999. University of Tennessee at Knoxville, Tennessee, USA.

Communicating Mobile Active Objects in Java

Françoise Baude, Denis Caromel, Fabrice Huet, and Julien Vayssière

OASIS Team, INRIA - CNRS - I3S
Univ. Nice Sophia-Antipolis
First.Last@inria.fr

Abstract. This paper investigates the design and implementation of mobile computations in Java. We discuss various issues encountered while building a Java library that allows *active objects* to migrate transparently from site to site, while still being able to communicate with each other. Several optimizations are introduced, and a set of benchmarks provides valuable figures about the cost of migration in Java: basic cost of migration with and without remote classloading, migration vs. standard remote method invocation in a typical information retrieval application. Our conclusion is that mobile computations are a viable alternative to remote method invocation for a large domain of Java applications that includes Web-based application.

1 Introduction

This paper is concerned with *mobile computations* [Car99]: the ability to start a computation on a site, suspend the execution of the computation at some point, migrate the computation to a remote site and resume its execution there. Sometimes called *mobile agents*, it is a far more complex problem than simply moving objects around as done in RMI or CORBA, or moving code alone as with Java applets.

A distinction is usually made between *strong migration* and *weak migration*. Strong migration means migrating a process by sending its memory image to a remote site: the current state of the stack, the value of the program counter, and of course all the objects reachable from the process. Moreover, the migration may occur preemptively; the process does not even need to know it has migrated. On the other hand, weak migration usually only involves the objects reachable from the process, and requires that the process has agreed to migrate (it is non preemptive). To the best of our knowledge, there exists no implementation of strong migration in Java that does not break the Java model or require user instrumentation of the code.

In this paper, we have taken the weak migration approach, and are trying to provide a flexible, extensible, yet efficient Java library for migration. As a result, we introduce a Java library for mobile computations allowing *active objects* to migrate transparently from site to site, while still being able to communicate with each other. It is built as an extension to the *ProActive PDC* library [CKV98].

Section 2 presents some related work in mobile agents technology and introduces the main features of the *ProActive PDC* library. Section 3 discusses and details the library for mobile computations, and presents some possible optimization techniques for mobile agents. Results from several benchmarks are presented in section 4, section 5 concludes.

2 Background on Mobility and Active Objects

2.1 Related Work

The best-known Java libraries for mobile agents are Aglets [Ven97] and Voyager [Obj99]. Both implement a form of weak migration, in the sense that an object must explicitly invoke a primitive in order to migrate, and this can only take place at points in the execution where a checkpointed state (either implicit or explicit) is reached. A *checkpoint* is a place in the code where the current state of the computation can be safely saved and restored.

For example, the `MoveTo` primitive of Voyager waits until all threads have completed. On the other hand, in Ajents [IC99], any object can be migrated while executing by interrupting its execution, moving the most recently checkpointed state of the object to a remote site and re-executing the method call using the checkpointed object state. Apart from the fact that whether checkpointing occurs lies in the hands of the user program, general and well-known issues related to checkpoint inconsistency due to rollback are kept unresolved.

Systems also differ in the way mobile objects interact: either by remote method call (Ajents, Voyager), or using a message-centric approach (Aglets). The default interaction mode is usually synchronous, even if some form of asynchronous communication is sometimes provided (in Aglets or Ajents for instance). Interacting synchronously simplifies the overall management of migration: while a remote method or message is handled, nothing else can happen to the two partners (and especially no migration), but this of course incurs a performance penalty.

Remote interaction is achieved transparently by a proxy which hides the effective location of the destination object to the caller. The proxy also often acts as a forwarder for locating the mobile object (see section 3.2) and is a convenient place for performing security-related actions.

Protocols for managing the migration of an object introduce several events, typically events such as *departure* or *arrival*, which can be customized by the user, like for example with Aglets. It is also possible to provide the mobile agent with an itinerary, i.e. a list of hosts to visit.

2.2 Asynchronous Active Objects

The results presented in this paper capitalize on research done over the last few years around the *ProActive PDC* library [CKV98]. *ProActive PDC* is a Java library for concurrent and distributed computing whose main features are transparent remote active objects, asynchronous two-way communications with transparent futures and high-level synchronization mechanisms. *ProActive PDC* is built on top of standard Java APIs (Java RMI [Sun98b], the Reflection API [Sun98a]). It does not require any modification to the standard Java execution environment, nor does it make use of a special compiler, preprocessor or modified virtual machine.

A distributed or concurrent application built using *ProActive PDC* is composed of a number of medium-grained entities called *active objects* which can informally be thought of as "active components". Each active object has its own thread of control and has the ability to decide in which order to serve incoming method calls. There are no shared passive objects (normal objects) between active objects, which is a very desirable property for implementing migration.

Method calls between active objects are always asynchronous with transparent *future objects* and synchronization is handled by a mechanism called *wait-by-necessity*.

As both active objects and future objects are type-compatible with the equivalent 'normal' objects, the programmer does not need to modify any code when reusing old code with ProActive. The only code that needs to be changed is the code that instanciates the objects we now want to be active. Here is a sample of code for creating an active object of type A:

```
class pA extends A implements Active{}
Object[] params={"foo", 7};
A a = (A) ProActive.newActive ("pA", params, node);
```

A full description of the library is outside the scope of this paper and can be found in [CKV98].

3 Communicating and Asynchronous Mobile Objects

Enhancing active objects with mobility is a nice feature to have but it is not enough: mobile active objects also need to be able to communicate with each other, regardless of where they are.

In this section we present the mechanism we have built into *ProActive PDC*.

3.1 Programmer Interface

The principle is to have a very simple and efficient (optimized) primitive to perform migration, and then to build various abstractions on top of it.

Primitives. Any active object has the ability to migrate and if it references some passive objects, they will also migrate to the new location. That is, we not only migrate an active object but also its complete subsystem. Since we rely on the serialization to send the object on the network, the active object must implement the Serializable interface. Many different migration primitives are available, all with the same general form :

```
public static void migrateTo(...) throws ...
```

Notice that we are only able to perform weak migration since we do not have access to the execution stack of the JVM. Hence, a migration call must be encapsulated into a method of the mobile object so that in the end the call comes from the object itself. This call never returns because, strictly speaking, the object is not present anymore and so the execution must stop.

There are two different ways to move an active object in *ProActive PDC*. The first one consists in migrating to a remote host. In order to do so, the remote host must have a running Java object called a *Node*, a kind of daemon in charge of receiving, restarting active objects and keeping trace of locally accessible active objects.

The other way to migrate is to join another active object on a remote host, even if the address of this host is unknown. Indeed, only a reference to the remote object is needed. However it can not be guaranteed that after the migration the two objects will be on the same host, the referenced object having possibly migrated. An example of a very simple mobile object can be found in example 3.1. For the sake of clarity, this code has been simplified and exception handling is not shown.

Exemple 3.1 SimpleAgent

```
public class SimpleAgent implements Active, Serializable {

  public void moveToHost(String t)
    {
       ProActive.migrateTo(t);
    }

  public void joinFriend(Object friend)
    {
       ProActive.migrateTo(friend);
    }

  public ReturnType foo(CallType p)
    {
       ...
    }
}
```

Higher Level Abstractions. With this primitive for migrating active objects from host to host, we can build an API on top of it in order to implement autonomous active objects, also known as mobile agents in the distributed computing literature.

Itinerary. We want an autonomous object to be able to successively visit all the sites on a list and perform a possibly different action on each of them (see [KZ97]). Such a list does not need to be known at compile-time and can be dynamically modified at runtime in order to react to changing environmental conditions.

Any active mobile object in *ProActive PDC* can have an itinerary, which is a list of <*destination, action*> pairs, where *destination* is the host to migrate to and *action* is the name of the method to execute on arrival. Therefore, the method to execute on arrival at a host differs from host to host and can be modified dynamically.

Automatic Execution. In some Java libraries for mobile agents, such as Aglets [Ven97], only one predefined hard-wired method to be called on arrival is allowed. This is an annoying limitation, since no decision on which method to call on arrival can be made at runtime.

The approach we have chosen makes use of reflection techniques in order to bring more flexibility in choosing the method to execute on arrival. Setting the method to execute on arrival is done through this call:

ProActive.onArrival(String)

There are two limitations on the signature of the method to be executed on arrival. First, the method cannot return any value since it is called by an internal mechanism and there is nobody to return the result to. Second, the method cannot have any parameter. Remember that this method will be executed when the mobile object arrives on a new site, so there might be a delay between the moment the method is set and the moment it is called. The value of the parameters cannot be known for sure, and so, proving properties on the system can be difficult. Therefore, the signature of the method to execute on arrival must be:

void myMethodToExecuteOnArrival()

This is not in any case a limitation. The method can call other methods and access attributes of the object as would do any other method.
There is also the event corresponding to the start of a migration:

ProActive.onDeparture(String)

The method specified with onDeparture() could be used to do some clean-up once the mobile object has left the local host.

3.2 Rationale and Discussion on Implementation Techniques

Implementing mobility features into *ProActive PDC* raised a number of interesting issues. One of them is how do we make sure that a mobile active object always remains reachable, even if it never stops jumping from one host to the next ?

In this section we survey two of the most popular solutions to the *location problem*. The first solution is based on a *location server*, chain of *forwarders*. Other solutions exist as the one described in [Bru99] that uses a distributed two-phase transaction and a sophisticated reference-tracking mechanism.

Location Server. The location server responsability is track the location of each mobile active object: Every time an active object migrates, it sends its new location to the location server it belongs to. As a result of the active object leaving a host, all the references pointing to its previous location become invalid. There exists different strategies for updating those dangling references with the new location, one of them being lazy: when an object tries to send a message to a mobile active object using a reference that is no longer valid, the call fails and a mechanism transparently queries the location server for the new location of the active object, updates the reference accordingly and re-issues the call (see figures 1 and 2).

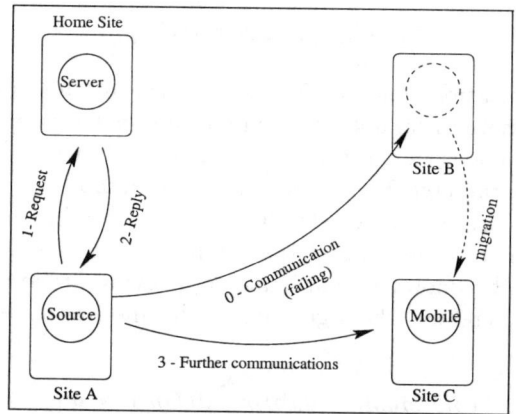

Fig. 1. Localisation - the caller side

Being a centralized solution, it is very sensitive to network or hardware failures. Standard techniques for fault-tolerance in distributed systems could be put to use here, such as using a hierarchy of possibly replicated location servers instead of a single server. Nevertheless, this solution is costly, difficult to administer and would certainly not scale well.

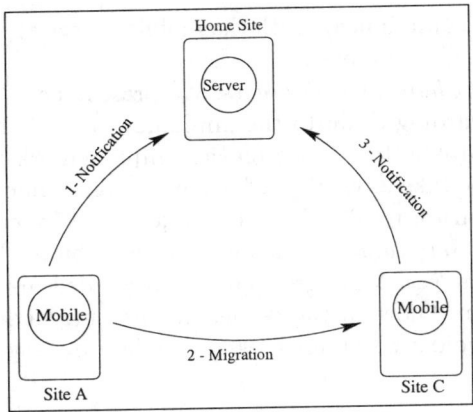

Fig. 2. Localisation - the mobile object side

Forwarders.

An alternative solution is to use *forwarders* [Fow85]: knowing the actual location of a mobile object is not needed in order to communicate with it; rather, what really matters is to make sure that the mobile object will receive the message we send to it.
To do so, a chain of references is built, each element of the chain being a *forwarder* object left by the mobile object when it leaves a host and that points to the next location of the mobile object [Obj99]. When a message is sent, it follows the chain until it reaches the actual mobile object. This appears to be the solution of choice for several systems [KZ97].
In *ProActive PDC*, for efficiency purpose, it is the active object that is turned into a forwarder object at the moment it leaves a host. This mechanism is fully transparent to the caller because the forwarder has exactly the same type as the mobile object. Moreover, the same mechanism is used for sending asynchronous replies to active objects, which means that an active object can migrate with some of the future objects in its subsystem still in the *awaited* state.

Forwarders can be considered as a distributed solution to the location problem, as opposed to the previous centralized solution. Moreover, contrary to the location server, in the absence of network or host failure, a message will finally reach any mobile object even if it never stops migrating. However, this solution also suffers from a number of drawbacks.
First, some elements of the chain may become temporarily or permanently unreachable because of a network partition isolating some elements from the rest of the chain or just because a single machine in the chain goes down; it would destroy all the forwarder located on it[1]. The chain is then broken, and it be-

---
[1] Except if the forwarders are persistent objects.

comes impossible to communicate with the mobile object at the end of the chain, although it is still well and alive.

Shortcutting the Chain of Forwarders. We present here a technique to keep the length of a forwarding chain to the minimum to limit the consequence of a break, as done in [Obj99]. It is based on the simple remark that the shorter the chain, the better. We take advantage of two-ways communication, i.e invocation of methods that return a result. When the object that executes the remote call receives a request, it *immediately* sends its new location to the caller if and only if it does not know it yet. Messages that are forwarded are marked so that an object receiving them knows if the sender has its current location or an older one. Thus we can avoid sending unnecessary update messages.

4 Benchmarks

Cost of one Migration. The first benchmark has been conducted to evaluate the cost of the migration between different computers.

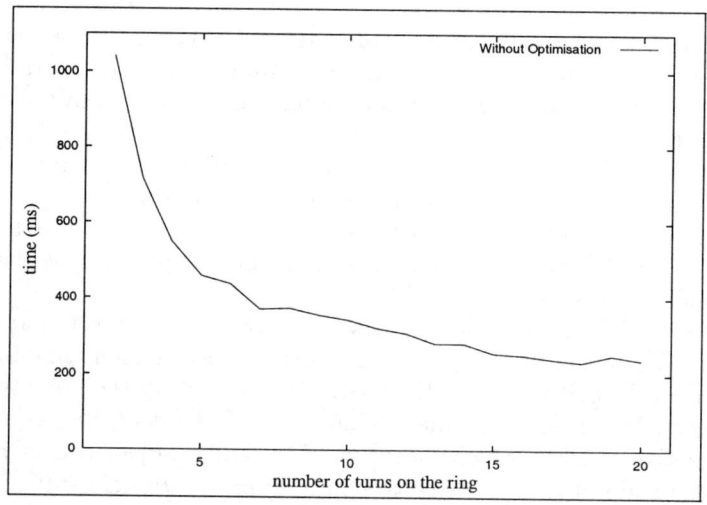

Fig. 3. Cost of one migration

Our test network was made of 2 Sparc-10, one Pentium-Pro, and a Bi-PentiumII. The program consisted in a mobile object moving from one host to another, always in the same order and always ending on the one it first started. It is equivalent to having the mobile object make one or more turns on a virtual ring of computers. The object is initially out of the ring, and so has to migrate first before actually starting its journey. We have measured the time taken to go through the whole itinerary, including the initial migration. The start-up time (first creation of all objects) has not been taken into account. Figure 3, shows the average time of one migration against the number of turns on the ring.

As time goes, the average cost of a migration decreases. This can easily be explained by the fact that on the first round on the ring, all the JVMs must load the class corresponding to the mobile object: when benchmarking mobile objects system, one should not forget that there is always a start-up time if the host is visited for the first time, even if all the class files are available locally or through NFS.

We can infer that the average migration time should be around 200ms on a LAN for some given network conditions. Keep in mind that these results are heavily dependent on the computating power.

Information Retrieval. The second benchmark has been designed to test a common application of mobile code: information retrieval. A text file representing the directory listing of a FTP server [2] is available on a remote host and we want to perform a search on this file to find all lines containing a certain keyword (i.e we are looking for a file). Without code mobility, we must first download the file (using a remote call on the active object that manages the file) and then perform the search; a mobile object would simply migrate to the remote host, perform the search and only bring the resulting lines.

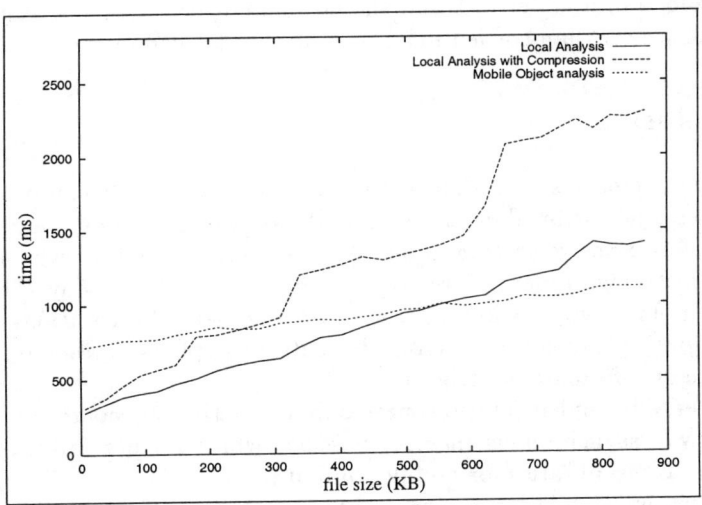

Fig. 4. Execution time of a file search

The purpose of this benchmark is to find the threshold above which migration becomes a better solution than local analysis of the downloaded file. Two

[2] ftp.lip6.fr

versions have been designed, both using the same methods for reading the file from disk and analysing it. In order to experiment with potential optimisation in the remote method call version, we have implemented "on the fly" compression of the downloaded file with the Java compression facilities (ZipInputStream). The test has been conducted on a 100 Mbits LAN with a Bi-PentiumII and a Bi-PentiumIII.

Figure 4 shows the result of 3 experiments, one with a mobile object, and two using local analysis. In the latter case, the file has been downloaded with and without compression to measure the possible gain.
Contrary to what was expected, compression of the file didn't decrease the time taken to perform the search. Even worse, the difference increases as the file size grows. Because the computers are on a high bandwidth LAN, the cost of compressing data on the fly is greater than the time saved on transmission. Also, we can notice that at file sizes 180, 350 and 650 KB, the time suddenly increases. We believe this is due to the algorithms used to compress the data and to related memory managment operations.
The mobile object program shows a remarquable scability since its execution time increases only slightly with the size of the file. Most of the time is spent performing migration (one to the remote host, and one back to the home host) as we can see from the time taken with a small file. Even on a high bandwidth and low latency LAN, mobile objects can be useful in many ways. First they reduce the load on the network and the hosts, and second their execution time can be much lower than conventional search, even for common sized files.

5 Conclusion

In this paper we have shown that mobile computation in Java can be efficiently implemented without breaking the standard Java execution environment in any way. The cost of one migration in our system is around 200 ms which is of the same order of magnitude as a remote call across the Web today. This means that using mobile computations instead of remote method invocations for client-server computing can quickly deliver benefits in application domains such as e-commerce or information retrieval.

Moreover, our implementation enables mobile active objects to communicate using two-way asynchronous message-passing, which further improves performances over other libraries for mobile computations.

Our future work will be aimed at studying mechanisms for fault-tolerance and security in mobile computation systems. The *ProActive PDC* library, its extension for mobile computations discussed here and the code for all the examples are freely available for download at http://www.inria.fr/oasis/proactive.

References

[Bru99] E Bruneton. Indirection-Free Referencing for Mobile Components. In *Proc. of the 1999 International Conference on Parallel and Distributed Processing Techniques and Applications (PDPTA'99)*, Madeira Island, Portugal, April 1999.

[Car99] Luca Cardelli. Abstractions for mobile computation. *Secure Internet Programming: Security Issues for Mobile and Distributed Objects*, LNCS 1603:51–94, 1999.

[CKV98] D. Caromel, W. Klauser, and J. Vayssiere. Towards Seamless Computing and Metacomputing in Java. *Concurrency Practice and Experience*, 10(11–13):1043–1061, November 1998.

[Eck98] Bruce Eckel. *Thinking in Java*. Prentice Hall, 1998.

[Fow85] Robert Joseph Fowler. *Decentralized Object Finding Using Forwarding Addresses*. PhD thesis, University of Washington, 1985.

[HP99] B. Haumacher and M. Philippsen. More efficient object serialization. In *Parallel and Distributed Processing, International Workshop on Java for Parallel and Distributed Computing*, pages 718–732, San Juan, Puerto Rico, April 1999. Springer-Verlag. LNCS 1586.

[IC99] M. Izatt and P. Chan. Ajents: Towards an Environment for Parallel, Distributed and Mobile Java Applications. In *Proc. of the 1999 Java Grande Conference*. ACM, 1999.

[KZ97] K. Kiniry and D. Zimmerman. A hands-on look at java mobile agents. *IEEE Internet Computing*, 1(4):21–30, July/August 1997.

[Obj99] ObjectSpace, Inc. ObjectSpace Voyager http://www.objectspace.com/developers/voyager/index.html, 1999.

[Sun98a] Sun Microsystems. Java core reflection, 1998. http://java.sun.com/products/jdk/1.2/docs/guide/reflection/index.html.

[Sun98b] Sun Microsystems. Java remote method invocation specification, October 1998. ftp://ftp.javasoft.com/docs/jdk1.2/rmi-spec-JDK1.2.pdf.

[Ven97] Bill Venners. Under the hood: The architecture of aglets. *JavaWorld: IDG's magazine for the Java community*, 2(4), April 1997.

A Service-Based Agent Framework for Distributed Symbolic Computation

Ralf-Dieter Schimkat*, Wolfgang Blochinger, Carsten Sinz**, Michael Friedrich, and Wolfgang Küchlin

Symbolic Computation Group, WSI for Computer Science
University of Tübingen, Sand 13, D-72076 Tübingen, Germany
http://www-sr.informatik.uni-tuebingen.de

Abstract. We present OKEANOS, a distributed service-based agent framework implemented in Java, in which agents can act autonomously and make use of stationary services. Each agent's behaviour can be controlled individually by a rule-based knowledge component, and cooperation between agents is supported through the exchange of messages at common meeting points (*agent lounges*). We suggest this general scheme as a new parallelization paradigm for Symbolic Computation, and demonstrate its applicability by an agent-based parallel implementation of a satisfiability (SAT) checker.

1 Introduction

Symbolic Computation comprises Computer Algebra on the one hand, and Computational Logic on the other hand. It is increasingly acknowledged that Symbolic Computation will play an essential role in future problem solving environments (PSEs). Where it is theoretically applicable and practically feasible, it yields answers with the quality of mathematical proofs. Moreover, answers may be given at very high levels of abstraction, containing symbolic parameters.

Due to the very high level of abstraction at which Symbolic Computation takes place, it is characterized by high computational demands and highly irregular and data dependent control flows. At the same time, there is practically no hardware support (e.g. for big integer arithmetic). However, there is a great potential for parallelization. Parallel Symbolic Computation is therefore an interesting research topic. Since automatic parallelization of Symbolic Computation algorithms is rather difficult due to highly dynamic data structures and irregular control flow, our approach is to investigate middleware architectures and associated programming paradigms which will support human programmers in parallelizing their sequential code base. As we move towards more processors and even less homogeneous and larger scale networks, it is time to explore more flexible and loosely coupled parallel architectures.

In this paper we explore the use of agent based middleware with a parallelization paradigm that is based on a stricter separation of the sequential code base from aspects

---

* Supported by *debis Systemhaus Industry*.
** Partially supported by *Deutsche Forschungsgemeinschaft (DFG)* under grant Ku 966/4-1, and partially supported by *debis Systemhaus Industry*.

dealing with the parallelization, such as making parallelization decisions, communicating, and distributing and balancing computational load. Our OKEANOS middleware provides an infrastructure for *mobile agents* which may access computational *services* and may communicate by passing messages in KQML [5], using Remote Method Invocation only as a general transport layer. The agents themselves are implemented in Java [9] and may contain rule-based knowledge interpreted by the Java expert system shell Jess [6], and they are supported with distributed information about the computational load provided by agent-related middleware services.

The agent based style of parallelization is much less synchronized than a client / server style. In our agent based approach, the sequential code is left largely intact, and the parallelization aspects are factored out to the level of agents. In this sense it bears some resemblance to Aspect-oriented programming [14].

We evaluate the OKEANOS system with a parallel satisfiability (SAT) checker for Boolean logic. Although theoretically intractable, practical advances have opened up important applications to SAT checkers, such as hardware verification or scheduling problems [1, 12]. Due to the large volumes of data in industrial applications, any significant advance in the speed of SAT checking is likely to open up new classes of practical problems. Our implementation is the computational core of an industrial system which we are developing for DaimlerChrysler AG. Its application will be in checking the consistency of symbolic product model data bases for trucks and passenger cars [15].

2 OKEANOS Agent Framework

2.1 Introduction

OKEANOS is an agent-based middleware framework for processing and distributing complex computational tasks. The primary goals of OKEANOS are to hide the process of task distribution within a computer network and to offer high-level interfaces for integrating distributed symbolic applications. The complexity of how, where, and why, the computational processes are running should be hidden behind uniform middleware-related interfaces. The notion of a service within OKEANOS establishes a middleware infrastructure where each participant's behaviour, i.e. a problem solver agent, is characterized either by making use of a collection of available services, or by offering a certain service itself. The trading and consuming of services in OKEANOS is done by exchanging language-neutral messages between participants and service providers.

In order to meet the challenges of scalability in large computer networks such as the Internet, the service management itself is strictly decoupled from the utilization of services. Thus, whereas all participants in OKEANOS can communicate asynchronously with each other or with service providers by exchanging messages, a distinct management process is responsible for keeping the service infrastructure consistent. It is up to the middleware to provide appropriate communication facilities as the *facilitators* [7] for exchanging messages between participants in a general manner.

While OKEANOS is completely implemented in Java [9], the implementation of potential service providers is not necessarily bound to Java. For example, they can be integrated into OKEANOS either by using the Java Native Interface (JNI) or by using a client/server like communication mechanism. In the latter one, the service provider would act as a manager which forwards service requests to a remote server. These kind of requests can be transferred by applying the paradigm of Remote Procedure Call or by using ordinary socket-based communication styles. In Section 3 a symbolic computing engine, a propositional satisfiability checker, is integrated into OKEANOS as a calculation service by interfacing to its C++ implementation using JNI. Thus, OKEANOS forms the core component of a heterogeneous, distributed infrastructure which is characterized by services implemented in a wide range of programming languages. Moreover, by choosing Java as the implementation language, all OKEANOS components can be used uniformly on several platforms, as described in Section 3.5.

In the following sections we give a description of the main design principles of OKEANOS.

2.2 Framework Components

In OKEANOS there are two basic components: lounges and agents. A lounge is located at a host machine within a computer network like the Internet, providing a service-centered processing environment for agents. It offers the computing infrastructure for agents on top of the underlying host machine. A lounge supports the management of various kinds of services like a directory and messaging service. A new lounge is added to OKEANOS by registering itself to a well-known registry server called observer. The observer maintains a table of all registered and active lounges. By registering at the observer, the new lounge receives a list of callback entries of all other lounges. It is up to the directory services to propagate the new lounge and to synchronize its directory entries.

To minimize the complexity of an agent-based middleware framework, there is a strict distinction between two kinds of agents in OKEANOS: stationary and mobile agents. A stationary agent is located at one single lounge which it can not leave. Its primary task is to implement and to provide services to other agents like querying the lounge about locally available and registered application services. The complete communication infrastructure in OKEANOS is designed as a general service which is managed by stationary agents called portal agents. In contrast to stationary agents, mobile agents (MA) are characterized by their potential to move among lounges transparently by using the appropriate communication service available at the local lounge. Since the communication infrastructure is hidden behind the notion of a service, the agent logic of MAs has only to deal with the core application-related logic which facilitates the development of MAs as discussed in Section 3.3 and as illustrated in Figure 1a. When moving between lounges, MAs keep their state and behaviour. The lounge at the destination host is responsible for restarting the transferred MA correctly.

Lounges and agents are designed as framework components to establish a high-level, semi-complete software architecture which can be specialized to produce custom applications [4]. They form an abstract set of classes and interfaces among them which, taken together, set up a generic architecture for a family of related applications [10, 11]. In order to design an agent-based distributed application in OKEANOS, the application requirements have to be clearly identified in order to distinguish between such functionalities which can be addressed at the framework level, i.e. at the service level of

A Service-Based Agent Framework for Symbolic Computation

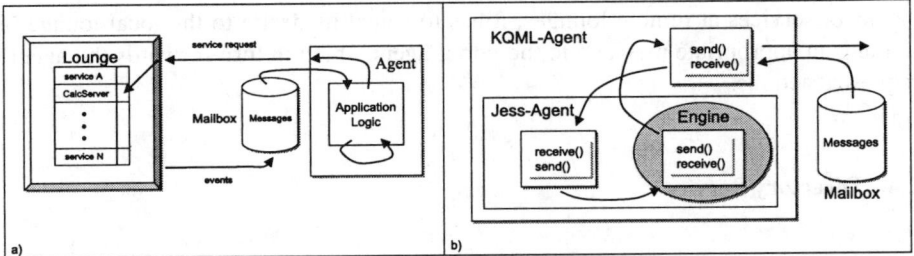

Fig. 1. a) Interaction between application logic and services located at lounges; b) Flow of control within a Jess-based agent

OKEANOS, and the application level. The benefit of shifting requirements to the more abstract framework level is in facilitating the design and implementation of the remaining application logic. Furthermore it increases the reuse by other applications which in turn eases their development. Among the patterns discussed in [13] are the *Layered Agent*, *Broker*, and *Reasoner* pattern which form the basis for the design of agents in OKEANOS.

2.3 Communication Concepts

Since interaction between agents can exhibit a complex structure, the paradigm of remote method invocation is too restrictive. Therefore communication within OKEANOS is characterized by the concept of message passing using the Knowledge Query Manipulation Language (KQML) [5]. KQML provides a general message container for possibly different types of message contents. It communicates dedicated properties about the transferred content of a message, rather than communicating within some special language. A mailbox is assigned to each agent which is basically a buffer for incoming and outgoing messages, as depicted in Figure 1a. The mailbox supports asynchronous communication between agents and lounges since it does not interfere directly with the agent's flow of control. Each agent is responsible for the management of its mailbox by itself. In OKEANOS, Remote Method Invocation (RMI) serves as a transport layer for messages which are sent between lounges. Since the communication interface between lounges does not change over time regardless of the number of lounges and type of messages, no additional communication-related code has to be provided [18], i.e. such as compiled static interfaces like stubs and skeletons. Instead, KQML messages are used as generic interfaces to initiate remote agent requests and to transfer mobile agents.

Using messages as the exclusive communication medium provides an extensible and adaptable approach to an uniform management of a possibly wide range of different agent communication interfaces, as discussed in [18]. As far as implementation issues are concerned we need only one communication interface within the entire distributed system which reduces potential incompatibilities between agents in OKEANOS.

Nevertheless, in OKEANOS other communication paradigms can be integrated transparently and made available to the entire system by wrapping them into a service. The interaction of a MA is restricted to the local lounge. If it wishes to communicate with agents or services at remote lounges, it has to signal its desire to the local lounge by sending an appropriate message to the portal agent which in turn forwards the agent's requested actions.

2.4 Directory Service

The directory service in OKEANOS is designed as a distributed bulletin board for announcing application services. It is kept up to date by using so-called Updater Agents (UA). An UA is small in code size (about 1 KByte). It updates the directory of each lounge. From a lounge's point of view, UAs are handled with high priority since their process of updating each directory is very short in time.

One directory service is assigned to each lounge. There are two types of directory entries: global and local entries. A local entry describes a service which is offered only locally by a service provider at the respective lounge. A global entry specifies a service which is available at the respective lounge and can be accessed by other lounges and agents, respectively. The global part of the directory service is replicated at each lounge in order to minimize remote communication overhead: An agent's service request can be processed locally without remote communication to other lounges. Therefore it caches globally available services at each lounge and speeds up the lookup of services by reducing the overhead of remote communication tremendously. This caching mechanism aids the scalability of the number of lounges and agents in OKEANOS because of the decreased remote communication as opposed to a more central directory server.

The global part of the directory service serves as a snapshot of the state of all services of the distributed system. It is contacted by agents to make application-related decisions which need to have the most up-to-date information of the overall distributed system. The updating mechanism in OKEANOS is characterized by its simplicity and robustness. However, there is a small window of inconsistency of each globally replicated directory entry since the service update is processed in a decentralized manner avoiding a central manager for coordinating and propagating the updates appropriately. To eliminate potential bottlenecks in large-scale systems, the design of OKEANOS does not insist on the overall consistency of the distributed directory services at all times. If an MA contacts such a service at a remote lounge which is actually not available any more, it simply has to adjust its behaviour to the new state of the system. For example, this MA can go back to the lounge it has come from. By this, one has a tradeoff between a strict and a weak consistent point of view of the distributed directory service. Since OKEANOS is a service-centric system with autonomous MAs acting on their own, it is not predictable where and how agents interact with each other [1]. The primary goal of the distribution of the global directory entries is to provide a service for registering and propagating all types of services in a simple, robust and scalable manner.

---

[1] In Section 3.5 each performance measurement for a given number of particating lounges ends up with a slightly different runtime because of the unpredictable behaviour of the agents and the non-deterministic nature of the SAT problem.

2.5 Intelligent Agents

Agents in OKEANOS are either pure KQML-agents or Jess-agents. Jess [6] is a rule engine and scripting environment written entirely in Java. Agents based on Jess have the capacity to reason, using knowledge which is supplied in the form of declarative rules. Jess-agents are integrated into OKEANOS in a natural way in that declarative rules are wrapped into KQML messages and vice versa, as shown in Figure 1b. A Jess-agent consists of two parts. The KQML part of the agent is responsible for managing incoming and outgoing messages and for transforming KQML messages into Jess-based declarative rules. Then, after the Jess engine has interpreted the incoming rules, new facts are added (`assert`) to the knowledge base, or existing facts are retracted (`retract`) from it, depending on the content of the incoming messages. The resulting knowledge base of a Jess-agent determines its state and is characterized by its dynamic change. It also depends directly on the processing environment, i.e. the lounge where it is anchored. Therefore, integrating Jess into OKEANOS provides an infrastructure for developing agents which have facilities related to the area of artificial intelligence.

Jess-agents encourage the design and implementation of autonomous MA which adapt to changes of the location of lounges, of the global availability of services, and of application-related knowledge in a rule-based fashion. In contrast to Jess-agents, pure KQML-agents pursue an object-oriented mechanism for managing their dynamic behaviour accordingly. This resulting implementation is characterized by vast quantities of agent code which is hard to maintain and to adjust. Nevertheless, the increased code size of such an agent contradicts the scalability and bandwidth issues of large-scale networks. Since the memory footprint of a Jess-agent in OKEANOS is rather small, it is suitable for designing autonomous agents for the World Wide Web.

2.6 Small Memory Footprints

A further design goal of agents in OKEANOS is to facilitate their use within large-scale distributed systems such as the Internet. Therefore agents and their processing environments, the lounges, have to be small in code size in order to minimize the network bandwidth consumption. Additionally, they require less processor time to be serialised. As far as agents are concerned, Jess-agents fulfill the requirement of minimal code size because most of the application logic is specified as declarative rules. For example, the Jess-based agents described in Section 3.3 contain the entire application logic for a distributed symbolic computation and are as small as 25 KByte. It is adequate to transmit only the rules and facts, since an instance of the Jess engine is provided by each lounge. An agent's serialized code size is even further reduced if the participating lounges use sophisticated compression techniques for transferring agents[2]. As far as a lounge is concerned, the size depends on the number of services which are made globally available. The size of the serialized lounge is between 400 KByte and 600 KByte.

---

[2] For example, customized socket factories of the Remote Method Invocation facility of Java enables such a data compression by applying several compression formats, such as GZIP or ZIP.

3 SAT: An Application from Symbolic Computation

We consider as an application from the realm of Symbolic Computation the well-known satisfiability problem for Boolean formulae (SAT). Important applications of the SAT algorithm include cryptography [16], planning and scheduling [12], model checking for hardware verification [1], checking formal assembly conditions for motor-cars [15], and finite mathematics (e.g. quasigroup problems [19]). The computational complexity of these problems can vary considerably, where the hardest practical problems can require thousands of hours of running time. So parallelization is an important issue.

The SAT problem asks whether or not a Boolean formula has a model, or, alternatively, whether or not a set C of Boolean constraints has a solution. Usually the constraints are kept in conjunctive normal form (CNF). Each constraint is then also called a clause and consists of a set of literals, where a literal is a variable or its negation. A clause containing exactly one literal is called a unit clause. A solution assigns to each variable a value (either TRUE or FALSE), such that in each clause at least one literal becomes true.

3.1 The Davis-Putnam Algorithm

Basically, by trying all possible variable assignments one after the other, one finally finds a solution to a given SAT-instance, provided that such a solution exists. The *Davis-Putnam* (DP) algorithm [3] performs an optimized search by extending partial variable assignments, and by simplifying the resulting subproblems by applying two operations known as *unit propagation* (consisting of *unit subsumption* and *unit resolution*) and *pure literal deletion* (see Figure 2).

```
boolean dp(ClauseSet C)
{
    while ( C contains a unit clause {L} ) {
        delete clauses containing L from C;      // unit-subsumption
        delete L̄ from all clauses in C;          // unit-resolution
    }
    if ( ∅ ∈ S ) return FALSE;                   // empty clause?
    pureLiteralDeletion();
    if ( C = ∅ ) return TRUE;                    // no clauses?
    choose a literal L occurring in C;           // case-splitting
    if ( dp(C ∪ {L}) ) return TRUE;              // first branch
    else if ( dp(C ∪ {L̄}) ) return TRUE;         // second branch
    else return FALSE;
}
```

Fig. 2. The Davis-Putnam Algorithm

In the following, we associate with each run of the DP algorithm a search tree, which is a finite binary tree generated by the recursive calls of the case splitting step.

A Service-Based Agent Framework for Symbolic Computation

The nodes of the tree represent executions of the DP algorithm with a fixed input clause set C. We will label the outgoing edges of each node with the literal L resp. $\bar{L}$ which is added to C to generate the new subproblem.

3.2 Parallel SAT Checking Using Agents

This section deals with the basic concepts of parallel satisfiability checking with the Davis-Putnam algorithm using an agent approach. Section 3.3 provides detailed information about the realization of a parallel distributed SAT prover within the OKEANOS agent framework.

Overview of the Parallel Execution Process. For the parallel execution of the Davis-Putnam algorithm the search space has to be divided into mutually disjoint portions to be treated in parallel. However, static generation of balanced subproblems is not feasible, since it is impossible to predict the extent of the problem reduction delivered by the unit propagation step in advance. Consequently, when parallelizing the Davis-Putnam algorithm we have do deal with considerably different and completely unpredictable run-times of the subproblems.

We adopt a search space splitting technique presented in [19] which is based on the notion of a *guiding path*. A guiding path describes the current state of the search process. More precisely, a guiding path is a path in the search tree from the root to the current node, with additional labels attached to the edges. Each level of the tree where a case splitting literal is added to clause set C, i.e. each (recursive) call to the DP procedure, corresponds to an entry in the guiding path, and each entry consists in turn of the following information:

1. The literal L which was selected at the corresponding level.
2. A flag indicating whether or not backtracking has already been done for that level; we use B to indicate backtracking and N to state that no backtracking is required.

Each entry in the guiding path with flag B set is a potential candidate for a search space division. The whole subtree rooted at the node corresponding to this entry may be examined by another independent agent.

The guiding-path approach allows dynamic problem decomposition, as at any point of time during the search any agent may decide to further split its portion of the search space. Moreover, the selected literals coincide with the selections of the sequential version. Thus, approved literal selection strategies may be carried over to the parallel agent-based version of the DP algorithm.

Implementation of Search Space Splitting. To allow search space splitting, we have modified the DP algorithm to accept a guiding path object as an additional input parameter. A call to the extended DP algorithm with a non-empty guiding path makes the case-splitting literals to be chosen as indicated by the path element of the corresponding level instead of by querying the literal chooser. Additionally, in all levels backtracking, i.e. the second recursive call to DP, is only performed when the corresponding flag is set B.

When a search-space split is requested, the computation is asynchronously stopped by the agent and the actual guiding path P is used to build two new paths P_1 and P_2. Then a new agent is started with guiding path P_1, and the interrupted agent continues work with a modified guiding path P_2.

The computation starts with one agent to which the whole search space is assigned. The agents are notified about changes concerning the availability of processing capacity. Every time a free processor is found, a split is performed. This happens, for example, when an agent has completed the search in the assigned subtree without finding a model. As mentioned above, due to the nature of the SAT problem the size of a subtree cannot be predicted in advance. In turn this leads to an unpredictable splitting behavior. Dealing with this dynamic evolution of parallelism is the major challenge in parallelizing SAT provers.

3.3 Realization of the SAT Prover in OKEANOS

The implementation of the parallel SAT prover in OKEANOS is basically made up of four types of agents, which are discussed in greater detail in the following paragraphs.

DP Service Agent. Agents of class *DPServiceAgent* are stationary agents that actually perform the search in a subtree by executing the DP algorithm. The core Davis-Putnam algorithm is a legacy application implemented in C++. It is integrated as a native library module available in every lounge of the system. *DPServiceAgents* can use the methods of the native code via the Java Native Interface. The C++ implementation is wrapped by the *DPServiceAgents* and thus can act as a service provider for the *CalcService* to all OKEANOS agents. The *CalcService* is only registered as long as local processing capacity is available.

DP Master Agent. Each search process is initiated by a stationary agent of class *DPMasterAgent*, which first creates an initial agent of type *DistributorAgent*, and then sets it up by sending several KQML performatives. These messages contain the input clauses and the guiding path, which in this case is empty. Thus, the whole search space is assigned to the first agent.

During its lifetime, the *DPMasterAgent* waits for replies from *DistributorAgents* containing their partial results. As soon as a model is found, the whole search is terminated. Until then, the master agent periodically checks for *CalcServices* that are currently occupied and thus indicate ongoing search. If no such *CalcService* exists, all work is done and consequently the whole search space has been traversed without finding a model. In this case the set of input clauses is not satisfiable and the computation is completed. As *DistributorAgents* keep track of which part of the search tree they have handled or given to other agents, the *DPMasterAgent* can finally detect missing agents (e.g. due to a crashed lounge) and restart a partial search when necessary in order to complete the whole search.

Distributor Agent. Agents of class *DistributorAgent* are mobile agents that are responsible for solving a given SAT problem. They are moving around, looking for a service

A Service-Based Agent Framework for Symbolic Computation

that is capable of solving such a kind of problem (*CalcService* in our case). They compete with other agents for these special resources. The task of finding a suitable service is supported by an expert system built into every *DistributorAgent*. As soon as a *DistributorAgent* has gained access to a *CalcService*, it places a request for calculation, and waits until this request is granted. Then the input clauses and guiding path are passed to the *CalcService*. The *CalcService* reports the result of the computation as soon as it is available, which is then taken back to the *DPMasterAgent* by the *DistributorAgent*. Access to the *CalcService* is always managed and synchronized via a *DPServiceAgent*, which acts as a service provider within OKEANOS for this service.

Strategy Service Agent. At each lounge where a *DPServiceAgent* resides, there is another stationary agent called *StrategyServiceAgent* which offers a *StrategyService*. *DistributorAgents* may consult this service about whether or not they should split their search space. The *StrategyService* can be accomodated to the specific environment of the lounge, such as the processor speed or kind of network connection. In case a decision is made to perform a split, the following steps are taken:

1. The *DistributorAgent* requests the *DPServiceAgent* to asynchronously stop the calculation and return the current guiding path as result.
2. The *DistributorAgent* splits the guiding path and passes one of these guiding paths together with the clause list to a newly created *DistributorAgent*.
3. The new *DistributorAgent* is sent off to a lounge with an unoccupied *CalcService*.
4. Finally, the interrupted *CalcService* is restarted with the second guiding path generated in step 2.

3.4 Benefits of the Agents Approach

The key benefit of using the described approach is the service-oriented character of OKEANOS agents. Each computation-related task such as the lookup of the calculation service, the determination of split times and the right strategy for splitting is encapsulated within different kinds of services. Thus there are strategy and calculation services provided by each lounge. Since each of these services is only attached to one lounge, they can be tailored to the specific lounge's environment, such as the type and location of the host machine. For example each lounge can have its own dedicated strategy service for carrying out the best splits at this lounge. However, agents are not bound to use the provided services at all. They also can work out splits completely autonomously without involving such a local service provider.

In contrast to a central master process in a classical master slave parallelization technique, our proposed agent approach is more flexible because by employing autonomous agents reaction to load changes is possible in a decentralized manner. Basically, parallelism in the agent approach is achieved through the process of migration of agents within a pool of compute-service providers.

3.5 Empirical Results

For the realization of performance measurements we have selected three benchmarks (dubois20, dubois23 and dubois26) from the publicly available [3] DIMACS benchmark suite for SAT provers. The chosen benchmarks have different sequential running times and all exhibit a highly irregular search splitting behaviour.

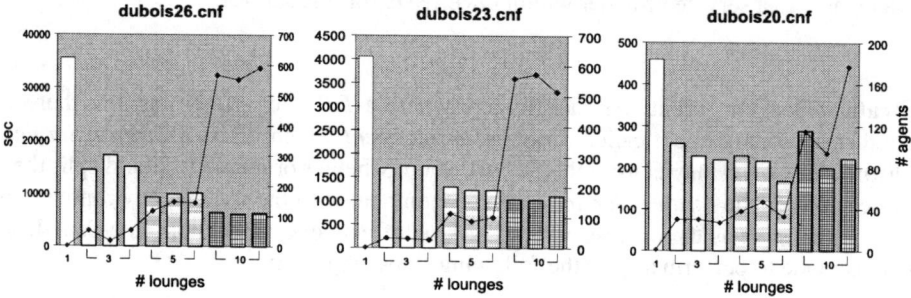

Fig. 3. Results of the Performance Measurements

The measurements where carried out on a heterogeneous computing pool consisting of machines of the following types: Two Sun Ultra E450 with 4 processors each at 400MHz running under Solaris 7, and up to four PentiumII PCs at 400MHz running under Windows NT 4.0. For all measurements the wall clock time in seconds was taken.

Figure 3 shows the results for 3, 5 and 10 lounges and relates them to the sequential running times. For each number of lounges the times of three program runs are shown. Figure 3 shows also the total number of agents that were involved during the computation. Each lounge provided a simple strategy service which suggests to initiate a split and sends a distributor agent as soon as a calculation service within OKEANOS is made available. This strategy did not take into account the overall number of lounges and number of calculation services. Each performance measurement for a given number of participating lounges ends up with a slightly different runtime because of the unpredictable behaviour of the agents and the non-deterministic nature of the SAT problem.

Generally, the overall speedup is well suited for parallelizing large search spaces (dubois26.cnf, seq. running time about 10 hours). If the sequential running time is rather short (dubois20.cnf, seq. running time about 8 minutes), the performance gain of our agent approach is less evident. When adding more lounges to the distributed computation infrastructure, the number of generated agents increases accordingly.

Our future work will focus on more sophisticated splitting strategies in order to adapt the number of generated agents more appropriately to the current number of globally available computing services. The main goal is to find adequate, cooperative strategies which match the dynamic and unpredictable character of SAT problems.

---

[3] ftp://dimacs.rutgers.edu/pub/challenge/sat/benchmarks/cnf/

4 Related Work

JavaParty [17] provides transparent remote Java objects and remote threads using pre-compiling techniques. *Java// (ProActive)* [2] is a 100% Java library for seamless cross-paradigm high performance computing. Both environments are targeted towards high performance computing in pure Java. In contrast, our approach uses Java for the parallelization infrastructure, while the actual algorithm is provided as a native code module and is accessible as a service.

MATS [8] is a mobile agent system for distributed processing. It is based on collections of agents which form teams to solve distributed tasks. The user is may be forced to structure the problem in such a way that it can be easily broken into a set of co-operating tasks, whereas in OKEANOS it is the responsibility of the autonomous agent to figure out an appropriate plan to distribute the tasks. The granularity of distribution in OKEANOS is directly related to each agent's strategy which is determined by applying techniques from the area of artificial intelligence.

PSATO [19] is a distributed/parallel prover for propositional satisfiability for a network of workstations. In contrast to our work, a master slave model is applied, where a central master is responsible for the division of the search space and for assigning the subtasks to the slaves.

5 Conclusion

In this paper, we have described a distributed infrastructure that provides mobility and application-level services to software agents. Its design as a framework enables the conceptual and operational reuse of services in a generic manner to support the scalability issues for agents in large computer networks. The agent framework supports asynchronous communication and uses message passing via an open and standardized message format. By integrating techniques from the area of artificial intelligence, the complexity of the design and implementation of autonomous mobile agents for distributing tasks is reduced tremendously. We have implemented the service agent framework in Java to illustrate its feasibility. A distributed symbolic computation is then used to show the benefits.

It turns out that a service framework for mobile agents is very suitable for distributed, symbolic computing environments as it allows to benefit from its framework services in a general acessible manner and provides and extensible computing platform over wide area networks.

References

1. A. Biere, A. Cimatti, E. Clarke, and Y. Zhu. Symbolic Model Checking without BDDs. In *Tools and Algorithms for the Analysis and Construction of Systems (TACAS'99)*, number 1579 in LNCS. Springer-Verlag, 1999.
2. D. Caromel, W. Klauser, and J. Vayssiere. Towards Seamless Computing and Metacomputing in Java. *Concurrency: Practice and Experience*, 10(11–13):1043–1061, 1998.

3. M. Davis and H. Putnam. A Computing Procedure for Quantification Theory. In *Journal of the ACM*, volume 7, pages 201–215, 1960.
4. M. Fayad and D. Schmidt. Object-Oriented Application Frameworks. *Communications of the ACM*, 40(10), October 1997.
5. T. Finn, Y. Labrou, and J. Mayfield. KQML as an Agent Communication Language. In J.M. Bradshaw, editor, *Software Agents*, pages 291–316. MIT Press, 1997.
6. E.J. Friedman-Hill. Jess, The Java Expert System Shell. Available at the URL: http://herzberg.ca.sandia.gov/jess/, 1999.
7. M.R. Genesereth. An Agent-Based Framework for Interoperability. In J.M. Bradshaw, editor, *Software Agents*, pages 317–345. MIT Press, 1997.
8. M. Ghanea-Hercock, J.C. Collis, and D.T. Ndumu. Heterogenous Mobile Agents for Distributed Processing. In *Proceedings of the Third International Conference on Autonomous Agents (Agents '99)*, May 1999. (Workshop on Agent-based Highperformance Computing, Seattle, USA).
9. J. Gosling and K. Arnold. *The Java Programming Language*. Addison-Wesley, Reading, Massachusetts, 1996.
10. R. Johnson and B. Foote. Designing Reusable Classes. *Object-Oriented Programming*, 1(2):22–35, 1988.
11. R. Johnson and V. Russo. Reusing Object-Oriented Design. Technical Report 91-1996, University of Illinois, 1991.
12. H. Kautz and B. Selman. Pushing the Envelope: Planning, Propositional Logic, and Stochastic Search. In *Proceedings of the Thirteenth National Conference on Artificial Intelligence (AAAI-96)*, 1996.
13. E.A. Kendall and M.T. Malkoun. The Layered Agent Patterns. Available at the URL: http://www.cse.rmit.edu.au/~rdsek/, 1997.
14. G. Kiczales, J. Lamping, A. Mendhekar, C. Maeda, J.-M. Loingtier C. Lopes, and J. Irwin. Aspect-Oriented Programming. Technical Report SPL97-008 P9710042, XEROX Palo Alto Res. Center, February 1997.
15. W. Küchlin and C. Sinz. Proving Consistency Assertions for Automotive Product Data Management. In I. P. Gent and T. Walsh, editors, *Journal of Automated Reasoning*, volume 24, pages 145–163. Kluwer Academic Publishers, Feb. 2000.
16. F. Massacci and L. Marraro. Logical Cryptoanalysis as a SAT Problem. In I. P. Gent and T. Walsh, editors, *Journal of Automated Reasoning*, volume 24, pages 165–203. Kluwer Academic Publishers, Feb. 2000.
17. M. Philippsen and M. Zenger. JavaParty – Transparent Remote Objects in Java. *Concurrency: Practice and Experience*, 9(11):1225–1242, 1997.
18. R. Schimkat, S. Müller, W. Küchlin, and R. Krautter. A Lightweight, Message-Oriented Application Server for the WWW. In *ACM 2000 Symposium on Applied Computing*, Como, Italy, March 2000. Association for Computing Machinery.
19. H. Zhang, M. P. Bonacina, and J. Hsiang. PSATO: A Distributed Propositional Prover and its Application to Quasigroup Problems. In *Journal of Symbolic Computation*, volume 21, pages 543–560. Academic Press, 1996.

Performance Analysis of Java Using Petri Nets

Omer F. Rana and Matthew S. Shields

Department of Computer Science
University of Wales Cardiff, POBox 916
Cardiff CF24 3XF, UK
{o.f.rana,m.s.shields}@cs.cf.ac.uk

Abstract. Understanding and improving Java performance is an important objective, for both application and tool developers. Current efforts towards developing benchmarks have played a crucial role towards this objective, particularly in identifying general trends that can be observed through such a study. An alternative mechanism for categorising Java performance is investigated here, based on developing a performance model of Java execution. The presented model is based on stochastic Petri nets and aimed at evaluation performance of method calls and class loading in Java programs. The primary objective of this paper is to use the Java profiler to determine distributions of execution time for method executions, and use this as a first step in analysing Java performance using stochastic Petri net models.
Keywords: Performance Analysis and Prediction, Object Oriented Petri nets.

1 Introduction

Object-orientation and concurrency are useful features of the Java programming paradigm, and it is important to understand the interaction between objects, threads, class loaders etc, to understand Java performance. We describe a way to analyse Java program execution, making use of compiled byte code and the Java source – and differ from existing work which makes use of either byte code execution or source profiling, but not both. Existing efforts on developing benchmarks which range from kernel codes to applications, complement this work, as does the work on creating formal models of Java semantics. A formalism based on Algebraic nets and object oriented Petri nets [5] to model Java programs is presented. The proposed formalism can also be used as a software engineering tool, to analyse synchronisation in multi-threaded Java applications, making use of monitors and locks, for instance. Hence, the proposed benefits of the approach are to: (1) enable developers to analyse program execution, to support the discovery of execution bottlenecks; (2) to develop a formal model of Java execution semantics, as a software engineering aid. Our emphasis is on analytic models, which can be supported by quantitative results obtained through simulation/benchmarking. Previous work which has an impact on this area is described in section 2. The proposed scheme for modelling Java is presented in section 3, which can subsequently be simulated on the GreatSPN [7] Petri net simulator, to derive metrics such as total execution times, structural properties of the Petri net such as Place and Transition invariants, and conditions that could lead to deadlock.

2 Previous Work

Various authors have looked at Java semantics, and modelling of both executable semantics [1], where formal semantics are provided for a large subset of Java (including inheritance, primitive types, classes, instance variables, methods, interfaces, overloading etc) for the Centaur system, and language semantics (for soundness of the type system) [17], [19]. The language semantics are considered both at an abstract level [9], without reference to a particular Java Virtual Machine (JVM), and also at the byte code level for a particular JVM [4]. The motivation for these analysis approaches have ranged from deriving proof of correctness of the Java type system (and eventually the Java language in the longer term), security [18] aspects associated with given byte code programs during dynamic class loading, and analysis of byte code after network transmission to ensure consistency [11, 10]. Our motivation is different, in that we develop a formal model of Java execution to aid performance analysis. On the other hand, existing efforts under the Java Grande initiative, described in [15], [6], are mainly aimed at developing benchmarks for bytecode analysis, and often the associate Java source code is hard to correlate with the benchmark. With benchmarks, it is also difficult to generalise results when the underlying operating environment changes, and one for which a benchmark is not available. Augmenting results of these benchmarks with formal models can support a more detailed investigation of Java performance, and more importantly, could lead to a better understanding of deficiencies in the Java language for high performance applications.

2.1 Reasons for Using Petri Nets

We use a Petri net (PN) formalism as theoretical results concerning PNs are plentiful, and they have a sound mathematical foundation [8]. Also, numerous tools are available for a quantitative analysis of such discrete event based systems. Other techniques for representing performance include queuing networks, finite state automata, process algebras and temporal logics. We make use of PNs primarily because of the wide number of tools available for analysing systems modelled by PNs, the modular way in which PN can be hierarchically constructed, and the ability to model stochastic and determinsitic variables within the same model. Furthermore, various PN simulators provide visual feedback to the performance engineer, and the representation is easier to use and analyse. However, it is important to realise that PNs are primarily used to derive a Markov chain, which is subsequently analysed for understanding properties of the system being analysed.

The basic PN model comprises two components: *places* and *transitions*, connected together via arcs to model system behaviour. A PN model may be extended by introducing the notion of time, leading to *timed Petri nets* for a quantitative analysis. When random variables are used in specifying the time behaviour, a timed PN is called a *stochastic PN* (SPN). It has been shown that SPNs are, under certain conditions, isomorphic to homogeneous Markov chains. By analysing metrics of the Markov chain (such as the steady state probability distribution) it is possible to investigate the behaviour of the underlying system being modelled by the PN. A PN consists of a *structural* part and a *dynamic* part. The structural part is a bipartite and oriented graph $S = (P, T, A, r, s)$, where P is the set (finite) of places, T the set (finite) of transitions: these two sets form the edges of the graph. $A \subset P \times T \cup T \times P$ is the set of arcs. $r : P \times T \to \mathbb{N}$ is the value of an arc going from a transition to a place and $s : T \times P \to \mathbb{N}$ is the value of an arc going from a place to a transition. The dynamic part of a PN involves the change in markings over time. The firing of a transition changes the state of the PN from marking $\mathcal{M}$ into $\mathcal{M}'$ as follows:

$$\mathcal{M}'(p) = \begin{cases} \mathcal{M}(p) - k_1 \times r(p,t) \ if \ p \in {}^\bullet t \\ \mathcal{M}(p) + k_2 \times s(p,t) \ if \ p \in t^\bullet \\ \mathcal{M}(p) \ otherwise \end{cases}$$

where k_1 and k_2 represent multiplicative factors associated with each arc in *(A)*. See [16] for more details. Object-oriented Petri nets have been introduced and used by various researchers, for providing a more intuitive way of describing dynamic systems. Various proposals and simulators have been developed, such as CO-OPN/2 [5], LOOPN [14], CLOWN [3], PN-TOX [13] and others. Two general themes emerge in making use of Objects and Petri nets, namely "Objects inside Petri nets" and "Petri nets inside Objects", as described in [2]. Most existing applications of these approaches are aimed at modelling systems, rather than software. This work differs in modelling Java execution and object interaction semantics.

3 Modelling Java

During the executing of a Java program, a sequence of operations take place involving the creation, use and update of objects and threads, namely:

- When running a Java application, a JVM is initialised and set-up on the target architecture, which includes the setting up of a default class loader;
- the class loader creates an instance of the identified class, and runs the `main` method in it. Class loading involves locating the `getLoader` method, and then calling `getResource` to invoke the `java.lang.ClassLoader.loadClass` method;
- subsequently, either new classes are loaded as needed, based on class dependencies in the executable (called "lazy" instantiation), or all required classes are loaded together (called "eager" instantiation);

– once a class has been loaded within the JVM, the class loader does not need to be invoked again, unless the class is subsequently modified. In this study, we do not consider optimisation such as HotSpot or JIT compilers, which make use of pre-compiled native code in subsequent invocations of a previously used (or often used) method. The presented approach is therefore a first step, and such optimisations will be considered in subsequent efforts.

Each method call has an associated cost, which can change dynamically, based on the operating environment within which the program runs. The use of the **synchronized** keyword to provide mutually exclusive access to shared resources will involve additional execution cost, to first associate a lock when an instance of the class is created, to block any subsequent threads making a call, and then to release the lock. The locking strategy in Java is recursive, implying that if a thread has acquired the lock on an object via a **synchronized** method, that method may itself call another **synchronized** method from the same object without having to re-aquire the lock. Furthermore, if the execution of the **synchronized** method is not completed, due to an abort operation or an exception, an unlock is automatically performed. A simplified model of Java execution is being used as a first step, and in subsequent models we will incorporate specialised PN blocks for modelling **wait-notify** and other synchronisation primitives.

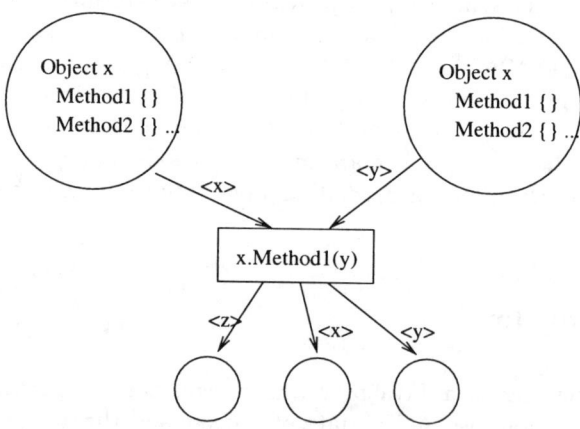

Fig. 1. *Object inside Petri net*

Using OPNs, the set 'P' represents a collection of place holders for objects, where each object is represented by a token. Each transition in set 'T' models the execution cost of a method call. Such a cost may be obtained from profiling code, or may be modelled as an exponentially distributed random variable, for instance. Synchronisation and locks within objects, and the inner behaviour of objects, are modelled using the Petri net in Objects (PNO) approach. Figure 2 illustrates the modelling of the **synchronized** method, where the thick bars represent timed transitions modelling the period to acquire and release a lock, tokens express data flow between methods of an object, and places represent entry points to method which may be invoked by other objects. We can therefore combine OPNs representing calls between methods on different objects, with PNOs representing calls to methods within an object, or to **synchronized** methods to construct a single call graph based PN for Java, which is annotated with performance information from benchmarks or, for analysis purposes, with exponentially distributed random variables to measure bounds on execution time. Recursive calls to **synchronized** methods can be modelled by combining PN blocks modelling such method calls. The combined Java program is then represented by a composition of OPNs and PNOs.

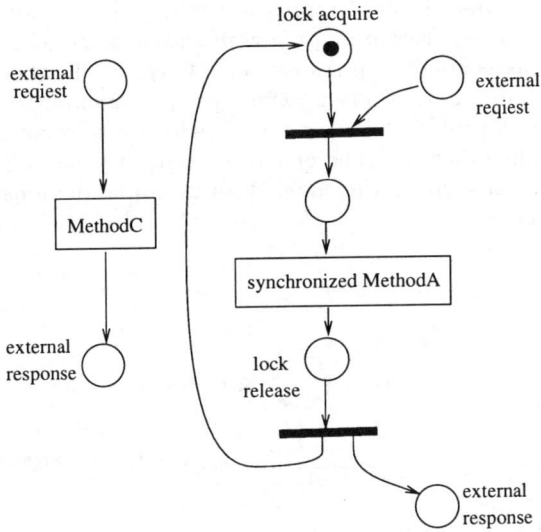

Fig. 2. *Petri net inside Object*

Consider the program in Code Segment 1, which converts a hostname into an IP address, and contains a single class.

```
import java.net.*;

class translator {
 public static void main (String args[]) {
   try {
    InetAddress[] addresses
       = InetAddress.getAllByName("www.cs.cf.ac.uk");
      for (int i = 0; i < addresses.length; i++) {
       System.out.println(addresses[i]); } }
    catch (UnknownHostException e) {
     System.out.println("Could not find www.cs.cf.ac.uk");
   } } }
```

Code Segment 1

The program contains a single class, which invokes the **main** method. This results in execution of the **try** block which instantiates an instance of the **InetAddress** class. The associated PN model is illustrated in figure 3. Each place has an associated firing delay α which can either be measured or modelled as a random variable. Places $P0$ and $P4$ represent the 'start' and 'end' places respectively, and represent the initialisation of the PN when a token is delivered to place $P0$, and a termination of the PN when a token eventually comes to place $P4$. **Exceptions** and **try-catch** blocks are modelled as a fork within the PN, corresponding to whether the specified exception is generated or not. The **for** loop is not directly modelled in the PN, although this can be refined in subsequent version of the PN if necessary. Currently, loops and other language constructs are subsumed in the execution time of the subsequent transition – $T4$ in this case. At a coarser level of representation, a single transition models the execution time of the **System.out.println(addresses[i])** method, identified as parameter $\alpha 4$, associated with transition $T4$. We profile the code using either the **java -prof** command, or **java -Xrunhprof:cpu=samples,format=a** command from Java2. A profile file is generated, which can be analysed with shareware such as "ProfilerViewer". The generated output shows CPU time spent on particular method calls within the code. Both examples demonstarted use "lazy" instantiation of classes.

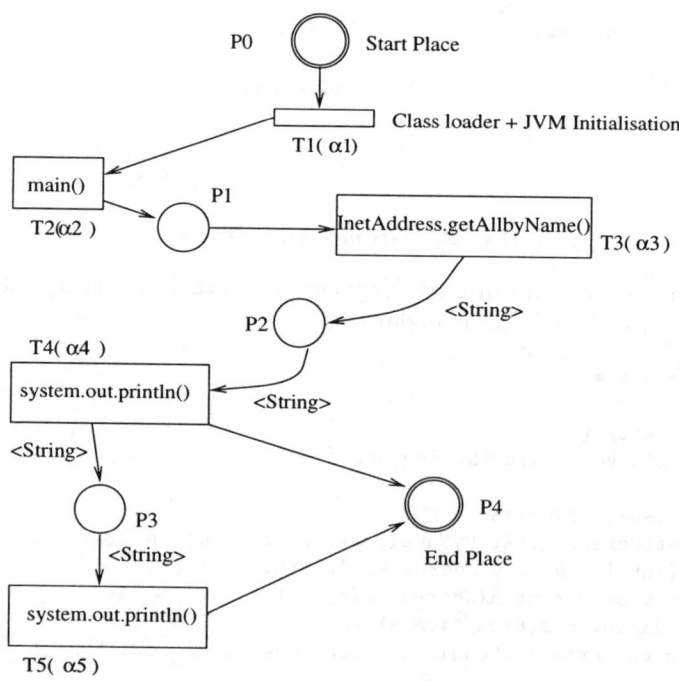

Fig. 3. *Petri net for Code Segment 1*

As an alternative example, we consider the DHPC_ThreadBench benchmark obtainable from [12]. We use the profiling tool in Java2, to determine time spent per method call by the JVM, the output is illustrated in Code Segment 2. The output shows a stack trace depth of 6. The self column is an estimate of the percentage of time a particular stack trace is active. By analysing method calls as identified by the trace, and adding up references to similar method calls, we can deduce a single number for a particular method call by analysing and following references in the stack trace.

```
CPU SAMPLES BEGIN (25630 samples, 3156 ticks, depth 6)
rank   self  accum  method
1    804.53% 100.00% TestThread.run            ()V
2      1.11% 100.00% DHPC_ThreadBench.main     ([Ljava/lang/String;)V
3      1.08% 100.00% DHPC_ThreadBench.JGFrun   ()V
4      0.35% 100.00% java/net/URLClassLoader$1.run ()Ljava/lang/Object;
5      0.29% 100.00% sun/misc/URLClassPath.getResource    (Ljava/lang/String;Z)Lsun/misc/Resource;
6      0.22% 100.00% java/net/URLClassLoader.findClass    (Ljava/lang/String;)Ljava/lang/Class;
7      0.22% 100.00% sun/misc/URLClassPath.getLoader      (I)Lsun/misc/URLClassPath$Loader;
8      0.19% 100.00% java/security/Security.access$0      ()V
9      0.19% 100.00% java/security/Security.initialize    ()V
10     0.19% 100.00% java/security/Security.<clinit>      ()V
11     0.19% 100.00% java/security/Security$1.run         ()Ljava/lang/Object;
12     0.16% 100.00% java/util/Properties.load            (Ljava/io/InputStream;)V
13     0.13% 100.00% java/lang/ClassLoader.loadClass      (Ljava/lang/String;Z)Ljava/lang/Class;
14     0.13% 100.00% java/util/jar/JarFile.getManifest    ()Ljava/util/jar/Manifest;
```

Code Segment 2:
Output of java -Xrunhprof:cpu=samples,depth=6,format=a

In this particular example, external procedures are being invoked, such as methods from the JGFUtil library, which forms part of the import declarations in the Java source. These import declarations should also form part of the PN model that is constructed, but here are treated as external method calls, that will have their own PN model. Figure 4 illustrates the PN model for the second example, where external calls are modelled by PN blocks which can be refined when profiling calls to external methods. In the first example we only make use of PNO, whereas in the second example, we can utilise both. An outer PN models the overall control flow within the DHPC_ThreadBench class, while inner or OPNs can model a more detailed structure of an inner class or method. The overall control loop contains only the class loader starting the DHPC_ThreadBench.main() method modelled as transition $T2$. A transition is represented in further details with an additional PN, as illustrated to the right of transition $T2$. From the profiler, we can associate a parameter with transition $T2$, which is 1.11% of the total execution time, and as most of the time is spent in the TestThread.main() method, we can similarly scale the execution time for transition $T3$ and subsequently $T14$. Hence, execution times can be specified as ratios of the total run time, rather than as exact values for a particular execution. Multiple runs of a benchmark, and subsequent analysis with a profiler, can help construct a better associated PN model.

3.1 Composition and Associating Costs with Transitions

A larger program can be analysed by dividing it into a collection of PN blocks, as illustrated in figure 4 where a single program is divided into blocks. Specialised

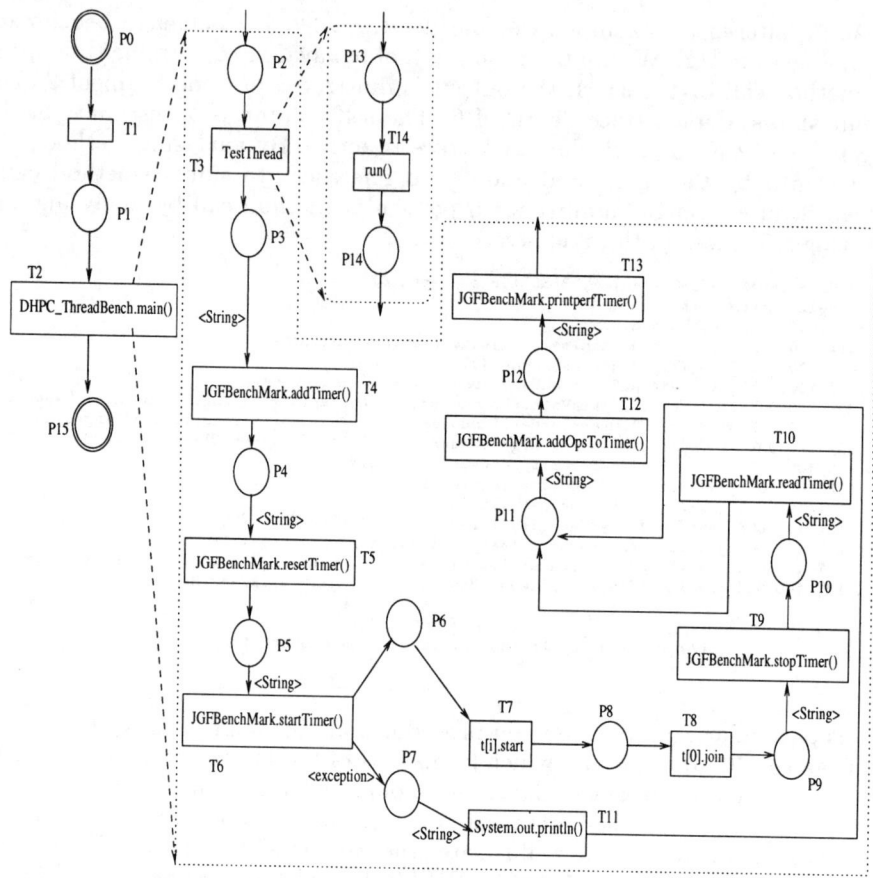

Fig. 4. Petri net for example 2

PN blocks can be provided for language constructs, such as `wait-notify` and `notifyAll`, for instance, and these blocks can be inserted into the model when they are encountered in the source. Each of the blocks can be individually profiled, and an execution cost be associated for the corresponding transition in the composed PN.

Hence, each transition in the PN model has an associated cost, which constitutes the execution cost for a a single or multiple methods. This cost can either be calculated using a profiling technique as mentioned above, where a profiler in Java2, or a third party profiler, is used to measure execution times for particular method calls. These times are then associated with deterministic transitions, and a PN simulator can be used to analyse the execution of a particular program. Alternatively, execution times can be modelled as stochastic variables, to be associated with particular transitions. Subsequent analysis of relationships between these variables is used to analyse the execution semantics of the Java program. The stochastic variables may be calculated in a number of ways:

- accounting for the execution environment, such as operating system, execution platform, caching strategies supported, amongst others. Each of these criteria can be modeled in varying details, and results from current literature may be used to develop these models;
- using a profiler to calculate the execution time of a program, and then associating variables with the firing delay of each transition based on these times. Note that the execution profiling technique outlined in Code Segment 2 also provides details about various methods that are not part of the source code. These have to be compounded to identify a single value that can be associated with a transition, based on the source code

Some preliminary results based on utilising performance figures for the code outlined in Code Segment 1, and a PN model in figure 3 are illustrated in figure 5. The figure shows the CPU samples required to execute a given method within Code Segment 1 (based on a particular transition), and the costs associated with the corresonding transition, calculated as (100-CPU Cycles)/100. These costs or firing rates are normalised to lie between 0 and 1, and are based on the premise that the greater the number of CPU cycles required to execute a method, the lower the firing rate.

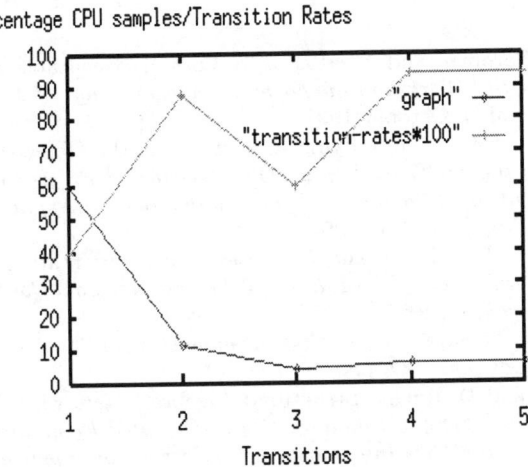

Fig. 5. *Comparisons - comparing execution times based on model, and actual times obtained from the profiler*

4 Discussion and Conclusion

We provide a PN based approach for modelling Java execution semantics, as a first step to calculating bounds on Java performance. Developing formal models can help a developer better understand the performance of Java programs, and consequently, could use code transformation techniques (such as inlining) to improve bottlenecks identified by the model. We look at a standard and a `synchronized` method call in Java, and also provide the ability to model a combination of the two by composition of PN blocks representing these method calls. The results presented here are a first step, and subsequent analysis will consider the use of other language primitives, such as `notify-all` primitives, modelled as PN blocks. At present, we manually extract information from the profiler generated text file, and are currently working on a parser to deduce CPU cycles for method executions from this file. We can automatically construct a PN model from a Java source program, and simulate the resulting model in GreatSPN [7].

The principal contributions of this paper are to introduce formal methods for analysing Java execution, to support performance analysis. We feel such a study is crucial for analysing bottlenecks in Java program execution, and to enable a developer to better understand Java execution semantics, and facilitate what-if scenarios. Furthermore, benchmarking is an important activity, but must be complemented with a formal model of Java execution semantics, and one which is not specific to particular Java APIs or applications, and can generalise across the various uses of Java.

References

1. I. Attali, D. Caromel, and M. Russo. A Formal Executable Semantics for Java. *Proceedings of the Workshop on Formal Underpinnings of Java, at OOPSLA98, Vancouver, Canada*, October 1998.
2. R. Bastide. Approaches in unifying Petri nets and the Object-Oriented Approach. *Proceedings of the 1st Workshop on Object-Oriented Programming and Models of Concurrency, at the 16th International Conference on Application and Theory of Petri nets, Turin, Italy*, July 1995.
3. E. Battiston, F. De Cindio, and G. Mauri. Class Orientation and Inheritance in Modular Algebraic Nets. *Proceedings of the International Conference on Systems, Man and Cybernetics*, 1993.
4. P. Bertelsen. Java Byte Code Semantics, 1999. See web site at http://www.dina.kvl.dk/~pmb/.
5. O. Biberstein and D. Buchs. Structured Algebraic Nets with Object-Orientation. *Proceedings of the 1st Workshop on Object-Oriented Programming and Models of Concurrency, at the 16th International Conference on Application and Theory of Petri nets, Turin, Italy*, July 1995.
6. J. M. Bull, L. A. Smith, M. D. Westhead, D. S. Henty, and R. A. Davey. A Methodology for Benchmarking Java Grande Applications. *Proceedings of the ACM Java Grande conference, San Francisco, USA*, June 1999.
7. G. Chiola, G. Franceschinis, R. Gaeta, and M. Ribaudo. *GreatSPN2.0*. Dipartimento di Informatica, Università di Torino (Italy), 1997.
8. R. David and H. Alla. Petri nets for modeling of dynamic systems - a survey. *Automatica (Elsevier Science Limited)*, 30(2):175-202, 1994.
9. S. Drossopoulou and S. Eisenbach. Java is Type Safe – Probably. *Proceedings of ECOOP97, LNCS 1241*, Springer Verlag, 1997.
10. S. Freund. The Costs and Benefits of Java Bytecode Subroutines. *Proceedings of the Workshop on Formal Underpinnings of Java, at OOPSLA98, Vancouver, Canada*, October 1998.

11. S. N. Freund and J. C. Mitchell. A Type System for Object Initialization in the Java Bytecode Language. *Proceedings of the ACM Conference on Object Oriented Programming: Languages, Systems, and Applications*, October 1998.
12. K. Hawick, P. Coddington, and J. Mathew. Java Grande Benchmarking, 1999. See web site at
http://www.dhpc.adelaide.edu.au/projects/javagrande/benchmarks/.
13. T. Holvoet and P. Verbaeten. PN-TOX: a Paradigm and Development Environment for Object Concurrency Specifications. *Proceedings of the 1st Workshop on Object-Oriented Programming and Models of Concurrency, at the 16th International Conference on Application and Theory of Petri nets, Turin, Italy*, July 1995.
14. C. A. Lakos and C. D. Keen. LOOPN – Language for Object Oriented Petri nets. *Proceedings of the SCS Multiconference on Object-Oriented Simulation*, 1991.
15. J. A. Mathew, P. D. Coddington, and K. A. Hawick. Analysis and Development of Java Grande Benchmarks. *Proceedings of the ACM Java Grande conference, San Francisco, USA*, June 1999.
16. T. Murata. Petri nets: Properties, analysis and applications. In *Proceedings of the IEEE*, April 1989.
17. T. Nipkow and D. Von Oheimb. Java Light is Type Safe – Definately. *Proceedings of the 25th ACM Symposium on Principles of Programming Languages*, 1998.
18. J. Posegga and H. Vogt. Java Bytecode Verification Using Model Checking. *Proceedings of the Workshop on Formal Underpinnings of Java, at OOPSLA98, Vancouver, Canada*, October 1998.
19. R. Stata and M. Abadi. A Type System for Java Bytecode Subroutines. *Proceedings of the 25th ACM Symposium on Principles of Programming Languages*, 1998.

A Framework for Exploiting Object Parallelism in Distributed Systems

Chen Wang and Yong Meng Teo

Department of Computer Science
National University of Singapore
3 Science Drive 2
Singapore 117543
teoym@comp.nus.edu.sg

Abstract. To support parallel computing in a distributed object-based computing platform, a uniform high performance distributed object architecture layer is necessary. In this paper, we propose a distributed object-based framework called DoHPC to support parallel computing on distributed object architectures. We present the use of dependence analysis technique to exploit *intra-object parallelism* and an *interoperability model* for supporting distributed parallel objects. Experimental results on a Fujitsu AP3000 UltraSPARC workstation cluster computer show that with intra-object parallel computation speedup efficiency is greater than 90% and with overhead of less than 10% for large problems. In addition, the interoperability model improves speedup by 20%.

1 Introduction

Distributed object technology has attracted much attention recently from high performance computing researchers. The compositional and interoperable features embodied in distributed object technology allow the development of high performance computing software that meets the requirements of open systems [9]. However, performance results reported show that the efficiency of applications encapsulated with CORBA is always far lower than programmed directly in C/C++ [4,5]. To achieve high performance in distributed object computing, one key issue is to optimize the distributed object architecture. Current distributed object architectures contain three main layers. A *component framework layer* such as OMG's Component Model, Sun's Java Beans and Microsoft's ActiveX. A *distributed architecture layer* such as OMG's CORBA, Sun's Java RMI, and Microsoft's DCOM. The *process communication layer* such as OMG's GIOP/IIOP and OSF's DCE. Current research concentrates mainly on optimizing the event delivery mechanism of ORB and in improving the efficiency of event service in CORBA. This includes various implementations such as multi-threaded ORB [6], high performance and real-time ORB [11,14], optimized IDL (Interface Definition Language) compiler [2], and environment sensitive event delivery mechanism [13].

Besides, there is some research that focus on making the component framework layer suitable for parallel computing to facilitate extending parallel computing over wide-area networks. In general, this is achieved by implementing the component framework layer based on parallel programming languages or mechanisms. Examples include LSA [3] that is based on HPC++, and in POOMA [10] the key mechanism to ensure reusability is the field class in which data-parallel operations are encapsulated. The main reasons for using parallel languages or mechanisms are: (a) Current distributed object systems do not support these parallel programming languages and the underlying parallel mechanism; (b) The efficiency of current distributed architecture layer does not meet the requirement for many parallel applications. To support parallel computing more efficiently on distributed architectures, we propose a uniform distributed architecture layer called DoHPC. Difference in remote invocation models for parallel computing and in distributed computing hinders efficiency when their components must interoperate. For example, HPC++ based on IIOP in LSA must be efficiently implemented to facilitate interoperation with external components that are defined using the OMG's Component Model.

In this paper, we propose a distributed architecture layer framework that aims to achieve high performance while keeping the interoperability with other components. Our major contributions include a method to exploit implicit parallelism in distributed objects through dependence analysis at compile-time, and an interoperability model to support distributed parallel objects that reduce internal message transfer. Section 2 presents the architecture of DoHPC. Sections 3 and 4 discuss DoHPC supports for intra-object parallelism and the interoperability model for distributed parallel objects respectively. Performance analysis is discussed in section 5. In section 6, we discuss related works on optimizing distributed architecture and comparing them with DoHPC. Section 7 summarizes our conclusion.

2 Overview of DoHPC

In DoHPC, the active object concept is fully supported. A distributed object is an active object in DoHPC, which is a service-providing entity as well as an autonomous entity that could also run its own program. DoHPC supports two categories of parallelism in distributed objects: **inter-object parallelism** and **intra-object parallelism** using the following mechanisms:

1. To exploit **inter-object parallelism**, a mechanism is provided for client to invoke methods of supplier asynchronously.
2. To exploit **intra-object parallelism**, we provide a mechanism for the supplier to serve more than one request in parallel and to dispatch some of these requests to idle processors.
3. A mechanism to support the **interoperability of distributed parallel objects** efficiently. Current interoperability model of distributed object cannot harness the parallelism among distributed parallel object very well [7]. This can be regarded as a special case of inter-object parallelism.

As shown in figure 1, DoHPC consists of *three* parts: client, supplier and a runtime system. A *client* consists of client applications and several distributed objects' implementation stubs. The *supplier* consists of a *Method Queue*, a *Deciding Controller*, an *Event Service mechanism*, a *Transaction Controller*, *Port Information* and *Runtime System*. The method invocation queue (Method Queue) stores all client requests for the supplier. The *Deciding Controller* accesses the Method Queue and dispatches methods to processors with lighter workload. The *Event Service* is used to "push" message from supplier to clients and the *Transaction Controller* executes methods of a supplier object. To improve the efficiency of a parallel supplier, the *Port Information* mechanism ensures that all parallel executing methods can interoperate with clients and other parallel objects directly. The *runtime system* provides common services such as synchronous/asynchronous message passing, name service, distributed shared-memory service and object schedule service, etc.

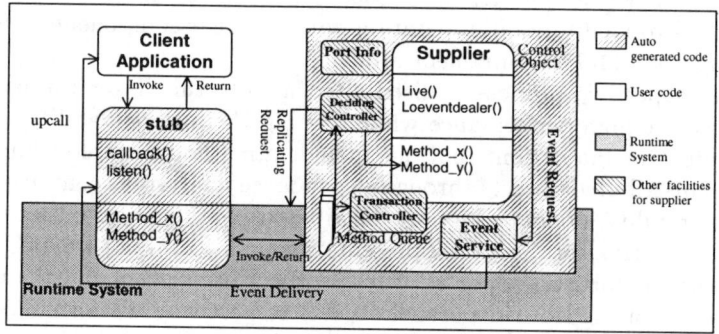

Fig. 1. Architecture of DoHPC

3 Support for Intra-Object Parallelism

Intra-object parallelism is exploited implicitly by automatically dispatching clients' requests to different processors. Methods belonging to an object that have no dependence are executed in parallel. Inter-method dependence in DoHPC is determined during compilation.

3.1 Inter-Method Dependence Analysis

Figure 2 shows the inter-method dependence analysis process. Firstly, the *dependence analyzer* obtains the method names from an *IDL file*, starts the analysis by invoking the analyzer for a specific programming language and generates a new IDL file. In the new IDL file, a method declaration consists of its method name, return type and parameter list but also a list of all dependent methods. Secondly, an *enhanced IDL compiler* automatically generates the *dependence table*, *supplier skeletons* and *client stubs*. The dependence table together with the implementing skeleton of a distributed object is used by the Deciding Controller to schedule methods for execution. To obtain the inter-method dependence, we require the summary information for each method, namely **flow sensitive information** and **flow insensitive information**. Computation of the flow sensitive information is a NP problem when variables are permitted to have alias [1].

Our Dependence Analyzer currently computes only the flow insensitive information. Let p and q denote methods in the interface declaration of a distributed

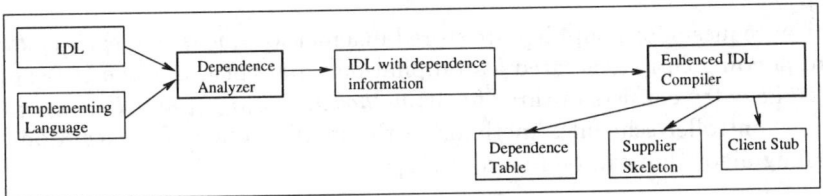

Fig. 2. Inter-Method Dependence Analysis

object. W_p, W_q and R_p, R_q denote sets of all fields p, q write and read in the superclass, superinterface and implementing class of this interface. In general, if $(R_p \cap W_q) \cup (W_p \cap R_q) \neq \Phi$, then p depends on q or q depends on p, otherwise no dependence exists between p and q. This is called **strong dependence**. It is necessary to relax this condition to increase parallelism. Assume p is ahead of q in the Method Queue, then

1. If $W_p \cap R_q \neq \Phi$, the dependence is called **true dependence**. Thus, p, q must be executed in sequence to ensure the correctness of the result and these two methods cannot be parallelized.
2. If $R_p \cap W_q \neq \Phi$ and $W_p \cap R_q = \Phi$, the dependence is called **pseudo-dependence** and the two methods can be executed in parallel. To ensure correct computation, shared variables must be localized in the processes running p and q. After p completes, these variables should be merged. In DoHPC, the localization is achieved through replication of an object.
3. If $W_p \cap W_q \neq \Phi$ and $W_p \cap R_q = \Phi$, the dependence is called **pseudo-dependence**, also called **output dependence**. The two methods p, q can be executed in parallel and those shared variables q holds should be merged after the execution of p is completed.

For illustration, we discuss DoHPC dependence analyzer for the Java programming language. The read/write set, S, for a Java Interface comprises of three parts: *all fields inherited from its superclass, all fields inherited from its super interfaces*, and *all fields of itself*. Private fields of superclass or super interface are excluded in the read/write set. Methods of a distributed object that are accessible to its clients include all public methods defined in its interface and all public methods it inherited. For each method p, the analyzer generates two sets R_p, W_p containing the flow insensitive information. There are twelve assignment operators in Java. If v, where $v \in S$, appears on the left of these operators in method p then $W_p = W_p \cup \{v\}$ else v is on the right and $R_p = R_p \cup \{v\}$. Assume that the alias of v exists, such as,

```
SOMECLASS obj = new SOMECLASS();
SOMECLASS another_obj = obj;
```

Operations on *another_obj* are in fact performed on *obj*. In this case, *another_obj* is called an alias of *obj*. The DoHPC analyzer maps aliases to the original variable first, then merge all read/write sets of these aliases to the flow insensitive information of their original variables. For cases whereby there are function calls in the source code of method p, such as, *foo(arg1,arg2,...,argn)*, the flow insensitive information of p can be obtained with the algorithm presented in [12].

Client requests for a supplier are stored in a method queue. For each *method_i*, a corresponding entry generated at compilation time is placed in the Dependence Table. The entry consists of three fields: *method_i*, W_{method_i} and R_{method_i}. The deciding controller schedules methods in the method queue for execution. The scheduling algorithm can be found in [12].

4 Support for Interoperability of Distributed Parallel Objects

The efficient interoperation among objects and with particular client's remains a problem. Since the IDL is unable to describe how its methods are scheduled, each call has to be forwarded to those processors on which the object is running. This increases the communication overhead. We propose an interoperability model that allows distributed parallel objects to interoperate more efficiently.

To generalize the characteristics of *data parallelism* and *task parallelism*, we introduce task parallel and data parallel object in DoHPC. Two classes *PObj* and *DSeq*, define the templates for these two kinds of objects. *PObj* provides an abstracted interface of distributed object's interoperation. *DSeq* describes the common interface of distributed data in a data parallel program, and can be used as the base class for implementing data parallel object.

4.1 Supporting Task Parallel Distributed Objects

Figure 3(a) shows the structure of *PObj*. Darken arrows represent normal channels that a distributed object used to map network message to its methods, while hollow arrows represent channels for interoperation between parallel programs. A port class defined in *PObj* contains the mapping information of methods already dispatched by *PObj*'s scheduler (see [12] for details of the *port* and *PObj* classes).

The address portion consists of three parts: global process communicating address (host name and process_id), object name, and method name. This definition contains input as well as return value, and is implemented as a duplex port. It can be used to represent the call and reply message when performing remote method invocation. To minimize extra synchronization overhead, we restrict concurrent execution on one processor by closing the port to all other requests before the current active client release it. Only the scheduler can spawn parallel executing object entities.

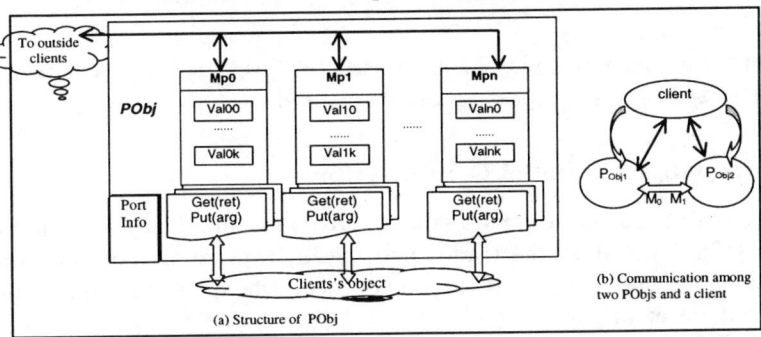

Fig. 3. Interoperability Model for Task Parallel Distributed Objects

Client requests without port operation is treated as common object method invocations. The client stub automatically converts client requests with port operation into two phases: **port query** and **method invocation**. The former fetches the runtime information of the requested method while the latter send request directly to the corresponding port.

4.2 Supporting Data Parallel Distributed Objects

The distributed object implementation using the data parallel model provides various transportation methods to support distributed data interoperation. However, we can consider distributed data as an object too. Distributed data objects can be viewed as task parallel objects that are distributed among several nodes and provide data services to client applications. A major difference is that a *DSeq* object provides many array operation and data distribution functions. In addition, methods of *DSeq* are not dynamically dispatched by the scheduler, but dispatched to different processors during object construction (through dispatch) and provide services to data parallel applications. *DSeq* provides data distribution information through *get_part_info()*. The method called Partition in *DSeq* allows client to define data partition ratio, for example, prt.strsplit="1:2:2:4" denotes partitioning these data into four blocks as 1:2:2:4 and distributing these blocks into four processors. Data partition by row or column is supported, for example, part.rowsplit = BLOCK denotes putting data sequentially on each nodes by row, with the length of each row defined by BLOCK. Each data block is assigned a port for listening to the data request of applications and send related data to them. It also incorporates an interface to the scheduler, so that the client can dispatch data to it through the scheduler. The declaration of *DSeq*'s interface can be found in [12]. This class overrides the subindex operator, so that parallel statements can get the port number of a data portion transparently when accessing the data resource, and complete data operations through port access.

5 Performance Analysis

The test environment is a Fujitsu AP3000 workstation cluster consisting of 143MHz SUN UltraSPARC processors running the Solaris operating system.

Processors are connected via a 200MB/sec bi-directional Torus network. The Fujitsu PVM/AP message-passing library is used.

5.1 Intra-Object Parallel Computation

Detail results of inter-object parallelism and remote method invocation can be found in [12]. To analyze the trade-off between transparency of obtaining parallelism and the performance of the intra-object scheduler, we implemented the Mandelbrot set example.

In the implementation of the Mandelbrot class, we use a public method to compute the vertical lines with fix horizontal distance in a Mandelbrot set. The method places the computed result into a matrix. Obviously, the method has access dependence with itself because two simultaneous invocations write their results to the same matrix. Let l denotes the fix distance and assume a client invokes the method l times with different starting vertical line to get the whole Mandelbrot set, and the scheduler dispatches all these requests to execute in parallel on idle processors. Since the supplier is duplicated during dispatching, the access dependence will not affect the achieving of parallelism. The maximum iteration number is 1024.

We implemented the Mandelbrot set as a parallel program and as a DoHPC program, and collected the elapsed times for different plane sizes and varying the number of processors from 2(P2) to 16(P16). The percentage differences in

	Plane Size	Speedup				Efficiency			
		P2	P4	P8	P16	P2	P4	P8	P16
Parallel	200 × 200	1.92	3.74	7.59	13.32	0.96	0.94	0.95	0.83
	400 × 400	1.92	3.89	7.70	15.05	0.96	0.97	0.96	0.94
	800 × 800	1.96	3.98	7.71	15.59	0.98	0.99	0.96	0.97
DoHPC	200 × 200	1.88	3.65	7.28	12.17	0.94	0.91	0.91	0.76
	400 × 400	1.90	3.82	7.38	13.89	0.95	0.96	0.92	0.87
	800 × 800	1.94	3.91	7.50	14.65	0.97	0.98	0.94	0.92

Table 1. Mandelbrot Set – Speedup and Efficiency

elapsed time show that DoHPC is problem size scalable, e.g., for 16 processors, the overhead reduces from 9.3%(200 × 200) to 6.4%(800 × 800). Table 1 shows that speedup efficiency of DoHPC for the Mandelbrot set with large problem size is greater than 90%. DoHPC's overheads for achieving transparent intra-object parallelism include scheduling, object replication, state recovery and in the merging of results. The results indicate that the intra-object parallelism mechanism is efficient and is scalable.

5.2 Interoperability of Distributed Parallel Objects

To evaluate the performance of interoperability among distributed parallel objects, we use a 2D-FFT (2 Dimension Fast Fourier Transform) parallel algorithm.

We compare three different implementations of the 2D-FFT algorithm. Assume we have eight images with each image represented by a 512 × 512 matrix, and using eight processors as the execution platform. In the *first case*, we use one distributed parallel object and dispatch the execution processes of its 1D-FFT method belonging to this object to the eight processors for execution. In the *second case*, we use two distributed parallel objects and divide the processors into two groups, each containing four processors. Computation starts with the first object computing the 1D-FFT of rows in parallel, and on completion, the object transposes the matrix and transfer it to the second object. Subsequently, the second object completes the computation of the matrix in parallel. In the meantime, the first object starts computation on the next image (matrix). For the two distributed parallel objects to interoperate, we adapted the CORBA interoperability model of communicating through the scheduler. The *third case* is similar to the second but implements the DoHPC interoperability model distributed parallel objects. The interoperation of two distributed objects is analogous to a 2-stage pipeline, and overlaps communication with computation. Table 2 shows

Number of Processors	Schemes	Elapsed Time(ms)		Speedup	
		8 images	16 images	8 images	16 images
4	one object	157051	308606	3.14	3.23
	two distributed objects – CORBA	156549	301333	3.15	3.31
	two distributed objects – DoHPC	152333	291136	3.24	3.43
8	one object	84394	174513	5.85	5.72
	two distributed objects – CORBA	83002	166163	5.94	6.01
	two distributed objects – DoHPC	77925	148925	6.33	6.70
16	one object	48293	90294	10.23	11.05
	two distributed objects – CORBA	47486	92682	10.40	10.77
	two distributed objects – DoHPC	41876	84030	11.78	11.87

Table 2. 2D-FFT – Elapsed Time and Speedup

the elapsed time and speedup for 4 to 16 (divided into 2 groups of 8 processors each) processors.

As the dispatching of rows of data to processors can incur a high communication overhead, the performance result of the 2D-FFT is better with two distributed objects (overlaps in computation and communication) than using one distributed object. However, when the number of processors reaches 16, speedup with CORBA style interoperability model is even lower than with one object (16 images). In this case, processors used by the second object is blocked waiting for data from the scheduler, and hinders the smooth operation of the pipelined 2D-FFT algorithm. In DoHPC, the interoperability model achieves higher speedup in this situation by ensuring that these processors receive data from the first object directly.

One of the factors that influence speedup is the degree of problem parallelism relative to the available machine parallelism. For a fixed problem size and increasing machine parallelism, the speedup varies from 79-86% on 4 processors, 71-84% on 8 processors and 64-74% on 16 processors. However, DoHPC's interoperability model improves speedup efficiency by up to 20% over the one object implementation. Data-dispatching overhead is reduced by communicating through port in DoHPC.

6 Related Works

Much research has been carried out to optimize the underlying layer of distributed object architecture, such as improving the efficiency of current distributed object's communication [6, 2], extending the real-time and QoS features [11, 14] and supporting the mobile environment in a distributed object framework [13]. Some frameworks support the composition of parallel component, such as LSA [3], POOMA [10], PARDIS [8], etc.

DoHPC differs Multi-threaded ORB in two aspects. Firstly, DoHPC provided partial concurrency control through inter-method analysis, which makes the achieving of parallelism in a distributed object more transparent to its clients. Secondly, the implementation of multi-threaded ORB is based on parallel computers with shared memory, such as SMP parallel computer. DoHPC is designed for both shared- and distributed-memory systems.

PARDIS is a distributed system that aims to provide interoperability between distributed, heterogeneous, and data-parallel objects [7]. DoHPC differs from PARDIS in supporting the interoperability of both SPMD and MPMD parallel objects through a uniform interoperating model.

LSA provided a visual programming environment to compose parallel components. Components transfer data from one to another through ports. The port object in LSA could establish a data channel differing from the stub/skeleton channel in common distributed object architecture. However, the port of LSA cannot transfer data from one parallel thread in an object to another object directly. This function is supported in DoHPC.

7 Conclusions

In this paper, we present a distributed object-based framework for parallel computing called DoHPC to exploit both intra- and inter-object parallelism in distributed object architecture. The framework emphasizes both performance and compositional features in distributed object architecture. The implicit intra-object parallelism is exploited through inter-method dependence analysis. The proposed interoperability model exploits both task and data parallelism. Performance results obtained on a cluster of UltraSPARC workstations show that implicit intra-object parallelism can be efficiently exploited and are scalable. Support for interoperability of distributed parallel objects improves speedup by 20%.

The functionality of DoHPC can be extended in the following directions. Firstly, the dependence analyzer can be enhanced to estimate the computing complexity of programs to improve on load balancing. Secondly, we can use parallel compilation techniques to partition compute- intensive methods to exploit intra-method parallelism for parallel execution. Thirdly, more efficient method of distributing shared data among processors is necessary for data that has to be replicated many times. This reduces the scheduling cost and enhances the scalability of the scheduler. Finally, the data parallel mechanism can be optimized for various network environments. For example, in a wide-area network with different types of connection mode, how to make use of them in the data parallel mechanism efficiently is important for the overall efficiency of a parallel application.

References

1. J. Barth, A Practical Inter-procedural Data flow Analysis Algorithm, *Communications of the ACM*, Vol. 21, No. 9, pp. 721-736, 1978.
2. E. Eide, K. Frei et.al, Flick: A Flexible, Optimizing IDL Compiler, *ACM SIGPLAN Notices*, Vol. 32, No. 5, May 1997.
3. D. Gannon, R. Bramley et.al, Developing Component Architectures for Distributed Scientific Problem Solving, *IEEE Computational Science & Engineering*, Vol. 5, No. 2, April-June 1998.
4. A. Gokhale and D. C. Schmidt, Measuring the Performance of Communication Middleware on High-Speed Networks. In *Proceeding of SIGCOMM'96*, pp. 306-317, Stanford, CA, August 1996.
5. A. Gokhale and D. C. Schmidt, The Performance of the CORBA Dynamic Invocation Interface and Dynamic Skeleton Interface over High-Speed ATM Networks. In *Proceedings of GLOBECOM'96*, pp. 50-56, London, England, November 1996.
6. A. Gokhale and D. C. Schmidt, Evaluating Latency and Scalability of CORBA Over High-Speed ATM Networks., In *Proceedings of the International Conference on Distributed Computing Systems*, Baltimore, Maryland, May 1997.
7. K.Keahey, A Model of Interaction for Parallel Objects in a Heterogeneous Distributed Environment, *Technical report*, ftp://ftp.cs.indiana.edu/pub/techreports/TR467.ps.Z, Sept, 1996
8. K. Keahey, Dennis Gannon, PARDIS: A Parallel Approach to CORBA, Technical report, ftp://ftp.cs.indiana.edu/pub/techreports/TR475.ps.Z, February 1997.
9. O. Nierstrasz, A Tour of Hybrid-A Language for Programming with Active Objects, *Advances in Object-Oriented Software Engineering*, ed. D. Mandrioli and B. Meyer, Prentice-Hall, 1992.
10. J. Reynders, P. J. Hinker et.al, POOMA: A Framework for Scientific Simulation on Parallel Architectures, available in: http://www.acl.lanl.gov/pooma/documentation.
11. D.C. Schmidt, S. Mungee et.al, Alleviating Priority Inversion and Nondeterminism in Real-Time CORBA ORB Core Architectures, In *Proceedings of the Fourth IEEE Real-Time Technology and Applications Symposium*, San Francisco, December 1997.
12. C. Wang and Y. M. Teo, Supporting Parallel Computing on a Distributed Object Architecture, *Technical Report*, School of Computing, National University of Singapore, February 2000.
13. G. Welling, B.R. Badinath, Mobjects: Programming Support for Environment Directed Application Policies in Mobile Computing, *ECOOP95*,1995.
14. V.F. Wolfe, L.C. DiPippo, et.al, Real-time CORBA, In *Proceedings of the Third IEEE Real-Time Technology and Applications Symposium*, Montréal, Canada, IEEE, June 1997.

Clustering SMP Nodes with the ATOLL Network: A Look into the Future of System Area Networks

Lars Rzymianowicz, Mathias Waack, Ulrich Brüning,
Markus Fischer, Jörg Kluge and Patrick Schulz

Dept. of Computer Engineering, University of Mannheim, Germany
{larsrzy,mathias,ulrich,mfischer,joerg,patrick}@atoll-net.de
http://www.atoll-net.de

Abstract. Though several System Area Networks (SAN) are available to enhance the communication performance of a cluster, none of them has taken yet into account the special needs of clusters built out of SMP machines. Since single-CPU computers will be more and more replaced by small-scale SMP systems in the near future, an appropriate cluster interconnect should be adopted to the needs of multi-CPU nodes.
In this paper, we investigate the properties of the ATOLL SAN[1], which offers with its four replicated network interfaces (NI), the integrated 8x8 switch and its four links the possibility to boost the total communication performance and message throughput inside a SMP node. As the chip design is in its final stage at the time of writing, we use detailed simulations of all hard- and software components to precisely predict the performance of the ATOLL SAN and show that ATOLL is capable of serving efficiently several message transfers at a time while eliminating the need to multiplex the network interface in software.

1 Introduction

Cluster Computing tries to deliver supercomputer performance through interconnecting powerful workstations or PCs with high performance networks. As the number of nodes steadily increases, a consequence of more scalable software and higher demand for computing power by applications, issues like maintainance effort, power consumption and simpleness of administration become major factors when building a cluster. This leads to the current trend to use two- or four-way SMP machines as node architecture instead of single-CPU systems, since it is much easier to deal with 8 quad-CPU nodes than with 32 single-CPU PCs.

Another positive effect is the use of fast shared memory-based communication inside a node, while only messages between different nodes cross the network. This can be accomplished through a communication library, which transparently directs communication requests to the appropriate interface, like the SCore system of the Japanese RWCP project [1] or BIP-SMP [2]. Communication between

[1] ATOLL is jointly developed with Siemens PC Systems GmbH, Augsburg, Germany

processes on the same node takes place in SCore v3.x by using shared memory regions, whereas calls with destinations on remote nodes are served by using the network. Another solution to program clusters of SMP nodes is to mix both programming models (shared memory and message passing), e.g. Pthread and MPI calls [3].

But as the number of CPUs inside a single node steadily increases, a single Network Interface Card (NIC) quickly becomes the major bottleneck [4]. The NIC has to be multiplexed between all CPUs of the node, significantly reducing message throughput of the system. This bottleneck can be solved by adding more NICs to each node, but this solution increases the costs of the system to an unacceptable level in most cases.

The ATOLL SAN addresses the bottleneck problem by simply replicating NIs and a very aggressive integration of network components into a single ASIC. Besides offering four independent NIs to the node, the chip also integrates an 8x8 crossbar switch, together with four bytewide links. This unique *Network on a Chip* design is able to satisfy the communication needs of SMP nodes at a yet unmatched cost level.

The rest of this paper is organized as follows: After briefly introducing the ATOLL SAN in Section 2, the simulation environment and performance prediction method are described in Section 3. Detailed performance numbers are then presented in Section 4, before in Section 5 the results are summarized.

2 The ATOLL SAN

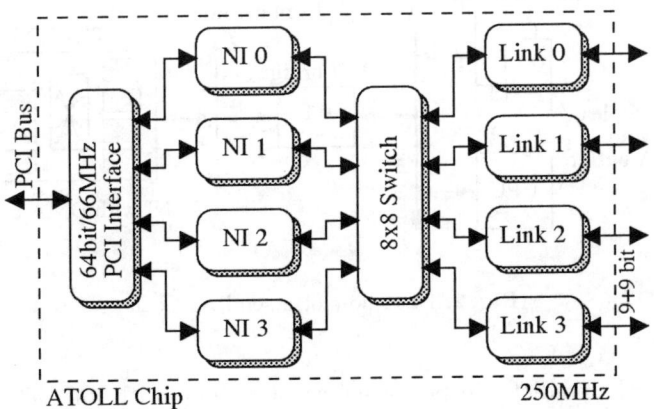

Fig. 1. The architecture of the ATOLL SAN ASIC

The ATOLL network [5] is a new SAN, which enables users to build cost-efficient clusters out of single-CPU, and especially, SMP nodes. Its unique architecture, as depicted in Fig.1, of four replicated NIs and their tight integration, together with four link interfaces and an 8x8 crossbar switch, into one single ASIC is an ideal base for building clusters of SMPs. The integrated switch eliminates the need for any external switching hardware. With its four link connectors out of each network adapter, topologies like a 2D torus or grid can be formed by simply connecting each of the four links to the adjacent nodes.

Furthermore, with its four independent NIs the ATOLL PCI card offers the possibility to serve four concurrent open calls on network devices by applications without the need to multiplex them. Inside a quad-CPU node e.g., all processors can send messages simultaneously, even though the PCI bus must be shared by all NIs. By implementing support for the high-end version of the PCI bus (64 bit, 66 MHz), the I/O bottleneck is not removed but significantly softened. With each of its eight unidirectional links offering a peak data rate of 250 Mbyte/s, the ATOLL NIC will be the only available SAN interface capable of saturating the high-end PCI bus, as shown later on.

3 The Simulation Environment

Prior to the ability to benchmark the real implementation, it is desirable to predict the performance of a design by extracting performance numbers from a simulation model. This permits the validation of the expected performance and gives future users first outlooks about the applicability of the design. This section describes the simulation environment used to gather performance data of a node equipped with the ATOLL NIC.

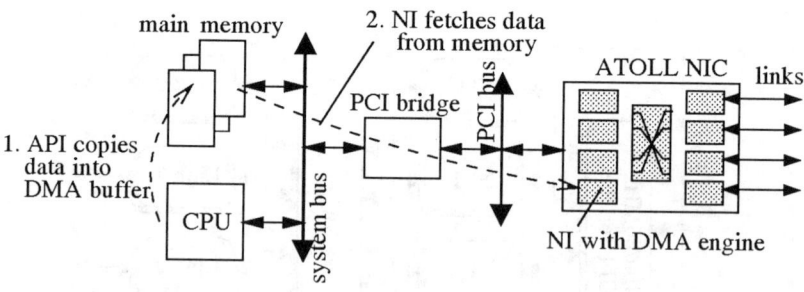

Fig. 2. Data flow of a send API call

To correctly predict the performance of a node with an ATOLL card, all parts of a communication operation must be included in the simulation. For this evaluation, we assume that a user issues function calls of the ATOLL API, which offers a low user-level interface to the ATOLL SAN. We also focus on the performance of small-sized messages (32–4096 bytes), since this is the most interesting category.

So first of all, on the software level the performance of the ATOLL API send/receive calls has to be measured. On the system level, the main memory and the PCI bridge have to be taken into account. And finally, the NIC hardware itself needs to be simulated. Fig.2 shows all data movements involved on the sending side, with the receiving one being analogic to it.

3.1 Software Simulation

To develop the driver software and the low-level API for the ATOLL NIC prior to the first running prototypes, a software simulator was implemented to act like the ATOLL NIC. This task is simplified through the fact, that most interactions between software and the NIC are memory-mapped. This enabled the implementation of most software components without a real ATOLL PCI card. In detail, a send call of the ATOLL API includes the following steps:

1. various checks (valid destination, enough buffer space available, etc.)
2. copy data within the user space into the DMA buffer area
3. assemble send descriptor and enqueue it into the descriptor table
4. trigger message transfer by storing the new descriptor table pointer

Except the data copy, all remaining steps are independent from the message size. Therefore, this part of the API call can be measured once and simply added to the specific memcpy performance. We measured an average value of 0.5 μs for this overhead.

Table 1. memcpy performance for data transfer into(out of) DMA buffer area

msg size (byte)	send: user $\to$ DMA (μs)	receive: DMA $\to$ user (μs)
32	0.3	0.2
256	1.3	0.9
512	1.4	1.1
1024	2.3	1.6
4096	11.3	10.9

To finetune the API calls for high performance, send and receive calls have to be treated differently. More specific, the DMA buffer areas have to be allocated with different memory types to avoid a significant loss of performance. Suppose, the DMA buffer for sending data is allocated as normal write-back memory. The CPU would copy data to be sent now only into the cache, since DMA buffers are not shared within the ATOLL API. When the DMA engine inside a NI now tries to read the data from the DMA buffer, each access to a new cache line would result in an expensive invalidation and write-back cycle. This would significantly slow down the data transfer. To avoid this behavior, it has to be made sure that the data is written immediately to main memory. Therefore, the send buffer area is declared as 'write-combining', which is a new memory type introduced with the P6 family of Intel processors [6]. Several store instructions to consecutive addresses get assembled in a write buffer and are written to memory in a single burst, resulting in superior bandwidth. For the receiving side, the normal cachable memory type is sufficient. To measure the `memcpy` performance from the DMA buffer area into the final destination, only a single `memcpy` call was timed, since only with the first access to this region a read miss and a following cache line fill are triggered. This corresponds exactly to the real behavior of a running ATOLL system.

All measurements were made on a Dual Pentium III 500 MHz system with a 100 MHz Front Side Bus and the Intel 440BX AGPset chipset, running Linux. An internal CPU-cycle counter was utilized to measure precise execution times [7]. Tab.1 depicts the times measured to copy data within user space to/from the DMA buffer area.

3.2 Hardware Simulation

The ATOLL ASIC is implemented as Verilog HDL[2] RTL model. To verify its correct behavior, extensive simulations have been performed. Special care has been taken to validate the PCI compliance of the chip using PCI Bus Functional Models (BFM). The simulation environment, as depicted in Fig.3, is composed of the following units:

1. two ATOLL chips connected via their four links
2. each ATOLL chip interfaces to a dedicated PCI bus
3. a PCI Master BFM is used on each of the two PCI busses to initiate data transfers and generally acts as CPU of the host system
4. a PCI Slave BFM is used to model system memory

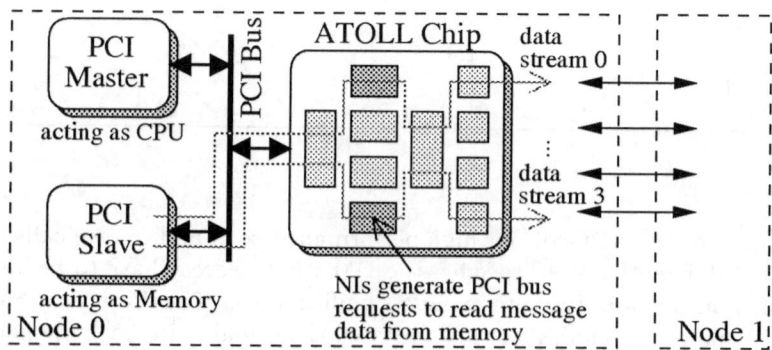

Fig. 3. Simulating two nodes connected by an ATOLL network

The PCI busses are running 64 bit transfers at 66 MHz, thus offering a peak bandwidth of 528 Mbyte/s. Two components of a real system are not modeled in

---

[2] The Verilog Hardware Description Language (HDL) is the leading programming language used to implement hardware on various levels of abstractions

this environment: the system bus and the PCI bridge. The system bus normally offers much more bandwidth than the I/O bus, so that contention between data transfers initiated by the ATOLL NIC and the remaining memory accesses occurs on a rare basis. Therefore, we decided to ignore the impact of system bus contention in this evaluation.

However, the PCI bridge adds some latency to data transfers between the NIC and main memory. Before the requested data is available, the PCI bus must be arbitrated, and the PCI bridge must forward the address to the system bus. It is difficult to measure a concrete value at this point, since performance varies significantly between different chipset implementations. So we decided to add a fixed *bridge penalty* $t_{bridge} = 200\,ns$ (≈ 12 PCI cycles) each time the ATOLL NIC initiates a PCI transfer to read/write data from/to main memory.

4 Performance Measurements

One-way latency and sustained bandwidth have been measured by simulating a constant stream of messages with various sizes. In the following, the specific formulas are listed to compute both values:

$$latency = APIcall_{snd} + (n_{snd} \times t_{bridge}) + t_{atoll} + (n_{rcv} \times t_{bridge}) + APIcall_{rcv} \quad (1)$$

where $APIcall_{snd/rcv}$ corresponds to the sum of the `memcpy` latency and the fix API call overhead, and $n_{snd/rcv}$ to the number of PCI (burst) transfers. t_{atoll} is the complete message transfer time through the ATOLL network taken from the hardware simulation, from the start of the DMA engine inside the NI to the PCI write cycle of the last data at the receiver.

$$bandwidth_{snd} = \frac{size_{message}}{(T_{start}^{n+1} - T_{start}^{n}) + (n_{snd} \times t_{bridge})} \quad (2)$$

where T_{start}^{n} corresponds to the time the ATOLL NIC starts sending message n by fetching the first data from memory. Since the hardware simulation does not include the PCI *bridge penalty*, it is added for each PCI transfer initiated by the NIC. So roughly, the bandwidth is measured as the data rate, at which the ATOLL NIC is able to feed messages into the network.

4.1 Single Sender Performance

First of all, the single sender performance was simulated. It is assumed, that one single process utilizes a single NI of the four offered by the ATOLL NIC to communicate with other nodes. Tab.2 depicts the parameters extracted from simulation runs.

Regarding latency, one can expect one-way API latency starting at about $5\,\mu s$ for a single NI in use. Most parameters show a sub-linear scaling with respect to message size. This is a result of the higher utilization of all components along the

Table 2. Parameters for a single NI in use

msg size (byte)	n_{snd}	t_{atoll} (μs)	n_{rcv}	$T^{n+1}_{start} - T^{n}_{start}$ (μs)	latency (μs)	BW (Mbyte/s)
32	3	1.6	3	0.9	4.3	20.3
256	3	2.7	6	1.8	7.7	101.7
512	4	3.8	10	2.8	10.1	135.6
1024	6	6.0	18	5.1	15.7	155.0
4096	18	19.6	66	18.7	59.6	175.1

data path with increasing message size. E.g., the PCI interface is able to fetch up to 256 bytes of data in a single burst (if supported by the bridge), which can be observed in the fact, that for both 32 and 256 byte messages, the NIC only needs 3 memory accesses to fetch the data (1 descriptor, 1 routing data, 1 payload).

Numbers extracted from the simulation for the achieved bandwidth (BW) are very promising. With 175.1 Mbyte/s for 4 Kbyte messages about 70% of the peak link bandwidth are reached. The characteristic $N_{1/2}$ value is around 400 bytes, a remarkable number.

4.2 Multiple Sender Performance

One unique feature of the ATOLL NIC is its ability to serve up to four communicating processes simultaneously. So simulations have been made with one, two and with all four NIs in use. This puts a lot of pressure on the PCI interface, which has to multiplex between all data streams. Since its implementation tries to serve all active NIs in a fair manner, measured parameters have been almost the same for all NIs. So Tab.3 and Tab.4 list the numbers of one specific NI, which are representative for all other NIs.

Table 3. Parameters for two NIs in use

msg size (byte)	n_{snd}	t_{atoll} (μs)	n_{rcv}	$T^{n+1}_{start} - T^{n}_{start}$ (μs)	latency (μs)	single NI BW (Mbyte/s)	acc. BW (Mbyte/s)
32	3	1.9	3	1.1	4.6	17.9	35.8
256	3	3.4	5	2.4	8.2	81.3	162.6
512	4	5.0	9	3.7	11.1	108.5	217.0
1024	6	7.4	14	6.0	16.3	135.6	271.2
4096	18	20.4	36	19.1	54.4	172.0	344.0

With two sending processes only minor performance changes can be observed. While most parameters show slightly increasing values due to resource conflicts, some also get better with more pressure on the NIC. E.g. the number of PCI transfers on the receiving side dramatically shrinks for larger messages. With only one data stream, the data FIFOs inside the PCI interface often run empty, which results in PCI transfers that are ended and restarted repeatedly. With the PCI interface switching between both data streams, more data gets assembled in FIFOs and, thus, longer burst cycles can be made. For 4 Kbyte messages, latency is even 5 μs less than for a single NI, and bandwidth drops only by 2%. By adding the numbers for both NIs in use (acc. BW), one is able to obtain a data rate of more than 340 Mbyte/s out of a single node.

Table 4. Parameters for all four NIs in use

msg size (byte)	n_{snd}	t_{atoll} (μs)	n_{rcv}	$T^{n+1}_{start} - T^{n}_{start}$ (μs)	latency (μs)	single NI BW (Mbyte/s)	acc. BW (Mbyte/s)
32	3	2.9	3	2.2	5.6	10.8	43.2
256	3	4.9	6	3.9	9.9	54.2	216.8
512	4	7.3	8	6.1	13.2	70.7	282.8
1024	6	12.9	14	10.6	21.8	82.7	330.8
4096	18	39.4	32	36.8	72.6	96.6	386.4

With all four NIs simultaneously sending data, the conflicts for the PCI bus increase significantly. Latency is still below 10 μs for small messages, but for larger ones the message data travels twice as long as for two senders along the data path. Bandwidth is almost halved, if one compares the numbers for two and four senders. This indicates, that two NIs in use can almost saturate an ATOLL-equipped node system.

But accumulating the bandwidths of all NIs results in almost 400 Mbyte/s for 4 Kbyte messages, a number yet unmatched by any single SAN NIC.

4.3 Comparison with Other SANs

Even though the numbers reported are taken from simulations, they should come close to the real performance of the ATOLL SAN, since we have tried to model most components involved in a communication call very accurately.

The single sender performance of ATOLL is very close to competitive products, even often exceeds their numbers. E.g. Myricom, Inc. reports bandwidths of more than 140 Mbyte/s for their user-level GM layer on top of Myrinet 64 bit/66 MHz SAN interfaces [8]. One-way latency of less than 5 μs on Myrinet were measured for the BIP message layer [9]. Other SANs like ServerNet II or Giganet offer similar latencies, but provide only a peak link bandwidth of 125 Mbyte/s.

Comparing ATOLLs multiple-sender performance is quite difficult, since no other SAN NIC offers multiple NIs. One would have to use multiple NICs with a software layer capable of managing them to obtain the same functionality. The relative high costs of todays SAN solutions (\$ 1000+ for NICs, \$ 200+ per switch port) make this approach appear very uneconomic.

The only SAN able to compete with ATOLLs accumulated bandwidth of 400 Mbyte/s is QsNet from Quadrics Supercomputers World Ltd. With its byte-wide links operating at 400 MHz, bandwidths of more than 210 Mbyte/s were measured. This value is expected to increase up to 300 Mbyte/s with better PCI implementations available [10]. Though prices for QsNet are not available, they should be significantly higher than the ones expected for ATOLL, due to the large amount of on-board SRAM (32 Mbyte).

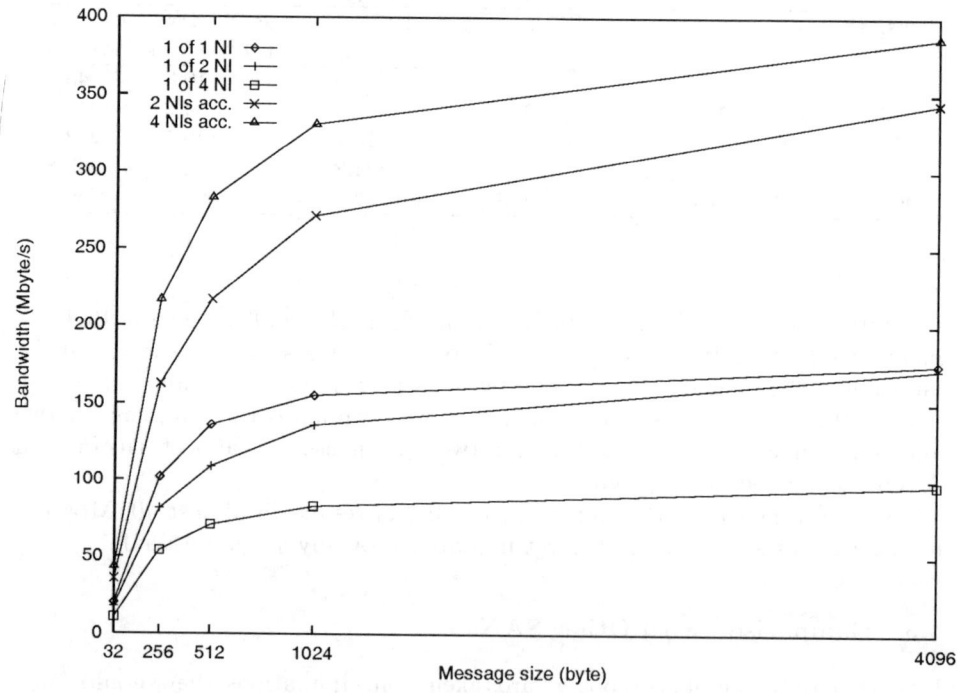

Fig. 4. Single-NI and accumulated bandwidth for 1, 2 and all 4 NIs in use

Since Cluster Computing has become so attractive because of its better price/performance ratio, special care has been taken to develop with ATOLL a low-cost version of a SAN NIC. With its single chip solution and the elimination of separate switching hardware and on-board memory, costs have been reduced to a minimum. Though a detailed price is not fixed yet, the commercial price for a single ATOLL NIC should be in the range of \$ 600. So overall, ATOLL should simply offer the best performance at the lowest price.

5   Results and Future Work

With this detailed evaluation and performance prediction we have shown that the ATOLL PCI card will be an attractive solution to build clusters. Not only that its single sender performance is highly competitive. Its accumulated performance of multiple parallel NIs inside a single NIC gives users the ability to fully exploit the potential of clustered SMP nodes. With the ATOLL SAN, even for a 64 bit/66 MHz version of the PCI bus, the network is not anymore the performance bottleneck. No other current SAN NIC is able to deliver a data throughput exceeding 400 Mbyte/s. Fig.4 displays all measured bandwidth numbers.

As soon as the first prototype cards are running, we will run benchmarks to validate the numbers presented in this publication. Future work will then concentrate on an in-depth evaluation of the ATOLL SAN and suggestions to further improve the hardware.

References

1. Ishikawa Y., Tezuka H., Hori A., Sumimoto S., Takahashi T., O'Carroll, and Harada H.: RWC PC Cluster II and SCore Cluster System Software – High Performance Linux Cluster. Proceedings of the 5th Annual Linux Expo, (1999) 55–62
2. Geoffray P., Prylli L., and Tourancheau B.: BIP-SMP: High Performance Message Passing over a Cluster of Commodity SMPs. Proceedings of the ACM/IEEE SC99 Conference, Portland, Oregon, (1999)
3. May J., and de Supinski B.: Experience with Mixed MPI/Threaded Programming Models. Proceedings of the Intl. Conference on Parallel and Distributed Processing Techniques and Applications (PDPTA), Las Vegas, NV, **Vol. VI** (1999) 2907–2912
4. Lumetta S., Mainwaring A., and Culler D.: Multi-protocol Active Messages on a Cluster of SMPs. Proceedings of the ACM/IEEE Supercomputing Conference, San Jose, CA, (1997)
5. Rzymianowicz L., Brüning U., Kluge J., Schulz P., and Waack M.: ATOLL: A Network on a Chip. Proceedings of the Intl. Conference on Parallel and Distributed Processing Techniques and Applications (PDPTA), Las Vegas, NV, **Vol. V** (1999) 2307–2313
6. Intel Corp.: Intel Architecture Software Developer's Manual. **Vol. 3**, System Programming Guide, (1997)
7. Gooch R.: Linux Documentation I've written: Performance Monitoring Interface Control. WWW: http://www.atnf.csiro.au/ rgooch/linux/docs/index.html, (1999)
8. Myricom, Inc.: Performance Measurements: GM API with 64-bit Myrinet/PCI Interfaces. WWW: http://www.myri.com/myrinet/performance, (1999)
9. Prylli L., and Tourancheau B.: BIP: A New Protocol Designed for High Performance Networking on Myrinet. Lecture Notes in Computer Science (LNCS), **Vol. 1388** (1998)
10. Quadrics Supercomputers World Ltd.: QsNet High Performance Interconnect. WWW: http://www.quadrics.com/web/public/fliers/qsnet.pdf, (1999)

An Architecture for Using Multiple Communication Devices in a MPI Library[1]

Hernâni Pedroso and João Gabriel Silva

Dept. Engenharia Informática/CISUC
Universidade de Coimbra – Polo II
3030-397 Coimbra
Portugal
{hernani,jgabriel}@dei.uc.pt

Abstract. Every year new intercommunication technologies emerge and the parallel computing libraries have to evolve to use these new technologies. Most of the existing libraries have the communication medium dependent code embedded in the core and rely on it for the management of the environment. This makes it difficult to adapt the libraries to those emerging technologies. A new architecture, implemented in the WMPI library, cleanly separates the communication device dependent code from the library core, making that adaptation much easier. Third party institutions (technology vendors and other research institutions) can thus easily adapt WMPI for any technology. Additionally, the architecture enables the concurrent use of any number of communication media by the WMPI library. The freedom is also given to the developer of the communication device dependent code to use or not dedicated threads for sending and/or receiving. This paper describes the main characteristics and rationale of that new architecture.

1 Introduction

One of the drawbacks of constructing a PC cluster for high performance computing used to be the lack of interconnection networks that could present high bandwidth and low latency between nodes. Most of the PC clusters use TCP over Ethernet and Fast Ethernet networks, which are not optimized for message exchange performance but to reliability and low cost. As the computational power of the PCs grew, the network became the bottleneck of the cluster. Aware of this fact, the hardware vendors have created new technologies that improve the message passing performance between the computer nodes. VIA [1] has recently presented great acceptance, although SCI [2], Myrinet [3] and Gigabit Ethernet are also widely used. Other proprietary interconnection mechanisms are used in dedicated parallel machines.

---

[1] This work was partially supported by the Portuguese Ministry of Science and Technology through the R&D Unit 326/94 (CISUC) and the project PRAXIS XXI 2/2.1/TIT/1625/95 named ParQuantum.

The parallel computing libraries have to follow the improvements of the intercommunication technologies. However, most of the available implementations strongly depend on the communication protocol through which they communicate. To make a release for a new technology deep changes are necessary in the core of the library. This slows the pace of new releases. To overcome this drawback it is necessary to separate from the core of the library the code that depends on each communication device.

A new architecture to achieve that objective is presented here. The base idea was to ease the development of the communication device dependent code, thus allowing everyone, even without the access to the core library's source code, to adapt it to virtually any communication medium. This is very important since e.g. the technology vendors can develop their own adaptation code and speed up the release of the library for their products. This new architecture is already implemented in the current version of the WMPI library.

WMPI (Windows Message Passing Interface) [4,5] was the first full implementation of the MPI standard [6] for Windows operating systems. It was originally based on MPICH [7] and ever since has received several optimizations to improve its performance and reliability. In a recent study, which evaluated implementations of MPI for Windows NT environment [8], WMPI was considered the best freely available implementation. In addition, the study concluded that WMPI rivals with other commercial implementations in performance and functionality. The new structure of the WMPI core is also driven by the need of making deep changes in the library for the implementation of the MPI-2's Dynamic Process Creation chapter [9]. A full MPI-2 implementation is under development and will be released in the near future.

2 The WMPI Internal Design

The first versions of WMPI were strongly based on the MPICH implementation. Hence, WMPI inherited a highly structured implementation that aimed for portability and performance. MPICH uses an Abstract Device Interface (ADI) [10] to mediate the interaction between the library core, that is independent of the communication protocol used, and the code that depends on it. In this paper we will call to this protocol dependent code "MPI-device". On the other hand, in this paper "communication device" is used to refer to the interface, offered by the layer below the MPI library, to some underlying communication protocol or media. That interface can be offered either by the operating system, by some other communication library or by the hardware itself.

Although the MPICH structure is well fitted for portability, the support for simultaneous communication devices is absent. Moreover, the MPI-devices are responsible for the management of part of the environment. Hence, they have a quite complex architecture. The most common implementation of MPICH uses, within an MPI-device for the TCP protocol and shared memory, an implementation of p4 [11], that is responsible for setting up the environment and managing the communication between processes. In fact, p4 is by itself a complex library for parallel computing.

The WMPI inherited this characteristic and used an implementation of p4 for Windows environments. In order to increase the performance of the system, the p4 code became increasingly dependent on the communication devices and deep changes would be necessary to port WMPI for different communication media. This was slowing the release of WMPI for new technologies.

Hence, we decided to design a new architecture for WMPI that removed from the MPI-device functionality all tasks that are independent from the particular communication device (Figure 1). Contrary to MPICH, the environment management is completely performed in the core of the WMPI library, which we call the WMPI Management Layer (WML), and not in the MPI-devices. These are only used when the WML does need to perform a communication device dependent operation. The interface between the MPI-Devices and the WML is called Multiple Device Interface (MDI).

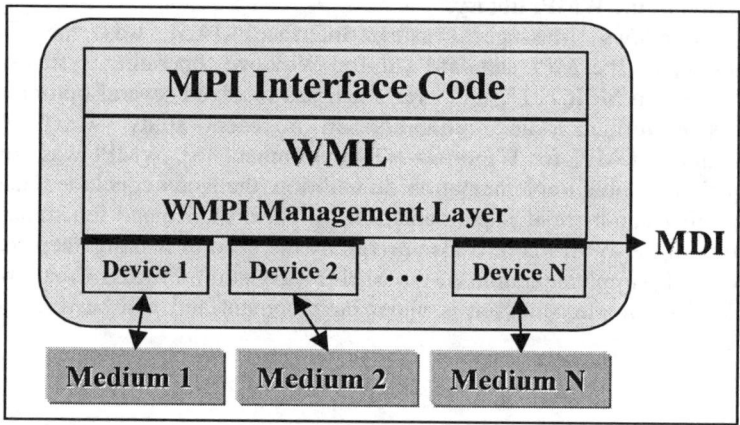

Fig. 1. WMPI layer scheme.

The MPI-devices are independent DLLs that are loaded by each process at the startup of the MPI environment. When the WML initializes the MPI-devices, these return a structure that contains pointers to the functions and some shared variables that implement the MDI interface. Also during initialization, the WML passes to the MPI-devices some pointers to functions of the WML ("upcalls"), for the MPI-devices to use. The decision of sharing data and having upcalls was quite hard, since it makes the MPI-device implementation more difficult, but it significantly improves the performance and scalability of the system.

Each communication device uses different information to represent the address of a certain communication end-point or a connection. Since this implementation does not want to restrict in any way the kind of communication protocols that can be used, we decided to keep the address information in the MPI-devices, because we considered it not to be possible to arrive at any general format. The WML never handles the addresses directly, only pointers to buffers (opaque to the WML) containing those addresses. Only the MPI-devices know and process the buffers contents.

Using Multiple MPI-Devices Simultaneously

It is common to find more than one type of communication media in a cluster. The most common configuration is, probably, shared memory and TCP/IP. However, other configurations are possible and some of them may use more than two different communication media. The WML has the capability to interact with any number of different MPI-devices. Within each process, the WML, according to a cluster configuration file, associates a MPI-device to each machine of the cluster.

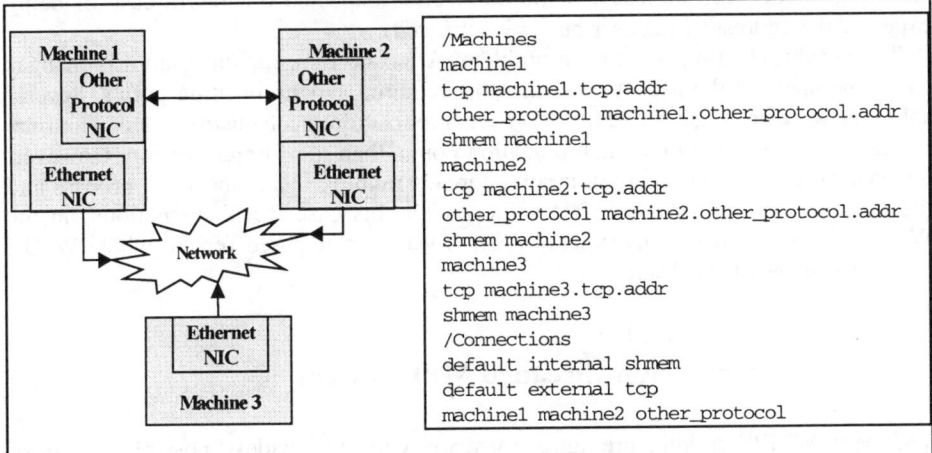

Fig. 2. Example of a cluster definition.

The user, through a text file, defines the configuration of the cluster. Figure 2 presents an example of such a file. The configuration file is divided in two parts; one defines the machines that constitute the cluster and the other the MPI-devices used to exchange information between the machines. For each machine in the cluster, the user has to specify which MPI-devices the processes running on that machine use and the address of the machine within each communication protocol. In the second part, the user has to indicate which are the MPI-devices that processes of one machine should use to communicate with the other machines. It is possible to set a default MPI-device for internal communication (within the same machine) and for external communication (with remote machines). The user also has the possibility to set specific MPI-devices for the communication between two specific machines.

When a process starts it loads the DLLs that implement the MPI-devices available in the machine where it is running. Using the connections section of the configuration file, the processes associate each machine of the cluster (hence all the processes running on that machine) with a MPI-device. When it is necessary to send or receive data from a process the WML finds the MPI-devices' structure that is associated with the machine where the other process is running and uses the function pointers that it specifies.

Performance

Such a modular and flexible architecture usually presents performance problems. The memory copies and synchronization are the most important factors in performance. This new architecture does not use more memory copies than the previous version. If a message arrives at a process when the corresponding receive has already been issued, the MPI-devices are able to directly place the data into the user-level buffer. This is accomplished by using an upcall function of the WML that searches the receiving queues and indicates if the message is expected or not (in case of being expected the address of the user buffer is retrieved).

The synchronization was also a problem. After the first implementations, changes had to be made to the MDI to diminish the required synchronization points. Several solutions were considered and different synchronization mechanisms tested. The final result contains some more synchronization points than the former version. However, this synchronization points only involve threads running inside the same process and the operating system provides fast solutions for this case ("critical sections" in the Win32 API). Moreover, this synchronization was also required for the whole WMPI library to become thread safe.

3 Threaded and Non-threaded MPI-Devices

The new WMPI architecture aims to work with the widest possible range of communication media. While gathering information about the several existing technologies, we noticed that some required the received information to be removed from the medium as soon as possible. Technologies fitting in this category normally use a message queue from where the reading process gets the messages. This message queue may overflow and refuse further messages if it gets full. This may happen when a big message or a big number of small messages arrive and they are not removed from the queue quickly enough. Applications using this type of architecture normally have a thread to remove the information from the queue to the application memory. Since in MPI the user threads cannot be expected to be always available to remove data from the communication devices, because their main objective is performing user level computations, each MPI-device must have its own dedicated thread for receiving. On the other hand, some communication devices do not require a receiving thread. Imposing a thread that is not mandatory causes an added latency, due to context switching, when receiving messages through the MPI-device and in the global system. It was thus decided that the MDI should not require the existence of a thread in MPI-devices.

For the case when that thread does exist, the WML and MDI were designed to take full advantage of its presence. Hence, when a receiving thread receives a message, it uses the WML upcalls to know if the message is expected or not. If expected, the thread fully completes the receive. Otherwise, the message is stored in the WML receive queue, where the user finds it when he/she finally issues the corresponding MPI Receive.

If a device does not have a receiving thread, the messages remain in the communication medium until the WML needs to receive a message through that communication device. When this happens, the user thread that issues the MPI call goes into the MPI-device code (passing through the WML) and executes the same steps a dedicated receiving thread would. It receives all the available messages and returns to the WML. If the expected message was not in the set of received messages, the thread waits until another message arrives at the MPI-device and executes again the MPI-device's code for receiving.

It is possible to use MPI-devices, without any threads, that use polling to wait for new messages. In this case the MPI-device uses the user thread that issued the MPI function call to poll inside the receive function. However, it is required that the MPI-device frees the thread to let it check for incoming messages in the other MPI-devices. The implementation of the MPI-device can be more or less aggressive in the usage of the thread for pooling, depending on the time it takes before letting the user thread go check for messages in other devices. This should be a configurable parameter, which would let the user choose it according to the number of other MPI-devices and the probability of receiving messages from other MPI-devices.

4 The Multiple Device Interface

The MDI primitives can be grouped in four categories: environment, process creation, connection management and communication.

Environment

The environment functions are used to initialize and finalize the MPI-devices. When a WMPI process starts, it loads each MPI-device's DLL and calls an initializing function. This function should create all the internal structures necessary for the operation of the MPI-device. It is expected that after the execution of this function, the MPI-device be ready to exchange data or open connections. The initialization function is also responsible for returning a structure that contains the address of the MPI-device's functions that implement the interface, along with other internal data.

When the user calls the MPI_Finalize function the MPI environment is dismantled, at this point all the devices have to end and the allocated resources be freed. The WML calls an end function of each MPI-device. This function should stop the MPI-device and free any allocated resources.

Whenever a MPI process gets a fatal error, the abort sequence is started. It is necessary to abort all the processes of the MPI environment. The WML uses the MPI-devices to broadcast an abort message through all the processes. It is expected that the abort message be sent immediately with the highest priority possible. When receiving an abort message, the MPI-devices must immediately inform the WML layer. It is not required to the MPI-devices to end the environment, they just notify the WML, which tries to free all the allocated resources (including all MPI-devices).

Process Creation

When setting up the environment it is necessary to create processes in remote machines. The MPI library requires some mechanism to remotely start the processes. The only communication medium that the WML knows to be available to contact the other machine is the MPI-device specified in the cluster's configuration file. Hence, the library has to rely upon the MPI-devices to create new processes. Each MPI-device has a function to create local processes and another to create remote processes. It was decided to use different functions for each of these two cases, because MPI-devices usually have different mechanisms for each situation.

Connections Management

The WML is connection oriented. A connection is opened between each pair of processes and is maintained while the processes are connected (have a common communicator or group). Although the WML uses connections, the MPI-devices can emulate them over a connectionless protocol (e.g. UDP). The MPI-devices just have to pass to the WML some connection handle, which remains associated with the address of the remote process in the device. The interface contains functions to open and close connections. Notice that connections to the same address can be opened and closed several times, since the MPI-2 Process Creation and Management chapter allows two MPI computations to connect and disconnect.

Communication

Communication functions can be divided into two separate groups:

Send

The functionality that must be provided by the MPI-devices to send messages is very similar to the basic functionality of MPI. The MPI-devices have two different functions, one is blocking (similar to the MPI_Send) and the other is non-blocking (similar to the MPI_Isend). When the blocking send function is called, it blocks until the message is completely sent. When the function returns, the memory buffer, where the data to be sent was placed, can be reused without any impact on what arrives at the target process.

The non-blocking send allows to optimize the send operation by scheduling it to the best opportunity. The function should return immediately to let the user thread continue its computation. However, the user or the WML cannot manipulate the buffer contents until the device notifies that the message was actually sent. When the WML makes a call to the MPI-device's non-blocking send function, it passes to the MPI-device a message handle. When the MPI-device completes the send, it uses that handle to notify the WML of that event, through a specific upcall function. This way it is possible to have the user thread working while the thread of the MPI-device performs the communication. When the user thread wishes to know if the message was already sent, it just verifies the state of the data in the WML layer, by means of the MPI_Wait or MPI_Test functions. To speedup the notification of the completion of a send and avoid some synchronization, the WML provided the MPI-device with a variable, together with the message handle, that should be set to true as soon as possible after the message is sent. The user thread will test this variable prior to issuing a synchronization call that waits for the notification from the MPI-device.

Receive

It is required by the MPI standard that communication channels have FIFO behavior. This means that MPI-devices must process messages in arrival order and that the communication devices are FIFO channels.

Each message has a header that contains the information that identifies it in the MPI environment (e.g. context, tag, source, etc). For each message, the MPI-device uses an upcall function to directly access the WML receiving queues and verify, using the header information, if the message is expected or not.

If the message is expected, the upcall function retrieves the address of the user buffer where the data is expected. This way it is possible to receive the data directly from the communication medium into the user buffer. When the message has been copied to that buffer, another upcall is issued to notify the WML.

If the message is not expected the upcall function places an entry in the unexpected queue. In this case, the upcall does not retrieve a memory address for the data destination. The MPI-device is responsible for managing an intermediate buffer for unexpected messages. This allows the MPI-device to optimize the reception of unexpected messages. Having some pre-allocated memory or leaving the data in the communication device (which might be the case in a shared memory device) can speedup the reception. Additionally, the MPI-device has to place a handle of the message data in a variable received from the upcall function. When the user issues a receive for a message that has already been received, the WML call a function of the MDI-device, where the message to be copied is identified by that handle. In this case it is the user thread that makes the work of completing the receive, since it would be blocked anyway and two context switches can be saved.

5 Developing New MPI-Devices for WMPI

Compared with the former WMPI version, the development of a new MPI-device is much simpler. The separation between modules allows the developers to focus only on the part being developed. Moreover the reduced set of actions required, which form a set of twelve specific functions, are much smaller and simpler code. The new TCP and shared memory MPI-devices have one order of magnitude less code lines than the ones that integrated the former WMPI version. Each of the twelve functions performs a well-defined action, which is extensively specified in a document available in the WMPI web site [5].

Since each MPI-device is a separated DLL it is easy to test each DLL separately. Moreover, since the WMPI dynamically loads the MPI-devices' DLLs at startup according to the cluster configuration file, it is possible to quickly test the new MPI-device under the WMPI environment. This permits testing and tuning the MPI-device in the real environment without any change to the WMPI core or even knowing how the internals work.

The developed MPI-devices can be used even if new versions of the WMPI core are released, since the MDI is expected to be maintained. In addition the release of a new version of a MPI-device can be done completely asynchronously with a new WMPI release, since the users only have to update the MPI-device's DLL.

6 Related Work

The other MPI implementations known to the authors only support one MPI-device at a time. While many can use simultaneously shared memory with another protocol like TCP or VIA, each combination requires a different MPI-device. The ability of having one MPI-device for shared memory, another for TCP and so on, and mixing them at will at run-time, is unique to WMPI.

The MPICH implementation, that was used as the basis for many other implementations because it was the first open implementation of the MPI 1.0 standard, uses an internal structure that assigns important tasks in the management of the environment to the MPI-devices. Hence, it is difficult to construct new MPI-devices, an endeavor that requires strong knowledge of the library core. The MPI/Pro [12], MP-MPICH [13], FM-MPI [14], MPICH-NT [15] implementations are all based on MPICH and suffer from the same restrictions of MPICH regarding MPI-devices.

The ability of loading the necessary MPI-devices at startup time according to a configuration is also absent in all these other libraries.

7 Conclusions

Most of the available MPI libraries require a very significant amount of work when adding support for a new communication medium. This is due to the dependence between the MPI management code and the communication access code. In fact, part of the management functionality is performed by technology dependent code. The new architecture described in this paper frees the dependent code from core management activities and makes the library core totally independent of the communication media. A reduced set of operations has been identified as communication device dependent. The code that interacts with the communication device must implement all of these operations. A generic interface (Multiple Device Interface - MDI) is specified and is used by the WMPI Management Layer to perform the necessary actions on the communication device. This way the code that accesses the medium is completely independent from the library core.

Communication medium vendors or any other research institutions can develop their own device implementation for the WMPI library. This should diminish the time to have the WMPI available for new technologies and to have better devices. The complete specification of the MDI can be found at the WMPI web site [5].

Another very important characteristic of the described architecture is its capability of supporting an unlimited number of simultaneous MPI-devices. The user just has to specify in a simple text configuration file which devices to use and they get dynamically loaded at start-up time.

MPI-devices can be threaded and non-threaded, both types being supported in an efficient way and being able to co-exist at run-time. If threaded, the MPI-devices work asynchronously with the user threads, allowing them to compute while communicating with other processes. If the process has more than one MPI-device, the MPI-devices work concurrently, since the access to the WML data, through the upcalls, is thread safe.

Last but not least, the whole WMPI library is now thread safe.

References

1. Virtual Interface Architecture Interface Specification: Version 1.0 http://www.viarch.org (December 1997).
2. Hellwagner, H., Reinefeld, A.: SCI: Scalable Coherent Interface – Architecture and Software for High-Performance Compute Clusters. Lecture Notes in Computer Science Vol. 1734, Springer, ISBN 3-540-66696-6 (1999).
3. Boden, N., Cohen, D., Felderman, R., Kulawik, A., Seitz, C., Seizovic, J. and Su, W.: Myrinet: A Gigabit-per-second Local Area Network. IEEE Micro, pp. 29-36 (February 1996).
4. Marinho, J. and Silva, J.G.: WMPI – Message Passing Interface for Win32 Clusters. Proc. of 5th European PVM/MPI User's Group Meeting, pp.113-120 (September 1998).
5. WMPI Homepage – http://dsg.dei.uc.pt/wmpi
6. Message Passing Interface Forum: MPI: A message-passing interface standard. International Journal of Supercomputer Applications, 8(3/4):165-414 (1994).
7. Gropp, W., Lusk, E., Doss, N. and Skejellum, A.: A High-Performance, Portable Implementation of the MPI Message Passing Interface Standard. Parallel Computing Vol. 22, No. 6, (September 1996).
8. Baker, M: MPI on NT: The Current Status and Performance of the Available Environments. NHSE Review, Volume 4, No 1 (September 1999).
9. Message Passing Interface Forum: MPI-2: Extensions to the Message-Passing Interface. (June 1997), available at http://www.mpi-forum.org.
10. Groop, W., Lusk, E.: MPICH ADI Implementation Reference Manual – DRAFT. ANL-000, Argonne National Laboratory, Mathematics and Computer Science Division (August 1995).
11. Butler, R. and Lusk, E.: Monitors, messages and clusters: The p4 parallel programming system. Parallel Computing, 20:547-564 (April 1994).
12. MPI Software Technology, Inc – http://www.mpi-softtech.com
13. MP-MPICH: Multiple Platform MPICH - http://www.lfbs.rwth-aachen.de/~joachim/MP-MPICH.html
14. Lauria, M., Pakin, S., Chien, A.: Efficient Layering for High Speed Communication: The MPI over Fast Messages (FM) Experience. Cluster Computing, HPDC7 special issue (1999)
15. MPICH – A Portable MPI Implementation - http://www-unix.mcs.anl.gov/mpi/mpich/

Results of the One-year Cluster Pilot Project

Kimmo Koski, Jussi Heikonen, Jari Miettinen, Hannes Niemi, Juha
Ruokolainen, Pekka Tolvanen, Jussi Mäki, and Jussi Rahola

CSC - Scientific Computing Ltd.
P.O.Box 405
FIN-02101 Espoo, Finland
email{Kimmo.Koski, Jussi.Heikonen, Jari.Miettinen, Hannes.Niemi,
Juha.Ruokolainen, Pekka.Tolvanen}@csc.fi
http://www.csc.fi

Abstract. During recent years clustered systems using off-the-shelf processors and standard Ethernet networks have been increasingly popular. The motivation has been primarily the cheap price of systems, but also the rapid development of standard processors. So-called Beowulf systems have spread around the world. This development is further accelerated by the LINUX-boom which provides an ideal and free operating system for these clusters.

In the end of 1998 a pilot project was started at CSC in order to study PC clusters for parallel computing. The aim was to gather experience about performance, bottlenecks and quality of programming tools. We will present in this paper our experiences in building a 32-processor PC cluster and running parallel applications on it.

The initial problems in getting the systems to work properly and tuned represented a fair workload. Data communication bottlenecks turned out to be a major performance issue. A small number of parallel scientific applications, mostly running on Cray T3E or other parallel systems, was transferred to the PC cluster environment in order to get porting experience. A large set of benchmarks was run on the cluster, in addition to the benchmarking of the ported applications.

1 Introduction

As the widespread use of so-called Beowulf cluster computer systems expands, criticism of expensive supercomputer systems has grown. For a computer center providing mainly centralized supercomputer systems, such as CSC, it has become necessary to further study the usability of clustered systems. A pilot project of PC clusters was hen established to determine whether clusters could provide a solution to the increasing need for supercomputer resources in the Finnish research community.

The pilot project was started in the end of 1998 by contacting a number of university laboratories running parallel applications. The main target was to port their codes to a pilot cluster and compare the performance with traditional supercomputers. In addition, practical experience with the work load required to build a cluster was of interest.

A test system of sixteen dual-processor computing nodes and one server node was purchased and installed. This pilot system was at the disposal of the project from February 1999 until January 2000. During the time the cluster was operational, a large amount of activities took place: from benchmark runs to porting and optimizing parallel applications.

The paper describes experiences in building and running a PC cluster. A number of application benchmark results (CFD, physics, biocomputing, others) are included and compared with more tightly coupled systems.

2 Cluster environment

The LINUX cluster which was used during the project consisted of sixteen dual-processor Dell Precision 410 workstations and one Dell PowerEdge 2300 front-end server. The systems were attached together with a standard Ethernet using a BayStack 450-24T switch. The processor was a Pentium II with a 400 MHz frequency. Each computing node had 128 MB central memory and a 4 GB local SCSI disk, except the front-end system in which the amount of memory was 512 MB and disk 18 GB.

The systems were running LINUX RedHat 5.2 and later 6.0 with a kernel level of 2.2.5. Multiple compilers were used, such as public domain gnu (C, C++, Fortran 77) compilers and commercial Portland Group and Absoft compilers (Fortran77, Fortran90). The MPI implementation used MPICH from Argonne National Laboratories with a standard p4 (TCP/IP) communication layer.

The communication network was based on switched 100 Mbit/s Ethernet. The switch had 24 ports from which 16 ports were used for the computing nodes and one for the server. The fast Ethernet was based on a 3Com chip and embedded in the mother board of each system. The largest parallel jobs were run with 32 processors using both processors from all the systems.

The cluster had a dedicated private network. The login server was connected both to the public local area network and the private cluster network. Design was such that users did not need to login to the nodes directly. The login server was used with a script automatically starting the jobs in the selected nodes.

3 Building the computing environment

When building the cluster environment the two tasks that proved most time-consum were getting the programming environment (compilers, debuggers etc.) to work and tuning the communication network. The physical installation of hardware and operating systems consumed less than 10% of the time. The breakdown for various areas is illustrated in Fig. 1.

Most of the initial problems were related to the network tuning. Since the nodes were dual-processor systems, there were issues in the multi-level hierarchy

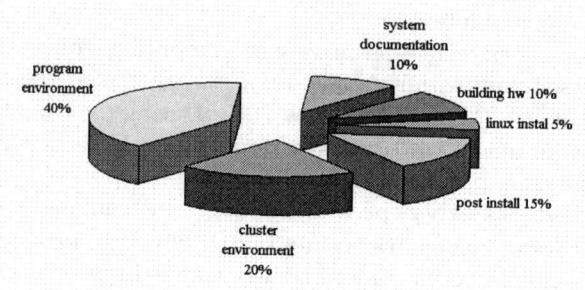

Fig. 1. Workload for different areas in building the computing environment.

of communication between processors. LINUX kernel level updates and network tuning solved a part of the problem, even though there still seemed to be some level of problems in shared memory multiprocessor support.

Testing a set of compilers and programming tools was another major issue. The selection of available software tools, at least in the spring of 1999, was quite limited. A few C, C++ and Fortran77/90 compilers were selected (Absoft, GNU, Portland Group). Other tools, especially parallel debuggers, were mostly inadequate or not available at all. MPICH was available and worked nicely.

Using the cluster together with the research groups located in various universities in different geographical locations was a challenge for the system maintenance. Since there were no adequate queuing systems in the beginning, users had to reserve a slot in a physical calendar. During the summer of 1999 PBS (Portable Batch System) was selected as a queuing system and installed into the cluster.

Suitable monitoring and maintenance tools were not available. For that reason a group of people from the project searched and evaluated the available public domain and commercial tools. Product development was carried out by the research partners from Abo Akademi, who designed a specific cluster management and monitoring system.

Despite these initial building problems the project team got the cluster system up and running. The programming environment was adequate to port and tune the selected application codes. Also, a set of benchmarks could be run after a reasonable amount of results was obtained.

4 Performance

Benchmarks varied from the low level single CPU and subsystem tests including parallel communication to the real parallel application tests. In the applications some reference tests, such as NAS Parallel benchmarks, were used as well as a number of customer applications, for example weather code, ab-initio physics and molecular dynamics simulations.

The following benchmarks were run:

1. Single CPU tests (CSCSuite2b, matrix multiplication, vector kernels, STREAMS)
2. MPI communication benchmarks (point-to-point, synchronization, global communication)
3. NAS Parallel benchmarks (subset: Multigrid, FFT 3D)
4. Application tests (FINGER, HIRLAM, PARCAS, VASP)

The single processor performance has been tested at CSC during the last several years by CSCSUITE2b benchmark set which consists of 14 different codes. Four of the codes are scalar intensive, four are vector dominant and six are mixed scalar-vector code. The benchmark set provides also a single number (TOTAL) which is a geometric mean of the individual benchmark results. The CSCSUITE2b results are presented in Fig. 2.

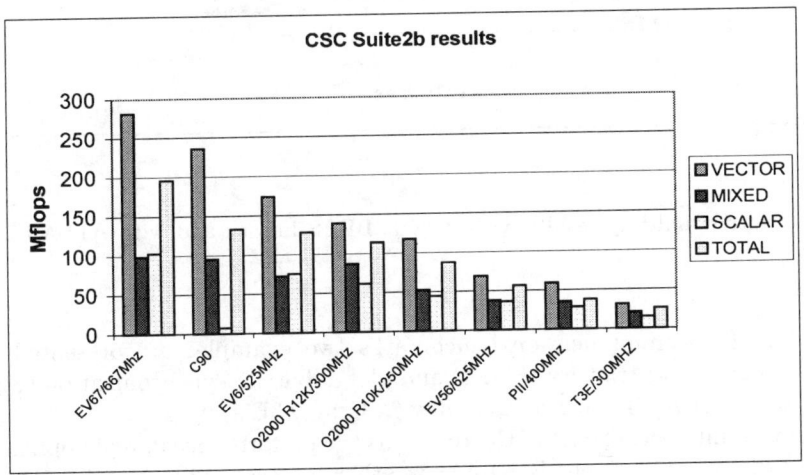

Fig. 2. CSCSUITE2b results

To measure the achievable peak performance of the Pentium processor various matrix multiplication tests were used. First, a typical Fortran loop based on non-optimised code was used, along with the default BLAS library from netlib. Second, an optimised library was used (netlib ATLAS).

The default option from [11] gives a peak performance of about 100 Mflop/s for small 40x40 matrix size. For large 1000x1000 matrixes it rates only 35 Mflop/s. The significant drop in performance for large matrixes means that the overall performance is not satisfactory. Also, the compilers do not recognize the loop structure and do not seem to know how to implement the block accesses to the matrix data.

The best BLAS3 performance is given with the ATLAS [12] optimised library (298 Mflop/s for large 1000x1000 matrixes). The library optimises itself automatically during the compile time. The ATLAS performs well for small matrixes, e.g. the 10x10 matrix runs with 222 Mflop/s (Fig. 3).

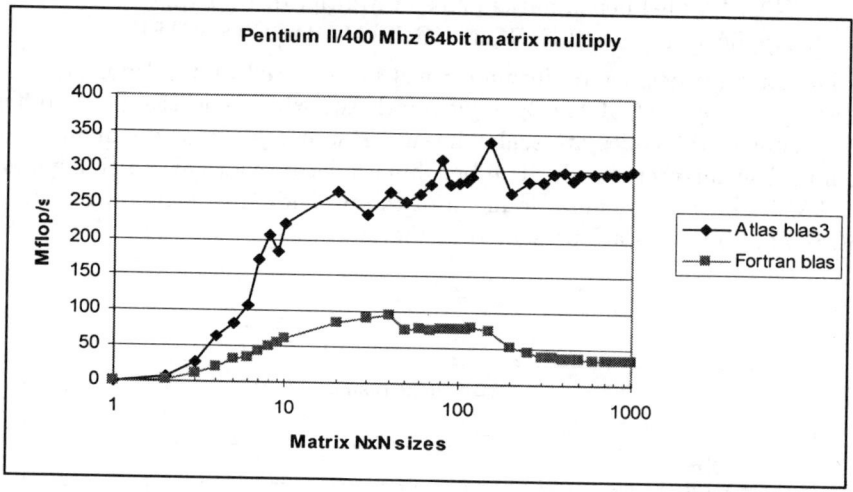

Fig. 3. Matrix multiply results with default BLAS library and with ATLAS BLAS3 library

From MPI communication benchmarks two examples are presented: MPI point-to-point data transfer (Fig. 4) and global barrier synchronization (Fig. 5). The PC cluster results are compared with Cray T3E system.

Another interesting part of the tests was the performance in real applications. Three examples are given: HIRLAM weather forecast code (Fig. 6), PARCAS molecular dynamics code (Fig. 7) and VASP ab-initio physics code (Fig. 8). The results of PC cluster are compared with results from various systems, such as Cray T3E, C90, SGI Origin and different Compaq systems.

The complete benchmark report is available from [8] and [10].

5 Parallel applications

A number of parallel applications were ported and tuned to the cluster. The main codes were:
  1. FINFLO CFD code from the Helsinki University of Technology Laboratory of Applied Thermodynamics and Laboratory of Aerodynamics

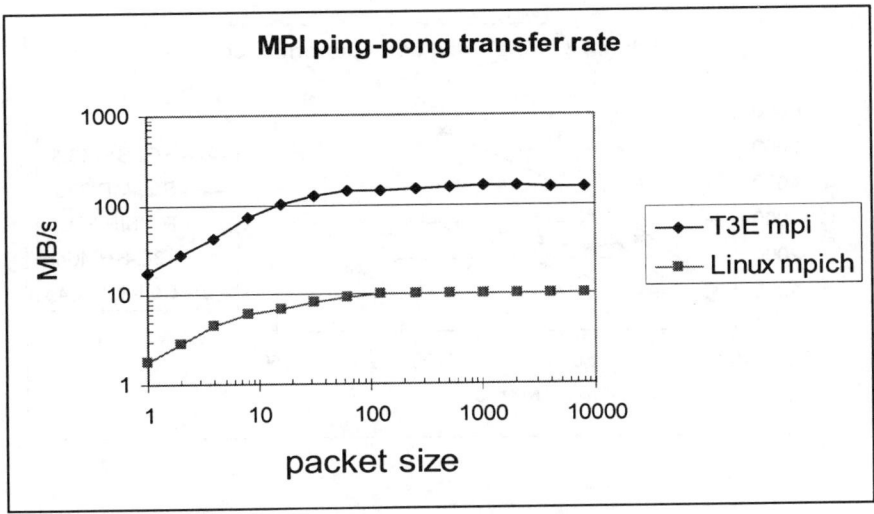

Fig. 4. MPI ping-pong transfer rate

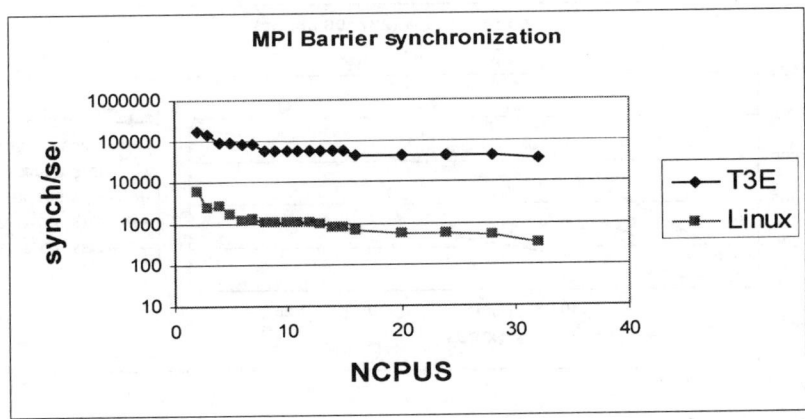

Fig. 5. MPI global barrier synchronization

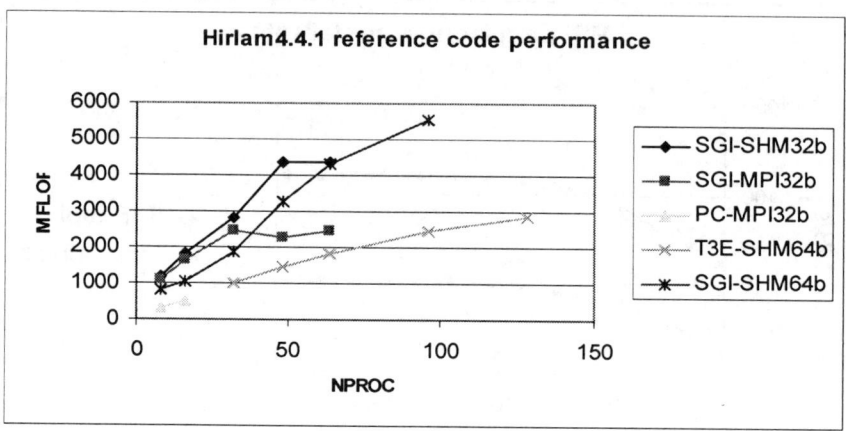

Fig. 6. HIRLAM 4.4.1 weather forecast code (MPI and SHMEM libraries, 32- and 64-bit versions)

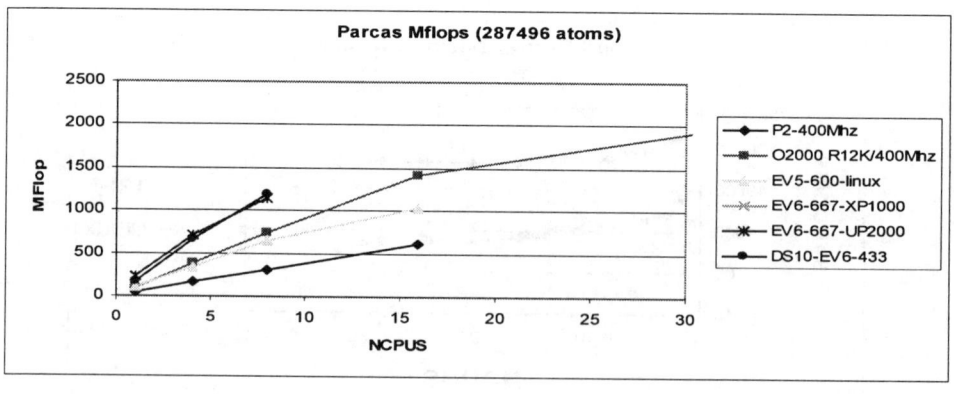

Fig. 7. PARCAS molecular dynamics performance with various different systems

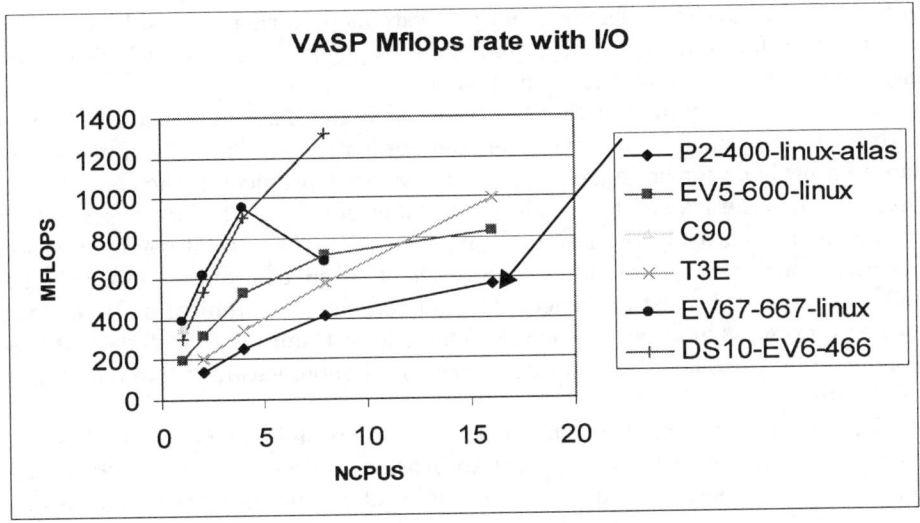

Fig. 8. VASP ab-initio code

2. FINGER Car-Parrinello code from the University of Oulu Laboratory of Chemistry
3. GENFIT genetic algorithm code from the Åbo Akademi Department of Biochemistry and Pharmacy
4. ELMER multi-physical tool from CSC - Scientific Computing Ltd.

Most of the codes had already a MPI version available which made the porting fairly easy. The more challenging task was to optimize the performance of the codes.

A relatively high computation/communication ratio in FINFLO and GENFIT made them suitable candidates for Ethernet-based clusters. ELMER and FINGER were more problematic, for example due to the algorithm design or the large amount of communication between processors. Especially for FINGER, a fast network with low latency would be needed to gain high sustained performance. Additional information can be found from [10] and [9].

6 Conclusions

The tests during the project have shown that PC clusters are well suited for a limited number of parallel applications, especially when communication between the processors is not very intense. For these problems, clusters can provide cost efficient computing capacity. On the other hand, there are a lot of applications in which using clusters does not make much sense - at least with a low performance network such as Ethernet.

In PC systems the single CPU performance is comparative to previous generation RISC processor performance. The price/performance ratio is usually better in the PC class systems. The memory subsystems performance can, however, be a bottleneck for some types of applications (especially for those which require large data set accesses and are tuned to vector systems).

The standard communication hardware, switched Fast Ethernet 100 Mbit/s with TCP/IP, provides an entry level communication system. The transfer rate can be a problem for large datasets but the biggest problem in the communication system is the scalability for a larger number of CPUs. The lack of scalability is due to the relatively big latency compared with the special communication hardware, such as Myrinet (latency in order of 10 microseconds compared with 150 microseconds in Fast Ethernet). The long barrier synchronization time of the cluster represents another bottleneck. The rule of thumb is that if the application communicates less than 10 times a second it should scale well in a standard PC cluster.

The results of the application benchmarks are promising. For example scalability of the cluster in PARCAS until 16 processors (one processor used of each sixteen nodes) is shown to be good. VASP results scale reasonably until 8 processors. It can be estimated that a number of real applications exist who can cost-effectively use Ethernet-based PC clusters.

Shared memory nodes create another challenge for the PC clusters. In PC nodes a memory subsystem does not usually scale for two CPUs. Also the operating system does not seem to share the communication hardware as efficiently as it should. This means that for communication intensive applications the regular Fast Ethernet shared memory nodes are not very useful.

Based on the performance measurements and practical experience of running the cluster, we conclude that this kind of system can not efficiently replace the more tightly coupled systems, such as SGI Origin 2000 or IBM SP, at least with a heterogeneous application environment. PC clusters and RISC based clusters should be seen as complementary solutions to the traditional supercomputers in stead of competing solutions. A small and carefully chosen set of parallel software could be run cost-efficiently on a cluster.

7 Future issues

In the future, alternative communication hardware should be investigated. Both the academic implementations such as Purdue PAPERS, Genova GAMMA and NIST M-VIA, and commercial communication hardware such as Myrinet or SCI from Dolphin/SCALI should be tested.

Faster networks, such as Myrinet or Giganet, could possibly overcome the performance problems in communication networks. However, the high price of these networks raises the cluster costs closer to the traditional supercomputers. Accurate benchmark results of fast networks are unfortunately not available in the scope of this project, but this is a research topic for the near future.

A better model should be investigated for performance scalability analysis. This model should take into the account the effect of global communications. The modeling should include both the system architecture and the application.

A number of interesting areas need to be studied further. For example barrier synchronization mechanisms, such as PAPERS hardware, can solve some of the bottlenecks. In system software further development is needed, for example in the area of parallel debugging. An interesting product is for example Totalview with a LINUX cluster version available in the beginning of the year 2000.

Due to the positive experience in this project, we will build a large cluster for production work. The goal is to get a system with 64-128 processors (Intel or RISC) and a fast network in production in the spring of year 2000.

Acknowledgements

CSC's PC cluster pilot has been a success story of collaboration between many partners. The following university research groups have actively participated in the project:

1. University of Jyväskylä: Center for Mathematical and Computational Modeling
2. Åbo Akademi: Department of Computer Science and Department of Biochemistry and Pharmacy
3. Lappeenranta University of Technology: Laboratory of Telecommunications
4. Helsinki University of Technology: Laboratory of Applied Thermodynamics and Laboratory of Aerodynamics

References

1. Pfister, G. F.: In search of Clusters. ISBN 0-13-899709-8.
2. Aznar, G.: LINUX HOWTO Index http://www.linuxdoc.org/HOWTO/HOWTO-INDEX.html
3. Gropp, W. and Lusk, E.: Installation Guide to mpich, a Portable Implementation of MPI.
4. Gropp, W., Lusk, E., Doss, N., Skjellum, A.: A High-Performance, Portable Implementation of the MPI Message Passing Interface Standard.
5. The home page of the Beowulf project at CESDIS http://beowulf.gsfc.nasa.gov/
6. Buyya, R. (ed.): High Performance Cluster Computing: Architectures and Systems, Vol. 1 and 2, Prentice Hall PTR, New Jersey, USA 1999.
7. Hill, J., Warren, M. and Goda, P.: I'm Not Going to Pay a Lot for This Super Computer. Linux Journal 1997. http://www2.linuxjournal.com/cgi-bin/frames.pl/lj-issues/issue45/2392.html
8. Koski, K. et al.: PC Cluster Project, Final Report. CSC - Scientific Computing Ltd. report series. (To be published in March 2000).
9. Kaurinkoski, P., Rautaheimo, P., Siikonen, T., Koski, K.: Conference Presentation submitted to Parallel CFD 2000. Trondheim, Norway (May 2000).
10. http://www.csc.fi/metacomputer/pckluster/
11. http://www.netlib.org/blas
12. http://www.netlib.org/atlas

Clusters and Grids for Distributed and Parallel Knowledge Discovery

Mario Cannataro

ISI-CNR, Via P. Bucci, 41/c,
87036 Rende, Italy
cannataro@si.deis.unical.it

Abstract. Parallel and Distributed Knowledge Discovery (PDKD) is emerging as a possible killer application for clusters and grids of computers. The need to process large volumes of data and the availability of parallel data mining algorithms, makes it possible to exploit the increasing computational power of clusters at low costs. On the other side, grid computing is an emerging "standard" to develop and deploy distributed, high performance applications over geographic networks, in different domains, and in particular for data intensive applications. This paper proposes an approach to integrate cluster of computers within a grid infrastructure to use them, enriched by specific data mining services, as the deployment platform for high performance distributed data mining and knowledge discovery.

1 Introduction

Knowledge Discovery in Database (KDD) is a general term to indicate the semi-automatic analysis of large volumes of data to find "useful" knowledge, based on the use of techniques and algorithms collectively named data mining (DM) [3]. KDD is a complex process both in time and space, so it can benefit from the use of parallel algorithms that can both speedup the process and improve the accuracy of results [2].

The unprecedented rate at which data is being produced by many fields of human activity, such as retail, banking, e-commerce, sensor data collection, physics and nuclear simulation and experimentation, and the increasing need to analyze and use them in effective way, poses new challenges to the designers and users of KDD and DM systems. These systems should be able:

- to face with large, high dimensional data sets, that very often are geographically distributed;
- to analyze different type of data, where DBMS managed data could be only a short percentage of all data (e.g., mining of unstructured or semi-structured web contents);
- to be used in an easy way, on the basis of high level DM programming models and output specification languages;
- to offer intuitive, easy to use interfaces, including advanced immersive interfaces;
- to cope with security and privacy of data.

In the latest years some KDD systems have been designed implemented on traditional parallel computing platforms to speedup their execution time and scaleup performance with respect to the problem size. However, KDD systems should make use of emerging techniques and advances in parallel and distributed computational environments, such as *cluster computing* and *grid computing*, and high performance networks as well. Very often the task of parallelizing algorithms or their porting over parallel platforms, is not trivial and not cheap, so together with speedup, important goals (perhaps long term) are architecture independence and portability. This architectural abstraction objective could be taken into account in the design of data mining algorithms together with the goal of interoperability among data mining tools. Efforts are made in the field of distributed data mining, facing with data distribution but also with networks details [7,8,9]. A language-based approach in the design of DM and KDD systems, offering a set of data mining primitives and data types, could both simplify the design of algorithms and allows for architecture transparency.

The previous problems cannot be faced only from an architectural point of view, and in fact many efforts are devoted to discover new algorithms and to integrate different scientific and theoretical knowledge into existing ones (e.g. scaling data mining solution to linear algebra or global optimization) [10,11].

This paper, addresses the main requirements of new Parallel and Distributed Knowledge Discovery (PDKD) systems [5] from an architectural point of view, with regards to their implementation and use on cluster computers and grids of geographically distributed clusters. Section 2 gives the basic concepts of cluster computers, grid computing and PDKD. Section 3 presents major architectural requirements. Finally, Section 4 contains conclusions and outlines future work.

2 Background

2.1 Clusters and Grids

Cluster computing is a new area of distributed computing, based on the availability of commodity computers, networks and middleware software [17,18]. Cluster computers are composed from tens to thousands of commodity, stand-alone computers connected by a high performance network, working together as a single computing resource. Each node of a cluster is a complete computer, hosting its own operating system and software – moreover, nodes can be heterogeneous with regard to software, hardware, and configuration. Basic cluster aspects are:
- Single System Image, e.g., the possibility to manage the systems as it was composed by a single node.
- High level communication mechanisms.
- Process allocation, migration and load balancing.
- Support to parallel and distributed computing.
- High availability.
- Heterogeneity transparency (at various levels).
- Parallel I/O and File Systems.

In some sense, clusters comprise functionality of both parallel and distributed systems. Currently, the use of cluster computers is increasing in many application domains, and efforts are made at different level, from the operating system to the middleware, to enhance their functionality and manageability. From a programmer perspective, cluster computing is a cheap and so affordable way to conduct parallel computation. In the PDKD domain, clusters are often used "only" to execute parallel DM algorithms. Recent researches address the problem of integrate clusters within grids [12,13] to obtain worldwide high performance computing environments.

In the last few years, the Web is becoming a powerful infrastructure for analyzing distributed data. More and more application deployment and communication platforms (Java, CORBA) have been developed on it, allowing new transactional and distributed services. To scale-up this trend to massive data analysis and distributed problem solving over geographical wide networks, "the grid" architecture, allowing secure and effective sharing of computing resources and networks, is emerging.

The grid term refers to middleware software allowing transparent remote access to distributed instrumentation and data, yet providing security, resource management, access management, accounting, and other services necessary for applications, users, and resource managers to operate effectively (e.g., Globus, Legion).

Grids have successfully been used especially in computational science to solve large scale experiments and simulations. Currently, grids are used to share computer, supercomputer, dedicated hardware, databases, and clusters of computers, in their various flavors. Today, a need for a *standard* grid architecture framework is emerging as a necessary step to extend the scope of utilization to different scientific domains, comprising data intensive applications. The Integrated Grid Architecture [12] is a promising effort in this directions. It comprises the following layers:

- A set of resources (possibly empty) that can be used to *grid-enable* basic computing or network resources. They can allow the implementation of basic communication functionality (e.g. Quality of Service), resource reservation, allocation and monitoring, and so on.
- A set of *grid services* (middleware), i.e. instrumentation-independent and application-independent services. Among those, authentication, authorization and accounting, remote data location, service level agreement.
- A set of *application toolkits*, providing more specific services and resources for the particular domain, e.g., parallel and distributed computing, remote data access and selection, algorithm selection (problem solving).
- At the top we find the *grid-aware* applications, i.e., applications designed on the top of grid services.

Nowadays, many international projects are developing components for the grid architecture [15, 16] and these efforts will play a key role in the future of high performance computing.

2.2 Parallel and Distributed Knowledge Discovery and Data Mining

PDKD is the application of the KDD techniques to distributed, large, possibly heterogeneous, volumes of data that are residing over computing nodes distributed on a geographic area. As mentioned before, several parallel algorithms for single data mining tasks such as classification, clustering and association have been designed in the past years. However, it lacks a proposal for integrated environments that use novel computing platforms to PDKD environments that integrate different sources, models, and tools. Significant examples where PDKD environments could be used are:
- large organizations that need to mine their distributed data;

- mining of data owned by different organization;
- mining of distributed, geographic wide, sensor originated data;
- mining of data using different techniques in different sites to get and compare output in parallel.

The basic aspect of a PDKD systems is the distributed mining of the data, that can be performed according to the following schemes:

- *Move Data*. The data residing over remote computing nodes is selected and than transmitted to a central node (possibly a cluster) for processing.
- *Move Models*. Each node processes the data locally, and send the predictive model to another node for further processing.
- *Move Results*: Each node processes the data locally until a result is obtained, and send it to another node for further processing.

Some distributed data mining models and systems have been proposed [19, 20, 21] supporting a combination of these schemes. However, the architecture and communication infrastructure aspects are often approached from an application point of view. In the simplest cases, the data are owned by the same organization, and their formats and availability is known. In a more complex case, concerning multi-owned, heterogeneous data, some of the distributed data mining requirements could be filled by the combination of grid services and cluster computers resources.

Combining the results and models emerging in these apparently far areas, such as cluster computing, grid computing and distributed data mining, could empower PDKD and allow a new class of applications to be developed. Each of these areas could leverage on the counterpart efforts reducing costs, and domain scientists could be free to concentrate over specific problems, benefiting by this architectural independence. Furthermore, PDKD could benefits from the efforts that are in progress to define standards for the data mining process such as the CRISP-DM process model that aims to provide a standard process structure for carrying out data mining and the predictive modeling markup language (PMML) specification for dealing with different input and output data formats [1].

3 Basic Architecture to Support PDKD Systems

In this Section we give some basic directions for the integration of KDD in the framework of grid of clusters. The definition of an architecture for PDKD can be approached from different points of view. Here, using a practical approach, the goal is the definition of basic mechanisms and approaches to allow the data mining of geographically distributed data, using cluster of computers interconnected by grids, leveraging general purpose and specific grid services.

The main problems that should be faced are:

- the definition of the middleware to *grid-enable* cluster computers, yet preserving the cluster autonomy and hiding as much as possible its internal organization and management;
- the integration of specific distributed knowledge discovery requirements in the grid architecture, and finally
- the design of *grid-aware* distributed knowledge discovery systems, using basic and domain specific grid services.

The overall architecture, shown in figure 1, is composed by two main components:
- a set of grid-enabled nodes declaring their availability to do some PDKD computation that are connected by,
- a grid infrastructure, offering basic grid-services (authentication, data location, service level negotiation, etc.) and PDKD services.

Here the hypothesis is that the grid continues to serve pre-existing connected nodes. Each node can be a generic computer or a cluster of processors capable of parallel data mining computation. Moreover, a generic node can be involved in the PDKD environment in different ways. It can start a PDKD process, or it just owns data available for distributed data mining, or it can be charged of the execution of a data mining algorithm.

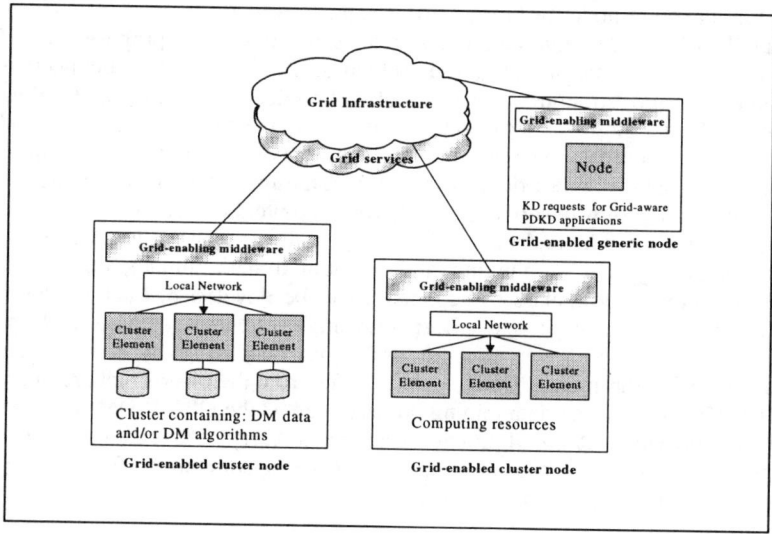

Fig. 1. Grid of cluster computers to solve PDKD problems

3.1 Grid-Enabling Clusters

Grid-enabling a cluster of computers means that:
1. the grid is aware of cluster resources (computing power, data and application) availability;
2. the cluster is able to access grid-services (i.e. to search the grid, to receive, negotiate, accept/reject and serve *grid requests*);

The mutual recognition between cluster and grid is based on a grid-enabling middleware, that should be present on the cluster computer. The goal of this software is to:
- Announce the cluster presence, such as configuration details, and its availability to share resources (data for DM analysis, algorithms for DM execution, simply computing power) and to accept external program code (e.g. Java DM agents). The degree of resources availability (type of cluster, kind of nodes, library, etc) and

their security requirements, are care of the cluster, but should adhere to minimal requirements. This task can be accomplished by using directory services as LDAP or by using grid information service, as in Globus [15].
- Authenticate grid-requests, and, on the basis of service level agreement policy, to accept them - a specific PDKD requirement involves the protection of data to be mined, based on a hierarchy of authorization levels. As previously said, basic security services can be directly offered by the grid. A possible schema, using Public Key Technology, could be based on interoperable Trusted Third Parties, which implement "roaming of security services", as in GSM wireless systems.
- Accept and serve grid-requests. To maintain cluster autonomy, its public description should suffice a requester to choose the cluster as a participant to PDKD, but should prevent access and management of internal resources (scheduling, resource reservation and so on). A resource allocation plan could be negotiated and agreed between parts (e.g., min-max service level). We want underline the importance of cluster autonomy in a heterogeneous multi-owned grid of clusters, where the cluster should be transparent when it shows data and resource availability, but should be *opaque* with respect to the way a task is internally processed.

3.2 Knowledge Discovery Grid-Services

The main services that the grid computing environment must offer to specifically support the knowledge discovery process are:
1. Search of input data for a given KDD goal; different level of details should be allowed (e.g., data warehouse and OLAP aggregation could be useful for). For this task a specialized search engine based on mobile agents could be provided.
2. Selection of useful data (the owner should provide some detailed data description and eventually it can provide data conversion routines).
3. Selection of Data Strategy, i.e. how the DM can choose to access and manage relevant data (major strategies are Move Data, Move Results and Move Models, to which correspond different processing strategies). Moreover, to cope with congested networks, priority-based transmission protocols could be used guaranteeing the real-time delivery of (approximate) data [22].
4. Search of useful discovery tools and algorithms among those that are available through the grid.
5. Selection of Task strategy, i.e. how the DM can choose to coordinate or not data mining algorithms over several nodes (*independent learning* vs *coordinated learning*), enabling multiple and collaborative use of different data mining models that can run on different nodes (clusters or processors).
6. Map a data mining computation to one or more nodes that autonomously schedule and start the computation. Moreover, it should be possible both to search and access distributed DM algorithms libraries, and to send portable object code to clusters not owning proper DM algorithms, using recognized platforms as CORBA, Java and enhanced systems [14].
7. Collect and present the output of the data mining process that has been executed in a parallel and distributed way. Moreover, the user should be able to analyze results using advanced visual interface, such as immersive virtual environments [8]

These basic PDKD services must be coupled and integrated with more general services that support user authorization, authentication, accounting, logging and monitoring.

3.3 Grid-aware PDKD systems

Grid-aware PDKD systems could be described using a language-based approach, combining DM models and algorithms specified using PMML or CRISP-DM [1], with rewriting techniques as used in CORBA. A "grid PDKD program" should:
- specify data set names and search for possible useful data available over the grid;
- describe PDKD computation (data need, application description, process structure, data access pattern, parallelism degree, etc);
- use the grid-services to dynamically match program requirements with the available resources.

At some extend, grid-aware PDKD systems are *not cluster-aware*, in the sense that they negotiate services, maintaining cluster autonomy. The use of the basic and specific grid services, allows architecture independence, but requires the development of middleware software. However, the Internet architecture is an important reference for various aspects.

An important aspect to be considered is the impact over performance of high latency and (often) low bandwidth of grids that involves both general grid applications and PDKD computations. Latency and propagation delay make synchronous communications inefficient and impractical. Other than using asynchronous communications or sending multiple messages in parallel, latency can be hidden by overlapping computation and communication as much as possible. It could be obtained designing applications able to advance computation as long as data are received (data driven).

Although the most intuitive (not easy) approach to this problem is to develop communication layers able to face with these constraints [23, 15], a more comprehensive approach could cope with them at the programming level [24]. A grid application should use communication techniques which are enough expressive to describe it and allow the grid support to dynamically select the best communication and transmission strategy, on the basis of the current situation (available resources and bandwidth). Moreover, the system should trade-off parameters such as degree of parallelism and process granularity, that affect the communication patterns. In other words, the constraints of the grid environments, other than faced at the various levels of the grid architecture, could be accounted in a more integrated way allowing the coexistence of different programming models and execution strategies. These could be dynamically chosen at run time by the grid support. For example, a portion of an application which is allocated over a SMP cluster could use shared-memory communication techniques, whereas another portion, remotely started, could use enhanced communication techniques, improve computation and communication overlapping, or use topology-aware communications.

4 Conclusions and future work

Cluster computing, grid infrastructure, and data mining are three key areas of computer science and they will play a paramount role in the next years. A big challenge for the future is to integrate them for the implementation of high performance knowledge discovery systems that will improve industry and business processes. In fact, distributed mining of large, geographically dispersed, multi-owned, heterogeneous data is not more an option, but a need to solve large scale, inter-country problems.

In this paper we proposed a basic framework to integrate cluster of computers within a grid infrastructure to use them as the deployment platform for high performance distributed data mining and knowledge discovery. The paper, addressed the main architectural requirements of PDKD systems with regards to their implementation and use on cluster computers and grids of geographically distributed clusters. To give a more complete and detailed specification of a PDKD framework will be necessary to further study how each layer that composes the architecture will be implemented, which current and novel tools can be utilized for, and which detailed services should be offered. These issues constitute our future work along the direction of effective exploitation of the computing power of novel architectures for high performance knowledge discovery.

Acknowledgements

I wish to thank the anonymous referees for the useful comments and Domenico Talia for his help on an early version of the manuscript.

References

1. G. Piatesky-Shapiro, The data mining Industry coming of age, *IEEE Intelligent Systems*, pp. 32-34, november/december 1999
2. A. Freitas, S. Levington, *Mining Very Large Databases with Parallel Processing*, Kluwer, 1998.
3. M.J.A. Michael, J.A. Berry, *Data Mining Techniques*, John Wiley & Sons, 1997.
4. D. Abramson, From PC Clusters to a Global Computational Grid, 1st IEEE Workshop on Cluster Computing (IWCC99), Melbourne, 1999.
5. R. Moore, Collection-Based Data Management, Workshop on Large-Scale Parallel, KDD Systems (KDD99), San Diego, CA, 1999.
6. S. Bailey, E. Creel, R. Grossman, S. Gutti, H. Sivakumar, A high performance implementation of the data space transfer protocol (DSTP), Workshop on Large-Scale Parallel, KDD Systems (KDD99), San Diego, CA, 1999.
7. U. Dayal, Large-Scale Data Mining Applications: Requirements and Architectures, Workshop on Large-Scale Parallel KDD Systems (KDD99), San Diego, CA, 1999.
8. G. Williams, Integrated Delivery of Large-Scale Data Mining Systems, Workshop on Large-Scale Parallel KDD Systems (KDD99), San Diego, CA, 1999.

9. R. Grossman, S. Kasif, R. Moore, D. Rocke, J. Ullman, Data Mining Research: Opportunities and Challenges, A report on three NFS Workshops on Mining Large, Massive and Distributed Data, available at http://www.ncdm.uic.edu/m3d-finalreport.htm
10. B. Grossman and Yike Guo, Communicating Data Mining: Issues and Challenges in Wide Area Distributed Data Mining, Workshop on Large-Scale Parallel KDD Systems (KDD99), San Diego, CA, 1999.
11. V. Kumar, Large-Scale Data Mining: Where is it Headed? , Workshop on Large-Scale Parallel KDD Systems (KDD99), San Diego, CA, 1999.
12. Building the Grid: An Integrated Services and Toolkit Architecture for Next-Generation Networked Applications, Working Draft,http://www.gridforum.org/building_the_grid.htm.
13. Foster and C. Kesselman, editors, The Grid: Blueprint for a New Computing Infrastructure, Morgan Kaufmann Publishers, 1999.
14. Foster, G. H. Thiruvathukal, S. Tuecke, Technologies for Ubiquitous Supercomputing: A Java Interface to the Nexus Communication System, *Concurrency: Practice and Experience*, special issue edited by G. C. Fox, June 1997.
15. The Globus project, available at http//www.globus.org.
16. The Nimrod project, available at http//www.dgs.monah.edu/~davida/nimrod.html.
17. Rajkumar Buyya (editor), High Performance Cluster Computing: Architectures and Systems, Prentice Hall PTR, NJ, USA, 1999.
18. M. Baker, editor, Cluster Computing White Paper, http://www.dcs.port.ac.uk/~mab/tfcc/WhitePaper/.
19. R. L. Grossman, S. Kasif, D. Mon, A. Ramu and B. Malhi, The Preliminary Design of Papyrus: A System for High Performance, Distributed Data Mining over Clusters, Meta-Clusters and Super-Clusters, Proceedings of the KDD-98 Workshop on Distributed Data Mining, AAAI, 1999.
20. S. Stolfo, A. L. Prodromis, P.K. Chan, JAM: Java Agents for Meta-Learning over Distributed Databases, Proc. of the 3rd Int. Conf. On Knowledge Discovery and data Miing, AAAI Press, CA, 1997.
21. Y. Guo et al., Meta Learning for parallel Data Mining, in Proc. o the 7th Parallel Computing Workshop, 1997.
22. Albanese, M. Cannataro, P. Rullo, D. Saccà, Transmitting Datacubes over Congested Networks, Proc. of the IEEE International Conference on Coding and Transmission (ITCC2000), Las Vegas, 2000 (to appear).
23. Foster, I., A Grid-Enabled MPI: Message Passing in Heterogeneous Distributed Computing Systems, Proc. of the SC98 Conference, Orlando, USA, Nov. 7-13, 1998.
24. DiNucci, D. "The Role and Requirements of a Grid Programming Model", available at http://www.elepar.com/GPMWG/gpm.1.ps

Author Index

Adorni, G., 555
Afsarmanesh, H., 163, 507
Albada, G.D. van, 249, 395
Alda, W., 588
Alkindi, A.M., 280
Aloisio, G., 32
Aoki, K.F., 237
Arbab, F., 197

Baiardi, F., 71
Baude, F., 633
Beliën, A.J.C, 119
Benabdelkader, A., 163
Blochinger, W., 644
Bogdanov, A.V., 592
Böszörményi, L., 543
Breidler, K., 517
Brezany, P., 323
Brooke, J., 22
Brüning, U., 678
Bubak, M., 373
Buntine, M. A., 535

Cafaro, M., 32
Camarinha-Matos, L.M., 149
Cannataro, M., 708
Caromel, D., 633
Cela, J.M., 572
Charoy, F., 227
Chen, W., 668
Cherif, N., 580
Chi-Huing, C., 527
Chiti, S., 71
Cinque, L., 333
Clinckemaillie, J., 439
Coddington, P.D., 535
Costen, F., 22
Courson, M., 497
Cremonini, M., 187
Czerwinski, P., 323

Dandamudi, S.P., 427
Danelutto, M., 385, 407
Daoudi, E.M., 576, 580
De Florio, V., 313
Deconinck, G., 313

Dominguez, A., 454
Dydejczyk, A., 588

Estévez, A., 551

Falabella, P., 32
Fischer, J., 560
Fischer, M., 678
Frank, J., 99
Friedrich, M., 644
Fröhlich, A.A., 417

Gabriel Silva, J., 688
Gabriel, E., 22
Galbas, H.G., 439
Garatani, K., 445
Garita, C., 163
Gevorkyan, A.S., 592
Goedbloed, J.P., 61, 119
González, J.A., 551
Graham, P.J., 477
Grigori, D., 227
Grigoryan, A.G., 592
Gronek, P., 588
Grove, D.A., 535
Guo, Y., 207
Gustavson, F., 629

Halderen, A.W. van, 507
Hawick, K.A., 41, 363, 535
Heikonen, J., 698
Hertzberger, L.O., 163, 507
Holst, B. van der, 119
Hruz, T., 547
Huet, F., 633

Jaâra, E.M., 580
James, H.A., 41, 363, 535
Jensen, T., 217
Jeong, C.S., 12, 81
Jialin, W., 568
Jo, S.U., 81
Julià, A., 572
Jun, W., 618

Kaletas, E.C., 163, 507
Karl, W., 270
Keppens, R., 61
Kerbyson, D.J., 280
Kesselman, C., 32
Khanlarov G.O., 584
Kim, H.D., 12
Kim, K., 618
Kitowski, J., 543, 558
Kluge, J., 678
Kobler, R., 597
Kolp, O., 439
Kopeć, M., 588
Koppler, R., 597
Kosch, H., 517, 543
Koski, K., 698
Kranzlmueller, D., 597
Kremenetsky, M., 487
Kruk, D., 558
Küchlin, W., 644
Kurzyniec, D., 373
Kuwabara, T., 564

Lancaster, D.J., 3
Lauwereins, R., 313
Lecomber, D., 51
Lee, D.T., 237
Liao, C., 487
Lim, A., 527
Limniotes, T.A., 177
Linden, F. van der, 507
Lloyd, A.D., 477
Ludwig, T., 261
Lukianov, G.A., 584
Łuszczek, P., 373

Mäki, J., 698
Magdoń-Maksymowicz, M.S., 588
Magolu monga Made, M., 89
Maksymowicz, A.Z., 588
Malashonok, D.Yu., 584
Manoharan, S., 353
Marçais, G., 497
Maresca, M., 343
Marongiu, A., 333
Mastronardo, F., 333
Matsumoto, R., 564
Meijster, A., 109
Merkl, D., 136
Meziane, A., 576

Miettinen, J., 698
Migliardi, M., 622
Ming-Chin, L., 217
Ming-Chun, C., 217
Mink, A., 497
Miyaji, S., 564
Mohamed El Hadj, Y.O., 576
Morales, D.G., 551
Mori, P., 71
Mourkousis, G., 467
Müller, M., 22

Nakamura, H., 445
Nakamura, K., 564
Nazief, B.A.A., 395
Nicolopoulos, D., 454
Niemi, H., 698
Nikolow, D., 517
Nool, M., 61, 119
Nudd, G.R., 280

Okuda, H., 445
Omicini, A., 187
Ord, S., 22
Otfinowski, J., 543

Paker, Y., 568
Palazzari, P., 333
Panizzi, E., 539
Pantoja-Lima, C., 149
Papadimitriou, G.I., 610
Papadopoulos, G.A., 177, 197
Papaefstathiou, E., 280
Parkinson, D., 568
Pasquarelli, A., 547
Pedroso, H., 688
Pérez, M., 551
Pickles, S., 22
Plattner, B., 531
Ploeg, A. van der, 119
Podlipnig, S., 517
Pogoda, M., 517
Pomportsis, A.S., 610
Posey, S., 487
Priano, F.H., 551
Protonotarios, M., 467
Pucci, C., 407

Qingping, G., 568

Author Index

Rahola, J., 698
Rana, O.F., 605, 657
Rauber, A., 136
Reeve, J.S., 3
Resch, M., 22
Ricci L., 71
Rod Blais, J.A., 127
Roda, J.L., 551
Rudgyard, M., 51
Ruey-Kai, S., 217
Ruf, L., 531
Ruokolainen, J., 698
Ryu, S.H., 12
Rzymianowicz, L., 678

Sacco, G., 539
Santoso, J, 395
Schimkat, R.D., 644
Schreiber, A., 560
Schröder-Preikschat, W., 417
Schulz, M., 270
Schulz, P., 678
See-Mu, K., 353
Shamonin, D.P., 460
Shesheng, Z., 568
Shields, M.S., 657
Shyan-Ming, Y., 217
Sinz, C., 644
Skaf-Molli, H., 227
Sloan, T.M., 477
Sloot, P.M.A., 249, 395
Słota, R., 517, 543
Smyllie, K., 477
Soleimany, C., 427
Song, Y.M., 81
Spinnato, P., 249
Stankova, E.N., 592
Steihaug, T., 601
Strietzel, M., 560
Sunderam, V., 261, 622
Swietanowski, A., 323

Tan, C.J.K., 127

Tao, J., 270
Thole, C.A., 439
Tichy, W.F., 300
Tientcheu, G.P., 417
Tolvanen, P., 698
Tomsich, Ph., 136
Traverse, B., 497
Trinitis, J., 261

Vakali, A.I., 610
van der Vorst, H.A., 89
Varvarigou, T., 467
Vayssière, J., 633
Vinter, B., 614
Vlachoutsis, S., 439
Volkert, J., 597
Vuik, C., 99
Vyridis, I., 467

Waśniewski, J., 629
Waack, M., 678
Walter, T., 531
Wendel, P., 207
Werner-Kytölä, O., 300
Williams, R., 32
Winslett, M., 323
Wismüller, R., 261
Wójcik, P., 543
Wubs, F.W., 109

Xiang, L., 527

Yagawa, G., 445
Yalçınkaya, Y., 601
Yao-Jin, H., 217
Yong, M.T., 668
Yue-Shan, C., 217

Zakharov, V.V., 584
Zambonelli, F., 187
Zavanella, A., 290
Zegeling, P.A., 61
Zingirian, N., 343
Zudilova, E.V., 460

Lecture Notes in Computer Science

For information about Vols. 1–1725
please contact your bookseller or Springer-Verlag

Vol. 1726: V. Varadharajan, Y. Mu (Eds.), Information and Communication Security. Proceedings, 1999. XI, 325 pages. 1999.

Vol. 1727: P.P. Chen, D.W. Embley, J. Kouloumdjian, S.W. Liddle, J.F. Roddick (Eds.), Advances in Conceptual Modeling. Proceedings, 1999. XI, 389 pages. 1999.

Vol. 1728: J. Akoka, M. Bouzeghoub, I. Comyn-Wattiau, E. Métais (Eds.), Conceptual Modeling – ER '99. Proceedings, 1999. XIV, 540 pages. 1999.

Vol. 1729: M. Mambo, Y. Zheng (Eds.), Information Security. Proceedings, 1999. IX, 277 pages. 1999.

Vol. 1730: M. Gelfond, N. Leone, G. Pfeifer (Eds.), Logic Programming and Nonmonotonic Reasoning. Proceedings, 1999. XI, 391 pages. 1999. (Subseries LNAI).

Vol. 1731: J. Kratochvíl (Ed.), Graph Drawing. Proceedings, 1999. XIII, 422 pages. 1999.

Vol. 1732: S. Matsuoka, R.R. Oldehoeft, M. Tholburn (Eds.), Computing in Object-Oriented Parallel Environments. Proceedings, 1999. VIII, 205 pages. 1999.

Vol. 1733: H. Nakashima, C. Zhang (Eds.), Approaches to Intelligent Agents. Proceedings, 1999. XII, 241 pages. 1999. (Subseries LNAI).

Vol. 1734: H. Hellwagner, A. Reinefeld (Eds.), SCI: Scalable Coherent Interface. XXI, 490 pages. 1999.

Vol. 1564: M. Vazirgiannis, Interactive Multimedia Documents. XIII, 161 pages. 1999.

Vol. 1591: D.J. Duke, I. Herman, M.S. Marshall, PREMO: A Framework for Multimedia Middleware. XII, 254 pages. 1999.

Vol. 1624: J. A. Padget (Ed.), Collaboration between Human and Artificial Societies. XIV, 301 pages. 1999. (Subseries LNAI).

Vol. 1635: X. Tu, Artificial Animals for Computer Animation. XIV, 172 pages. 1999.

Vol. 1646: B. Westfechtel, Models and Tools for Managing Development Processes. XIV, 418 pages. 1999.

Vol. 1735: J.W. Amtrup, Incremental Speech Translation. XV, 200 pages. 1999. (Subseries LNAI).

Vol. 1736: L. Rizzo, S. Fdida (Eds.): Networked Group Communication. Proceedings, 1999. XIII, 339 pages. 1999.

Vol. 1737: P. Agouris, A. Stefanidis (Eds.), Integrated Spatial Databases. Proceedings, 1999. X, 317 pages. 1999.

Vol. 1738: C. Pandu Rangan, V. Raman, R. Ramanujam (Eds.), Foundations of Software Technology and Theoretical Computer Science. Proceedings, 1999. XII, 452 pages. 1999.

Vol. 1739: A. Braffort, R. Gherbi, S. Gibet, J. Richardson, D. Teil (Eds.), Gesture-Based Communication in Human-Computer Interaction. Proceedings, 1999. XI, 333 pages. 1999. (Subseries LNAI).

Vol. 1740: R. Baumgart (Ed.): Secure Networking – CQRE [Secure] '99. Proceedings, 1999. IX, 261 pages. 1999.

Vol. 1741: A. Aggarwal, C. Pandu Rangan (Eds.), Algorithms and Computation. Proceedings, 1999. XIII, 448 pages. 1999.

Vol. 1742: P.S. Thiagarajan, R. Yap (Eds.), Advances in Computing Science – ASIAN'99. Proceedings, 1999. XI, 397 pages. 1999.

Vol. 1743: A. Moreira, S. Demeyer (Eds.), Object-Oriented Technology. Proceedings, 1999. XVII, 389 pages. 1999.

Vol. 1744: S. Staab, Extracting Degree Information from Texts. X; 187 pages. 1999. (Subseries LNAI).

Vol. 1745: P. Banerjee, V.K. Prasanna, B.P. Sinha (Eds.), High Performance Computing – HiPC'99. Proceedings, 1999. XXII, 412 pages. 1999.

Vol. 1746: M. Walker (Ed.), Cryptography and Coding. Proceedings, 1999. IX, 313 pages. 1999.

Vol. 1747: N. Foo (Ed.), Adavanced Topics in Artificial Intelligence. Proceedings, 1999. XV, 500 pages. 1999. (Subseries LNAI).

Vol. 1748: H.V. Leong, W.-C. Lee, B. Li, L. Yin (Eds.), Mobile Data Access. Proceedings, 1999. X, 245 pages. 1999.

Vol. 1749: L. C.-K. Hui, D.L. Lee (Eds.), Internet Applications. Proceedings, 1999. XX, 518 pages. 1999.

Vol. 1750: D.E. Knuth, MMIXware. VIII, 550 pages. 1999.

Vol. 1751: H. Imai, Y. Zheng (Eds.), Public Key Cryptography. Proceedings, 2000. XI, 485 pages. 2000.

Vol. 1752: S. Krakowiak, S. Shrivastava (Eds.), Advances in Distributed Systems. VIII, 509 pages. 2000.

Vol. 1753: E. Pontelli, V. Santos Costa (Eds.), Practical Aspects of Declarative Languages. Proceedings, 2000. X, 327 pages. 2000.

Vol. 1754: J. Väänänen (Ed.), Generalized Quantifiers and Computation. Proceedings, 1997. VII, 139 pages. 1999.

Vol. 1755: D. Bjørner, M. Broy, A.V. Zamulin (Eds.), Perspectives of System Informatics. Proceedings, 1999. XII, 540 pages. 2000.

Vol. 1757: N.R. Jennings, Y. Lespérance (Eds.), Intelligent Agents VI. Proceedings, 1999. XII, 380 pages. 2000. (Subseries LNAI).

Vol. 1758: H. Heys, C. Adams (Eds.), Selected Areas in Cryptography. Proceedings, 1999. VIII, 243 pages. 2000.

Vol. 1759: M.J. Zaki, C.-T. Ho (Eds.), Large-Scale Parallel Data Mining. VIII, 261 pages. 2000. (Subseries LNAI).

Vol. 1762: K.-D. Schewe, B. Thalheim (Eds.), Foundations of Information and Knowledge Systems. Proceedings, 2000. X, 305 pages. 2000.

Vol. 1763: J. Akiyama, M. Kano, M. Urabe (Eds.), Discrete and Computational Geometry. Proceedings, 1998. VIII, 333 pages. 2000.

Vol. 1764: H. Ehrig, G. Engels, H.-J. Kreowski, G. Rozenberg (Eds.), Theory and Application of Graph Transformations. Proceedings, 1998. IX, 490 pages. 2000.

Vol. 1765: T. Ishida, K. Isbister (Eds.), Digital Cities. IX, 444 pages. 2000.

Vol. 1767: G. Bongiovanni, G. Gambosi, R. Petreschi (Eds.), Algorithms and Complexity. Proceedings, 2000. VIII, 317 pages. 2000.

Vol. 1768: A. Pfitzmann (Ed.), Information Hiding. Proceedings, 1999. IX, 492 pages. 2000.

Vol. 1769: G. Haring, C. Lindemann, M. Reiser (Eds.), Performance Evaluation: Origins and Directions. X, 529 pages. 2000.

Vol. 1770: H. Reichel, S. Tison (Eds.), STACS 2000. Proceedings, 2000. XIV, 662 pages. 2000.

Vol. 1771: P. Lambrix, Part-Whole Reasoning in an Object-Centered Framework. XII, 195 pages. 2000. (Subseries LNAI).

Vol. 1772: M. Beetz, Concurrent Ractive Plans. XVI, 213 pages. 2000. (Subseries LNAI).

Vol. 1773: G. Saake, K. Schwarz, C. Türker (Eds.), Transactions and Database Dynamics. Proceedings, 1999. VIII, 247 pages. 2000.

Vol. 1774: J. Delgado, G.D. Stamoulis, A. Mullery, D. Prevedourou, K. Start (Eds.), Telecommunications and IT Convergence Towards Service E-volution. Proceedings, 2000. XIII, 350 pages. 2000.

Vol. 1775: M. Thielscher, Challenges for Action Theories. XIII, 138 pages. 2000. (Subseries LNAI).

Vol. 1776: G.H. Gonnet, D. Panario, A. Viola (Eds.), LATIN 2000: Theoretical Informatics. Proceedings, 2000. XIV, 484 pages. 2000.

Vol. 1777: C. Zaniolo, P.C. Lockemann, M.H. Scholl, T. Grust (Eds.), Advances in Database Technology – EDBT 2000. Proceedings, 2000. XII, 540 pages. 2000.

Vol. 1778: S. Wermter, R. Sun (Eds.), Hybrid Neural Systems. IX, 403 pages. 2000. (Subseries LNAI).

Vol. 1780: R. Conradi (Ed.), Software Process Technology. Proceedings, 2000. IX, 249 pages. 2000.

Vol. 1781: D.A. Watt (Ed.), Compiler Construction. Proceedings, 2000. X, 295 pages. 2000.

Vol. 1782: G. Smolka (Ed.), Programming Languages and Systems. Proceedings, 2000. XIII, 429 pages. 2000.

Vol. 1783: T. Maibaum (Ed.), Fundamental Approaches to Software Engineering. Proceedings, 2000. XIII, 375 pages. 2000.

Vol. 1784: J. Tiuryn (Ed.), Foundations of Software Science and Computation Structures. Proceedings, 2000. X, 391 pages. 2000.

Vol. 1785: S. Graf, M. Schwartzbach (Eds.), Tools and Algorithms for the Construction and Analysis of Systems. Proceedings, 2000. XIV, 552 pages. 2000.

Vol. 1786: B.H. Haverkort, H.C. Bohnenkamp, C.U. Smith (Eds.), Computer Performance Evaluation. Proceedings, 2000. XIV, 383 pages. 2000.

Vol. 1787: J. Song (Ed.), Information Security and Cryptology – ICISC'99. Proceedings, 1999. XI, 279 pages. 2000.

Vol. 1790: N. Lynch, B.H. Krogh (Eds.), Hybrid Systems: Computation and Control. Proceedings, 2000. XII, 465 pages. 2000.

Vol. 1792: E. Lamma, P. Mello (Eds.), AI*IA 99: Advances in Artificial Intelligence. Proceedings, 1999. XI, 392 pages. 2000. (Subseries LNAI).

Vol. 1793: O. Cairo, L.E. Sucar, F.J. Cantu (Eds.), MICAI 2000: Advances in Artificial Intelligence. Proceedings, 2000. XIV, 750 pages. 2000. (Subseries LNAI).

Vol. 1795: J. Sventek, G. Coulson (Eds.), Middleware 2000. Proceedings, 2000. XI, 436 pages. 2000.

Vol. 1794: H. Kirchner, C. Ringeissen (Eds.), Frontiers of Combining Systems. Proceedings, 2000. X, 291 pages. 2000. (Subseries LNAI).

Vol. 1796: B. Christianson, B. Crispo, J.A. Malcolm, M. Roe (Eds.), Security Protocols. Proceedings, 1999. XII, 229 pages. 2000.

Vol. 1800: J. Rolim et al. (Eds.), Parallel and Distributed Processing. Proceedings, 2000. XXIII, 1311 pages. 2000.

Vol. 1801: J. Miller, A. Thompson, P. Thomson, T.C. Fogarty (Eds.), Evolvable Systems: From Biology to Hardware. Proceedings, 2000. X, 286 pages. 2000.

Vol. 1802: R. Poli, W. Banzhaf, W.B. Langdon, J. Miller, P. Nordin, T.C. Fogarty (Eds.), Genetic Programming. Proceedings, 2000. X, 361 pages. 2000.

Vol. 1803: S. Cagnoni et al. (Eds.), Real-World Applications and Evolutionary Computing. Proceedings, 2000. XII, 396 pages. 2000.

Vol. 1805: T. Terano, H. Liu, A.L.P. Chen (Eds.), Knowledge Discovery and Data Mining. Proceedings, 2000. XIV, 460 pages. 2000. (Subseries LNAI).

Vol. 1806: W. van der Aalst, J. Desel, A. Oberweis (Eds.), Business Process Management. VIII, 391 pages. 2000.

Vol. 1807: B. Preneel (Ed.), Advances in Cryptology – EUROCRYPT 2000. Proceedings, 2000. XVIII, 608 pages. 2000.

Vol. 1811: S.W. Lee, H.. Bülthoff, T. Poggio (Eds.), Biologically Motivated Computer Vision. Proceedings, 2000. XIV, 656 pages. 2000.

Vol. 1815: G. Pujolle, H. Perros, S. Fdida, U. Körner, I. Stavrakakis (Eds.), Networking 2000 – Broadband Communications, High Performance Networking, and Performance of Communication Networks. Proceedings, 2000. XX, 981 pages. 2000.

Vol. 1816: T. Rus (Ed.), Algebraic Methodology and Software Technology. Proceedings, 2000. XI, 545 pages. 2000.

Vol. 1818: C.G. Omidyar (Ed.), Mobile and Wireless Communications Networks. Proceedings, 2000. VIII, 187 pages. 2000.

Vol. 1823: M. Bubak, H. Afsarmanesh, R. Williams, B. Hertzberger (Eds.), High Performance Computing and Networking. Proceedings, 2000. XVIII, 719 pages. 2000.